Polymerized Ionic Liquids

Smart Materials

Series editors:
Hans-Jörg Schneider, *Saarland University, Germany*
Mohsen Shahinpoor, *University of Maine, USA*

Titles in this series:

1: Janus Particle Synthesis, Self-Assembly and Applications
2: Smart Materials for Drug Delivery: Volume 1
3: Smart Materials for Drug Delivery: Volume 2
4: Materials Design Inspired by Nature
5: Responsive Photonic Nanostructures: Smart Nanoscale Optical Materials
6: Magnetorheology: Advances and Applications
7: Functional Nanometer-Sized Clusters of Transition Metals: Synthesis, Properties and Applications
8: Mechanochromic Fluorescent Materials: Phenomena, Materials and Applications
9: Cell Surface Engineering: Fabrication of Functional Nanoshells
10: Biointerfaces: Where Material Meets Biology
11: Semiconductor Nanowires: From Next-Generation Electronics to Sustainable Energy
12: Supramolecular Materials for Opto-Electronics
13: Photocured Materials
14: Chemoresponsive Materials: Stimulation by Chemical and Biological Signals
15: Functional Metallosupramolecular Materials
16: Bio-Synthetic Hybrid Materials and Bionanoparticles: A Biological Chemical Approach Towards Material Science
17: Ionic Polymer Metal Composites (IPMCs): Smart Multi-Functional Materials and Artificial Muscles Volume 1
18: Ionic Polymer Metal Composites (IPMCs): Smart Multi-Functional Materials and Artificial Muscles Volume 2
19: Conducting Polymers: Bioinspired Intelligent Materials and Devices
20: Smart Materials for Advanced Environmental Applications
21: Self-cleaning Coatings: Structure, Fabrication and Application
22: Functional Polymer Composites with Nanoclays
23: Bioactive Glasses: Fundamentals, Technology and Applications
24: Smart Materials for Tissue Engineering: Fundamental Principles

25: Smart Materials for Tissue Engineering: Applications
26: Biobased Smart Polyurethane Nanocomposites: From Synthesis to Applications
27: Inorganic Two-dimensional Nanomaterials: Fundamental Understanding, Characterizations and Energy Applications
28: Ionic Liquid Devices
29: Polymerized Ionic Liquids

How to obtain future titles on publication:
A standing order plan is available for this series. A standing order will bring delivery of each new volume immediately on publication.

For further information please contact:
Book Sales Department, Royal Society of Chemistry, Thomas Graham House, Science Park, Milton Road, Cambridge, CB4 0WF, UK
Telephone: +44 (0)1223 420066, Fax: +44 (0)1223 420247
Email: booksales@rsc.org
Visit our website at www.rsc.org/books

Polymerized Ionic Liquids

Edited by

Ali Eftekhari
Ulster University, Belfast, UK
Email: eftekhari@elchem.org

ROYAL SOCIETY OF CHEMISTRY

THE QUEEN'S AWARDS FOR ENTERPRISE: INTERNATIONAL TRADE 2013

Smart Materials No. 29

Print ISBN: 978-1-78262-960-3
PDF eISBN: 978-1-78801-053-5
EPUB eISBN: 978-1-78801-221-8
ISSN: 2046-0066

A catalogue record for this book is available from the British Library

The Royal Society of Chemistry is a charity, registered in England and Wales, Number 207890, and a company incorporated in England by Royal Charter (Registered No. RC000524), registered office: Burlington House, Piccadilly, London W1J 0BA, UK, Telephone: +44 (0) 207 4378 6556.

For further information see our web site at www.rsc.org

Printed in the United Kingdom by CPI Group (UK) Ltd, Croydon, CR0 4YY, UK

Foreword

The reader holds a reference book in hand, published as a new addition to the *Smart Materials* series, which provides a great overview of recent progress in this swiftly growing field in the form of chapters, written by well-known experts in their respective research areas. The volume is edited by Professor Ali Eftekhari, a distinguished chemist and materials scientist, and a renowned expert in ionic liquids, nanotechnology, electrochemistry and electroactive materials.

In order to place the objectives of this monograph, let us first consider molten salts, low temperature ionic liquids (or "designer" media), and polymeric ionic liquids (as functional materials) as different laps in the knowledge and technology chain resulting in the current state-of-the-art of the subject discussed in this book. Ionic compounds that have a low melting point, such as (the arbitrarily chosen) 100 °C, are usually classified as ionic liquids. When compared with salts like NaCl (with a melting point of 801 °C) the distances between oppositely charged ions in the structure of ionic liquids is (substantially) larger, the bonding coordination is poorer, and the charges are delocalized. Due to this charge delocalization (in π-electron bonds) and to the presence of at least one organic constituent, and related to conformational freedom, the formation of stable crystal structures in ionic liquids is often prevented. Fine-tuning of the chemical structure of the organic constituents allows one to control the properties of ionic liquids, which endows these systems with a "designer" characteristic. The fine tuning of the structure, and thus properties, provides engineered materials with controlled solubility, dissolving characteristics, viscosity, vapor pressure, and electrical conductivity. Rapid growth of research into these low-melting point ionic liquids is also related to their manifold applications as dispersing agents (solvents), electrolytes, enhancing agents of chemical process efficiency and process safety,

Smart Materials No. 29
Polymerized Ionic Liquids
Edited by Ali Eftekhari

Published by the Royal Society of Chemistry, www.rsc.org

and adhesive-sealant barriers, the latter being related to their low vapor pressure. But reduced toxicity and non-flammability in combination with low vapor pressure further enhance the scope of their applications, *e.g.* in green chemistry. Although ionic liquids were first reported in 1914 by Walden, they did not trigger substantial interest until recent decades. This is also due to quickly emerging applications, as processes that were feasible only at high temperatures have become now accessible due to the use of ionic liquids at low temperatures opening new technologies in organic chemistry and energy-related fields.

Naturally, the field of ionic liquids has been covered by some excellent reviews and books. As a great introduction, the volume published by the Royal Society of Chemistry, authored by M. Freemantle, T. Welton, and R. D. Rogers, can be mentioned.[1] An excellent review published in the scientific journal literature by Plechkova and Seddon[2] provides great insights into "Applications of ionic liquids in the chemical industry" from the historical beginnings to recent times.

Incorporation of polymerizable groups in the structures of the ionic liquid constituents yields polymerizable ionic liquids, which are the subject of a very active area of contemporary materials research. The emergence of polymeric ionic liquids – or poly(ionic liquid)s (PILs) – as functional materials has opened novel applications in polymer electrolytes, dispersants and solvents, microporous membranes, green chemistry, and electroactive polymers. One of the early reviews by M. A. Firestone *et al.* still provides excellent insights into the emergence of poly(ionic liquid)s.[3]

Of course, in physical chemistry of macromolecules, charged polyions have been a subject of intensive research, not least due to the charged nature of many biological macromolecules. One should however keep in mind that in polyelectrolyte systems there is a dissolving liquid and counterions in a supporting electrolyte, thus the liquid phase is a solvent–ion-charged polymer system. Strictly, for poly(ionic liquid)s no solvents are needed.

Swift progress in research and applications of poly(ionic liquid)s has more than justified the publication of subsequent additional reviews, such as ref. 4 by M. Antonietti and colleagues.

In this present volume published by the Royal Society of Chemistry, the fundamental concepts of and the basic differences between ionic liquids and poly(ionic liquid)s, as well as detailed accounts on technological progress, are provided in an excellently balanced fashion. Swift progress and growth of interest, and the still relatively sparse book literature more than justify the publication of the current volume edited by Professor Eftekhari. In this book, in eighteen chapters, leading researchers give a comprehensive overview of the state-of-the-art (and often even go "beyond the edge") of the field.

The book begins with a great summary by J. Mays and co-authors discussing the use of ionic liquids as polymerization media in the "green synthesis" of polymers. The advantages and the reduced environmental footprint of the use of designer ionic liquids as solvents are presented. The second chapter is also related to materials synthesis, but with a focus on porous, charged materials (with pore sizes across the length scales) using ionic and poly(ionic) liquids. The authors, A. Dani, S. Bordiga *et al.*, describe the fundamentals of using ionic liquid functionalities in the assembly, charging and reorganization of functional materials, and making hybrid and responsive functional materials. A nice section in this chapter discusses porosity evaluation approaches and contrasts micro-mesoporous structures. Olga Kuzmina in Chapter 3 very nicely describes cationic and anionic polyion-containing poly(ionic liquid) structures, and provides very informative tables of the various poly(ionic liquid) cations–anions, the corresponding counterions, and some material properties, like conductivity. The next section by H. Ohno and co-author draws the reader's attention to stimulus responsive PILs, with a focus on hydrophobic/hydrophilic transitions and the thermodynamic phase behavior of PIL–solvent systems. Structure–property relations and how these influence thermodynamic phases (including LCST-type phase transitions) are explained. In the next chapter, Mikkola and co-authors continue the discussion on responsive behavior, in particular regarding polarity-switchable media of molecular liquids. Such systems possess great promise in designer chemical transformations. Still along the highly promising line of research with a focus on electro-responsive fluids with charged constituents, Jianbo Yin and co-workers provide a very nice section introducing the principles and the potential applications of PILs as novel fluidic systems in electrorheology, including molecularly aligned anisotropic fluids. Chapter 7 provides a discussion on preparing nanogels from PILs, authored by Y. B. Xiong *et al.* A facile one-step cross-linking copolymerization is used here to obtain thermo- and multiple-responsive nanogels with potential use *e.g.* in the capture and release of molecular cargos. The next section by H. Randriamahazaka and co-authors takes the reader beyond the more conventional stimuli and offers a discussion on redox-active PIL functional systems. Regarding applications, employing a redox trigger opens the range of uses encompassing devices with electrochemical switching, supercapacitors, electrolytes for use in batteries, fuel cells, sensors, and actuators. Next, H. Lin and co-workers focus on "doping" of polymers with PILs for morphology control to regulate *e.g.* gas separation of membranes, while Y. J. Li and co-author in the subsequent chapter show how polymer/IL "composites" can be used to tailor anti-static and dielectric performance. Ionogels and hydrogels have gained widespread use as biomaterials. In the next Section S. S. Silva and R. L. Reis focus on polymeric hydrogels processed in ionic liquids, and provide examples of biomaterials obtained *via* this route. Regarding biomaterials, Kadokawa then shows how the functionality of polysaccharides can be enhanced by ionic liquids. In this part, the particularly important use of ionic liquids to dissolve polymers, including cellulose and chitin that cannot be dissolved, or are notoriously difficult to dissolve in other solvents, is nicely documented. Torresi *et al.* return, in their contribution, to the added versatility regarding the functionality of ionic liquids and PILs, with a particular focus on electrochromic and gas storage applications. Zopf *et al.* in Chapter 14 show an exciting idea of how ionic liquid gels can be made and used as wearable electrochemical storage devices, such as batteries and supercapacitors. Next, McIntosh and co-authors also focus on wearable devices, but the application in their case is related to sensors. Here, ionic liquids provide enhanced analyte sensing specificity, and sensor performance improvement. They point out that although a large number of ionic liquids have been explored, there is still a lot of space for further research into task-specific selection of ionic liquids for particular use in designer sensors. Chapter 16 then provides an overview of ionic gels, electrochemical ionic actuators, ionic metal–polymer composites, and related responsive materials, authored by A. Maziz, C. Bergaud and co-worker. The focus here is primarily on mechanical actuation. The penultimate chapter shows yet another exciting application area of using PILs to capture CO_2, which is a very important focus field in combatting climate change. Finally, the last contribution written by C. P. Lee and K. C. Ho introduces the reader to the highly interesting potential application of PILs as electrolytes in dye-sensitized solar cells. This contribution also offers a good general introduction to employing solar radiation as a sustainable source of energy.

The unique collection of chapters in this book, showing the state-of-the-art, from the basics, *via* sketching the new trends, to introducing emerging possibilities of ionic liquids in combination with polymers, or of poly(ionic liquid)s, with a very strong focus on responsive materials, is a very essential read for chemists, materials scientists, physicists and engineers.

G. J. Vancso

References

1. M. Freemantle, T. Welton and R. D. Rogers, *An Introduction to Ionic Liquids*, RSC Publishing, 2009, ISBN-13: 978–1847551610.
2. N. V. Plechkova and K. R. Seddon, Applications of ionic liquids in the chemical industry, *Chem. Soc. Rev.*, 2008, **37**(1), 123–150.
3. O. Green, S. Grubjesic, S. Lee and M. A. Firestone, The design of polymeric ionic liquids for the preparation of functional materials, *Polym. Rev.*, 2009, **49**(4), 339–360.
4. J. Yuan, D. Mecerreyes and M. Antonietti, Poly(ionic liquid)s: An Update, *Prog. Polym. Sci.*, 2013, **38**(7), 1009–1036.

Preface

Polymerized ionic liquids (PILs) are a new member of the family of ionic polymers, and are of both fundamental and practical interest. The classic ionic polymers have a partial ionicity, while PILs are entirely ionic. The flagship of ionic liquids is Nafion, which is a commercial product that has been widely utilized since the 1970s. Owing to its high cost and reduced flexibility in design, a simple approach in studies of PILs is to find alternatives to Nafion in various applications employing Nafion membranes. However, the applicability of PILs goes beyond the conventional applications of ionic polymers, as they have introduced numerous new opportunities.

Despite their potential, PILs have not yet attracted the attention they deserve because of the diverse range of research studies, which are not adequately linked. Researchers from quite different disciplines are now tuning PILs to meet the requirements of different applications by focusing on various aspects of PILs, whether as flexible polymers or ionic materials. Even the terminology of this field is not straightforward, and it is not easy to find all relevant works using common searching methods in the literature databases. The present book aims to provide a convenient collection of different research topics in this field. While we attempted to avoid overlap between the chapters, each chapter has its own introduction to the field to preserve the independence of the individual chapters, since some researchers may be interested in specific chapters. On the other hand, many readers will just find a particular chapter when browsing online.

My dear friend, Prof. G. Julius Vancso, has provided a good overview of the book in his foreword, and I do not repeat it here. Owing to page limits, the goal was not to offer a comprehensive volume on PILs, but to highlight the emerging potential in this area of research. In fact, we hope to attract the attention of researchers from quite different disciplines to strengthen the networking connections in this field.

Smart Materials No. 29
Polymerized Ionic Liquids
Edited by Ali Eftekhari

Published by the Royal Society of Chemistry, www.rsc.org

PILs are not just a class of polymers, but also an alternative to ionic liquids where the liquidity is not the key factor. In a sister book entitled *Ionic Liquid Devices*, the emerging potential of ILs has been reviewed. There are many overlaps between the areas of interest in these two sister books. In other words, polymerization is a practical approach for controlling the ionicity of ILs. Many obstacles in the practical development of ILs in various applications are because of the intrinsic ionicity, which is ideally controllable. By fixing one side of the IL ionicity as the polymer moiety, it is much easier to control the ionicity. Owing to the ionic nature of ILs, PILs can be seen as an analog to the lattice structure of ionic solids. Although there is no such unit cell, the architecture of this ionic network can be designed. In applications involving solid-state diffusion, the similarity is tangible when designing the diffusion channels.

The present book is the result of invaluable contributions from the individual authors, who closely worked with me to reach a satisfactory integrity for such an emerging field. However, the sincere efforts of the people backstage should not be forgotten. It would be a long list to provide a formal acknowledgement, but I would like to especially thank Dr Leanne Marle who was with me throughout this process from the inception of the project. When we were facing an obstacle, she always came up with brilliant solutions to keep the project going. In addition, I would like to thank Prof. Mohsen Shahinpoor, the Editor of RSC Smart Materials, as the present book project is actually the outcome of a friendly chat with him. Overall, it was an absolute pleasure for me to closely work with the enthusiastic team at the Royal Society of Chemistry.

Ali Eftekhari

Contents

Chapter 1 Polymerization in Ionic Liquids 1
Nikhil K. Singha, Kunlun Hong and Jimmy W. Mays

1.1 Introduction 1
1.2 ILs in Conventional Free Radical Polymerization 2
1.3 ILs in Controlled Free Radical Polymerizations 6
1.4 ILs in Ionic Polymerizations and Group Transfer Polymerizations 9
1.5 Ring Opening Polymerization 10
1.6 Transition Metal-catalyzed Polymerizations 11
1.7 Electrochemical Polymerization 12
1.8 Step-growth Polymerization 12
1.9 Enzymatic Polymerization 12
1.10 ILs as Solvents for Grafting from Biopolymers 13
1.11 Other Applications of ILs in Polymerization 13
1.12 Conclusions and Future Prospects 14
Acknowledgements 15
References 15

Chapter 2 Porous Ionic Liquid Materials 23
Alessandro Dani, Valentina Crocellà, Giulio Latini and Silvia Bordiga

2.1 Introduction 23
2.1.1 General Definition of Porosity 24

Smart Materials No. 29
Polymerized Ionic Liquids
Edited by Ali Eftekhari

Published by the Royal Society of Chemistry, www.rsc.org

2.2 Porosity and Ionic Liquids 28
2.2.1 Porous Poly(Ionic Liquid)s 30
2.2.2 Supported Porous Ionic Liquids 51
2.2.3 Hybrid Porous IL Materials 63
2.2.4 Supramolecular Ionic Liquids 77
References 77

Chapter 3 Cationic and Anionic Polymerized Ionic Liquids: Properties for Applications 83
Olga Kuzmina

3.1 Introduction 83
3.2 Comparison of the Properties of Cationic *vs.* Anionic PILs 84
3.2.1 Electroconductivity 84
3.2.2 CO_2 Sorption 95
3.2.3 Sensors 103
3.2.4 Thermoresponsive Materials 105
3.3 Summary and Future Directions of Research 109
References 109

Chapter 4 Switchable Hydrophobicity and Hydrophilicity 117
Yuki Kohno and Hiroyuki Ohno

4.1 Introduction 117
4.2 Ionic Liquids with Switchable Hydrophobicity and Hydrophilicity Depending on the Temperature 118
4.3 Thermoresponsive Poly(ionic liquid)s with Switchable Hydrophobicity/Hydrophilicity 125
4.4 Potential Applications of Thermoresponsive Ionic Liquid-based Materials 134
4.5 Conclusion 139
Acknowledgements 140
References 140

Chapter 5 Switchable Polarity Liquids 143
S. G. Khokarale, I. Anugwom, P. Mäki-Arvela, P. Virtanen and J.-P. Mikkola

5.1 Introduction 143
5.2 Preparation and Characterization 145
5.2.1 Two-component Switchable Polarity Solvent/Ionic Liquids for CO_2/SO_2 Capture 145
5.2.2 Switchable Ionic Liquids from DBU, Alcohols and CO_2 146

5.2.3 Switchable Ionic Liquids from TMG, Alcohols and CO_2 147
5.2.4 CO_2 Release and Recyclability of SPSs 149
5.2.5 DBU Bicarbonate 153
5.2.6 One-component Switchable Polarity Solvents/Ionic Liquids for CO_2 Capture 155
5.2.7 Regeneration of Silylamines from ILs 161
5.2.8 Recyclability of Silylamine SPSs/ILs 161
5.3 Applications 162
5.3.1 CO_2 Capture with Switchable Ionic Liquids 162
5.3.2 Fractionation of Alga with Switchable Ionic Liquids 166
5.3.3 Wood Fractionation 167
5.3.4 Switchable Ionic Liquids as Reaction and Separation Media 174
5.4 Conclusions 176
Acknowledgements 176
References 176

Chapter 6 Stimuli Responsive Smart Fluids Based on Ionic Liquids and Poly(ionic liquid)s 180
Jianbo Yin, Qi Lei, Yuezhen Dong and Xiaopeng Zhao

6.1 Introduction 180
6.2 Electro/magneto-responsive Smart Fluids 182
6.3 Electro-responsive Electrorheological Fluids 184
6.3.1 Electrorheological Fluids Based on Ionic Liquids 187
6.3.2 Electrorheological Fluids Based on Poly(ionic liquid)s 191
6.4 Summary 198
Acknowledgements 199
References 199

Chapter 7 Thermo-responsive Poly(ionic liquid) Nanogels Prepared *via* One-step Cross-linking Copolymerization 202
J. R. Yu, Y. Zuo and Y. B. Xiong

7.1 Introduction 202
7.2 Thermo-responsive Systems Comprising ILs 204
7.3 Thermo-responsive PIL Nanogels Prepared *via* One-step Cross-linking Copolymerization 206
7.4 Summary and Outlook 222
References 222

Chapter 8 Redox-active Immobilized Ionic Liquids and Polymer Ionic Liquids 225
Thuan-Nguyen Pham-Truong, Jalal Ghilane and Hyacinthe Randriamahazaka

8.1 Introduction 225
8.2 Electrochemistry in Ionic Liquids 226
8.2.1 Electrodeposition of Metals, Metal Alloys and Semiconductors 226
8.2.2 Electropolymerization of Conducting Polymers 227
8.3 Redox-active Ionic Liquid 228
8.4 Immobilization of Ionic Liquids and Redox-active Ionic Liquids 230
8.4.1 Introduction 230
8.4.2 Approaches to form Thin Layers of Ionic Liquids 231
8.5 Approaches for Polymer Ionic Liquids 238
8.5.1 Synthetic Route and Structure of PILs 238
8.5.2 Physicochemical Properties 241
8.6 Applications of Poly(ionic liquid)s 243
8.6.1 Nanostructuration 243
8.6.2 Switchable Devices 245
8.6.3 Energy Applications 247
8.6.4 Sensors 252
8.7 Concluding Remarks 255
References 256

Chapter 9 Doping Polymers with Ionic Liquids to Manipulate Their Morphology and Membrane Gas Separation Properties 262
Xianda Hou, Junyi Liu, Hien Nguyen and Haiqing Lin

9.1 Introduction 262
9.2 Background 264
9.3 Effect of IL Doping on the T_g of Blends 265
9.3.1 T_g Depression and Modeling Using the Gordon–Taylor Equation 265
9.3.2 Estimation of the T_g for ILs 266
9.4 Effect of IL Doping on Polymer Crystallization 267
9.4.1 Effect of IL Doping on T_m Depression 267
9.4.2 Effect of IL Doping on Polymer Crystallinity 268
9.4.3 Effect of ILs Doping on Dissolution of Cellulose Acetate 268
9.5 Effect of IL Doping on Gas Permeation Properties 270
9.5.1 Gas Solubility in ILs 270

9.5.2 Effect of IL Doping on Gas Solubility in Polymer/IL Blends 271
9.5.3 Effect of IL Doping on Gas Diffusivity in Polymer/IL Blends 272
9.5.4 Effect of IL Doping on Gas Separation Properties 274
9.6 Conclusion 275
Acknowledgements and Disclaimer 276
References 276

Chapter 10 Ionic Liquid-modified Poly(Vinylidene Fluoride): from High Performance Anti-static Materials to Flexible Dielectric Materials 280
C. Y. Xing and Y. J. Li

10.1 Introduction 280
10.2 Anti-static PVDF/IL Composites 281
10.2.1 Anti-static Miscible PVDF/IL Films 281
10.2.2 Anti-static PVDF/IL Nanofibrous Films 284
10.2.3 Anti-static PVDF/IL–CNT Nanocomposites 286
10.3 Dielectric PVDF/IL Composites 288
10.3.1 Formation of PVDF-*g*-IL Films 288
10.3.2 Dielectric PVDF/IL Nanostructured Composites 290
10.3.3 Block-like Copolymers of PVDF-*g*-IL Chains and Their Microphase Separation Behaviours 291
10.3.4 Dielectric PVDF/IL–CB Nanocomposites 296
10.4 Conclusion and Outlook 300
References 300

Chapter 11 Ionic Liquids as Tools in the Production of Smart Polymeric Hydrogels 304
S. S. Silva and R. L. Reis

11.1 Introduction 304
11.2 Polymeric Hydrogels Using Ionic Liquids 305
11.2.1 Agarose 306
11.2.2 Cellulose 307
11.2.3 Chitin and Chitosan 308
11.2.4 Silk Fibroin 310
11.2.5 Xanthan Gum 311
11.3 Smart Polymeric Hydrogels 311
11.4 Conclusions 314
List of Abbreviations 315
Acknowledgements 315
References 315

Chapter 12 Preparation of Functional Polysaccharides and Related Materials Combined with Ionic Liquids **319**
J. Kadokawa

12.1 Introduction 319
12.2 Polysaccharide Ion Gels 321
12.2.1 Ion Gels of Abundant Polysaccharides with Ionic Liquids 321
12.2.2 Ion Gels of Hydrocolloid Polysaccharides with Ionic Liquids 327
12.3 Polysaccharide–Polymeric Ionic Liquid Composite Materials 332
12.3.1 Polymeric Ionic Liquids 332
12.3.2 Preparation of Polysaccharide Films Reinforced by Polymeric Ionic Liquids 334
12.3.3 Preparation of Polysaccharide–Polymeric Ionic Liquid Composites 335
12.4 Conclusion 338
Acknowledgements 338
References 339

Chapter 13 Tailoring Transport Properties Aiming for Versatile Ionic Liquids and Poly(Ionic Liquids) for Electrochromic and Gas Capture Applications **342**
Roberto M. Torresi, Cintia M. Corrêa, Tânia M. Benedetti and Vitor L. Martins

13.1 Introduction 342
13.2 Physicochemical Properties of ILs and PILs and the Effect on Transport 346
13.2.1 Density of ILs 347
13.2.2 Transport Properties of ILs and PILs 348
13.3 Ionic Liquids, Polymeric Ionic Liquids and Electrochromism 357
13.4 Transport of Gases by Ionic Liquids and Poly(ionic liquid)s: CO_2 Separation 363
13.4.1 Why Can Ionic Liquids Selectively Dissolve CO_2? 364
13.4.2 Factors Affecting CO_2 Solubility in Ionic Liquids 365
13.4.3 Supported Ionic Liquids for CO_2 Separation 368
13.5 Concluding Remarks 372
Acknowledgements 373
References 373

Chapter 14 Wearable Energy Storage Based on Ionic Liquid Gels **381**
Stephanie F. Zopf, Anthony J. D'Angelo, Huan Qin and Matthew J. Panzer

14.1 Introduction 381
14.1.1 Wearable Technology 381
14.1.2 Energy Storage for Wearable Applications 382
14.2 Ionic Liquid Gels for Energy Storage 384
14.2.1 Ionic Liquid Gels Overview 384
14.2.2 Ionic Liquid Gel Electrolytes for Battery Applications 389
14.2.3 Ionic Liquid Gel Electrolytes for Supercapacitors 394
14.3 Fabrication Techniques for Ionic Liquid Gel Integration into Wearable Systems 395
14.3.1 Device Assembly Techniques 395
14.3.2 Considerations for Integrating Ionic Liquid Gel-based Energy Storage into Wearable Systems 404
14.4 Conclusions 408
References 410

Chapter 15 Ionic Liquids in Wearable Chemical Sensors **416**
S. A. Goodchild, M. R. Sambrook and A. J. S. McIntosh

15.1 Introduction 416
15.2 Sensing with Wearable Technologies 417
15.3 The Benefits of Ionic Liquids for Use in Wearable Chemical Sensors 418
15.4 Exploiting the Selective Solvation of Ionic Liquids in Sensor Systems 419
15.4.1 Towards Selective Sampling Using Ionic Liquid Solvents 420
15.4.2 Improved Selectivity and Specificity of Sensing Strategies Achieved Using Ionic Liquids 423
15.5 Progression of Ionic Liquid Sensors Towards Viable Wearable Sensor Options 427
15.5.1 Optical Systems 428
15.5.2 Electrochemical Sensors 430
15.5.3 Skin-worn Chemical Sensors 438
15.5.4 *In situ* Environmental Detection Using Paper-based Sensors 443
15.5.5 Environmental Detection of Vapours 446
15.6 Prospects for the Future of Ionic Liquids in Smart Chemical Sensors 446
References 448

Chapter 16 Ionic Electrochemical Actuators **456**
A. Maziz, A. Simaite and C. Bergaud

16.1 Introduction 456
16.1.1 Ionic Gels 457
16.1.2 Ionic Polymer–Metal Composites 458
16.1.3 Carbon Nanotubes 460
16.1.4 Conducting Polymers 461
16.2 Volume Change in Ionic Conducting Polymers 461
16.3 Synthesis of Conducting Polymers 462
16.4 Ionic Electromechanical Actuators Based on Conducting Polymers 465
16.4.1 Actuators Immersed in an Electrolyte: Linear Deformation 466
16.4.2 Bilayer Bending Actuators 468
16.4.3 Trilayer Bending Actuators: Use of Ionic Liquids as Electrolytes for Air Working Actuators 468
16.4.4 Creeping Effects 473
16.5 Interfacing and Actuation 474
16.6 Applications 475
16.7 Conclusions and Challenges 479
References 481

Chapter 17 Capturing CO_2 with Poly(Ionic Liquid)s **489**
S. Einloft, F. L. Bernard and F. Dalla Vecchia

17.1 Introduction 489
17.2 Carbon Capture Technologies 490
17.3 Ionic Liquids (ILs) 492
17.4 Poly(Ionic Liquid)s 493
17.4.1 Poly(Ionic Liquid) Syntheses 493
17.5 Performance of PILs Synthesized by Direct Radical Polymerization of IL Monomers in CO_2 Capture and Separation 498
17.5.1 The Effect of the Cation, Anion and Backbone Structure on CO_2 Sorption 499
17.6 Performance of PILs Synthesized by Condensation Polymerization and Polymer Modification in CO_2 Capture and Separation 505
17.7 Composites (PIL–ILs) 509
Acknowledgements 511
References 512

Chapter 18 Ionic Liquid-based Polymers and Crystals for Dye-sensitized Solar Cells **515**
Chuan-Pei Lee and Kuo-Chuan Ho

18.1 Introduction to Solar Energy & Dye-sensitized Solar Cells 515
18.2 Toward All/Quasi-solid-state Dye-sensitized Solar Cells *via* Ionic Liquid Electrolytes 517
18.2.1 Polymeric Ionic Liquids for Solid-state Dye-sensitized Solar Cells 518
18.2.2 Ionic Liquid Crystals for Solid-state Dye-sensitized Solar Cells 522
18.3 Summary 527
Acknowledgement 528
References 528

Subject Index **531**

CHAPTER 1

Polymerization in Ionic Liquids

NIKHIL K. SINGHA[a], KUNLUN HONG[b] AND JIMMY W. MAYS*[c]

[a]Rubber Technology Centre, Indian Institute of Technology, Kharagpur 721302, India; [b]Center for Nanophase Materials Sciences, Oak Ridge National Laboratory, Oak Ridge, TN 37831, USA; [c]Department of Chemistry, University of Tennessee, Knoxville, TN 37996, USA
*E-mail: jimmymays@utk.edu

1.1 Introduction

Most liquids used as solvents are composed of neutral molecules. In contrast, ionic liquids (ILs) are salts in the liquid state at ambient or near ambient temperatures. This room temperature ionic liquid state is often achieved by choosing ion pairs where one is organic and has a delocalized charge, or by choosing bulky asymmetric substituents. This causes the ions to be poorly coordinated, resulting in low melting temperatures. In principle, literally millions of ionic liquids with an exceptionally wide range of properties can be produced. This has led to ILs being considered as designer solvents or task specific solvents, where their extremely low vapor pressures offer potential to minimize pollution associated with volatile organic compound (VOC) solvents through recycling.[1]

Some examples of common IL cations and anions and their general characteristics are shown in Figure 1.1. Imidazolium- and pyridinium-based ILs feature delocalized cations, whereas quaternary ammonium cations are asymmetrically substituted. A very wide range of properties, including viscosity, hydrophobicity or hydrophilicity, thermal and chemical stability,

Smart Materials No. 29
Polymerized Ionic Liquids
Edited by Ali Eftekhari

Published by the Royal Society of Chemistry, www.rsc.org

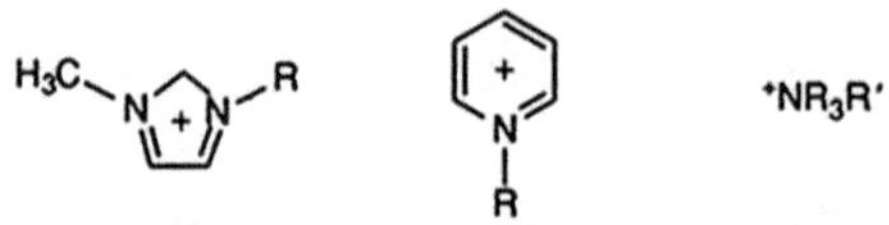

			$^{+}NR_3R'$
Cl^-	Easy to synthesize, hydrophilic	Easy to synthesize, hydrophilic	Inexpensive, hydrophilic
NO_3^-	Anion can produce reactive RTIL	Hydrophilic	Inexpensive, hydrophilic
PF_6^-	High thermal stability	Less thermally stable	Dopant RTIL to decrease cost
$N(CF_3SO_2)_2^-$	Low viscosity, hydrophobic, expensive	Hydrophobic	Low viscosity

Figure 1.1 Some common IL cations and anions and their properties.

melting point, flammability and cost, can be tuned by judicious selection of the cation and anion.

Probably the first room temperature IL, ethylammonium nitrate (m.p. 12 °C), was reported by Walden[2] in 1914. However, there was little interest in ILs until the mid-1970s when they attracted attention as electrolytes for batteries.[3,4] Nowadays, ILs are being intensely investigated in a wide range of applications, including use as solvents in a wide range of chemical processes.[5] ILs have been used as solvents for Diels–Alder reactions,[6] hydrogenations,[7] alkylations,[8] Friedel–Crafts reactions,[9] Heck reactions,[10] Suzuki couplings,[11,12] metathesis reactions,[13,14] and many others.

To the best of our knowledge, the first polymerization in ILs was reported in 1990 by Carlin *et al.*[15] They reported that $TiCl_4$ and $AlEthylCl_2$ in $AlCl_3$/1-ethyl-3-methylimmidazolium chloride ([EMIM]Cl) could polymerize ethylene in low yields. Subsequent work by the same group, where $TiCl_4$ was replaced by Cp_2TiCl_2, gave higher yields.[16] This pioneering work, using ILs as a reaction medium for polymerization, has inspired numerous researchers over the past quarter of a century to investigate a wide range of different types of polymerization in ILs. While much of this work was inspired by the "green" aspects of ionic liquids (very low vapor pressure and potential for recycling), it quickly became apparent that chemistry could often proceed differently (faster polymerization rates, higher molecular weights, enhanced yields, *etc.*) in ILs. Providing a review of the field of polymerization in ionic liquids, with particular attention to developments over the past several years, is the subject of this chapter. The reader is referred to earlier reviews in this field for additional details on work in this area.[17–23]

1.2 ILs in Conventional Free Radical Polymerization

Free radical polymerization, because of its compatibility with a wide range of monomers having different types of functional groups, is one of the most widely used polymerization techniques. Free radical solution polymerization is of great commercial importance as dilution of radical polymerizations with a solvent, typically a VOC, is effective in controlling viscosity and the exotherm accompanying polymerization. Hong *et al.*[24] noted large increases in the rate of polymerization and much higher molecular weights for free radical polymerization of methyl methacrylate (MMA) in $[BMIM]PF_6$ as compared to polymerizations carried out under identical conditions in VOCs. These effects were attributed at least in part to the high viscosity of the polymerization medium. A "diffusion-controlled termination" mechanism was proposed to explain the decreased rate of chain termination in these viscous systems. A decrease in termination rate could explain a simultaneous increase in the rate of polymerization and molecular weight.[24] In contrast to variations in the rate of polymerization and molecular weight, the polymers synthesized in RTILs have similar glass transition temperatures and microstructures as compared to those obtained in benzene or in bulk, based upon thermal analysis and ^{13}C-NMR experiments.[25] Since then, other groups[26–28] have reported similar behavior, high molecular weights and rapid polymerization rates, for MMA and other methacrylates in $[BMIM]PF_6$. Harrison *et al.*[29,30] used pulse laser polymerization (PLP) techniques to polymerize MMA in $[BMIM]PF_6$. They found that both the propagation and termination rates, k_p and k_t, respectively, were strongly affected by the presence of $[BMIM]PF_6$. They attributed the increase in propagation rate to the high polarity of the ionic liquid solution, which reduces the activation energy of propagation *via* charge–transfer interactions. The termination rate decrease was attributed to the increased viscosity of the polymerization medium. Both the increase of k_p and the decrease of k_t combine to account for a ten-fold increase in the overall rate of polymerization. Subsequent PLP studies[31,32] have reported k_t to be decreased by an order of magnitude and k_p to be increased by a factor of 4 for free radical polymerization in ILs.

Polenz and co-workers recently studied the polymerization of MMA co-initiated by imine bases and found that the reaction was accelerated greatly by even a trace amount of IL.[33] They demonstrated that the polymerization proceeded *via* a free radical mechanism, with the addition of IL decreasing the activation energy of polymerization and increasing the rate of polymerization. These effects were attributed to interactions between the IL and the imine base. Cheng *et al.*[34] studied free radical polymerization of acrylonitrile in $[BMIM]BF_4$ using AIBN as an initiator. This team found that ionic liquids are excellent media for obtaining high molecular weight polymers. They attribute this finding to the low chain transfer constants for ILs and their ability to stabilize growing radical chain ends. Puttick *et al.*[35] used NMR to investigate nanoscale domains formed in dialkylimidazolium ILs. When polymerizing MMA in this type of IL, it was shown that the reactants and intermediates have different affinities for nanodomains that form within the IL. Segregation of different species within these domains accounts for the unusually high polymerization rates and increased molecular weights.

Many early studies on free radical polymerization in ILs used $[BMIM]PF_6$ due to its commercial availability and ease of synthesis. However, $[BMIM]PF_6$ can hydrolyze to generate toxic HF, and it is desirable to use more benign ILs in polymer systems. Zhang and co-workers[36] measured the viscosity and polarity of many different ILs containing a wide range of cations and anions, and attempted to correlate these physical properties with their behavior in the free radical polymerization of styrene and MMA. This study, which limited polymer conversion in order to avoid the Trommsdorf effect, showed no trends between the viscosity or polarity of the IL and the polymerization rate or molecular weight of the polymer. Strehmel and co-workers[37] studied free radical polymerization of *n*-butylmethacrylate in a wide range of ILs, imidazolium, pyridinium, and alkyl ammonium salts. They reported that the use of ILs results in higher molecular weights, which can even exceed those obtained in bulk radical polymerization. They also found increased glass transition temperatures (in contrast with prior studies) and reported improved thermal stability, although the microstructure (tacticity) reportedly did not change. Their results show that increased molecular weight is a general phenomenon observed for free radical polymerization in ILs. They also found that high molecular weights are favored by use of high viscosity imidazolium salts or by conducting polymerizations close to the melting temperature of the IL. They hypothesized that this behavior is due to locally ordered structures in the ILs, which favor propagation (their hypothesis is supported by later work, ref. 35). This team later studied copolymerization of nonpolar and zwitterionic methacrylates in ILs[38] as well as the use of 1-alkyl-3-methylimidazolium tosylates as solvents for polymerization of *n*-butylmethacrylate and styrene, where high yields and high molecular weights were obtained.[39] Styrene polymerization by iron-containing imidazolium-based ILs was recently reported.[40] The ILs showed high catalytic activity, with 2-3 catalytic species being present as evidenced by very broad multimodal molecular weight distributions.

IL gels for ionic transport and separations were synthesized by free radical copolymerization of 2-acrylamido-2-methyl-1-propanesulfonic acid and bisacrylamide in 1-ethyl-3-methylimidazolium ethylsulfate.[41] Improved grafting of polystyrene onto carbon black was reported by *in situ* free radical polymerization of styrene in the presence of carbon black. The improvement was believed to be due to the high viscosity of the ionic liquid solvent.[42] New ILs featuring hydrolytically stable anions have been studied.[43] ILs featuring four anions, tris(pentafluoroethyl)trifluorophosphate, trifluoromethanesulfonate, trifluoroacetate, and tetracyanoborate, were used as solvents in free radical polymerization, and the nature of the anion had a significant effect on the polymerization process.[43] Free radical polymerization of vinyl fluoride was recently reported to be enhanced through polymerization in ILs.[44] ILs containing alkylimidazolium cations and fluorinated anions gave poor conversions but polymerization in the presence of the lithium salts of the anions gave high conversions and high molecular weights at low pressures.

The use of ILs in free radical photopolymerization has been investigated by Dietlin and co-workers.[45] The efficiency of four Type I photoinitiators in ILs was investigated by kinetics, spectroscopic and photolysis studies. Polymerizations initiated by three of the photoinitiators were faster in ILs, but in the fourth case there was a significant decrease in reaction rate. This result underlines the need to match the photoinitiator to the ionic liquid.[45]

Addition of lithium bistriflimide RTIL to 1-vinylimidazole significantly improved both the monomer conversion and photopolymerization rate, likely due to coordination between the imidazole and the Li^+ cation.[46] Photopolymerization of PEG-methacrylate was investigated in [BMIM]BF_4 and 1-ethyl-3-methylimidazolium trifluoroacetate in order to probe the effect of the IL on the propagation and termination rates.[47] The use of [BMIM]BF_4 gave a greater increase in k_p due to its stronger Coulombic cation–anion attractions and the lower H-bond accepting ability of the BF_4^- anion. The reductions in k_t were believed to be caused by the high viscosities of the systems explored.

The insoluble nature of some polymers in ILs creates the opportunity to make diblock copolymers by simple sequential monomer addition. Zhang and co-workers[48] synthesized PS-*b*-PMMA by sequential addition of styrene, then MMA, in [BMIM]PF_6 through conventional free radical polymerization using benzoyl peroxide (BPO) as an initiator. The PS first block gradually precipitated out when the conversion reached around 50% due to the insolubility of PS in [BMIM]PF_6. The collapsed PS coils wrapped around the macroradicals, resulting in extended lifetimes due to diminished termination. Unreacted styrene was removed using a high vacuum line. After adding MMA, a diblock copolymer was formed at room temperature, although the re-initiation was not 100% (see Figure 1.2). Reversing the polymerization sequence resulted in a PMMA homopolymer.

The effect of IL solvents on reactivity ratios and sequence length distributions for statistical free radical copolymerization has been extensively studied. The calculated reactivity ratios for St and MMA (rSt = 0.381 ± 0.02 and rMMA = 0.464 ± 0.02) in [BMIM]PF_6 are significantly different from those (rSt = 0.54 ± 0.04 and rMMA = 0.50 ± 0.04) in benzene at 60 °C[49]. The "boot-strap" model,[50]

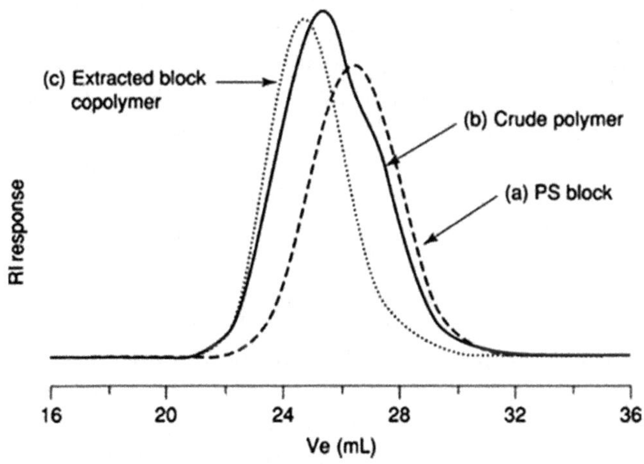

Figure 1.2 Size exclusion chromatograms for (a) the polystyrene first block, (b) the crude PS-b-PMMA product, and (c) purified PS-b-PMMA after extraction with acetonitrile and cyclohexane. Reproduced with permission from ref. 48. Copyright (2002) American Chemical Society.

polarity of the solvents, interaction between solvent and monomers, viscosity, and system heterogeneity all possibly contribute to the different reactivity ratios observed in ILs.

1.3 ILs in Controlled Free Radical Polymerizations

Over the past 25 years there has been tremendous interest, primarily in academia, in controlled radical polymerizations. These systems are sometimes called "living radical polymerizations", but they are not truly living systems in the manner that some anionic polymerizations are living (the latter can be truly termination-free). Controlled radical polymerizations employ an equilibrium between dormant chain ends and active radical chain ends, with the equilibrium heavily in favor of the dormant species, in order to minimize termination in the system, which occurs through radical coupling and/or disproportionation. The result is the ability to target precise molecular weights and obtain polymers with relatively narrow molecular weight distributions, and the ability to synthesize block copolymers by sequential monomer addition. There are three main ways of achieving this equilibrium, nitroxide mediated polymerization (NMP), atom transfer radical polymerization (ATRP), and reversible addition–fragmentation chain transfer (RAFT) polymerization. These polymerization techniques have been thoroughly reviewed.[51]

The first study of a controlled radical polymerization was carried out by Carmichael and co-workers.[52] They conducted ATRP of MMA using ethyl-2-bromoisobutyrate as an initiator and CuBr/N-propyl-2-pyridylmethanimine as the catalyst in [BMIM]PF_6. The reaction was found to be faster as compared to the same reaction carried out in VOCs, with narrow polydispersity indices (1.30–1.43). By extraction with toluene, the polymer was separated from the polymerization solution. The copper catalyst remained in the IL phase, which allowed catalyst recycling. The use of the RTIL medium also eliminated the need for post-purification to eliminate residual copper salts, which entails passing the polymer solution through purification columns. Carmichael and co-workers[52] attributed this polymerization behavior to the increased polarity of ionic liquids, because a similar increase in the rate has been observed with other polar/coordinating solvents.

Biedron and Kubisa[53] reported ATRP in [BMIM]PF_6 for various alkyl acrylates (methyl, butyl, hexyl, and dodecyl). The resulting polymers ranged from soluble to insoluble in this IL depending on the length of the alkyl substituent. For the heterogeneous systems, the alkyl acrylate formed an upper monomer phase while the CuBr/pentamethyldiethylenetriamine (PMDETA) catalyst remained in the lower IL phase. Methyl acrylate (MA) and poly(methyl acrylate) (PMA) are miscible with [BMIM]PF_6 and form a homogeneous polymerization reaction mixture where all reactions proceed in one phase. For the three other acrylates, the growing macromolecular chains react with the monomer at the interface but reside predominantly in the monomer phase during polymerization. All the polymerizations were reasonably well-controlled. Heterogeneous ATRP facilitates easy separation of polyacrylate from the IL phase after the reaction, with less copper catalyst contamination and reduction of side reactions due to the absence of catalyst in the upper monomer phase, as compared to bulk ATRP. It was discovered by this same group that a chiral IL (1-(R-(+)-2′-methylbutyl)-3-methylimidazolium hexafluorophosphate) could regulate the stereoregularity of the polymer.[54]

Sarbu and Matyjaszewski[55] used both iron and copper halide catalyst systems for ATRP in a range of 1-butyl-3-methylimidazolium ionic liquids. In iron-mediated ATRP, no added ligand was necessary for controlled polymerization of MMA although both the initiation and polymerization rates were low. Systems without organic ligands were effective only in a phosphonate ionic liquid for copper-mediated ATRP of MMA, and a ligand was also necessary in ionic liquids with halide or carbonate anions. ATRP in ionic liquids is characterized by low initiation efficiency. This can be attributed to the high concentration of the catalyst in the IL phase, thus generating a high concentration of free radicals which can undergo termination. The initiator efficiency was improved by employing macroinitiators, which show a greatly reduced tendency to diffuse into the IL phase. An advantage in these systems is that the catalyst can be readily regenerated after polymerization.

RAFT polymerization of MMA in [BMIM]PF_6, [BMIM]BF_4, and [DMIM]BF_4 (1-dodecyl-3-methylimidazolium tetrafluoroborate) has been reported.[56,57] Reverse ATRP of MMA was conducted in [BMIM]PF_6 using 2,2′-azobisisobutyronitrile (AIBN)/$CuCl_2$/2,2′-bipyridine initiation, which is a system that remains homogeneous throughout the reaction. The high viscosity of the IL may lead to a cage effect, causing termination of radicals formed by decomposition of AIBN before they can initiate polymerization, thus accounting for the low initiation efficiency. The Cu catalyst is soluble in the IL, and thus less catalyst is needed to effectively mediate the ATRP process in ILs than in other solvents.

ILs have proven to be excellent solvents for organotellurium-mediated living radical polymerization (TERP) of various monomers.[58] Polymerization in ILs significantly increased the rate of polymerization. For MMA as the monomer, essentially complete polymerization was achieved in <1 h with polydispersity indices lower than 1.1. The high level of control was attributed to the faster degenerative chain transfer reaction, which critically controls the polydispersity in TERP. [BMIM]PF_6 was used as a solvent in supplemental activator and reducing agent (SARA) for ATRP of methyl acrylate, allowing for very fast polymerizations yielding products with very low polydispersities and tunable chain end functionalities.[59] ILs have also been employed as solvents in ATRP using activators regenerated by electron transfer (ARGET).[60,61]

Biedron and Kubisa made PBA-*b*-PMA and PMA-*b*-PBA block copolymers (PBA is poly(butyl acrylate)) with different addition sequences of monomers using ATRP in [BMIM]PF_6.[62] When conversion of the first monomer, methyl acrylate, exceeded 70%, the diblock copolymer contained a considerable amount of PMA homopolymer. However, when methyl acrylate was added to "living" PBA in [BMIM]PF_6, a diblock containing no detectable homopolymer was formed. Ma *et al.*[56,57] described the synthesis of block copolymers where styrene was polymerized using chlorine-end-capped PMMA as a macroinitiator through reverse ATRP in [BMIM]PF_6, [BMIM]BF_4, and [C_{12}MIM]BF_4.

Nguyen and Shimm[63] reported IL-mediated synthesis and self-assembly of poly(ethylene glycol) (PEG)–PS diblock copolymers in [BMIM]PF_6 using ATRP. The same group[64] also reported IL-mediated synthesis of poly(2-hydroxyethyl methacrylate-*block*-methyl methacrylate)/Fe_3O_4 core–shell nanocomposites by surface-initiated ATRP in [BMIM]PF_6. A thermally regulated phase-separating catalyst system for ATRP has been developed based on a thermo-regulated IL[65]. The system allows the catalyst to be easily recovered and re-used multiple times without significant loss of activity.

Recently, Wang *et al.*[66] studied photo-induced ATRP of MMA in an IL using 2-bromoisobutyrate as the initiator, $CuBr_2$/pentamethyldiethylenetriamine as the catalyst, and TiO_2 as the photocatalyst. Well-controlled polymerizations were achieved and the resulting PMMA had low polydispersity indices. The polymerizations could be manipulated by turning the light on or off. A very thorough investigation of photoinduced ATRP of several acrylates in a variety of ILs was reported by Davis, Haddleton, and their co-workers.[67] Many of the polymerizations were well-controlled and gave products of predictable molecular weights and low polydispersities in quantitative yields within 30 min. All of the polymerizations were much faster than the corresponding polymerizations in DMSO, while maintaining high levels of end group fidelity.

In terms of NMP in ILs, Zhang *et al.*[68] employed both benzoyl peroxide (BPO) + 2,2,6,6-tetramethylpiperidine 1-oxyl (TEMPO) and the universal initiator system (2,2,5-trimethyl-3-(1-phenylethoxy)-4-phenyl-3-azahexane (TMPPAH)) for NMP of MMA and styrene. While polymers were obtained, the polymerizations were not controlled with either monomer, possibly due to the low diffusion rates of the mediating radicals and/or slow degradation of free TEMPO at elevated temperatures in the presence of [BMIM]PF_6. Ryan and co-workers[69] demonstrated the first controlled NMP of MA in 50% v/v [HMIM]PF_6 initiated by AIBN + 4-oxo-2,2,6,6-tetramethyl-1-piperidinyl-N-oxyl (4-oxo-TEMPO) at higher temperature (~150 °C). The polymerization rates in the IL were greater than for reactions in anisole. In both studies, thermal polymerization occurred because of the high temperatures used. Later, the BlocBuilder alkoxyamine was used for NMP of MMA at 90 °C[70], with a high level of control, and greater overall polymerization rate and conversion. Both are consistent with an increase in k_p and a decrease in k_t.

Other types of controlled free radical polymerization, such as RAFT and charge transfer polymerization, have also been reported using ILs as the solvent.[71–75] Zhang and Zhu[76] recently described the use of ILs as media for the preparation of vesicles *via* polymerization-induced self-assembly. RAFT was used in dispersion polymerization with styrene, BMA, and HEMA using a PEG-RAFT agent. The block copolymers produced spontaneously self-assembled into vesicles. Singha *et al.*[77] reported the use of RAFT polymerization in IL to prepare tailor-made poly(furfuryl methacrylate). The polymerization was extremely fast, yielding well-defined functional polymers that were

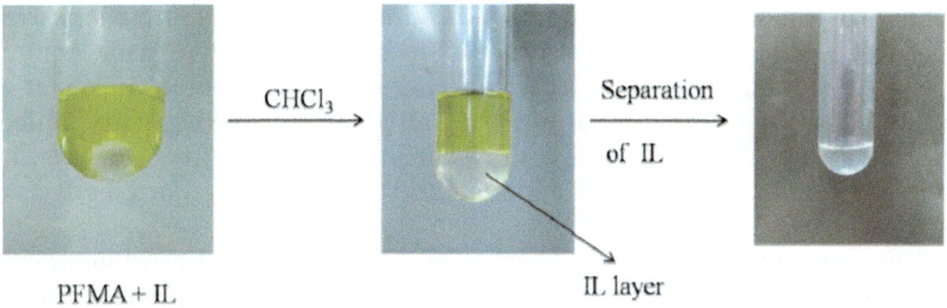

Figure 1.3 Recovery and reuse of the IL used in the RAFT polymerization of furfuryl methacrylate. Reproduced from ref. 77 with permission from The Royal Society of Chemistry.

capable of reversible crosslinking with maleimide-functionalized POSS *via* Diels–Alder and Reverse Diels–Alder reactions. The IL was readily recovered from the polymerization mixture for recycling (Figure 1.3). Use of dispersion RAFT polymerization in ILs to prepare block copolymer nano-objects has been reported recently.[78]

1.4 ILs in Ionic Polymerizations and Group Transfer Polymerizations

Living anionic polymerization usually gives much narrower molecular weight distribution polymers as compared to those obtained by controlled or sometimes called "living" free radical polymerizations. Anionic polymerizations can proceed to complete conversion with far fewer side reactions, such as termination and chain transfer, as compared to any known radical polymerization process. The ability to quantitatively introduce terminal functional groups by using selective termination agents is an additional advantage of anionic polymerization. However, anionic polymerization requires rigorous purification of reagents and rigorous synthesis conditions if living polymerization is to be achieved.[79] In addition, judicious choices of initiators and reaction conditions are required.

To date there has been little success in conducting living anionic polymerizations in ILs. Biedron and Kubisa[80] first described anionic polymerization of MMA in an IL. Their work involved *n*-BuLi in [BMIM]BF_4 at 0 °C as an initiator, and they obtained low molecular weight polymers with high polydispersity and end-capped with an imidazolium moiety. Kokubo and Watanabe[81] investigated *n*-BuLi and 1,1-diphenylhexyllithium as initiators for MMA polymerization in [BMIM]PF_6 at 0 °C. As with the prior reaction, low yields of polymer with high polydispersity were obtained, likely due to the relatively high temperature employed and the reaction of the anionic centers with the imidazolium cation. Anionic polymerization of styrene in hydrophobic ionic liquids was reported by Vijayaraghavan *et al.*[82] with low to moderate yields of polydisperse polymers obtained, indicating that the polymerizations were not living. Diaz de Leon and Cordova[83] recently carried out living anionic polymerization of butadiene in cyclohexane with IL present at molar ratios of 8–20 relative to the *n*-BuLi initiator. Fairly narrow polydispersity indices of ~1.25 were obtained along with molecular weight control. However, no advantage of using the IL was apparent and there was no reported effect on polydiene microstructure.

Vijayaraghavan and MacFarlane[84] were the first to report cationic polymerization in an IL, specifically styrene in *N*-butyl-*N*-methylpyrrolidinium bis(trifluoromethanesulfonyl)amide ($[P_{14}][Tf_2N]$) using bis(oxalato)boric acid as an initiator. The polymerization proceeded in a controlled manner at 60 °C to yield products with low molecular weights (1300–1700) and moderate polydispersity indices (1.3–1.5). The IL was believed to promote the dissociation of the acid and initiate polymerization. Biedron and Kubisa[85] cationically polymerized styrene in $[BMIM]PF_6$ using 1-phenetyl chloride/$TiCl_4$ as an initiator. Polymers were obtained but in the presence of chain transfer, thus there was limited control over molecular weight and the products exhibited broad polydispersity. Recently Han *et al.*[86] studied cationic polymerization in $[BMIM]PF_6$ using cumylchloride as an initiator with various co-initiators. Compared with polymerizations in VOC solvents, the polymerization in IL proceeded under milder exothermic conditions. The mechanism of styrene polymerization in $[BMIM]PF_6$ was discussed. New initiators allowing rapid cationic polymerization of styrene in ILs have been reported.[87] Cationic polymerization of *p*-methylstyrene in various ILs has been reported very recently.[88] Controlled polymerization was obtained in $[Bmim][NTf_2]$ using a $CumOH/BF_3OET_2$ initiator at −25 °C in the presence of 2,6-di-*tert*-butylpyridine, and the polymerization mechanism was proposed.

Cationic polymerization of isobutyl vinyl ether in ILs has also been reported recently by two groups.[89,90] Wu *et al.*[89] studied polymerization in $[Omim][BF_4]$ using various co-initiators with IBVE-HCl as an initiator. The polymerizations were faster than in VOCs under similar conditions, resulting in higher molecular weights and conversions. However, the polymerizations in $[Omim][BF_4]$ were not well-controlled due to beta proton elimination. Yoshimitsu *et al.*,[90] using $[Omim][NTf_2]$ as the IL, 1-(isobutoxy)ethyl acetate/$TiCl_4$ as an initiator, and 2,6-di-*tert*-butylpyridine as a proton trap, achieved controlled polymerization of IBVE and polymers with fairly low polydispersity indices. Polymerization was much faster than in CH_2Cl_2. Group transfer polymerization in ILs has also been reported.[91,92]

1.5 Ring Opening Polymerization

A very useful ring-opening metathesis polymerization of norbornene in a biphasic medium (1-butyl-2,3-dimethylimidazolium hexafluorophosphate ($[BDMIM]PF_6$)/toluene) has been reported.[93] The catalyst resides in the IL phase, and the toluene phase dissolves the polymer product. The IL and catalyst could be recycled six times giving quantitative conversion and without significant loss of catalytic activity. Biedron *et al.*[94] reported cationic ring opening polymerization of 3-ethyl-3-hydroxymethyloxetane in [BMIM][BF_4] using BF_3–Et_2O initiation. The use of an IL has the advantage of allowing high reaction temperatures (up to 180 °C), reducing intermolecular hydrogen bonding that leads to the formation of aggregates. Conversely, the use of ILs can facilitate reduced reaction temperature, avoiding side reactions that occur at higher temperatures. Polycarbonate made by ring opening polymerization of ethylene carbonate undergoes decarboxylation at high temperatures of 180–200 °C[95]. In acidic ILs, however, the reaction can take place at much lower temperatures of 100–120 °C depending on the IL used.

ILs have been used in the microwave-assisted polymerization of ε-caprolactone.[96] Phosphonium-based ILs containing different ions were used to modify layered double hydroxides (LDHs) followed by microwave irradiation. LDHs with intercalated IL anions were exfoliated due to caprolactone polymerization within 3 min. Imidazolium IL alcohols have been employed as initiators for ring opening polymerization of ε-caprolactone, as well as formation of its block copolymers.[97]

1.6 Transition Metal-catalyzed Polymerizations

Poly(phenyl acetylene) (PPA) has been obtained by Rh(I)-catalyzed phenylacetylene polymerization in ILs including *n*-butylpyridinium tetrafluoroborate ([bupy]BF_4) and [BMIM]PF_4.[98] The catalyst was (diene)Rh(acac) or [(diene)$RhCl_2]_2$, while the co-catalyst was triethylamine. The polymer was separated from the IL by extraction with toluene or by filtration after adding methanol into the IL to form a suspension of PPA. The PPA obtained had a molecular weight of 55 000 to 200 000. Rogers *et al.*[99] used a mixed solvent, 1-hexylpyridinium bis(trifluoromethanesulfonyl) imide ([C_6Pyr][NTf_2])/methanol, for palladium-catalyzed alternating copolymerization of carbon monoxide and styrene. The IL and catalyst could be recycled, and the yields and molecular weights were higher than those when the reaction was carried out in methanol alone. Catalyst stability and propagation rate were improved due to the inhibited chain transfer and catalyst decomposition in the IL. Biphasic IL systems for metallocene-catalyzed polymerization of ethylene have been described.[100,101]

Ionic liquid-modified silica has been used in combination with metallocenes for ethylene polymerization.[102] Yim *et al.*[103] reported metallocene catalysts supported on aminosilane and IL-functionalized silica for ethylene polymerization. Ethylene polymerization using vanadium catalysts supported on silica in the presence of pyridinium ILs has also been reported.[104] The resulting PE was a linear polymer of very high molecular weight.

Metathesis step-growth polymerization in ILs was carried out, where the high boiling point of the IL allowed low pressures to be achieved at high temperatures.[105] Short reaction times yielded high molecular weight polymers. Acyclic diene metathesis polymerization (ADMET) and Suzuki coupling polymerization have also been performed in ILs.[106,107] A short review on transition metal-catalyzed oligomerization and polymerization in ILs has recently been published.[108]

1.7 Electrochemical Polymerization

Electrochemical polymerization is a major method employed in the synthesis of conducting polymers for applications in energy storage, electrochromic devices, and light-emitting diodes. This process has advantages over chemical synthesis, including faster reaction times, generation of the polymer in the doped state, and facile control over film thickness. Naudin *et al.*[109] have used ILs in electrochemical polymerization of poly(3-(4-fluorophenyl)thiophene) (PFPT). The electrochemical behavior was similar to that observed in common non-aqueous electrolytes, however some IL residue remained in the formed film. ILs were used as solvents in electrochemical synthesis of poly(3-octyl thiophene), resulting in an excellent yield of 99%.[110] Other kinds of conjugated polymers, such as polypyrrole, polyaniline, and unsubstituted polythiophene, have also been synthesized in ILs.[111–123]

1.8 Step-growth Polymerization

Most step-growth polymerizations are carried out in solvent-free or near solvent-free systems because of their sluggish polymerization kinetics and the need to go to elevated temperatures in order to achieve high conversions and high molecular weights. However, polyimides and polyamides, important materials due to their exceptional thermal and mechanical properties, are sometimes synthesized in solution at high temperatures. Vygodskii and co-workers[124] studied their step-growth polymerization in various ILs, and high molecular weight polyimides and polyamides were obtained without the addition of a catalyst. Subsequently, ILs have been shown to be effective solvents and catalysts for a wide range of step growth polymerizations, including polyamides,[125,126] polyesters,[127,128] polyethers,[129–131] poly(ether ketones),[132] poly(ether sulfones),[133] polyurethanes,[134] poly(amino acids),[135] and poly(amide imides).[136,137] Recently a sustainable process for preparation of high-performance thin film composite aromatic polyamide membranes, using ILs as the reaction medium, was reported.[138]

1.9 Enzymatic Polymerization

Enzymatic polymerization in IL media has attracted increasing attention in recent years due to the growing interest in polymers from renewable resources. The first such report was Candida Antarctica lipase (lipase CA)-catalyzed formation of biodegradable polyesters in $[BMIM]PF_6$ and $[BMIM]BF_4$.[139] Ring-opening polymerization of ε-caprolactone and polycondensation of dicarboxylic acid diesters with 1,4-butanediol were explored. The resulting higher molecular weight products and improved conversions show the potential for green polymer chemistry *via* a combination of non-toxic enzyme catalysts, mild reaction conditions, and potentially environmentally benign IL solvents. Further studies on enzymatic polymerization in ILs have been applied to lactones,[140–143] lactides,[140,144] and water soluble polyanilines.[145] Very recently the enzymatic polymerization of phenols catalyzed by chloroperoxidase in the presence of ILs has been described.[146] The results show that the presence of even low levels of IL improved the efficiency of the process, and a mechanism for the polymerization was also proposed.

1.10 ILs as Solvents for Grafting from Biopolymers

Because of their ability to dissolve cellulose and other biopolymers, ILs have enjoyed extensive recent utilization in the synthesis of graft copolymers based on cellulose,[147–158] chitosan,[159–161] soy protein isolate,[162] xylan,[163–165] and biomass.[166] A very wide range of grafting chemistries have been employed and some recent short reviews of these developments are available.[167,168] IL technology also enables facile synthesis of familiar cellulose derivatives, such as cellulose acetate.[169]

1.11 Other Applications of ILs in Polymerization

Very recently, the use of ILs in emulsion polymerization processes has been explored. Polyaniline doped with decylbenzenesulfonic acid having different morphologies and compositions was made by inverse emulsion polymerization.[170] Polymerization was carried out in toluene using ammonium peroxydisulfate in the presence of imidazolium ILs. The influence of ILs on the morphology and particle size was investigated by dynamic light scattering and scanning electron microscopy. The presence of certain ILs has been shown to have a strong effect on the polymerization rate in conventional emulsion polymerization.[171] Addition of $[Omim]PF_6$ had a marked effect on the reaction rate at low levels when AINB was used as an initiator, but not when hydrogen peroxide was used. The use of ILs in microemulsions as templates for the synthesis of nanostructured polymer materials has been reviewed.[172]

Molecularly imprinted polymers with increased affinity for oleanic acid were prepared using IL as a pyrogen.[173] The type of macromolecular crowding agent used and the nature of the IL had a strong impact on the imprinting properties of the molecularly imprinted polymers. The resulting template could purify oleanic acid from crude extracts efficiently with 72% recovery.

Jet fuel paraffins have been produced by low temperature polymerization of gaseous olefins using ILs.[174] Paraffins with the desired 8–15 carbons were produced using $[Bmim][A_{12}Cl_7]$ at ambient pressure. The influence of the reaction conditions on the polymerization was investigated.

ILs have been employed for depolymerization as well. Mouawia and co-workers[175] described the use of olefin metathesis reactions in ILs to controllably depolymerize natural rubber with low polydispersity telechelic polymers as the product. The process (Figure 1.4) was successfully applied to waste automobile tires and is thus extremely promising for utilization of a major polymer waste product. Iannone *et al.*[176] employed ILs and ZnO

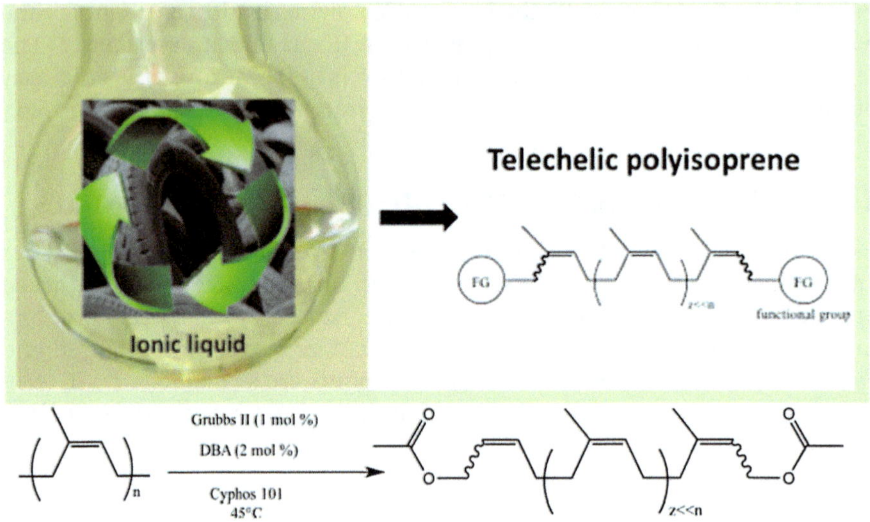

Figure 1.4 Recovery of oligomeric polyisoprene from tire waste by using a Grubbs catalyst in an IL. Reproduced with permission from ref. 175. Copyright (2017) American Chemical Society.

nanoparticles in the presence of various nucleophiles (amines, alcohols, water, *etc.*) to depolymerize polycarbonate, leading to complete recovery of the bisphenol A (BPA) monomer. When polycarbonate was treated with glycerol, only BPA and glycerol carbonate, useful as a fuel additive and industrial solvent, were obtained.

The combination of ILs and microwave irradiation as tools in polymer synthesis has been reviewed recently by Tarasova *et al.*[177] These processes were demonstrated to exhibit high resource efficiency and show great promise as energy efficient processes for polymer synthesis in the future. The use of ILs to incorporate novel multi-responsive character into polymer systems by doping, grafting, and polymerization has been reviewed.[178,179]

1.12 Conclusions and Future Prospects

The use of ILs as media for polymerization has seen tremendous growth over the past 15 or so years. ILs have now been employed advantageously in nearly every major type of polymerization. While initial interest in using ILs as solvents was to replace VOC solvents and reduce pollution, their capacity for altering the fundamental chemistry of various reactions is now obvious. ILs have been shown to give a combination of very fast polymerization and very high molecular weight products in conventional free radical polymerization. ILs have also been demonstrated to accelerate controlled radical polymerizations and emulsion polymerizations. The use of ILs to create trapped radicals has facilitated simple production of block copolymers using conventional free radical polymerization. ILs have shown their utility even in step-growth polymerizations by allowing reactions to be carried out at high temperatures due to their very low vapor pressures. ILs also provide the capacity for incorporating biological species (proteins, enzymes, *etc.*) into polymer systems in a non-denaturing environment. Essentially all types of polymer production processing have now been advantageously affected by the use of ILs. ILs are now also showing their potential as media for the recycling of polymer waste *via* depolymerization processes. In particular, their use in the recycling of automobile tire waste is tackling a critical environmental need.

Acknowledgements

KH acknowledges support from the US Department of Energy, Basic Energy Sciences, for support of the Center for Nanophase Materials Sciences at Oak Ridge National Laboratory.

References

1. T. Welton, *Chem. Rev.*, 1999, **99**, 2071.
2. P. Walden, *Bull. Acad. Imp. Sci. St.-Petersbourg*, 1914, **8**, 405.
3. H. L. Chum, V. R. Koch, L. L. Miller and R. A. Osteryoung, *J. Am. Chem. Soc.*, 1975, **97**, 3264.
4. J. S. Wilkes, J. A. Levisky, R. A. Wilson and C. L. Hussey, *Inorg. Chem.*, 1982, **21**, 1263.
5. M. Freemantle, *An Introduction to Ionic Liquids*, RSC Publishing, Cambridge, 2009.
6. T. Fischer, A. Sethi, T. Welton and J. Woolf, *Tetrahedron Lett.*, 1999, **40**, 793.
7. Y. Chauvin, L. Mussmann and H. Olivier, *Angew. Chem., Int. Ed.*, 1995, **34**, 2698.
8. M. Badri, J. J. Brunet and R. Perron, *Tetrahedron Lett.*, 1992, **33**, 4435.
9. A. Stark, B. L. MacLean and R. D. Singer, *J. Chem. Soc., Dalton Trans.*, 1999, 63.
10. D. E. Kaufmann, M. Nouroozian and H. Henze, *Synlett*, 1996, 1091.
11. C. J. Mathews, P. J. Smith and T. Welton, *Chem. Commun.*, 2000, 1249.
12. R. Rajagopal, D. V. Jarikote and K. V. Srinivasan, *Chem. Commun.*, 2002, 616.
13. Q. Yao and Y. Zhang, *Angew. Chem., Int. Ed.*, 2003, **42**, 3395.
14. N. Audic, H. Clavier, M. Mauduit and J. C. Guillemin, *J. Am. Chem. Soc.*, 2003, **125**, 9248.
15. R. T. Carlin, R. A. Osteryoung, J. S. Wilkes and J. Rovang, *Inorg. Chem.*, 1990, **29**, 3003.
16. R. T. Carlin and J. S. Wilkes, *J. Mol. Catal.*, 1990, **63**, 125.
17. P. Kubisa, *Prog. Polym. Sci.*, 2004, **29**, 3.
18. V. I. Parvulescu and C. Hardacre, *Chem. Revs.*, 2007, **107**, 2615.
19. M. R. Xie, H. J. Han, L. Ding and J. X. Shi, *Polym. Rev.*, 2009, **49**, 315.

20. J. M. Lu, F. Yan and J. Texter, *Prog. Polym. Sci.*, 2009, **34**, 431.
21. P. Kubisa, *Prog. Polym. Sci.*, 2009, **34**, 1333.
22. V. Strehmel, *Chem. Ing. Tech.*, 2011, **83**, 1443.
23. H. Zhang, P. K. Behera, N. K. Singha, K. Hong and J. W. Mays, Polymerization in Ionic Liquids, *Encyclopedia of Polymer Science and Technology*, Wiley, New York, 2015, pp. 1–19.
24. K. Hong, H. Zhang, J. W. Mays, A. N. Visser, C. S. Brazel, J. D. Holbrey, W. M. Reichert and R. D. Rogers, *Chem. Commun.*, 2002, 1368.
25. H. Zhang, L. Bu, M. Li, K. Hong, A. E. Visser, R. D. Rogers and J. W. Mays, in *Ionic Liquids: Industrial Application for Green Chemistry*, ed. R. D. Rogers and K. R. Seddon, ACS Symposium, 2001.
26. M. G. Benton and C. S. Brazel, *Polym. Int.*, 2004, **53**, 1113.
27. P. Wang, Y. Liu, W. Su and L. Liu, in *Advanced Materials and Processes III*, ed. S. Li, W. Li and J. Li, pts 1 and 2, Trans Tech Publications Ltd, Zurich, 2013, vol. 395–396, pp. 411–14.
28. E. Andrzejewska, A. Dembna, J. Bledniak and I. Stepniak, *Przem. Chem.*, 2013, **92**, 1570.
29. S. Harrison, S. R. MacKenzie and D. M. Haddleton, *Chem. Commun.*, 2002, 2850.
30. S. Harrison, S. R. MacKenzie and D. M. Haddleton, *Macromolecules*, 2003, **36**, 5072.
31. J. Barth, M. Buback, G. Schmidt-Naake and I. Woecht, *Polymer*, 2009, **50**, 5708.
32. S. Beuermann, *Macromol. Rapid Commun.*, 2009, **30**, 1066.
33. I. Polenz, I. Tchernook, J. Friedrich, F. G. Schmidt and S. Spange, *Macromol. Chem. Phys.*, 2015, **216**, 386.
34. A. L. Cheng, Y. Zhang, T. Zhao and H. Wang, *Macromol. Symp.*, 2004, **216**, 9.
35. S. Puttick, A. L. Davis, K. Butler, L. Lambert, J. El Harfi, D. J. Irvine, A. K. Whitaker, K. J. Thurecht and P. Licence, *Chem. Sci.*, 2011, **2**, 1810.
36. H. Zhang, K. Hong and J. W. Mays, in *Ionic Liquids in Polymer Systems: Solvents, Additives, and Novel Applications*, ed. R. Rogers and C. S. Brazel, ACS Symposium Series No. 913, American Chemical Society Press, Washington, DC, 2005.
37. V. Strehmel, A. Laschewsky, H. Wetzel and E. Gornitz, *Macromolecules*, 2006, **39**, 923.
38. V. Strehmel, *Macromol. Symp.*, 2007, **254**, 25.
39. V. Strehmel, E. Reynaud, H. Wetzel, E. Gornitz and A. Laschewsky, *Macromol. Symp.*, 2008, **275**, 242.
40. T. S. Rodrigues, F. Machado, P. M. Lalli, M. N. Eberlin and B. A. D. Neto, *Catal. Commun.*, 2015, **63**, 66.
41. J. F. Stanzione, R. E. Jenson, P. J. Costanzo and G. R. Palmese, *ACS Appl. Mater. Interfaces*, 2012, **4**, 6142.
42. X. L. Wu, J. H. Qiu, P. Liu, E. Sakai and L. Lei, *J. Polym. Res.*, 2013, **20**(6), 167.

43. Y. S. VygodskiI, D. A. Sapoznikov, A. S. Shaplov, E. I. Lozinskaya, N. V. Ignat'ev, M. Schulte, P. S. Vlasov and I. A. Malyshkina, *Polym. J.*, 2011, **43**, 126.
44. B. D. Mather, N. M. Reinartz and M. B. Shiflett, *Polymer*, 2016, **82**, 295.
45. C. Dietlin, M. Podgorska-Golubska and E. Andrzejewska, *J. Photochem. Photobiol., A*, 2014, **281**, 8.
46. J. W. Whitley, W. J. Home, S. P. O. Danielsen, M. S. Shannon, J. E. Marshall, S. H. Hayward, C. J. Gaddis and J. E. Bara, *Eur. Polym. J.*, 2014, **60**, 92.
47. E. Andrzejewska and A. Dembna, *Polimery*, 2014, **59**, 459.
48. H. Zhang, K. Hong and J. W. Mays, *Macromolecules*, 2002, **35**, 5738.
49. H. Zhang, K. Hong, M. Jablonsky and J. W. Mays, *Chem. Commun.*, 2003, 1356.
50. H. Harwood, *J. Macromol. Sci. Macromol. Symp.*, 1987, **10/11**, 331.
51. (a) C. J. Hawker, A. W. Bosman and E. Harth, *Chem. Rev.*, 2001, **101**, 3661; (b) M. Kamigaito, T. Ando and M. Sawamoto, *Chem. Rev.*, 2001, **101**, 3689; (c) K. Matyjaszewski and J. Xia, *Chem. Rev.*, 2001, **101**, 2921; (d) *Progress in Controlled Radical Polymerization: Mechanisms and Techniques*, ed. K. Matyjaszewski, B. Sumerlin and N. V. Tsarevsky, American Chemical Society, Washington D.C., 2012; (e) G. Moad, E. Rizzardo and S. H. Thang, *Polymer*, 2008, **49**, 1079; (f) G. Moad, E. Rizzardo and S. H. Thang, *Chem.–Asian J.*, 2013, **8**, 1634.
52. A. J. Carmicheal, D. M. Haddleton, S. A. F. Bon and K. R. Seddon, *Chem. Commun.*, 2000, 1237.
53. T. Biedron and P. Kubisa, *Macromol. Rapid Commun.*, 2001, **22**, 1237.
54. T. Biedron and P. Kubisa, *Polym. Int.*, 2003, **52**, 1584.
55. T. Sarbu and K. Matyjaszewski, *Macromol. Chem. Phys.*, 2001, **202**, 3379.
56. H. Ma, X. Wan, X. Chen and Q. Zhou, *J. Polym. Sci., Part A: Polym. Chem.*, 2003, **41**, 143.
57. H. Ma, X. Wan, X. Chen and Q. Zhou, *Polymer*, 2003, **44**, 5311.
58. S. Feng, W. J. Xu, K. Nakanishi and S. Yamago, *ACS Macro Lett.*, 2012, **1**, 146.
59. J. P. Mendes, F. Banco, C. M. R. Abreu, P. V. Mendonca, A. V. Popov, T. Guliashvili, A. C. Serra and J. F. J. Coelho, *ACS Macro Lett.*, 2014, **3**, 544.
60. H. Chen, C. H. Wang, D. L. Liu, M. Wang and C. N. Ji, *J. Appl. Polym. Sci.*, 2011, **122**, 3298.
61. X. Y. Du, J. L. Pan, M. T. Chen, L. F. Zhang, Z. P. Cheng and X. L. Zhu, *Chem. Commun.*, 2011, **50**, 9266.
62. T. Biedron and P. Kubisa, *J. Polym. Sci., Part A: Polym. Chem.*, 2002, **40**, 2799.
63. V. H. Nguyen and J. J. Shimm, *Colloid Polym. Sci.*, 2015, **293**, 617.
64. V. C. Tran, V. H. Nguyen, D. Tuma and J. J. Shimm, *Colloid Polym. Sci.*, 2016, **294**, 777.
65. X. Y. Du, J. L. Pan, M. T. Chen, L. F. Zhang, Z. P. Cheng and X. L. Zhu, *Chem. Commun.*, 2014, **50**, 9266.

66. G. X. Wang, M. Lu, Z. H. Hou, C. A. Yang, E. X. Liang, L. C. Liu, H. Wu, X. L. Li and Y. X. Xu, *J. Polym. Res.*, 2015, **22**, 60.
67. A. Anastasaki, V. Nikolaou, G. Nurumbetov, N. P. Truong, G. S. Pappas, N. G. Engelis, J. F. Quinn, M. R. Whittaker, T. P. Davis and D. M. Haddleton, *Macromolecules*, 2015, **48**, 5140.
68. H. Zhang, K. Hong and J. W. Mays, *Polym. Bull.*, 2004, **52**, 9.
69. J. Ryan, F. Aldabbagh, P. B. Zetterlund and B. Yamada, *Macromol. Rapid Commun.*, 2004, **25**, 930.
70. S. Brusseau, O. Boyron, C. Schikaneder, C. C. Santini and B. Charleux, *Macromolecules*, 2011, **44**, 215.
71. S. Perrier, T. P. Davis, A. J. Carmichael and D. M. Haddleton, *Chem. Commun.*, 2002, 2226.
72. S. Perrier, T. P. Davis, A. J. Carmichael and D. M. Haddleton, *Eur. Polym. J.*, 2003, **39**, 417.
73. R. Vijayaraghavan and D. R. MacFarlane, *Aust. J. Chem.*, 2004, **57**, 129.
74. L. Ban, X. Han, X. H. Wang, Y. P. Huang and Z. S. Liu, *Anal. Bioanal. Chem.*, 2013, **405**, 8597.
75. J. Ma, H. Chen, M. Zhang and M. M. Yu, *J. Polym. Sci., Part A: Polym. Chem.*, 2012, **50**, 609.
76. Q. Zhang and S. P. Zhu, *ACS Macro Lett.*, 2015, **4**, 755.
77. N. K. Singha, N. B. Pramanik, P. K. Behera, A. Chakrabarty and J. W. Mays, *Green Chem.*, 2016, **18**, 6115.
78. H. Zhou, C. G. Liu, C. Q Gao, Y. Q. Qu, K. Y. Shi and W. Q. Zhang, *J. Polym. Sci., Part A: Polym. Chem.*, 2016, **54**, 1517.
79. D. Uhrig and J. W. Mays, *J. Polym. Sci., Part A: Polym. Chem.*, 2005, **43**, 6179.
80. T. Biedron and P. Kubisa, *J. Polym. Sci., Part A: Polym. Chem.*, 2007, **45**, 4168.
81. H. Kokubo and M. Watanabe, *Polym. Adv. Technol.*, 2008, **19**, 1441.
82. R. Vijayaraghavan, J. M. Pringle and D. R. McFarlane, *Eur. Polym. J.*, 2008, **44**, 1758.
83. R. Diaz de Leon and M. T. A. Cordova, *Macromol. Symp.*, 2013, **325–326**, 194.
84. R. Vijayaraghavan and D. R. MacFarlane, *Chem. Commun.*, 2004, 700.
85. T. Biedron and P. Kubisa, *J. Polym. Sci., Part A: Polym. Chem.*, 2004, **42**, 3230.
86. L. Han, Y. B. Wu, Y. Dan, H. Wang, X. Q. Zhang, X. L. Wei, W. L. Guo and S. X. Li, *RSC Adv.*, 2016, **6**, 105322.
87. R. Vijayaraghavan and D. R. MacFarlane, *Sci. China: Chem.*, 2012, **55**, 1671.
88. Q. Y. Zhang, W. L. Guo, Y. B. Wu, L. F. Gong, W. li, X. N. Li, S. X. Li, Y. W. Shang, D. Yang and H. Wang, *Polym. Chem.*, 2016, **7**, 5099.
89. Y. B. Wu, L. Han, X. Q. Zhang, J. Mao, L. F. Gong, W. L. Guo, K. Gu and S. X. Li, *Polym. Chem.*, 2015, **6**, 2560.
90. H. Yoshimitsu, A. Kanazawa, S. Kanaoka and S. Aoshima, *J. Polym. Sci., Part A: Polym. Chem.*, 2016, **54**, 1774.

91. R. Vijayaraghavan and D. R. MacFarlane, *Chem. Commun.*, 2005, **9**, 1149.
92. V. Strehmel and V. Senkowski, *J. Polym. Sci., Part A: Polym. Chem.*, 2015, **53**, 2849.
93. S. Csihony, C. Fischmeister, C. Bruneau, I. T. Horvath and P. H. Dixneuf, *New J. Chem.*, 2002, **26**, 1667.
94. T. Biedron, M. Bednarek and P. Kubisa, *Macromol. Rapid Commun.*, 2004, **25**, 878.
95. J. Kadokawa, Y. Iwasaki and H. Tagaya, *Macromol. Rapid Commun.*, 2002, **23**, 757.
96. J. Kredatusova, H. Benes, S. Livi, O. Pop-Georgievski, P. Ecorchard, S. Abbrent, E. Pavlova and D. Bogdal, *Polymer*, 2016, **100**, 86.
97. U. H. Choi, A. Mittal, T. L. Price, R. H. Colby and H. W. Gibson, *Macromol. Chem. Phys.*, 2016, **217**, 1270.
98. P. Mastrorilli, C. F. Nobile, V. Gallo, G. P. Suranna and G. Farinola, *J. Mol. Catal. A: Chem.*, 2002, **184**, 73.
99. M. A. Klingshirn, G. A. Broker, J. D. Holbrey, K. H. Shaughnessy and R. D. Rogers, *Chem. Commun.*, 2002, 1394.
100. W. Ochedzan-Siodlak and K. Dziubek, *Chem. Eng. Process.*, 2013, **72**, 74.
101. W. Ochedzan-Siodlak, K. Dziubek, K. Czaja, S. Rabiej and R. Szatanik, *J. Polym. Res.*, 2014, **21**, 558.
102. W. Ochedzan-Siodlak and K. Dziubek, *Appl. Catal., A*, 2014, **484**, 134.
103. J. H. Yim, J. S. Lee and Y. S. Ko, *Polymer (Korea)*, 2015, **39**, 169.
104. W. Ochedzan-Siodlak, A. Bihun, A. Olszowy, M. Rajfur, T. Jesionowski and K. Siwinska-Stefanska, *Polym. Int.*, 2016, **65**, 1089.
105. C. Simocko, Y. Yang, T. M. Swager and K. B. Wagener, *ACS Macro Lett.*, 2013, **2**, 1061.
106. J. Zhao, D. R. Wang, B. Autenrieth and M. R. Buchmeiser, *Macromol. Rapid Commun.*, 2015, **36**, 190.
107. Z. A. Page, F. Liu, T. P. Russell and T. Emrick, *Chem. Sci.*, 2014, **5**, 2368.
108. A. M. Trzeciak, Ionic Liquids (ILs) in Organometallic Catalysis, *Topics in Organometallic Chemistry*, 2015, vol. 51, pp. 307–322.
109. E. Naudin, H. A. Ho, S. Branchaud, L. Breau and D. Belanger, *J. Phys. Chem. B*, 2002, **106**, 10585.
110. T. J. Park, Y. S. Kim, E. Kan and S. H. Lee, *RSC Adv.*, 2015, **5**, 25590.
111. K. Sekiguchi, M. Atobe and T. Fuchigami, *J. Electroanal. Chem.*, 2003, **557**, 1.
112. J. M. Pringle, J. Efthimiadis, P. C. Howlett, J. Efthimiadis, D. R. MacFarlane, A. B. Chaplin, S. B. Hall, D. L. Officer, G. G. Wallace and M. Forsyth, *Polymer*, 2004, **45**, 1447.
113. J. H. Shi, C. H. Yang, G. Y. Gao and Y. F. Li, *Chin. J. Chem. Phys.*, 2004, **17**, 503.
114. S. Esir, S. Shifiua, R. Topkaya and A. Baykal, *J. Inorg. Organomet. Polym. Mater.*, 2013, **23**, 1368.
115. X. L. Li, Y. F. Liu, W. Guo, J. J. Chen, W. X. He and F. F. Peng, *Electrochim. Acta*, 2014, **135**, 550.

116. J. Arjomandi, D. Nematollahi and A. Amani, *J. Appl. Polym. Sci.*, 2014, **131**, 40094.
117. S. Lagoutte, P. H. Aubert, M. Pinault, T. V. Francois, M. Mayne-L'Hermite and C. Chevrot, *Electrochim. Acta*, 2014, **130**, 754.
118. C. Dalmolin, S. R. Biaggio, N. Bocchi and R. C. Rocha, *Mater. Chem. Phys.*, 2014, **147**, 99.
119. M. E. Abdelhamid, G. A. Snook and A. P. O'Mullane, *ChemPlusChem*, 2015, **80**, 74.
120. S. Shafiu, B. Unal and A. Baykal, *J. Inorg. Organomet. Polym. Mater.*, 2013, **23**, 1335.
121. B. Y. Lu, S. M. Zhang, L. Q. Qin, S. Chen, S. J. Zhen and J. K. Xu, *Electrochim. Acta*, 2013, **106**, 201.
122. F. X. Zou, X. X. Yu, J. Zhang, N. Cheng and X. R. Huang, *Synth. Met.*, 2015, **204**, 76.
123. C. Dalmolin, S. R. Biaggio, N. Bocchi and R. C. Rocha, *Mater. Chem. Phys.*, 2014, **147**, 99.
124. Y. S. Vygodskii, E. I. Lozinskaya and A. S. Shaplov, *Macromol. Rapid Commun.*, 2002, **23**, 676.
125. M. Taghavi, M. Ghaemy, S. M. A. Nasab and M. Hassanzadeh, *Polymer*, 2013, **54**, 3828.
126. M. Taghavi, R. Alizadeh and M. Ghaemy, *RSC Adv.*, 2015, **5**, 9582.
127. J. Y. Dou, Z. P. Liu, K. Mahmood and Y. T. Zhao, *Polym. Int.*, 2012, **61**, 1470.
128. N. Guo, D. W. Hu, H. Wang, R. M. Wang and Y. B. Xiong, *Polym. Bull.*, 2013, **70**, 3031.
129. J. Wang and Z. P. Liu, *Chin. Sci. Bull.*, 2013, **58**, 1262.
130. S. D. Zhang, A. Feret, H. Lefebvre, M. Tessier and A. Fredet, *Chem. Commun.*, 2011, **47**, 11092.
131. H. Q. N. Gunaratne, C. R. Langrick, A. V. Puga, K. R. Seddon and K. Whiston, *Green Chem.*, 2013, **15**, 1166.
132. Y. Kanetaka, S. Yamazaki and K. Kimura, *Macromolecules*, 2016, **49**, 1252.
133. J. Wang, Y. Wu and Z. P. Liu, *Chin. J. Polym. Sci.*, 2016, **34**, 981.
134. F. Rafiemanzelat and E. Abdollahi, *J. Appl. Polym. Sci.*, 2012, **123**, 1015.
135. S. D. Zhang, L. D. Goncalves, H. Lefenvre, M. Tessier, B. Rousseau and A. Fradet, *ACS Macro Lett.*, 2012, **1**, 1079.
136. S. Mallakpour and M. Dinari, *Polym. Bull.*, 2013, **3**, 1049.
137. S. Mallakpour and A. Zadehnazari, *Adv. Polym. Technol.*, 2013, **32**, 21333.
138. H. Marien, L. Bellings, S. Hermans and I. J. F. Vankelecom, *ChemSusChem*, 2016, **9**, 1101.
139. H. Uyama, T. Takamoto and S. Kobayashi, *Polym. J.*, 2002, **34**, 94.
140. U. Piotrowska and M. Sobczak, *Molecules*, 2015, **20**, 1.
141. C. Y. Wu, Z. Y. Zhang, F. He and R. X. Zhuo, *Biotechnol. Lett.*, 2013, **35**, 879.
142. J. G. Yuan, Y. P. Dai, Y. Y. Yu, P. Wang, Q. Wang and X. R. Fan, *Eng. Life Sci.*, 2016, **16**, 371.

143. E. Champagne, S. Strandman and X. X. Zhu, *Macromol. Rapid Commun.*, 2016, **37**, 1986.
144. M. Mena, A. Lopez-Luna, K. Shirai, A. Tecante, M. Gimeno and E. Barzana, *Bioprocess Biosyst. Eng.*, 2013, **36**, 383.
145. J. Zhang, F. X. Zou, X. X. Yu, X. R. Huang and Y. B. Qu, *Colloid Polym. Sci.*, 2014, **292**, 2549.
146. S. J. Wang, L. X. Liu, Y. C. Jiang, M. C. Hu, S. N. Li and Q. G. Zhai, *Chem. Res. Chin. Univ.*, 2016, **37**, 1733.
147. Y. Z. Guo, X. H. Wang, Z. G. Shen, X. C. Shu and R. C. Sun, *Carbohydr. Polym.*, 2013, **92**, 77.
148. A. Hufendiek, V. Trouillet, M. A. R. Meier and C. Barner-Kowollik, *Biomacromolecules*, 2014, **15**, 2563.
149. Y. Shen, L. Dai, M. F. Li, J. He, B. Wang and R. C. Sun, *Cellul. Chem. Technol.*, 2014, **48**, 653.
150. X. Y. Wei, G. Chang, J. H. li, F. Wang, L. H. Cui, T. K. Fu and L. X. Kong, *J. Polym. Res.*, 2014, **21**, 535.
151. D. H. Liu, W. Luo, Y. Z. Wei, C. X. Lin, C. Y. Du and M. H. Liu, *Asian J. Chem.*, 2014, **26**, 6295.
152. S. J. Lee, H. S. Lee, S. W. Jeong, H. C. Kim, S. G. Lee and T. H. Oh, *J. Appl. Polym. Sci.*, 2015, **132**, 41331.
153. Y. D. Liu, X. S. Jin, X. S. Zhang, M. M. Han and S. X. Ji, *Carbohydr. Polym.*, 2015, **117**, 312.
154. R. Kakuchi, M. Yamaguchi, T. Endo, Y. Shibata, K. Ninomiya, T. Ikai, K. Maeda and K. Takahashi, *RSC Adv.*, 2015, **5**, 72071.
155. E. J. Tang, K. D. Du, X. Y. Feng, M. Yuan, S. J. Liu and D. S. Zhao, *Eur. Polym. J.*, 2015, **66**, 228.
156. I. Zhang, Y. Z. Guo, J. H. Zhou, G. W. Sun, Y. Han and X. H. Wang, *Bioresources*, 2016, **11**, 2698.
157. L. P. Hinner, J. L. Wissner, A. Beurer, B. A. Nebel and B. Hauer, *Green Chem.*, 2016, **18**, 6099.
158. P. Chmielarz, *eXPRESS Polym. Lett.*, 2017, **11**, 140.
159. Z. D. Wang, L. C. Zheng, C. C. Li, D. Zhang, Y. N. Xiao, G. H. Guan and W. X. Zhu, *Carbohydr. Polym.*, 2013, **94**, 505.
160. C. X. Lin, D. H. Liu, W. Luo, Y. F. Liu, M. S. Q. Zhu, X. J. Li and M. H. Liu, *J. Appl. Polym. Sci.*, 2015, 132.
161. J. Kadokawa, *Pure Appl. Chem.*, 2016, **88**, 621.
162. S. F. Deng, J. Cheng, X. J. Guo, L. Jiang and J. W. Zhang, *J. Polym. Environ.*, 2014, **22**, 17.
163. X. Q. Zhang, M. J. Chen, C. F. Liu, A. P. Zhang and R. C. Sun, *Molecules*, 2015, **20**, 6033.
164. X. Q. Zhang, M. J. Chen, H. H. Wang, C. F. Liu, A. P. Zhang and R. C. Sun, *Ind. Eng. Chem. Res.*, 2015, **54**, 6282.
165. X. Q. Zhang, H. H. Wang, X. X. Wen, A. P. Zhang, X. Y. Wang, L. X. Zhong, C. F. Liu and R. C. Sun, *Macromol. Mater. Eng.*, 2016, **301**, 287.
166. A. Llevot, P. K. Dannecker, M. von Czapiewski, L. C. Over, Z. Soyler and M. A. R. Meier, *Chem.–Eur. J.*, 2016, **22**, 11509.

167. X. D. Yang, C. D. Qian, Y. Li and T. D. Li, *React. Funct. Polym.*, 2016, **100**, 181.
168. M. Isik, H. Sardon and D. Mecerreyes, *Int. J. Mol. Sci.*, 2014, **15**, 11922.
169. O. Jogunola, V. Eta, M. Hedenstrom, O. Sundman, T. Salmi and J. P. Mikkola, *Carbohydr. Polym.*, 2016, **135**, 341.
170. L. F. Calheiros, B. G. Soares, G. M. O. Barra and S. Livi, *Mater. Chem. Phys.*, 2016, **179**, 194.
171. B. J. Hu, Y. Y. Kong, R. M. H. Zheng, J. Dong, K. L. Choy and H. H. Zhao, *Faraday Discuss.*, 2016, **190**, 487.
172. C. Yuan, J. N. Guo, Z. H. Si and F. Yan, *Polym. Chem.*, 2015, **6**, 4059.
173. C. Zhang, F. Li, S. X. Wang, Z. S. Liu and H. A. Aisa, *Anal. Methods*, 2015, **7**, 10256.
174. P. W. Jiang, X. P. Wu, L. J. Zhu, F. Jin, J. X. Liu, T. Y. Xia, T. J. Wang and Q. X. Li, *Energy Convers. Manage.*, 2016, **120**, 338.
175. A. Mouawia, A. Nourry, A. C. Gaumont, J. F. Pilard and I. Dez, *ACS Sustainable Chem. Eng.*, 2017, **5**, 696.
176. F. Iannone, M. Casiello, A. Monopoli, P. Cotugno, M. C. Sportelli, R. A. Picca, N. Cioffi, M. M. Dell'Anna and A. Nacci, *J. Mol. Catal. A: Chem.*, 2017, **426**, 107.
177. N. Tarasova, A. Zanin, K. Burdakov and P. Sobolev, *Polym. Adv. Technol.*, 2015, **26**, 687.
178. Y. J. Li, C. C. Zhang, Y. L. Zhou, Y. X. Dong and W. Y. Chen, *Eur. Polym. J.*, 2015, **69**, 441.
179. S. Livi, J. Duchet-Rumeau, J. F. Gerard and T. N. Pham, *Macromol. Chem. Phys.*, 2015, **216**, 359.

CHAPTER 2

Porous Ionic Liquid Materials

ALESSANDRO DANI*[a], VALENTINA CROCELLÀ[b], GIULIO LATINI[b,c] AND SILVIA BORDIGA*[b]

[a]Department of Colloid Chemistry, Max Planck Institute of Colloids and Interfaces, Am Mühlenberg 1, 14476 Potsdam, Germany; [b]Department of Chemistry, NIS and INSTM Reference Centre, University of Turin, Via Quarello 15, 10135 Torino, Italy; [c]Department of Applied Science and Technology (DISAT), Politecnico Di Torino, C.so Duca degli Abruzzi 24, 10129 Torino, Italy
*E-mail: silvia.bordiga@unito.it; Alessandro.Dani@mpikg.mpg.de

2.1 Introduction

The development of novel synthetic routes for the design of new porous materials is probably one of the most relevant achievements in chemistry and materials science of the last few decades. As a matter of fact, from their first synthetic attempts, porous materials became a hot research topic. The design at the nanoscale level of these materials gives rise to a high surface/weight ratio, opening up a route to a wide variety of potential applications in the fields of gas storage, adsorption and catalysis. In fact, the increased specific surface area of a porous system multiplies the interaction possibilities between the material surface and the possible guest molecules, increasing the fraction of available active sites and so enhancing the adsorption/catalytic properties with respect to the equivalent non-porous counterpart. Moreover, the theoretically unlimited combinations of available chemical structures allows materials to be tailored for many different well-defined applications.

Smart Materials No. 29
Polymerized Ionic Liquids
Edited by Ali Eftekhari

Published by the Royal Society of Chemistry, www.rsc.org

Nowadays, several methodologies are available to create porous materials, each of them suitable for generating a tailored porous structure in terms of pore size and volume. The present goal in the field of porous materials is to introduce, inside the porous structure, functional groups that make the material able to meet the requirements of a desired application. Ionic liquids (ILs) are functional molecules that have already proved to have a broad range of applications in different fields, mainly due to the huge number of available chemical structures and the possibility to combine different anions and cations. Therefore, the combination of ionic liquids and porous materials gave rise to several novel IL-functionalized porous materials, with a broad range of chemical structures, porosity features and potential applications.

The aim of the present chapter is to give the reader a thorough overview of all the different families of ionic liquid-functionalized porous materials, with remarkable properties from a scientific point of view. In particular, the present chapter includes a description of the porous poly(ionic liquid)s, which, thanks to their fully organic structure, can be obtained by means of all the common methodologies to make porous polymers. This chapter also describes the materials obtained by supporting ionic liquid moieties on pre-formed porous structures of various kinds. A section is also devoted to a new concept of hybrid materials, where the ionic liquid moiety is an integrated part of a metal–organic framework (MOF) or an ordered mesoporous silica. These hybrid materials have a well-defined structure with long-range order. In the case of ionic liquid metal–organic frameworks (MOF–ILs), the material usually exhibits a microporous structure, whereas silica–ILs are generally mesoporous and, for this reason, they are also called periodic mesoporous organosilica (PMOs).

This chapter considers all kinds of ILs, in particular imidazolium, pyridinium and phosphonium ILs, the most common cations used for the development of porous materials; however, other examples of cations are also reported.

A general overview regarding the state-of-the-art of the various methodologies to create ionic liquid-functionalized porous materials will be presented to the reader. Many synthesis examples will be reported along with some information on the porous structure of the resulting material and its potential applications. Furthermore, combining information from different studies, it will allow the definition of general principles suitable for the various classes of materials, and it will give the reader a more conscious view of the potential of porous ionic liquids, and the synthetic strategies to adopt in order to obtain a porous ionic liquid that meets the requirements of the desired application.[1,2]

2.1.1 General Definition of Porosity

Porosity may generally be defined as the fraction of the apparent specific volume of the sample that is attributed to the pores. Porosity is a highly valued structural trait of solid materials for applications such as gas storage and separation or catalysis.

Nanoporous materials refer to a class of porous materials having pore diameters from 100 nm down to around 1 nm. The International Union of Pure and Applied Chemistry (IUPAC) classifies pores according to their size (IUPAC recommendation, 1985[3]: (i) pores with widths exceeding about 50 nm are called macropores; (ii) pores of widths between 2 nm and 50 nm are called mesopores; (iii) pores with widths not exceeding about 2 nm are called micropores). In the range of micropores, it is useful to distinguish between narrow micropores (ultramicropores) of approximate width <0.7 nm and wide micropores (supermicropores).

In the characterization of porous materials, inert gas (commonly N_2) physical adsorption is very useful and efficient in measuring pore-related information such as specific surface area (SSA), pore volume, and pore size distribution. Adsorption isotherms are the main tool for the study of adsorption in solid adsorbents.

Physical adsorption isotherms can be classified into six types (Figure 2.1) according to the IUPAC definition. Type I isotherms (I(a) and I(b)) are given by microporous solids with small external surfaces. In this type of isotherm, the adsorbed amount of N_2 approaches a limiting value governed by the accessible micropore volume rather than by the internal surface area. Micropore filling and therefore high uptakes are observed at relatively low pressures, because of the narrow pore width and the high adsorption potential. Type II isotherms are obtained in the case of nonporous or macroporous adsorbent,

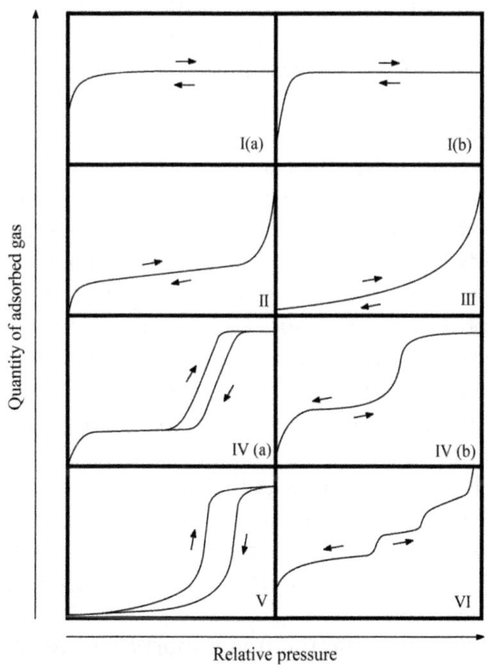

Figure 2.1 Classification of adsorption isotherms.

where unrestricted monolayer–multilayer adsorption occurs. The inflection point or knee of the isotherm is called point B and usually corresponds to the completion of the monolayer. Type III isotherms do not exhibit a point B and therefore there is no identifiable monolayer formation. This indicates that the attractive adsorbate–adsorbent interactions are relatively weak and that the adsorbate–adsorbate interactions play an important role. Type IV isotherms (IV(a) and IV(b)) are typical for mesoporous materials. In this case, the initial monolayer–multilayer adsorption is followed by pore condensation or capillary condensation (*i.e.* the phenomenon in which a gas condenses to a liquid-like phase in a pore at a pressure lower than the saturation pressure of the liquid itself), which can be accompanied by hysteresis, when the pore width is above a specific width, depending on the adsorbate and on the temperature (for example, in the case of nitrogen adsorption in cylindrical pores at 77 K, hysteresis can appear for pores wider than 4 nm). At least six different main shapes of hysteresis loops are recognized as reported in ref. 4. Each of these six hysteresis loops is related to particular features of the pore structure. Type V isotherms exhibit pore condensation and hysteresis but, in this case, weak adsorbent–adsorbate attractive interactions are responsible for the shape of the isotherm in the low p/p_0 range (as for Type III isotherms). Finally, the peculiar stepwise Type VI isotherms are representative of layer-by-layer adsorption on a highly-uniform nonporous surface.

The application of the BET equation is the most popular approach for the calculation of the specific surface area (SSA) of a solid material. This method is based on an over-simplified model of physisorption (the theory of multilayer adsorption).[5] In specific controlled conditions, the BET area of a nonporous, macroporous or mesoporous solid can be considered as the "probe accessible area", *i.e.* the area available for the adsorption of the specified adsorbate. The BET method is usually applied to materials possessing well-defined Type II and Type IV isotherms. In the case of microporous systems (*i.e.*, with Type I isotherms or combinations of Type I and II or Type I and IV isotherms) extreme caution is needed. In fact, in the presence of microporous solids, the range of linearity of the BET plot may be very difficult to identify due to the difficulty in separating the processes of monolayer/multilayer adsorption and micropore filling. For this reason, the BET area derived from a Type I isotherm must not be considered as a realistic "probe accessible surface area" but as an apparent surface area.

In the case of microporous materials with physisorption isotherms of Type I that can be described by the Langmuir equation,[5] which was developed on the assumption that adsorption was limited to at most one monolayer, the Langmuir equation is often applied for the determination of the specific surface area. However, despite the often-observed good fit of the Langmuir equation to the experimental data, the obtained surface area results do not reflect a true surface area.

At the beginning of this chapter, it is important to recall a specific feature of porous polymer networks. These "soft" materials often present peculiar adsorption/desorption isotherms with respect to stiff microporous inorganic materials like, for example, zeolites, as reported in Figure 2.2. In fact, the softness of polymeric materials in comparison to inorganic materials can lead to swelling effects.

In particular, the adsorption/desorption isotherms of polymeric materials exhibit two distinct characteristics derived from swelling effects: (i) the N_2 loading does not reach a plateau after micropore filling, and (ii) a significant hysteresis is present at low relative pressures, in which the desorption branch of the isotherm does not close to the adsorption branch. Both of these effects are derived from the elastic deformations of the polymeric matrix due to polymer relaxations occurring during nitrogen adsorption that increase the amount of available free volume.[6–8] The swelling effect typical of porous polymers can be described taking into account a similar process, observed in ordinary (*i.e.* nonporous) polymer films and membranes in high-pressure gas permeation experiments, called dual-mode adsorption. The dual-mode adsorption model assumes the existence of a fractional free volume of pores within polymer films, which can be filled by gas molecules at first (pocket filling), and of an additional Henry adsorption within the film, which is instead proportional to the gas pressure (swelling).[6,9]

The deformation–relaxation phenomenon is schematically depicted in Figure 2.3: part (I) represents the as-made polymeric networks containing residual solvent or gas molecules, part (II) displays the networks after evacuation and removal of the solvent/gas molecules that appear contracted due to the deformation of chains and bonds and part (III) represents the polymeric network that can relax again by the uptake of gas molecules (PI1), as long as this process is not restricted by intermolecular interactions such as hydrogen bonding (PA1).[6]

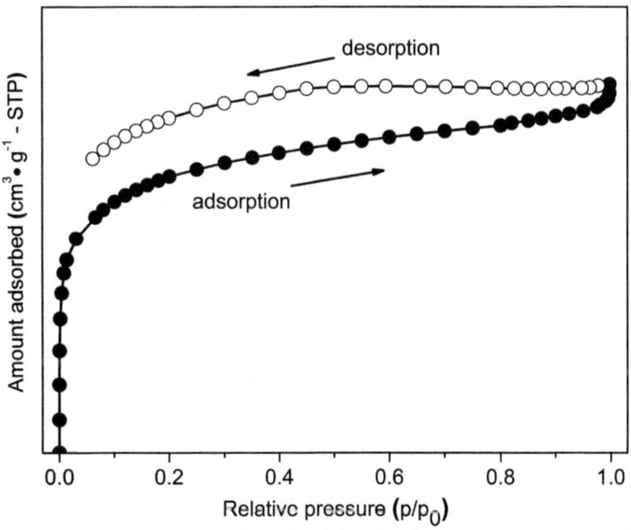

Figure 2.2 Typical low temperature N_2 adsorption/desorption isotherms of microporous materials possessing swelling capacity.

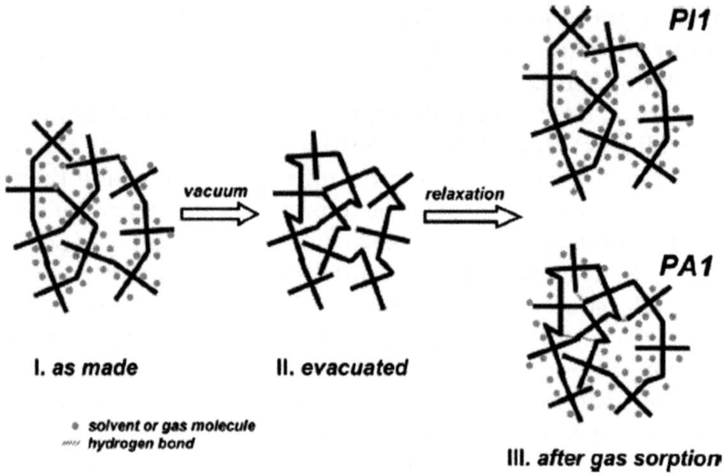

Figure 2.3 Schematic drawing of the suggested deformation–relaxation process during evacuation and gas adsorption. Reproduced with permission from ref. 6, Copyright 2008, American Chemical Society.

Due to the relaxation of the polymer altering the local structure and free volume within the framework during gas adsorption, the porosity of these materials can be better described by the desorption branches of the isotherms. In fact, a fraction of pores is available for N_2 adsorption only at high values of relative pressure, being the swelling proportional to the gas pressure.

Although nitrogen adsorption is, in general, an extremely versatile tool in the analysis of rigid and well-defined porous structures (such as MOFs, zeolites, and porous silica) nitrogen suffers from several drawbacks when employed as an adsorbate for the analysis of the textural properties of microporous organic materials. In fact, precisely due to the possible swelling effects, it is questionable whether the N_2 isotherms really reflect the equilibrium state.

The textural properties of this type of material can be derived from CO_2 adsorption at 273 K. At this temperature, the mobility of the chains would allow the micropores to open at lower partial pressure.

2.2 Porosity and Ionic Liquids

The field of porous ionic liquids can be divided into three main classes, depending on the material structures. The first class arises from the direct synthesis of fully organic porous polymer networks, which is pursued by means of various stratagems and techniques. The second class includes polymer-supported ionic liquid moieties on pre-formed porous materials, *i.e.* porous silica or metal–organic frameworks (MOFs). The third class, instead, can be considered as a combination of the first and second ones, where the organic ionic liquid moieties become a structural part of the organic–inorganic hybrid porous ionic materials. In this section, we describe the most common synthetic methods employed to develop porous ionic liquids belonging to all three classes, and we briefly describe the structural and chemical properties of the resulting porous ionic liquids.

As expected, the three classes of porous ionic liquids have different advantages and drawbacks. Since these three material classes are very broad, it is difficult to extrapolate common behaviours to compare them. The materials belonging to the first class are created by polymer synthesis, which is intrinsically more tunable. Depending on the synthetic method, materials with different pore sizes and SSAs can be achieved ranging from microporous to macroporous. The materials belonging to the second and third classes are modifications of previously synthesized porous materials, bearing ionic liquid moieties along the structures. In these cases, the pore size and SSA of the resulting porous ionic liquids are strictly related to those of the supporting porous materials or their non-ionic analogue; thus, the final porous ionic liquids can be either micro-, meso- or macroporous.

The chemical properties of porous ionic liquids mainly depend on the type of ionic liquid moiety employed, whereas they are not directly related to the "class" to which they belong, which only determines the porous structure of the system and the accessibility of the ionic liquid sites.

There are a variety of synthetic strategies used to fabricate a desired porous structure using ionic liquids, and they all have some common parameters to take into account in order to achieve a stable porous structure. A porous material can be created using a single linear polymer, however these structures are not stable and the pore structure can immediately collapse.[10,11] In this case, the simple addition of a solvent is sufficient to dissolve the linear polymer and permanently break the pore structure. In order to stabilize the porous structure, it is then strictly necessary to use a cross-linking agent or to perform post-functionalization on linear ionic polymers to introduce electrostatic or covalent cross-links upon pore structure formation. In addition, the structural rigidity of the ionic liquid building block, and of the material structure in general, is crucial to determine the pore size and the SSA of the resulting material. If the ionic moiety bears one methylene bridge ($-CH_2-$) along the alkyl chain or other flexible parts, this confers some degree of rotation to the ionic liquid moieties or to the polymeric chains, leading to a partial collapse of the porous structure. The ionic liquid moieties of the polymer chains are permanently charged entities; therefore, they are subject to mutual columbic attraction between the cations and the anions of different chains, which, in the case of chain mobility, induces the collapse of the polymeric chains to a bulky structure. As a general rule, if there are flexible linkages in the ionic porous structure, the SSA of the resulting porous material is usually below 200 $m^2\ g^{-1}$ and only meso/macro-porous structures are obtained. In contrast, if the ionic liquid moiety does not have methylene bridges along its alkyl chain or other flexible bonds, but instead possesses a rather stiff structure (such as a fully conjugated system), microporous materials with SSAs in the range of 200–1000 $m^2 g^{-1}$ can be obtained. The rigidity of the monomer does not affect the materials of the "second class" where the ionic liquid is supported on a pre-formed porous material. In such a case, the SSA and the pore size are strictly related to the supporting material, and only the size of the supported ionic liquid affects the structural properties of the final ionic material, regardless of its structural flexibility.

2.2.1 Porous Poly(Ionic Liquid)s

The "first class" of porous ionic liquids involves the generation of polymerized ionic liquids, also known as poly(ionic liquid)s (PILs). These polymers, bearing ionic liquid functionalities, have many useful properties, which have been discovered in the last few decades.[12–14] Although PILs bear permanent charges along the polymeric backbone, they can be processed and polymerized as normal non-ionic polymers. This fact allows the creation of porous PILs by means of the common techniques used to introduce porosity into non-ionic polymers. Porous PILs can be obtained by means of two main methodologies. The first one makes use of a templating agent, both hard and soft. The second one (called a templating-free methodology) allows a porous polymer to be obtained directly by means of free-radical polymerization or other direct synthesis techniques that make use of various C–C bond couplings. A further templating-free methodology for the generation of porous PILs involves the electrostatic cross-linking of a pre-formed PIL using inter-polyelectrolyte complexation. In this process, simultaneous phase separation and cross-linking generate porosity in the system while stabilizing the final structure of the material.

Table 2.1 lists the main advantages and drawbacks of the more common synthetic routes that lead to porous PILs. All these techniques will be extensively described in the following sections.

2.2.1.1 Templating Methodologies

Templating methodologies for creating porous PILs are essentially molding or casting techniques that lead to the inverse replication of the template structure.[15,16] This structure can be pre-formed, as in the case of hard templating (also called direct templating), or generated *in situ* during the synthesis of the porous PILs, as in the case of soft templating (also called indirect templating).

A further distinction can be made according to the raw material, which can be either a polymer or a monomer. The former is infiltrated into the void of the template or coated onto its surface, whereas the latter is polymerized directly inside the template.[1] Both these raw materials can be employed in the hard templating route, whereas soft templating is usually performed starting from monomers.

Table 2.1 List of the main methodologies to generate porous PILs, and their advantages and drawbacks.

Methodologies	Advantages	Drawbacks
Templating methodologies		
Hard (direct)	Easy synthesis of mesoporous and macroporous networks with well-defined pore size	Difficulty in removing the sacrificial template No microporous structures are achievable
Soft (indirect)	Mesoporous structures with a well-defined architecture can be obtained	Difficult control of monomer–template interactions No microporous structures are achievable
Templating-free methodologies		
Direct synthesis	High SSA and microporous structures are achievable An ordered crystalline network is achievable with peculiar reactions	Monomers with strict defined requirements need to be synthesized Challenging synthesis of the network
Free radical polymerization	Simple synthetic strategy leading to polymers with a broad range of porosity Upscale is possible	Difficulty in fine tuning the porous properties
Polyelectrolyte complexation	Simple synthetic strategy for meso/macroporous materials The tuning of the pore size is possible by changing the polyelectrolyte Porous membranes are achievable	Materials not stable in high ionic strength environments No micropores are achievable

Obviously, the starting template structure affects the final pore structure of the resulting PIL. A template with a well-defined porous structure is recommended in order to precisely tune the PIL replica porous structure. Templating methodologies allow macroporous and mesoporous materials to be achieved, but they do not allow microporous materials to be obtained. Microporosity, in fact, arises from inefficient packing of the polymer chains and, therefore, a good template for micropores in PILs should be small enough to intercalate between the polymers chains; for that reason, it should have molecular dimensions. In this specific case, porosity can be generated by a porogenic solvent. We will further discuss this concept in Section 2.2.1.2.1: Free Radical Polymerization.

The choice of an appropriate templating methodology to create porous PILs implies some important preliminary considerations. First, a good compatibility between the raw material (either the monomer or the polymer) and the template is necessary to achieve a good interaction and, as a consequence, proper filling of the template voids with the raw material itself. In some cases of hard templating synthesis, the surface of the template needs to be modified to make it compatible with the raw material, or to introduce reaction sites for the growth of the PILs. The template structure should be stable enough to tolerate the polymerization conditions and it should be easily removed after the reaction by means of solvent extraction or chemical etching, avoiding critical conditions that could damage the just formed porous PILs. Cross-linked PILs are more likely to survive the template removal process and avoid pore collapse because their robust structure is able to contrast the high interfacial energy and strong capillary action involved during template removal.

2.2.1.1.1 Hard Templating. In hard templating methodologies, the selected templating structure is pre-formed and it is usually constituted by nanostructured porous silicas. The general synthetic route is as follows: a cross-linker monomer, liquid or in solution, and a polymerization initiator (if required) are infiltrated into the voids of the solid template, and then polymerization is triggered and subsequently the template is removed by chemical etching to produce the final porous PILs. This very intuitive technique has been used to obtain porous PILs with different pore sizes and shapes depending, obviously, on the employed templating agent. To the best of our knowledge, hard templating synthesis using PILs as a raw material has never been reported in the literature, whereas many studies describe the use of this technique for the synthesis of porous PILs starting from cross-linked ionic liquid monomers.

Only two types of hard templating agent have been used to produce porous PILs: silica and iron oxide. Silica is easy to obtain in different nanostructured shapes and, thanks to its polar surface, it can easily interact with the charged ionic liquid moiety, permitting good permeation of the monomer within the nanostructured silica. Furthermore, silica is easy to remove by means of an etching step with diluted HF solution or NaOH solution. Imidazolium and pyridinium PILs are rather stable in both basic and acidic conditions, unless the NaOH concentration does not exceed 1 M. An issue can arise during silica etching with HF solutions because the reaction of HF with the siliceous framework produces $[SiF_6]^{2-}$ anions, which can partially substitute the IL anions and remain as a leftover byproduct in the final material.

The first attempt to synthesize porous PILs *via* hard templating was performed by Li and co-workers.[17] They used uniform silica nanoparticles assembled into monodisperse spherical colloidal crystals as a template, in order to get inverse opal porous PILs. The resulting materials showed remarkable optical properties, which depended on the PIL anion. Following the same methodology, and by applying some variations to the silica nanoparticles employed as the template and to the PILs, Li's group were able to obtain porous PILs applicable as films for electro-optic switching and optical sensors for different anions and environmental humidity.[17–19]

Highly-ordered mesoporous SBA-15 has also been used as a templating agent to produce porous PILs (see Figure 2.4a).[20] In this case, divinylbenzene was used as a cross-linker for 1-benzyl-3-vinylimidazolium bromide in order to stabilize the final nanorod structure. Since divinylbenzene is strongly hydrophobic, the hydroxyl groups at the surface of the silica were protected by trimethylsilyl groups in order to facilitate the diffusion of the monomer inside the SBA-15 channels. Thanks to the resulting ordered porous structure of the PIL, the BET SSA is as high as 289 m^2 g^{-1}, and the pore size distribution shows a narrow maximum at 2.4 nm, which matches the wall thickness of SBA-15.[20] Randomly ordered silica nanoparticles were also reported as a porous PIL templating agent. In this case, the silica particles are not efficiently packed together, minimizing the empty spaces; therefore, it is important to tune the template/PIL ratio in order to obtain a porous structure with fully interconnected pores, thus allowing easy accessibility of guest molecules to all the available surface area. In this work, two different porous PILs were obtained. The first one was synthesized by free radical polymerization of a cross-linker ionic monomer, and the resulting mesoporous material exhibited remarkable CO_2 adsorption properties.[21] The second porous PIL was obtained by condensation polymerization of a trivalent alkyl bromide monomer and a series of bipyridine monomers. The condensation polymerization led to a mesoporous pyridinium PIL useful for supporting gold nanoparticles for catalytic applications.[22]

Recently, Fe_3O_4 nanoparticles were also used as a novel templating agent to introduce macroporosity into PILs. Fe_3O_4 nanoparticles are cheap and easy to synthesize, but it is necessary to functionalize their surface with 3-(trimethoxysilyl)propyl methacrylate monomer in order to favour the growth of the PIL on their surface. This templating agent can be easily removed by means of HCl/H_2O/ethanol solutions and ultrasonic treatment. The resulting macroporous PIL was exploited for biodiesel production, taking advantage of the combination of the sulfonic acid group on the imidazolium alkyl chain and the hydrogen sulfate anion of the PIL.

2.2.1.1.2 Soft Templating. Soft templating methodologies rely on the use of soft matter, mainly in the form of self-assembled block copolymers, as a templating agent. Briefly, self-assembly occurs when supramolecular organization takes place in a dispersion of a block-copolymer consisting of two or more chemically immiscible homopolymers that are covalently linked together. The phase separation of this block-copolymer from the solvent generates an ordered aggregate, which minimizes the contact energy between the segments of different polarity. The phase separation is restricted to the nanometer scale because there is a covalent bond between the two chain segments leading to the creation of a nanostructured template.

The self-assembly of the templating agent takes place inside the polymerization mixture. For this reason, it is crucial to balance the interaction forces between the templating agent and the monomer to drive the polymerization around the template avoiding close conjunction of the organic compounds

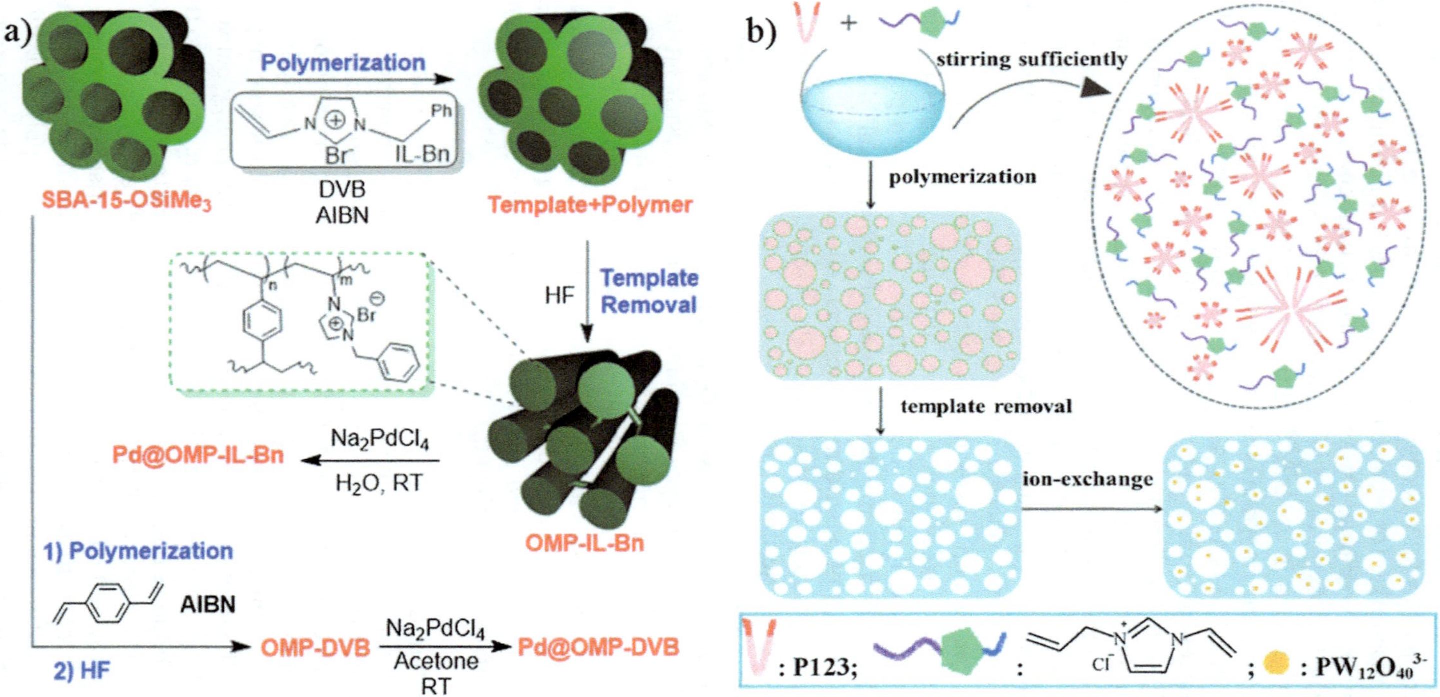

Figure 2.4 Section (a): synthesis scheme of a porous PIL starting from the co-polymerization of an imidazolium monomer and DVB, using SBA-15 as a hard templating agent. Reproduced from B. Karimi, M. R. Marefat, M. Hasannia, P. F. Akhavan, F. Mansouri, Z. Artelli, F. Mohammadi, H. Vali, *ChemCatChem*, John Wiley and Sons, Copyright © 2016 WILEY-VCH Verlag GmbH & Co. KGaA, Weinheim.[20] Section (b): synthesis of a porous PIL starting from a cross-linker imidazolium monomer using P123 as a soft templating agent. Reproduced from ref. 23 with permission from The Royal Society of Chemistry.

and so permitting the removal of the template by means of solvent extraction, without causing the porous structure to collapse. The self-assembled template structure also needs to be strong enough to support the stiffening induced by the cross-linking polymerization. The most common block-copolymers used as templates to generate porous material are commercially known as P123 (EO_{20}–PO_{70}–EO_{20}) and F127 (EO_{106}–PO_{70}–EO_{106}), which, due to the different lengths of their poly(ethylene glycol) and poly(propylene glycol) chains, are able to generate different micelle arrangements. Porous materials generated *via* soft templating techniques usually possess an ordered mesoporous structure.

A soft templating methodology has been exploited to create porous PILs using P123 as a templating agent and 1-allyl-3-vinylimidazolium chloride as a cross-linking monomer *via* free radical polymerization (see Figure 2.4b).[23] After the polymerization, extraction with ethanol leads to the formation of a hierarchically structured meso-macroporous PIL with a BET SSA as high as 144 $m^2\ g^{-1}$. The hierarchical porosity arises from the broad range of dimensions of the templating micelles. The size of the pores and the continuous pore structure of this material permit anion exchange of the PIL with the phosphotungstic anion making the material exploitable as a heterogenous catalyst in the H_2O_2-mediated epoxidation of *cis*-cyclooctene.[23]

2.2.1.2 *Templating-free Methodologies*

Templating free methodologies allow porous polymers to be achieved without the use of sacrificial templates. The three main templating-free techniques are (i) free radical polymerization, (ii) direct synthesis and (iii) polyelectrolyte complexation. The first two are classical polymerization techniques to obtain cross-linked polymer networks that, thanks to particular strategies, lead to the generation of porous structures. In principle, in both of these methods the raw monomer needs to be carefully designed to be able to create a porous polymer network. The monomer should be multitopic in order to cross-link the network, and stiff enough to sustain the high capillary force of both micro- and mesopores and to avoid the collapse of the network into a bulky structure. In particular, the monomer must possess a rigid and sterically hindered structure that helps the inefficient packing of the polymer chains leading to the formation of micropores. The main advantage of these two polymerization techniques is that they allow the generation of microporous and mesoporous materials with rather high specific surface areas.

In the case of free radical polymerization, the starting monomers are usually cheap and easy to synthesize. Moreover, the polymer synthesis is usually easy to upscale. In contrast, in the case of direct synthesis, the monomers require more tricky synthesis steps, due to the need to generate a multitopic molecule that can react in carbon coupling reactions to create a network. However, an additional advantage of the direct synthesis route is that good design of the starting monomers leads to fine tuning of the polymer porous structure in terms of pore size and distribution.

Polyelectrolyte complexation uses a completely different approach. In this synthetic route, a linear PIL is electrostatically cross-linked with another polyelectrolyte or a small organic molecule. In this way, macroporosity, and sometimes mesoporosity, can be introduced by processing a pre-existing polymer. An important advantage of this technique is its extreme flexibility, due to the possibility to work with several different PILs and cross-linking agents. By means of this route, it is possible to obtain PIL membranes useful for various applications. The main drawback of this technique is the impossibility to generate microporosity because the phase separation mechanism that induces the porosity only works in the meso/macro scale range.

2.2.1.2.1 Free Radical Polymerization. Free radical polymerization is one of the most common and versatile techniques to obtain polymers, and by means of some proper modifications and expedients, this synthetic route can be employed to produce porous polymers.[1] In accordance with all the other methodologies, the structure of the resulting porous polymer must be stabilized by means of an appropriate cross-linking agent, which is able to generate a polymer network.

Free radical polymerization derives from the chain reaction of vinyl groups to generate an alkyl chain, which is intrinsically very flexible. Using this polymerization technique, the porosity of the polymer arises from both a microphase separation mechanism of the polymer chains from the reaction mixture, and from inefficient packing of the polymer chains due to the steric hindrance of the rigid monomers. The employed solvent can also act as a porogenic agent, inducing pore creation from the phase separation occurring during the polymerization process. This situation occurs if the solvent is, at the same time, a rather good solvent for the raw monomers and a poor solvent for the polymer chains. This polymerization methodology is rather flexible and leads to PILs with various porous structures and pore size distributions ranging from micro- to macropores.

PIL porosity is mostly governed by the chemical structure of the starting IL monomer and its interaction with the reaction solvent. In order to obtain a cross-linked polymer network, a starting monomer possessing at least two double bonds needs to be employed, which can be an IL cross-linker or even a non-ionic one. The first studies on porous PILs involved the co-polymerization of various IL monomers using divinylbenzene (DVB) as a non-ionic cross-linker. DVB is a rather usual cross-linker, commonly used for the production of porous polystyrene–divinylbenzene (PS–DVB) resins. The structural rigidity of DVB leads to inefficient packing of the polymeric chains, opening up a route to generate microporosity by means of free-radical polymerization.

Ding *et al.* reported the first study in this field, working on the co-polymerization of 3-butyl-1-vinylimidazolium chloride with DVB.[24] The reaction was performed in chloroform, which acts as the only porogenic solvent. SEM images revealed a rough surface of the PIL microparticles; however, no direct evidence was provided to estimate the SSA. This material was employed to catalyze the coupling between carbon dioxide and various epoxides in order to form cyclic carbonates.[24]

Two subsequent studies, performed using DVB as a cross-linker for PILs, extensively investigated the effect of imidazolium alkyl chains of different length, the employed solvents and the reaction temperature on the resulting porous PILs, with the aim to acquire information on the porous structure. The co-polymerization reaction of different imidazolium monomers was run both under reflux and in an autoclave, in the latter case exceeding the boiling point of the reaction solvent and generating micro-mesoporous PIL monoliths with a BET SSA ranging from 100 to 900 $m^2 g^{-1}$, depending on the reaction parameters (see Figure 2.5a).[25,26] An analogous synthetic pathway involves the synthesis of the corresponding non-ionic copolymer between 1-vinylimidazole and DVB. Subsequently, the polymer is alkylated to introduce the imidazolium moiety into the porous structure. The final results in terms of SSA and pore size were similar to those of the PILs made from ionic liquid monomers. However, this synthetic strategy allows the properties of the non-ionic porous polymer to be compared directly with those of the porous PIL, revealing the specific influence of the ionic liquid moiety in various catalytic reactions (see Figure 2.5b).[27–29]

A step forward in this field was the employment of IL cross-linkers, instead of non-ionic cross-linkers such as DVB, which are functional and structural at the same time. For this purpose, bis-vinylimidazolium salt monomers with alkyl chains of different lengths and 1,3-bis(4-vinylbenzyl)imidazolium chloride were exploited in an attempt to obtain porous PILs from free-radical polymerization.[30] The syntheses performed with IL cross-linker monomers clearly evidence that the first mandatory condition to obtain porous PILs is the rigidity of the monomer structure, and if this condition is satisfied, the synthesis parameters have a minor influence on the porosity of the final material. For this reason, careful choice of the starting monomer and fine-tuning of the synthesis conditions are very important. The polymerization of 1,3-bis(4-vinylbenzyl)imidazolium chloride (see the structure in Figure 2.6, part b), performed in reflux of ethanol, leads to PIL microparticles of less than 200 nm and a BET SSA of only 24 $m^2 g^{-1}$, probably arising from the inter-particle voids.[30] In this case, even if the monomer has two relatively rigid styrene groups, the methylene bridges of the monomer give enough flexibility to the resulting polymer chains to cause the collapse of the network into bulky nanoparticles.

Better results in terms of SSA were obtained in the case of bis-vinylimidazolium salt monomers with variable lengths of the alkyl chain connecting the two imidazolium moieties (see the general structure in Figure 2.6, part a). In particular, two studies deal with bis-vinylimidazolium salt polymerization under ionothermal conditions in the presence of various ionic liquids as solvents and possibly water as a co-solvent.[31,32] The IL employed as a solvent can act as porogenic agent and, thanks to the high affinity between the IL solvent and the IL monomer, very small pores can be obtained.[31] On the other hand, if water is used as a co-solvent, it supposedly forces the IL solvent

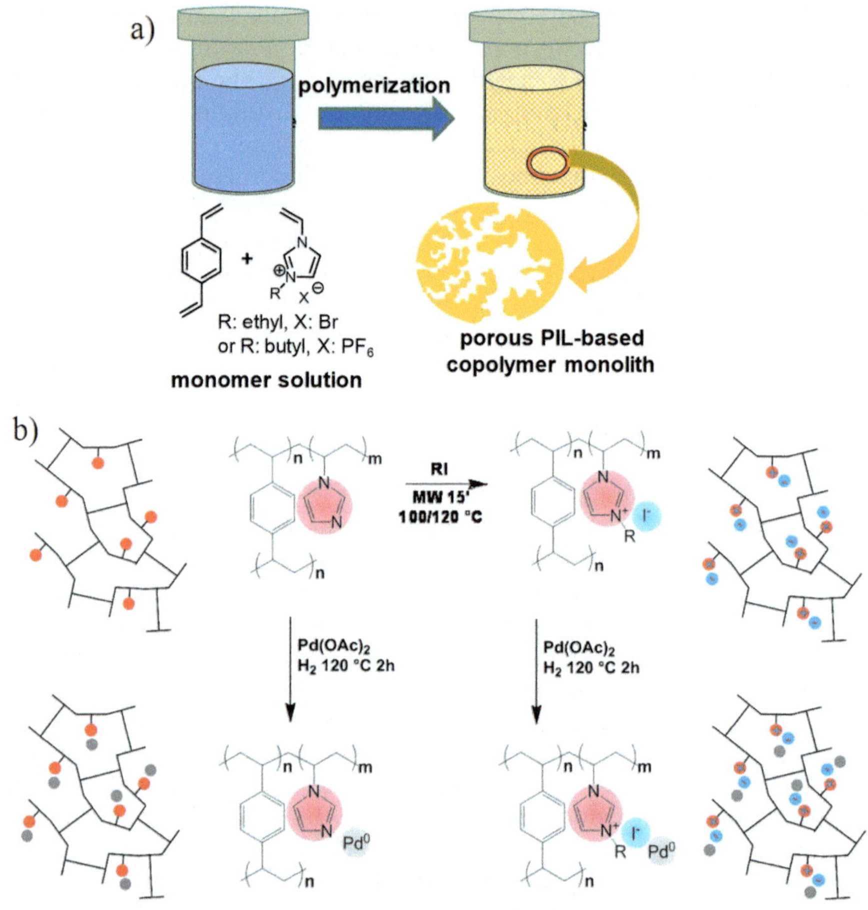

Figure 2.5 Section (a): scheme of the solvothermal synthesis of various porous PILs performed in an autoclave using an imidazolium starting monomer and DVB as a cross-linking agent. Reprinted from *Polymer*, 55 (16), D. Kuzmicz, P. Coupillaud, Y. Men, J. Vignolle, G. Vendraminetto, M. Ambrodhi, D. Taton, J. Yuan, Functional mesoporous poly(ionic liquid)-based copolymer monoliths: From synthesis to catalysis and microporous carbon production, 3423–3430, Copyright 2014 with permission from Elsevier.[25] Section (b): synthesis scheme of a porous polymer obtained from imidazole and DVB and its following alkylation step that leads to various porous PILs. These porous polymers were used to host Pd nanoparticles in order to study how the ionic environment influences the catalytic properties. Reproduced with permission from ref. 28 (Copyright 2015, American Chemical Society).

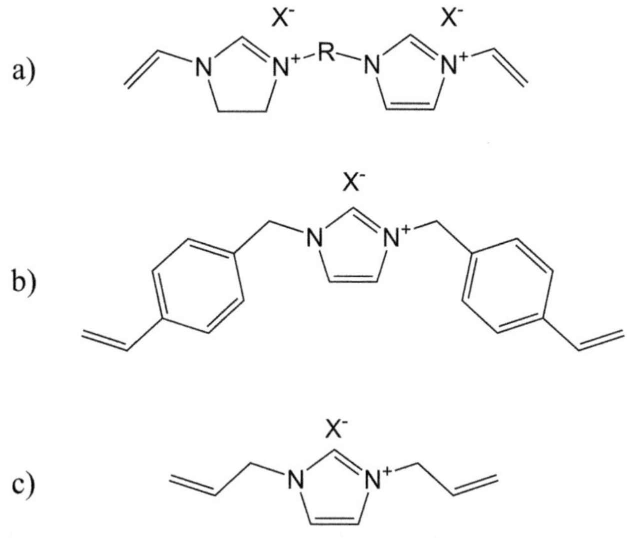

Figure 2.6 Structures of the most common ionic liquid monomers employed for the synthesis of PILs.

to aggregate in an ordered fashion creating a nano-structural organization, which is able to act as a soft templating agent for mesopore formation, rather than being just a porogenic solvent.[32] The bis-vinylimidazolium salt monomers with two or four carbons in the alkyl chain usually generate porous networks with SSAs below 90 $m^2 g^{-1}$, mainly due to the flexibility of the resulting polymer structures. The best results were obtained in the case of bis-vinylimidazolium with an alkyl chain of only one methylene group. In such a case, the structure of the monomer is very rigid, and by means of an appropriate IL/water solvent mixture, a hierarchically structured meso-macroporous PIL monolith with a BET SSA as high as 205 $m^2 g^{-1}$ is formed.[32]

Very recently, 1,3-diallylimidazolium salts (see the structure in Figure 2.6, part c) were co-polymerized with tris(4-vinylphenyl)phosphine. Both of these monomers can act as a cross-linker and by means of hydrothermal free-radical polymerization in DMF they were able to form a micro-mesoporous PIL with a SSA as high as 591 $m^2 g^{-1}$, arising from the very rigid structure of the phosphine monomer.[33]

Yan and Texter reported a synthetic strategy for obtaining PIL copolymers that can be tuned between a hydrogel and a porous structure by simply changing the anions of the ionic liquid moiety. The starting monomer, 1-(2-acryloyloxyun-decyl)-3-methylimidazolium tetrafluoroborate, is a surfactant and was polymerized in a H_2O/1-propanol/methyl methacrylate mixture, with a variable percentage of ethylene glycol dimethacrylate as a cross-linking agent. When the co-polymer is exchanged to a PF_6^- anion, the structure can switch from a bulky morphology, if the material is immersed in aqueous DMSO, to a porous morphology of highly interconnected macropores, if the co-polymer is immersed in pure water. This morphology shift is ascribable to the swelling and deswelling of the co-polymer depending on the solvent polarity.[34]

2.2.1.2.2 Direct Synthesis of Polymeric Networks. The direct synthesis of porous PILs is a flourishing method in the field of porous PIL synthesis, since it opens up a route to obtain microporous materials, avoiding the use of a templating agent (with its challenging etching step) and directly generating the porous network, usually in just one step. In order to introduce microporosity into organic materials, it is important to prevent efficient packing of the polymer chains that irremediably leads to a bulky non-porous material. This is generally true for all polymers, but it is particularly relevant in the case of charged polymer backbones, because their coulombic interactions cause the collapse of the polymer chains together. As in the case of free radical polymerization, the resulting polymer network should be stiff enough to avoid the collapse of the pores, which arises from bending and twisting chains, and to counteract the high capillary pressure and surface energy present in the microporous network. A contorted polymer chain with a high degree of cross-linking or connectivity is then helpful to stabilize the micropore structure and to minimize the swelling of the material in the presence of solvents or gas molecules. In contrast with the free radical polymerization technique, direct synthesis mainly uses condensation polymerization, or other C–C coupling methods, to build up the network between tritopic or tetratopic monomers, which are sterically hindered in order to introduce inefficient packing of the resulting polymer chains. It is worth noting that, due to the difficulty in finely controlling the ionic network synthesis, a small fraction of mesopores is usually present in addition to a large fraction of micropores. The removal of the solvent is generally the critical step of the direct synthesis route. In fact, the solvent is able to keep the polymer structure in a swollen state with open pores and, during its removal, the residual mobility of the polymer chain can lead to the collapse of the pore network. Hence, to avoid this possibility, solvent removal is usually performed *via* sublimation employing supercritical carbon dioxide drying or freezer drying from water. In this way, thanks to the low temperature of these processes, the mobility of the polymer chains is limited and the porous structure is preserved.

According to their definition, PILs are polymers obtained from the polymerization of ionic liquid monomers. This definition is in good agreement with free radical polymerization, but does not match the direct synthesis of porous PILs. In fact, in the direct synthesis route, the ionic liquid moiety can be directly generated during the formation of the polymer network, without passing through an isolable ionic liquid. Thus, the porous ionic liquids obtained from direct synthesis are rarely indicated as PILs, whereas they are more properly termed porous cationic/anionic polymers. Despite their official definition, it is worth including these materials in this chapter, because they still have the ionic liquid moiety in their structure (giving them the chemical properties of porous PILs), combined with the advantage of a microporous texture. A wide variety of direct synthesis approaches have been reported in the literature, and, therefore, it is very difficult to find some common character among all the different syntheses. In order to give a brief schematic overview, we will initially discuss the ionic polymers obtained starting from a pre-formed ionic monomer, and then the materials generated starting from a non-ionic monomer and in which the ionic moiety is generated *in situ* during the synthesis of the polymer network.

Direct synthesis starting from an ionic monomer requires the synthesis of a rigid multitopic salt based on B^-, P^+ or N^+ ions. A polymer network is obtained by connecting the ionic monomers by means of various kinds of bond formation, such as Yamamoto-type Ullmann cross-coupling, Sonogashira–Hagihara cross-coupling, Schiff base reaction or Friedel–Crafts reaction. Sometimes, it is also necessary to employ a non-ionic multitopic monomer to obtain the cross-links necessary for the creation of the polymer network.

The first microporous anionic borate network was prepared by taking advantage of the Sonogashira–Hagihara coupling of tetrakis(4-bromo-2,3,5,6-tetrafluorophenyl)borate salt with 1,3,5-triethynylbenzene. The resulting anionic network shows a BET SSA of 761 $m^2\ g^{-1}$ and a pore volume of 0.54 $cm^3\ g^{-1}$.[35] Similar borate microporous networks were also obtained starting from analogous multitopic ethynyl tectons and tetrakis(4-bromophenyl)borate salts.[36] Phosphonium microporous polymer networks were obtained for the first time starting from tetrakis(4-chlorophenyl)phosphonium bromide, self-cross-linked by means of a Yamamoto-type cross-coupling. The resulting materials exhibit BET SSAs ranging from 650 to 980 $m^2\ g^{-1}$, depending on the halogen anion.[37] Friedel–Crafts catalysis was also used to react various phenyl phosphonium salts in the presence of benzene, in order to obtain hypercross-linked meso/microporous polymers with BET SSAs ranging from 770 to 1168 $m^2\ g^{-1}$, depending on the starting phosphonium salt.[38]

Various rigid imidazolium ditopic salts were synthesized by means of the Debus–Radziszewski imidazolium synthesis, which leads to an imidazolium ring directly conjugated to a functionalized phenyl ring. The imidazolium monomers formed with this synthesis only possess a double functionality; therefore, in order to generate a network, they need to react with a multitopic monomer. Son and co-workers report the first example of this polymer synthesis taking advantage of the Sonogashira coupling of 1,3-bis(2,6-diisopropyl-4-iodophenyl)imidazolium chloride and tetrakis(4-ethynylphenyl)methane, obtaining a microporous material with a BET SSA of 620 $m^2\ g^{-1}$ (see Figure 2.7a).[39] Using the same concept, but through Suzuki–Miyaura cross-coupling and Yamamoto reaction, similar microporous imidazolium networks were also obtained (see Figure 2.7b and c).[40,41]

A ditopic pyridinium ionic monomer was also obtained from the quaternarization of 4,4′-bipyridine. Subsequently, a microporous cationic polymer was generated by the Sonogashira–Hagihara cross-coupling of the just formed 1,1′-bis-(4-iodophenyl)-[4,4′-bipyridine]-1,1′diium salts with different anions (Cl^-, BF_4^- and PF_6^-) and tetrakis(4-ethynylphenyl)methane (see Figure 2.8a). The resulting microporous polymers exhibit BET SSAs in the

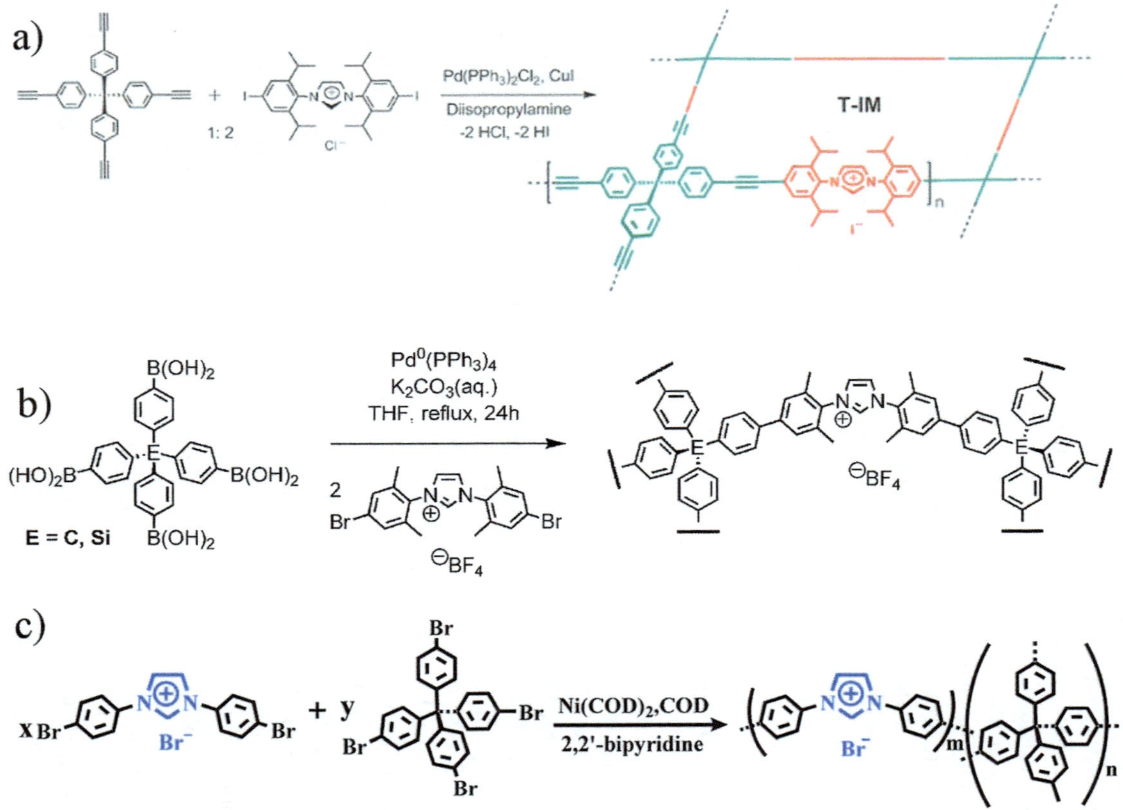

Figure 2.7 Main strategies for the direct synthesis of porous PILs starting from bifunctional imidazolium ionic liquid monomers. The materials are obtained by means of Sonogashira coupling (section a), Suzuki–Miyaura cross-coupling (section b) and Yamamoto reaction (section c). Adapted from ref. 39, ref. 40 and ref. 41 with permission from The Royal Society of Chemistry.

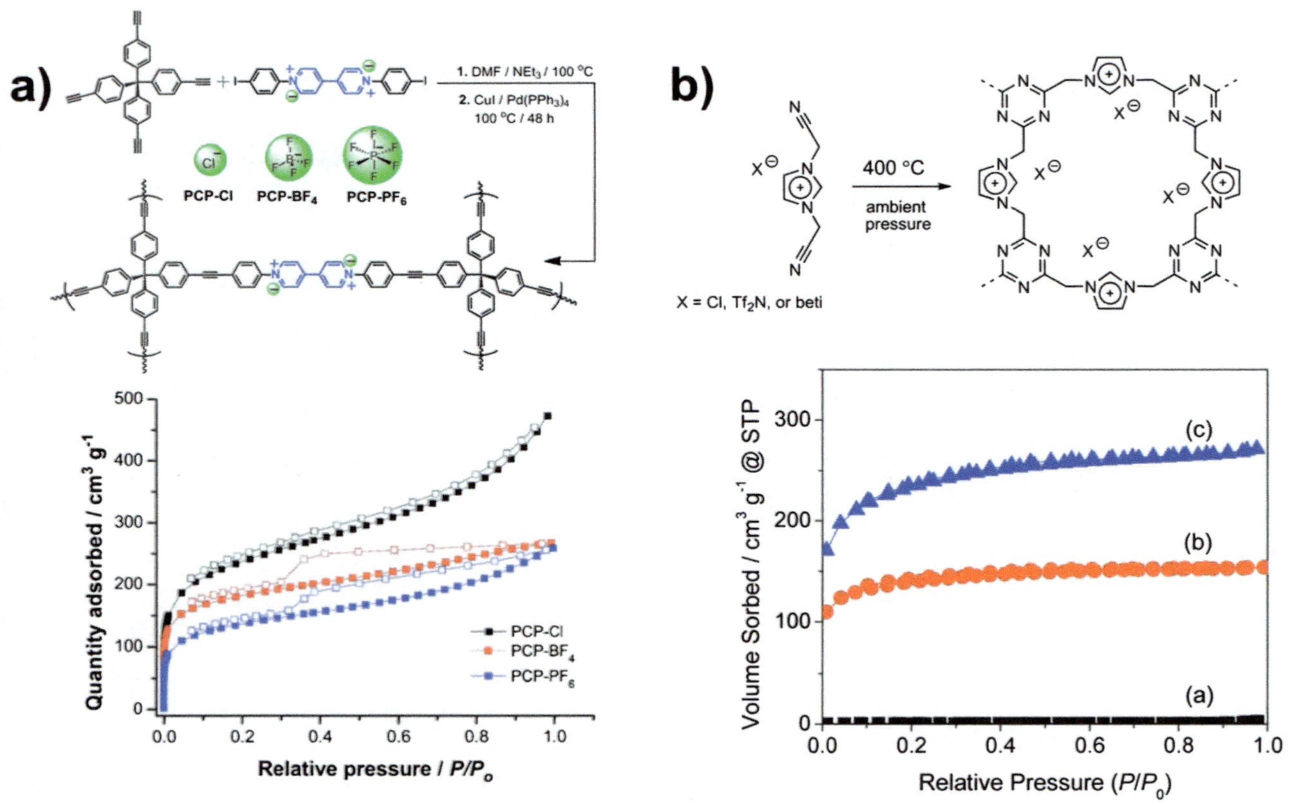

Figure 2.8 Section (a): direct formation of porous PILs by means of Sonogashira coupling starting form a pyridinium bifunctional monomer. The SSA of the sample is inversely proportional to the size of the anion, as clearly visible from the N_2 adsorption isotherms at 77 K. Adapted from ref. 42 with permission from The Royal Society of Chemistry. Section (b): formation of porous PILs by ionothermal cyclotrimerization reaction of imidazolium nitrile moieties. Also in this case, the N_2 adsorption isotherms at 77 K show that the SSA inversely depends on the size of the anion. Adapted with permission from ref. 43, Copyright 2009, American Chemical Society.

433–755 $m^2 g^{-1}$ range, depending on the nature of the anion.[42] By means of an innovative approach, Dai *et al.* used the ionothermal cyclotrimerization reaction of nitrile to cross-link a cyano-functionalized imidazolium ionic liquid and connect the cations forming a polymer network based on 1,3,5-triazine (see Figure 2.8b).[43] The negligible vapor pressure of the ionic liquid was the key point for the development of the porous network, since the cyclotrimerization reaction was performed at 400 °C. The results reported in this study claim that the anion of the ionic liquid can act as a templating agent. In fact, different anions lead to porous PILs with very different BET SSAs, ranging from 2 $m^2 g^{-1}$ in the presence of chloride to 814 $m^2 g^{-1}$ in the case of bis(pentafluoroethylsulfonyl)imide.[43]

Another example of a quaternary ammonium network comes from the Schiff base reaction between ethidium bromide and 1,3,5-triformylphloroglucinol (see Figure 2.9).[44] The Schiff base reaction, under proper conditions, is reversible, and long reaction times lead to the thermodynamic equilibrium structure of the porous polymer network, *i.e.* an ordered porous

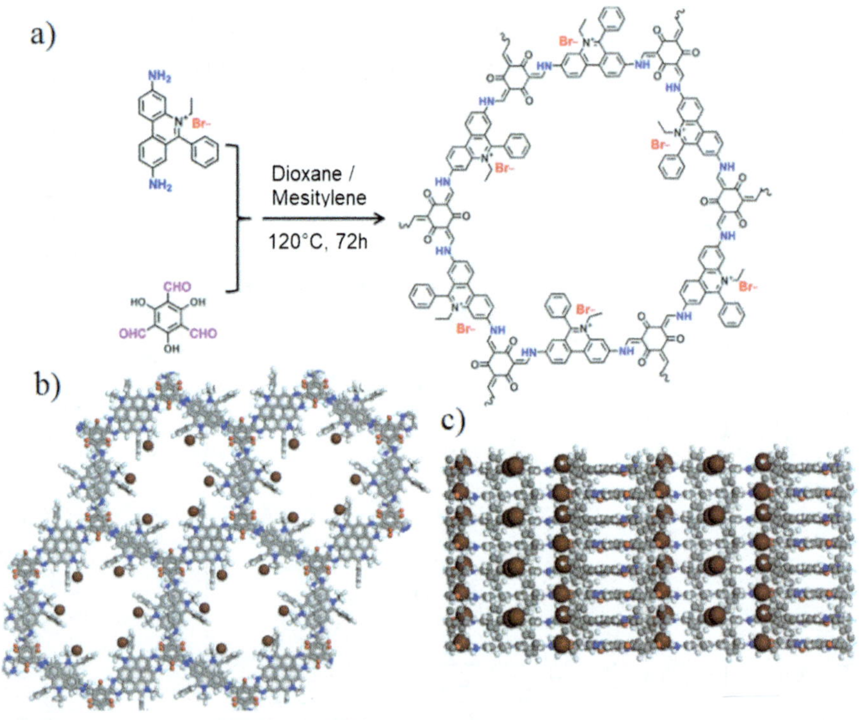

Figure 2.9 Section (a): the synthesis scheme of a porous covalent organic framework obtained from the Schiff base reaction of ethidium bromide (bearing a quaternary nitrogen) and 1,3,5-triformylphloroglucinol. The obtained material structure is ordered as illustrated in the pictures in sections (b) and (c). Reproduced with permission from ref. 44, Copyright 2016, American Chemical Society.

structure. This peculiar ordered porous structure makes this polymer definable as a covalent organic framework (COF). If this material is obtained bearing fluorine as the anion, the BET SSA shows a value of 1002 $m^2\ g^{-1}$, progressively decreasing in the case of materials bearing other halide anions (Cl^-, Br^- or I^-) to values of 954, 774 and 616 $m^2\ g^{-1}$, respectively.[44] Two different COF ionic polymers were obtained using triaminoguanidinium halides and 1,1-bis(4-formylphenyl)-4,4′-bipyridinium dichloride as starting quaternary ammonium sources.[45,46]

Porous ionic networks may also be synthesized starting from non-ionic monomers, by creating the ionic moiety and the porous network simultaneously. A recent example, reported by Dani *et al.*, takes advantage of the click-based Debus– Radziszewski imidazolium synthesis to directly cross-link the amino group of tetrakis(4-aminophenyl)methane (see Figure 2.10a).[47] This reaction generates an imidazolium moiety that connects the tectons together. The resulting cationic polymer bears the acetate anion, which was subsequently exchanged with other anions, such as PF_6^-, Tf_2N^-, BF_4^- and TfO^-. All the synthesized polymers are microporous and their BET SSAs range from 176 to 426 $m^2\ g^{-1}$, depending on the anion present in the network.[47] All the porous ionic networks synthesized by Dani *et al.* exhibit peculiar isotherms for both N_2 adsorption at 77 K and CO_2 adsorption at 298 K (see Figure 2.10b and c, respectively), in which the desorption branch of the isotherms is not close to the adsorption branch even at low relative pressures. As reported in the introduction section of this chapter, these particular adsorption/desorption isotherms are indicative of swelling effects, derived from the elastic deformations of the polymeric network during the gas adsorption process. The SSA strictly depends on the size of the anion as evidenced by the N_2 isotherms at 77 K (see Figure 2.10b). The presence of the imidazolium ionic liquid moiety, together with the microporous structure, makes these PILs promising materials for CO_2 capture. In fact, they exhibit excellent behavior towards carbon dioxide adsorption, either at 298 K and 1 bar, with a CO_2 loading above 1 mmol g^{-1} (see Figure 2.10c), or even more at 273 K and 1 bar, where the loading was 2 mmol g^{-1}. To the best of our knowledge, the above reported values are the highest in the field of imidazolium PILs.

Two similar microporous PILs were obtained *via* a two-step synthesis of the imidazolium ring, passing through the polymer bearing the diimine and closing the imidazolium ring just in a successive step.[48,49]

Another way to directly create porous ionic polymers, starting from a non-ionic monomer, is the alkylation of a multifunctional nitrogen monomer using a multivalent alkylating agent to generate the porous network, as reported by Chen *et al.*[50] This method was employed to alkylate the free electronic doublet of nitrogen present in various organic linkers, based either on imidazole, pyridine, amines or 1,4-diazabicyclo[2.2.2]octane. All these molecules are ditopic and they are alkylated using octakis(chloromethyl)silsesquioxane, which is a multitopic monomer. The present study is a clear example of how the rigidity of the ionic liquid moiety affects the BET SSA of the final polymer network. In fact, the higher the flexibility of the linker structure,

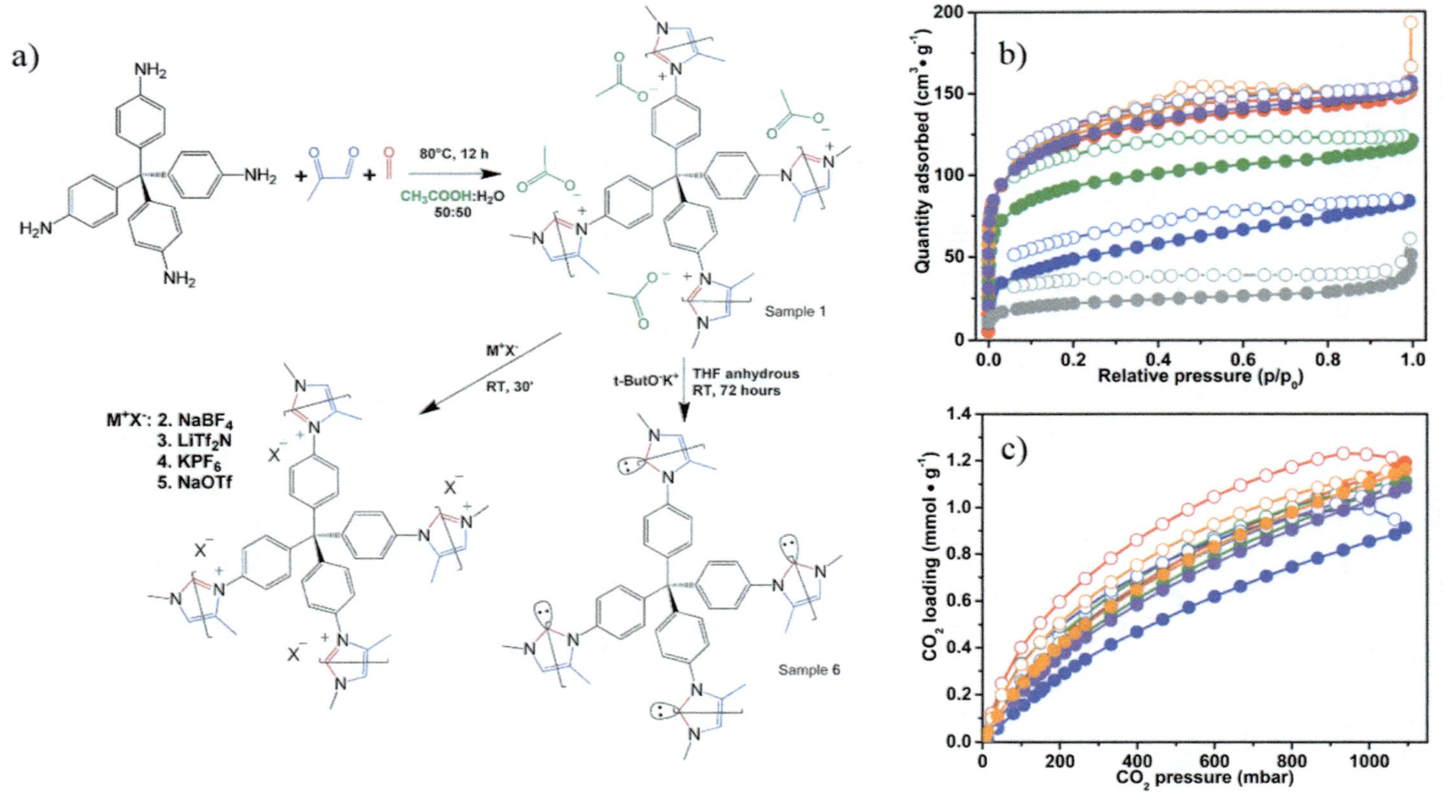

Figure 2.10 Section (a): synthesis scheme of porous PILs obtained by Debus–Radziszewski click reaction to directly create the imidazolium moiety during the synthesis of the porous network. The porous PILs are obtained with different anions and the carbene moiety was also introduced on the imidazolium ring. Section (b): N_2 adsorption/desorption isotherms at 77 K. Section (c): carbon dioxide adsorption isotherms at 298 K. Adapted from ref. 47 with permission from The Royal Society of Chemistry.

the lower the surface area of the final material, ranging from 1025 $m^2\ g^{-1}$ for the 4,4′-bipyridine linker to 26 $m^2\ g^{-1}$ for the 1,3-bis(4-pyridyl)propane linker, which possesses a propyl alkyl chain within its structure.[50]

Nitrogen alkylation was also performed on *N*-methylimidazole using various bifunctional benzyl halides *via* a one pot synthesis, in which the so-formed ionic molecules were directly hypercross-linked using the Friedel–Craft reaction. The obtained porous cationic polymers exhibit mostly microporous structures, with also a small fraction of both meso- and macropores. The BET SSAs are rather high, ranging from 530 to 926 $m^2\ g^{-1}$, depending on the starting benzyl halide.[51] Zhu and co-workers recently reported an innovative porous cationic polymer created by the condensation of 4-pyridinylboronic acid and cyanuric chloride (see Figure 2.11).[52] In this synthesis, the quaternarization of the pyridine moiety to form a pyridinium salt and the trimerization of the boronic acid occur simultaneously, generating a

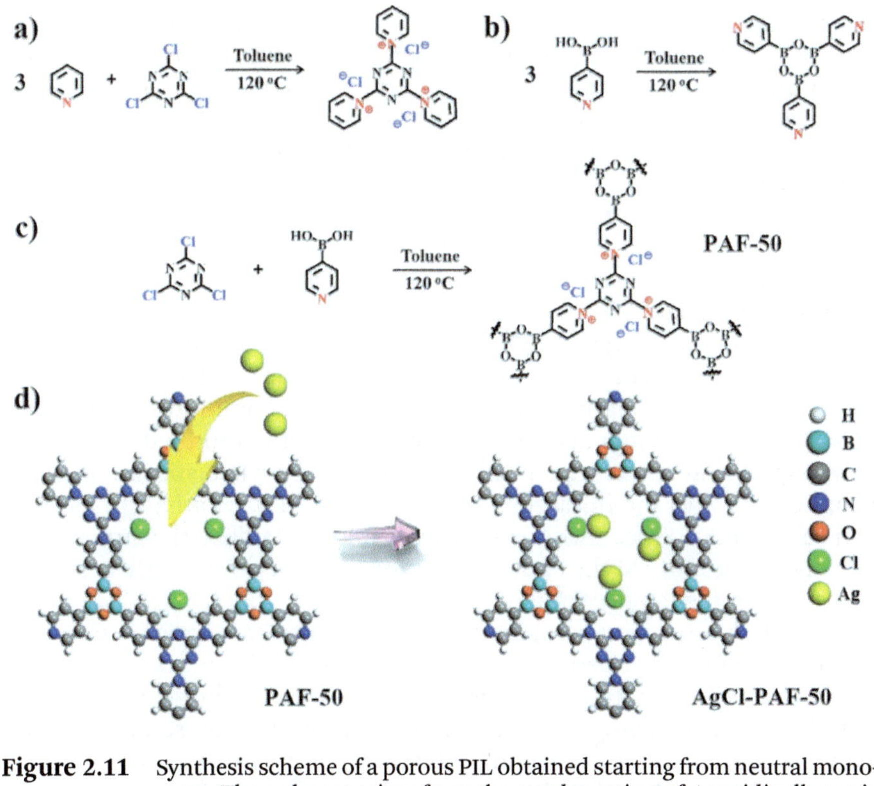

Figure 2.11 Synthesis scheme of a porous PIL obtained starting from neutral monomers. The polymer arises from the condensation of 4-pyridinylboronic acid and cyanuric chloride. The formation of the pyridinium moiety and network formation occur simultaneously. These porous PILs can be obtained bearing various halide anions and they are used as hosts for silver atoms. Adapted from Y. Yuan, F. Sun, F. Zhang, H. Ren, M. Guo, K. Cai, X. Jing, X. Gao, G. Zhu, *Advanced Materials*, John Wiley and Sons, Copyright © 2013 WILEY-VCH Verlag GmbH & Co. KGaA, Weinheim.[52]

microporous polymer network with a BET SSA of 384 $m^2 g^{-1}$. In a subsequent study, the chlorine anion of this material was exchanged with other halides anions, leading to fine tuning of the pore size and making this material useful for the separation of various gas mixtures.[53]

2.2.1.2.3 Polyelectrolyte Complexation. Previously, we introduced several methods to create porous PILs exploiting free radical and condensation polymerization. Both these methodologies are employed to generate covalently cross-linked polymer networks. An alternative way to cross-link polymer chains is to take advantage of the inter-polyelectrolyte complexation that takes place between polyelectrolytes bearing opposite charges along their backbones. PILs are polyelectrolytes bearing IL moieties as permanent charges (usually positive) and, therefore, an electrostatic cross-link can be made between a PIL and a negatively charged polyelectrolyte like neutralized poly(acrylic acid) (PAA) or a multivalent acid molecule. The generation of the electrostatic cross-link leads to a polymer network and may also induce phase separation of the network from the solvent. This phase separation, occurring at the nano/microscale level, produces porosity inside the structure of the material, while the simultaneous formation of a cross-linked network is able to stabilize the as-formed porous structure.

Taking advantage of the so-called "acidic blending process", followed by deprotonation, Zhao and co-workers were able to generate a series of meso- and/or macroporous PILs with various applications.[54] In a typical synthetic process, a hydrophobic PIL is solubilized in DMF (or another non-protic solvent), together with a multivalent carboxylic acid or with PAA. The acid group is in the protonated form and, since DMF is an aprotic solvent, no acid–base equilibrium takes place; therefore, no complexation is possible allowing the "acidic blending process". The PIL/acid solution is then added dropwise into an organic solvent containing a small percentage of ammonia (commonly 0.5 wt%). The ammonia acts as an activating agent deprotonating the carboxylic groups and, subsequently, the formed carboxylate species undergo complexation with the imidazolium cations of the PIL creating the electrostatic cross-links. Since the cross-links are able to stabilize the structure of the polymer network at the early stage of the phase separation, the resulting pore size is in the range of micro/mesopores.

If inter-polyelectrolyte complexation is applied to conventional polyelectrolytes, the porous properties of the resulting materials are not noteworthy. In fact, due to their charges, conventional polyelectrolytes are all water-based and, therefore, the complexation is performed in aqueous solution, *i.e.* in the presence of strong interacting forces. In contrast to the conventional polyelectrolytes, PILs, with their peculiar property of being a permanently charged polymer with a tunable polarity that depends on the anion–cation pair, opened up a route to obtain porous polyelectrolytes by means of inter-polyelectrolyte complexation. In fact, in the case of PILs, the polymer cross-linking occurs in organic media, where the set of interaction forces is completely different. In particular, the hydrophobic interaction is null, the hydrogen bonds are weakened or canceled and the columbic interactions have a different character. This situation allows the formation of a micro/mesoporous structure of the polymer complex.[2]

The first porous PILs obtained using this technique were synthesized starting from a homogeneous DMF solution of poly(3-cyanomethyl-1-vinylimidazolium) bearing various anions (Tf_2N^-, PF_6^-, BF_4^-, Br^-) and PAA, and were precipitated in ethanol containing 0.5 wt% NH_3 (see Figure 2.12, left side).[55] The polyelectrolyte complex precipitates as a powder consisting of aggregated nanoparticles of 30–50 nm diameter, as pointed out by the SEM image reported in Figure 2.12 (left side). The nanoparticles exhibit an internal micro/mesoporous structure, as highlighted by the N_2 isotherms at 77 K (see Figure 2.12, left side), conferring to the material a BET SSA as high as 310 $m^2\ g^{-1}$ and a total pore volume of 0.98 $cm^3\ g^{-1}$.[55] This synthetic route has been demonstrated to be reliable and applicable for different PIL/multi-polyacid combinations. Other porous PILs, with a similar porous structure, were reported starting from PIL/multi-carboxylic acid, PIL-*co*-PAA random copolymers, and a zwitterionic PIL homopolymer with a carboxylic group attached to the imidazolium alkyl chain.[56–58]

The inter-polyelectrolyte complexation of PILs can be also employed to generate membranes covering a broad range of shapes, porosities, chemical structures and applications (see Figure 2.12, right side).[59,60] In a typical synthesis, the solution of PIL and multi-carboxylic acid/polyacid is cast onto a glass plate and dried to evaporate the solvent. Then the as obtained film is immersed in an aqueous ammonia solution. This solution is able to penetrate the cast mixture of the two intimately mixed components, because the acid component is water soluble. The deprotonation is triggered when NH_3 is put in contact with the carboxylic acid group, which is able to complex with the PIL chains generating the electrostatic cross-links of the polymer chains.[59] Quasi-solid state chemistry is involved in the membrane synthesis mechanism with respect to the porous PIL synthesis, since only one of the two intimately mixed components is readily soluble in the ammonia solution. Therefore, the phase separation step occurs at a slightly later stage leading to a final macroporous membrane, with a pore size ranging between 30 and 100 nm.

As the membrane is cast onto a glass surface, the ammonia solution penetrates inside the film from top to bottom, creating a gradient of ammonia concentration, which, in turn, generates a gradient of cross-linking density along the membrane cross-section. Thanks to this peculiar property, the asymmetric swelling of the membrane forces the structure to fold in the time scale of seconds in the presence of organic solvent, either in solution or in the gas phase. After removal of the organic solvent, the starting structure is restored, allowing these membranes to work as a perfect actuator.[60,61] The pore size of these membranes can be simply tuned by changing the multivalent carboxylic acid used as the multi-anion.[62]

Interestingly, it is possible to tune the membrane pore structure by changing the redox potential in a membrane made from a redox-responsive PIL. Vancso and co-workers reported the synthesis of a macroporous membrane starting from a poly(ferrocenylsilane)-based PIL and PAA.[63] The immersion of the membrane in aqueous ammonia deprotonates the PAA triggering the

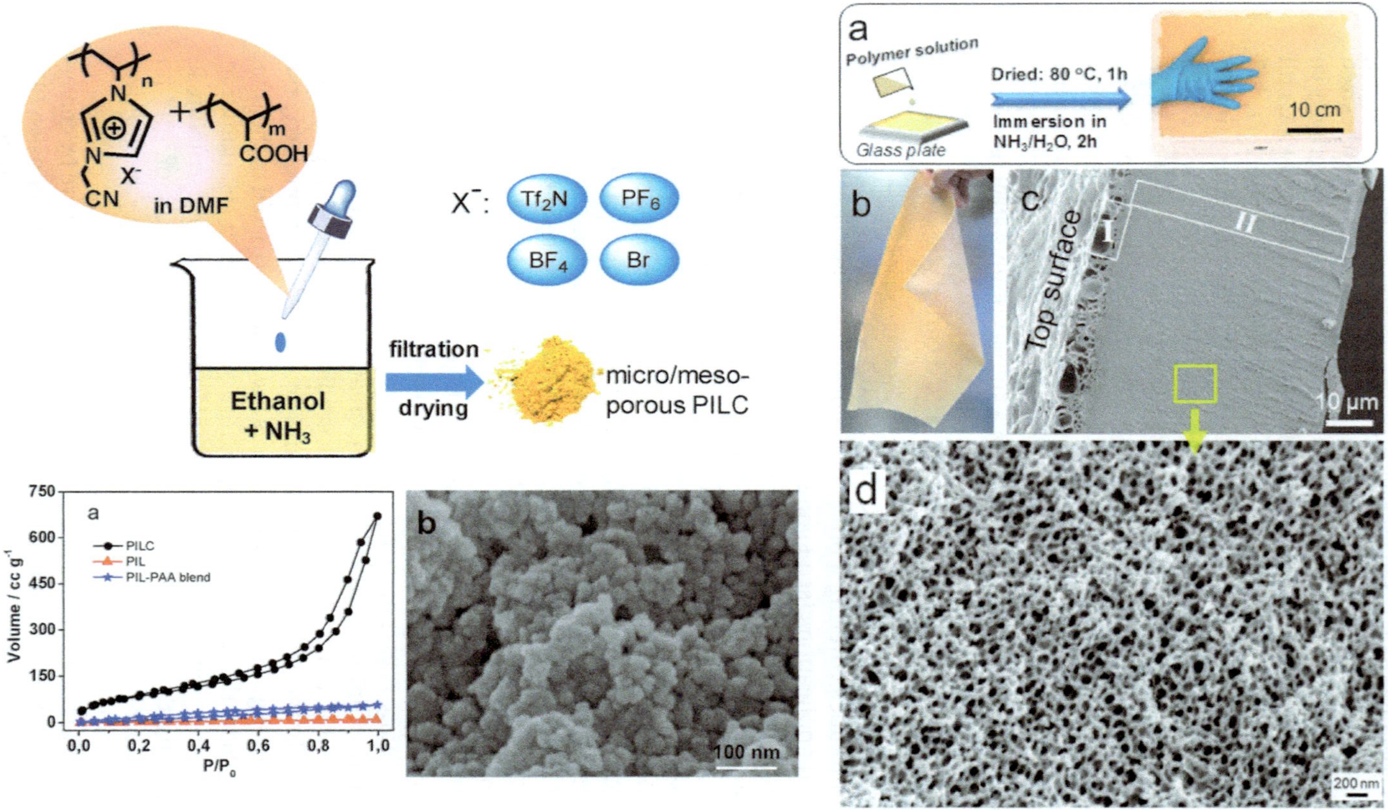

Figure 2.12 Left side: synthesis scheme, N_2 adsorption/desorption isotherms at 77 K (a) and SEM image (b) of a porous PIL obtained *via* polyelectrolyte complexation of various imidazolium PILs and PAA. Adapted with permission from ref. 55, Copyright 2012 American Chemical Society. Right side: (a) scheme of the preparation of a nanoporous polyelectrolyte membrane from a solution mixture of a cationic poly(ionic liquid) and poly(acrylic acid) in DMF; (b) photograph of a 35 cm × 25 cm free-standing membrane; (c and d) representative SEM images of the cross-sectional area of the as-prepared membrane. Adapted with permission from ref. 59. Copyright 2013 American Chemical Society.

polyelectrolyte complexation (see Figure 2.13). The resulting membrane exhibits an interconnected pore structure and the opening and closing of the pores is regulated by the redox state of the system. In particular, if the ferrocenyl iron is in the oxidized form, the density of the open pores is higher. The ability of the pore system to breathe on command is useful to tune the permeability of the membrane and, as a consequence, to regulate the water flux across it.[63] A similar pore switching process, regulated by the solvent in contact with the membrane, was reported for a PIL/multi-carboxylic acid realized with a hydrophilic PIL.[64] In this specific case, the PIL chains swell in the presence of water blocking the membrane pores, but when the membrane is in contact with isopropanol, the hydrophilic PIL structure de-swells and the pores open again.[64] Membranes formed by pyridinium PILs have also been synthesized,[64,65] thus highlighting the extreme flexibility of this synthetic route to generate porous polyelectrolyte membranes.

Since electrostatic cross-linking does not lead to porous PIL membranes that are stable in high ionic strength environments, a method was recently developed to introduce covalent cross-links in addition to the existing electrostatic ones. A membrane with a residual dicyanamide anion and a nitrile alkyl chain dangling from the imidazolium PIL was annealed at 200 °C, in order to create 1,3,5-triazine rings able to covalently cross-link the PIL chains (see Figure 2.14).[66] The starting porous structure of the membrane is retained after the annealing, and the covalently cross-linked membrane has proven to be stable in various salt solutions, *i.e.* in highly ionic solution environments, as demonstrated by the SEM image reported in Figure 2.14b, in which it is evident that the porous structure of the membrane is unaltered in terms of pore size and distribution after the salt treatment.[66]

2.2.2 Supported Porous Ionic Liquids

Ionic liquids (ILs) are usually very viscous liquids, and in some cases (when a symmetric plane is present in both the cation and the anion), they are solid at room temperature. Due to their high viscosity in the liquid phase, ionic liquids are clearly not able to generate porous materials, and, at the same time, if they are in the solid state, they crystalize in ordered bulky reticles similar to all inorganic salts. The synthetic strategies introduced in the previous sections allowed the creation of new porous materials bearing ionic liquid functionalities from scratch. A different approach to create porous materials with ionic liquid moieties is to immobilize ionic liquids on a previously generated porous material. The materials belonging to this class are called supported porous ionic liquids (SPILs).

The immobilization of the ILs on a porous support was extensively studied in the last few years, not only with the aim of creating new porous materials bearing IL functionalities, but also to improve the handling of the ILs and to enhance the properties of the final composite. Both polymeric and inorganic porous materials, amorphous or ordered, have been studied as potential supports for ILs. Selvam *et al.*[67] classified the supported ionic liquids (SILs) into three main categories: supported ionic liquid catalysts (SILCs), solid catalysts

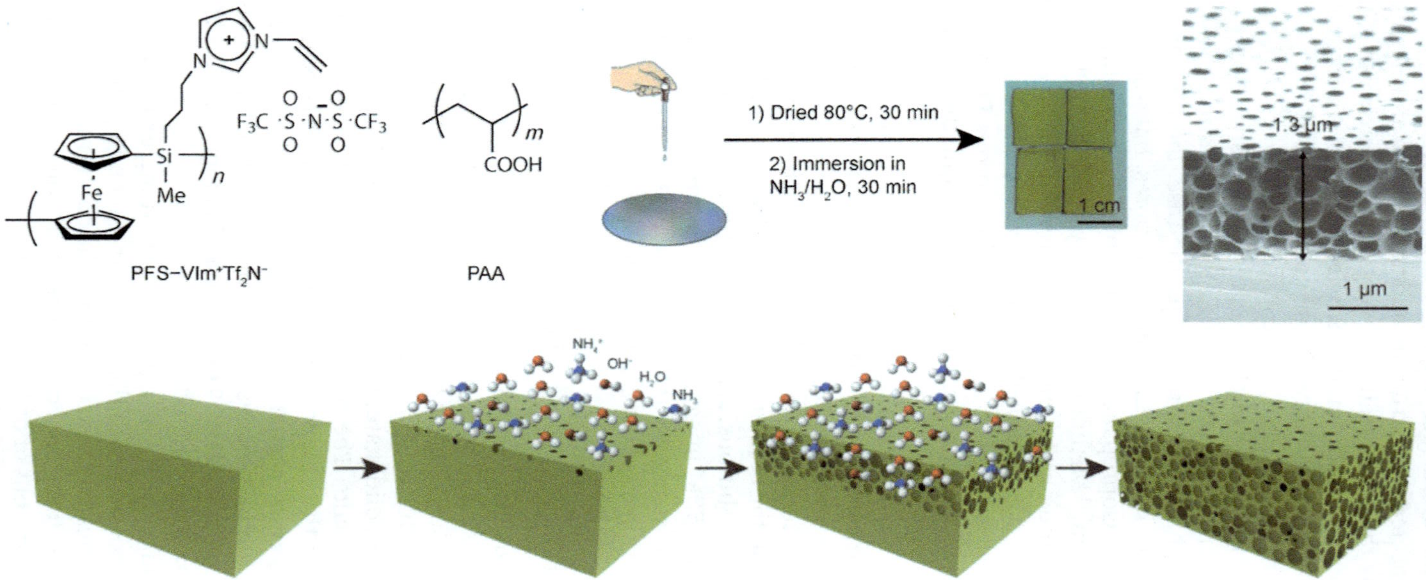

Figure 2.13 Synthesis scheme of a polyelectrolyte complexation membrane obtained from the complexation of a poly(ferrocenylsilane)-based PIL with PAA. The immersion of the membrane in aqueous ammonia deprotonates the PAA triggering the polyelectrolyte complexation. The resulting membrane exhibits a macroporous structure. Adapted from K. Zhang X. Feng, X. Sui, M. A. Hempenius, G. J. Vancso, *Angewandte Chemie International Edition*, John Wiley and Sons, Copyright © 2014 WILEY-VCH Verlag GmbH & Co. KGaA, Weinheim.[63]

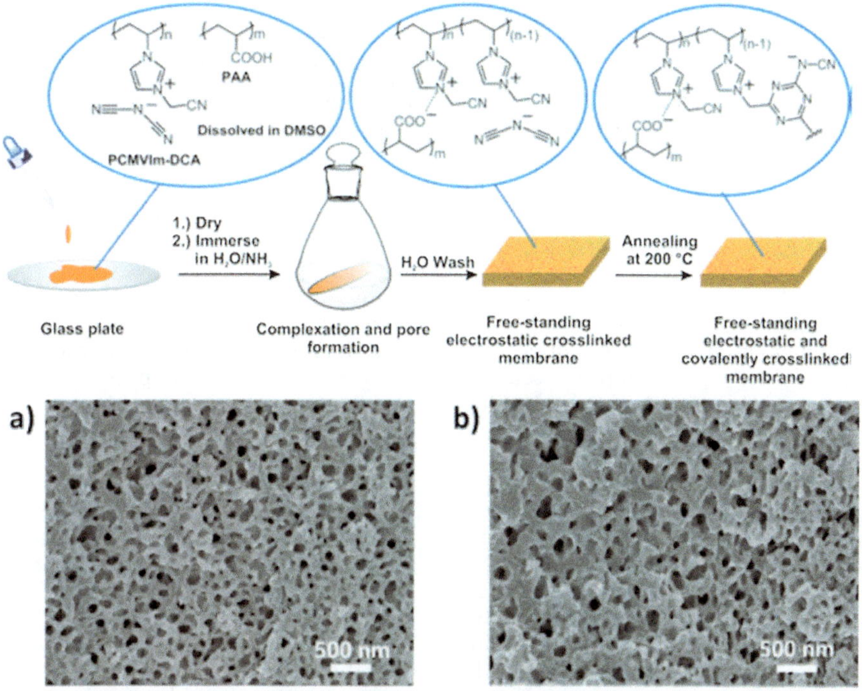

Figure 2.14 Synthesis scheme of a macroporous covalently cross-linked imidazolium PIL membrane obtained *via* polyelectrolyte complexation of an imidazolium PIL containing dicyanamide as the anion and PAA. The as-formed membrane is stable in salt solutions with high ionic strengths. The SEM image in part (a) represents the membrane before treatment in salt solution, whereas the one in part (b) represents the covalently cross-linked membrane after 72 hours of treatment in 1 M $LiPF_6$ solution. Adapted with permission from ref. 66. Copyright 2016 American Chemical Society.

with ionic liquids (SCILs), and supported ionic liquid phases (SILPs), along with many others subcategories.

In our discussion, we classify SPILs into several classes depending on the starting porous material. In the present section, we will deal with a large variety of possible supports, including polymers, metals, porous carbons, metal–organic frameworks (MOFs) and inorganic oxides, with a special focus on silicas. Further classification can be done considering the binding method of the ionic liquid with the support material. In fact, in general, the IL can be connected to the support in two possible ways: (i) by physical adsorption or (ii) by chemical bonding with the species present at the surface of the support. From a general point of view, physically immobilized SPILs may suffer from leaching problems; however, they usually have the advantage that very simple synthetic routes can be employed. On the other hand, in the case of chemically immobilized SPILs, the leaching problems are limited but their synthetic procedures involve chemical reactions that are not always easily achievable.

Table 2.2 Summary of the immobilization techniques.

Method	Description	Advantages	Drawbacks
Impregnation/ encapsulation	The porous support is dispersed in an IL solution. Capillary forces constrain the IL inside the pores	✓ Simple ✓ Always achievable	× SSA and pore volume decrease of the porous support × IL leaching problems
Incipient wetness impregnation	An IL solution is dropped on a dried porous support until it appears wet. The IL remains inside the pores due to capillarity	✓Easy ✓Always achievable	× SSA and pore volume decrease of the porous support × IL leaching problems
Co-condensation	The IL species enter the structure simultaneously with the synthesis of the support, such that the IL is embedded in the structure of the support	✓One step synthesis ✓No IL leaching problems ✓ The IL is an integrated part of the main structure ✓The IL directly faces the open pores of the material	× Special requirement of the IL molecules (must be able to link to the support structure and must be compatible with the reaction conditions of the support)
Grafting	The IL is covalently bonded to the surface of the support by chemical reaction on specific reactive sites	✓No IL leaching problems ✓Good control of the amount of IL species at the surface	× Requirement of specific IL structures
Surface modification	Chemically active species at the surface of the support are modified to introduce the ILs	✓No IL leaching problems ✓Good control of the amount of IL species at the surface	× Not always achievable × Active moieties on the surface of the porous material are mandatory
Ship-In-Bottle synthesis	The IL is synthetized inside the support: the reactants are able to penetrate inside the support and, once the IL forms, it is stuck inside the support	✓Novel approach ✓No leaching problems	× Special requirements regarding the support pore size and the dimensions of the reactant and final IL molecules

The immobilization of an IL on a support occurs following some general methods. The most common immobilization strategies are listed in Table 2.2, together with a brief description of the procedures and both their main advantages and drawbacks. The impregnation, encapsulation and incipient wetness impregnation methods generally lead to physical immobilization of the IL moiety. On the other hand, the grafting and surface modification techniques necessarily involve a chemical immobilization. The ship-in-bottle synthesis, instead, is a special synthetic route in which the support acts as the container for the synthesis of the IL, which is eventually trapped inside the pores of the support. In the following sections, we will try to provide a general discussion regarding SPILs, considering ILs supported on different porous systems.

2.2.2.1 Ionic Liquids on Polymeric Supports

Different from the poly(ionic liquid)s already described in Section 2.1, polymer supported ionic liquids (PSILs) are non-ionic porous polymers that undergo surface functionalization in order to support ionic liquid moieties of various kinds.

PSILs have been widely studied in recent years. In particular, systems based on polystyrene (PS) and PS-modified resins, divinylbenzene (DVB)-cross-linked PSILs, PEG-1000 and chitosan-based PSILs, were used to produce catalysts for a certain number of important catalytic reactions.[67] In addition, supported ionic liquid membranes (SILMs), using nylon, cross-linked nylon, polyimide or polymer films as supporting polymers, belong to the family of PSILs. SILMs have found applications in catalytic membrane reactors and in gas separation.[67]

In the context of PSILs, one of the first and easy synthetic attempts exploited an impregnation technique to produce an IL-impregnated resin, which was used as a fixed bed column for a butyric acid separation process.[68] A commercial macroporous Amberlite resin (XAD-1180N, based on a styrene–divinylbenzene copolymer) was immersed in an ammonium IL (trialkylmethylammonium bis-(2,2,4-trimethylpentyl)phosphinate ($[C_nC_nC_nC_1N]^+[BTMPP]^-$)) solution. The IL was obtained *via* a metathesis reaction between Aliquat 336 (trialkylmethylammonium chloride, $[C_nC_nC_nC_1N]^+[Cl]^-$, mostly containing C8 alkyls) and Cyanex 272 (bis-(2,4,4-trimethylpentyl)phosphinic acid, H[BTMPP]). The final supported material possesses an average particle size of 553 μm and an IL/support mass ratio ($g\ g^{-1}$) of 1.28.

2.2.2.2 Ionic Liquids on Metal Catalysts

According to Selvam *et al.*[67] we can refer to a material as a solid catalyst with an ionic liquid layer (SCILL) if the IL support is catalytically active. SCILLs are porous heterogeneous catalysts coated with an IL layer, which is able to enhance the catalytic properties of the support itself. An example of the synthesis of a solid catalyst with an ionic liquid layer was reported by Kernchen *et al.*, who prepared a SCILL following an impregnation method.[69] In this synthetic procedure, a commercial Ni-catalyst (BET SSA 105 $m^2\ g^{-1}$)

was coated with different loadings of an IL (1-butyl-3-methylimidazolium-*n*-octylsulfate ([BMIM][*n*-$C_8H_{17}OSO_3$])) up to 20 wt%. The procedure was repeated until the desired loading of the IL was reached. The catalyst characterization pointed out that the higher the IL loading, the lower the catalyst SSA and pore volume. A complete coating of the catalyst surface was achieved with a 10 wt% loading of IL, but under these conditions, the catalyst exhibited a SSA of 50 $m^2\ g^{-1}$. The aim of the functionalization was to increase the selectivity of the hydrogenation reaction of the cyclooctadiene. From the reactivity point of view, the IL coating was robust and no detectable leaching problems of the IL were observed. Obviously, the reduced SSA decreased the activity of the SCILL with respect to the pristine catalyst, but in this case, the presence of the IL layer increased the selectivity of the catalyst in the hydrogenation reaction of cyclooctadiene (COD) to cyclooctene (COE) and cyclooctane.

2.2.2.3 Ionic Liquids on Carbon-based Supports

Carbon-based materials are probably the most used supports for chemical applications. Their very high surface areas, strong chemical stability and easy availability, and the tunability of their properties and morphology, are the main strengths of this class of materials.

In the case of carbon-based supports, the most straightforward way to generate a SPIL is the incipient wetness impregnation technique. Rufete-Beneite *et al.*[70] studied the impregnation of different activated carbon materials, including spherical activated carbon nanoparticles, powder activated carbon, and multiwalled carbon nanotubes (MWCNTs), with BET SSAs ranging from 176 $m^2\ g^{-1}$ to 1968 $m^2\ g^{-1}$. The purpose of this work was to establish which properties of the supports (*e.g.*, porosity (volume and type), surface area, oxygen surface chemistry and morphology) determine the IL adsorption capacity and the stability of the supported IL phase. Rufete-Beneite and co-workers focused their studies on the filling of the support pores, evaluating both the way the pores were filled and the filling degree of the porosity, and tested the stability of the different impregnated carbon materials in water. The impregnation technique used is quite simple: a solution of acetone and 1-butyl-3-methylimidazolium hexafluorophosphate ($[bmim]^+[PF_6]^-$) was dropped onto activated carbon and dried in an Ar atmosphere until the carbon material cannot be dried any further. As described in the paper, all the carbon materials are compatible with the employed impregnation method allowing 88% loading of the total pore volume with the IL phase. The presence of the IL inside the pores nullifies the total pore volume of the carbon support because the porosity is both filled and blocked by the IL.

The same technique was also employed by Mikkola *et al.* for the immobilization of Pd-catalyst nanoparticles together with various IL layers on a high-surface area, structurally active carbon cloth (see Figure 2.15).[71,72] In the present case, the catalyst precursor was dissolved in the impregnating solution together with the IL.

Of all the carbon-based materials, the most unstable supports after impregnation are the carbon nanotubes (CNTs). As described by Rufete-Beneite

Figure 2.15 Pd nanoparticles dispersed on an ionic liquid layer supported on an active carbon phase. Reproduced from *Topics in Catalysis*, Supported Ionic Liquid Catalysts (SILCA) for Preparation of Organic Chemicals, 53 (15), 2010, 1096–1103, P. Virtanen, T. O. Salmi, J. P. Mikkola, © Springer Science+Business Media, LLC 2010, with permission from Springer.[72]

et al.,[70] MWCNTs exhibit the highest amount of blocked porosity due to the particularity of the pore structure in which the tubular channels are probably not filled, or filled just in the initial part and, moreover, undergo almost complete leaching of the loaded IL (around 92%). This behavior could be ascribed to the low interaction strength between the surface of the carbon material and the IL itself, which, therefore, does not enter inside the inner cavities of the nanotubes. The creation of a covalent bond between the IL species and the nanotube surface could help to prevent the leaching of the IL from the inner cavities of the CNTs. This possibility is described in the work of Fu and co-wokers,[73] in which the authors report on a facile ion exchange strategy to functionalize single-walled carbon nanotubes (SWCNTs) with a fluorescent imidazolium salt. In particular, they compare two different methods, to obtain fluorescent SWCNTs, chemical grafting on the one hand and physical grinding on the other. The grafting method starts from COOK-functionalized SWCNTs and exploits ion exchange to link the cationic species of the IL to the COO^- group, whereas, in the case of physical grinding, a part of the IL chemically links directly to the CNT surface (see Figure 2.16).

2.2.2.4 *Ionic Liquids on Porous Metal–Organic Frameworks*

Porous metal–organic frameworks (MOFs) are crystalline materials where metal clusters coordinate with organic linkers to form a microporous structure. Due to their very high SSAs and the tunability of the organic linker, porous MOFs are considered to be very promising materials for heterogenous catalysis. The properties of MOFs can be improved, similar to other materials, by introducing specific IL functionalities into their structure.

Since the present section just deals with supported IL systems, a dedicated Section (2.2.3.1) will be devoted to materials in which the IL is directly incorporated into the chemical structure of the MOF.

The encapsulation of an IL inside a MOF structure can be easily achieved through the dispersion of the MOF in an IL solution. Luo *et al.*[74] evaluated the effect of different solvents (H_2O, ethanol, DMF) on the encapsulation procedure of an amino-functionalized basic IL (ABIL) with a hydroxyl anion in

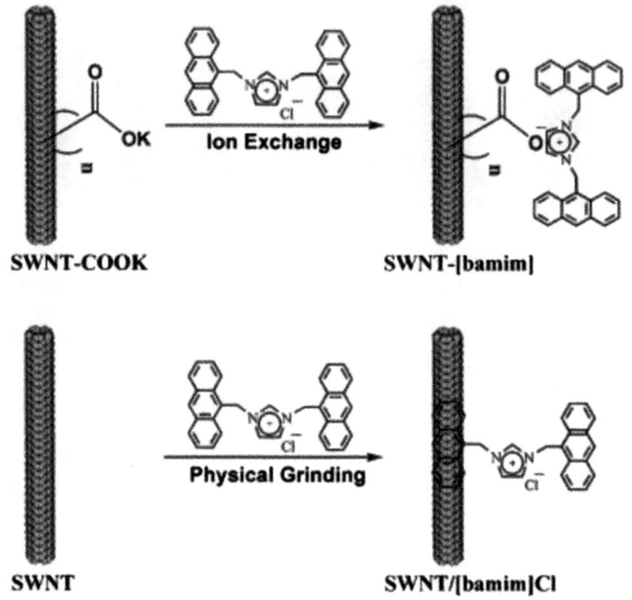

Figure 2.16 Preparation of functionalized SWNTs *via* ion exchange or physical grinding. Reproduced from C. Fu, L. Meng, Q. Lu, Z. Fei, P. J. Dyson, *Advanced Functional Materials*, John Wiley and Sons, Copyright © 2008 WILEY-VCH Verlag GmbH & Co. KGaA, Weinheim.[73]

the structure and the copper-based MOF HKUST-1 (BET SSA: 1158 m^2 g^{-1}). It was found that the chemical behaviour of the ABIL, dissolved in these solvents, had a profound impact on the catalytic performance of this MOF. The authors discovered that the ABIL penetrates inside the cavities of the MOF, where the Cu metal centers can coordinate with the ammine of the IL. Furthermore, they pointed out how the choice of the solvent critically affects the ABIL/HKUST-1 structure and the final IL loading. In particular, considering the above-mentioned solvents, it was found that: (i) the use of a DMF solution slightly decreases the SSA and allows the highest IL loading (BET SSA: 941 m^2 g^{-1}, ABIL loading: 1.56 mmol g^{-1}). However, in this case, the loading data could be inaccurate because the DMF molecules can be retained in the MOF instead of the ABIL molecules. (ii) The employment of an ethanol solution reduces the IL loading and, at the same time, introduces mesoporosity into the final material with a decrease of the SSA (BET SSA: 338 m^2 g^{-1}, ABIL loading: 1.07 mmol g^{-1}). (iii) The use of a water solution allows a high loading of the IL, almost destroying the microstructure of the HKUST-1 (BET SSA: 52 m^2 g^{-1}, ABIL loading: 1.45 mmol g^{-1}). A possible reason for this collapse could be derived from the strong alkaline environment (pH = 10–11) created by the ABIL/water solution. In fact, at these pH values, the ABIL can be entirely ionized and form free ions (alkyl-imidazolium and hydroxyl anions) that can extract the Cu atoms from the HKUST-1 framework, which collapses due to the host–guest interaction.

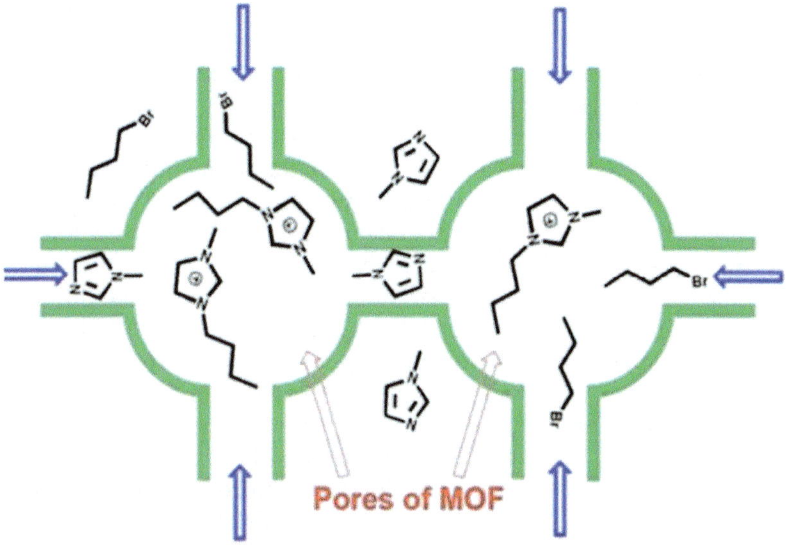

Figure 2.17 Schematic diagram of "Ship-In-Bottle" synthesis of ILs. Reproduced from ref. 75 with permission from The Royal Society of Chemistry.

A new concept to introduce IL species inside MOF structures was proposed by Khan *et al.*,[75] by means of the so called "*Ship-In-Bottle*" (*SIB*) approach. SIB synthesis employs a MIL-101 porous framework as the host (the bottle) to obtain the desired guest product (the ship), that is technically or sterically difficult to remove because the synthesized IL is larger than the pore openings of the host framework. Differently, smaller IL precursors can easily move through the host structure and react inside the cages, to form the larger IL which is trapped inside the cavities. In this case, the synthesis of the IL takes place inside the pores of the MOF as represented in Figure 2.17, following a dry media reaction.[75] Briefly, the MIL-101 MOF is dispersed in *N*-methylimidazole, and then 1-bromobutane is added. The synthesis works both inside and outside of the MOF structure and the excess of both reactants and IL deposited outside is removed by washing with ethanol. As a consequence of the presence of the IL inside the MOF-matrix, the SSA slightly decreases from 2956 $cm^2 g^{-1}$ to 2071 $cm^2 g^{-1}$. In the SIB method, the IL molecules are trapped inside the cavities of the MOF avoiding leaching-related problems.

2.2.2.5 *Ionic Liquids on Inorganic Oxide Supports*

The inorganic oxides include a great variety of compounds that, in principle, can support ILs. Silica is probably the most popular and the most studied support.

The common techniques employed to immobilize an IL on a silica support are grafting, incipient wetness impregnation, so-called co-condensation and self-assembly. Valkenberg *et al.* illustrated these three possible methodologies to immobilize highly acidic ILs.[76] Although the paper was dedicated to a specific class of IL, the illustrated methods can be generally used for functionalization of silica supports.[77] In this section, we will describe the immobilization of ILs on silica supports by both grafting and incipient wetness impregnation, which can be considered post-synthetic functionalization methods to support an IL on a pre-formed material. In contrast, the co-condensation and self-assembly methodologies, employed to directly generate porous IL–silica hybrid materials, will be discussed in Section 2.2.3.2.

The grafting method aims to create a covalent bond between the ionic liquid and the surface silanol groups of the support. A general reaction scheme is represented in Figure 2.18 as reported by Valkenberg *et al.*[76] This synthesis involves the suspension of the dried support (MCM-41) in dry toluene, the addition of the IL (1-(triethoxysilyl)propyl-3-methylimidazolium chloride) and the following mixture reaction at 90 °C. The toluene and the formed ethanol are then outgassed, the solid is dried under high vacuum and the excess of unreacted IL is removed by extraction in dichloromethane. The grafting technique is a versatile procedure since it obviously allows the incorporation of different IL moieties, just after the functionalization of the IL with the alkyloxysilyl group. The literature reports the grafting of ILs based on imidazolium,[78,79] pyridinium[79] and phosphonium.[80]

The described methodologies can be ideally suitable for any kind of silica structure: amorphous silica, silica nanoparticles, as well as ordered mesoporous silica (such as SBA-15, MCM-41, or HMS). It is worth noting that both the impregnation and grafting procedures do not modify the structure of the silica support, whereas the co-condensation technique affects the formation of the final structure of the IL–silica hybrid material.

The supporting of an IL on an inorganic oxide is not only limited to the silica systems. A large variety of less common oxidic supports can also be employed (such as alumina, titania and zirconia), as reported by De Castro and co-workers.[81] This work describes the preparation of supported 1-butyl-3-methyl-imidazolium tetrachloroaluminate ([bmim]$AlCl_4^-$) (commonly referred to as AL-IL) on a wide variety of oxides, including silicon oxide (FK-700: amorphous silica; SP-18: Aerosil silica; T350: silica powder), aluminum oxide (PURAL SB: alumina powder, average particle size 45 nm),

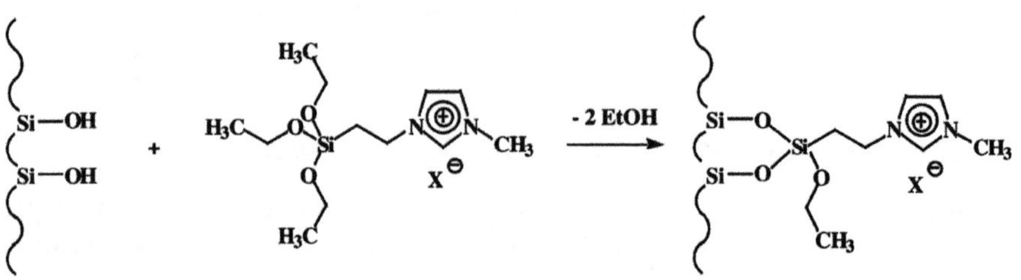

Figure 2.18 Immobilization of an IL on a silica surface *via* a grafting method. Adapted from ref. 76 with permission from The Royal Society of Chemistry.

titanium oxide (P25: anatase/rutile powder, average particle size 21 nm), zirconium oxide and H-beta zeolite.

These supported AL-ILs were prepared *via* incipient wetness impregnation, adding the AL-IL dropwise onto the supports, which were previously calcined and dried, until they change appearance from dry to wet. The excess of IL is then removed with boiling dichloromethane in a Soxhlet extraction setup.

The authors found that, as a consequence of the impregnation process, all the supports undergo a reduction of the SSA, as reported in the Table of Figure 2.19, directly related to the AL-IL loading (see the graph on the right side of Figure 2.19).

After impregnation, almost all the supports undergo a reduction of the SSA and of the pore size. Only zirconia and titania supports appear to be almost unmodified after the impregnation procedure; however, the IL loading on these two systems is quite low, as pointed out by the loss of IL after Soxhlet extraction reported in Figure 2.19. In fact, it is evident that both titania and zirconia supports almost do not retain the IL.[81] Supported AL-ILs with the $AlCl_4^-$ species as a counterion can also be synthesized by adding $AlCl_3$ to an IL bearing the Cl^- anion, to obtain supported AL-ILs with catalytic applications in Friedel–Crafts reactions.[82]

The immobilization of the AL-IL on the support surface could involve the formation of a chemical bond between the aluminum atom of the IL $AlCl_4^-$ anion and the surface hydroxyl groups of the supports (see Figure 2.20). This idea could justify the low retention on titania and zirconia, which lack hydroxyl groups at the surface, and the formation of HCl during the immobilization.

A summary of all the supports and immobilization techniques described in Section 2.2 is provided in Table 2.3.

2.2.2.6 *Porous Liquids*

One last case that, in some way, is quite far from the SPIL concept, being closer to the idea of a "real" porous IL, was reported by Dai and co-workers.[83] They created an ammonium porous liquid *via* surface engineering of micro-porous hollow silica spheres with suitable ionic corona and canopy species by a two-step synthetic procedure (see Figure 2.21). At first, they synthesized hollow silica (HS) spheres with microporous shells acting as molecular sieves by blocking species larger than 1.9 nm. In order to prevent the self-filling of the cavities, a positively charged ammonium organosilane (OS) moiety, used as the corona for the modification, was bound on the inner surface of the HS nano-spheres, termed OS@HS.

Then, the chloride counter-ion, balancing the positively charged corona, was replaced with a negative poly(ethylene glycol)-tailed sulfonate (PEGS) canopy. The final result is an optically transparent hollow silica porous ionic liquid (HS-liquid) stable at room temperature. The porosity of the silica nanospheres is maintained through the synthetic path as proved by N_2 adsorption measurements collected on both the HS and the OS@HS system and by the TEM images of the HS-liquid.

Textural Properties of the Supports Used and the Supported AL-ILs*

Materials	Description	Surface area (m^2/g)	Average pore diameter (Å)
T-350	SiO_2	168	156
T-350/AL-IL ($N=0.60$)*	SiO_2/AL-IL ($N=0.60$)	150	147
Pural SB	Al_2O_3	201	75
Pural SB ($N=0.60$)	Al_2O_3/AL-IL ($N=0.60$)	167	73.2
FK700	SiO_2	391	59.6
FK700/AL-IL ($N=0.60$)*	SiO_2/AL-IL ($N=0.60$)	306	58.7
SP18	SiO_2	254	234
SP18/AL-IL ($N=0.60$)	SiO_2 AL-IL ($N=0.60$)	203	222
H-beta (Si/Al = 13.8)	SiO_2/Al_2O_3	328	—[a]
H-beta/AL-IL ($N=0.60$)*	SiO_2/Al_2O_3 AL-IL ($N=0.60$)	306	—[a]
TiO_2	TiO_2	29.3	214
TiO_2/AL-IL ($N=0.60$)*	TiO_2/AL-IL ($N=0.60$)*	25.2	214
ZrO_2	ZrO_2	31.3	222
ZrO_2/AL-IL ($N=0.60$)	ZrO_2/AL-IL ($N=0.60$)	31.5	202

[a] H-beta is a microporous material. Adsorption isotherms only allow calculation of the size of mesopores.

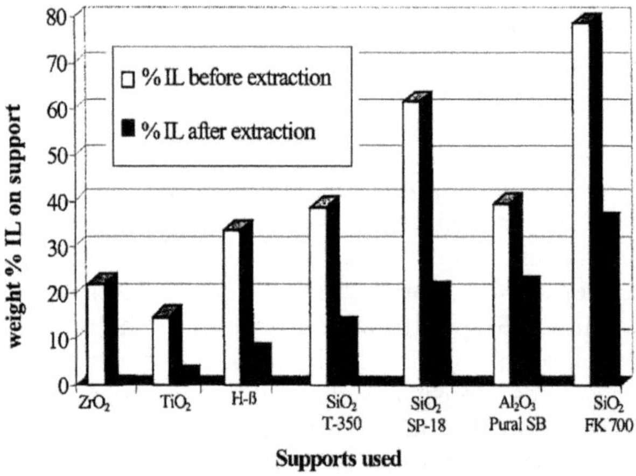

Figure 2.19 Upper part: SSA of inorganic supports before and after IL impregnation. The material names are described in ref. 81. Lower part: percentage of immobilized AL-IL on different supports before and after extraction. Reprinted from *Journal of Catalysis*, 196 (1), C. DeCastro, E. Sauvage, M. H. Valkenberg, W. F. Hölderich, Immobilised Ionic Liquids as Lewis Acid Catalysts for the Alkylation of Aromatic Compounds with Dodecene, 86–94, Copyright 2000 with permission from Elsevier.[81]

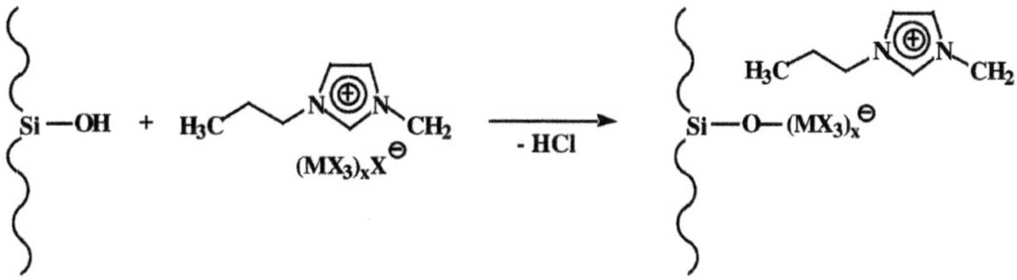

Figure 2.20 Immobilization of an IL on a silica surface *via* anion binding. Adapted from ref. 76, with permission from The Royal Society of Chemistry.

Table 2.3 Summary of the supports and immobilization techniques dealt with in the present section.

Type of support	Technique	Support (M) = macroporous (m) = mesoporous (μ) = microporous	Ref.
Polymer	Impregnation	Amberlite (M)	68
Metal catalyst	Impregnation	Ni-catalyst (m)	69
Carbon	Incipient wetness impregnation	Activated carbon, spherical NPs, powder, MWCNTs (m) (μ)	70
	Incipient wetness impregnation for catalyst immobilization	Active carbon cloth	71
	Grafting	COOK-functionalized SWCNTs (m) (μ)	73
	Grinding	SWCNTs (m) (μ)	73
MOF	Encapsulation	HKUST-1 (μ)	74
	SIB synthesis	MIL-101 (μ)	75
Inorganic oxide	Incipient wetness impregnation	SiO_2 (m), Al_2O_3 (m), TiO_2 (m), ZrO_2 (m), H-beta zeolite (μ)	81
	Grafting	SiO_2 (m)	76

2.2.3 Hybrid Porous IL Materials

Porous ionic liquid materials can be made from fully organic building blocks, as illustrated in Section 2.2.1 devoted to porous PILs, or they can be obtained by supporting the ionic liquid moieties on different pre-formed porous materials (organic, inorganic or hybrid). If the ionic liquids are supported on a pre-formed porous inorganic material, they could be described as hybrid materials (*e.g.* Sections 2.2.2.4 and 2.2.2.5). Conversely, in this section of the chapter, we consider hybrid IL materials to be ones where the ionic liquid is an integrated part of the chemical structure and is directly connected to the inorganic part, being an essential constituent of the overall material. This is the case when the ionic liquid is a linker for a metal–organic framework (MOF), or the organic moiety of an organosilica precursor for a periodic mesoporous organosilica (PMO).

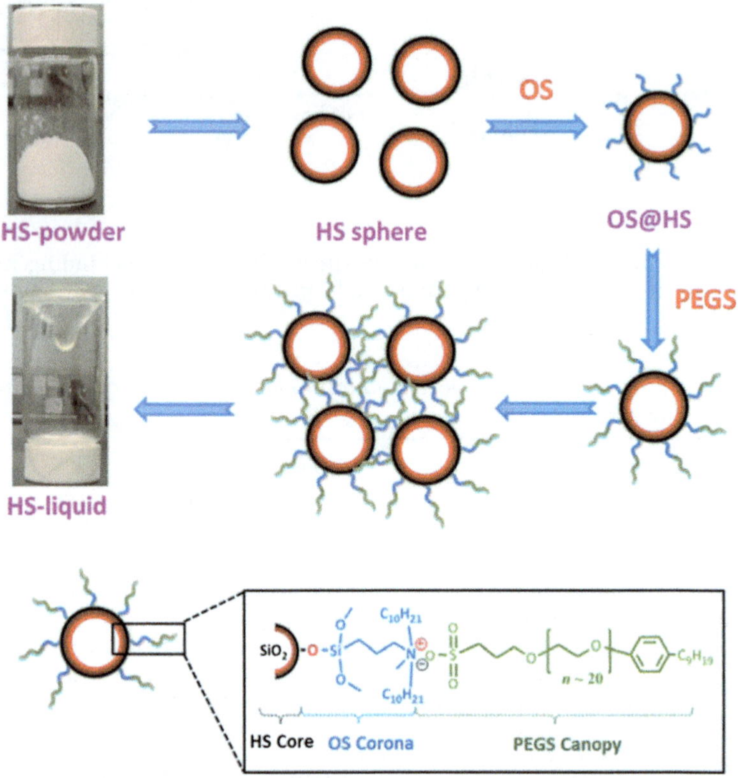

Figure 2.21 Two-step strategy for porous liquid fabrication. Reproduced from J. Zhang, S. H. Chai, Z. A. Qiao, S. M. Mahurin, J. Chen, Y. Fang, S. Wan, K. Nelson, P. Zhang, S. Dai, *Angewandte Chemie International Edition*, John Wiley and Sons, Copyright © 20XX WILEY-VCH Verlag GmbH & Co. KGaA, Weinheim.[83]

2.2.3.1 MOF–ILs

Metal–organic frameworks (MOFs) are a particular kind of coordination polymer. The term MOF was introduced for the first time in 1995 by Yaghi[84] and it refers to: "a crystalline compound consisting of metal ions or clusters coordinated to often rigid organic molecules to form one-, two-, or three-dimensional structures that can be porous".[85] This definition differs from the one of coordination polymers because it is very specific for materials that have high crystallinity and well-defined final structure morphology.[86] Often self-assembling synthesis gives rise to microporous structures with particularly high SSAs. Their extraordinary degree of variability in terms of inorganic and organic building blocks make these MOFs suitable in the fields of gas adsorption and separation, and catalysis.[87–89] The functionality of the MOFs can be ascribed to the organic linker or to the metal cluster, which is generated during the synthesis of the material.

If opportunely designed, azolium-containing ionic liquids can be employed as organic linkers for MOFs, opening up a route to new hybrid materials (MOF–ILs) that incorporate an ionic liquid moiety in the microporous structure of a MOF. The ionic liquid moiety leads to increased functionality of the framework and it also simultaneously generates an ionic environment within the MOF pores. However, the ionic environment increases the difficulty in completely removing the synthesis solvent from the MOF pores, due to the strong interaction occurring between the ion pairs and the guest molecules. Nevertheless, the ionic liquid moieties enhance the interaction strength with gas molecules having high quadrupole moment and polarizability, such as carbon dioxide, leading to an increased uptake. Carbon dioxide adsorption properties are improved when the MOF–IL combines an open metal site on the metal nodes and an ionic liquid anion with high affinity for the guest molecule.[90,91] Moreover the ionic environment of MOF–ILs makes these materials suitable for proton conductivity.[92] MOF–ILs are microporous materials that, due to the porous crystalline structure of the framework, are able to maximize the exposition of the ionic liquid moieties to the open pores of the framework.

Some MOFs are relevant materials in the field of heterogeneous catalysis, mainly due to their easy recyclability and permanent microporosity.[88] Their catalytic activity can arise from an open metal site on the metal cluster, a metal complex tethered on a Lewis base functional linker, or an active catalyst encapsulated inside the MOF pores.[93]

The use of ionic liquid linkers broadens the range of available catalytic sites. In fact, the azolium moiety can act as a catalyst and, moreover, it is also a valuable precursor of stable NHC carbene species, which are well-known for their Lewis base catalytic properties and as ligands for transition metals.[94,95] The MOF–IL structures, due to their structural rigidity and the presence of the azolium linker, allow the generation of an NHC carbene in the heterogeneous phase, bypassing the dimerization process that these species can undergo in solution.[96]

The azolium linkers for MOF–ILs include imidazolium, triazolium or benzimidazolium moieties; however, the large majority contain imidazolium linkers. Table 2.4 reports the most common imidazolium liquid linkers used for MOF synthesis, together with the relative metal nodes and the reference study.[93] As expected, the presence of permanent charge on the organic linker affects the final MOF structure. A comparative study, considering triazole and triazolium-based linkers aims to demonstrate the differences in the MOF topology and concatenation for materials obtained using an ionic liquid linker.[97]

The azolium moiety can be integrated in the core of the ligand or can be appended to a common linker. In the first case, the presence of the imidazolium in the linker structure gives rise to a bent molecule with an angle of 137° as shown in Figure 2.22. This bent geometry generates unusual MOF topologies which are often highly interpenetrated (three-fold or four-fold), drastically diminishing the porosity and the available surface area.[98,99] In contrast, if the imidazolium moiety is appended to a common organic linker, it protrudes

Table 2.4 Azolium linkers used for the synthesis of MOFs with their respective metal nodes and references.

Entry	Structure of the IL linker		Metal nodes	Ref.
1	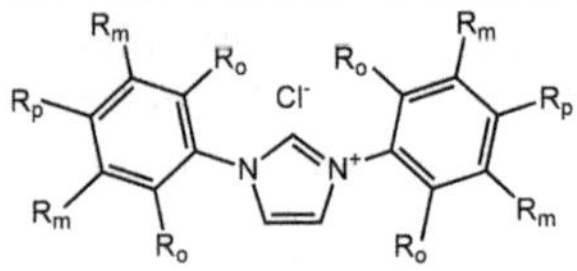	$R_{o1,2}$: H; $R_{m1,2}$: H; R_p: COOH	Zn(II), Cd(II)	92 and 104
2		$R_{o1,2}$: $-CH_3$; $R_{m1,2}$: H; R_p: COOH	Cu(II), Zn(II)	105
3		$R_{o1,2}$: -iPr; $R_{m1,2}$: H; R_p: COOH	Cu(II), Cd(I), Ce(III)	106
4		$R_{o1,2}$: H; $R_{m1,2}$: COOH; R_p: H	Zn(II)	91
5		$R_{o1,2}$: $-CH_3$; $R_{m1,2}$: COOH; R_p: H	Cu(II)	90
6		$R_{o1,2}$: H; R_{m1}: H; R_{m2}: COOH; R_p: H	Zn(II)	104
7	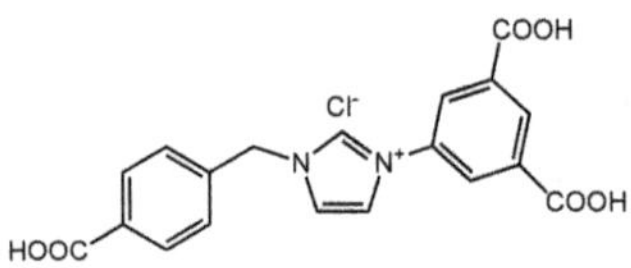		Cu(II)	107
8	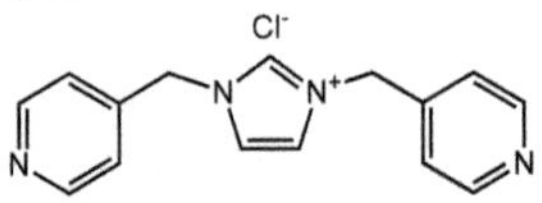		Ag(I), Cu(I)	110
9	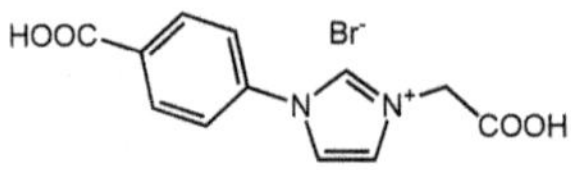		Cu(II), Zn(II), Cd(II), Mn(II), Co(II), Ni(II)	103
10	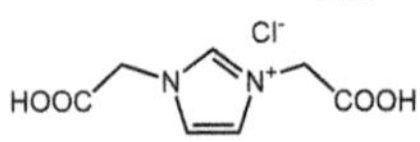		Pb(II)	101
11	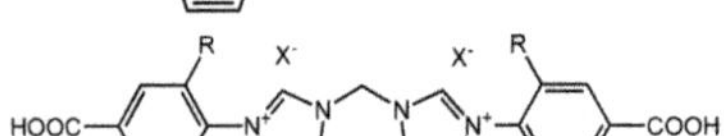	R: H; X: Cl R: $-CH_3$; X: Br	Cu(II) Zn(II)	108 109
12	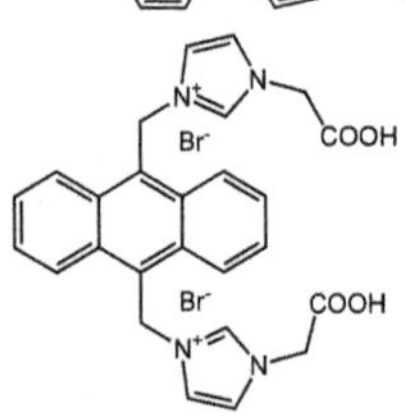		Cu(II), Zn(II)	111
13	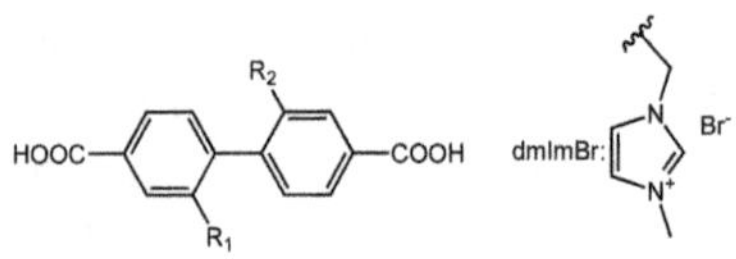	R_1: H; R_2: dmImBr R_1, R_2: dmImBr	Cu(II), Zn(II) Cu(II)	100 100
14	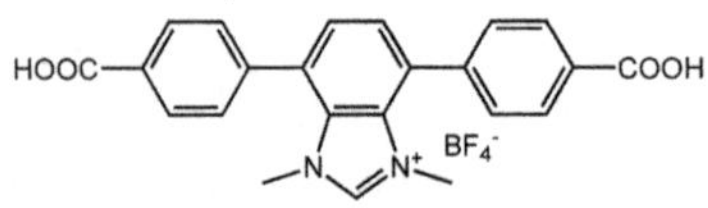		Zn(II)	112

from the linker and generates a steric hindrance during the synthesis of the framework, avoiding the interpenetration (see Table 2.4, entry 13).

Roberts and co-workers demonstrated that one dangling imidazolium from a biphenyl dicarboxylate linker is not sufficient to avoid the formation of a catenated system for a MOF created using copper as the metal node. The same study also reveals that if two imidazolium rings are dangling from the linker a non-catenated structure is achievable (Figure 2.23).[100] Since the core of this linker is analogous to other common linkers employed in MOF synthesis, the resulting material with just one dangling imidazolium is isostructural with MOF 9 (IRMOF-9), while the one obtained with two dangling imidazoliums is isostructural with MOF 10 (IRMOF-10). An innovative strategy to obtain non-interpenetrated MOF–ILs, starting from an imidazolium core linker, is the combined use of a tetracarboxylate imidazolium linker with different pyridine-based linkers that form a porous three-dimensional bipillared-layer structure when complexed with zinc.[91]

In order to study how the rigidity of the imidazolium linker affects the final structure of the MOF–IL, different studies have been performed with more flexible linkers bearing methylene bonds between the imidazolium and the carboxylic group. The flexibility of the spacers that contain a methylene bridge allows the linkers to bend and rotate when they coordinate to metal centers, often causing structural diversity. As an example, 1,3-bis(carboxymethyl)imidazolium chloride was used as a linker to form a lead-based MOF, with a chiral

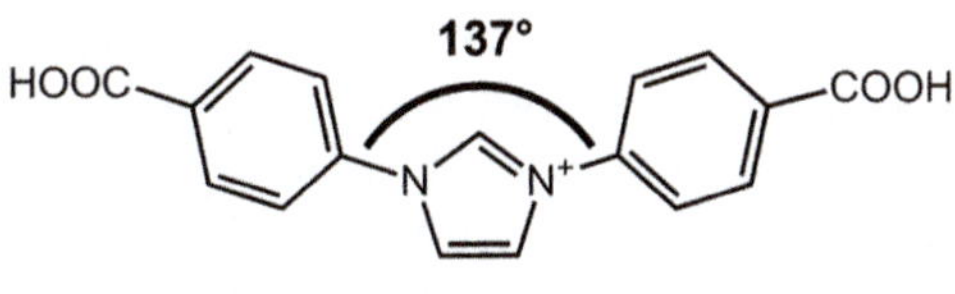

Figure 2.22 Chemical structure of 1,3-bis(4-carboxyphenyl)imidazolium (H_2L + linker). This is a typical example of an imidazolium-based linker for the synthesis of MOFs. The imidazolium leads to a bend of 137° in the molecule, which generates unusual MOF topologies.

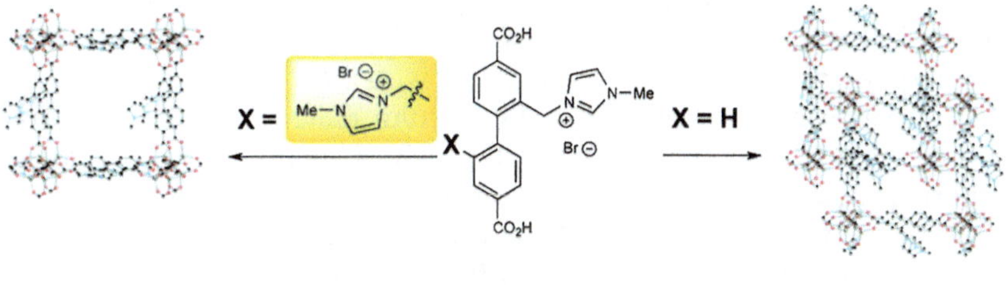

Figure 2.23 Structure of a MOF biphenyl dicarboxylate-based linker with appended methyl imidazolium salts. In the case of one dangling imidazolium, the linker can form an Me-based MOF with an interpenetrated structure. However, when two dangling imidazolium rings are present in the linker structure, the steric hindrance gives rise to a non-interpenetrated Me-based MOF. Reproduced with permission from ref. 100. Copyright 2011 American Chemical Society.

double interpenetrated framework (Table 2.4, entry 10).[101] The same linker was used to obtain several different MOF topologies when the complexation was performed with lanthanum and neodymium.[102] A semi-rigid 1-(carboxymethyl)-3-(4-carboxyphenyl) imidazolium bromide bearing only one methylene bridge in the structure was used to create a balance between rigidity and flexibility of the azolium ligand, resulting in other different MOF topologies depending on the metal in the nodes (Table 2.4, entry 9).[103]

A large number of studies have been devoted to producing MOFs using imidazolium linkers made from Debus–Radziszewski imidazolium synthesis,[90–92,104–106] the same reaction used to obtain the monomers for the direct synthesis of imidazolium porous cationic polymers, but, in this case, the starting amine is functionalized with one or two carboxylic groups, in order to obtain an ionic liquid able to act as a ligand. This reaction is flexible enough to work with various alkyl group substituents on the phenyl ring useful to introduce steric hindrance inside the linker. The azolium linkers, used for the MOF synthesis, are listed in Table 2.4, entries 1–6. More flexible ionic liquid monomers can be obtained by quaternarization of a pre-formed 1-substituted imidazole. The quaternarization step, performed using an alkyl halide, introduces a flexible methylene bridge in the structure of the linker. The resulting linkers have, as a substituent of the imidazolium ring, multiple combinations of mono-carboxyphenyl, di-carboxyphenyl and carboxymethyl (Table 2.4, entries 7, 9 and 10).[101,103,107] If dibromomethane is used as the alkylating agent for 1-(carboxyphenyl)imidazole, it generates a rigid bis-imidazolium linker (Table 2.4, entry 11).[108,109] The majority of the ionic linkers bear carboxylic acids as sources of free electron doublets, but it is also possible to generate a pyridine-functionalized imidazolium ligand by reacting picolyl chloride and imidazole in the presence of NaH (Table 2.4, entry 8).[110] A luminescent imidazolium ligand is obtained by functionalizing anthracene with two carboxymethylimidazoles. These luminescent linkers can be inserted into a copper and Zn-based MOF (Table 2.4, entry 12).[111] A different approach, developed by Oisaki *et al.*, uses a synthetic step to create 4,7-bis(4-carboxylphenyl)-1,3-dimethylbenzimidazium tetrafluoroborate, a linear linker based on benzimidazolium (Table 2.4, entry 14). In this case, the ionic moiety is part of the ligand core structure; however, its linear rigid structure leads to a Zn-based MOF called IRMOF-76, a non-interpenetrated framework isoreticular with MOF-5.[112]

Several MOFs can be synthesized by azolium linkers; nevertheless, not all of them show permanent porosity. Obviously, a successful porous MOF–IL synthesis procedure has to avoid channel collapse during the final solvent removal step, and the blockage of the channels due to excessive framework interpenetration.[91] It is also necessary to consider that if the MOF presents a porous structure, the porosity will not be accessible to the surface area gas probes (for example N_2). In fact, the anion of the ionic liquid, with its solvent coordination sphere, can prevent the N_2 probe molecules from entering inside the pores. A thorough outgassing of the material, for 12 to 24 hours at a temperature ranging from 50 to 200 °C, is necessary before its available SSA can be probed. The outgassing temperature is a very important parameter: it should be high enough to allow the complete removal of the solvent, and, at the same time, not too high to avoid the thermal decomposition of the material. Complete solvent removal from the framework can be a challenging step, if the solvent inside the pores has a high boiling point and a low vapor pressure, like for example DMF. In this case, solvent exchange (with a solvent like methanol that is easier to remove) can be necessary before the final removal step.[107] After the outgassing procedure the MOF XRD pattern has to be checked, since this procedure can lead to a framework structure change.[105]

A Zn-based porous MOF–IL was synthesized using 1,3,-bis(3,5-dicarboxyphenyl)imidazolium as a linker to form a 2D layered structure, pillared by three different pyridine additional linkers, which create a 3D porous MOF with a non-interpenetrated structure.[91] The employed pyridine linkers are: 4,4′-bipyridine, 1,2-di(4-pyridyl)ethylene, and 1,4-bis(4-pyridyl)benzene. These linkers generate three different MOFs with a BET SSA of 181, 235, and 425 $m^2\ g^{-1}$, respectively. It is clear how the increasing length of the pyridine linkers generates MOFs with a larger pore size and higher SSAs, as shown in Figure 2.24a.[91] The interpenetrated structure of a Zn-based MOF containing 1,3-bis(4-carboxyphenyl)imidazolium as the linker, drastically decreases the available BET SSA to a value of 102 $m^2\ g^{-1}$.[92,113] Starting from a similar linker, 1,3-bis(4-carboxy-2,6-dimethylphenyl)imidazolium, two MOFs were created with copper and zinc in the metal nodes, respectively. The BET SSA of the MOF obtained with copper is 170 $m^2\ g^{-1}$; in contrast, the MOF made with zinc exhibited no significant SSA. The outgassing step preceding the SSA measurement was responsible for the structural modification of these MOFs. For this reason, dye adsorption/desorption experiments were performed to test the accessibility of the pores, without a preliminary solvent outgassing step. In accordance with the N_2 adsorption experiments, the Zn-based MOF shows

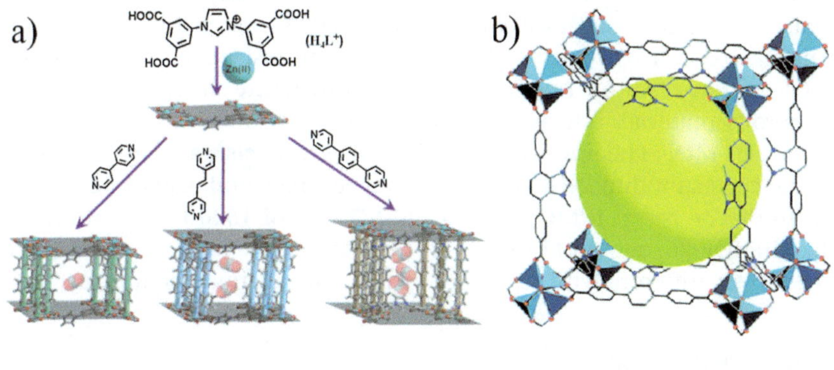

Figure 2.24 Section (a): the structure of a Zn-based MOF obtained with a tetravalent imidazolium linker and three different divalent pyridine linkers. Adapted with permission from ref. 91. Copyright 2014 American Chemical Society. The lengths of the pyridine linkers are directly proportional to the pore size of the resulting MOF. Section (b): a porous Me-based MOF obtained with a linear azolium linker, which leads to a material with a conventional MOF topology. Adapted with permission from ref. 112. Copyright 2010 American Chemical Society.

no dye adsorption; in contrast, the Cu-based MOF readily adsorbs Nile blue, methyl red and fluorescein.[105] The dye adsorption experiment was also useful to probe the accessibility of the pores for a Zn-based MOF synthesized using 1,1′-methylenebis-(3-(4-carboxy-2-methylphenyl))-imidazolium as the linker. In this case, the N_2 adsorption at 77 K was not used to explore the porous structure of the material; however, the accessibility of the MOF channels, calculated to be 3.3 nm, was successfully probed by the uptake of Congo red, Oil red, Sudan Black B and Methylene Blue.[109]

It is worth noting that some MOFs show negligible N_2 uptake at 77 K, but they exhibit significant uptake for CO_2 or CH_4 at 273 or 298 K.[114,115] The gas uptake does not depend only on the pore size, but also on the interaction occurring between the host MOF and the guest molecules. CO_2 and CH_4 are more polarizable molecules with respect to N_2, and thus they can easily bind to the ionic moiety present in the imidazolium-based linker. In particular, two studies, reporting MOFs with imidazolium linkers, deal with this behavior. Lee and co-workers described a Cu-based MOF with *N*,*N*′-bis(2,6-dimethyl-3,5-dicarboxyphenyl)imidazolium as an ionic linker. The N_2 adsorption at 77 K was negligible for this MOF, whereas the CO_2 adsorption at 273 K and 1 atm was significant (1.8 mmol g^{-1}). The SSAs, based on nonlocal density functional theory (NLDFT), derived from these CO_2 adsorption isotherms, range from 300 to 520 m^2 g^{-1} depending on the outgassing conditions. This MOF also exhibits an important CO_2 over CH_4 selectivity.[90] Wang *et al.* reported a Zn-based MOF obtained from *N*-(3,5-dicarboxyphenyl)-*N*′-(4-carboxyphenyl)imidazolium chloride. Also for this material, the N_2 adsorption at 77 K was negligible, whereas the CO_2 adsorption at 195 K and 1 atm (2.45 mmol g^{-1}) was high enough to allow the calculation of a BET SSA of 222 m^2 g^{-1}. This MOF shows enhanced selectivity for CO_2, H_2, and CH_4 with respect to N_2, mainly due to the lower polarizability of the N_2 molecule. The different imidazolium counter anions present in the MOF can also modulate the adsorption properties of this framework.[107]

Aiming to obtain a porous MOF, Oisaki *et al.* synthesized the above described IRMOF-76. The resulting non-interpenetrated structure should have given rise to a porous material, but, unfortunately, this work does not report the SSA value of this MOF. The linker employed in this synthesis was also transformed into the corresponding NHC-PdI_2(pyridine) complex and, subsequently, used to form the Zn-based IRMOF-77, with a BET SSA of 1590 m^2 g^{-1}. IRMOF-77, unlike IRMOF-76, does not bear an ionic linker and the results in terms of SSA could be very different with respect to its ionic liquid linker analogous.[112]

2.2.3.2 Silica–ILs

The immobilization of ILs on a pre-formed porous silica support by means of incipient wetness impregnation and grafting techniques (post-synthetic functionalization methods) was already discussed in Section 2.2.2.5. Briefly, mesoporous silica with a periodic structure is produced by the condensation of a silica precursor, usually TEOS (tetraethyl orthosilicate), *via* a soft-templating methodology and, subsequently, the silica support can be impregnated with an IL solution or grafted with the desired IL groups (Figure 2.25, path I). A different approach could be to generate a porous silica–IL hybrid material (hereafter referred to as a silica–IL) in a single synthetic step. In this latter case, a particular silica precursor, functionalized with an IL moiety, is employed in the synthesis of the hybrid material. The co-condensation and the self-assembly methodologies are the two main techniques used to create porous silica–ILs. The first technique, the so-called co-condensation method, involves the simultaneous condensation of a mixture of a common silica precursor (such as TEOS) and a bifunctional or mono-functional (*i.e.* containing two or one silylated groups, $-Si(OR)_3$) organosilica precursor, which in our case is an IL-functionalized silica precursor (Figure 2.25, path II). The second technique is the self-condensation of an organosilica precursor with at least two silylated groups that leads to a mesoporous material with the IL moiety embedded in the silica structure (Figure 2.25, path III). As for all the synthetic routes to obtain ordered mesoporous silica materials, both these synthesis techniques employ a soft-templating methodology to generate mesoporosity in the final material.[116]

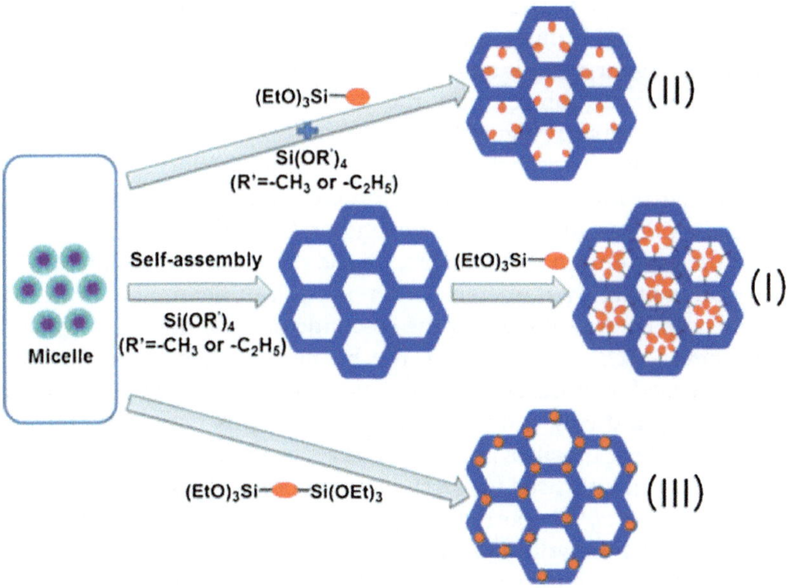

Figure 2.25 Synthetic pathways to produce IL-functionalized silica and silica–IL hybrid materials. The red dots represent the organic moiety (in this case IL) covalently bound to the silica precursor, which can be bi-functional or mono-functional. Path I represent post synthetic grafting, path II co-condensation and path III self-condensation of a bifunctional organosilica moiety. Adapted from *Korean Journal of Chemical Engineering*, Periodic mesoporous organosilica (PMO) for catalytic applications, 31 (10), 2014, 1707–1719, S. S. Park, M. S. Moorthy, C. S. Ha, © Korean Institute of Chemical Engineers, Seoul, Korea 2014 with permission of Springer.[116]

Nowadays, different organosilica precursors, with various organic moieties covalently bound to the silica monomer, are available. Among all the varieties of organic moieties, the ionic liquids represent a well-defined reality at the current state-of-the-art.[116–119] The silica–IL precursors can bear one or more silyl groups, originating from mono-functional or multi-functional precursors. The mono-functional precursors can be exploited in both the co-condensation and the grafting techniques, as they need the support of a conventional silica network to create a mesoporous material. The IL moieties, synthesized starting from mono-functional precursors, hang from the silica network. The multi-functional precursors usually undergo co-condensation, but they are also capable of self-condensation giving rise to porous IL–organosilicas. In this last case, the IL moiety is bonded through two silyl functionalities, and, therefore, is embedded inside the periodic crystalline structure of the final mesoporous silica–IL. All these materials belong to the class of periodic mesoporous organosilicas (PMOs).

2.2.3.2.1 Silica–ILs from Co-condensation. The first step of the co-condensation technique to create PMOs is the synthesis of the mono- or multi-functional organosilica precursors. The organosilica precursor is then co-condensated in the presence of the inorganic precursor (TEOS), a proper solvent and a soft-templating agent, usually an anionic, a cationic, or a non-ionic surfactant. After the co-condensation reaction, the surfactant is removed together with the unreacted precursors. Figure 2.26 reports an example of co-condensation between 1-(triethoxysilylpropyl)-3-methylimidazolium chloride and TEOS. The PMOs obtained from the co-condensation technique exhibit an ionic liquid loading and a porous structure that depend on the ratio between the organosilica and the conventional precursors, on the nature of the templating agent and on other synthetic parameters. Different multi-functional IL organosilica precursors have been exploited for the co-condensation methodology, aiming to obtain new PMOs. Among them, the bi-silylated IL–organosilica precursors, mostly based on imidazolium, bis-imidazolium, and bipyridinium (Table 2.4, entries 1-6), are very common precursors for this type of synthesis. No mono-functional organosilica precursor was reported for the co-condensation, because they are mainly used for grafting on pre-formed porous silica.

In 2001, Alvaro *et al.*[120] reported the synthesis of an ordered MCM-41 system incorporating a 4,4′-bipyridinium ionic liquid, produced by co-condensation of a bi-silylated bipyridinium precursor (Table 2.4, entry 1) in combination with TEOS, using cetyltrimethylammonium bromide (CTABr) as a templating agent.

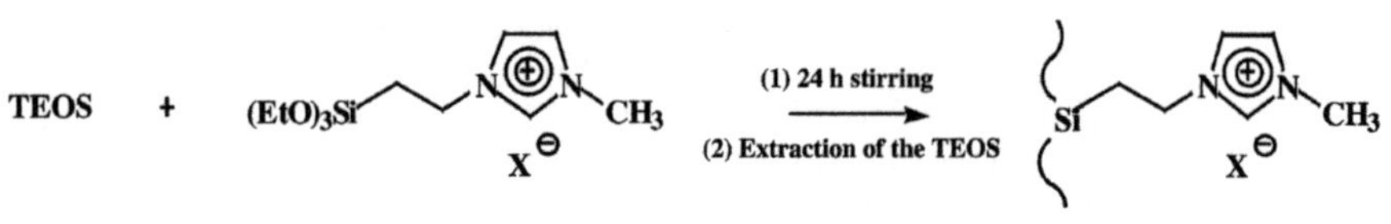

Figure 2.26 General synthetic pathway for the co-condensation of TEOS and an organosilica precursor. Adapted from ref. 76 with permission from The Royal Society of Chemistry.

The obtained solid showed a BET SSA of 930 $m^2 g^{-1}$, a pore size of 3.8 nm and a total pore volume of 0.88 $cm^3 g^{-1}$. The bipyridinium moiety is strongly bound to the silica walls, and no traces of unreacted IL organosilica precursor were found during the final removal of the CTABr. In addition, the bipyridinium moiety endowed the final PMO with a peculiar electron accepting ability, as well as photo- and thermo-chromic responses typical of viologen derivatives. In the following year, the same research group synthesized another MCM-41-like PMO material, with *trans*-1,2-bis(4-pyridyl)ethylene (Table 2.4, entry 2) incorporated inside the silica–IL walls.[121] The peculiarity of this material is that, upon UV irradiation, the C=C bond changes its configuration from trans to cis, as shown in Figure 2.27. This configuration transformation affects the textural properties of the material. After the UV irradiation, the SSA and the pore volume increase from 350 to 473 $cm^2 g^{-1}$ and from 0.28 to 0.38 $cm^3 g^{-1}$, respectively, while the pore size decreases from 39.8 to 36.5 Å.

The co-condensation of a bis-imidazolium-based organosilica precursor (1,1′-di(3-propyltrimethoxysilane)-3,3′-propylenediimidazolium dichloride, Table 2.4, entry 3) and TEOS, in the presence of P123 as a templating agent, produced a silica–IL porous material, as reported by Zheng *et al.*[122] Subsequently, copper(II) chloride was reacted with the chloride anion in order to generate the $[CuCl_4]^{2-}$ complex, which is the final anion present in the silica–IL system. The silica mesoporous structure appeared to be stable to the anion exchange, retaining the hexagonally ordered shape, whereas the BET SSA and the pore volume decreased from 596 to 391 $m^2 g^{-1}$ and from 1.22 to 0.46 $cm^3 g^{-1}$, respectively. Moreover, the average pore size decreased from 77.6 to 38.4 Å. This effect is ascribable to the presence of the large copper complex inside the silica–IL pores. The obtained PMO was used as a heterogeneous catalyst

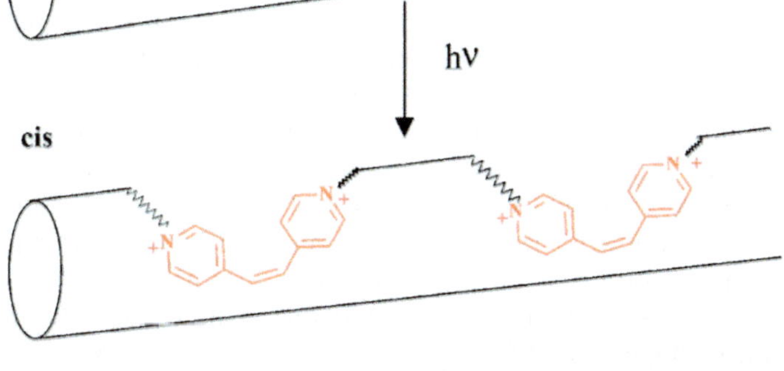

Figure 2.27 Structural changes inside a porous pyridinium-based silica–IL upon UV irradiation. Reproduced from ref. 121 with permission from The Royal Society of Chemistry.

in the decomposition of cyclohexyl hydroperoxide and showed improved stability and negligible metal leaching due to the electrostatic binding between the bisimidazolium dication and the $[CuCl_4]^{2-}$ complex.[122]

In analogy to the corresponding IL–MOF linker, a fully conjugated imidazolium precursor was synthesized.[118] The 1,3-bis(4-triethoxysilyl-2,6-dimethylphenyl)-imidazolium (Table 2.4, entry 4) has two phenyl groups directly conjugated to the imidazolium ring, both bearing a triethoxysilyl group, useful for the creation of the silica network. The co-condensation of this organosilica precursor, using P123 as the soft-templating agent, leads to an ordered silica–IL nanostructure with surprisingly high porosity, with a BET SSA as high as 1170 $m^2\ g^{-1}$. The pore volume of the material is 1.20 $cm^3\ g^{-1}$, and the average pore size is 54 Å.[118]

2.2.3.2.2 Silica–ILs from Self-condensation. The self-condensation methodology for the synthesis of PMOs avoids the use of TEOS in the synthesis of the organosilica system. By means of this methodology, the organosilica precursor is the only building block employed to obtain the PMO. The challenge of this synthetic methodology is to achieve a mesoporous structure using more flexible organosilica precursors. The main advantage of this technique is the enhanced loading of the IL moiety in the final material. The organosilica precursors are both structural and functional components, opening up a route to the next generation of PMOs. The di-functional organosilica precursors currently developed for the generation of PMOs by self-condensation are mainly based on imidazolium, di-imidazolium and ammonium ionic liquids. The final structure of the PMOs strongly depends on the synthetic medium, on the nature of the templating agent and on the structure of the organosilica precursor (see Table 2.5).

The self-condensation of 1,3-bis(3-triethoxysilylpropyl)-imidazolium iodide (Table 2.5, entry 5) in the presence of 1-hexadecane-3-methylimidazolium bromide as a templating agent produces a mesoporous PMO with a BET SSA of 452 $m^2\ g^{-1}$ and a pore volume of 0.48 $cm^3\ g^{-1}$.[123] The structure of the final material is characterized by the presence of wormholes instead of the usual ordered hexagonal structure. This silica–IL hybrid material works as an anion exchange resin for perrhenate adsorption.

A recent study by Dinker and co-workers[124] compares the self-condensation and the co-condensation of an organosilica precursor, constituted by a PEG chain connecting two imidazolium rings, both functionalized with a triethoxysilylpropyl group (Table 2.5, entry 6). The co-condensation is performed by using TEOS as an auxiliary precursor, and CTAB (a cationic surfactant) as a soft-templating agent for both techniques. A general scheme of this syntheses is reported in Figure 2.28. In both cases, the resulting PMOs show a hexagonal pore arrangement and a very similar porous structure. The co-condensation route leads to a BET SSA of 443 $m^2\ g^{-1}$ and a pore volume of 0.47 $cm^3\ g^{-1}$, whereas self-condensation leads to a BET SSA of 581 $m^2\ g^{-1}$ and a pore volume of 0.62 $cm^3\ g^{-1}$.

The first study reporting an organosilica precursor based on ammonium ILs dates back to 2010.[119] Two different tetraalkylammonium precursors were

Table 2.5 Summary of PMO precursors, the resulting structure, and the synthetic conditions.

Entry	PMO IL-precursors	Structure[a]	Synth. Path[b]	Templating agent	Ref.
1	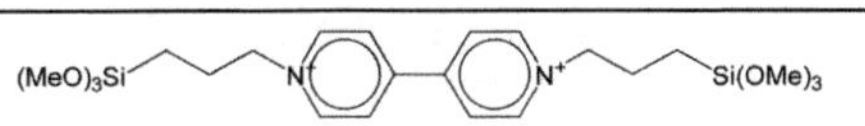	H	C	CTAB	120
2	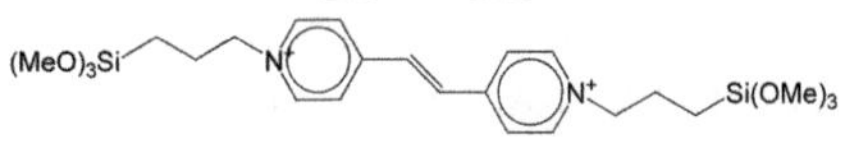	H	C	CTAB	121
3	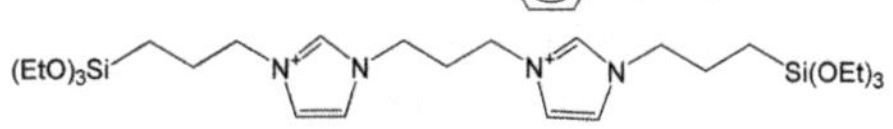	H	C	P123	122
4	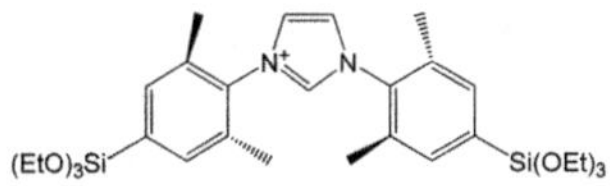	H	C	P123	118
5	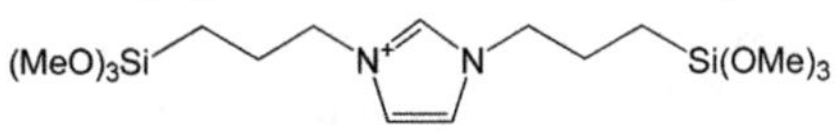	W	S	$[C_{16}MIM]$ [Br]	123
6	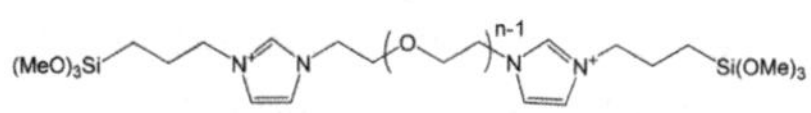	H	Both	CTAB	124
7	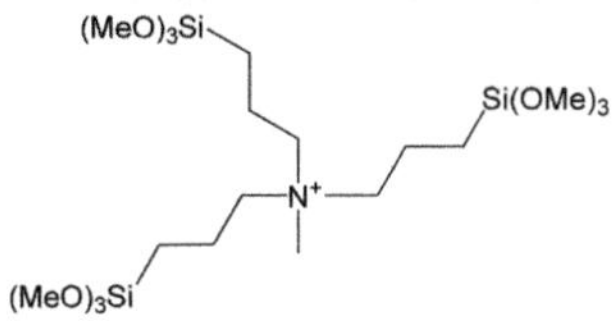	W	S	Anionic surfactants	119
8	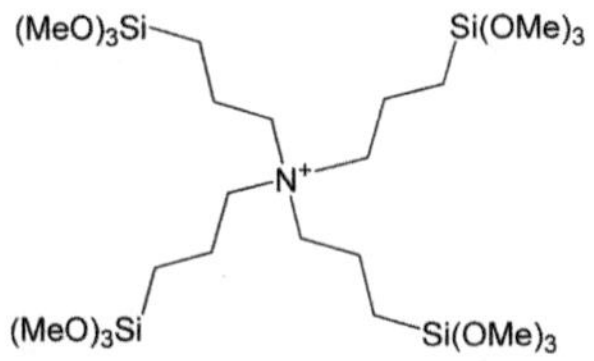	W	S	Anionic surfactants	119
9	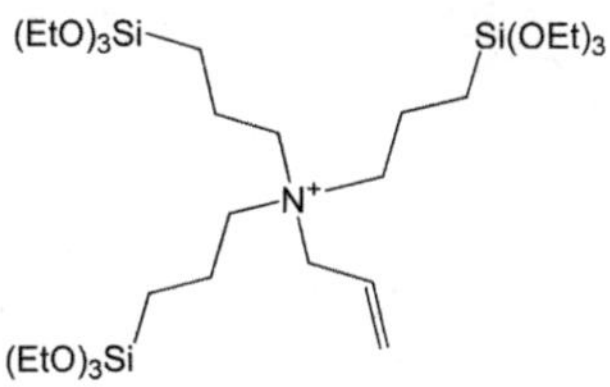	W	S	Anionic surfactants	117
10	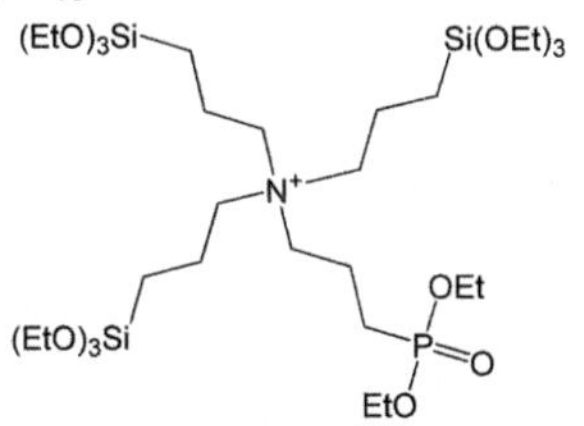	W	S	Anionic surfactants	117

[a]H: hexagonal ordered, W: worm-like.
[b]C: co-condensation, S: self-condensation.

synthesized, methyl-(tris(3-trimethoxysilyl)propyl) ammonium iodide (Table 2.5, entry 7) and tetrakis(3-(trimethoxysilyl)propyl)ammonium iodide (Table 2.5, entry 8), with three or four silylated groups on the alkyl chains, respectively. These precursors were self-condensated in the presence of sodium hexadecyl sulfate (an anionic surfactant), generating ammonium-based PMOs. In both cases, a mesoporous structure was obtained. In particular, the PMO obtained from the tri-functional precursor (Table 2.5, entry 7) exhibits a hexagonal arrangement, with a BET SSA of 575 $m^2\ g^{-1}$, a pore volume of 0.42 $cm^3\ g^{-1}$, and a pore size of 22 Å. In contrast, the PMO obtained from the tetra-functional precursor (Table 2.5, entry 8) shows a more disordered "worm-like" structure, with a definitely higher BET SSA (910 $m^2\ g^{-1}$), a pore volume of 0.54 $cm^3\ g^{-1}$, and a pore size of 20 Å. El Hankari and co-workers extended the number of available ammonium-based PMO precursors.[117] The newly synthesized *N*,*N*,*N*-tris(3-(triethoxysilyl)propyl)prop-2-en-1-ammonium bromide (Table 2.5, entry 9), and 3-(diethoxyphosphoryl)-*N*,*N*,*N*-tris(3-(trimethoxysilyl)propyl)propan-1-ammonium bromide (Table 2.5, entry 10) precursors were self-condensed in the presence of three different kinds of surfactants. It is worth highlighting that in the case of cationic and non-ionic surfactants, CTAB and P123, respectively, a non-structured material was obtained.

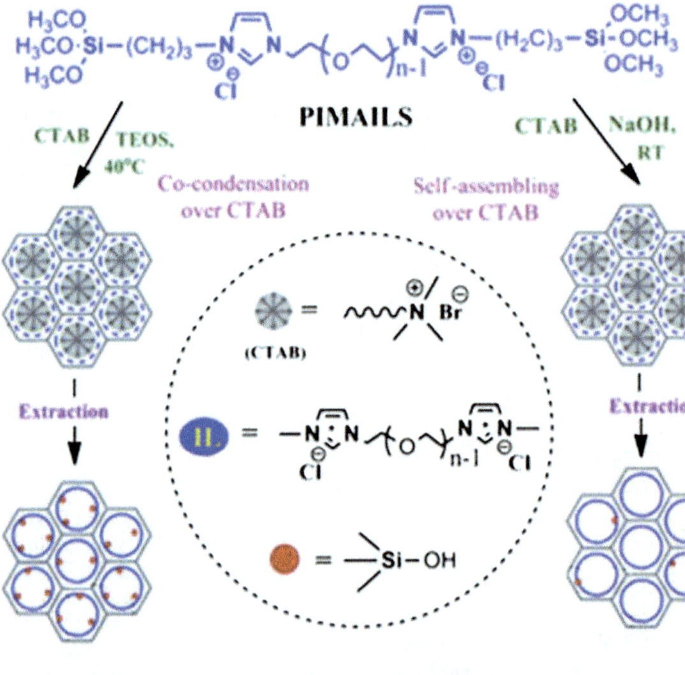

Figure 2.28 Synthesis scheme of the formation of two different PMOs obtained *via* co-condensation and self-assembly. The same organosilica precursor is employed. Reprinted from *Microporous and Mesoporous Materials*, 230, M. K. Dinker, P. S. Kulkarni, Insight into the PEG-linked bis-imidazolium bridged framework of mesoporous organosilicas as ion exchangers, 145–153, Copyright 2016 with permission from Elsevier.[124]

In contrast, the use of an SHS-anionic surfactant produced two PMOs with "worm-like" regular structures. The allyl-functionalized precursor (Table 2.5, entry 9) led to a PMO with a BET SSA as high as 640 $m^2 g^{-1}$, a pore volume of 0.38 $cm^3 g^{-1}$ and a pore size below 20 Å. The diethoxyphosphoryl-functionalized precursor (Table 2.5, entry 10) generates a PMO with a BET SSA of 567 $m^2 g^{-1}$, a pore volume of 0.33 $cm^3 g^{-1}$ and a pore size below 20 Å.

2.2.4 Supramolecular Ionic Liquids

Similarly to MOFs, a fully organic network can be created by electrostatic complexation of multivalent ionic species. These materials are called supramolecular ionic networks and they were developed for the first time by Wathier and Grinstaff, combining multicationic and multianionic molecules.[125] These authors created a supramolecular network, exploiting the electrostatic interaction between a tetra-alkyl phosphonium dication and ethylenediaminetetraacetate (a multivalent acid). The as obtained network was in the form of a viscous liquid salt and, unfortunately, no porosity information was reported.[125] A subsequent study by Mecerreyes and co-workers investigated the formation and the rheological properties of supramolecular ionic networks based on citric acid and di-imidazolium, di-pyridinium and di-pyrrolidinium dications. The final products were obtained in the form of non-crystalline soft solids and, also in this case, the porosity was not investigated.[126] Recently, Hei *et al.* synthesized a supramolecular porous ionic network based on imidazolium and triazinonide. They reacted 2,4,6-triimidazolyl-1,3,5-triazine with cyanuric chloride under very mild conditions, generating an orange precipitate consisting of a supramolecular network with a BET SSA as high as 263 $m^2 g^{-1}$ and a total pore volume of 0.75 $cm^3 g^{-2}$. The N_2 adsorption isotherm at 77 K pointed out the formation of a meso-macroporous supramolecular network. A thorough structural analysis evidenced that the reaction of the two starting molecules led to the quaternarization of the imidazole moiety to form one or more positively charged imidazolium rings. This di-cationic molecule could spontaneously hydrolyze introducing negative charges along the 1,3,5-triazine ring. This combination of positively and negatively charged molecules was arranged into a supramolecular network, which due to the rigidity of the conjugated building blocks was intrinsically porous.[127]

References

1. D. Wu, F. Xu, B. Sun, R. Fu, H. He and K. Matyjaszewski, *Chem. Rev.*, 2012, **112**, 3959–4015.
2. J.-K. Sun, M. Antonietti and J. Yuan, *Chem. Soc. Rev.*, 2016, **45**, 6627–6656.
3. K. S. W. Sing, D. H. Everett, R. A. W. Haul, L. Moscou, R. A. Pierotti, J. Rouquerol and T. Siemieniewska, *Pure Appl. Chem.*, 1985, **57**, 603–619.
4. M. Thommes, K. Kaneko, A. V. Neimark, J. P. Olivier, F. Rodriguez-Reinoso, J. Rouquerol and K. S. W. Sing, *Pure Appl. Chem.*, 2015, **87**, 1051–1069.
5. S. J. Gregg and K. S. W. Sing, *Adsorption, Surface Area, and Porosity*, Academic Press, London and New York, 2nd edn, 1982, vol. 1, pp. 42–60.

6. J. Weber, M. Antonietti and A. Thomas, *Macromolecules*, 2008, **41**, 2880–2885.
7. J. Weber, J. Schmidt, A. Thomas and W. Böhlmann, *Langmuir*, 2010, **26**, 15650–15656.
8. K. E. Hart, J. M. Springmeier, N. B. McKeown and C. M. Colina, *Phys. Chem. Chem. Phys.*, 2013, **15**, 20161–20169.
9. Y. Tsujita, *Prog. Polym. Sci.*, 2003, **28**, 1377–1401.
10. M. Steinhart, J. H. Wendorff, A. Greiner, R. B. Wehrspohn, K. Nielsch, J. Schilling, J. Choi and U. Gösele, *Science*, 2002, **296**, 1997.
11. M. Steinhart, R. B. Wehrspohn, U. Gösele and J. H. Wendorff, *Angew. Chem., Int. Ed.*, 2004, **43**, 1334–1344.
12. J. Yuan and M. Antonietti, *Polymer*, 2011, **52**, 1469–1482.
13. J. Yuan, D. Mecerreyes and M. Antonietti, *Prog. Polym. Sci.*, 2013, **38**, 1009–1036.
14. D. Mecerreyes, *Prog. Polym. Sci.*, 2011, **36**, 1629–1648.
15. A. Thomas, F. Goettmann and M. Antonietti, *Chem. Mater.*, 2008, **20**, 738–755.
16. A. H. Lu and F. Schüth, *Adv. Mater.*, 2006, **18**, 1793–1805.
17. X. Hu, J. Huang, W. Zhang, M. Li, C. Tao and G. Li, *Adv. Mater.*, 2008, **20**, 4074–4078.
18. J. Huang, C.-a. Tao, Q. An, W. Zhang, Y. Wu, X. Li, D. Shen and G. Li, *Chem. Commun.*, 2010, **46**, 967–969.
19. J. Huang, C.-a. Tao, Q. An, C. Lin, X. Li, D. Xu, Y. Wu, X. Li, D. Shen and G. Li, *Chem. Commun.*, 2010, **46**, 4103–4105.
20. B. Karimi, M. R. Marefat, M. Hasannia, P. F. Akhavan, F. Mansouri, Z. Artelli, F. Mohammadi and H. Vali, *ChemCatChem*, 2016, **8**, 2508–2515.
21. A. Wilke, J. Yuan, M. Antonietti and J. Weber, *ACS Macro Lett.*, 2012, **1**, 1028–1031.
22. P. Zhang, Z.-A. Qiao, X. Jiang, G. M. Veith and S. Dai, *Nano Lett.*, 2015, **15**, 823–828.
23. C. Gao, G. Chen, X. Wang, J. Li, Y. Zhou and J. Wang, *Chem. Commun.*, 2015, **51**, 4969–4972.
24. Y. Xie, Z. Zhang, T. Jiang, J. He, B. Han, T. Wu and K. Ding, *Angew. Chem., Int. Ed.*, 2007, **46**, 7255–7258.
25. D. Kuzmicz, P. Coupillaud, Y. Men, J. Vignolle, G. Vendraminetto, M. Ambrogi, D. Taton and J. Yuan, *Polymer*, 2014, **55**, 3423–3430.
26. X. Feng, C. Gao, Z. Guo, Y. Zhou and J. Wang, *RSC Adv.*, 2014, **4**, 23389–23395.
27. A. Dani, E. Groppo, C. Barolo, J. G. Vitillo and S. Bordiga, *J. Mater. Chem. A*, 2015, **3**, 8508–8518.
28. A. Dani, V. Crocellà, L. Maddalena, C. Barolo, S. Bordiga and E. Groppo, *J. Phys. Chem. C*, 2016, **120**, 1683–1692.
29. F. Liu, L. Wang, Q. Sun, L. Zhu, X. Meng and F.-S. Xiao, *J. Am. Chem. Soc.*, 2012, **134**, 16948–16950.
30. S. Ghazali-Esfahani, H. Song, E. Paunescu, F. D. Bobbink, H. Liu, Z. Fei, G. Laurenczy, M. Bagherzadeh, N. Yan and P. J. Dyson, *Green Chem.*, 2013, **15**, 1584–1589.

31. I. Azcune, I. García, P. M. Carrasco, A. Genua, M. Tanczyk, M. Jaschik, K. Warmuzinski, G. Cabañero and I. Odriozola, *ChemSusChem*, 2014, **7**, 3407–3412.
32. X. Wang, Y. Zhou, Z. Guo, G. Chen, J. Li, Y. Shi, Y. Liu and J. Wang, *Chem. Sci.*, 2015, **6**, 6916–6924.
33. W. Wang, C. Li, L. Yan, Y. Wang, M. Jiang and Y. Ding, *ACS Catal.*, 2016, **6**, 6091–6100.
34. F. Yan and J. Texter, *Angew. Chem., Int. Ed.*, 2007, **46**, 2440–2443.
35. S. Fischer, J. Schmidt, P. Strauch and A. Thomas, *Angew. Chem., Int. Ed.*, 2013, **52**, 12174–12178.
36. Z. Yan, Y. Yuan, Y. Tian, D. Zhang and G. Zhu, *Angew. Chem., Int. Ed.*, 2015, **54**, 12733–12737.
37. Q. Zhang, S. Zhang and S. Li, *Macromolecules*, 2012, **45**, 2981–2988.
38. J. Wang, J. G. Wei Yang, G. Yi and Y. Zhang, *Chem. Commun.*, 2015, **51**, 15708–15711.
39. H. C. Cho, H. S. Lee, J. Chun, S. M. Lee, H. J. Kim and S. U. Son, *Chem. Commun.*, 2011, **47**, 917–919.
40. M. Rose, A. Notzon, M. Heitbaum, G. Nickerl, S. Paasch, E. Brunner, F. Glorius and S. Kaskel, *Chem. Commun.*, 2011, **47**, 4814–4816.
41. Y. Su, X. Li, Y. Wang, H. Zhong and R. Wang, *Dalton Trans.*, 2016, **45**, 16896–16903.
42. O. Buyukcakir, S. H. Je, D. S. Choi, S. N. Talapaneni, Y. Seo, Y. Jung, K. Polychronopoulou and A. Coskun, *Chem. Commun.*, 2016, **52**, 934–937.
43. J. S. Lee, H. Luo, G. A. Baker and S. Dai, *Chem. Mater.*, 2009, **21**, 4756–4758.
44. H. Ma, B. Liu, B. Li, L. Zhang, Y.-G. Li, H.-Q. Tan, H.-Y. Zang and G. Zhu, *J. Am. Chem. Soc.*, 2016, **138**, 5897–5903.
45. S. Mitra, S. Kandambeth, B. P. Biswal, A. M. Khayum, C. K. Choudhury, M. Mehta, G. Kaur, S. Banerjee, A. Prabhune, S. Verma, S. Roy, U. K. Kharul and R. Banerjee, *J. Am. Chem. Soc.*, 2016, **138**, 2823–2828.
46. S.-B. Yu, H. Lyu, J. Tian, H. Wang, D.-W. Zhang, Y. Liu and Z.-T. Li, *Polym. Chem.*, 2016, **7**, 3392–3397.
47. A. Dani, V. Crocella, C. Magistris, V. Santoro, J. Yuan and S. Bordiga, *J. Mater. Chem. A*, 2017, **120**, 1683–1692.
48. S. N. Talapaneni, O. Buyukcakir, S. H. Je, S. Srinivasan, Y. Seo, K. Polychronopoulou and A. Coskun, *Chem. Mater.*, 2015, **27**, 6818–6826.
49. K. Thiel, R. Zehbe, J. Roeser, P. Strauch, S. Enthaler and A. Thomas, *Polym. Chem.*, 2013, **4**, 1848–1856.
50. G. Chen, Y. Zhou, X. Wang, J. Li, S. Xue, Y. Liu, Q. Wang and J. Wang, *Sci. Rep.*, 2015, **5**, 11236.
51. J. Wang, W. Sng, G. Yi and Y. Zhang, *Chem. Commun.*, 2015, **51**, 12076–12079.
52. Y. Yuan, F. Sun, F. Zhang, H. Ren, M. Guo, K. Cai, X. Jing, X. Gao and G. Zhu, *Adv. Mater.*, 2013, **25**, 6619–6624.
53. Y. Yuan, F. Sun, L. Li, P. Cui and G. Zhu, *Nat. Commun.*, 2014, **5**, 4260.
54. Q. Zhao, Q. F. An, Y. Ji, J. Qian and C. Gao, *J. Membr. Sci.*, 2011, **379**, 19–45.

55. Q. Zhao, P. Zhang, M. Antonietti and J. Yuan, *J. Am. Chem. Soc.*, 2012, **134**, 11852–11855.
56. Q. Zhao, S. Soll, M. Antonietti and J. Yuan, *Polym. Chem.*, 2013, **4**, 2432–2435.
57. S. Soll, Q. Zhao, J. Weber and J. Yuan, *Chem. Mater.*, 2013, **25**, 3003–3010.
58. S. Soll, P. Zhang, Q. Zhao, Y. Wang and J. Yuan, *Polym. Chem.*, 2013, **4**, 5048–5051.
59. Q. Zhao, M. Yin, A. P. Zhang, S. Prescher, M. Antonietti and J. Yuan, *J. Am. Chem. Soc.*, 2013, **135**, 5549–5552.
60. Q. Zhao, J. W. C. Dunlop, X. Qiu, F. Huang, Z. Zhang, J. Heyda, J. Dzubiella, M. Antonietti and J. Yuan, *Nat. Commun.*, 2014, **5**, 4293.
61. Q. Zhao, J. Heyda, J. Dzubiella, K. Täuber, J. W. C. Dunlop and J. Yuan, *Adv. Mater.*, 2015, **27**, 2913–2917.
62. K. Täuber, Q. Zhao, M. Antonietti and J. Yuan, *ACS Macro Lett.*, 2015, **4**, 39–42.
63. K. Zhang, X. Feng, X. Sui, M. A. Hempenius and G. J. Vancso, *Angew. Chem., Int. Ed.*, 2014, **53**, 13789–13793.
64. K. Täuber, A. Zimathies and J. Yuan, *Macromol. Rapid Commun.*, 2015, **36**, 2176–2180.
65. K. Tauber, B. Lepenies and J. Yuan, *Polym. Chem.*, 2015, **6**, 4855–4858.
66. K. Täuber, A. Dani and J. Yuan, *ACS Macro Lett.*, 2017, **6**, 1–5.
67. T. Selvam, A. Machoke and W. Schwieger, *Appl. Catal., A*, 2012, **445–446**, 92–101.
68. M. Blahušiak, Š. Schlosser and J. Marták, *React. Funct. Polym.*, 2011, **71**, 736–744.
69. U. Kernchen, B. Etzold, W. Korth and A. Jess, *Chem. Eng. Technol.*, 2007, **30**, 985–994.
70. M. Rufete-Beneite, M. C. Román-Martínez and A. Linares-Solano, *Carbon*, 2014, **77**, 947–957.
71. J.-P. T. Mikkola, P. P. Virtanen, K. Kordás, H. Karhu and T. O. Salmi, *Appl. Catal., A*, 2007, **328**, 68–76.
72. P. Virtanen, T. O. Salmi and J.-P. Mikkola, *Top. Catal.*, 2010, **53**, 1096–1103.
73. C. Fu, L. Meng, Q. Lu, Z. Fei and P. J. Dyson, *Adv. Funct. Mater.*, 2008, **18**, 857–864.
74. Q.-x. Luo, B.-w. An, M. Ji, S.-E. Park, C. Hao and Y.-q. Li, *J. Porous Mater.*, 2015, **22**, 247–259.
75. N. A. Khan, Z. Hasan and S. H. Jhung, *Chem. Commun.*, 2016, **52**, 2561–2564.
76. M. H. Valkenberg, C. deCastro and W. F. Holderich, *Green Chem.*, 2002, **4**, 88–93.
77. F. Hoffmann, M. Cornelius, J. Morell and M. Fröba, *Angew. Chem., Int. Ed.*, 2006, **45**, 3216–3251.
78. B. Xin and J. Hao, *Chem. Soc. Rev.*, 2014, **43**, 7171–7187.
79. R. Skoda-Foldes, *Molecules*, 2014, **19**, 8840–8884.
80. J. M. Zhu, F. Xin, Y. C. Sun and X. C. Dong, *Theor. Found. Chem. Eng.*, 2014, **48**, 787–792.

81. C. De Castro, E. Sauvage, M. H. Valkenberg and W. F. Hölderich, *J. Catal.*, 2000, **196**, 86–94.
82. M. J. Earle, K. R. Seddon, C. J. Adams and G. Roberts, *Chem. Commun.*, 1998, 2097–2098.
83. J. Zhang, S.-H. Chai, Z.-A. Qiao, S. M. Mahurin, J. Chen, Y. Fang, S. Wan, K. Nelson, P. Zhang and S. Dai, *Angew. Chem., Int. Ed.*, 2015, **54**, 932–936.
84. O. M. Yaghi and H. Li, *J. Am. Chem. Soc.*, 1995, **117**, 10401–10402.
85. K. Biradha, A. Ramanan and J. J. Vittal, *Cryst. Growth Des.*, 2009, **9**, 2969–2970.
86. J. L. C. Rowsell and O. M. Yaghi, *Microporous Mesoporous Mater.*, 2004, **73**, 3–14.
87. K. Sumida, D. L. Rogow, J. A. Mason, T. M. McDonald, E. D. Bloch, Z. R. Herm, T.-H. Bae and J. R. Long, *Chem. Rev.*, 2012, **112**, 724–781.
88. J. Lee, O. K. Farha, J. Roberts, K. A. Scheidt, S. T. Nguyen and J. T. Hupp, *Chem. Soc. Rev.*, 2009, **38**, 1450–1459.
89. J. D. Evans, C. J. Sumby and C. J. Doonan, *Chem. Soc. Rev.*, 2014, **43**, 5933–5951.
90. J. Y. Lee, J. M. Roberts, O. K. Farha, A. A. Sarjeant, K. A. Scheidt and J. T. Hupp, *Inorg. Chem.*, 2009, **48**, 9971–9973.
91. S. Sen, S. Neogi, A. Aijaz, Q. Xu and P. K. Bharadwaj, *Inorg. Chem.*, 2014, **53**, 7591–7598.
92. S. Sen, N. N. Nair, T. Yamada, H. Kitagawa and P. K. Bharadwaj, *J. Am. Chem. Soc.*, 2012, **134**, 19432–19437.
93. C. I. Ezugwu, N. A. Kabir, M. Yusubov and F. Verpoort, *Coord. Chem. Rev.*, 2016, **307**(pt 2), 188–210.
94. S. P. Nolan, *N-Heterocyclic Carbenes in Synthesis*, Wiley-VCH Verlag GmbH & Co. KGaA, 2006, p. 304.
95. W. A. Herrmann, *Angew. Chem., Int. Ed.*, 2002, **41**, 1290–1309.
96. R. H. Crabtree, *Chem. Rev.*, 2015, **115**, 127–150.
97. A. Burgun, C. J. Doonan and C. J. Sumby, *Aust. J. Chem.*, 2013, **66**, 409–418.
98. S. Sen, T. Yamada, H. Kitagawa and P. K. Bharadwaj, *Cryst. Growth Des.*, 2014, **14**, 1240–1244.
99. A. Burgun, R. S. Crees, M. L. Cole, C. J. Doonan and C. J. Sumby, *Chem. Commun.*, 2014, **50**, 11760–11763.
100. J. M. Roberts, O. K. Farha, A. A. Sarjeant, J. T. Hupp and K. A. Scheidt, *Cryst. Growth Des.*, 2011, **11**, 4747–4750.
101. X.-W. Wang, L. Han, T.-J. Cai, Y.-Q. Zheng, J.-Z. Chen and Q. Deng, *Cryst. Growth Des.*, 2007, **7**, 1027–1030.
102. X.-C. Chai, Y.-Q. Sun, R. Lei, Y.-P. Chen, S. Zhang, Y.-N. Cao and H.-H. Zhang, *Cryst. Growth Des.*, 2010, **10**, 658–668.
103. P. Suresh, C. N. Babu and G. Prabusankar, *Polyhedron*, 2015, **89**, 322–329.
104. R. S. Crees, M. L. Cole, L. R. Hanton and C. J. Sumby, *Inorg. Chem.*, 2010, **49**, 1712–1719.
105. G. Nickerl, A. Notzon, M. Heitbaum, I. Senkovska, F. Glorius and S. Kaskel, *Cryst. Growth Des.*, 2013, **13**, 198–203.

106. J. Chun, I. G. Jung, H. J. Kim, M. Park, M. S. Lah and S. U. Son, *Inorg. Chem.*, 2009, **48**, 6353–6355.
107. S. Wang, Q. Yang, J. Zhang, X. Zhang, C. Zhao, L. Jiang and C.-Y. Su, *Inorg. Chem.*, 2013, **52**, 4198–4204.
108. G.-Q. Kong, X. Xu, C. Zou and C.-D. Wu, *Chem. Commun.*, 2011, **47**, 11005–11007.
109. G.-Q. Kong, S. Ou, C. Zou and C.-D. Wu, *J. Am. Chem. Soc.*, 2012, **134**, 19851–19857.
110. I. T. Siraj and M. D. Spicer, *Int. J. Chem. Eng. Appl.*, 2013, **4**, 199–203.
111. P. Suresh, S. Radhakrishnan, C. Naga Babu, A. Sathyanarayana, N. Sampath and G. Prabusankar, *Dalton Trans.*, 2013, **42**, 10838–10846.
112. K. Oisaki, Q. Li, H. Furukawa, A. U. Czaja and O. M. Yaghi, *J. Am. Chem. Soc.*, 2010, **132**, 9262–9264.
113. C. I. Ezugwu, B. Mousavi, M. A. Asrafa, A. Mehta, H. Vardhan and F. Verpoort, *Catal. Sci. Technol.*, 2016, **6**, 2050–2054.
114. L. Pan, K. M. Adams, H. E. Hernandez, X. Wang, C. Zheng, Y. Hattori and K. Kaneko, *J. Am. Chem. Soc.*, 2003, **125**, 3062–3067.
115. Y.-S. Bae, O. K. Farha, J. T. Hupp and R. Q. Snurr, *J. Mater. Chem.*, 2009, **19**, 2131–2134.
116. S. S. Park, M. S. Moorthy and C.-S. Ha, *Korean J. Chem. Eng.*, 2014, **31**, 1707–1719.
117. S. El Hankari, A. Bouhaouss and P. Hesemann, *Microporous Mesoporous Mater.*, 2013, **180**, 196–208.
118. T. P. Nguyen, P. Hesemann, P. Gaveau and J. J. E. Moreau, *J. Mater. Chem.*, 2009, **19**, 4164–4171.
119. T. P. Nguyen, P. Hesemann, T. M. Linh Tran and J. J. E. Moreau, *J. Mater. Chem.*, 2010, **20**, 3910–3917.
120. M. Alvaro, B. Ferrer, V. Fornes and H. Garcia, *Chem. Commun.*, 2001, 2546–2547.
121. M. Alvaro, B. Ferrer, H. Garcia and F. Rey, *Chem. Commun.*, 2002, 2012–2013.
122. X. Zheng, M. Wang, Z. Sun, C. Chen, J. Ma and J. Xu, *Catal. Commun.*, 2012, **29**, 149–152.
123. B. Lee, H.-J. Im, H. Luo, E. W. Hagaman and S. Dai, *Langmuir*, 2005, **21**, 5372–5376.
124. M. K. Dinker and P. S. Kulkarni, *Microporous Mesoporous Mater.*, 2016, **230**, 145–153.
125. M. Wathier and M. W. Grinstaff, *J. Am. Chem. Soc.*, 2008, **130**, 9648–9649.
126. M. A. Aboudzadeh, M. E. Munoz, A. Santamaria and D. Mecerreyes, *RSC Adv.*, 2013, **3**, 8677–8682.
127. Z.-H. Hei, G.-L. Song, C.-Y. Zhao, W. Fan and M.-H. Huang, *RSC Adv.*, 2016, **6**, 92443–92448.

CHAPTER 3

Cationic and Anionic Polymerized Ionic Liquids: Properties for Applications

OLGA KUZMINA

Department of Chemistry, Imperial College London, South Kensington Campus, Exhibition Road, SW7 2AZ, London, UK
*E-mail: okuzmina@imperial.ac.uk

3.1 Introduction

Over the last two decades, a class of organic salts with large cations and usually small anions, liquid at temperatures below 100 °C, has been investigated due to their unique properties and possible applications. However, the possibility to polymerise these ionic liquids (ILs) to create a new type of polymer, polymeric ionic liquids (PILs), was not considered until more recent years.[1] As an emerging interdisciplinary topic among polymer chemistry and physics, materials science, catalysis, separation, analytical chemistry, and electrochemistry, PILs are attracting increasing interest.[2] PILs are mainly considered as a product that combines the unique electroconductive and thermostable properties of ILs and mechanical strength of solid materials. Despite ILs generally being liquids over a wide temperature range, most PILs are amorphous solids or gel-like materials, often with low glass transition temperatures.[3] Currently known PILs have started to be investigated as a new class of polyelectrolytes. The advantages of polymeric electrolytes, such

Smart Materials No. 29
Polymerized Ionic Liquids
Edited by Ali Eftekhari

Published by the Royal Society of Chemistry, www.rsc.org

as easy handling, no-leakage of electrolyte, higher energy density, flexible geometry and improved safety, have resulted in increasing interest in polymer research for use in energy applications, especially batteries, separately or in combination with Li-ion.[4]

The potential of PILs has not yet been fully explored, but can also include such applications as electrochemistry, preparation of polymeric surfactants, hybrid nanocomposites, carbon precursors, anion-sensitive smart materials, CO_2-sorbents, bio- and gas-sensors (because of their ability to conduct electricity), and catalysts.[3]

Depending on the location of the polymerisable units, PILs can be mainly classified as cationic (polymerisable unit on a cation), anionic (polymerisable unit on an anion) or zwitterionic (backbone containing both cations and anions).[3] Cationic ionic liquids based on imidazolium, pyrrolidinium, pyridinium and ammonium have received the most attention from researchers, due to their ease of synthesis and the wide range of available non-polymerisable anions, such as tetrafluoroborate, hexafluorophosphate, triflates and halides. In this chapter, the properties of PILs regarding their applications in different fields are described in terms of the effect of the structure of the anions and cations of the PILs.

3.2 Comparison of the Properties of Cationic *vs.* Anionic PILs

3.2.1 Electroconductivity

One of the major areas of potential application for PILs is as polyelectrolytes. Polyelectrolytes are defined as polymers carrying either positively or negatively charged ionisable groups.[5] The dependence between ionic conductivity and glass transition temperature, common for polyelectrolytes, was found to be replicated for PILs.[6]

In the last few years, large numbers of cations based on imidazolium (vinyl, styrene, (meth)acrylate and (meth)acrylamide, ethylene glycol, vinyl ether and norbornene) as well as on tetraalkyl ammonium, pyridinium, pyrrolidinium guanidinium and piperidinium have been studied for charge transfer in PILs. These cations were modified by the inclusion of different spacers between the backbone and charged species, and combined with a wide selection of anions and even doped with other polymers, ILs, cross-linkers *etc.* As correctly noted by Mecerreyes,[3] one combination of cation–anion pair in PILs that works better for one type of electrochemical device, may not necessarily work in others. This complicates the comparison of PILs used for electrochemical devices. Recently, an overview of the influence of different factors including the structure of PILs was reported.[7] Below, a deeper, broader and updated comparison of the structural dependence of ion conductivity for cationic and anionic PILs is summarised (Tables 3.1 and 3.2). Poly(1,2,3-triazolium)s are not included in these tables. Their synthesis and conductivity ranging from 10^{-11} to 10^{-5} S cm^{-1} at close to room temperature (with the highest conductivities reported for PILs containing the NTf_2^- anion) were recently reported in great detail.[8,9]

As can be seen from Table 3.1, among the systems based on neat PILs, imidazolium cations contribute to higher conductivity reaching 10^{-4} S cm^{-1} (entries 10, 15, 19, 23, 24 and 33b). The anion also plays a crucial role in defining the conductivity of ILs. For imidazolium cationic PILs, the highest conductivity in most cases was found for the NTf_2^- anion. PF_6^-, BF_4^-, Cl^-, $B(CN)_4^-$ and OTf^- anions gave lower conductivities when compared to NTf_2^--containing cationic imidazolium PILs (Table 3.1, entries 2–4, 6, 18, 19, 22, 24 and 26). In some cases,

Table 3.1 Structural dependence of ion conductivity (δ) for cationic PILs.

Entry	PIL cation	PIL anion	Highest reported δ, S cm^{-1}	Comments	Ref.
1	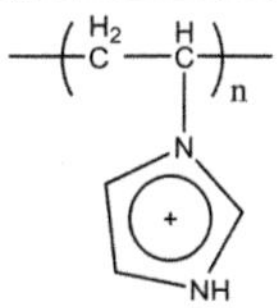	BF_4^-	10^{-9} at 25 °C	Conductivity was improved by adding Li^+ salts	90
2	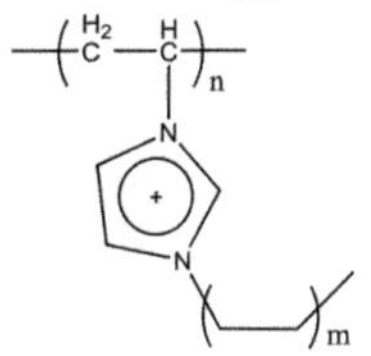 $m = 1, 3,$ or 7	Br^- BF_4^- OTf^- NTf_2^-	10^{-5} at 85 °C for the NTf_2^- PIL with $m = 3$.	Conductivity increased in the order: $m = 7 < 1 < 3$, with the NTf_2^- anion showing the highest values	11
3	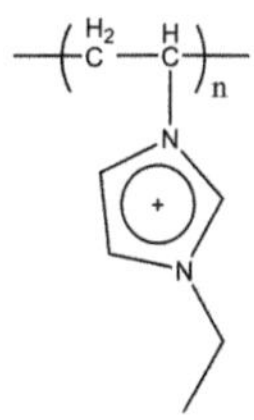	OTf^- NTf_2^- $N(CN)_2^-$	1.83×10^{-8} 7.65×10^{-8} 1.51×10^{-5} at 10^6 Hz, 25 °C	Conductivity decreased when the PILs were cross-linked with acrylates	91
4	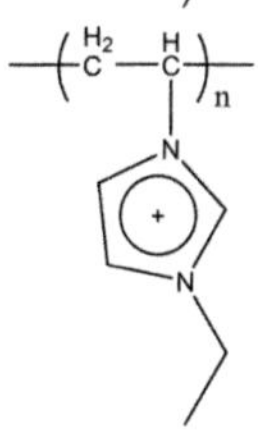	Br^- BF_4^- NTf_2^-	6×10^{-6} 2.5×10^{-6} 10^{-5}	Conductivity increased upon the addition of IL	92
5	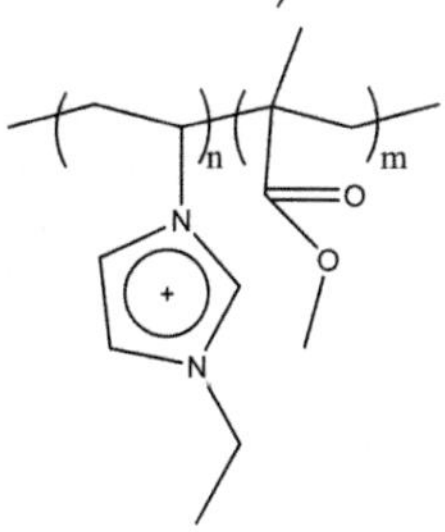 $n : m = 6 : 4$	NTf_2^-	6.23×10^{-9} at 10^6 Hz, 25 °C	Conductivity increased upon the addition of IL	93

(continued)

Table 3.1 (*continued*)

Entry	PIL cation	PIL anion	Highest reported δ, S cm^{-1}	Comments	Ref.
6	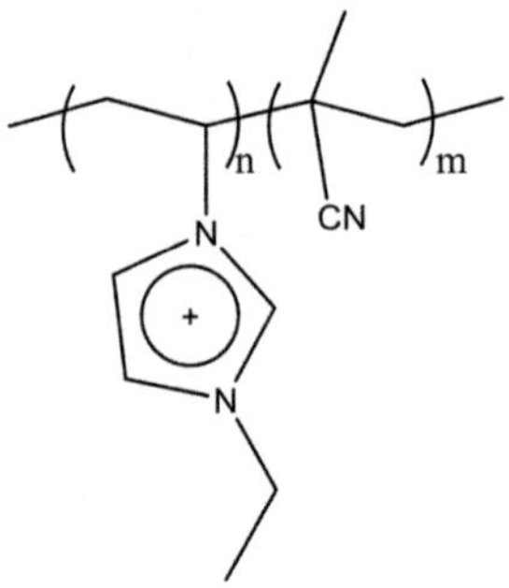	$N(CN)_2^-$	1.24×10^{-8} at 10^6 Hz, 25 °C	Conductivity increased upon the addition of IL	93
7	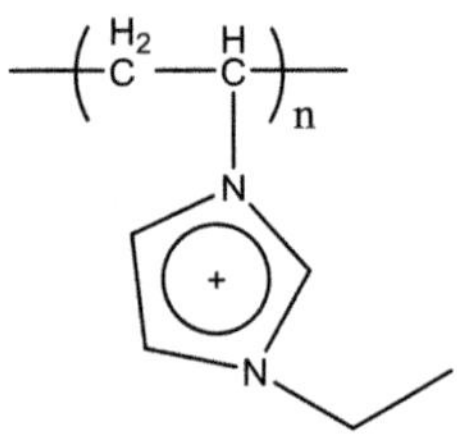	NTf_2^-	1.89×10^{-5} at 25 °C	As a mixture with 1,2-dimethyl-3-ethoxyethyl imidazolium bis(trifluoromethanesulfonyl) imide and Li^+ salt	94
8	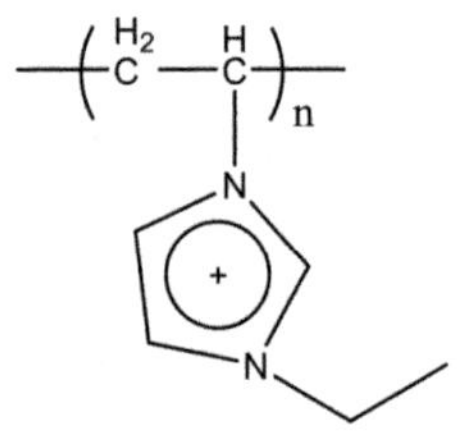	NTf_2^-	1×10^{-4} at 25 °C	As a conducting membrane including PEO, $LiNTf_2$ and free-radical photoinitiator	95
9	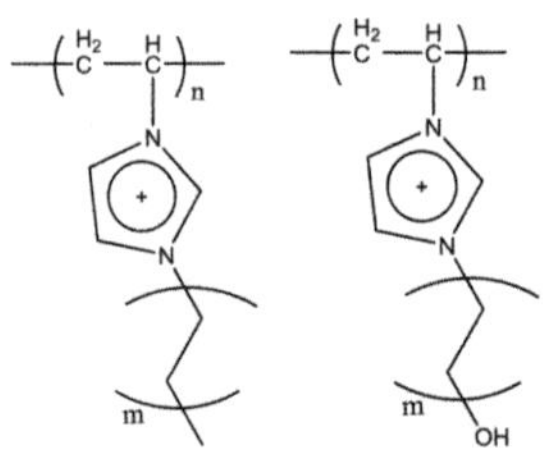 m = 2, 3, 6, or 8	NTf_2^-	10^{-6} at 30 °C for the hydroxyl-containing polymer with $m = 8$	Conductivity increased with the introduction of an OH group	10
10	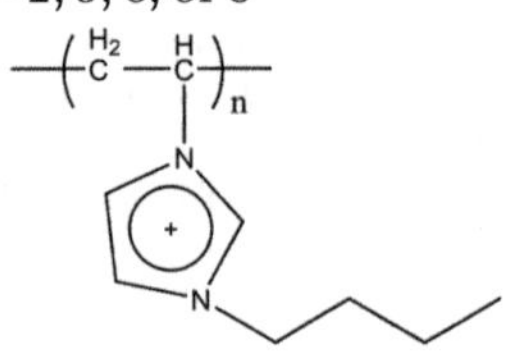	Br^-	6.8×10^{-4}	1-Bromo-butane was incorporated onto the polymer membrane	96

11	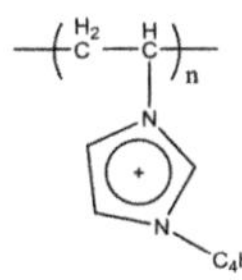	NTf_2^-	2.92×10^{-3}	25 wt% in the ionic liquid electrolyte under simulated AM 1.5 solar spectrum illumination at 100 mW cm^{-2}	97
	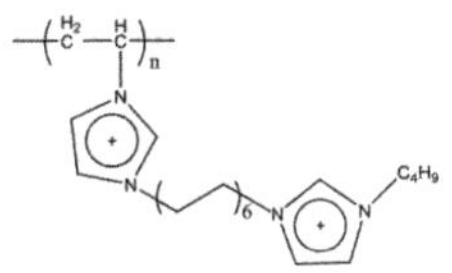	$2 \times NTf_2^-$	5.83×10^{-3}		
12	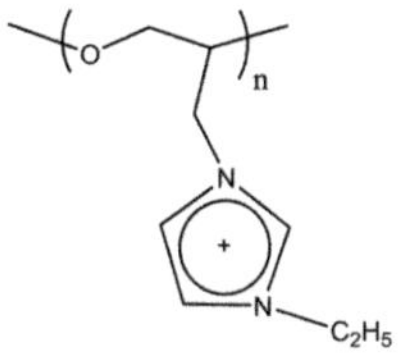	NTf_2^-	5.3×10^{-6} at 30 °C	—	98
13	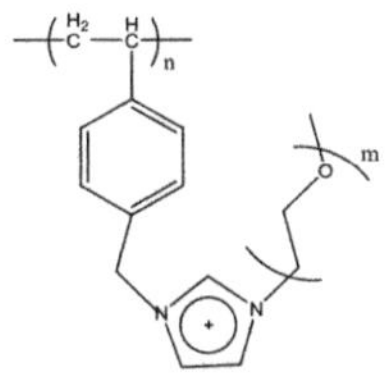 $m = 1, 2, 3, 8,12,$ or 17	NTf_2^-	1×10^{-4} at 25 °C	Highest conductivity for $m = 8$	99
14	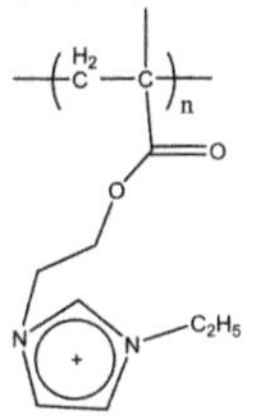	I^-	above 1×10^{-3} at 20 °C	Plasticizers, I_2 and polyacrylonitrile were incorporated	17
15	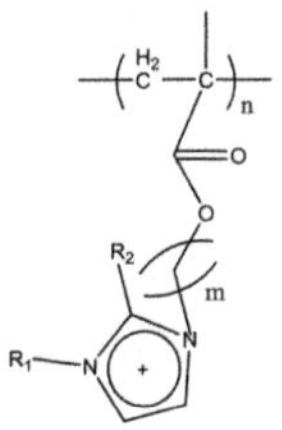 R1 = CH_3, C_2H_5, or C_4H_9; R_2 = H, or CH_3; $n = 2, 3, 6, 9,$ or 12	NTf_2^-	1.37×10^{-4} for $R_1 = C_2H_5$, R_2 = H	Conductivity change: $m = 6 > 2 > 3 > 9 > 12$ R_1 = E > M > B > EM	12
16	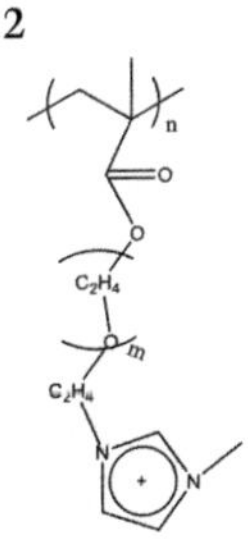	NTf_2^- $B(CN)_4^-$	10^{-5} 3×10^{-6}	—	100

(*continued*)

Table 3.1 (*continued*)

Entry	PIL cation	PIL anion	Highest reported δ, S cm^{-1}	Comments	Ref.
17	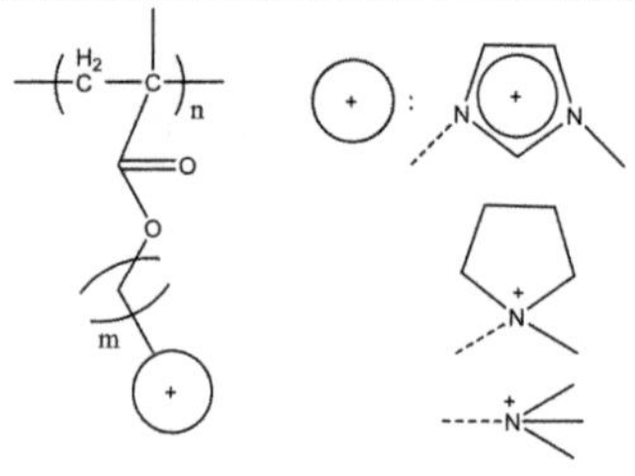 $m = 2, 3,$ or 11	NTf_2^-	10^{-10}–10^{-6}	PILs with imidazolium cations were elastic, the others were brittle	18,24
18	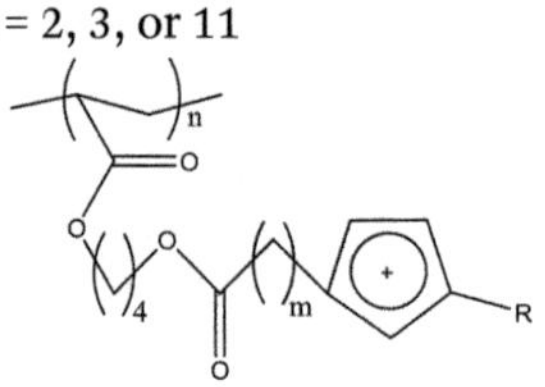 $m = 5,$ or 10 $R = C_4H_9$, or $(CH_2CH_2O)_2CH_3$	NTf_2^- PF_6^-	10^{-5} 10^{-7}–10^{-6}	$m = 10$ $m = 5$	101
19	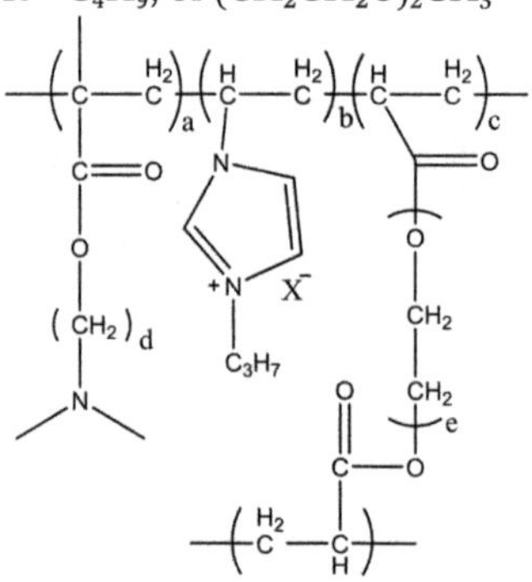	BF_4^- PF_6^- NTf_2^-	6.52 ± 0.04 5.22 ± 0.05 3.12 ± 0.03 $\times 10^{-4}$ S cm^{-1} at 30 °C	PIL-NTf_2 PIL-BF_4 PIL-PF_6 Conductivity increases after CO_2 bubbling	102
20	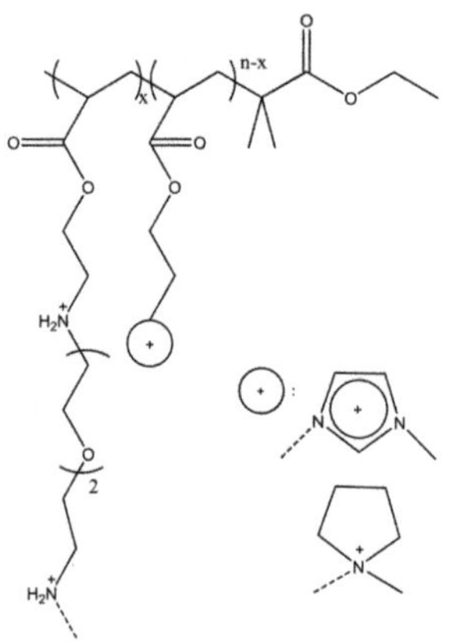	$3 \times NTf_2^-$	up to 10^{-3}	Depending on the amount of cross-linker or IL	16
21	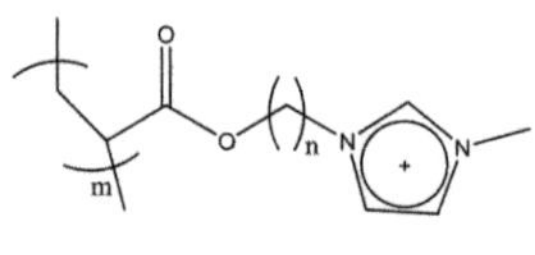	$H_3PO_4^-$	1.95×10^{-13} 4.43×10^{-11} 10^{-6}	$n = 2, x = 0$ $n = 6, x = 0$ $n = 6, x = 2$	103

22	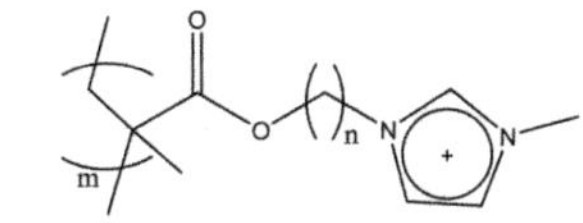 ($n = 2, 6, 7$, or 10)	Br^-	(3.32–3.74) $\times 10^{-5}$	Conductivity improved by 10^3 times upon the addition of $[C_2C_1im][BF_4]$	104
	$n = 3$	NTf_2^-	3.1×10^{-8}		105
23	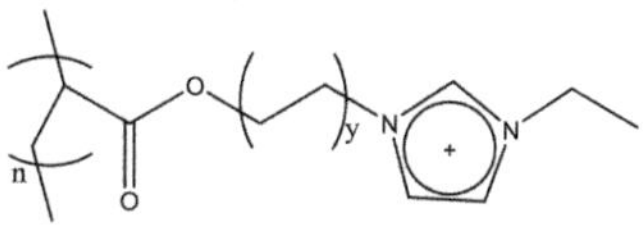 $y = 1, 1.5, 3$, or 5	NTf_2^-	10^{-4}	Conductivity increases with the elongation of the alkyl spacer	6
24	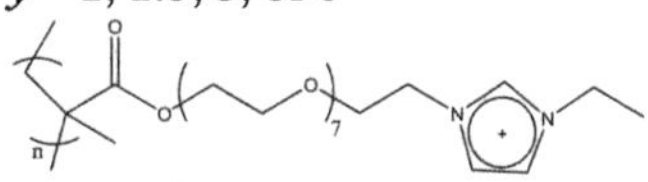	NTf_2^- Cl^-	1.2×10^{-4} 10^{-6}	at 30 °C	106
25	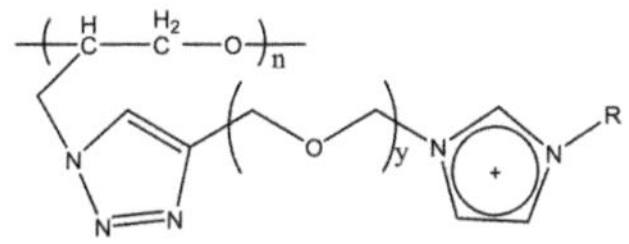 $y = 2$, or 3 R = CH_3, C_2H_5, C_3H_7, or C_4H_9.	NTf_2^-	1.5×10^{-5}	at 30 °C	107
26	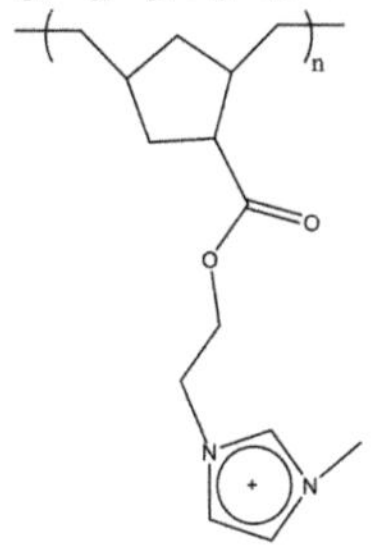	NTf_2^- PF_6^- BF_4^-	2.91×10^{-7} 1.09×10^{-8} 1.21×10^{-9} at 10^3 Hz	Conductivity of the polymer coating	108
27	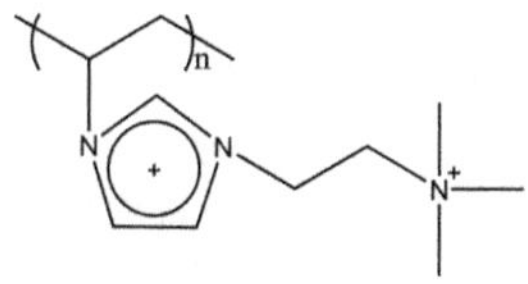	$2 \times NTf_2^-$	10^{-4}	With 1,2-dimethyl-3-ethoxyethyl imidazolium bis(trifluoromethane-sulfonyl)imide (IM(2o2)-11NTf$_2$) ionic liquid and $LiNTf_2$ salt in different weight ratios	109
28	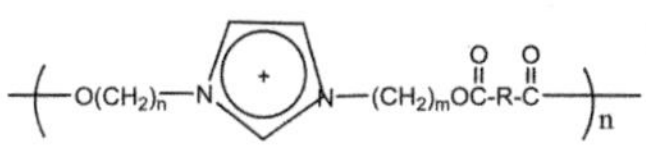 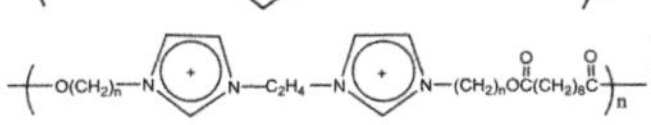 $n, m = 6, 11$. R = $(CH_2)_2$, $(CH_2)_4$, $(CH_2)_8$, or n-Ph	NTf_2^- PF_6^-	Up to 3.2×10^{-5}	The highest conductivities were obtained for monocationic PILs with the NTf_2^- anion and R = $(CH_2)_8$	14
29	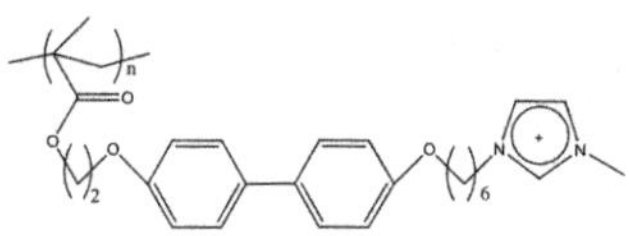	BF_4^-	10^{-6}	PILs of layered conductive structure	110

(*continued*)

Table 3.1 *(continued)*

Entry	PIL cation	PIL anion	Highest reported δ, S cm^{-1}	Comments	Ref.
30	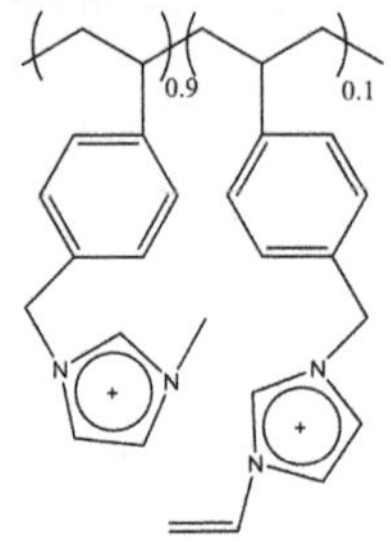	NTf_2^- OTf^- NO_3^-	≈10^{-2}	As a mixture with ILs based on $[C_2C_1im]^+$, $[C_4C_1im]^+$ and $[NH_3]^+$	15
31	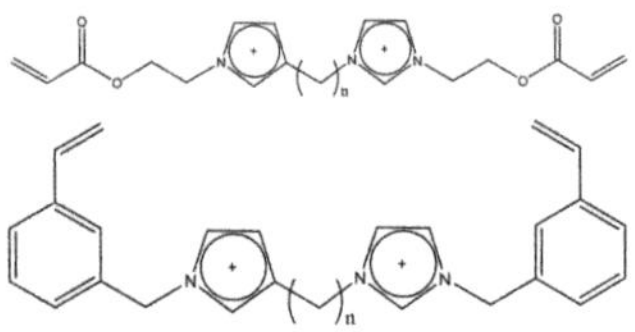 $n = 2$ or 6	2 × NTf_2^-	1.36×10^{-4} (at 50 °C)	With cross-linkers	111
32	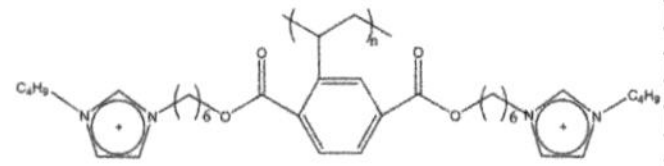	Br^-, BF_4^-, PF_6^-, NTf_2^-	8.5×10^{-4} S cm^{-1} for BF_4	LC phase was formed	13
33	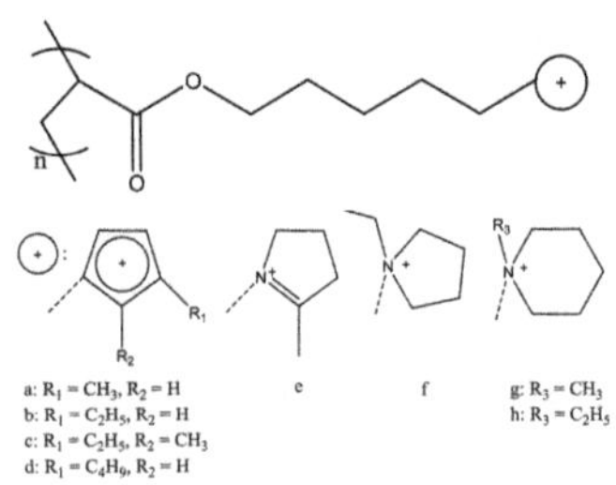	NTf_2^-	a: 4.4×10^{-5} b: 1.4×10^{-4} c: 8.5×10^{-6} d: 4.1×10^{-5} e: 2.4×10^{-5} f: 2.1×10^{-5} g: 6.2×10^{-6} h: 2.4×10^{-5} at 30 °C	F S F F S S F F F-fluid, S-solid	19
34	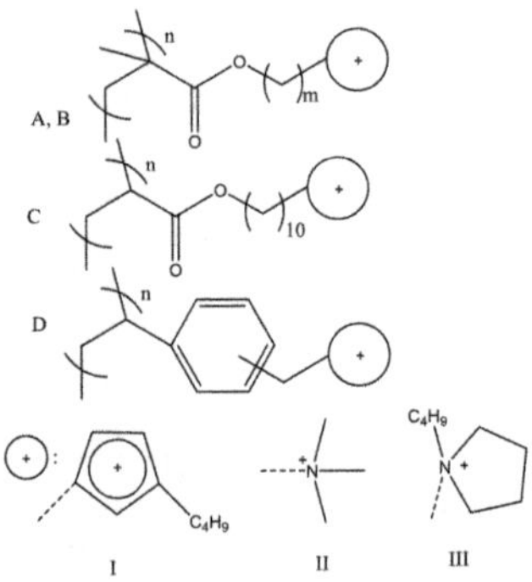 $m = 2$ (A), or 10 (B)	Br^-	A-I: 1.0×10^{-2} A-II: 2.1×10^{-2} A-III: 1.0×10^{-2} B-I: 6.9×10^{-3} B-II: 1.3×10^{-2} B-III: 3.0×10^{-2} C-I: 6.3×10^{-3} C-II: 7.3×10^{-3} C-III: 7.9×10^{-3} D-I: 1.0×10^{-2} D-II: 2.7×10^{-3} D-III: 1.5×10^{-2}	At 60 °C and 90% RH Conductivity	20

35	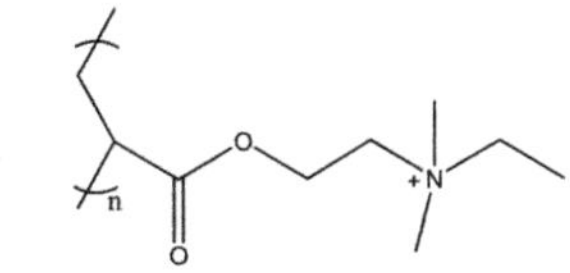	NTf_2^-	4.9×10^{-4} at 60 °C	As electrolyte PIL–IL–$LiNTf_2$–SiO containing 25 wt% of IL	112
36	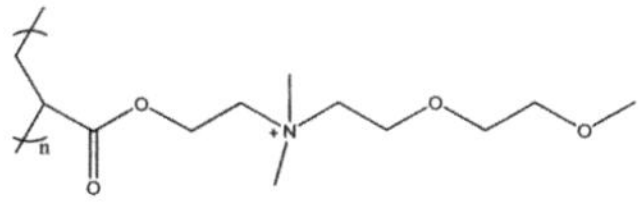	NTf_2^- FSI$^-$	4×10^{-6} 2×10^{-6} at 30 °C	Conductivity varied with Li^+ addition	113
37	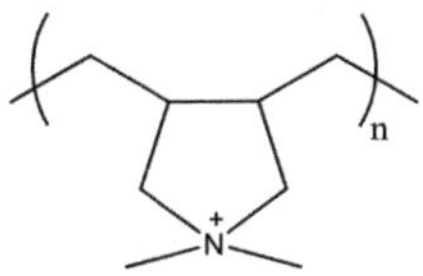	NTf_2^-	10^{-5} at 25 °C	Conductivity was improved upon addition of IL	21,22
			10^{-6}	As a composite with 2 wt% Li^+ salt. Conductivity improved upon introduction of succinonitrile	114
38	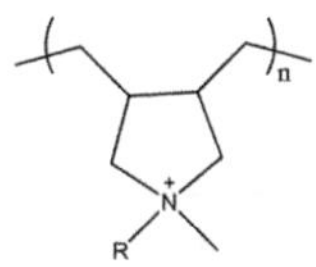 R: 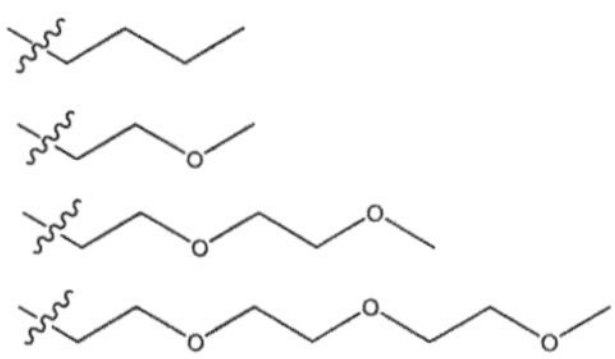	NTf_2^-	10^{-3}–10^{-2} at 25 °C	As a blend with IL. Conductivity increases with addition of long chain PEG	23
39	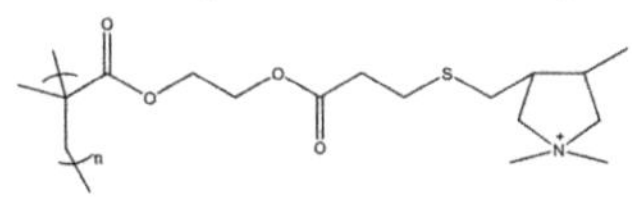	NTf_2^-	8.2×10^{-9}	Conductivity increased by copolymerisation of ILMs with PEGDM and PEGM	115
	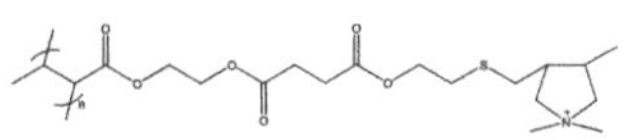		6.7×10^{-8}	—	105
	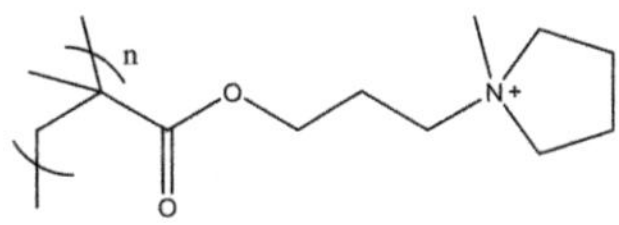		7.8×10^{-9} (3.3×10^{-10})		
40	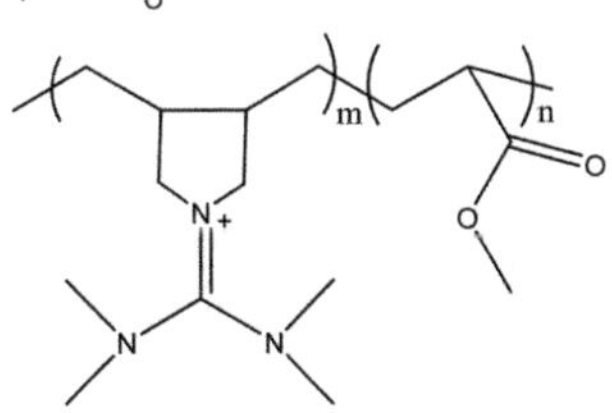	NTf_2^-	4×10^{-6}	As a polymer membrane containing Li^+ salt, IL and nano-size SiO_2	116

(*continued*)

Table 3.1 (*continued*)

Entry	PIL cation	PIL anion	Highest reported δ, S cm^{-1}	Comments	Ref.
41	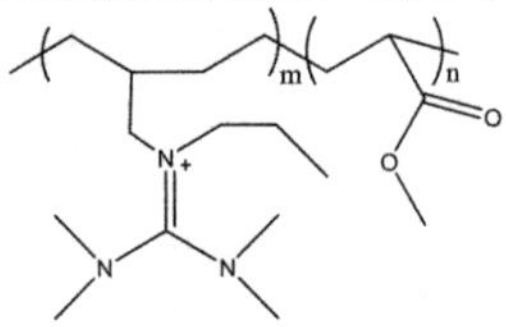	NTf_2^-	10^{-7}–10^{-6} depending on the IL content	As a polymer membrane containing Li^+ salt, IL and nano-size SiO_2	117
42	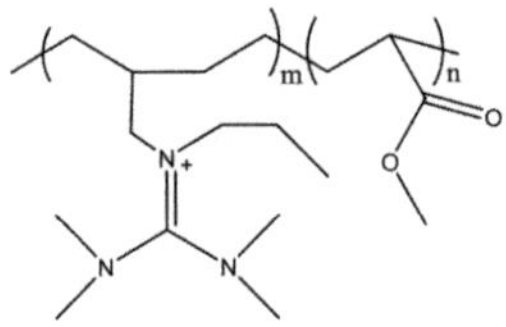	ClO_4^- PF_6^- NTf_2^- BF_4^-	10^{-7}–10^{-3} depending on the Li^+ salt content	Contains the corresponding Li^+ salt. On average, the conductivity increases in the order: $ClO_4^- < PF_6^- < NTf_2^- < BF_4^-$	118

Table 3.2 Structural dependence of ion conductivity for anionic PILs.

Entry	PIL anion	PIL cation	Highest reported δ, S cm^{-1}	Ref.
1	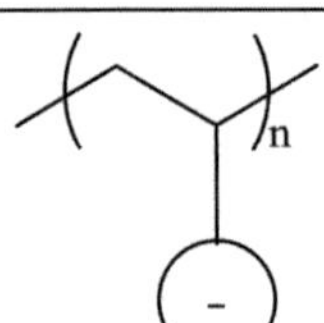 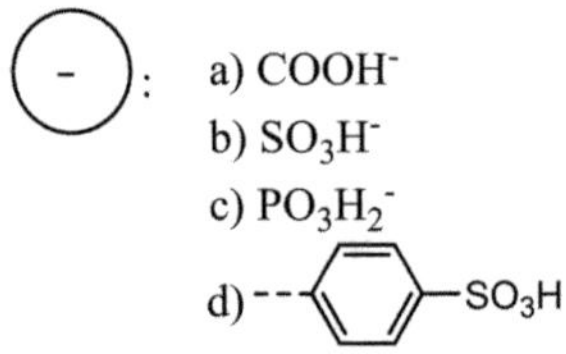	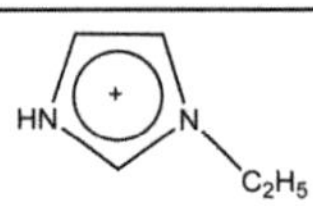	(a) 1.2×10^{-6} (b) 1.1×10^{-4} (c) 2.9×10^{-5} (d) 1.1×10^{-8}	6
2	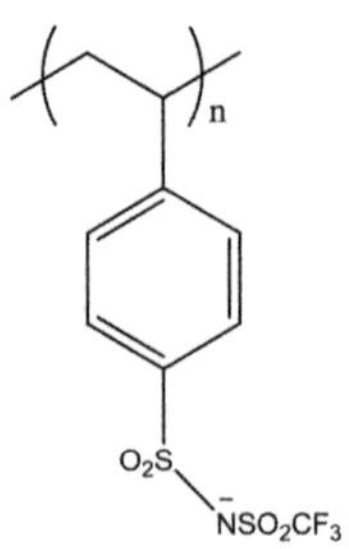	(a) 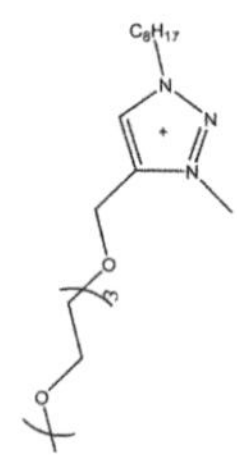	(a) 8.4×10^{-7}	119
		(b) 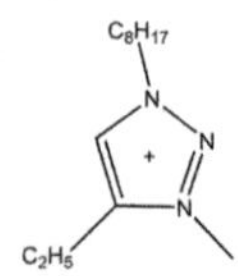	(b) 7.8×10^{-8}	

3	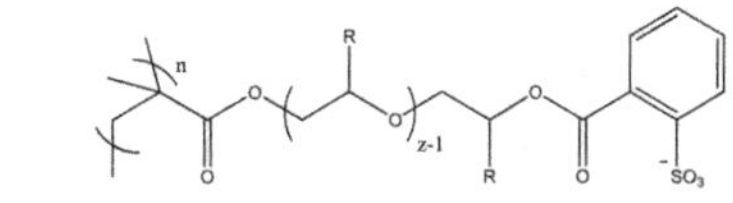 R = H, or CH_3; z = 0, 2, 8, or 9	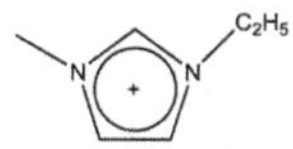	up to 4×10^{-6}	6
4	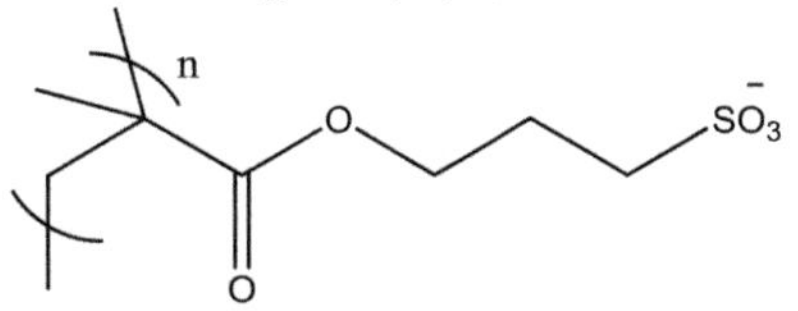	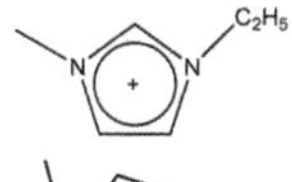	3.2×10^{-8} (6.3×10^{-5})	105 6
		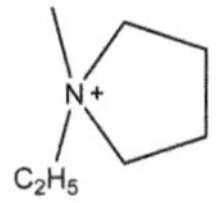	3.3×10^{-10}	105
5	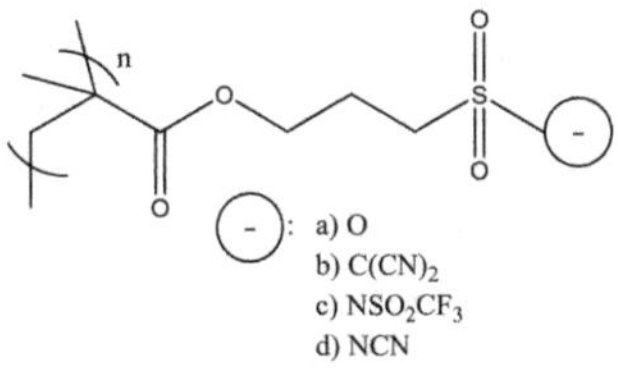	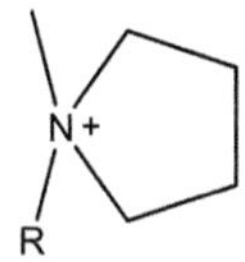 $R_1 = C_2H_5$, $R_2 = C_4H_9$	a-R_1: 3.3×10^{-10} a-R_2: 4.9×10^{-10} b-R_2: 3.1×10^{-8} c-R_2: 2.0×10^{-8} d-R_2: 1.7×10^{-7}	24,120
		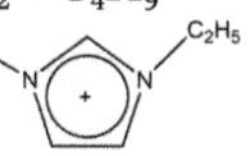	3.2×10^{-8}	
6	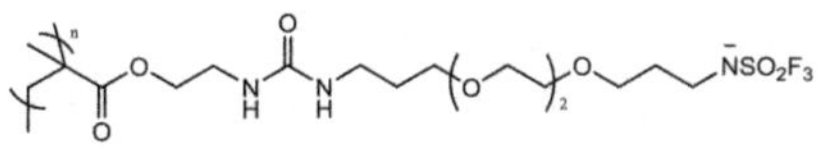	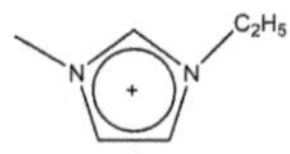	6.5×10^{-4} at 20 °C	121
7	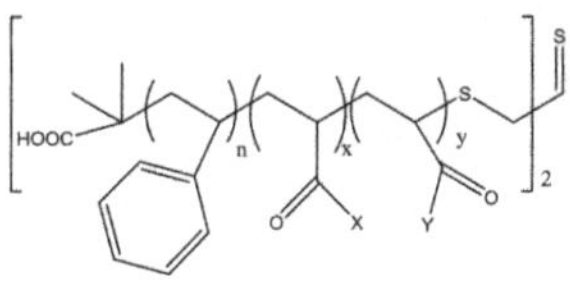 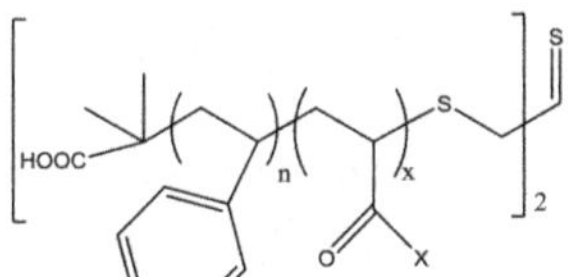 X, Y = O, OC_4H_9	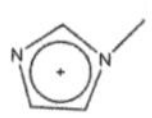	10^{-5}–10^{-4}	25

however, imidazolium PILs containing PF_6^- and BF_4^- anions showed very close values of conductivity, outperforming NTf_2^- (Table 3.1, entry 19). It was also found that cationic imidazolium PILs with Br^- and $N(CN)_2^-$ anions may have higher conductivity values as compared to those with NTf_2^- (Table 3.1, entries 3 and 22). Separately, alkyl chains on imidazolium cations have a great effect on the PIL conductivity. When the imidazolium cation was directly attached to the polymeric chain, the highest conductivity was found for PILs containing the ethyl side chain and the conductivity was improved by the introduction of a hydroxyl group at the end of this side chain.[10] Another study showed that the conductivity of the polymerised 1-butyl-3-vinylimidazolium cation was higher than that of the 1-ethyl-substituted one, agreeing, however, that the elongation of the side chain to 7 carbon atoms led to a decrease in conductivity.[11] For poly(methacryloxyimidazolium) PILs, the conductivity decreased in the order $C_2H_5 > CH_3 > C_4H_7$, with the lowest conductivity exhibited by PILs with bi-substituted 1-ethyl-2-methylimidazolium cations.[12,13] The length of the side chain affects the aggregate state of the obtained monomer and polymer, which also affects the mobility of the ionic species and charge transfer.[13]

The suggestion that the mobility of the cation influences the conductivity of the PILs was raised a long time ago.[6] Since then, it has been shown that the conductivity of PILs can be improved to a certain degree by providing some flexibility to the polymerised cation. For poly(methacryloxyimidazolium) PILs, the conductivity decreased in the following order of the length of hydrocarbon spacer between the polyacrylic chain and the imidazolium cation: 6 > 2 > 3 > 9 > 12.[12] Similar results were obtained in other studies, proving that the conductivity can be increased by elongating the hydrocarbon spacer from 1 to 5 carbons.[6] This confirms that a moderate length of the spacer between the polymer chain and the charged species improves the conductive properties of PILs; however, when the spacer is too large the molecule's steric configuration is changed, which leads to dramatically decreased conductivity. The inclusion of bulky side polymeric units between vinylimidazolium cations can potentially decrease the packing density of the PIL, resulting in a lower T_g (Table 3.1, entries 5, 6, 19, 15 and 26). This approach, however, did not result in higher conductivities of the synthesised PILs. The same unsatisfying results were obtained for dicationic imidazolium-based PILs.[14] For example, the conductivity of polymerised mono-imidazolium-C_6-sebacate-C_6 bis(trifluoromethanesulfonyl) imide was found to be 3.1×10^{-5}, while for its polymerised dicationic analogue the conductivity was only 7.8×10^{-6} S cm^{-1}.[14]

It is still common to add non-polymerised ILs, cross-linkers and corresponding Li^+ salts to PILs to improve their conductivity. In this case, the conductivity of such a mixture can reach 10^{-3}–10^{-2} S cm^{-1}.[15–17]

Pyrrolidinium-based PILs seem to be promising conductive materials with conductivity values comparable to imidazolium- and *N*,*N*,*N*-trimethylammonium-based PILs.[18–20] PILs containing pyrrolidinium pendant units with bis(trifluoromethanesulfonyl) imide can show conductivity values as high as 10^{-5} S cm^{-1} at room temperature.[21,22] Their blends with non-polymerisable ILs can reach conductivities of 10^{-2} S cm^{-1} depending on the length of the side chain of the pyrrolidinium cation.[23]

Recent work provided a broad comparison of synthesised cationic and anionic PILs across multiple properties and compared the influence of polymer molecular weight on conductivity.[24] For poly(1-[3-(methacryloyloxy)propyl]-3-methylimidazolium bis(trifluoromethanesulfonyl)imide), the conductivity decreased with increased molecular mass of the PIL and varied around 3.1×10^{-8} S cm^{-1} at molecular weights larger than 4×10^5.

Other cationic PILs reported to date do not show improved conductivity and are reported at the end of Table 3.1.

As shown in Table 3.2, there is much less information available on the conductivity of anionic PILs. A study by Ohno *et al.* from more than a decade ago showed the potential for anionic PILs to act as highly conductive materials.[6] In this study, different polymerised organic acids were combined with 1-ethylimidazolium cations. The highest conductivity of 1.1×10^{-4} was achieved for polysulfonic PIL. However, the conductivity of this anionic PIL was not improved by the introduction of an ethylenoxide spacer between the cation and polymeric chain, although this previously worked for cationic PILs (Table 3.2, entries 1 and 4). However, for bulky PILs based on styrenesulfonic acid and 1-ethylimidazolium cations, the conductivity increased with elongation of the polyether spacer reaching 10^{-6}.[6] Bulky anionic PILs with 1-ethylimidazolium cations were also studied in combination with 30% non-polymerised $[C_2C_1m][OTf]$ showing conductivities in a moderately good range of 10^{-5}–10^{-4} S cm^{-1} depending on the side chains on the cation.[25]

An interesting comparison was done by Shaplov *et al.*[24] When anionic and cationic PILs with the correspondingly matching species of anion and cation are compared, it appears that polyanions show higher conductivities than the corresponding polycations. For example, poly(1-[3-(methacryloyloxy)propyl]-1-methylpyrrolidinium bis[(trifluoromethanesulfonyl)imide]) has a conductivity value of 7.8×10^{-9} S cm^{-1}, while the corresponding polyanion poly(1-methyl-1-butylpyrrolidinium 1-[3-(methacryloyloxy)propyl]-sulfonylimide) has a conductivity of 2.0×10^{-8} S cm^{-1}.[24]

Bearing in mind the available information on the ion conductivity of cationic and anionic PILs, the following conclusions can be drawn:

1. PILs bearing the 1-ethylimidazolium species in either the cation or anion show higher conductivities among the studied PILs.
2. Bis(trifluoromethanesulfonyl) imide anion-containing PILs overperform most of the other studied anionic PILs under similar conditions.
3. To a certain degree, increasing the distance between the polymer backbone and the ionic species, *via* introduction of a polyethylene spacer, increases the ionic conductivities of both polycations and polyanions.
4. Polyanions overperform polycations with equal length of the bridging group and similar structures of ions.

3.2.2 CO_2 Sorption

The change in electroconductivity of PILs upon absorption of certain gases makes them promising materials for CO_2 detection, sorption and separation. Similar to the research on electroconductivity, much more attention has been paid to the study of the CO_2 absorption properties of cationic PILs than anionic PILs. Moreover, there is no consistency in the manner in which the information is reported: many studies were performed without precise quantification of the absorbed values, preventing wider comparison of the performance of the polyanions and polycations. For example, the sensitivity of a series of polymerisable tetraalkylammonium ionic liquids based on the [2-(methacryloyloxy)ethyl]dimethylheptylammonium cation and bis(trifluoromethylsulfonyl)imide, nonafluoro-1-butanesulfonate, dodecylbenzenesulfonate, heptadecafluorooctanesulfonate, and heptadecafluoroundecanoate anions, as well as their corresponding homo- and copolymers was studied. Rapid frequency responses with no memory-effect were observed, but the sorption values were not reported.[26,27] Furthermore, the efficient fixation of CO_2 into cyclic carbonates catalysed by hydroxyl-functionalised poly(ionic liquids)[28–30] and the effect of the structure of polybenzene imdazoles[31–34] with rigid polyaromatic groups[35] on permeation and selectivity was reported but not quantified, and this therefore complicates the comparison.

In 2009, Radosz and Shen first patented PILs as new materials for CO_2 separation and other applications, claiming that imidazolium-, phosphonium-, ammonium- and pyridinium-based polycations are efficient selective CO_2 sorbents.[36] Table 3.3 summarises the data available on the sorption capacity of some cationic and anionic PILs. To date, only two reports have been found on the application of polyanionic PILs in this field (Table 3.3, entries 19 and 20). When these anionic PILs are compared to bulky cationic PILs with similar structures (Table 3.3, entries 9 and 11), the similarity in their performance is obvious. Data from Table 3.3 also suggest that ammonium- and phosphonium-based anionic PILs outperform imidazolium-based ones. Among the phosphonium-based PILs more permeable to light gases and generally less selective were those with large and bulky cations like poly([P_{888}VB][NTf_2]).[37] Sorption of CO_2 by imidazolium anionic PILs was reduced with the introduction of a longer side chain on the imidazolium cation, but improved for the PILs with higher molecular weight.[38] It was confirmed that the sorption ability was increased in the following order: poly([2-(methacryloyloxy)ethyl]trimethylammonium nitrate) (0.066 mol fraction) < poly(vinylbenzyltriethylammonium nitrate) (0.133 mol fraction) < poly(α-(vinylbenzyl)trimethylammonium nitrate) (0.177 mol fraction).[39] It was suggested that longer alkyl side chains on the polycations may act as a steric obstruction that hinders CO_2 sorption. Therefore, long alkyl substituents on the cation (as well as cross-linking) would decrease the CO_2 sorption capacity of PILs.

The influence of other different polycations with the bis(trifluoromethylsulfonyl)imide anion on the solubility and selectivity of CO_2 was also studied and recently reported.[40] It was found that the solubility of CO_2 in these PILs strongly depends on the sorption properties of the polycation and decreases in the order: pyridinium > cholinium > pyrrolidinium > imidazolium > ammonium, with imidazolium-based PILs outperforming other cations in terms of the solubility of methane and nitrogen.[40] However, it can be seen from entry 3, Table 3.3, that ammonium-based polycation absorbs four times more CO_2 than the corresponding imidazolium PIL.[41] The same conclusion can be drawn from the comparison of entries 5–8, 12 and 13 of imidazolium- and ammonium-based PILs with a selection of anions.

The anions of cationic PILs also strongly affect their CO_2 sorption capacity. For example, for poly(α-(vinylbenzyl)trimethylammonium)-based PILs,

Table 3.3 Sorption capacity of some cationic and anionic PILs.

Entry	PIL cation	PIL anion	Highest reported CO_2 sorption, mol%	Comments	Ref.
1	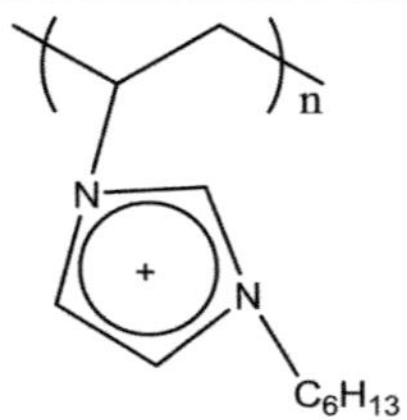	NTf_2^- Taurate	6.7–41.7	PIL-coated solid-phase micro-extraction fibres	122
2	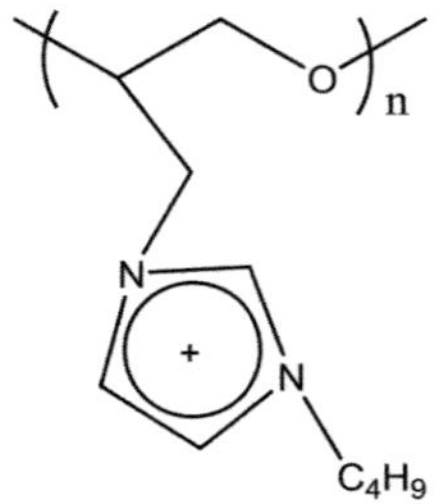	BF_4^-	1.06	—	123
3	(a) (b) 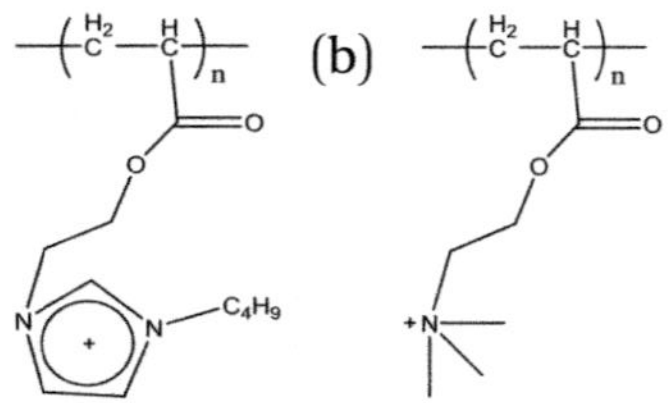	BF_4^-	(a) 1.9 (1.77) (b) 8.0	—	41 44
4	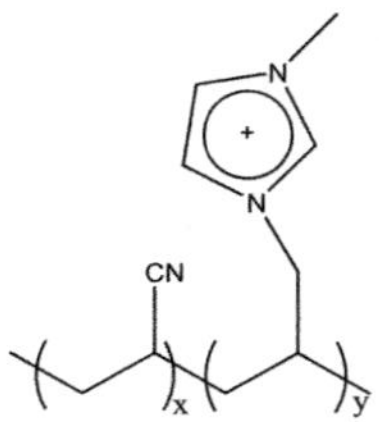	BF_4^- PF_6^-	49.18 53.40	PILs with high absorption were synthesised *via* the seed-swell method	46

(continued)

Table 3.3 (*continued*)

Entry	PIL cation	PIL anion	Highest reported CO_2 sorption, mol%	Comments	Ref.
5	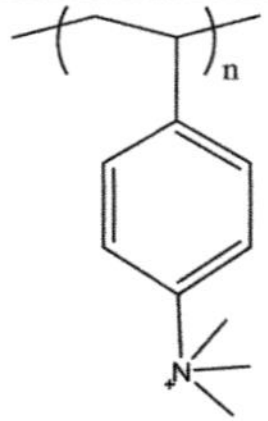	BF_4^-	47.94	PILs synthesised *via* free-radical polymerisation	46
6	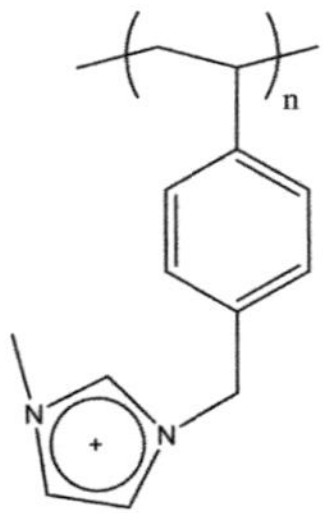	BF_4^-	3.05 (3.26)	(at 0.79 bar)	123,124
7	(a) 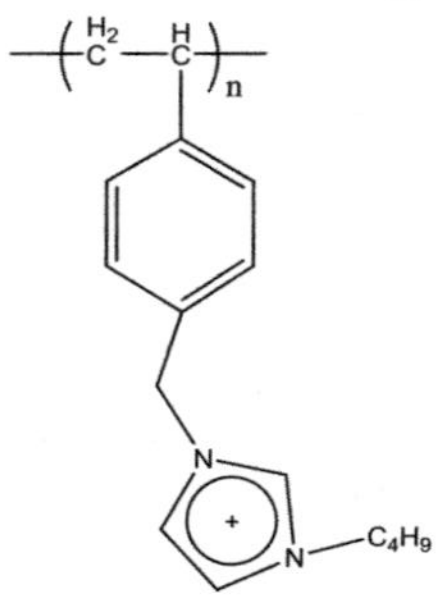 (b) 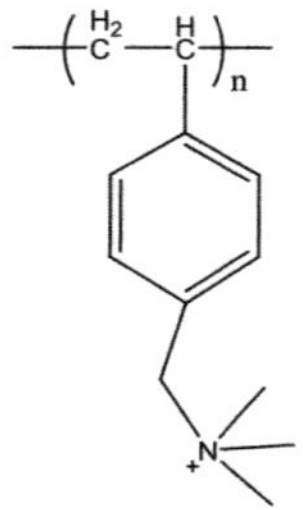	(1) BF_4^- (2) NTf_2^-	a-1: 2.2 a-2: 2.2 b-1: 10.1	—	41

8	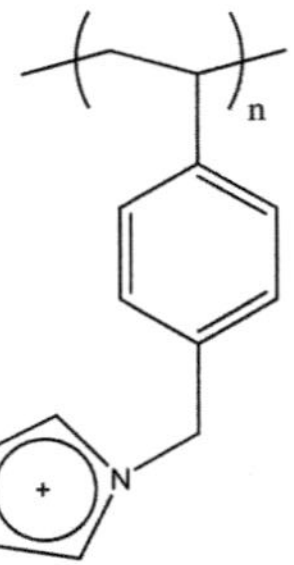	BF_4^- PF_6^-	2.22 2.75	—	44,45
9	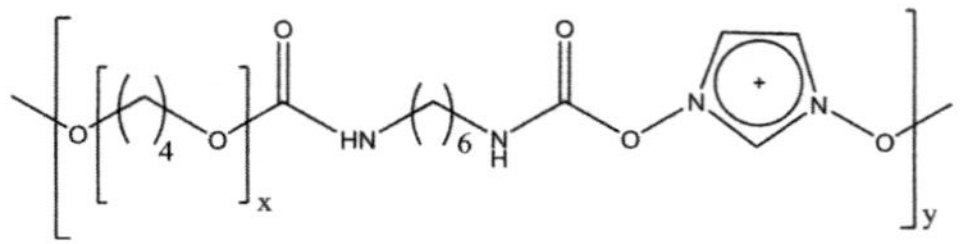 $x = 2000$	NTf_2^-	30	at 1 bar	38
10	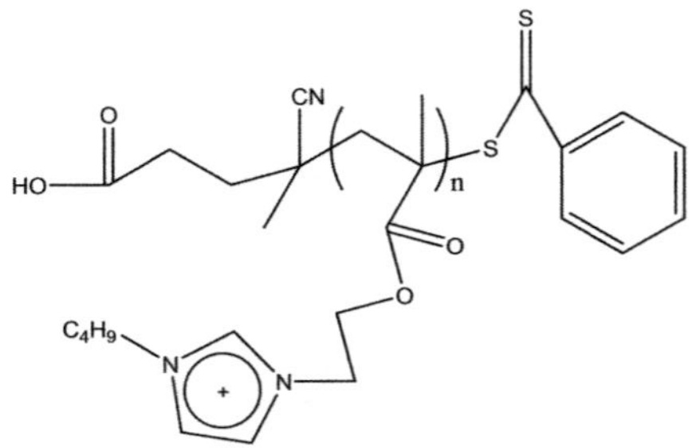	BF_4^- PF_6^- NTf_2^- Br^- OTf^- OAc^-	2.13 2.90 1.82 2.13 1.83 7.59	Manuscript analyses the influence of the PIL's Mr on CO_2 sorption	42
11	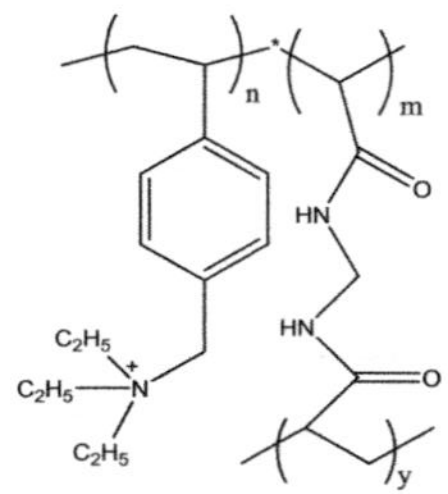	PF_6^-	11.72	—	125

(*continued*)

Table 3.3 (*continued*)

Entry	PIL cation	PIL anion	Highest reported CO_2 sorption, mol%	Comments	Ref.
12	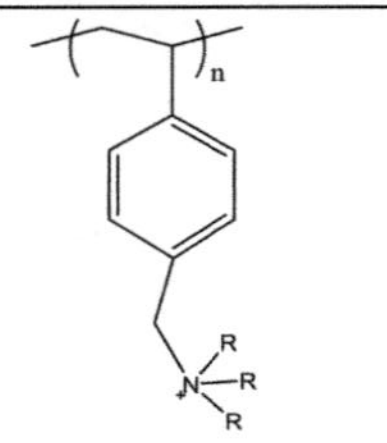 R = a) CH_3, b) C_2H_5, c) C_4H_9	BF_4^- PF_6^- NTf_2^- 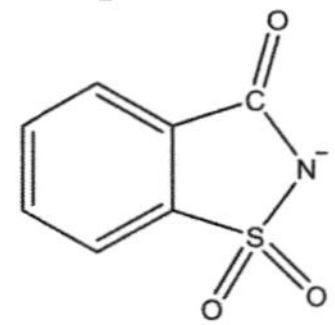	(a) 10.22 10.66 2.85 2.67	Sorption decreased with an increase of the R length	43
13	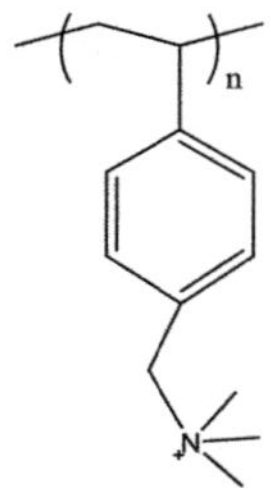	PF_6^-	85.51	As an ionogel	126
14	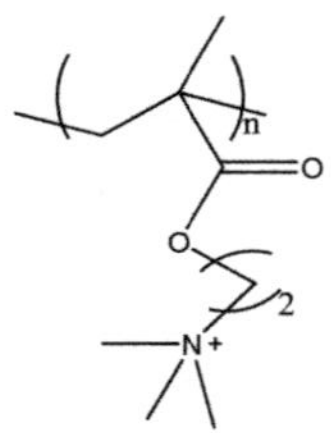	BF_4^-	7.99	—	43

15	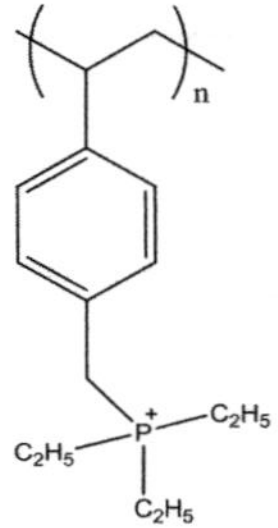	BF_4^-	4.62	at 0.79 bar	124
16	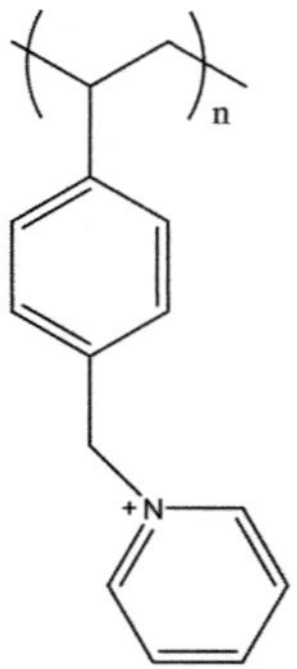	BF_4^-	4.35	at 0.8 bar	124
17	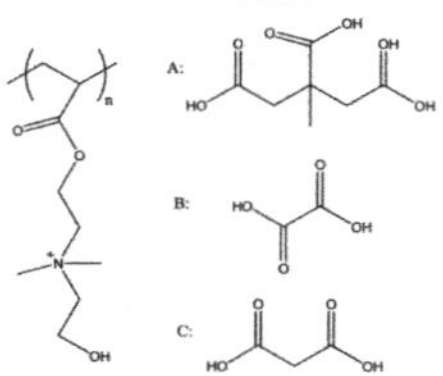	Br^-	A: ~4.2 B: ~5.0 C: ~0.9	Deep eutectic mixtures	127
18	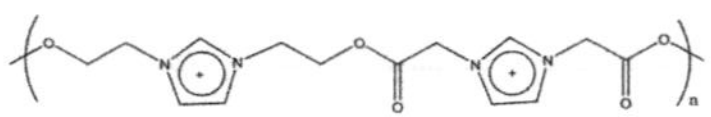	PF_6^- BF_4^-	4.2 4.5	—	47

(*continued*)

Table 3.3 (*continued*)

Entry	PIL cation	PIL anion	Highest reported CO_2 sorption, mol%	Comments	Ref.
19	(I) 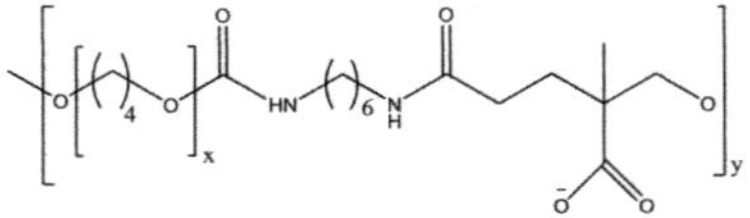 (II) 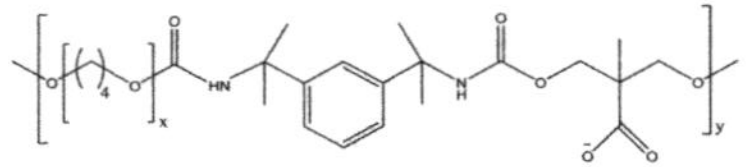 (a) $x = 1000$, (b) $x = 2000$	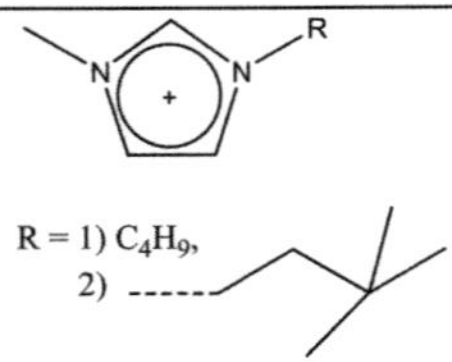	Ia-1: 3 Ib-1: 26 Ib-2: 16 IIa-1: 3 IIb-1: 28	at 1 bar	38
20	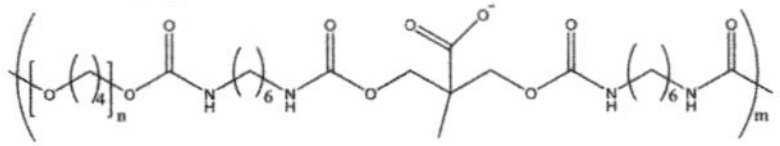	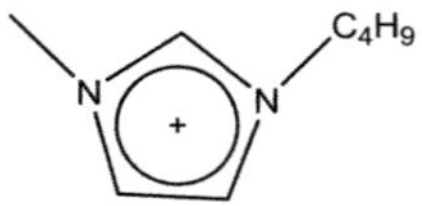	11.04	—	128
		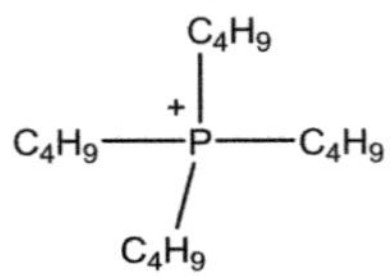	19.21		
		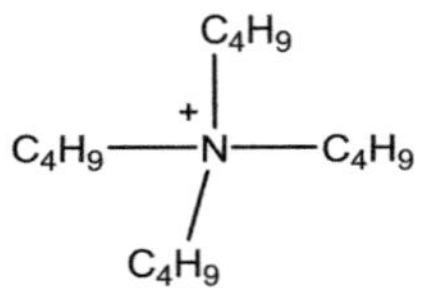	19.21		

the sorption of CO_2 decreases in the order of: BF_4^- (0.209 mol fraction) > Cl^- (0.188 mol fraction) > PF_6^- (0.180 mol fraction) > NO_3^- (0.177 mol fraction) > OTf^- (0.164 mol fraction), which is explained by the steric effect of the anioins.[39]

Privalova *et al.* (Table 3.3, entry 10) found the following decreasing trend of CO_2 sorption by PILs depending on the anion: OAc^- > PF_6^- > Br^- > BF_4^- > OTf^- > NTf_2^-.[42] Other reported studies have confirmed this general trend.[43–46] However, for some imidazolium-based PILs, the sorption capacity did not change when BF_4^- was replaced with NTf_2^-.[41] Another exception from Privalova's trend was found by Xiong *et al.* for dicationic imidazolium PILs, with the BF_4^- anion showing better sorption than with the NTf_2^- (Table 3.3, entry 18).[47]

As indicated by the comparison of PILs with different anions, unlike for ILs, a fluorine atom is not a decisive factor for PIL sorption of CO_2. However, amino acid PILs (AAPILs) with tethered amines in the anions were synthesised to allow chemisorption of CO_2. AAPILs with polymerised amino acid-anions showed much higher CO_2 sorption compared to other PILs, reaching 0.530 mol fraction at room temperature and 0.7 bar, which is more than twice as high as the value for non-AA PILs.[39]

3.2.3 Sensors

Electrochemical sensors and biosensors are another two broad areas of application of cationic PILs. To the author's knowledge, there is just one report on the application of an anionic PIL, *N*,*N*-didecyl-*N*-methyl-*d*-glucaminium poly(2-methyl-acrylic acid 2-[1-(3-{2-[2-(3-trifluoromethanesulfonylamino-propoxy)-ethoxy]-ethoxy}-propylamino)-vinylamino]-ethyl ester), (poly([DDMGlu][MNTf$_2$])), for the detection of furfural benzofuran, styrene oxide, *m*-toluidine, 2,4-dimethylaniline, 2-methoxy-5-methylaniline and furfural (Figure 3.1).[48] When compared to the other cationic PILs used in this study, poly([DDMGlu][MNTf$_2$]) fibers generally exhibited lower sensitivities, except for in the detection of benzofuran and furfural. The precision of the poly([DDMGlu][MNTf$_2$]) PIL ranged from 1.9 to 9.8%, while its limit of detection (LOD) ranged from 0.9 to 12 μg L^{-1}.[48]

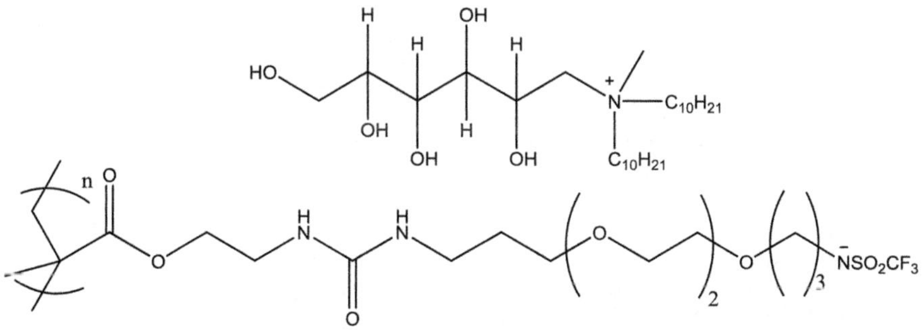

Figure 3.1 Anionic PIL used for the sensing of alkyl halides and aromatics.

The majority of the studied cationic PILs for sensing applications bear NTf_2^-, BF_4^-, PF_6^- and Cl^- anions. It is interesting to see the application of PILs for detecting food aromas, such as beer volatiles (Table 3.4, entry 6)[49] and coffee aroma.[50] Cationic PILs have also been studied for the detection of nitroaromatic explosives (fluorescent PILs),[51] as highly chemoresistive humidity sensors[52] and in other sensor applications described in the recent review by Guterman *et al.*[53] Most of them are applied as coatings or fiber sorbents in solid-phase microextraction (SPME), often coupled with gas chromatography (SPME-GC), and include various additives such as PPy, GO, GCE,[54] ZnCdHgSe with *N*-acetyl-L-cysteine,[55] poly(acrylic acid),[56] cross-linkers,[57] magnetic nanoparticles,[58] multiwalled carbon nanotubes,[59,60] polypyrrole nanotubes[61] and titanium.[62] Table 3.4 summarises the analytes and reported detection limits or uptake of the PILs used.

Table 3.4 Reported detection limits or uptake values (in brackets) of PILs for some analytes.

Entry	Analyte	Detection limit/(Uptake)	Ref.
1	Acrylamide	0.5 μg L^{-1} (10 μg L^{-1})	129,130
2	Alcohols, *n*-alkanes, phthalate esters and halogenated aromatic hydrocarbons	from 1 μg L^{-1}	131
3	Antihuman neuron specific enolase (anti-NSE)	0.20 ng L^{-1}	55
4	Ascorbic acid	5.0 μM	61
5	Aqueous solutions of estrogens	0.25 μg L^{-1}	57
	Bisphenol A	0.25 μg L^{-1}	
	Phthalate esters	0.2 μg L^{-1}	
6	Aqueous solutions of polycyclic aromatic hydrocarbons (PAHs), alkylphenols and parabens	up to 7 μg L^{-1}	63
7	1-Butanol, 2-butanol, isobutanol	values are not reported; the detection limit decreased in the order 1-butanol > 2-butanol > isobutanol	56
8	Beer volatiles	from 0.4 μg L^{-1}	49
9	Diethylstilbestrol and 17α-ethynylestradiol	1 μg L^{-1}	132
	Estrone and hexestrol	2 μg L^{-1}	
10	Dopamine	73.3 nM (180 nM)	54,61
11	Esters	from 2.5 μg L^{-1}	133
12	Furfural benzofuran, styrene oxide, *m*-toluidine, 2,4-dimethylaniline, 2-methoxy-5-methylaniline, furfural	0.3–3.3 μg L^{-1}	48
13	Glucose	10^{-6} M	134
14	Hexestrol	2.1×10^{-9} mol L^{-1}	135
15	Hydrogen bonding compounds	from 0.35 ng L^{-1}	62
16	*p*-Hydroxybenzoic acid, sorbic acid and cinnamic acid	from 12 ng L^{-1}	136
17	Ketamine	0.05 mg L^{-1}	137

(*continued*)

18	Organic solvents	from 0.3 μg L^{-1}	138
19	Pharmaceutical drugs, phenolics, and insecticides	from 0.2 to 2 μg L^{-1}	139
20	Phenolic compounds	from 2 μg L^{-1}	140
21	Phenols, phenol derivatives, and polycyclic aromatic hydrocarbons (PAHs)	from 135 μg L^{-1}	141
22	Polychlorinated biphenyls	low ng L^{-1} range	142
23	Polycyclic aromatic hydrocarbons (PAHs)	from 1 ng L^{-1}	60
24	PAHs and phthalate ester (PAEs)	from 5 ng L^{-1}	143
25	Simazine, atrazine, prometryn, metolachlor	0.88 μg L^{-1} 0.45 μg L^{-1} 0.35 μg L^{-1} 0.29 μg L^{-1}	58
26	Tryptophan	13–87 μmoles g^{-1}	144
27	Volatile compounds in cheeses	30 to 2000 μg L^{-1}	145
28	Volatile fatty acids (VFAs) and alcohols	PIL-Cl^-: from 0.02 μg L^{-1} PIL-NTf_2: from 0.2 μg L^{-1}	146

Unfortunately, the only provided comparison of the performance of some of the PIL sensors to the commercially available ones used to detect aqueous solutions of polycyclic aromatic hydrocarbons (PAHs), alkylphenols and parabens is still not satisfying.[63] On the basis of the published reports, it is too early to compare the selectivity, sensitivity and other sensing characteristics of cationic and anionic PILs. The author is looking forward to more research into the implementation of anionic PILs for sensing applications.

3.2.4 Thermoresponsive Materials

There are several thermoresponsive forms that are known for PILs, such as ion–anion responsive behaviour, inverse viscosity temperature dependence and the lower critical solution temperature (LCST) – all in solution or, for some examples, in thin films.[2]

Polystyrenesulfonate- and polysulfonate-based anionic PILs (including dianionic ones) with phosphonium cations were reported as ones showing LCST behaviour (Table 3.5).[64–68] As clearly shown for the example of poly[(1,8-octanediyl-bis(tri-*n*-butylphosphonium) 4-styrene) sulfonate], the temperature of the phase transition depends not only on the length of the side chain on the cation, but also on the PIL concentration.[65] PILs with phosphonium-based cations showed lower phase transition temperatures when compared to phosphonium-based anions. For example, for 10 wt% of poly[(1,8-octanediyl-bis(tri-*n*-butylphosphonium) 4-styrene) sulfonate], the temperature of the phase transition was reported to be 35 °C,[65] while for poly[(vinylbenzyltripentilphosphonium) chloride] it was recorded as 13 °C.[69] Okafuji *et al.* also showed that the temperature of the phase transition could be lowered by the addition of NaCl to the solution. They showed that

Table 3.5 Reported temperatures of the phase transitions (50% turbidity) for some solutions of anionic and cationic PILs.

PIL cation	PIL anion	Medium	T_c, °C (LCST/UCST)	Ref.
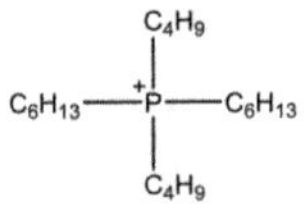	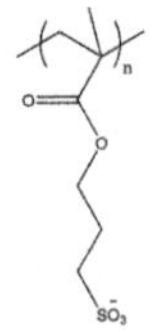	Water, blends with hydrophilic and hydrophobic anionic PILs	47.5 (~48)	66,67
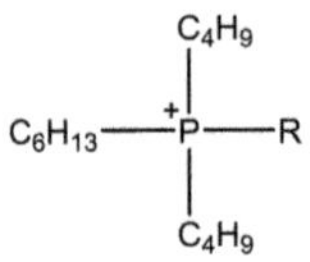 R=C_4H_9, C_6H_{13}	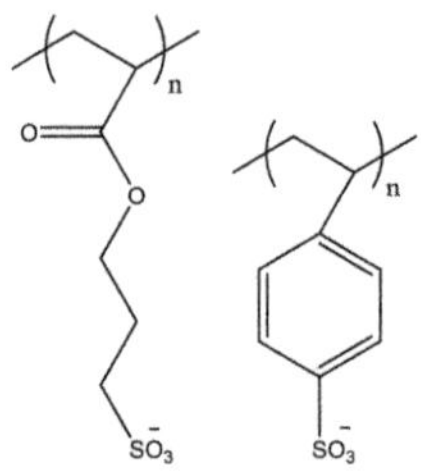	Water, PILs copolymerised with cross-linkers	40–53 (depending on the cross-linker concentration)	64
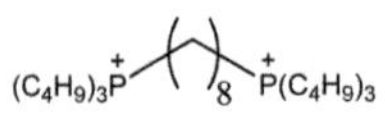	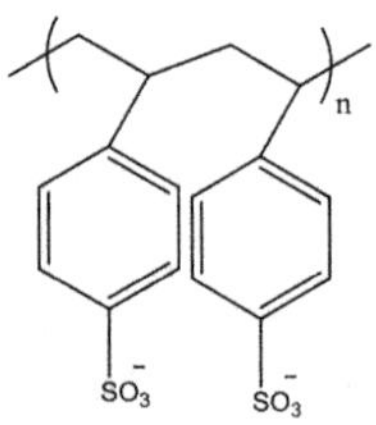	Water	35 (10 wt%) 38 (2 wt%) 51 (0.1 wt%)	65
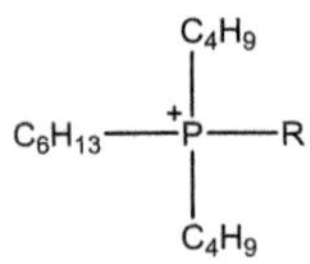 R=C_4H_9, C_6H_{13}	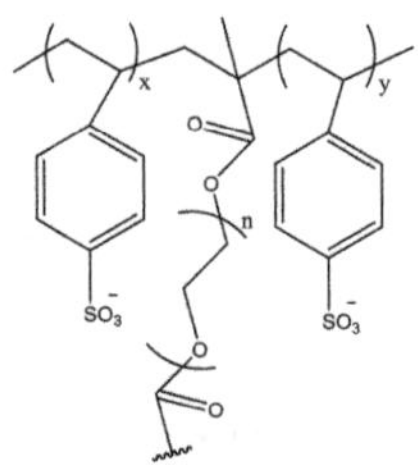	Water	R = P_{4444}^+: 58.7 R = P_{4446}^+: 44.6	68
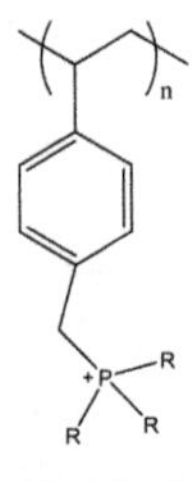 R= C_4H_9, C_5H_{11}, C_6H_{13}	Cl^-	Water	5–40 (depending on PIL and NaCl concentrations)	69

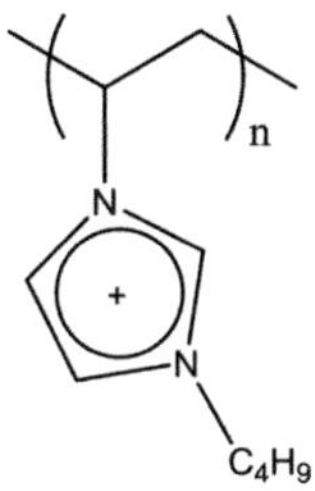	NTf_2^-	Cyclodextrins (CD)	48–62 (depending on the CD concentration)	70
(a) 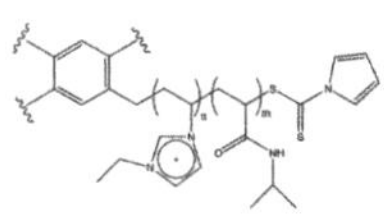 (b) 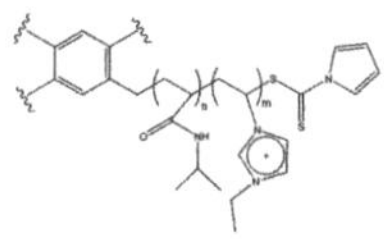	Br^-	Water	30–40 (depending on the polymerisation degree and the composition of the units in the star co-polymers)	71
(a) 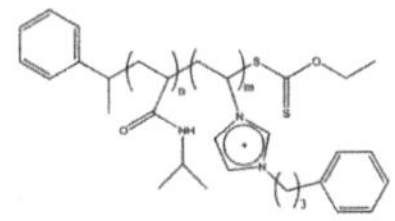 (b) 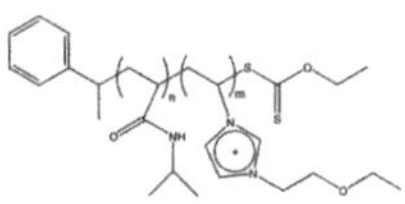 (c) 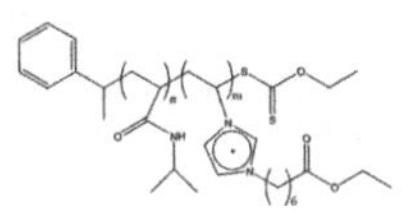	Br^-	Water	(a) 33–38 (b) 50 (c) no data	72
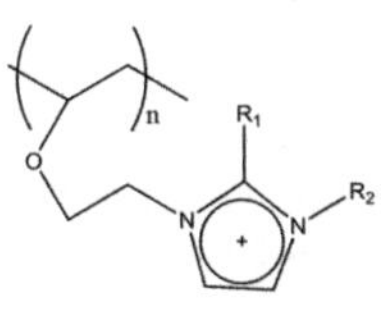 R_1= H, CH_3 R_2= CH_3, C_2H_5, C_4H_9, C_5H_{11}	Cl^- SbF_6^- BF_4^- NTf_2^-	Water and organic solvents	Varied from 0–45 (depending on type and concentration)	74
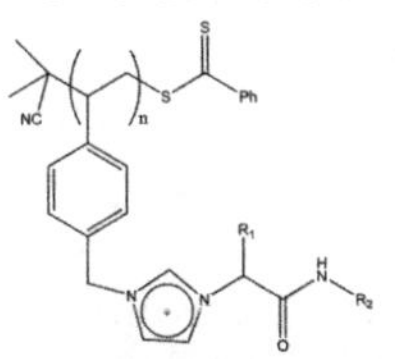 R_1 = (L)-valine or (L)-phenylalanine R_2 = benzyl or butyl	Cl^- NTF_2^-	Water	12.5–58.8 (depending on the composition and concentration)	75

(*continued*)

Table 3.5 (*continued*)

PIL cation	PIL anion	Medium	T_c, °C (LCST/UCST)	Ref.
(a) 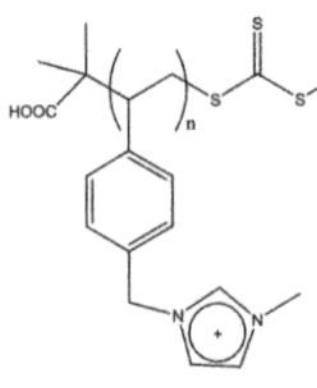 (b) 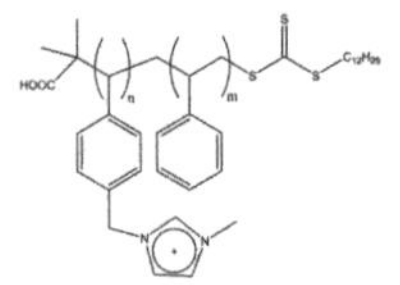	$BF4^-$	Methanol/water	(a) 20–35 at 0.1–3% PIL concentration (b) 39–45 depending on the polymerisation degree	147
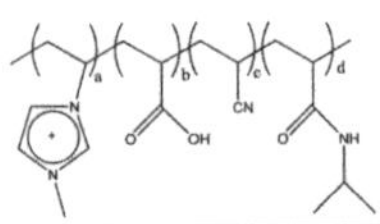	Br^- PF_6^- NTf_2^-		NTf_2^-: ~40	73

tripentylphosphonium-based PILs had lower phase transition temperatures in comparison to their tributyl and trihexylphosphonium analogues.[69] It appears that a "thermoresponsive segment" such as cyclodextrins,[70] or xanthate-type chain transfer agents (CTAs) such as poly(*N*-isopropylacrylamide) (poly(NIPAAm)),[71–73] are required for some imidazolium-based PILs to show phase transition behaviour. Yoshimitsu *et al.* copolymerised vinyl ethers and imidazoles obtaining cationic PILs exhibiting either LCST or UCST behaviours in water and some organic solvents. The character of the transition depended on the anion. PILs with hydrophilic anions, such as Cl^- and SbF_6^-, showed lower critical transition temperatures, while PILs with hydrophobic anions, such as BF_4^- and NTf_2^-, exhibited upper critical transition temperatures, in other words, a transition from an insoluble to a soluble state.[74] On the other hand, substitution of the hydrophilic Cl^- anion with the hydrophobic NTf_2^- anion led to a decrease in the LCST from 48.5 °C to 43.5 °C at 0.5 mg mL^{-1} for vinyl-based chiral-polymeric ionic liquids (CPILs).[75] Copolymerisation of vinyl cations with acrylic acid, acetonitrile and poly(NIPAAm) produced PILs that exhibited not only LCST/UCST behaviour, but also pH-sensitivity, and could be potentially used as soft actuators.[73] It would be interesting to research the properties of anionic PILs based on similar species and how their thermoresponsive behaviour correlates to that of their cationic analogues.

3.3 Summary and Future Directions of Research

As can be seen from this review, the vast majority of research is dedicated to cationic PILs. Although anionic PILs are generally more difficult to synthesise, and therefore currently less diverse than the cationic PILs, it would be interesting to compare the performance of anionic PILs in some of the other fields where cationic PILs show promising results, such as pyrolysis,[76–80] corrosion inhibition,[81] metal sorption for catalysis and sensor applications.[82–89] Perhaps it would be worthwhile to focus attention on the sensing and thermal properties of anionic PILs easily produced from acrylic acids and their salts, with a variety of IL cations. At present, the potential of this class of organic substances has not been fully explored and hopefully the high performance of anionic PILs as electroconductive materials will trigger the exploration of this type of polymer in other emerging applications.

References

1. N. Nishimura and H. Ohno, *Polymer (Guildf.)*, 2014, **55**, 3289.
2. J. Yuan, D. Mecerreyes and M. Antonietti, *Prog. Polym. Sci.*, 2013, **38**, 1009.
3. D. Mecerreyes, *Prog. Polym. Sci.*, 2011, **36**, 1629.
4. S. A. Manuel and K. S. Nahm, *Polymer (Guildf.)*, 2006, **47**, 5952.
5. A. V. Dobrynin and M. Rubinstein, *Prog. Polym. Sci.*, 2005, **30**, 1049.
6. H. Ohno, M. Yoshizawa and W. Ogihara, *Electrochim. Acta*, 2004, **50**, 255.
7. A. S. Shaplov, R. Marcilla and D. Mecerreyes, *Electrochim. Acta*, 2015, **175**, 18.
8. M. M. Obadia and E. Drockenmuller, *Chem. Commun. (Cambridge, U. K.)*, 2016, **52**, 2433.
9. G. Colliat-Dangus, M. M. Obadia, Y. S. Vygodskii, A. Serghei, A. S. Shaplov and E. Drockenmuller, *Polym. Chem.*, 2015, **6**, 4299.
10. M. H. Allen, S. Wang, S. T. Hemp, Y. Chen, L. A. Madsen, K. I. Winey and T. E. Long, *Macromolecules*, 2013, **46**, 3037.
11. M. D. Green, D. Salas-De La Cruz, Y. Ye, J. M. Layman, Y. A. Elabd, K. I. Winey and T. E. Long, *Macromol. Chem. Phys.*, 2011, **212**, 2522.
12. S. Washiro, M. Yoshizawa, H. Nakajima and H. Ohno, *Polymer (Guildf.)*, 2004, **45**, 1577.
13. L. Weng, J. J. Yan, H. Lou Xie, G. Q. Zhong, S. Q. Zhu, H. L. Zhang and E. Q. Chen, *J. Polym. Sci., Part A: Polym. Chem.*, 2013, **51**, 1912.
14. M. Lee, U. H. Choi, D. Salas-De La Cruz, A. Mittal, K. I. Winey, R. H. Colby and H. W. Gibson, *Adv. Funct. Mater.*, 2011, **21**, 708.
15. M. G. Cowan, A. M. Lopez, M. Masuda, Y. Kohno, W. M. McDanel, R. D. Noble and D. L. Gin, *Macromol. Rapid Commun.*, 2016, **37**, 1150.
16. H. Srour, M. Leocmach, V. Maffeis, A. Ghogia, S. Denis-Quanquin, S. Manneville, N. Taberlet, C. Andraud, C. Bucher and C. Monnereau, *Polym. Chem.*, 2016, **7**, 6608.

17. B. Yu, F. Zhou, C. Wang and W. Liu, *Eur. Polym. J.*, 2007, **43**, 2699.
18. A. S. Shaplov, E. I. Lozinskaya, D. O. Ponkratov, I. a. Malyshkina, F. Vidal, P.-H. Aubert, O. V. Okatova, G. M. Pavlov, L. I. Komarova, C. Wandrey and Y. S. Vygodskii, *Electrochim. Acta*, 2011, **57**, 74.
19. W. Ogihara, S. Washiro, H. Nakajima and H. Ohno, *Electrochim. Acta*, 2006, **51**, 2614.
20. K. M. Meek, J. R. Nykaza and Y. A. Elabd, *Macromolecules*, 2016, **49**, 3382.
21. A. L. Pont, R. Marcilla, I. De Meatza, H. Grande and D. Mecerreyes, *J. Power Sources*, 2009, **188**, 558.
22. G. B. Appetecchi, G. T. Kim, M. Montanino, M. Carewska, R. Marcilla, D. Mecerreyes and I. De Meatza, *J. Power Sources*, 2010, **195**, 3668.
23. M. Döbbelin, I. Azcune, M. Bedu, A. Ruiz De Luzuriaga, A. Genua, V. Jovanovski, G. Cabañero and I. Odriozola, *Chem. Mater.*, 2012, **24**, 1583.
24. A. S. Shaplov, D. O. Ponkratov, P. S. Vlasov, E. I. Lozinskaya, L. I. Komarova, I. A. Malyshkina, F. Vidal, G. T. M. Nguyen, M. Armand, C. Wandrey and Y. S. Vygodskii, *Polym. Sci., Ser. B*, 2013, **55**, 122.
25. E. Margaretta, G. B. Fahs, D. L. Inglefield, C. Jangu, D. Wang, J. R. Heflin, R. B. Moore and T. E. Long, *ACS Appl. Mater. Interfaces*, 2016, **8**, 1280.
26. P. G. Mineo, L. Livoti, M. Giannetto, A. Gulino, S. Lo Schiavo and P. Cardiano, *J. Mater. Chem.*, 2009, **19**, 8861.
27. P. G. Mineo, L. Livoti, S. Lo Schiavo and P. Cardiano, *Polym. Adv. Technol.*, 2012, **23**, 1511.
28. T.-Y. Shi, J.-Q. Wang, J. Sun, M.-H. Wang, W.-G. Cheng and S.-J. Zhang, *RSC Adv.*, 2013, **3**, 3726.
29. Z.-Z. Yang, Y. Zhao, G. Ji, H. Zhang, B. Yu, X. Gao and Z. Liu, *Green Chem.*, 2014, **16**, 3724.
30. Y. Zhang, S. Yin, S. Luo and C. T. Au, *Ind. Eng. Chem. Res.*, 2012, **51**, 3951.
31. A. S. Rewar, R. S. Bhavsar, K. Sreekumar and U. K. Kharul, *J. Membr. Sci.*, 2015, **481**, 19.
32. A. S. Rewar, S. V. Shaligram and U. K. Kharul, *J. Membr. Sci.*, 2016, **497**, 282.
33. R. S. Bhavsar, S. C. Kumbharkar and U. K. Kharul, *J. Membr. Sci.*, 2014, **470**, 494.
34. S. V. Shaligram, P. P. Wadgaonkar and U. K. Kharul, *J. Membr. Sci.*, 2015, **493**, 403.
35. S. V. Shaligram, A. S. Rewar, P. P. Wadgaonkar and U. K. Kharul, *Polymer*, 2016, **93**, 30.
36. M. Radosz and Y. Shen, US 20090266230 A1, 2009.
37. M. G. Cowan, M. Masuda, W. M. McDanel, Y. Kohno, D. L. Gin and R. D. Noble, *J. Membr. Sci.*, 2016, **498**, 408.
38. T. O. Magalhaes, A. S. Aquino, F. Dalla Vecchia, F. L. Bernard, M. Seferin, S. C. Menezes, R. Ligabue and S. Einloft, *RSC Adv.*, 2014, **4**, 18164.
39. M. S. Raja Shahrom, C. D. Wilfred and A. K. Z. Taha, *J. Mol. Liq.*, 2016, **219**, 306.
40. L. C. Tome, A. S. L. Gouveia, C. S. R. Freire, D. Mecerreyes and I. M. Marrucho, *J. Membr. Sci.*, 2015, **486**, 40.

41. J. Tang, H. Tang, W. Sun, H. Plancher, M. Radosz and Y. Shen, *Chem. Commun. (Cambridge, U. K.)*, 2005, 3325.
42. E. I. Privalova, E. Karjalainen, M. Nurmi, P. Mäki-Arvela, K. Eränen, H. Tenhu, D. Y. Murzin and J. P. Mikkola, *ChemSusChem*, 2013, **6**, 1500.
43. J. Tang, H. Tang, W. Sun, M. Radosz and Y. Shen, *Polymer (Guildf.)*, 2005, **46**, 12460.
44. J. Tang, W. Sun, H. Tang, M. Radosz and Y. Shen, *Macromolecules*, 2005, **38**, 2037.
45. H. Tang, J. Tang, S. Ding, M. Radosz and Y. Shen, *J. Polym. Sci., Part A: Polym. Chem.*, 2005, **43**, 1432.
46. J. M. Zhu, K. G. He, H. Zhang and F. Xin, *Adsorpt. Sci. Technol.*, 2012, **30**, 35.
47. Y. B. Xiong, H. Wang, Y. J. Wang and R. M. Wang, *Polym. Adv. Technol.*, 2012, **23**, 835.
48. T. D. Ho, M. D. Joshi, M. A. Silver and J. L. Anderson, *J. Chromatogr. A*, 2012, **1240**, 29.
49. J. González-Álvarez, D. Blanco-Gomis, P. Arias-Abrodo, J. Pello-Palma, N. Ríos-Lombardía, E. Busto, V. Gotor-Fernández and M. D. Gutiérrez-Álvarez, *J. Chromatogr. A*, 2013, **1305**, 35.
50. J. López-Darias, J. L. Anderson, V. Pino and A. M. Afonso, *Anal. Bioanal. Chem.*, 2011, **401**, 2965.
51. S. Shaligram, P. P. Wadgaonkar, U. K. Kharul, H. R. Nie, G. Sun, M. Zhang, M. Baumgarten, K. Mullen, S. Shanmugaraju, S. A. Joshi, P. S. Mukherjee, S. Toal, W. C. Trogler, D. Ownby, J. Belden, G. Lotufo, M. Lydy, D. Ownby, J. Belden, G. Lotufo, M. Lydy, S. Wallenborg, C. Bailey, J. Yang, T. Swager, J. Sylvia, J. Janni, J. Klein, K. Spencer, A. Hilmi, J. Luong, R. Luggar, M. Farquharson, J. Horrocks, R. Lacey, M. Krausa, K. Schorb, B. Zu, Y. Guo, X. Dou, D. McQuade, A. Pullen, T. Swager, K. Albert, N. Lewis, C. Schauer, G. Sotzing, S. Stitzel, T. Vaid, D. Walt, T. Naddo, Y. Che, W. Zhang, K. Balakrishnan, X. Yang, M. Yen, J. Zhao, J. Moore, L. Zang, H. Gang, N. Yan, J. Yang, H. Wang, L. Ding, S. Yin, Y. Fang, P. Beyazkilic, A. Yildirim, M. Bayindir, Y. Wang, A. La, Y. Ding, Y. Liu, Y. Lei, G. Demirel, B. Daglar, M. Bayindir, J. Goodpaster, J. Harrison, V. McGuffin, L. Guo, B. Zu, Z. Yang, H. Cao, X. Zheng, X. Dou, R. Manez, F. Sancenón, M. Shellaiah, Y. Wu, A. Singh, M. Raju, H. Lin, A. Ponnu, E. Anslyn, B. Gole, S. Shanmugaraju, A. Bar, P. Mukherjee, L. Desmonts, D. Reinhoudt, M. Crego-Calama, M. Li, R. Ho, Y. Lee, S. C. Kumbharkar, M. Islam, R. A. Potrekar, U. K. Kharul, R. S. Bhavsar, S. C. Kumbharkar, A. S. Rewar, U. K. Kharul, D. Kumar, K. Thomas, K. Kalyanasundaram, J. K. Thomas, H. Toiserkani, G. Yilmaz, Y. Yagci, L. Torun, I. A. Gorodetskaya, A. A. Gorodetsky, E. V. Vinogradova, R. H. Grubbs, F. M. Winnik, E. S. Forzani, D. Lu, M. J. Leright, A. D. Aguilar, F. Tsow, R. A. Iglesias, Q. Zhang, J. Lu, J. Li, N. Tao, A. Hakonen and S. Hulth, *J. Mater. Chem. A*, 2014, **2**, 13983.
52. L. Wang, X. Duan, W. Xie, Q. Li and T. Wang, *Chem. Commun.*, 2016, **52**, 8417.

53. R. Guterman, M. Ambrogi and J. Yuan, *Macromol. Rapid Commun.*, 2016, **37**, 1106.
54. H. Mao, J. Liang, H. Zhang, Q. Pei, D. Liu, S. Wu, Y. Zhang and X.-M. Song, *Biosens. Bioelectron.*, 2015, **70**, 289.
55. X. Yu, Y. Wang, X. Chen, K. Wu, D. Chen, M. Ma, Z. Huang, W. Wu and C. Li, *Anal. Chem.*, 2015, **87**, 4237.
56. Q. Zhao, J. Heyda, J. Dzubiella, K. Täuber, J. W. C. Dunlop and J. Yuan, *Adv. Mater.*, 2015, **27**, 2913.
57. J. Feng, M. Sun, Y. Bu and C. Luo, *Anal. Bioanal. Chem.*, 2015, **407**, 7025.
58. R. Zhang, P. Su, L. Yang and Y. Yang, *J. Sep. Sci.*, 2014, **37**, 1503.
59. J. Feng, M. Sun, L. Li, X. Wang, H. Duan and C. Luo, *Talanta*, 2014, **123**, 18.
60. C. Zhang and J. L. Anderson, *J. Chromatogr. A*, 2014, **1344**, 15.
61. H. Mao, H. Zhang, J. Liang, D. Liu, S. Wu, Y. Zhang, Y. Zhang, Q. Wu, G. Zhang and X.-M. Song, *J. Mater. Chem. B*, 2015, **3**, 5310.
62. J. Jia, X. Liang, L. Wang, Y. Guo, X. Liu and S. Jiang, *J. Chromatogr. A*, 2013, **1320**, 1.
63. J. López-Darias, V. Pino, Y. Meng, J. L. Anderson and A. M. Afonso, *J. Chromatogr. A*, 2010, **1217**, 7189.
64. B. Ziołkowski and D. Diamond, *Chem. Commun.*, 2013, **49**, 10308.
65. Y. Y. Men, H. Schlaad, A. Voelkel and J. Yuan, *Polym. Chem.*, 2014, **5**, 3719.
66. Y. Deguchi, Y. Kohno and H. Ohno, *Chem. Lett.*, 2015, **44**, 2014.
67. G. Wang and P. Wu, *Soft Matter*, 2015, **11**, 5253.
68. Y. Cai, R. Wang, W. B. Krantz, A. Fane and X. Hu, *RSC Adv.*, 2015, **5**, 97143.
69. A. Okafuji, Y. Kohno and H. Ohno, *Macromol. Rapid Commun.*, 2016, **37**, 1130.
70. S. Amajjahe and H. Ritter, *Macromolecules*, 2008, **41**, 3250.
71. H. Mori, Y. Ebina, R. Kambara and K. Nakabayashi, *Polym. J.*, 2012, **44**, 550.
72. H. Mori, M. Yanagi and T. Endo, *Macromolecules*, 2009, **42**, 8082.
73. F. Chen, J. Guo, D. Xu and F. Yan, *Polym. Chem.*, 2016, **7**, 1330.
74. H. Yoshimitsu, A. Kanazawa, S. Kanaoka and S. Aoshima, *Macromolecules*, 2012, **45**, 9427.
75. S. Montolio, L. Gonzáez, B. Altava, H. Tenhu, M. I. Burguete, E. García-Verdugo and S. V. Luis, *Chem. Commun.*, 2014, **50**, 10683.
76. S. Amajjahe and H. Ritter, *Macromol. Rapid Commun.*, 2009, **30**, 94.
77. J. Yuan, C. Giordano and M. Antonietti, *Chem. Mater.*, 2010, **22**, 5003.
78. X. Bo, J. Bai, J. Ju and L. Guo, *J. Power Sources*, 2011, **196**, 8360.
79. J. Yuan, A. G. Márquez, J. Reinacher, C. Giordano, J. Janek, M. Antonietti, K. K. R. Datta, V. V. Balasubramanian, K. Ariga, T. Mori, A. Vinu, J. P. Paraknowitsch, A. Thomas, M. Antonietti, J. P. Paraknowitsch, J. Zhang, D. Su, A. Thomas, M. Antonietti, W. Yang, T.-P. Fellinger, M. Antonietti, M. Terrones, P. Redlich, N. Grobert, S. Trasobares, W.-K. Hsu, H. Terrones, Y.-Q. Zhu, J. P. Hare, C. L. Reeves, A. K. Cheetham, M. Rühle,

H. W. Kroto, D. R. M. Walton, M. Terrones, H. Terrones, N. Grobert, W. K. Hsu, Y. Q. Zhu, J. P. Hare, H. W. Kroto, D. R. M. Walton, P. Kohler-Redlich, M. Ruhle, J. P. Zhang, A. K. Cheetham, M. Terrones, R. Kamalakaran, T. M. R. Seeger, M. Glerup, M. Castignolles, M. Holzinger, G. Hug, A. Loiseau, P. Bernier, A. G. Kudashov, A. V. Okotrub, L. G. Bulusheva, I. P. Asanov, Y. V. Shubin, N. F. Yudanov, L. I. Yudanova, V. S. Danilovich, O. G. Abrosimov, R. Sen, B. C. Satishkumar, A. Govindaraj, K. R. Harikumar, M. K. Renganathan, C. N. R. Rao, J. Liu, R. Czerw, D. L. Carroll, L. Zhi, T. Gorelik, R. Friedlein, J. Wu, U. Kolb, W. R. Salaneck, K. Müllen, A. Vinu, K. Ariga, T. Mori, T. Nakanishi, S. Hishita, D. Golberg, Y. Bando, R. V. Parthasarathy, K. L. N. Phani, C. R. Martin, J. Jang, J. H. Oh, T. Maiyalagan, B. Viswanathan, L. Zhi, T. Gorelik, J. Wu, U. Kolb, K. Müllen, Y. Xia, R. Mokaya, Y. Xia, R. Mokaya, F. Goettmann, M. Antonietti, R. Liu, D. Wu, X. Feng, K. Müllen, J. S. Lee, X. Wang, H. Luo, G. A. Baker, S. Dai, P. F. Fulvio, J. S. Lee, R. T. Mayes, X. Wang, S. M. Mahurin, S. Dai, J. Lu, F. Yan, J. Texter, O. Green, S. Grubjesic, S. Lee, M. A. Firestone, J. Yuan, M. Antonietti, F. Yan, J. Texter, J. Yuan, M. Antonietti, M. Graeser, M. Bognitzki, W. Massa, C. Pietzonka, A. Greiner, J. H. Wendorff, A. Greiner, J. H. Wendorff, H. Chen, Y. A. Elabd, A. Walther, A. Gröschel, A. H. E. Müller, A. Walther, J. Yuan, V. Abetz and A. H. E. Müller, *Polym. Chem.*, 2011, **2**, 1654.

80. C. Pan, L. Qiu, Y. Peng and F. Yan, *J. Mater. Chem.*, 2012, **22**, 13578.
81. M. Taghavikish, S. Subianto, N. K. Dutta, L. De Campo, J. P. Mata, C. Rehm and N. R. Choudhury, *ACS Omega*, 2016, **1**, 29–40.
82. K. Manojkumar, A. Sivaramakrishna and K. Vijayakrishna, *J. Nanoparticle Res.*, 2016, **18**, 1.
83. W. Zhang and J. Yuan, *Macromol. Rapid Commun.*, 2016, **37**, 1124.
84. Y. Chen, S. Cao, L. Zhang, C. Xi, X. Li, Z. Chen and G. Wang, *J. Chromatogr. A*, 2016, **1448**, 9.
85. M. J. Muldoon and C. M. Gordon, *J. Polym. Sci., Part A: Polym. Chem.*, 2004, **42**, 3865.
86. M. Döbbelin, V. Jovanovski, I. Llarena, L. J. Claros Marfil, G. Cabañero, J. Rodriguez and D. Mecerreyes, *Polym. Chem.*, 2011, **2**, 1275.
87. R. Gracia, K. Vijayakrishna and D. Mecerreyes, *React. Funct. Polym.*, 2014, **79**, 54.
88. M. T. Rahman, Z. Barikbin, A. Z. M. Badruddoza, P. S. Doyle and S. A. Khan, *Langmuir*, 2013, **29**, 9535.
89. H. Li, Q. Zhang, X. Liu, F. Chang, Y. Zhang, W. Xue and S. Yang, *Bioresour. Technol.*, 2013, **144**, 21.
90. M. Hirao, K. Ito and H. Ohno, *Electrochim. Acta*, 2000, **45**, 1291.
91. Y. S. Vygodskii, O. A. Mel'nik, A. S. Shaplov, E. I. Lozinskaya, I. A. Malyshkina and N. D. Gavrilova, *Polym. Sci., Ser. A*, 2007, **49**, 256.
92. R. Marcilla, F. Alcaide, H. Sardon, J. A. Pomposo, C. Pozo-Gonzalo and D. Mecerreyes, *Electrochem. Commun.*, 2006, **8**, 482.
93. J. Jagur-Grodzinski, *Polym. Adv. Technol.*, 2006, **17**, 395.
94. K. Yin, Z. Zhang, L. Yang and S. Hirano, *J. Power Sources*, 2014, **258**, 150.

95. J. R. Nair, L. Porcarelli, F. Bella and C. Gerbaldi, *ACS Appl. Mater. Interfaces*, 2015, 12961–12971.
96. T. J. Wood, W. C. E. Schofield, P. Lund, M. J. Larsen, J. P. S. Badyal and D. Dh, *Chem. Commun. (Cambridge, U. K.)*, 2012, **48**, 10201.
97. X. Chen, J. Zhao, J. Zhang, L. Qiu, D. Xu, H. Zhang, X. Han, B. Sun, G. Fu, Y. Zhang and F. Yan, *J. Mater. Chem.*, 2012, **22**, 18018.
98. H. Hu, W. Yuan, L. Lu, H. Zhao, Z. Jia and G. L. Baker, *J. Polym. Sci., Part A: Polym. Chem.*, 2014, **52**, 2104.
99. Z. Jia, W. Yuan, C. Sheng, H. Zhao, H. Hu and G. L. Baker, *J. Polym. Sci., Part A: Polym. Chem.*, 2015, **53**, 1339.
100. A. S. Shaplov, D. O. Ponkratov, P. H. Aubert, E. I. Lozinskaya, C. Plesse, A. Maziz, P. S. Vlasov, F. Vidal and Y. S. Vygodskii, *Polymer*, 2014, **55**, 3385.
101. M. Lee, U. H. Choi, R. H. Colby and H. W. Gibson, *Chem. Mater.*, 2010, **22**, 5814.
102. J. Zhang, D. Xu, J. Guo, Z. Sun, W. Qian and Y. Zhang, *Macromol. Rapid Commun.*, 2016, **37**, 1194.
103. H. Erdemi, Ü. Akbey and W. H. Meyer, *Solid State Ionics*, 2010, **181**, 1586.
104. K. Põhako-Esko, M. Timusk, K. Saal, R. Lõhmus, I. Kink and U. Mäeorg, *J. Mater. Res.*, 2013, **28**, 3086.
105. A. S. Shaplov, P. S. Vlasov, E. I. Lozinskaya, D. O. Ponkratov, I. A. Malyshkina, F. Vidal, O. V Okatova, G. M. Pavlov, C. Wandrey, A. Bhide, M. Sch and Y. S. Vygodskii, *Macromolecules*, 2011, **44**, 9792.
106. M. Yoshizawa and H. Ohno, *Chem. Lett.*, 1999, 889–890.
107. T. Ikeda, S. Moriyama and J. Kim, *J. Polym. Sci., Part A: Polym. Chem. Chem.*, 2016, **54**, 2896.
108. Y. S. Vygodskii, A. S. Shaplov, E. I. Lozinskaya, K. A. Lyssenko, D. G. Golovanov, I. A. Malyshkina, N. D. Gavrilova and M. R. Buchmeiser, *Macromol. Chem. Phys.*, 2008, **209**, 40.
109. K. Yin, Z. Zhang, X. Li, L. Yang, K. Tachibana and S. Hirano, *J. Mater. Chem. A*, 2015, **3**, 170.
110. K. Hoshino, M. Yoshio, T. Mukai, K. Kishimoto, H. Ohno and T. Kato, *J. Polym. Sci., Part A: Polym. Chem.*, 2003, **41**, 3486.
111. H. Nakajima and H. Ohno, *Polymer (Guildf.)*, 2005, **46**, 11499.
112. M. Li, B. Yang, L. Wang, Y. Zhang, Z. Zhang, S. Fang and Z. Zhang, *J. Membr. Sci.*, 2013, **447**, 222.
113. H. Zhang, C. Liu, L. Zheng, W. Feng, Z. Zhou and J. Nie, *Electrochim. Acta*, 2015, **159**, 93.
114. X. Li, Z. Zhang, S. Li, L. Yang and S. Hirano, *J. Power Sources*, 2016, **307**, 678.
115. A. S. Shaplov, P. S. Vlasov, E. I. Lozinskaya, O. A. Shishkan, D. O. Ponkratov, I. A. Malyshkina, F. Vidal, C. Wandrey, I. A. Godovikov and Y. S. Vygodskii, *Macromol. Chem. Phys.*, 2012, **213**, 1359.
116. M. Li, L. Yang, S. Fang, S. Dong, S. I. Hirano and K. Tachibana, *Polym. Int.*, 2012, **61**, 259.

117. M. Li, L. Yang, S. Fang, S. Dong, S. I. Hirano and K. Tachibana, *J. Power Sources*, 2011, **196**, 8662.
118. M. Li, L. Yang, S. Fang and S. Dong, *J. Membr. Sci.*, 2011, **366**, 245.
119. E. Drockenmuller, U. Claude and B. Lyon, *ACS Macro Lett.*, 2014, **3**, 658.
120. A. S. Shaplov, P. S. Vlasov, M. Armand, E. I. Lozinskaya, D. O. Ponkratov, I. a. Malyshkina, F. Vidal, O. V. Okatova, G. M. Pavlov, C. Wandrey, I. A. Godovikov and Y. S. Vygodskii, *Polym. Chem.*, 2011, **2**, 2609.
121. J. Juger, F. Meyer, F. Vidal, C. Chevrot and D. Teyssié, *Tetrahedron Lett.*, 2009, **50**, 128.
122. Q. Zhao, J. C. Wajert and J. L. Anderson, *Anal. Chem.*, 2010, **82**, 707.
123. J. Tang, H. Tang, W. Sun, M. Radosz and Y. Shen, *J. Polym. Sci., Part A: Polym. Chem.*, 2005, **43**, 5477.
124. J. Tang, Y. Shen, M. Radosz and W. Sun, *Ind. Eng. Chem. Res.*, 2009, **48**, 9113.
125. G. Yu, Q. Li, N. Li, Z. Man, C. Pu, C. Asumana and X. Chen, *Polym. Eng. Sci.*, 2014, **54**, 59.
126. S. Supasitmongkol and P. Styring, *Energy Environ. Sci.*, 2010, **3**, 1961.
127. M. Isik, S. Zulfiqar, F. Edhaim, F. Ruiperez, A. Rothenberger and D. Mecerreyes, *ACS Sustainable Chem. Eng.*, 2016, **4**, 7200.
128. F. L. Bernard, B. B. Polesso, F. W. Cobalchini, A. J. Donato, M. Seferin, R. Ligabue, V. V. Chaban, J. F. do Nascimento, F. Dalla Vecchia and S. Einloft, *Polymer (Guildf.)*, 2016, **102**, 199.
129. C. Cagliero, H. Nan, C. Bicchi and J. L. Anderson, *J. Chromatogr. A*, 2016, **1459**, 17.
130. C. Cagliero, T. D. Ho, C. Zhang, C. Bicchi and J. L. Anderson, *J. Chromatogr. A*, 2016, **1449**, 2.
131. J. Feng, M. Sun, Y. Bu and C. Luo, *J. Chromatogr. A*, 2015, **1393**, 8.
132. J. Feng, M. Sun, Y. Bu and C. Luo, *Anal. Bioanal. Chem.*, 2016, **408**, 1679.
133. F. Zhao, Y. Meng and J. L. Anderson, *J. Chromatogr. A*, 2008, **1208**, 1.
134. M. S.-P. López, D. Mecerreyes, E. López-Cabarcos and B. López-Ruiz, *Biosens. Bioelectron.*, 2006, **21**, 2320.
135. X. Chen, Y. Ma, D. Chen, M. Ma and C. Li, *Food Chem.*, 2015, **180**, 142.
136. L. Chen and X. Huang, *Anal. Chim. Acta*, 2016, **916**, 33.
137. F. Bianchi, F. Bisceglie, S. Dugheri, G. Arcangeli, V. Cupelli, E. del Borrello, L. Sidisky and M. Careri, *J. Chromatogr. A*, 2014, **1331**, 1.
138. C. M. Graham, Y. Meng, T. Ho and J. L. Anderson, *J. Sep. Sci.*, 2011, **34**, 340.
139. H. Yu, J. Merib and J. L. Anderson, *J. Chromatogr. A*, 2016, **1438**, 10.
140. M. Sun, Y. Bu, J. Feng and C. Luo, *J. Sep. Sci.*, 2016, **39**, 375.
141. M. Cordero-Vaca, M. J. Trujillo-Rodriguez, C. Zhang, V. Pino, J. L. Anderson and A. M. Afonso, *Anal. Bioanal. Chem.*, 2015, **407**, 4615.
142. M. D. Joshi, T. D. Ho, W. T. S. Cole and J. L. Anderson, *Talanta*, 2014, **118**, 172.
143. J. Feng, M. Sun, J. Li, X. Liu and S. Jiang, *J. Chromatogr. A*, 2012, **1227**, 54.

144. S. Hamdan, L. Moore, J. Lejeune, F. Hasan, T. K. Carlisle, J. E. Bara, D. L. Gin, A. L. LaFrate, R. D. Noble, D. A. Spivak and I. M. Warner, *J. Colloid Interface Sci.*, 2016, **463**, 29.
145. M. J. Trujillo-Rodríguez, H. Yu, W. T. S. Cole, T. D. Ho, V. Pino, J. L. Anderson and A. M. Afonso, *Talanta*, 2014, **121**, 153.
146. Y. Meng, V. Pino and J. L. Anderson, *Anal. Chim. Acta*, 2011, **687**, 141.
147. C. Liu, S. Wang, H. Zhou, C. Gao and W. Zhang, *J. Polym. Sci., Part A: Polym. Chem.*, 2016, **54**, 945.

CHAPTER 4

Switchable Hydrophobicity and Hydrophilicity

YUKI KOHNO[a] AND HIROYUKI OHNO*[b]

[a]National Institute of Advanced Industrial Science and Technology (AIST), 4-2-1, Nigatake, Miyagino-ku, Sendai 983-8551, Japan; [b]Department of Biotechnology, Tokyo University of Agriculture and Technology, Naka-cho, Koganei, Tokyo 184-8588, Japan

*E-mail: ohnoh@cc.tuat.ac.jp

4.1 Introduction

Stimuli-responsive materials that alter their properties in response to external stimuli (*e.g.*, temperature, pressure, and concentration of specific compounds) have attracted keen interest as "smart" materials. Thermoresponsive polymers capable of changing their hydrophobicity and hydrophilicity upon a slight change in temperature are well-studied examples of smart materials. For instance, poly(*N*-isopropylacrylamide) and some non-ionic polymers exhibit a lower critical solution temperature (LCST)-type phase change in water. Namely, these polymers are miscible with water at low temperature and become immiscible with water at higher temperatures. There were no examples of fully-charged materials that show an LCST-type phase transition in water until recently. The phase behaviour and physico-chemical properties of these interesting materials are discussed in this chapter.

In 2007, Fukumoto and Ohno reported the first example of an ionic liquid (IL) that shows an LCST-type phase change upon mixing with pure water,[1]

Smart Materials No. 29
Polymerized Ionic Liquids
Edited by Ali Eftekhari

Published by the Royal Society of Chemistry, www.rsc.org

which was the starting point of our studies on "smart ILs". This chapter aims to give an overview of the rapidly-growing research into IL-based *liquid-state* and *solid-state* materials that undergo the aforementioned phase change. Fundamental design criteria of thermoresponsive ILs and IL-derived polyelectrolytes showing highly temperature-sensitive and reversible phase changes are described in terms of the total hydrophilicity of both the cations and anions. In addition, potential (including expected) applications of these IL-based materials are summarised, which may not be accomplished by exploiting classical non-ionic thermoresponsive polymers.

4.2 Ionic Liquids with Switchable Hydrophobicity and Hydrophilicity Depending on the Temperature

Numerous studies on ILs, organic salts with very low melting temperatures, have recently been undertaken in interdisciplinary fields. In addition to the enticing properties of ILs (*e.g.*, vanishingly low vapour pressure, wide range of temperatures in the liquid state, high thermal/chemical stability, and high charge density), functionalising ILs is also possible upon suitable design of both cationic and anionic structures. Control of hydrophobicity and hydrophilicity is one of the key functions of ILs. Hydrophobic ILs, typically containing the bis(trifluoromethanesulfonyl)imide anion ($[Tf_2N]^-$), undergo phase separation with water, thus forming IL-based aqueous biphasic systems (ABSs). ABSs using ILs can also be obtained by mixing water-miscible ILs with salting-out agents such as kosmotropic inorganic salts. Most of these IL-based ABSs remain phase-separated regardless of external stimuli, such as temperature. Being totally different from the above-mentioned "stable" phase separation of IL/water mixtures, a thermoresponsive and reversible phase change between a homogeneous phase and two separate IL/water phases has been observed in specific IL/water mixtures.[2] Two types of thermoresponsive phase behaviour have been found in IL/water binary systems: upper critical solution temperature (UCST)-type phase change and LCST-type phase change. In the UCST-type phase transition, the mutual solubility of the IL and water increases upon heating. Many IL/water mixtures exhibit this type of phase behaviour. The opposite LCST-type phase change is observed when the mutual solubility of the IL and water increases upon cooling. Figure 4.1 displays representative phase diagrams expressing the thermoresponsive phase behaviour of reported IL/water mixtures. In many cases, the temperature at which the transparent solution turns turbid is determined as the phase separation temperature (T_c). The T_c value depends strongly on the amount of added water, and a convex curve is observed in the phase diagram when the mixture shows a UCST-type phase change (Figure 4.1, left). Upon cooling, the homogeneous mixture becomes turbid at a specific temperature on the convex curve, and settles in a phase-separated state below the curve. In the case of an LCST-type phase change, a concave curve is seen in the phase

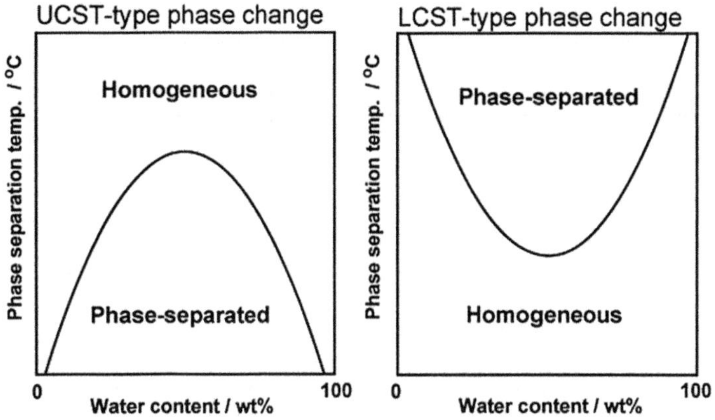

Figure 4.1 Typical phase diagrams of thermoresponsive IL/water mixtures.

diagram (Figure 4.1, right), and the mixture displays temperature-dependent behaviour opposite to UCST (*i.e.*, the homogeneous mixture is phase-separated above the concave curve). This chapter focuses on aqueous IL systems that exhibit a thermoresponsive hydrophobicity/hydrophilicity change. Other IL systems showing switchable hydrophobicity after mixing with different molecular solvents or driven by other stimuli are summarised elsewhere.[3,4]

Table 4.1 presents a summary of the reported IL/water mixed systems that show a thermoresponsive phase change. A UCST-type phase change was observed in several imidazolium-, ammonium-, and phosphonium-type ILs (**1–5**) after mixing with pure water or aqueous salt solutions.[5–10] This behaviour is understood from the viewpoint of the temperature–solubility relationship of salts. UCST behaviour has also been found in sulfuric acid-functionalised ILs (**6**).[11] On the contrary, LCST-type phase changes were scarcely observed in IL/water mixtures until recently. Ohno's group reported a series of phosphonium-type ILs with *N*-trifluoromethanesulfonyl-substituted amino acid anions (**7**).[1] These ILs indeed underwent an LCST-type phase change after mixing with water. When *N*-trifluoromethanesulfonyl-leucine was coupled with the tetra-*n*-butylphosphonium cation ($[P_{4444}]^+$), the resulting IL ($[P_{4444}]$[Tf-Leu]) was miscible with water at 22 °C, but converted to a liquid/liquid separated state after being kept at 25 °C for 10 min. The homogeneous mixture was obtained again upon cooling. The LCST-type phase change of the prepared IL/water mixture was therefore revealed to be highly temperature-sensitive and reversible. It should be noted here that the LCST behaviour itself had already been found in a few ammonium-type salt/water mixtures (**8–10**).[12–15] A ternary system composed of **11**–water–ethanol also showed LCST behaviour at a certain mixing ratio.[16]

Following intensive studies by Ohno and co-workers on the relation between IL structures and phase behaviour after mixing with water, several ILs were found to undergo the LCST-type phase change.[17] Figure 4.2 shows the phase behaviour of an aqueous solution containing **12** (IL concentration: 35 wt%).

Table 4.1 Thermoresponsive phase change of IL-based aqueous systems.

Entry	Structure	Solvent	Phase behaviour	Ref.
1	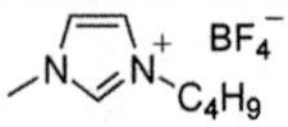	Water	UCST	5,6
2	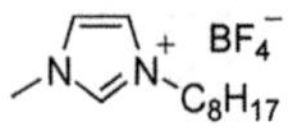	Water	UCST	7
3	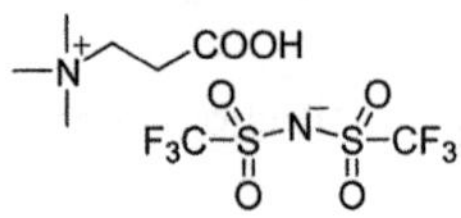	Water	UCST	8
4	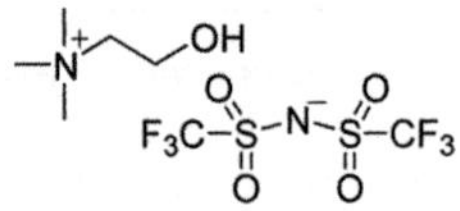	Water	UCST	9
5	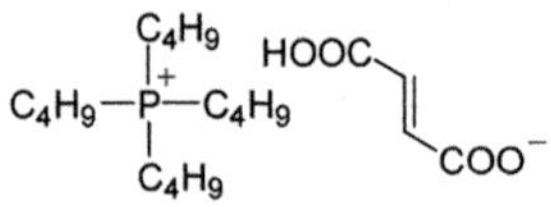	Water	UCST	10
6	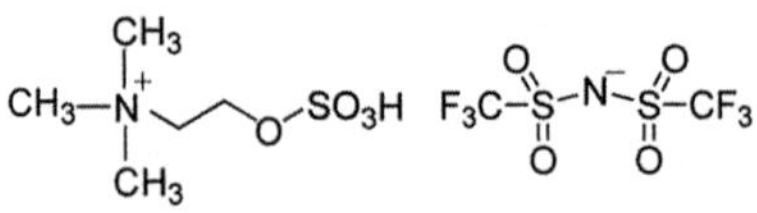	Aqueous salt solutions	UCST	11
7	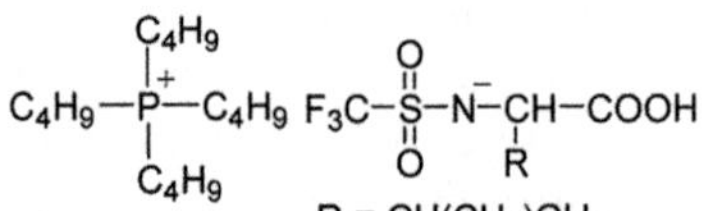 R = $CH(CH_3)CH_3$ R = $CH_2CH(CH_3)CH_3$ R = $CH(CH_3)CH_2CH_3$ R = $CH_2C_6H_5$	Water	LCST	1
8	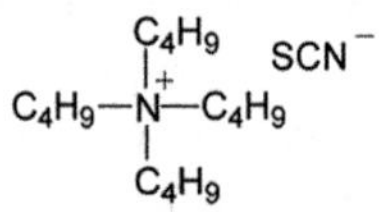	Water	UCST and LCST	12
9	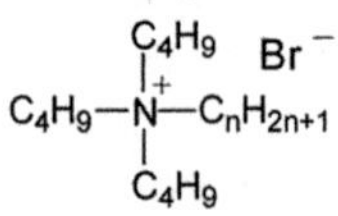 n = 12, 14, 16	Water	LCST	13
10	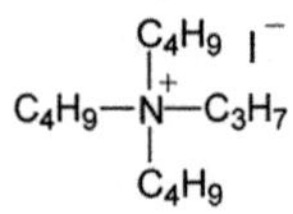	Water	UCST and LCST	14,15
11	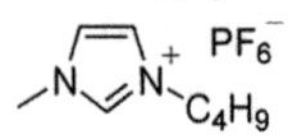	Water/ethanol	LCST	16
12	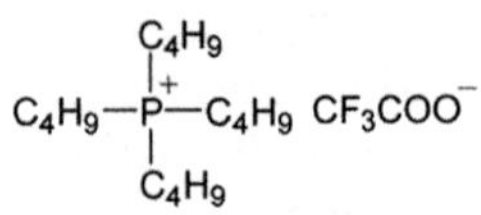	Water	LCST	17

13	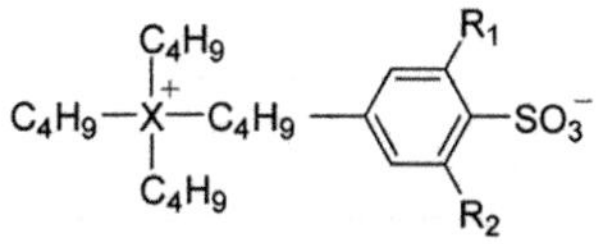 X = P, $R_1 = R_2 = H$ X = P, $R_1 = H$, $R_2 = CH_3$ X = P, $R_1 = R_2 = CH_3$ X = N, $R_1 = R_2 = CH_3$	Water	LCST	17
14	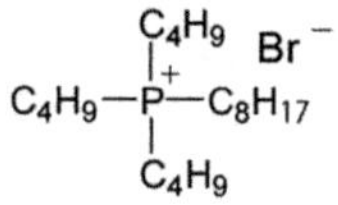	Water	LCST	17
15	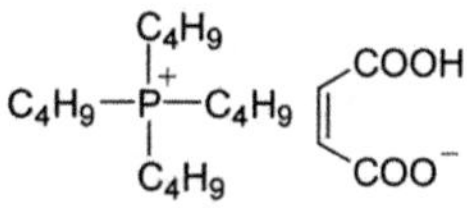	Water	LCST	10
16	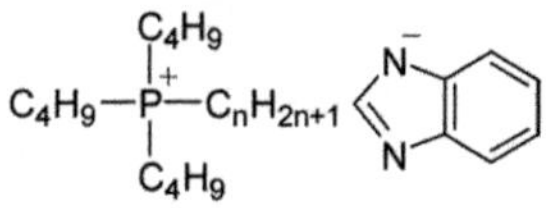 n = 4, 8	Water	LCST	18
17	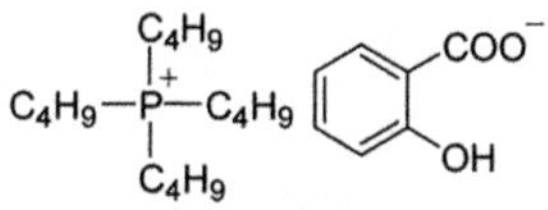	Water	LCST	19
18	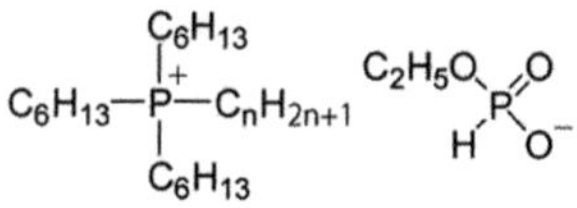 n = 6, 8	Water	LCST	20
19	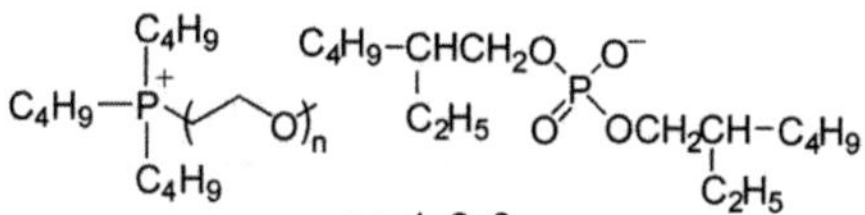 n = 1, 2, 3	Water	LCST	21
20	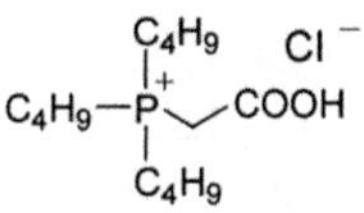	Aqueous salt solutions	LCST	22
21	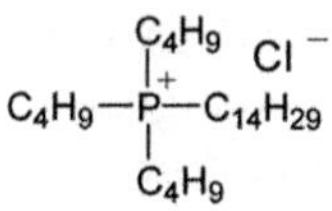	Aqueous salt solutions	LCST	23
22	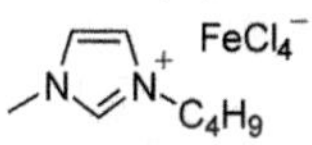	Water	LCST	24
23	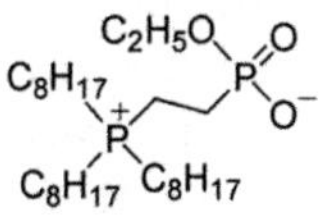	Water	LCST	25
24	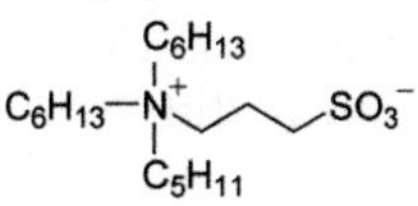	Water	LCST	26

(*continued*)

Table 4.1 *(continued)*

Entry	Structure	Solvent	Phase behaviour	Ref.
25	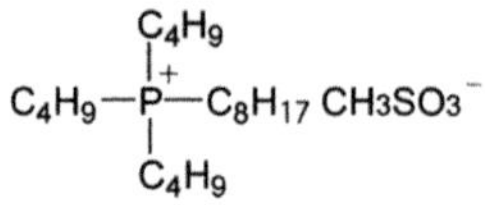	Water	Homogeneous	17
26	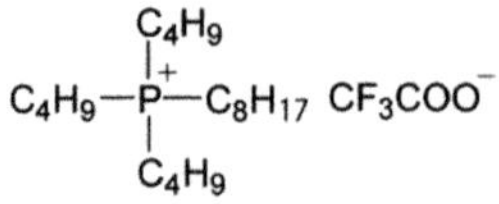	Water	Phase-separated	17
	Mixture of **25** and **26**	Water	LCST	17
27	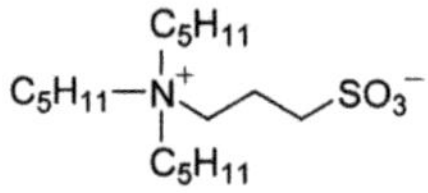	Water	Homogeneous	29
28	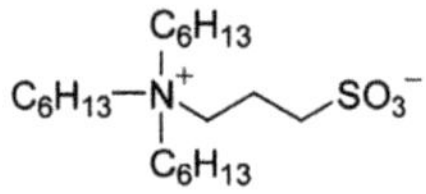	Water	Phase-separated	29
	Mixture of **27** and **28**	Water	UCST and LCST	29
	Mixture of **28** and HTfO	Water	LCST	26
29	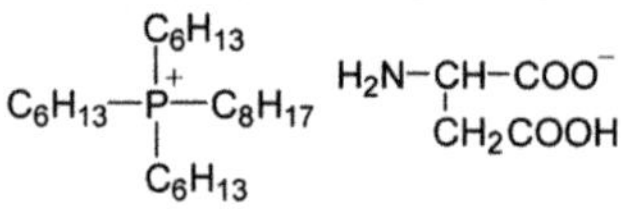	Water	Homogeneous	30
30	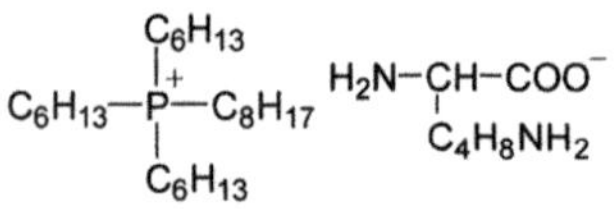	Water	Homogeneous	30
	Mixture of **29** and **30**	Water	LCST	30

This mixture was a homogeneous solution at 29 °C, and was phase-separated into an aqueous upper phase and an IL bottom phase above 34 °C. The separated phases were homogeneously mixed again below 29 °C. This LCST-type phase change suggests that other related ILs with various structures have the possibility to undergo such LCST-type phase changes.

In terms of cationic structures, alkylphosphonium cations were effective in preparing ILs showing the LCST-type phase transition in water (**7**, **12–21**).[1,17–23] The LCST-type phase change was also seen in an iron(III)-containing IL composed of the 1-butyl-3-methylimidazolium cation (**22**).[24] Of the anion species, various kinds of structure have been used to design ILs so far. Even highly hydrophilic anions (*e.g.*, halogenate, carboxylate, and phosphonate) can provide ILs with LCST behaviour if they are paired with adequately hydrophobic cations. It was also found that geometric isomerisation of anion structures offers significantly different phase behaviour. For example, an IL with the maleate anion (**15**) showed LCST behaviour, whereas its *trans*-isomer (**5**) underwent the inverse behaviour, *i.e.*, UCST.[10]

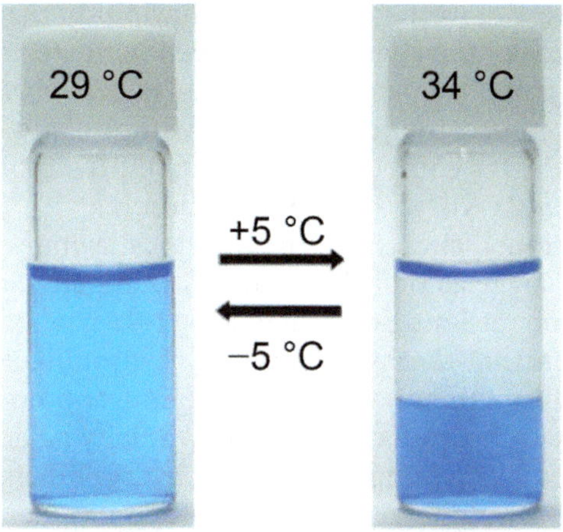

Figure 4.2 Temperature-sensitive LCST-type phase change of a **12**/water mixture. The IL phase was dyed with Coomassie Brilliant Blue. Reproduced from ref. 27 with permission from the PCCP Owner Societies.

Another interesting phase behaviour was observed in $[P_{4444}]^+$-based ILs containing benzoate derivatives as anions. While $[P_{4444}]$[Benzoate] was hydrophilic and freely miscible with water, a sufficiently hydrophobic IL with LCST behaviour was obtained after introduction of hydrophilic hydroxyl groups onto the *ortho*-position of the benzoate anion (**17**).[19] This is a very interesting result, indicating that the hydrophobicity of an IL can be increased by introducing a hydrophilic group onto the anion. Such striking phase behaviour should be derived from alternating the hydrated state of the anions *via* intra-/inter-molecular hydrogen bonding, or "proton-sharing" structures.[19] The LCST-type phase change was also found in phosphonium- or ammonium-type zwitterions (**23**, **24**),[25,26] in which both the cationic part and anionic part are covalently connected *via* an alkyl chain linker. An advanced property of aqueous zwitterion systems over IL-based ones is the ability to keep the ion pairs even in the presence of other ions. This will contribute to the desirable partition of ions in the media.

In the thermoresponsive IL-based mixed systems, the T_c value is totally dependent on the structure of the component ions. In the case of IL/water mixtures showing an LCST, the elevation of the T_c value was mainly seen in ILs composed of less hydrophobic ions. Upon comparison of a series of amino acid ILs (**7**), the T_c value decreased according to the extension of the alkyl chain length of the side groups on the amino acids. A similar trend was also observed in aqueous IL systems composed of benzenesulfonate derivatives as anions (**13**). Likewise, cationic structures also influence the T_c value, with an increase in the attached alkyl chains on the phosphonium cations lowering the T_c. In terms of the center atoms of cations, phosphonium-type ILs tend to show a lower T_c than ammonium-type ILs with the same alkyl chains. The added hydrophobicity/hydrophilicity effect of both cations and anions, therefore, governs the phase behaviour of the resulting ILs with water. The LCST-type phase change can be seen in the region between hydrophobic and hydrophilic ILs, when the phase behaviour of the IL/water mixtures was ordered in accordance with the hydrophilicity of the component ions. In other words, ILs composed of moderately hydrophilic ions have the potential to show an LCST-type phase transition.

The above-mentioned analysis of the phase behaviour of various IL/water mixtures strongly suggests that the total hydrophilicity/hydrophobicity balance of the component ions dominates the LCST-type phase change. Here we should mention that the hydrophobicity/hydrophilicity balance is generally unquantifiable with suitable physicochemical properties. To quantify the required range of hydrophobicity/hydrophilicity balance that allows for LCST behaviour, Ohno's group proposed the hydrophilicity index (HI) value, which represents the number of dissolved water molecules per ion pair in the phase separated IL-rich phase at 60 °C.[27] Figure 4.3 shows the T_c values of various ILs, including IL mixtures, as a function of their HI value. The T_c values of highly hydrophobic ILs that did not undergo an LCST-type phase change were tentatively set below 0 °C. As seen in Figure 4.3, there was a clear threshold value of HI after which ILs show the LCST-type phase change. When the HI of the ILs was less than 6, these ILs did not undergo any LCST response and were immiscible with water regardless of the temperature. The LCST-type phase transition took place when the HI value was 7 or more, and the T_c value showed higher values upon increasing HI. Figure 4.3 clearly shows that the number of water molecules per ion pair of 7 is critical to show the LCST-type phase change.[28]

In general, thermoresponsive IL/water mixtures can be obtained by careful design of the component ions to give an adequate hydrophobicity/

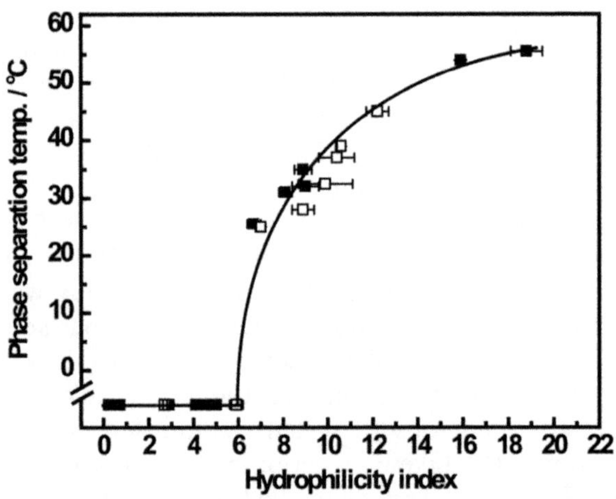

Figure 4.3 Relation between the phase separation temperature (T_c) and the hydrophilicity index (HI) value for various ILs (■) and IL mixtures (□). Reproduced from ref. 27 with permission from the PCCP Owner Societies.

hydrophilicity balance to result in LCST behaviour of the mixtures (*i.e.*, HI value higher than 7). On the other hand, mixing ILs was revealed to be another practical method to induce an LCST response in IL/water systems. More specifically, mixing relatively hydrophobic ILs was effective in making thermoresponsive ILs, providing that the prepared ILs are too hydrophilic to show an LCST-type phase transition. For example, when completely water-miscible **25** was mixed with water-immiscible **26**, the resulting mixtures showed LCST behaviour.[17] This result implies that mixing multiple ILs effectively influences the HI value to fulfill the requirement for LCST behaviour. This fine-tuning to convert the phase behaviour by mixing ILs was also observed in zwitterion/water mixtures, such as that composed of **27**/**28**[29] and **28**/trifluoromethanesulfonic acid (HTfO).[26]

Another interesting example for mixed IL systems is mixtures of amino acid-based ILs. Both **29** and **30** were miscible with water, but their mixtures started to show LCST behaviour at a suitable mixing ratio.[30] This unexpected behaviour was presumably derived from anion–anion interactions between the charged side chains on the amino acids. This presumption has recently been supported by MD simulations.[31] It should be noted here that not all ILs can acquire the LCST-type phase change by being mixed with suitable ILs. When a tributyl-*n*-octylphosphonium-based IL, $[P_{4448}][Tf_2N]$, was mixed with **25**, the resulting IL mixture behaved as a normal hydrophobic IL and did not show LCST behaviour.[17] This behaviour might be due to the partition of hydrophilic **25** between the aqueous phase and the $[P_{4448}][Tf_2N]$ phase, namely, **25** should be dissolved mainly in the aqueous phase.

4.3 Thermoresponsive Poly(ionic liquid)s with Switchable Hydrophobicity/Hydrophilicity

In addition to the design of *liquid-state* IL materials showing LCST behaviour, Ohno's group anticipated that it should not be difficult to fabricate *solid-state* IL-derived materials by suitable design of poly(ionic liquid)s (PILs), which also undergo this type of phase change after being mixed with water. PILs have recently been attracting much interest as a new subclass of solid-state polymeric materials. They are distinct from well-known polyelectrolytes because of their unique properties inherent to ILs. It was strongly suggested that several physico-chemical properties of the starting IL monomers, such as low glass transition temperature, high charge density, and inflammability, should be maintained even after polymerisation. Accordingly, it is assumed that when IL monomers show LCST-type phase behaviour in water, the resulting polymers should retain the LCST thermoresponsiveness. On the other hand, most polymer chemists might think that these thermoresponsive polymer materials are not new, since there are some examples of polymers showing LCST-type phase transition in an aqueous phase as mentioned above. We believe, however, that thermoresponsive and highly-charged polyelectrolytes, especially those derived from ILs, are unprecedented and expand the possibility of thermoresponsive materials in diverse fields upon considering the fascinating properties of the starting IL monomers.

Based on the HI value required for ILs to show an LCST-type phase transition as mentioned above, Kohno and Ohno started to design IL monomers that potentially exhibit LCST behaviour. As a preliminary study, ILs composed of $[P_{4444}]^+$ and benzenesulfonate derivatives were found to undergo an LCST-type phase transition after being mixed with water (**13**). An analogous styrenesulfonate ($[SS]^-$) was then chosen as a polymerisable anion candidate.[32,33] As counter cations, different phosphonium cations with different lengths of alkyl chain were used. These ILs were prepared by anion exchange reaction, and the as-prepared IL monomers were obtained as a white solid (**31**) or clear viscous liquids (**32, 33**). Next, the temperature-dependent phase behaviour of the IL monomers was investigated after adding an equal weight of water (50 wt%). Both **32** and **33** were scarcely soluble in water, and no homogeneous phase was obtained even after heating or cooling. In contrast, **31** formed a homogeneous mixture with water at low temperature. On gentle heating, the transparent solution became turbid and phase-separated, clearly indicating that **31** exhibited LCST behaviour. When **31** was mixed with an equal weight of water, the T_c value of the mixture was 34 °C. This value was almost the same as that of the analogous phosphonium-type IL (**13**) with two methyl groups on the benzenesulfonate anion.

The prepared IL monomers were then polymerised in water with 1.0 mol% of α,α′-azobis(isobutyronitrile) (AIBN) as a thermal polymerisation initiator to obtain poly(**31**) and poly(**32**).[32] Since both IL monomers were phase-separated at the temperature during polymerisation, the reaction proceeded with each monomer suspended in the mixture. Yuan and co-workers also prepared poly(**31**) using AIBN as a radical polymerisation initiator and ethanol as a solvent.[34] In this case, the monomer was homogeneously mixed in the solvent. The resulting poly(**31**) was mixed with pure water, and its thermal phase behaviour was visually confirmed as LCST behaviour. When an aqueous solution containing 10 wt% poly(**31**) was kept at 25 °C, the polyelectrolyte was homogeneously mixed, but it became turbid upon heating to 57 °C. The inhomogeneous mixture turned to a clear solution again upon cooling. The IL monomer **31**, therefore, retained its thermoresponsiveness even after polymerisation. This type of thermal response was first seen in PIL-based homopolymer/water mixtures. A large number of non-ionic polymers have already been reported to show an LCST-type phase transition with water and other organic solvents. However, a wholly-charged homopolyelectrolyte that reversibly shows an LCST-type phase transition with water has not been reported, especially, one which is derived from IL monomers with a unique set of the physico-chemical properties of ILs. To prepare polyelectrolytes showing an LCST in water, copolymerisation of a thermoresponsive non-ionic monomer with a charged co-monomer has frequently been used. These studies were previously summarised by us.[3]

The phase behaviour of an aqueous solution containing 10 wt% poly(**31**) was investigated by turbidity measurement.[32] On heating, the transmittance of the poly(**31**)/water mixture dropped to 90% at 57 °C (Figure 4.4, blue solid line). The transmittance of the solution sharply fell upon further heating, and reached less than 1% at 59 °C. The turbid solution became transparent again upon cooling (Figure 4.4, blue dotted line). As shown in Figure 4.4, there is a very small hysteresis in the solution turbidity. The reversible and sharp LCST behaviour of poly(**31**) has also been reported by other groups.[34,35] Furthermore, the T_c value of poly(**31**) could be readily decreased by copolymerising **31** with a relatively hydrophobic monomer such as **32** (Figure 4.4, red line). When 30 mol% of **32** was introduced into the copolymer, the resulting poly($\mathbf{31}_{0.7}$-*co*-$\mathbf{32}_{0.3}$)-type copolymerised PIL exhibited a T_c value approximately 20 °C lower than that for homopoly(**31**) at the same concentration. The shift in the T_c value clearly reflects the lowering of the total hydrophilicity of the component ions, which was consistent with the properties of thermoresponsive ILs (*i.e.*, mixing hydrophobic ILs lowers the T_c value). Clearly, there is a close relationship between **31** and poly(**31**) in terms of the thermoresponsiveness. Based on these studies mentioned above, it is strongly suggested that monomeric ILs showing an LCST-type phase change have the potential to maintain their thermoresponsiveness in the polymeric state. In addition, the copolymerisation of IL monomers is a facile and accurate potential strategy to control the T_c of polyelectrolytes in an aqueous solution. In the case of the famous poly(*N*-isopropylacrylamide) (PNIPAM) and other non-ionic polymers, copolymerisation with other monomers with different hydrophobicity was also found to be effective in controlling the phase behaviour. However, the sensitivity to temperature of the resulting copolymers often became sluggish. To maintain the thermal sensitivity, efforts have frequently been made to synthesise copolymers with very narrow molecular weight distribution by living or controlled radical polymerisation. On the other hand, the reported poly($\mathbf{31}_{0.7}$-*co*-$\mathbf{32}_{0.3}$)-type copolymer prepared by simple radical polymerisation

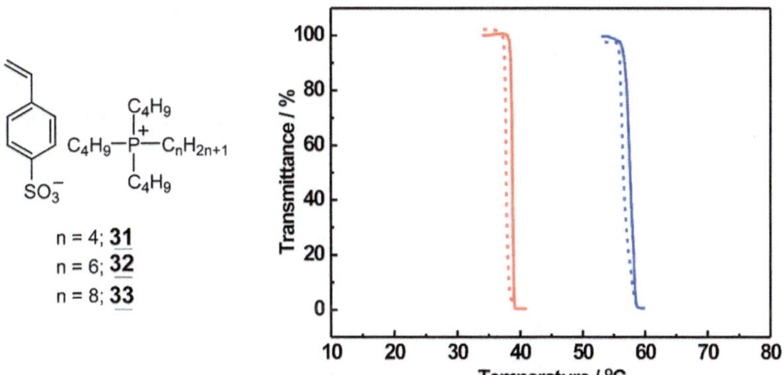

Figure 4.4 Left: structure of tetra-*n*-alkylphosphonium styrenesulfonate-type IL monomers (**31**-**33**); right: change in transmittance of poly(**31**) (blue line) and poly($\mathbf{31}_{0.7}$-*co*-$\mathbf{32}_{0.3}$) (red line) in water. Solid and dotted lines denote heating and cooling steps, respectively. Reproduced from ref. 2 with permission from the Royal Society of Chemistry.

displayed a very sharp phase transition. This highly temperature-sensitive and reversible phase change of the PILs is one of the advantages for practical applications.

The aforementioned studies on the phase behaviour of monomeric and polymeric **31** suggest that designing IL monomers with an LCST is a powerful method to obtain thermoresponsive polyelectrolytes showing the same phase behaviour in water. Soon after the discovery of poly(**31**), several monomeric ILs and PILs were designed and their temperature-responsive phase behaviour was investigated to seek thermoresponsive polyelectrolytes with various structures. We have summarised both the structure and phase behaviour of these IL monomers in Figure 4.5 and Table 4.2. As described above, an anionic IL monomer (**31**) underwent an LCST-type phase change in both monomeric and polymeric states, whereas **32** and **33** did not show LCST behaviour. Ohno and co-workers also reported other anionic 3-sulfopropyl-methacrylate ($[MC3S]^-$)-type IL monomers with different phosphonium cations (**34–36**).[36,37] The IL monomer **34** was miscible with water regardless of the temperature. On the other hand, both **35** and **36** showed LCST behaviour in the monomeric state. Upon comparing the phase behaviour of **31** and **34**, which contain the same cation, $[SS]^-$ is more hydrophobic than $[MC3S]^-$ in terms of total hydrophilicity of the component ions. These monomeric ILs were then polymerised by the same method to obtain poly(**31**), and the phase behaviour of the thus-polymerised PILs was studied. While poly(**34**) gave a homogeneous solution and poly(**36**) was phase-separated in water within the whole temperature range analysed, poly(**35**) was found to show an LCST-type phase change. Diamond's group reported a similar IL monomer, **37**, with the 3-sulfopropyl-acrylate anion, which showed LCST behaviour after polymerisation.[38] Although the phase behaviour of the monomer was not studied in detail, its T_c value should be higher than that for the analogous **35**. This is because the elimination of only one methyl group on the component ions raises the hydrophilicity of the corresponding ILs, resulting in an increase in the T_c value. Furthermore, a Gemini-type IL monomer (**38**)[39] and phosphonium/ammonium-type cationic IL monomers with alkylsulfonate anions (**39–44**)[40,41] were also reported, in which a handful of the corresponding

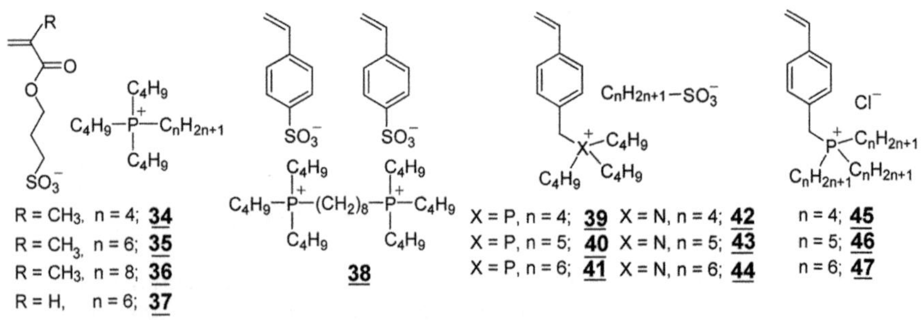

Figure 4.5 Structure of IL monomers prepared to seek thermoresponsive PILs showing LCST behaviour in water.

polyelectrolytes were found to show LCST behaviour. Highly hydrophilic chloride anion-based IL monomers also gave thermoresponsive polyelectrolytes when paired with adequately hydrophobic and polymerisable cations such as **45**, **46**, and **47**.[42] Poly(**46**) exhibited LCST behaviour in both pure water and aqueous sodium chloride solution. Temperature-dependent turbidity measurement is frequently used to determine the T_c values of PIL/water mixtures. In the case of PIL/water mixtures showing LCST behaviour, the total hydrophilicity of the component ions of the IL monomers dominantly influences the T_c value. This effect on the T_c value was more remarkably seen in PILs as compared to non-polymerisable IL/water mixtures. For instance, upon comparing ammonium-based poly(**43**) and phosphonium-based poly(**40**) with the same alkyl chains, poly(**43**) had a T_c value of 64 °C, which was 22 °C higher than that of poly(**40**).[41] This clearly shows that the hydrophobicity of the cation can also be changed by the center atom species. In addition, the elongation of one methyl group onto the anion of poly(**43**) significantly lowered the T_c value to 23 °C.

Table 4.2 clearly shows that most PILs inherit the thermoresponsive feature of the corresponding monomers. On the other hand, a few IL monomers showed different phase behaviour, and some of them did not show a thermal response after polymerisation. For example, IL monomer **36** showed LCST behaviour, but its polymer became more hydrophobic and was always phase-separated with water. These differences clearly showed that the hydrophobicity/hydrophilicity balance (or hydrated state) was somewhat susceptible to the molecular weight of the PILs. The T_c value also changed between the IL monomer and PIL. Ohno and co-workers reported that the T_c values of **31** and poly(**31**) at the same concentration (10 wt% in water) were 40 and 57 °C, respectively.[32]

Table 4.2 Phase behaviour of monomeric and polymerised ILs in water.

Entry	Phase behaviour (monomeric state)	Phase behaviour (polymeric state)	Entry	Phase behaviour (monomeric state)	Phase behaviour (polymeric state)
31	LCST	LCST	**40**	LCST	LCST
32	Phase-separated	Phase-separated	**41**	Phase-separated	Phase-separated
33	Phase-separated	Phase-separated	**42**	–[a]	Homogeneous
34	Homogeneous	Homogeneous	**43**	–[a]	LCST
35	LCST	LCST	**44**	–[a]	LCST
36	LCST	Phase-separated	**45**	LCST[b]	Homogeneous
37	–[a]	LCST	**46**	LCST	LCST
38	LCST	LCST	**47**	Phase-eparated	Phase-separated
39	LCST	Homogeneous			

[a]–: Data not shown.
[b]This behaviour was observed in aq. NaCl solution.

The same group also reported that the T_c values of monomeric and polymerised **35** were 44 and 40 °C (10 wt% in water).[36] Regarding such differences in the T_c values, valuable studies were performed by Aoshima and co-workers, who investigated the effect of number-average molecular weight on the UCST of imidazolium-based PILs with narrow molecular weight distributions.[43] The data demonstrated that the effect of molecular weight on the T_c values was small even in the low molecular weight region. The variation in the aqueous phase behaviour of monomeric/polymeric ILs should, therefore, depend on the monomer species. Further studies should be undertaken to analyse the effect of the molecular weight of various PILs on their phase behaviour with water.

The structure–property relationships of the as-prepared IL monomers and their polymers indicate that careful design of the component ions of the IL monomers is necessary to gain thermoresponsive PILs that exhibit the required hydrophilicity that allows for an LCST-type phase change. When component anions and cations are too hydrophilic (or too hydrophobic), the resulting PILs exhibit a stable homogeneous phase (or are phase-separated) regardless of the temperature. Instead, Ohno and co-workers reported a simple "blending" method to tune the LCST behaviour of PILs.[44] The advantage of this blending method is that it allows the acquisition of PILs whose LCST can be controlled by blending different components, while maintaining their high sensitivity toward temperature change. When thermoresponsive poly(**35**) was blended with hydrophobic poly(**36**) in the presence of water, the T_c value was lowered depending on the blending ratio of poly(**36**). The temperature-dependent turbidity measurement of the blended PILs demonstrated that the sharp phase transition was maintained even after blending these PILs, confirming its temperature-sensitive and reversible nature. The results mentioned here imply that the total hydrophilicity of blended PILs governs the phase behaviour, as already described in the case of non-polymerisable IL/water mixtures. It is therefore anticipated that LCST behaviour can also be realised by suitably blending hydrophobic and hydrophilic PILs, both of which do not singularly undergo an LCST-type phase change. To support this, hydrophobic poly(**36**) was blended with hydrophilic poly(**34**) with differing molar ratios, and their phase behaviour was analysed. It was observed that a few blended PILs with a certain blending ratio indeed underwent an LCST-type phase change in water. The T_c value was observed to be 45 °C, at which the mole fraction of hydrophobic poly(**36**) was 0.4, which was 14 °C higher than that of the poly(**35**)/poly(**36**) blend at the same mole fraction. This clearly suggests that the T_c value is a function of the total hydrophilicity of the ion species, and decreases as the fraction of hydrophobic cations increases. It should be remarked that the fine tuning of the LCST behaviour by blending polymerised ILs was also observed not only in PILs with alkylphosphonium cations, but also in PIL blends with highly hydrophilic, 1-ethyl-3-methylimidazolium ($[C_2mim]^+$)-based PILs. When an equimolar amount of poly($[C_2mim]$[MC3S]) was blended with poly(**36**), the resulting PIL blend exhibited LCST behaviour, and showed a T_c value of 64 °C. This high value of T_c reflects the hydrophilic nature of the $[C_2mim]^+$ cation. This result is of significant importance, since, given the structural diversity of *N*-alkylimidazolium-based ILs, it envisions a role of thermoresponsive PILs in practical applications.

Other factors influencing the T_c value include the concentration of the polymers and ionic strength. Even though the T_c values of some non-ionic polymers are almost unchanged by polymer concentration, thermoresponsive PILs have a wide range of T_c values depending on the polymer concentration. The reported LCST-type PIL/water mixtures showed lower T_c values upon increasing the polymer concentration. The opposite behaviour was observed in PIL/water mixtures showing UCST behaviour. In highly diluted solutions, some thermoresponsive PILs no longer showed an optically-detectable phase transition. The addition of suitable salts was also proved to be effective in controlling their T_c values. Yuan and co-workers reported that the T_c value of poly(**31**) increased dramatically upon adding KBr as a salt, but decreased following the addition of $[P_{4444}]$Br into the aqueous solution.[34] Ohno and co-workers reported that the T_c value of poly(**35**) became lower than that with pure water when phosphate buffer containing KH_2PO_4–K_2HPO_4 (pH = 7.0) was used instead of pure water.[36] The same trend was seen in the poly(**46**)/aqueous NaCl solution system.[42] Various factors should influence the T_c value after addition of the salts. These include the kosmotropicity of the salts, the fraction of the salts in the aqueous phase and separated polymer phase, and ion exchange reactions between the added salts and the component ions of the PILs.

A detailed analysis to comprehend the mechanism of the thermoresponsive behaviour is of great importance from a fundamental point of view. As several thermoresponsive ILs and poly(IL)s are now readily prepared in relatively large quantities, a handful of studies have been reported to analyse the LCST behaviour *via* spectroscopic and optical measurements. FT-IR spectroscopy was employed by Wu and co-workers to reveal the dynamic phase transition of both monomeric and polymerised ILs.[35,45,46] A few typical vibration modes for monomeric **31**/water mixtures (*e.g.*, the CH group of $[P_{4444}]^+$ and SO_3^- group of $[SS]^-$) suggested that the $[P_{4444}]^+$ moiety was aggregated and accordingly the hydrogen bonds between the water molecules and $[SS]^-$ were broken upon heating up to nearly the T_c value. These synergetic variations of both the $[P_{4444}]^+$ cation and $[SS]^-$ anion resulted in a randomly aggregated structure of globules during heating. In the case of poly(**31**)/water mixtures, however, the SO_3^- group was wrapped inside the aggregated globules, and the globules were surrounded by $[P_{4444}]^+$ cations. A repeating unit of $[SS]^-$ anions in poly(**32**), therefore, played a key role in the formation of aggregated globules. The morphological change of thermoresponsive PILs was reported by Ohno and co-workers using poly(**40**).[41] Combined with transmission electron microscopy (TEM), dynamic light scattering (DLS) and optical microscopy, poly(**40**) generated poly-dispersed nanospheres in water at a temperature below the T_c value, merging into unimodal microspheres upon heating above the T_c value. Gao and co-workers also studied the morphological change of non-polymerisable **12** with water by freeze-fracture transmission electron microscopy (FF-TEM).[47] From the FF-TEM measurements, **12** formed long-living spherical aggregates in water below the T_c value. The size of the aggregates was enlarged upon heating, and a remarkable increase in their size was seen near the T_c. The morphological transition of charged spherical aggregates in water at and around the T_c value should be a feature of both ILs and PILs. On the other hand, Nishikawa and co-workers proposed fuzzy and misty structures of **12**/water mixtures rather than distinct structures (*e.g.*, emulsions and micelles), as evidenced by small angle X-ray scattering experiments.[48] They pointed out that **12** was not completely separated from water even at the T_c value. As for non-polymerisable IL/water mixtures, increasing the temperature above the T_c value resulted in obvious liquid/liquid phase separation between the IL-rich phases and water-rich phase. Hydrated gel-like phases were instead seen in a few PILs upon heating the mixture far above the T_c. Such analytical studies will be further required to fully characterise the thermoresponsive phase behaviour of this new type of IL-based materials.

Since a handful of PIL linear polymers have shown LCST behaviour in water, various thermoresponsive solid-state materials can be fabricated from the above-mentioned IL monomers upon suitable choice of cross-linking methods. A few types of cross-linked thermoresponsive PIL materials have recently been reported. These include hydrogels,[37–39,42] membranes,[33] and microgels.[49] PIL-based hydrogels composed of **31**, **35**, **37**, **38**, and **46** were first prepared using different types of cross-linkers and initiators. Diamond and co-workers prepared PIL hydrogels derived from **31** and **37** *via* radical photo-polymerisation of these monomers using 2-hydroxy-2-methylpropiophenone as a radical photo-initiator.[38] Relatively high-molecular-weight cross-linkers, *e.g.*, poly(propylene glycol) diacrylate with an M_n of 800 and poly(ethylene glycol) diacrylate with an M_n of 700, were introduced to obtain stable and transparent hydrogels. Ohno and co-workers also prepared PIL hydrogels using both **35** and **36** as IL monomers and 1,14-tetradecanediol dimethacrylate as a cross-linker.[37] In this case, polymerisation of the monomers was undertaken at 80 °C using AIBN as a radical initiator. Yuan and co-workers reported Gemini-type dicationic phosphonium-based hydrogels made from **38**.[39] The hydrogel could be prepared by using divinylbenzene as a cross-linker, and AIBN as a radical initiator. These PIL hydrogels were revealed to maintain the LCST-type thermoresponsiveness; they shrunk and desorbed water with increasing temperature, while absorbing water with decreasing temperature. The T_c values of the thermoresponsive PIL hydrogels strongly depended on the structure of both the IL monomers and cross-linkers. Despite the very sharp phase transition of PIL linear polymers as evidenced by turbidity measurement and DSC measurement, the PIL hydrogels displayed a significant broadening in terms of the thermoresponsiveness. This might be due to the heterogeneity of the cross-linking density in the hydrogels and/or the decreased level of freedom of the component ions. In addition, some of the reported PIL hydrogels were fragile, and their mechanical strength was not sufficient to undergo reversible water absorption/desorption cycles.

Ohno and co-workers investigated the necessary conditions to prepare PIL hydrogels with sufficient mechanical properties to reversibly absorb/desorb water *via* LCST-type phase transition.[50] Through careful selection of both the initiators and cross-linkers, as well as the reaction procedures, a few mechanically-stable thermoresponsive PIL hydrogels were successfully prepared. One of the stable PIL hydrogels was prepared from an equimolar mixture of **35** and **36**, poly(ethylene glycol) dimethacrylate as a cross-linker, and 2,2′-azobis[2-(2-imidazolin-2-yl)propane] disulfate dehydrate as a polymerisation initiator. A certain amount of water was added to the mixture to completely homogenise them, and thermal polymerisation was carried out at 50 °C. Figure 4.6 shows a photograph of the as-prepared PIL hydrogel exhibiting thermoresponsive water absorption/desorption behaviour. The water-absorbed hydrogel was visually transparent at 25 °C (Figure 4.6, left), and it became turbid at the T_c (30 °C, as measured by DSC). A further increase in temperature up to 50 °C induced desorption of water from the PIL gel, and accordingly the gel started to shrink (Figure 4.6, right). The PIL hydrogel was then tested in a water absorption/desorption cycle test by soaking the gel in pure water at 5 °C for 2.5 h, and subsequently placing it at 50 °C for 30 min. This water absorption/desorption process was repeatedly carried out ten times, and it was found that this PIL hydrogel was stable without generation of any cracks, even after repeated cycles.

In addition to the design of PIL hydrogels, a new type of PIL membrane (PILM) showing LCST-type phase transition has recently been reported.[33] These PILMs were prepared from two IL monomers, **31** and an analogous but relatively hydrophobic IL monomer comprising a phosphonium cation with a longer alkyl chain (**33**). Free-standing, transparent, and flexible PILMs with a thickness of 150 μm were fabricated by radical photo-polymerisation of the two IL monomers in the presence of poly(ethylene glycol) diacrylate with an M_n of 700 as a cross-linker and 2-hydroxy-4′-(2-hydroxyethoxy)-2-methylpropiophenone as a radical photo-initiator (Figure 4.7, left). The water content dissolved in the PILMs increased with increasing molar fraction of the less hydrophobic **31** in the PIL, and the resulting poly($\mathbf{31}_{0.3}$-*co*-$\mathbf{33}_{0.7}$)-type PILM **1**

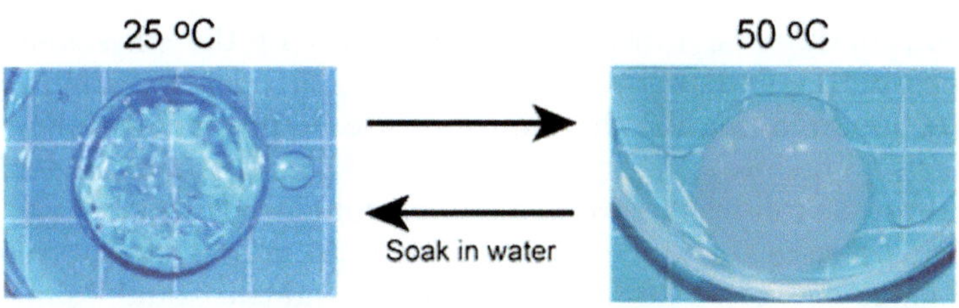

Figure 4.6 Thermoresponsive and reversible water absorption/desorption behaviour of a suitably designed PIL hydrogel composed of **35** and **36**. Reproduced from ref. 50 with permission from the Royal Society of Chemistry.

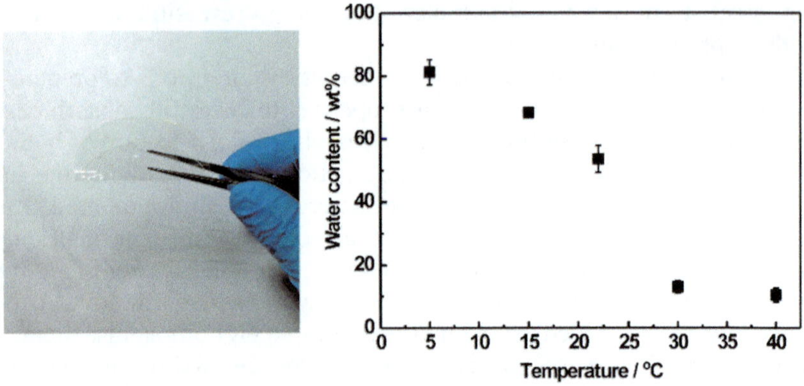

Figure 4.7 Left: photograph of the as-prepared PILM composed of **31** and **33**; right: temperature-dependent water content of PILM **1**. Reproduced from ref. 33 with permission from the Royal Society of Chemistry.

(*i.e.*, the molar fraction of **31** to **33** was 0.3) exhibited LCST behaviour as expected. The water content of PILM **1** fell exponentially with an increase in the temperature from 5 to 40 °C (Figure 4.7, right). Even a slight temperature change from 22 to 30 °C prompted a decrease in the water content from 54 to 14 wt%. This change in water content clearly indicates that PILM **1** is able to control the hydrated state *via* LCST behaviour. In the case of thermoresponsive PIL hydrogels as shown in Figure 4.6, the saturated water content was over 99 wt% and the PIL hydrogel desorbed water at higher temperatures, while a large amount of water remained in the hydrogel. This behaviour should be exploitable when treating a large amount of water. Very distinct from these PIL hydrogels, PILM **1** is able to control the hydrated state in a very narrow temperature range. This feature will enable affinity control of various molecules dissolved in water *via* LCST behaviour.

4.4 Potential Applications of Thermoresponsive Ionic Liquid-based Materials

Upon considering the intriguing combination of the physico-chemical properties of ILs and thermoresponsiveness, broad applications can be conceived by exploiting the thermoresponsive liquid-state or solid-state IL derivatives. Some research groups have already demonstrated the effectiveness of these thermoresponsive IL materials for specific applications. In the case of liquid-state IL/water mixtures with LCST responses, the reversible phase change between a homogeneous phase and two liquid/liquid phases can be controlled in a very narrow temperature range at around physiological temperature. This is particularly beneficial when it comes to extraction and separation of thermally unstable bio(macro)molecules, including proteins. Ohno's group first utilised the LCST-type phase transition of a

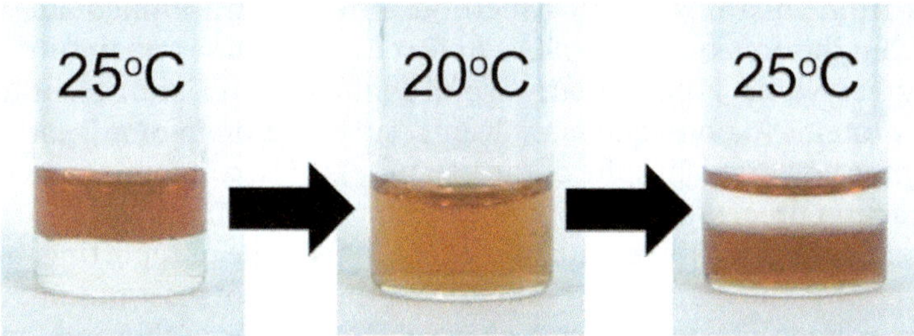

Figure 4.8 Extraction of Cyt. *c* from the aqueous upper phase to the IL bottom phase *via* the LCST-type phase change. Reproduced from ref. 2 with permission from the Royal Society of Chemistry.

$[P_{4444}]$[Tf-Leu]-type IL (7) for the extraction of water-soluble proteins.[51] An aqueous solution containing cytochrome *c* (Cyt. *c*), a typical water-soluble heme protein, was added to the IL at 25 °C (Figure 4.8, left). Once the solution was cooled down to 20 °C with gentle shaking, the separated phases immediately merged into a homogeneous solution (Figure 4.8, center). The phase separation was induced again by warming the mixture to 25 °C and incubating it until each phase became clear. As shown in Figure 4.8, right, Cyt. *c* was transferred from the aqueous upper phase to the IL bottom phase during this phase change. According to UV-visible spectroscopy, less than 0.1% of Cyt. *c* remained in the aqueous phase after the phase separation. The red colour at the top of the aqueous phase is merely a reflection of the liquid/liquid interface in the photo. Resonance Raman spectroscopic analysis confirmed that both the coordination state and spin state of heme were maintained even when dissolved in the IL. The distribution coefficient (*D*) value of several proteins between aqueous and IL phases was then measured spectroscopically. Typical heme proteins such as hemoglobin (Hb), myoglobin (Mb), and horse radish peroxidase (HRP) were used as well as several other proteins such as lysozyme (Lyz), chymotrypsin (Cht), laccase (Lac), and bovine serum albumin (BSA). For these proteins, the *D* value varied widely depending on the protein species. Lyz, Cyt. *c*, Cht, Hb, and BSA were extracted into the IL phase, whereas Lac and HRP remained in the aqueous phase even after the LCST-type phase change. Based on the difference in *D* value, selective extraction of proteins from protein mixtures was also carried out *via* an LCST-type phase change. As an example, an aqueous solution containing Cyt. *c* and HRP, which respectively have the highest and lowest *D* value, was mixed with the IL, and the LCST-type phase change was induced by changing the temperature. After the phase separation, SDS-PAGE analysis was employed to confirm the dissolved proteins in each phase. From SDS-PAGE, both the aqueous and IL phases exhibited a single band, but in different positions. Namely, only Cyt. *c* was dissolved in the IL phase, whereas HRP remained in the aqueous phase. This result clearly demonstrates the utility of thermoresponsive IL/water mixtures for the selective extraction of water-soluble proteins from mixtures. In this method, proteins are immediately transported from the aqueous to IL phase upon changing the temperature by only a few degrees, which is particularly useful compared to existing techniques to dissolve proteins in non-aqueous solvents, such as protein modification with amphiphilic polymers. In spite of the usefulness of IL/water mixtures for protein extraction processes, the concrete mechanism promoting the extraction of water-soluble proteins remained unclear at the time of reporting.

To determine the main factor influencing the protein distribution, Ohno's group examined the basic physico-chemical properties of thermoresponsive ILs after mixing with water. Based on the HI value of thermoresponsive ILs, certain amounts of water molecules were found in the separated IL phase even after the phase separation (*i.e.*, more than seven water molecules per ion pair). In turn, a small amount of the ILs should also be dissolved in the aqueous phase. To determine the concentration of ILs in both phases, a trimethylbenzenesulfonate ($[TMBS]^-$)-type thermoresponsive IL with the $[P_{4444}]^+$ cation (**13**) was used, and the temperature dependence of the concentration of the IL (C_{IL}/mol kg^{-1}) in both the aqueous and IL phases was determined from the absorbance of the $[TMBS]^-$ anion.[52] When $[P_{4444}]$[TMBS] was mixed with water to reach the C_{IL} of 0.80 mol kg^{-1}, the phase separation occurred at 31 °C. After storing the solution at 35 °C to obtain clear biphases, the C_{IL} value in each phase was measured. As expected, a relatively small amount of IL was detected in the upper aqueous phase, whose C_{IL} value was calculated to be 0.24 (mol kg^{-1}). The C_{IL} in the IL phase was 1.50 mol kg^{-1}, confirming that the IL bottom phase was of course composed mainly of the IL, but also contained water. The thermoresponsive $[P_{4444}]$[TMBS]/water mixture was, therefore, phase-separated into two phases with considerably different IL concentrations. It is highly possible that such difference in the C_{IL} value between the aqueous phase and the IL phase governs the solubility of proteins in both phases. Figure 4.9 shows the solubility of Cyt. *c* as a function of C_{IL}. Cyt. *c* dissolved fully in the mixture when C_{IL} was from 0 to 0.10 mol kg^{-1}; this solubility behaviour at lower IL concentrations can be comprehended as "salting-in". On the other hand, the "salting-out" profile was observed upon increasing the C_{IL}, where the solubility of Cyt. *c* dropped dramatically to less than 10% slightly above the C_{IL} of 0.10 mol kg^{-1}. Surprisingly, the salting-in phenomenon was observed again upon further increasing C_{IL} value. Cyt. *c* began to dissolve again in the mixture at a C_{IL} of 0.20 mol kg^{-1}, and finally became completely soluble when the C_{IL} was over 1.00 mol kg^{-1}. Upon comparing the data in Figure 4.9 with the C_{IL} in both phases, it was found that Cyt. *c* was negligibly soluble at the C_{IL} in the aqueous upper phase (Figure 4.9, **W**). On the other hand, the solubility of Cyt. *c* was significantly high at the C_{IL} in the IL bottom phase (Figure 4.9, **IL**). Given the correlation between C_{IL} and the solubility of Cyt. *c*, it is obvious that Cyt. *c* dissolves selectively in the IL phase containing a small amount of water after the phase change.

The extraction process for Cyt. *c* was then carried out using $[P_{4444}]$[TMBS] *via* the same method as in Figure 4.8, and Cyt. *c* was clearly transferred into the IL phase *via* LCST behaviour. The solubility of HRP in the $[P_{4444}]$[TMBS]/water

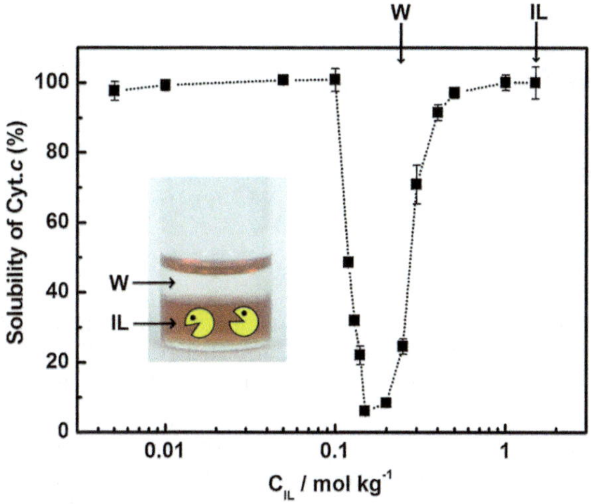

Figure 4.9 Effect of C_{IL} on the solubility of Cyt. *c*. W and IL respectively denote the C_{IL} of the separated aqueous phase, and the separated IL phase. Reproduced from ref. 52 with permission from CSIRO and the Australian Journal of Chemistry.

mixture was also determined, and only a salting-out profile was seen in the solubility–C_{IL} data. The solubility of HRP started to decrease at a C_{IL} of 1.00 mol kg^{-1}. HRP eventually remained in the aqueous phase, by considering the difference in the solubility of HRP in the aqueous phase and that in the IL phase. In general, the salting-out phenomenon of proteins occurs in concentrated aqueous salt solutions, and salting-in is observed in only dilute aqueous salt solutions. However, the salting-out profile of Cyt. *c* in the [P_{4444}][TMBS]/water mixture was found at a relatively low C_{IL}, followed by the salting-in of Cyt. *c*. These results suggest that both the salting-out and salting-in profiles of proteins depending on C_{IL} govern the distribution of proteins between the aqueous and IL phase. Ohno's group previously reported the enhanced solubility and stability of proteins in ILs containing a small amount of water, classified as hydrated ILs.[53] The unusual solubility of Cyt. *c* in condensed IL phases after phase separation with water should be due to the hydrated state of the separated ILs.

As liquid-state materials, thermoresponsive IL/water mixtures have been used for extraction of various compounds such as proteins[51,52,54] and metal ions.[8,9,11,21,23] These studies also inspired the potential application of thermoresponsive PILs as solid-state materials. One example of a way to apply such solid-state materials is for protein condensation processes. As shown in Figure 4.10, a chemically cross-linked PILM **1** (*vide supra*) was revealed to exhibit thermoresponsive and selective concentration of some water-soluble proteins without a significant loss of their higher-order structure.[33] The concentration amount of Cyt. *c* and MB exhibited high values, and increased when slightly increasing the temperature. On the other hand, HRP remained in the

Figure 4.10 Photographs of PILM **1** after immersion in aqueous solutions containing Cyt. *c* (left), Mb (middle), and HRP (right). Reproduced from ref. 33 with permission from the Royal Society of Chemistry.

aqueous phase regardless of the temperature change. Given the knowledge from liquid-state IL studies as mentioned above, the excellent selectivity in concentration amount for Cyt. *c* and HRP should result from the salting-out and salting-in profiles of the proteins in hydrated **PILM 1**. The new protein concentration process using the thermoresponsive behaviour of **PILM 1** has obvious benefits compared to existing polymer materials upon considering the selectivity based on the salting-out and salting-in profiles, as well as the unique set of properties derived from ILs. Further studies including detailed analysis of the hydrated states of PILMs with different component ions, and the correlation between the hydrated states and concentration capability of many proteins with different properties (*e.g.*, molecular weight, isoelectric point, and hydrophobicity) will offer various possibilities in bio-engineering fields.

In addition to the applications for extraction/separation of proteins, Ohno and co-workers proposed the utilisation of thermoresponsive PIL hydrogels for water purification, including desalination.[3,37] Considering the high charge density of PILs, there should be less chance for other ions to dissolve in the PIL phase from the aqueous phase. Accordingly, the thermoresponsive PIL-based hydrogels could be applicable for desalination processes where they reversibly absorb/desorb water molecules from salt water in response to a slight temperature change (Figure 4.11). In terms of applications for desalination, a handful of research groups started to investigate thermoresponsive PIL hydrogels as "smart" draw agents for forward osmosis (FO) processes. In FO processes, concentrated thermoresponsive materials are embedded onto semi-permeable membranes, and they absorb water from aqueous salt solutions driven through the membranes by osmotic pressure. After water absorption, the resulting diluted thermoresponsive materials are heated and undergo an LCST-type phase transition, promoting the desorption of water. In this context, thermoresponsive PIL-based materials are expected to act as alternatives to these thermoresponsive materials due to their highly charged characteristics and good thermal responses. Very recently, Hu and co-workers[55] and Liu and co-workers[56] individually reported FO desalination processes using thermoresponsive PIL hydrogels as smart draw agents. These pioneering studies highlight the potential of these PILs. However, we

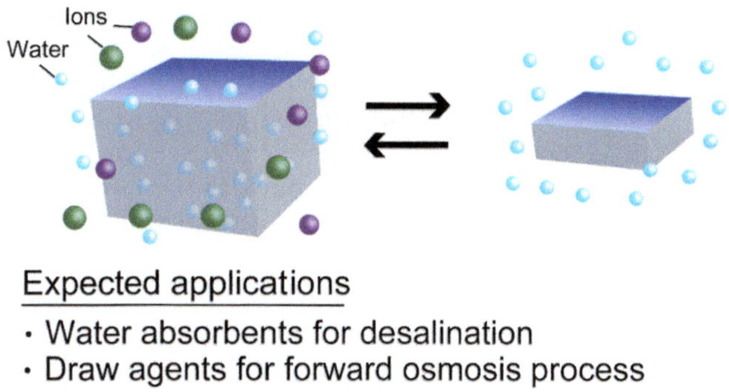

Figure 4.11 Potential applications of thermoresponsive PIL hydrogels as "smart" water absorbers. Reproduced from ref. 3 with permission from the Royal Society of Chemistry.

believe that many fundamental studies are still required for them to progress in diverse industrial processes. These include the effect of added salt species, salt concentration, and other organic/inorganic compounds on the phase behaviour and hydrated state of the thermoresponsive PIL hydrogels. The structure of the starting IL monomers should also influence the salt rejection performance. These factors are still elusive and have not been fully characterised so far. Recent studies by us demonstrated that the salting-out effect from adding highly kosmotropic NaCl dominantly dehydrates the phosphonium chloride-based PIL hydrogels derived from **46**, but the poor water absorption ability was easily improved by copolymerising the IL monomers with suitably hydrophilic IL monomer **45**, which is indeed a very valuable and practical strategy for designing thermoresponsive PIL hydrogels for desalination applications.[42]

4.5 Conclusion

Thermoresponsive IL-based materials with switchable hydrophobicity and hydrophilicity were thoroughly summarised. This is a smart way to control the hydrophobicity and hydrophilicity of ILs without disturbing the system. Since our systematic studies reveal that suitable hydrophobicity is a key parameter to obtain both liquid-state and solid-state IL materials that reversibly undergo LCST-type phase changes, introducing thermoresponsiveness into various ILs and PILs should be possible upon suitable design of the total hydrophilicity of the component ions. As a potential application, the extraction of target compounds such as proteins in the aqueous phase from the IL phase was successfully carried out. In addition, recent investigations demonstrated the potential of PIL-based solid-state materials for desalination applications. Now, switchable hydrophobicity/hydrophilicity can be regarded as one of the fascinating properties of ILs, combined with the well-known physico-chemical properties of classical ILs.

Acknowledgements

This work was mainly supported by a Grant-in-Aid for Scientific Research from the Japan Society for the Promotion of Science (KAKENHI, No. 26248049 (H. O.) and No. 16K17954 (Y. K.)). The authors also acknowledge all students in our lab involved in studies on thermoresponsive IL materials for their considerable efforts.

References

1. K. Fukumoto and H. Ohno, *Angew. Chem., Int. Ed.*, 2007, **46**, 1852.
2. Y. Kohno and H. Ohno, *Chem. Commun.*, 2012, **48**, 7119.
3. Y. Kohno, S. Saita, Y. Men, J. Yuan and H. Ohno, *Polym. Chem.*, 2015, **6**, 2163.
4. P. Jessop, S. M. Mercer and D. J. Heldebrant, *Energy Environ. Sci.*, 2012, **5**, 7240.
5. J. E. L. Dullius, P. A. Z. Suarez, S. Einloft, R. F. de Souza and J. Dupont, *Organometallics*, 1998, **17**, 815.
6. C. A. Cerdeiriña, J. Troncoso, C. P. Ramos, L. Romani, V. Najdanovic-Visak, H. J. R. Guedes, J. M. S. S. Esperança, Z. P. Visak, M. Nunes da Ponte and L. P. N. Rebelo, in *Ionic Liquids III A: Fundamentals, Progress, Challenges, and Opportunities*, ed. R. Rogers and K. R. Seddon, ACS Symp. Ser., 2005, ch. 13, vol. 901, p. 175.
7. P. J. Dyson, D. J. Ellis and T. Welton, *Can. J. Chem.*, 2001, **79**, 705.
8. P. Nockemann, B. Thijs, S. Pittois, J. Thoen, C. Glorieux, K. V. Hecke, L. V. Meervelt, B. Kirchner and K. Binnemans, *J. Phys. Chem. B*, 2006, **110**, 20978.
9. P. Nockemann, K. Binnemans, B. Thijs, T. N. Parac-Vogt, K. Merz, A.-V. Mudring, P. C. Menon, R. N. Rajesh, G. Cordoyiannis, J. Thoen, J. Leys and C. Glorieux, *J. Phys. Chem. B*, 2009, **113**, 1429.
10. Y. Fukaya, K. Sekikawa, K. Murata, N. Nakamura and H. Ohno, *Chem. Commun.*, 2007, 3089.
11. D. Dupont, E. Renders and K. Binnemans, *Chem. Commun.*, 2016, **52**, 4640.
12. H. Glasbrenner and H. Weingärtner, *J. Phys. Chem.*, 1989, **93**, 3378.
13. S. A. Buckingham, C. J. Garvey and G. G. Warr, *J. Phys. Chem.*, 1993, **97**, 10236.
14. H. Weingärtner, M. Kleemeier, S. Wiegand and W. Schröer, *J. Stat. Phys.*, 1995, **78**, 169.
15. M. Kleemeier, W. Schröer and H. Weingärtner, *J. Mol. Liq.*, 1997, **73–74**, 501.
16. V. Najdanovic-Visak, J. M. S. S. Esperança, L. P. N. Rebelo, M. N. da Ponte, H. J. R. Guedes, K. R. Seddon and J. Szydlowski, *Phys. Chem. Chem. Phys.*, 2002, **4**, 1701.
17. Y. Kohno, H. Arai, S. Saita and H. Ohno, *Aust. J. Chem.*, 2011, **64**, 1560.
18. Y. Tsuji and H. Ohno, *Chem. Lett.*, 2013, **42**, 527.

19. T. Ando, Y. Kohno, N. Nakamura and H. Ohno, *Chem. Commun.*, 2013, **49**, 10248.
20. Y. Fukaya and H. Ohno, *Phys. Chem. Chem. Phys.*, 2013, **15**, 4066.
21. D. Depuydt, L. Liu, C. Glorieux, W. Dehaen and K. Binnemans, *Chem. Commun.*, 2015, **51**, 14183.
22. D. Dupont, D. Depuydt and K. Binnemans, *J. Phys. Chem. B*, 2015, **119**, 6747.
23. B. Onghena, T. Opsomer and K. Binnemans, *Chem. Commun.*, 2015, **51**, 15932.
24. Z.-L. Xie and A. Taubert, *ChemPhysChem*, 2011, **12**, 364.
25. Y. Fukaya and H. Ohno, *Phys. Chem. Chem. Phys.*, 2013, **15**, 14941.
26. Y. Mieno, Y. Kohno, S. Saita and H. Ohno, *Chem.–Eur. J.*, 2016, **22**, 12262.
27. Y. Kohno and H. Ohno, *Phys. Chem. Chem. Phys.*, 2012, **14**, 5063.
28. H. Ohno, K. Fujita and Y. Kohno, *Phys. Chem. Chem. Phys.*, 2015, **17**, 14454.
29. S. Saita, Y. Mieno, Y. Kohno and H. Ohno, *Chem. Commun.*, 2014, **50**, 15450.
30. S. Saita, Y. Kohno, N. Nakamura and H. Ohno, *Chem. Commun.*, 2013, **49**, 8988.
31. Y. Zhao, H. Wang, Y. Pei, Z. Liu and J. Wang, *Phys. Chem. Chem. Phys.*, 2016, **18**, 23238.
32. Y. Kohno and H. Ohno, *Aust. J. Chem.*, 2012, **65**, 91.
33. Y. Kohno, D. L. Gin, R. D. Noble and H. Ohno, *Chem. Commun.*, 2016, **52**, 7497.
34. Y. Men, X.-H. Li, M. Antonietti and J. Yuan, *Polym. Chem.*, 2012, **3**, 871.
35. W. Li and P. Wu, *Polym. Chem.*, 2014, **5**, 5578.
36. Y. Kohno, Y. Deguchi and H. Ohno, *Chem. Commun.*, 2012, **48**, 11883.
37. Y. Deguchi, Y. Kohno and H. Ohno, *Aust. J. Chem.*, 2014, **67**, 1666.
38. B. Ziółkowski and D. Diamond, *Chem. Commun.*, 2013, **49**, 10308.
39. Y. Men, H. Schlaad, A. Voelkel and J. Yuan, *Polym. Chem.*, 2014, **5**, 3719.
40. Y. Men, H. Schlaad and J. Yuan, *ACS Macro Lett.*, 2013, **2**, 456.
41. Y. Kohno, Y. Deguchi, N. Inoue and H. Ohno, *Aust. J. Chem.*, 2013, **66**, 1393.
42. A. Okafuji, Y. Kohno and H. Ohno, *Macromol. Rapid Commun.*, 2016, **37**, 1130.
43. H. Yoshimitsu, A. Kanazawa, S. Kanaoka and S. Aoshima, *Macromolecules*, 2012, **45**, 9427.
44. Y. Deguchi, Y. Kohno and H. Ohno, *Chem. Lett.*, 2015, **44**, 238.
45. G. Wang and P. Wu, *Soft Matter*, 2015, **11**, 5253.
46. G. Wang and P. Wu, *Soft Matter*, 2016, **12**, 925.
47. R. Wang, W. Leng, Y. Gao and L. Yu, *RSC Adv.*, 2014, **4**, 14055.
48. A. Nitta, T. Morita, S. Saita, Y. Kohno, H. Ohno and K. Nishikawa, *Chem. Phys. Lett.*, 2015, **628**, 108.
49. S. Chen, Y. Peng, Q. Wu, A. Chang, A. Qu, J. Shen, J. Xie, Z. H. Farooqi and W. Wu, *Polym. Chem.*, 2016, 7, 5463.
50. Y. Deguchi, Y. Kohno and H. Ohno, *Chem. Commun.*, 2015, **51**, 9287.

51. Y. Kohno, S. Saita, K. Murata, N. Nakamura and H. Ohno, *Polym. Chem.*, 2011, **2**, 862.
52. Y. Kohno, N. Nakamura and H. Ohno, *Aust. J. Chem.*, 2012, **65**, 1548.
53. K. Fujita, D. R. MacFarlane, M. Forsyth, M. Yoshizawa-Fujita, K. Murata, N. Nakamura and H. Ohno, *Biomacromolecules*, 2007, **8**, 2080.
54. N. L. Mai and Y.-M. Koo, *Enzyme Microb. Technol.*, 2014, **63**, 34.
55. Y. Cai, R. Wang, W. B. Krantz, A. G. Fane and X. M. Hu, *RSC Adv.*, 2015, **5**, 97143.
56. X. Fan, H. Liu, Y. Gao, Z. Zou, V. S. J. Craig, G. Zhang and G. Liu, *Adv. Mater.*, 2016, **28**, 4156.

CHAPTER 5

Switchable Polarity Liquids

S. G. KHOKARALE[a], I. ANUGWOM[a,c], P. MÄKI-ARVELA[b], P. VIRTANEN[b] AND J.-P. MIKKOLA*[a,b]

[a]Technical Chemistry, Department of Chemistry, Chemical-Biological Centre, Umeå University, SE-90187 Umeå, Sweden; [b]Industrial Chemistry & Reaction Engineering, Process Chemistry Centre, Åbo Akademi University, FI-20500, Åbo-Turku, Finland; [c]Membrane Research Group, School of Engineering Science, Lappeenranta University of Technology, FI-53850, Lappeenranta, Finland
*E-mail: jpmikkol@abo.fi, jyri-pekka.mikkola@umu.se

5.1 Introduction

Solvents are an integral part and at the center of chemical transformations and induce the actual interactions of the reagents to produce the desired products. Many chemical processes, especially in organic synthesis, usually involve multiple steps (reactions, extractions, and/or separations) and each step is, in general, accompanied by various solvents with different properties. Further, the addition/removal of a solvent in each step gives rise to costs and environmental impacts associated with the particular process. In response to this, there is a great need to implement more sustainable solvent systems in which eventually one solvent can be used in more than one step. One can argue that, with a possible external trigger, a drastic change in the physico-chemical properties of the solvent can be induced and, consequently, the altered solvent can be used in the consecutive process steps. Also, after removing the trigger, if these physical properties actually are

Smart Materials No. 29
Polymerized Ionic Liquids
Edited by Ali Eftekhari

Published by the Royal Society of Chemistry, www.rsc.org

reversibly 'switchable' then it is more feasible to recover the original solvent back at the end of the process. These types of solvents that can reversibly and abruptly undergo a change in their inherent solvent properties are, therefore, typically referred to as 'switchable solvents'. The external 'trigger', such as a temperature or pressure change and/or the addition or removal of a gas, can often be exploited to switch the solvent from one form to another form – here one form is necessarily different from another one due to a change in one or more physical or chemical properties. Changes in properties such as viscosity, polarity, basicity, conductivity, solubilizing capacity or fluorophilicity are examples of these reversible transformations. Moderate changes in temperature and pressure are incapable of triggering significant changes in the properties of conventional solvents. In contrast, supercritical fluids[1] and CO_2/organic solvent mixtures[2] can be manipulated by pressure changes, but unfortunately only above 40 bar. Consequently, we need to design switchable solvents that are preferably capable of displaying dramatic changes in their physico-chemical properties reversibly under mild reaction conditions.

Amongst other properties, the polarity of a solvent demonstrates its important role in dissolving the reactants prior to their interactions upon chemical transformations, on the basis of the general rule 'like dissolves like'. In most organic transformations, the polarity of the reactants is usually not identical and, therefore, aprotic, dipolar solvents such as dimethyl sulfoxide are needed for the reaction to proceed. However, as the product is also often miscible in the same solvent system and due to the high boiling point of the solvent, it is difficult to proceed further in terms of the subsequent separation and purification of the product, as well as solvent recycling. Usually we can observe that in these cases the product separations proceed with the help of an anti-solvent strategy. This, nevertheless, increases the production costs and gives rise to environmental waste for the entire process. In these instances, the 'switchable polarity solvents (SPSs)' become more useful for facial separation of the product from the reaction mixture. In the SPSs, the external trigger switches the solvent from a lower polarity form to a higher polarity form. Such solvents could have applications in processes where a solvent of a certain level of polarity is needed for one step (such as a reaction) and a solvent of a different polarity is needed for a subsequent step (such as product extraction through phase separation or precipitation).

The polarity of the SPSs needs to be altered *via* structural modification of the solvent molecules to convert the non-polar or molecular form into polar or ionic moieties or *vice versa*. These structural modifications can be preceded by interacting the solvent molecules with external molecular moieties through chemical bonding. These external moieties can act as a 'trigger' to switch the polarity of the SPSs. Following a similar line of thought, Jessop *et al.* for the first time synthesized SPSs by exposing an equimolar mixture (1:1) of two non-ionic liquids, 1,8-diazabicyclo-[5.4.0]-undec-7-ene (DBU) and 1-hexanol, to gaseous carbon dioxide (CO_2), at normal pressure and at room temperature, thus triggering the conversion of the liquid mixture to a viscous ionic liquid (Figure 5.1a and b).[3] This liquid was further readily

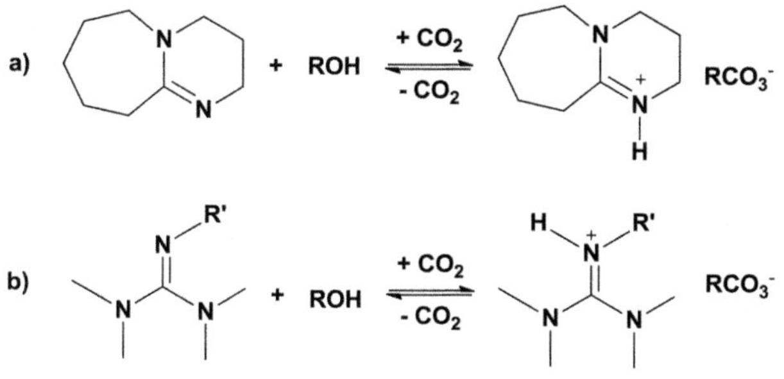

Figure 5.1 Formation of (a) a DBU–alcohol–CO_2 SPS, and (b) a DBU–alcohol–CO_2 SPS (R = alkyl group, R′ = H or butyl group).

converted back into a non-ionic liquid by bubbling N_2 through the liquid, at room temperature. They also observed that simply by heating the solution at 50 °C, the ionic liquid could be rapidly converted back to non-ionic moieties compared to the process of bubbling N_2 gas. All these transformations were monitored at the molecular level by nuclear magnetic resonance (^{1}H-NMR) spectroscopy, and by solvatochromic measurements of the polarity of the solvent before and after exposure to CO_2. In this instance, the CO_2 acted as a mild and easily available external trigger to synthesize SPSs from DBU and 1-hexanol. This groundbreaking work also opened the prospect of new and highly efficient pathways for carbon dioxide sequestration and storage. To date, several SPSs have been synthesized and characterized, and their use in various applications has been demonstrated. This chapter will summarize the synthesis of various types of SPSs and the corresponding characterizations to reveal their switchable nature.

5.2 Preparation and Characterization

5.2.1 Two-component Switchable Polarity Solvent/Ionic Liquids for CO_2/SO_2 Capture

Amidines, such as 1,8-diazabicyclo-[5.4.0]-undec-7-ene (here after referred to as DBU), and guanidines, such as 1,1,3,3-tetramethylguanidine or its derivatives (here after referred as TMG), are commonly used superbases in the synthesis of SPSs.[4–7] The SPSs from these superbases are prepared by bubbling CO_2 through an equimolar mixture of the superbase and an alcohol, at room temperature (Figure 5.1a and b). The formed SPSs are also often termed as DBU or TMG alkyl carbonates. For the demonstrated examples, the used alcohols ranged from lower alcohols to higher alcohols such as methanol, ethanol, 1-propanol, 1-butanol, 1-hexanol, 1-octanol, or 1-decanol.[5,6] In addition to pure alcohols, alkanol amines such as mono-ethanolamine can also be used in the synthesis of DBU-based SPSs.[8,9] Similar to CO_2-triggered synthesis, sulphur dioxide (SO_2)-mediated SPSs have also been synthesized and characterized.[10–12] It is worth mentioning here that the SPSs formed from the combination of DBU/TMG and alcohols with a CO_2 trigger possess characteristics such as ionic nature, reasonably high thermal stability, non-flammability, negligible vapor pressure, high ionic conductivity and the ability to solubilize both polar as well as non-polar species. The aforementioned properties are equivalent to the characteristics of liquid or solvent media termed as 'ionic liquids'. Hence, SPSs are also popularly known as switchable ionic liquids (SILs).[3,12] These physical and chemical properties of SPSs can be manipulated by selecting different alcohol/base pairs, as well as by chemically modifying the alcohol/base pairs.

5.2.2 Switchable Ionic Liquids from DBU, Alcohols and CO_2

It was observed that the synthesis of SPSs after bubbling CO_2 through a mixture of DBU and an alcohol is an exothermic reaction (Figure 5.1a). Also, the physical properties, such as melting point, viscosity, thermal stability *etc.*, of the obtained SPSs depend on the nature of the superbases and alcohols used in the synthesis. DBU alkyl carbonate salts, prepared by bubbling CO_2 through equimolar mixtures of DBU and ethanol, methanol or water, are solids at room temperature. On the other hand, those prepared from DBU and 1-propanol, 1-butanol, 1-hexanol, 1-octanol, or 1-decanol are viscous liquids at or near room temperature.[5] Considering the solid/semi-solid nature of the SPSs prepared from primary alcohols, the conversions were monitored gravimetrically, while in the case of secondary and tertiary alcohols, NMR spectroscopy analysis was used to verify the formation of the SPS. It was observed that the extent of conversion was a function of the alkyl chain length in the alcohols and it slowly decreased with an increase in the number of carbons in the alkyl chains.[5]

During the actual synthesis of DBU-based SPSs, both the viscosity and the ionic conductivity were influenced, indicating the formation of ionic/polar materials from non-ionic moieties. The viscosity of a DBU/1-propanol mixture under N_2 was 5.5 cP at 23 °C, but increased slowly during CO_2 bubbling. The conductivity of an equimolar (non-ionic) mixture of DBU-1/hexanol was 7 μS cm^{-1}, whereas the conductivity after CO_2 treatment was 187–189 S cm^{-1} after the formation of the ionic species. The change in the polarity, *i.e.* the formation of a polar compound from a non-polar mixture upon bubbling of CO_2, was also monitored by spectroscopic measurements such as ultraviolet (UV) spectroscopy using solvochromic dyes (*e.g.* Nile Red), where λ_{max} (the wavelength of maximum absorbance) was indicative of the solvent polarity.[5] The polarities of the physical equimolar mixtures of DBU and alcohols were slightly lower than that of pure DBU and were not significantly affected by the number of carbons in the alkyl chain of the alcohols. It was observed that the polarity of the equimolar mixture of DBU and alcohols linearly increased with periodical bubbling of CO_2 and the shift of the λ_{max} value to longer wavelengths, compared to a physical mixture of DBU and the alcohol in question. For example, the λ_{max} value for a DBU/MeOH mixture was 538 nm, while that for [DBUH][CH_3OCO_2] was 548 nm, after the formation of the SPS.[13] Similarly, for non-ionic mixtures of DBU/hexanol and for [DBUH][O_2CO-$(CH_2)_5CH_3$], the values were 536 and 543 nm, respectively. Based on Nile Red dye experiments, it was also observed that the polarity of both the molecular and ionic form depended on the length of the alkyl chain in the alcohol molecule. Similar to the polarity, the melting points of the SPSs were a function of the number of carbons in the linear alkyl chain. These observed melting points are only approximate because, as previously stated, the conversion slowly decreased with an increase in the length of the alkyl chain and the unreacted moieties could influence the measurement of the real melting point. The melting points of DBU/alcohol/CO_2 SPSs decreased from C_1 (methanol) to C_3 (propanol), while they increased again with a further increase in the alkyl chain length (*i.e.* from C_4 onwards) in the alcohol molecules.[5]

^1H (proton) and ^{13}C (carbon) NMR analysis revealed the formation of SPSs before and after the exposure of the DBU and alcohol mixture to CO_2.[3,5] In the case of the formation of SPSs from DBU and 1-hexanol, no significant changes were observed in the ^1H NMR spectra of the DBU/hexanol mixture and DBU/hexanol/CO_2, except in the oxygen-bound methylene of the hexyl group. In the sample, this peak shifted from the normal position for hexanol (3.58 ppm in $CDCl_3$) to 3.90 ppm. The ^{13}C NMR spectrum of $[DBUH][O_2C-(CH_2)_5CH_3]$ in $CDCl_3$ showed a carbonate carbon at 158.7 ppm, along with changes in the chemical shifts of the carbon atoms that are in the vicinity of the protonated nitrogen atom in the $[DBUH]^+$ cation. The composition of DBU methyl carbonate, $[DBUH][CH_3OCO_2]$, which was formed after exposing a molecular mixture of DBU and MeOH to CO_2, was also studied by NMR spectroscopy to confirm its chemical composition.[6] The stretching vibrations belonging to the –N–H (in the cation) and –C=O (in the anion) bonds were observed at 3117 cm^{-1} and 1648 cm^{-1}, respectively. In the FT-IR spectrum of DBU methyl carbonate, one could also see the formation of ionic moieties from the DBU, MeOH and CO_2 mixture. A SPS containing DBU and mono-ethanol amine (MEA) with CO_2 has also been synthesized and characterized by NMR spectroscopic analysis[8] (Figure 5.2). The ^1H and ^{13}C spectra of $[DBUH][O_2CO-(CH_2)_2-NH_2]$ revealed changes in the chemical shifts after the formation of an SPS compared to the physical mixture of DBU and MEA. A broad signal at 4.95 ppm in the ^1H NMR spectrum was observed for the exchangeable proton attached to the nitrogen atom in the $[DBUH]^+$ cation, while the ^{13}C NMR spectrum showed a chemical shift at 164.78 ppm for the carbonate carbon in the $[O_2CO-(CH_2)_2-NH_2]^-$ anion. The presence of the MEA carbonate anion was shown by long-range $^1H-^{13}C$ HMBC two-dimensional NMR spectroscopy. The correlation peaks observed at 3.00 ppm and 164.82 ppm, respectively, arise from the correlation between the protons at the 'a' position in the MEA molecule and the carbonate carbon in the anion of the synthesized SPS (Figure 5.2).

5.2.3 Switchable Ionic Liquids from TMG, Alcohols and CO_2

Like DBU, TMG and its derivatives (*e.g.* 2-butyl-1,1,3,3-tetramethylguanindine, TMBG) is also a frequently used superbase in SPS synthesis. Herein, a mixture of TMG/TMG derivatives and an alcohol exposed to CO_2 molecules results in TMG alkyl carbonates.[5] The TMBG alkyl carbonate salts prepared *via* bubbling

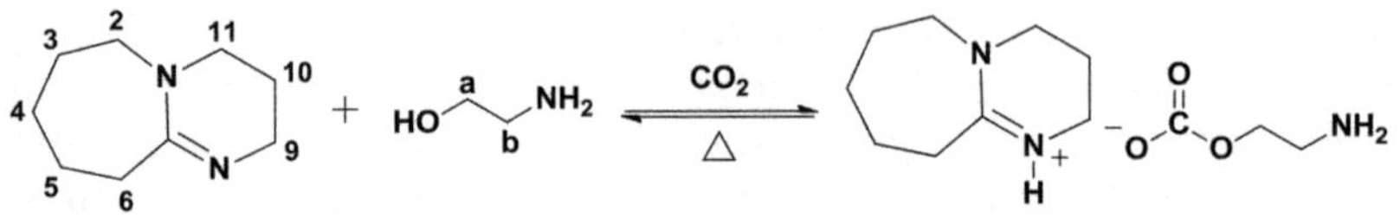

Figure 5.2 Formation of a DBU–monoethanol amine–CO_2 SPS.

of CO_2 through equimolar solutions of TMBG and methanol, 1-butanol, 1-hexanol, 1-octanol, or 1-dodecanol (Figure 5.1b) are also viscous liquids at room temperature. The melting points of the resultant SPSs with various alcohols were dependent on the alkyl group chain length and linearly increased with an increase in the alkyl chain length. Mixing TMBG with an equimolar amount of an alcohol results in a slight increase in the polarity, relative to TMBG alone. This behavior of TMGB was different from the DBU/alcohol physical mixtures, since the DBU/alcohol equimolar mixtures were less polar than either DBU alone or the alcohol alone. The authors assumed that this unusual behavior may be associated with the basicity and hydrogen bonding ability of DBU and alcohols, respectively. The formation of a hydrogen-bonded DBU/ROH adduct would effectively hide the basicity and hydrogen-bond donating ability of these liquids and, thereby, the equimolar mixture would give rise to a lower λ_{max} value than the amine or alcohol alone in UV spectroscopy measurements using solvochromic Nile Red dye. Similar to DBU/alcohol SPSs, in the case of TMGB/alcohol SPSs the λ_{max} values of TMG/alcohol/CO_2 were also higher than the values of TMGB and alcohol alone. In contrast to the irregular trend in the case of DBU/alcohol/CO_2 SPSs, the TMGB SPSs exhibited a linear increase in the melting points of the TMGB/alcohol/CO_2 SPSs with an increase in the number of carbon atoms in the alkyl chain of the alcohol. The formation of TMBG methyl carbonate was confirmed by ^{1}H NMR by an upfield shift in the aliphatic protons and by the collapse of the two N–Me peaks into one (two singlets at 2.53 and 2.62 ppm become one singlet at 3.1 ppm) not observed for neat TMBG molecule. The ^{13}C NMR spectra also illustrated the characteristic carbonate peak at 161 ppm. Also, a similar SPS was further characterized by elemental analysis and the analysis was consistent with the formation of the methyl carbonate anion and TBMG cation in a 1:1 ratio.

A previous study has shown that the ability of the bases/superbases to capture CO_2 depends on the type of base used upon SPS synthesis. TMG (the pK_a of the protonated base in MeCN is 23.3 [14,15]) is a weaker base than DBU (the pK_a of the protonated base in MeCN is 24.3 [16,17]). The ΔG ($\Delta G = \Delta H - T\Delta S$, where ΔG = Gibb's free energy, ΔH = change in enthalpy, ΔS = change in entropy and T = temperature) of formation of the SPS from TMG is positive, while the value in the case of its DBU counterpart is negative. This means that TMG has a less favorable ΔG of the reaction and, therefore, possesses a weaker ability to capture CO_2. The lower value of ΔG is not due to the reaction enthalpy, *i.e.* ΔH; in fact, the CO_2 captured by TMG and an alcohol has a more favorable reaction enthalpy compared to the case when DBU is selected. Nevertheless, it was observed that the resulting salt, [TMGH][$ROCO_2$], is capable

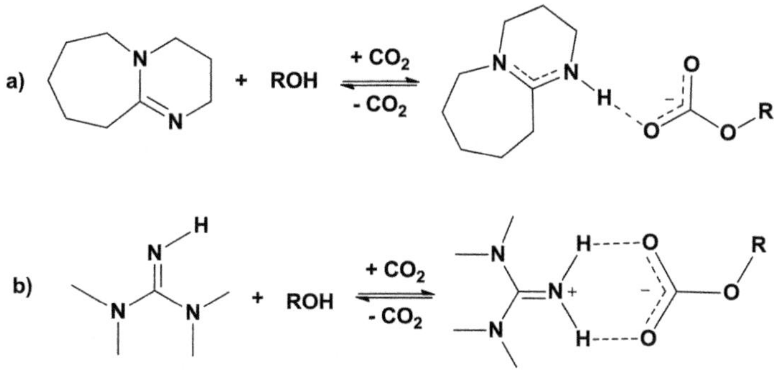

Figure 5.3 Hydrogen bonding of a cation with an anion for salts made from (a) DBU and (b) TMG with ROH and CO_2.

of more hydrogen bonding interactions than the DBU salt. Because of this excessive hydrogen bonding, the ΔS value became more negative, which is responsible for the positive ΔG and, consequently, the weak capture of CO_2 by TMG/alcohol mixtures. The alkyl carbonate salts of DBU give rise to fewer hydrogen-bonding interactions than the salt of TMG because $[TMGH]^+$ has two hydrogen-bond donor sites. As shown in Figure 5.3, $[TMGH]^+$ can form an entropically more favored 6-membered ring, resulting in strong hydrogen bonding.[7,18] On the basis of these observations, it was observed that the ability of a base/alcohol combination to trap CO_2 cannot be predicted or explained by the pK_a of the protonated base; instead, it is more dependent on the hydrogen-bonding contributions and the entropy of the reaction.

5.2.4 CO_2 Release and Recyclability of SPSs

5.2.4.1 CO_2 Release

The release of CO_2 from a classical aqueous MEA-based CO_2 capture system requires almost 60% of the electrical energy produced by the power plant for the thermal stripping of CO_2, which is not economical from an industrial point of view.[19] The large inefficiency of MEA systems is caused by the high specific heat capacity of water (4.18 J g^{-1} deg^{-1}).[20] Heldebrant *et al.* measured the rate of CO_2 evolution from CO_2-saturated SPSs by using an automatic burette system.[7] The SPS was loaded into the burette, which was further placed in a preheated oil bath, and upon stirring the CO_2 evolution was monitored. At 90 °C and with a stirring rate of 250 rpm, both $[DBUH]^+[ROCO_2]^-$ and $[TMGH]^+[ROCO_2]^-$ (R = 1-hexyl, 1-pentyl, or 1-butyl) showed 1st order kinetics in the CO_2 evolution rate with respect to the concentration of DBU or TMG alkyl carbonate salt (eqn (5.1)). The rate of stirring further altered the rate of CO_2 evolution, suggesting that the mass-transfer of CO_2 from the solution was the rate-determining step. The activation energy determined for CO_2 evolution from SPS systems was 23–33 kJ mol^{-1} based on an Arrhenius plot of the first order rate constant.

$$[BaseH]^+[RCO_3]^- \rightarrow Base + ROH + CO_2 \quad (5.1)$$

It was observed that the release of CO_2 from the SPS not only depended on the stirring rate but also on the desorption temperature used for CO_2 evolution. The SPS will not release CO_2 unless either mild heating or continuous flushing of an inert gas or air is applied through the liquid. It was observed that the amount of release of CO_2 was directly proportional to the applied temperature. In burette experiments without flushing of a gas through the liquid, $[DBUH]^+[ROCO_2]^-$ and $[TMGH]^+[ROCO_2]^-$ (R = 1-hexyl, 1-pentyl, or 1-butyl) released 0.25 equivalents of CO_2 when heated to 50 °C, 0.50 equivalents at 70 °C and up to 0.65 equivalents at 90 °C. The temperature switches at various temperatures were always accompanied by stabilization of a thermodynamic equilibrium between the gaseous and dissolved CO_2, whereby a temperature increase forces the equilibrium to shift towards the release of CO_2. Even though the specific heat capacities with respect to the release of CO_2 from DBU- and TMG-based switchable salts are not known, one can look at the study of the specific heat capacity of other commercially available ionic liquids such as 3-ethyl-1-methyl-imidazolium tetra-fluoroborate and 3-butyl-1-methylimidazolium tetra-fluoroborate (1.28 J $g^{-1}deg^{-1}$ and 1.66 J g^{-1} deg^{-1}, respectively[21]). One can therefore, assume that the specific heat capacity for the release of CO_2 may be between 1.2–2.0 J g^{-1} deg^{-1}. Hence, at least 50% less energy would be required to thermally strip CO_2 from SPSs than from existing aqueous alkanol amine systems, at comparable temperatures.

5.2.4.2 *Recyclability of SPSs*

From an industrial point of view, a CO_2 scrubbing system needs to be recyclable. In order to show the recyclability of their DBU-based SPSs, Heldebrant *et al.* performed consecutive capture and release cycles with the DBU:1-hexanol system using an automated gas burette system.[7] Initially, the CO_2-relaxed DBU/1-hexanol mixture was further exposed to fresh CO_2 for 5 min to re-synthesize the DBU hexyl carbonate salt. This salt was then again decarboxylated in the burette by plunging the stirred flask into a pre-heated oil bath at 90 °C and the CO_2 released was measured. After CO_2 release, the flask was cooled to room temperature and the DBU/1-hexanol mixture was again exposed to CO_2 for another five minutes. This process was repeated for five cycles. It was observed that an equivalent amount of CO_2 was released each time during consecutive CO_2 absorption–desorption cycles. This also suggests that no loss of activity occurred for the DBU/1-hexanol mixture during these cycles. The authors also explained that the CO_2 gas steam needs to be anhydrous, as if water is present then the more stable $[DBUH]^+[HCO_3]^-$ will be formed instead of $[DBUH]^+[ROCO_2]^-$, which requires 121 °C to release CO_2 and the challenges will be equivalent to the release of CO_2 from classical aqueous MEA systems.

5.2.4.3 *Switchable Ionic Liquids from DBU, Glycerol, and Acid Gas (CO_2 or SO_2)*

The synthesis of SPSs was also performed and characterized in the case of a DBU and glycerol mixture after exposure to acid gases such as CO_2 or SO_2 (Figure 5.4).[11] Anugwom *et al.* synthesized, for the first time, glycerol-containing SPSs by bubbling either CO_2 or SO_2 through a mixture containing 3:1 molar amounts of DBU and glycerol. The weights as well as the viscosities of the DBU/glycerol mixtures were increased upon bubbling of CO_2 or SO_2 compared to a physical mixture of DBU and glycerol, confirming the chemical interactions of DBU and glycerol with acid gases. Similar to primary alcohol-based SPSs, the formation of a polar compound from a non-polar mixture of DBU/glycerol after bubbling CO_2 or SO_2 was also monitored by using Nile Red as a solvatochromic dye. The DBU/glycerol mixture with Nile Red resulted in a red color, which is a common characteristic of Nile Red dye in non-polar solvents. After bubbling CO_2, the color changed from red to milky as the polarity of the liquid increased. In order to understand the concept of 'polarity switching', nitrogen was bubbled through the liquid and the red color again reappeared. This confirmed that CO_2 can be used as a trigger to change the polarity of the DBU/glycerol mixture. Furthermore, upon thermo-gravimetric measurements, it was observed that the formed SPS possessed a higher thermal stability compared to a physical mixture of DBU and glycerol. Amongst the ionic compounds formed with SO_2 and CO_2, the DBU–glycerol–SO_2 material demonstrated a higher thermal stability compared to DBU–glycerol–CO_2. This fact was found to depend on the Lewis acidic nature of the interacting acid gases. The strength of acid gas binding is directly proportional to the Lewis acidity of the acid gas, and thus the decomposition temperature of DBU–glycerol–SO_2 was higher compared to that of DBU–glycerol–CO_2 as SO_2 possesses a stronger Lewis acidic character than CO_2. Further evidence that an SPS was formed upon exposing the DBU/glycerol mixture to CO_2 or SO_2 was also further provided by NMR and FTIR analysis.

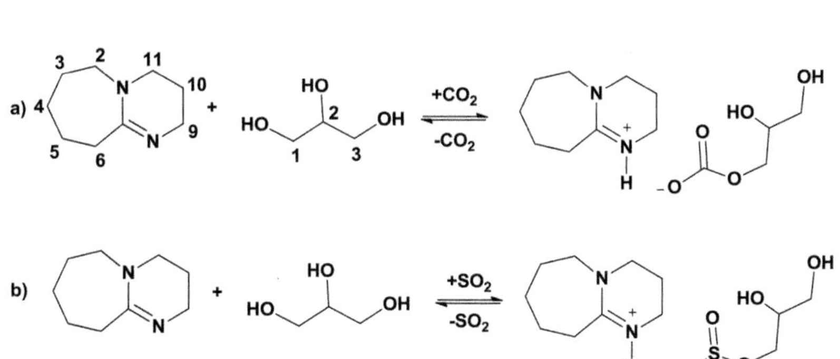

Figure 5.4 Formation of (a) a DBU–glycerol–CO_2 SPS, and (b) a DBU–glycerol–SO_2 SPS.

Both one (^{1}H and ^{13}C) and two (^{1}H–^{13}C correlation) dimensional NMR spectroscopic techniques were used to confirm the structures of the formed DBU glyceryl carbonate. The neat NMR spectra at 60 °C were acquired to avoid line broadening upon analysis. A capillary with $CDCl_3$ as an internal standard was used in the NMR tube during the analysis. The chemical shifts and signal assignments for the ^{1}H and ^{13}C NMR spectra of DBU–glycerol–CO_2 are depicted in Tables 5.1 and 5.2, respectively. As mentioned in Table 5.1, the ^{1}H NMR spectra did not display significant changes in the chemical shifts except for the broadening of the signals in the SPS compared to the physical

Table 5.1 ^{1}H NMR shifts and signal assignments in the spectra of DBU:glycerol (3 : 1) and DBU:glycerol (3 : 1):CO_2. The shifts are referenced to an external standard, DMSO (δ = 2.50 ppm). (The values in the Table are adapted from ref. 11 with permission from The Royal Society of Chemistry).

DBU:glycerol (3 : 1)			DBU:glycerol (3 : 1):CO_2		
H[a]	*nH*	δ, ppm	*H*[a]	*H*[a]	δ, ppm
DBU 3, 4 5	18	1.06–1.23	DBU 3, 4 5	18	1.00–1.21
DBU 10	6	1.28	DBU 10	6	1.27
DBU 6	6	1.89	DBU 6	6	1.93
DBU 9	6	2.71	DBU 9	6	2.68
DBU 2, 11	12	2.73–2.82	DBU 2, 11	12	2.74–2.87
Gly 1, 3	4	2.93–3.04	Gly 1, 2, 3	4	2.89–3.14
Gly 2	1	3.08			
NH, OH	3	5.34	NH, OH	3	6.73

[a]Atom numbering shown in Figure 5.4a.

Table 5.2 ^{13}C NMR shifts and signal assignments in the spectra of DBU:glycerol (3 : 1) and DBU:glycerol (3 : 1):CO_2. The shifts are referenced to an external standard, DMSO (δ = 39.50 ppm). (The values in the Table are adapted from ref. 11 with permission from The Royal Society of Chemistry).

DBU:glycerol (3 : 1)		DBU:glycerol (3 : 1):CO_2	
C[a]	δ, ppm	*H*[a]	δ, ppm
DBU 10	21.558	DBU 10	20.882
DBU 5	24.905	DBU 5	24.489
DBU 3	27.384	DBU 3	26.926
DBU 4	28.349	DBU 4	28.086
DBU 6	35.117	DBU 6	34.087
DBU 9	42.481	DBU 9	41.437
DBU 11	46.984	DBU 11	46.952
DBU 2	51.253	DBU 2	51.422
DBU 7	159.540[b]	DBU 7	160.454[b]
		–O–COO$^-$	156.513
GLY 1, 3	63.073[b]	GLY 1, 3	63.021[b]
GLY 2	71.375	GLY 2	71.430

[a]Atom numbering shown in Figure 5.4a.
[b]Strongly broadened signals.

mixture of DBU and glycerol. The significant broadening of the signals was observed mostly for the C–H protons of the glycerol ($\delta \approx 3.89$–3.14 ppm). Apart from that, the signal at $\delta \approx 6.73$ ppm, representing a mixed signal of the exchangeable protons (O–H of glycerol and N–H of protonated DBU), was also broadened and shifted downfield from $\delta \approx 5.34$ to 6.73 ppm. Similarly, after formation of ionic species of the DBU glyceryl carbonate, the signals in the ^{13}C NMR spectrum shifted between 0.0 and 1.2 ppm. The signals corresponding to the DBU carbons closest to the protonation site, such as C-6 and C-9, exhibited upfield shifts, while the signal for C-7 showed a downfield shift. The broadened signals of glycerol carbons at 63.021 and 71.430 ppm were observed to belong to both glycerol and carbonated glycerol. The broad signal of the carbonyl carbon of the formed DBU glyceryl carbonate was observed at 156.513 ppm. On the other hand, long-range ^{1}H/^{13}C correlation spectroscopy (HMBC) showed connectivity between the H-1 protons in glycerol and the carbonate carbon, as shown by the presence of a correlation signal situated at 3.18/157.34 (H/C) ppm. These observations from NMR spectroscopic analysis confirmed the formation of DBU glyceryl carbonate after exposure of the DBU and glycerol mixture to CO_2.

Glycerol gives rise to an IR band at 3286 cm^{-1}, while the SPS showed an IR band at 3255 cm^{-1} in the –OH stretching region. The broad signal obtained for the formed SPS in the region of 3000–3600 cm^{-1} was observed in response to the N–H stretching vibration, together with the unreacted OH-group. The asymmetric and symmetric C–O–C stretching vibrations were observed at 1274 and 1050 cm^{-1}, respectively, along with the characteristic C=O absorption band in the region of 1870–1540 cm^{-1}.

The possibility of switching the polarity of the DBU/glycerol reaction mixture with a CO_2 gas trigger was efficiently utilized for catalyst separation in the trans-esterification of soybean oil.[22–24] In this case, after complete conversion of the soybean oil to biodiesel (fatty acid methyl ester, FAME), the DBU was separated from FAME in the form of DBU glyceryl carbonate after bubbling CO_2 into the reaction mixture. More details about this process will be provided in upcoming chapters.

5.2.5 DBU Bicarbonate

Considering its highly basic nature, it may be possible for DBU to interact directly with molecular CO_2 without any presence of a proton donor such as an alcohol and form a [DBU:CO_2] zwitterionic adduct (Figure 5.5a). However, in practice, it has been observed that this is not possible and instead, DBU forms a DBU bicarbonate adduct in the presence of water and CO_2 (Figure 5.5b).[4] A stable zwitterionic adduct between DBU and CO_2 does not form either in the presence or the absence of water. Instead, it was observed that a white precipitate forms from the reaction of CO_2 with wet DBU (H_2O = 700 ppm) and this was confirmed by bubbling CO_2 through liquid DBU for 10 min. The initial clear non-viscous liquid turned into a white sludge with large amounts of a white solid precipitating out of solution. The formation

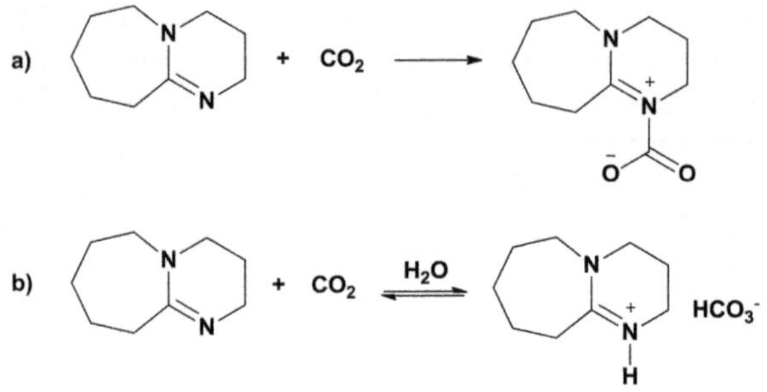

Figure 5.5 Formation of (a) a [DBU:CO_2] zwitterion adduct, and (b) a [DBUH][HCO_3] salt.

of polar moieties from non-ionic DBU molecules was further confirmed by solvochromic dye measurements, as well as ^{1}H and ^{13}C NMR and IR spectroscopy analysis.

The Reichardt's solvatochromic dye 'ET(30)' gave rise to a change in color from green to red when non-ionic DBU was converted to a white solid after reaction with CO_2. A similar experiment was performed for DBU in acetonitrile where the dye 'ET(30)' gave rise to a turquoise color, which further changed to pink through a purple color when the mixture was exposed to 1 bar of CO_2 gas. The IR analysis of the white solid precipitate showed characteristic stretching peaks at 1644 cm^{-1} and 835 cm^{-1} belonging to protonated DBU and out of place vibration of CO_2 in the bicarbonate anion. The ^{1}H NMR spectra of the white solid (in $CDCl_3$) showed a broad signal at 5.5 ppm belonging to a proton at the positively charged N atom of the DBU molecule. This is clear evidence of the formation of a DBU:CO_2 adduct through protonation of the amidine nitrogen. Furthermore, the ^{13}C NMR spectra demonstrated a shift in the case of the bridgehead carbon C7 (the atom numbering of the DBU molecule is given in Figure 5.4) that appeared at 162.6 ppm compared to 161.5 ppm for unreacted DBU. Still, ^{13}C-enriched CO_2 was bubbled through a DBU and water mixture (100 ppm) and the formation of a DBU:bicarbonate adduct was confirmed by NMR spectroscopy. In the ^{13}C NMR spectrum, a large peak at 159.0 ppm represents the DBU bridgehead carbon. A small new peak at 157.8 ppm belongs to the carbon atom in the $H^{13}CO_3^-$ species. Two-dimensional HMBC NMR analysis showed no evidence of coupling between the new peak (157.8 ppm) and any of the protons on DBU, a result that is consistent with the emergence of [DBUH][HCO_3] instead of a DBU:CO_2 zwitterionic adduct. Conductivity measurements were also performed in which anhydrous DBU (7 ppm water) in anhydrous acetonitrile showed a small increase in conductivity when CO_2 was bubbled through it. The reason is that less of the bicarbonate species forms as trace amounts of water are present. However, after an increase in the water concentration in the DBU:acetonitrile:CO_2 mixture, the conductivity increased further along with the formation of more bicarbonate species. The increase in the conductivity in the mixture further supported the hypothesis that [DBUH][HCO_3]

was formed instead of a DBU:CO_2 zwitterionic adduct, because the zwitterionic adduct would not give rise to an increased conductivity as it does not contain free ions.

5.2.6 One-component Switchable Polarity Solvents/Ionic Liquids for CO_2 Capture

The synthesis of ionic compounds through the chemisorption of CO_2 can also proceed through the interaction of a one-component system with molecular CO_2. This synthesis protocol is different from amidine- and guanidine-based SPS synthesis where alcohols generally help the amidine and guanidine to interact with acid gases to form ionic moieties. In this section, the synthesis and characterization of one component-based SPS systems based on silylamines are reviewed.

Silylamines such as 3-(trialkoxysilyl)propylamines (*e.g.* 3-(tri-methoxysilyl)propylamine and 3-(tri-ethoxysilyl)propylamine) and 3-(aminopropyl) trialkylsilanes (*e.g.* 3-(aminopropyl) triethylsilane, 3-(aminopropyl) tripropylsilane, and 3-(aminopropyl) trihexylsilane) have been used in the synthesis of one component-based SPSs/ILs after their chemical interaction with CO_2 (Figure 5.6).[25–29] Upon actual synthesis, two moles of non-ionic amines interact with CO_2 to form an ionic compound through two equilibrium sub-steps. The first equilibrium in Figure 5.6 is the reaction of one mole of silylamine with CO_2 to form the corresponding carbamic acid species. The second equilibrium represents the deprotonation of carbamic acid through

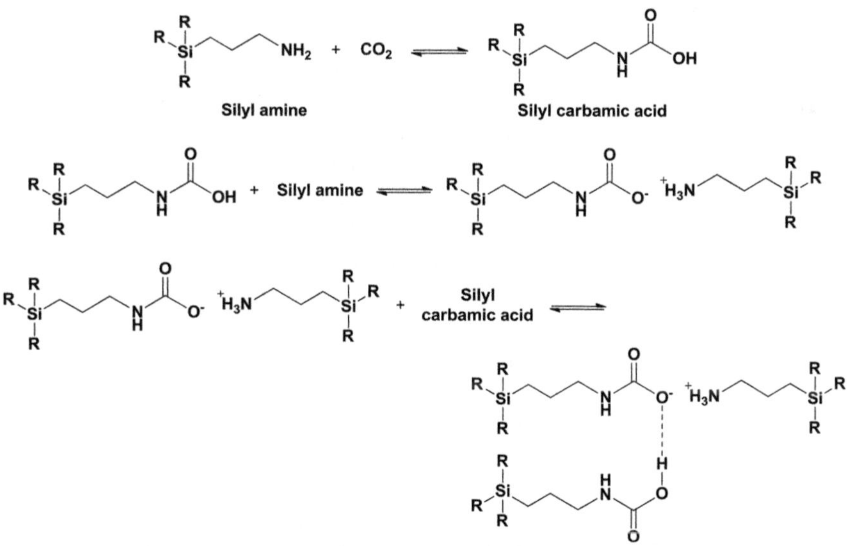

Figure 5.6 Equilibria for the reaction of CO_2 with silylamines (R = alkoxyl or alkyl group).

interaction with another silylamine molecule to yield ammonium–carbamate ion pairs.[25] It was also proposed that as the ion pair concentration increases, the free amine is no longer an effective reaction solvent and it is replaced by an increasing concentration of ionic species. As represented in the third equilibrium, these ionic species can be further stabilized by carbamic acid through hydrogen bonding.[26]

5.2.6.1 *(Trialkoxysilyl)propylamines and CO_2*

Alkoxysilylamines such as 3-(trimethoxysilyl)propyl amine and 3-(triethoxysilyl)propyl amine react with CO_2 under ambient conditions to form the corresponding ILs, such as 3-(trimethoxysilyl)propylammonium 3-(trimethoxysilyl)propylcarbamate (TMSAC) and 3-(triethoxysilyl)propylammonium 3-(triethoxysilyl)propylcarbamate (TESAC) (Figure 5.6).[25] The formation of an IL is an exothermic reaction accompanied by an increase in viscosity. TMSAC is gel-like, whereas TESAC is a viscous liquid, with the corresponding viscosity values of 2160 cP and 930 cP, respectively. The polarity switch from a molecular liquid to ionic species was confirmed by solvochromic measurements where the solvatochromic dye Nile Red was used. TMSA showed a 9.0 nm λ_{max} increase (from 528.1 nm to 537.1 nm) when reacting to form TMSAC, whereas TESA showed an 11.4 nm increase (from 522.6 to 534.0 nm) when forming TESAC. TMSA is more polar than TESA, but has a smaller polarity change window. The structure of the synthesized ILs was further evaluated by ^{1}H NMR, ^{13}C NMR, IR, and elemental analysis.

NMR analysis was carried out in $CDCl_3$ and the obtained ^{1}H NMR spectra showed that the signals related to the protons were shifted from the silylamine precursors (TMSA and TESA) with the appearance of hydrogen peaks for the hydrogens attached to nitrogen at 6 ppm ($-NH_3^+$). The ^{13}C NMR spectra also showed a shift in the carbon peaks from the precursors with the characteristic carbamate peak at 162.7 ppm. Additionally, the elemental analysis confirmed the formation of the final product accompanied by two moles of silylamine and one mole of CO_2. The molecular formulae for the TMSAC and TEASAC ILs from the elemental analysis were found to be $C_{13}H_{34}N_2O_8Si_2$ and $C_{19}H_{46}N_2O_8Si_2$, respectively. The FT-IR analysis demonstrated characteristic IR bands at 3400–2400 cm^{-1} ($-NH^+$ stretch), 1570 and 1471 cm^{-1} ($-CO_2$ asymmetric and symmetric stretch), 1071 cm^{-1} (Si–OR stretch), and 2940 and 2839 cm^{-1} (C–H stretch) for both of the synthesised ILs.

The switching ability of the TMAC and TEASAC ILs to their respective silylamines was demonstrated by thermogravimetric analysis (TGA), differential scanning calorimetry (DSC), and NMR. TGA indicated a mass loss of 13% at 88 °C and 9% at 125 °C in TMSAC and TESAC, respectively, corresponding to the loss of CO_2. However, the DSC isotherms showed that the CO_2 release window for TMSAC was between 75 and 175 °C, whereas for TESAC, it was between 50 and 150 °C. ^{13}C NMR analysis was further used to study the temperature-induced switching ability of the IL formed with the silylamine precursor TESA. The IL was heated at 120 °C for 2 hours to convert it back to

TESA. The carbamate peak at 162.3 ppm that appeared upon IL formation then disappeared upon heating. Therefore, these one-component systems offer elegant and simple processing opportunities for industrial applications upon the addition and removal of CO_2 alone.

5.2.6.2 *(Aminopropyl)trialkylsilane + CO_2*

Like 3-(trimethoxysilyl)propyl amine, 3-(aminopropyl)trialkylsilanes were also successfully used to produce a one-component silylamine-containing SPS/IL system for CO_2 capture. The formation of such solvents was supported by gravimetric and spectroscopic analysis.[26,27]

5.2.6.3 *ATR-FTIR Analysis*

Attenuated total reflectance-Fourier transform infrared (ATR-FTIR) measurements were used to measure both the chemical and physical absorption of CO_2 in 3-(aminopropyl) tripropylsilane (TPSA), 3-(aminopropyl)triethylsilane (TEtSA), 3-(aminopropyl)trihexylsilane (THSA) and 3-(aminobutyl)triethylsilane (αMe-TEtSA).[26] The IR spectrum of TPSA showed two distinct regions where characteristic peaks for C–H and N–H stretching were observed between 3100–2600 cm^{-1} and 1650–1590 cm^{-1}, respectively. Transformation of a molecular liquid to ionic species after CO_2 chemisorption was monitored by the appearance of a broad ammonium (NH_3^+) peak in the range of 2600–3100 cm^{-1} and an intense and slightly broad carbamate (NH–CO_2) peak at 1700 cm^{-1}. The theoretical mole ratio between CO_2 and silylamine is equal to 0.5; however, in practice, the overall experimental CO_2 capacities were higher due to physical absorption along with chemical absorption. Due to the formation of an IL, overlap of the C–H stretch and ammonium peaks occurred. Hence, the contribution of the molecular liquid C–H stretch, IL C–H stretch and ammonium peak, and the overlapped peaks in the range of 1740–1520 cm^{-1} (due to the N–H stretch of the molecular liquid and carbamate stretch), were considered to reveal the conversion of TPSA to an IL. In this regards the Beer–Lambert law was used to calculate the amount of CO_2 absorbed through the chemical interaction of CO_2 and TPSA. An analogous method was further also used for other silylamines, *viz*. TEtSA, THSA, and αMe-TEtSA.

The physical absorption capacities of TEtSA, TPSA, THSA, and αMe-TEtSA were measured by using ATR-FTIR analysis. Physically absorbed CO_2 showed an asymmetric stretch at approximately 2330 cm^{-1}. It was observed that the magnitude of the physical absorption of CO_2 was directly proportional to the applied CO_2 pressure. The absorbance or intensity of the asymmetric CO_2 stretch was linearly increased with an increase in the CO_2 equilibrium pressure from zero to 60 bar at 35 °C. At 1 bar of CO_2 pressure, the mole fraction of CO_2 in the IL was between 1.3×10^{-2} and 2.0×10^{-2} depending on the IL structure. THSA exhibited the highest CO_2 solubility ($X_{CO2} = 2.0 \times 10^{-2}$), while TEtSA, TPSA, and αMe-TEtSA have comparable but lower CO_2 physical absorptions ($X_{CO2} = 1.3 \times 10^{-2}$ to 1.7×10^{-2}).

5.2.6.4 Henry's Constant (H_{CO2}) Measurement

The physisorption capacity of the silylamine was further determined in terms of Henry's constant (H_{CO2}).[26] Henry's law states that the solubility of a gas in equilibrium with a liquid phase is directly proportional to the partial pressure of the gas in the vapor phase (eqn (5.2)).

$$y_{CO2}\,P = X_{CO2}H_{CO2} \tag{5.2}$$

$y_{CO2}\,P$ = partial pressure of CO_2 gas in the vapor phase, X_{CO2} = mole fraction of CO_2 and H_{CO2} = Henry's constant.

The Henry's law constants were determined from a straight-line fit (through the origin) and the data points below 30 bar. The average calculated Henry's constants for TEtSA, TPSA, THSA, and αMe-TEtSA were 77, 60, 50 and 67 bar, respectively. The CO_2 absorption capacities of these four silylamines were further determined by comparing the Henry's constant with the void volume (V_v) of the silylamine. Void volume is defined as the empty space between the molecules. This void volume was calculated by subtracting the theoretical van der Waals volume (V_{vdW}) from the experimentally determined molar volume (V_M)[28] (supporting information of ref. 26). A plot of H_{CO2} (bar, at 35 °C) *versus* V_v (cm^3 mol^{-1}) is given in Figure 5.7.

Also, Figure 5.7 shows that THSA exhibited a higher CO_2 absorption capacity as the longer alkyl chain on the silicon atom avoided the close packing of the molecules, hence providing a larger void space and greater CO_2 solubility compared to other silylamines. These results for the physical absorption of CO_2 were comparable to those obtained with ATR-FTIR analysis. Figure 5.8 depicts the overall CO_2 absorption capacities of silylamines obtained with both physical and chemical absorption. It was observed that the CO_2 absorption capacity of the silylamines was dependent on the length of the alkyl chain attached to the silicon atom in their composition.

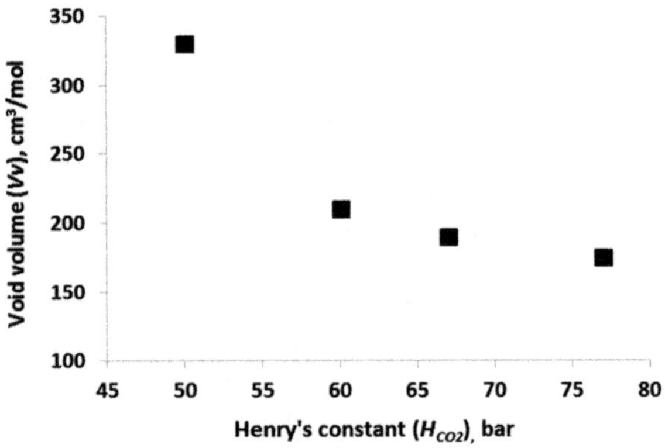

Figure 5.7 Henry's constants of silylamines at 35 °C *versus* the void volume (the values in the graph are adapted from ref. 26).

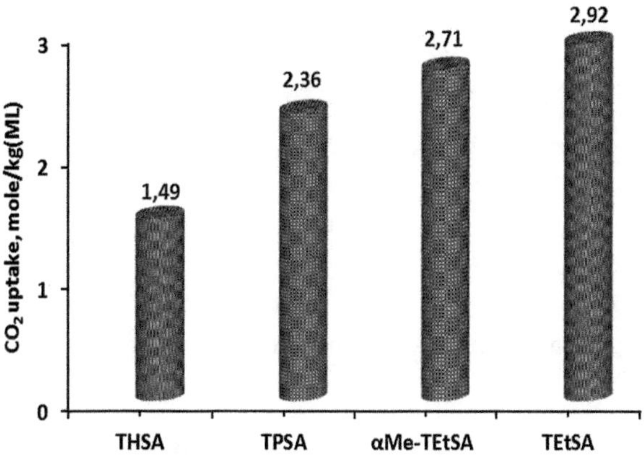

Figure 5.8 CO_2 absorption capacity of silylamines (the values in the graph are adapted from ref. 26).

Table 5.3 Silylamine CO_2 uptake. (Adapted with permission from ref. 27. Copyright 2013 American Chemical Society).

Silylamine	CO_2 uptake, mol CO_2 per mol amine[a]
3-(Aminopropyl) triethylsilane	0.63
3-(Aminopropyl) tripropylsilane	0.64
3-(Aminopropyl) trihexylsilane	0.67

[a]Measured by gravimetric analysis.

5.2.6.5 NMR Analysis

The NMR and FTIR analysis techniques provided qualitative as well as quantitative CO_2 absorption data for reversible silylamine ILs.[27] The conventional 2:1 amine to CO_2 stoichiometry would result in the capture of 0.50 moles of CO_2 per mole of amine. However, it was observed upon gravimetric measurements that the studied silylamines absorb 0.65 moles of CO_2 per mole of amine (Table 5.3). This shows that an additional 0.15 moles of CO_2 per mole of amine was absorbed through either physisorption or other types of interactions. It was observed that in accordance with Henry's law ($y_{CO2}P = x_{CO2}H_{CO2}$) under 1 bar of CO_2, the physical CO_2 uptake is approximately 0.01 moles of CO_2 per mole of amine. This means that the physical absorption of CO_2 at 1 bar along with the conventional chemical reaction gives a theoretical CO_2 capacity limited to 0.51 moles of CO_2 per mole of amine. Nevertheless, the studied silylamine ILs still outperformed this value and captured on average 0.65 moles instead of 0.51 moles of CO_2 per mole of amine. It was further proposed that considering the three equilibria observed in Figure 5.6, 2 moles of CO_2 per 3 moles of amine were absorbed, of which an additional 0.15 moles of CO_2 was captured through hydrogen bonding interactions of ammonium carbamate and carbamic acid.

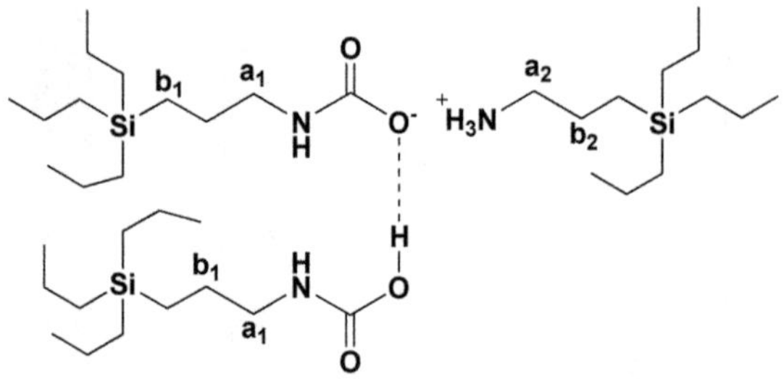

Figure 5.9 ^{13}C peak assignments for the 3-(aminopropyl) tripropylsilane reversible ionic liquid.

In support of the gravimetric analysis, ^{13}C NMR analysis of 3-(aminopropyl) tripropylsilane demonstrated that the integration of the values related to the a_2 and b_2 carbon of the nitrogen atom in the ammonium cation was observed as 1.00 and 1.12 ± 0 while the values belonging to the a_1 and b_2 carbon of the nitrogen atom in the carbamate anion were 1.69 ± 0.08 and 1.81 ± 0.11, respectively (Figure 5.9). The integrations observed in the carbamate anion were approximately 65% greater than those of the carbon atoms in the ammonium cations indicating the presence of more than one carbamate species in the molecular composition. This extra species was attributed as belonging to the carbamic acid that was hydrogen bonded to the carbamate anion. The FT-IR analysis also showed that during absorption of CO_2 in the 3-(aminopropyl) tripropylsilane IL, the carbonyl stretch at 1700 cm^{-1} related to carbamic acid was increased while the asymmetric CO_2^- stretch in the carbamate anion at 1575 cm^{-1} was decreased with an increase in the CO_2 pressure. This indicates an increase in the concentration of carbamic acid and a decrease in the concentration of carbamate.

Liotta *et al.* synthesized 13 types of silylamines with various lengths of the tether that connects the amine to the Si atom.[29] Moreover, variants changing the position of or with the addition of a single methyl group in the silylamine backbone or introducing unsaturation in the alkyl chain backbone were developed. Furthermore, the properties of the variants, such as viscosity, temperature of reversal, and enthalpy of regeneration, were studied in the CO_2 capture process. It was observed that for 3-(aminoalkyl) triethylsilane ILs (alkyl = methyl, ethyl, propyl or butyl), variation of the distance between the amine and the Si atom does not adversely affect the CO_2 capacity up to a propyl chain, but with a butyl chain, the absorption capacity decreased from 0.60 to 0.35.[29] The less prone chemisorption ability of the butyl chain-containing IL was due to its solid nature. It was also observed that the chemisorption ability of the ILs was influenced after introducing branching as well as unsaturation into the alkyl chain. Similar to the CO_2 absorption capacity, the effect of branching along the alkyl chain backbone, unsaturation of the propyl backbone, and the order of the amine (1^0 or 2^0) showed the strongest influence on the reversal temperature, enthalpy of regeneration and viscosity change during CO_2 uptake.

5.2.7 Regeneration of Silylamines from ILs

As mentioned previously, the regeneration of molecular moieties from SPSs is possible by thermal treatment. The release of CO_2 from the ILs of TEtSA, TPSA, THSA, and αMe-TEtSA and the subsequent transfer to the corresponding silylamines were also followed through heating and the progress of switching was monitored by differential scanning calorimetry (DSC).[26] The DSC thermograms provided information about the CO_2 release from the ILs and the subsequent evaporation of the silylamines at various temperatures. Table 5.4 summarizes the onset temperature for CO_2 release (T_{rev}), onset temperature of evaporation of the molecular liquid (T_{evap}) and regeneration enthalpies of the ILs ($\Delta H_{regeneration}$). The structure of the silylamines contributed significantly to the process of regeneration and it was observed that the regeneration temperature decreased with an increase in the alkyl chain length on the silicon atom. Also, the evaporation temperature of the silylamines increased with an increase in the substituent chain length. The enthalpy of regeneration ($\Delta H_{regeneration}$) upon the release of CO_2 of the studied ILs was found to be 87 ± 6 kJ mol^{-1}. This value is lower than the corresponding value obtained for the conventional aqueous MEA system (−72 kJ mol^{-1}) because the heat of regeneration for the aqueous system is significantly higher due to the energy loss associated with water heating.[19,30]

5.2.8 Recyclability of Silylamine SPSs/ILs

The recyclability of THSA in the process of consecutive CO_2 sorption measurement was studied.[26] In the actual process, THSA was reacted with CO_2 to form an IL, followed by heating of the IL at 100 °C for 1.5 hours to obtain a molecular liquid. This process was further continued for four more cycles. The progress of the formation of an IL from THSA and the further reversion of the formed IL to a molecular liquid was monitored by NMR, refractive index and gravimetric measurements. Figure 5.10 depicts the refractive index measurement of five consecutive cycles of CO_2 capture and release for the THSA system. It was observed that the representative THSA system showed constancy in all five consecutive CO_2 sorption measurements, where the THSA molecular liquid and IL showed a regular change in the refractive index values between 1.4590 (for THSA) and 1.4710 (for the IL), respectively.

Table 5.4 Onset temperature values for CO_2 release and silylamine evaporation. (Data taken from ref. 26).

Silylamine	T_{rev}, [°C]	T_{evap}, [°C]	$T_{evap} - T_{rev}$ [°C]	$\Delta H_{regeneration}$ [kJ mol_{CO2}^{-1}]
TEtSA	71 ± 3	168 ± 12	97	83 ± 6
TPSA	64 ± 2	180 ± 11	116	89 ± 3
THSA	51 ± 1	233 ± 1	182	81 ± 3
αMe-TEtSA	52 ± 2	150 ± 8	98	92 ± 6

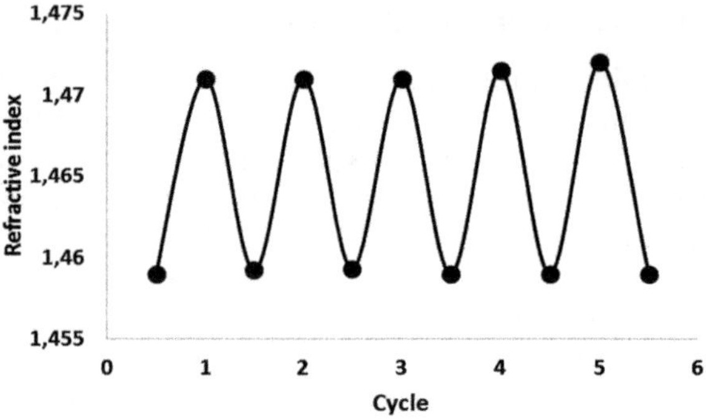

Figure 5.10 Recyclability of THSA monitored using refractive index measurements (the values in the graph are adapted from ref. 26).

5.3 Applications

5.3.1 CO_2 Capture with Switchable Ionic Liquids

Reutilization and capture of CO_2 is an important research topic. Currently, CO_2 is typically absorbed in monoethanolamine solutions. This process suffers, however, from solvent degradation, partial loss of amines and the fact that the solvent itself is corrosive.[31] Several alternative processes for CO_2 capture have been proposed, for example swing adsorption with metal–organic frameworks,[32] zeolites,[33] membranes,[34] chemical looping,[35] solvent room temperature ionic liquids[36,37] and recently also switchable ionic liquids.[7,12,38] In this section the alternative solvents for amines and, especially switchable ionic liquids, will be summarized. The main advantages of room temperature ionic liquids (RTILs) are their very low vapor pressures and relatively high stability. The efficiencies of RTILs, such as 1-ethyl-3-methylimidazolium or 1-butyl-3-methylimidazolium with different anions, for example hexafluorophosphonate[11] or tetrafluoroborate,[37] in absorbing CO_2 at atmospheric pressure are, however, only slightly improved, *i.e.* the molar ratio of CO_2 to nitrogen is less than 0.02 and they work better under elevated pressures.[39] In addition, RTILs are still rather expensive and also toxic limiting their application in CO_2 capture. Switchable ionic liquids, defined already in Section 5.2, are formed *in situ* during CO_2 capture from an alcohol or aminoalcohol and a base, an amidine, such as 1,8-diazabicyclo[5.4.0] udec-7-ene (DBU).[12] Aa an alternative to the use of (amino)alcohols, silylamines also form SILs with DBU and CO_2.[26] A necessity of using switchable ionic liquids in industry is also the requirement of their decomposition to recover CO_2 and reuse it. This back-switching occurs at a higher temperature and/or by bubbling an inert gas through the SIL. The literature is still very scarce regarding the application of SILs in CO_2 capture.[12,38] The main results of the application of SILs in CO_2 capture are summarized below. A comparative work of CO_2 capture and regeneration was performed with several (amino) alcohols together with a DBU-formed SIL during CO_2 capture.[12] Before

Table 5.5 Structures of (amino)alcohols used together with DBU in CO_2 absorption and the absorption capacity of CO_2 using a 5 ml liquid volume. Data taken from ref. 12.

DBU	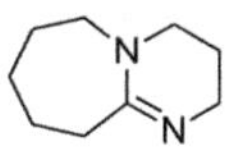	Mol CO_2 absorbed per mol alcohol	Viscosity [Pa s]	Degradation temperature of SIL [K]
1-Hexanol	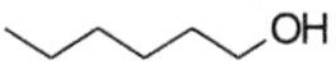	0.9	2.0	
4-Amino-1-butanol	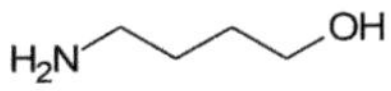	0.95	9.8	428
L-valinol	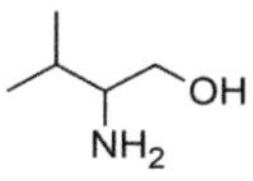	0.85	6.7	398
6-Amino-1-butanol	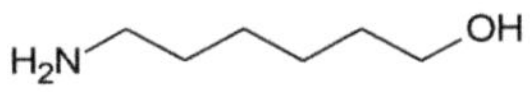	1.3	5.7	433
L-prolinol	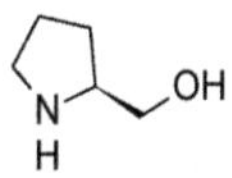	0.50	3.7	433

making a comparison of the CO_2 capturing efficiency of the novel aminoalcohol–DBU solvents, the method was validated against the CO_2 absorption capacities for hexanol–DBU,[40] L-prolinol and DBU[41] as well as for L-valinol and DBU.[41] The results showed that the most efficient SIL was DBU–1-aminohexanol (Table 5.5), followed by DBU–hexanol and DBU–4-aminohexanol and the lowest CO_2 absorption capacities were observed for DBU–L-valinol and DBU–L-prolinol.[12] If, however, a larger liquid phase volume, 10 ml, was used, the long chain aminoalcohol 6-amino-1-hexanol with DBU exhibited the highest CO_2 capturing efficiency (Figure 5.11),[38] which agrees with the literature data.[41] However, it can be seen from Figure 5.11 that the saturation of 6-amino-1-hexanol takes more time.

Mass transfer limitations can also cause a difference in CO_2 capturing efficiencies with different liquid volumes, for example, gas distribution, different residency times, gas hold-up and the fact that carbonate formation is more difficult on a larger scale. Generally, it can be stated that for secondary alcohols such as L-prolinol, the CO_2 capturing efficiency is lower than that obtained for primary alcohols. Charge distribution occurs in heterocyclic L-prolinol.

When SILs were formed, their viscosity increased typically very much opposite to the case when bubbling CO_2 through conventional ionic liquids, such as [C_4mim][BF_4] or [C_4mim][acetate]. Consequently, mass transfer limitations can become limiting steps also upon regeneration of the formed SILs. One alternative method to decrease the viscosity is to increase the temperature. It was already known that an increase in temperature from 298 K to 328 K can decrease the viscosity of DBU–CO_2–4-amino-1-butanol and DBU–CO_2–1-hexanol by a factor of 20 and 10, respectively. In addition, bigger volumes require longer absorption times for CO_2 saturation.

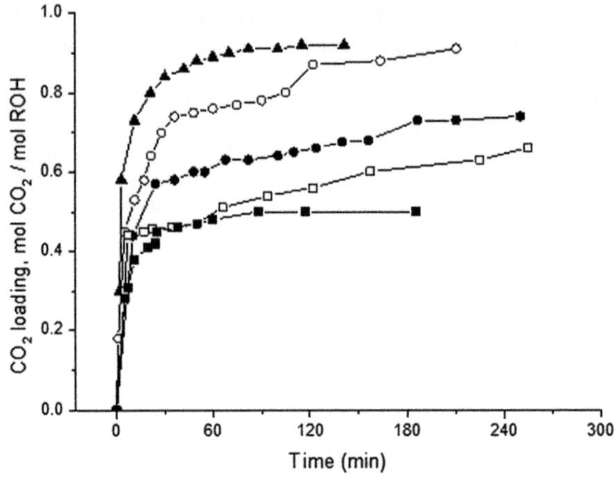

Figure 5.11 CO_2 capture by SILs based on equimolar amounts of DBU and different (amino)alcohols (V_L = 10 ml) at ambient temperature and pressure. Notation: (▲) 1-hexanol–DBU, (o) L-valinol–DBU, (•) 4-amino-1-butanol–DBU, (□) 6-amino-1-butanol–DBU and (■) L-prolinol–DBU. Adapted with permission from Privalova.[38]

Desorption of CO_2 from the formed SILs of (amino)alcohol–DBU–CO_2 has been investigated by different methods, such as desorption after regeneration at room temperature, desorption after regeneration at 380 K, desorption after regeneration at the degradation temperature of the SIL (see Table 5.1) and desorption in the presence of ultrasound.[12] It should be pointed out here that the high degradation temperatures of the SILs indicate that the ionic bonds are very strong.

The best regeneration ability was observed for the 1-hexanol–DBU–CO_2 system giving repeatable CO_2 absorption capacities using a larger 150 ml reactor in the case where evaporation of 1-hexanol was eliminated. The highest loading capacity of CO_2 after regeneration was obtained with DBU–CO_2–1-hexanol, and was about 70–75% with all other methods, but not with desorption after regeneration at room temperature in a small-scale reactor. On the other hand, for aminoalcohol–DBU–CO_2 systems, the CO_2 loading capacity in the second cycle in a small reactor showed quite large differences with different SILs varying from 54–67% of their initial value. Typically, the CO_2 loading efficiency in the subsequent cycle was not the same as in the first cycle in a small-scale reactor. In general, faster desorption could be achieved when increasing the regeneration temperature due to a decrease in the viscosity and the possibility to destroy the ionic bonds at higher temperatures. The change of viscosity at different temperatures for different (amino)alcohol–DBU–CO_2 SILs was also determined in ref. 12 showing that they decreased exponentially; for example, for 4-amino-alcohol–DBU–CO_2 the viscosity of 9.8 Pa s at 298 K changed to 0.3 Pa s at 327 K. A comparison of recycling with two conventional ionic liquids in CO_2 capture and desorption was also investigated. These results revealed that it was also possible to get nearly the same CO_2 loading as in the first cycle with [C_4mim][BF_4], whereas the CO_2 loading capacity for [C_4mim][acetate] in the second cycle was only about 20% of the one in the first cycle. The temperature dependency of conventional ionic liquids was also investigated for comparison, by measuring the viscosities of the two above-mentioned ionic liquids. In addition, the activation energy of viscous flow values both for SILs and conventional ionic liquids were determined by taking the logarithm of the viscosity. The activation energy for viscous flow values of the SILs were higher than those for the conventional ionic liquids, indicating that ions are moving more freely in conventional ionic liquids compared to SILs.

In conclusion, the study of SILs showed that it is possible to reversibly adsorb and desorb CO_2 from SILs and one candidate for further studies would be 6-amino-1-hexanol–DBU showing high CO_2 capturing capacity. The preferred regeneration strategy of the SILs would be to use higher temperatures facilitating the faster desorption of CO_2 due to the decreased viscosity.

Silylamines form with CO_2 ionic liquids, namely reversible ionic liquids (RevILs), such as ammonium carbamate,[26] the use of which has been demonstrated in CO_2 capture (Figure 5.12).

Different silylamines (Figure 5.13) were used in CO_2 capture by Rohan *et al.*[26] Since both physical and chemical absorption occurs, the overall absorption capacities of CO_2 were compared. The results showed that THSA

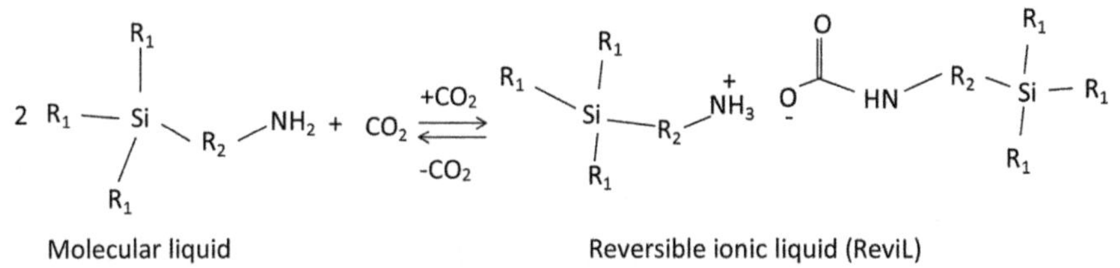

Figure 5.12 Reaction scheme for the formation of reversible ionic liquids.

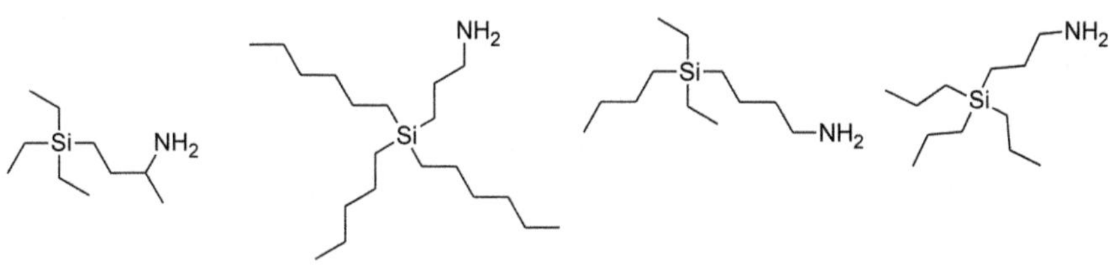

αMe-TEtSA

TPSA

TEtSA

THSA

Figure 5.13 Structures of reversible ionic liquids: 3-(aminopropyl)triethylsilane (TEtSA), 3-(aminopropyl)trihexylsilane (THSA), 3-(aminopropyl)tripropylsilane (TPSA) and 3-(aminobutyl)triethylsilane (αMe-TEtSA) used together with DBU for CO_2 capture. Data taken from ref. 39.

exhibited the lowest overall CO_2 absorption capacity, being 1.49 moles per kg of molecular solvent (ML), whereas the overall absorption capacities of CO_2 for TEtSA, αMe-TEtSA and TPSA were higher, being 2.92 moles per kg ML, 2.71 moles per kg ML and 2.36 moles per kg ML, respectively, at 35 °C. According to the stoichiometry, the chemical absorption capacity of CO_2 is only 0.5 moles per kg ML. On the other hand, the physical absorption of CO_2 was the highest for THSA at higher pressures due to the fact that when the hydrocarbon chain on the silicon atom is longer, it disrupts the packing ability of the RevIL creating a larger void space and, at the same time, facilitating a larger amount of CO_2 being dissolved.

5.3.2 Fractionation of Alga with Switchable Ionic Liquids

The possibility of fractionating the green marine alga *Ulva Rigida* with the switchable ionic liquid DBU–MEA–SO_2 has been investigated.[42] This SIL was very efficient in the dissolution of lignocellulosic biomass.[43] *Ulva Rigida* contains quite large amounts of carbohydrates, of which close to 40% of the main monomeric constituents are glucose and rhamnose. Fractionation of *Ulva Rigida* has only been investigated recently with SILs, but not with any conventional ionic liquids. On the other hand, the green alga *Chlorella* has been fractionated in ionic liquids, such as with [EMIM][Cl][44] and [EMIM][OAC].[45] In the former case, 90 wt% of sugars was obtained as a product when in the first step the alga were fractionated at 105 °C for 3 hours, followed by treatment of the biomass with concentrated HCl under the same conditions. In the latter case, [EMIM][OAc] was used as a solvent to fractionate lipids from *Chlorella* at 110 °C in 2 hours giving 219 mg per g of lipids.[19] In addition, ionic liquids have been used to extract isoprenoids from the cell wall of the green microalga *Botryococcus braunii*.[46]

A comparative fractionation study of *Ulva Rigida* with DBU–MEA–SO_2 SIL, a distillable ionic liquid (DIL), 1,1,3,3-tetramethylguanidium propionate, and a low viscosity ionic liquid (LVIL), 1,8-diazobicyclo-[5,4,0]-undec-7-ene-2,2,3,3,4,4,5,5-octafluoro-1-pentoxide, was performed.[15,47] The structures of the ionic liquids used for fractionation of *Ulva Rigida* are shown in Figure 5.14. The results indicated that opposite to the very promising results obtained in the fractionation of lignocellulosic material with DBU–MEA–SO_2, only 25 wt% of carbohydrates were solvated at 120 °C, whereas the DIL dissolved 67 wt% of carbohydrates, being thus a more suitable candidate

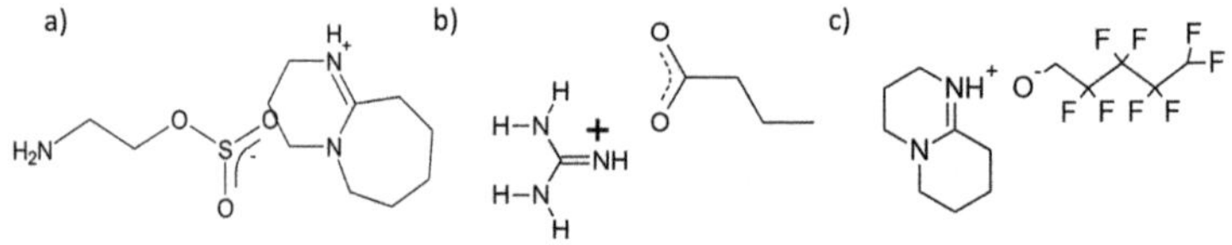

Figure 5.14 The structures of the ionic liquids used for the fractionation of *Ulva Rigida*. (a) DBU-MEA-SO2 (SIL), (b) 1,1,3,3-tetramethylguanidium propionate (DIL) and c) 1,8-diazobicyclo-[5,4,0]-undec-7-ene-2,2,3,3,4,4,5,5,-octafluoro-1-pentoxide (LVIL).

for carbohydrate dissolution from *Ulva Rigida*. On the other hand, the low viscosity ionic liquid formed from DBU and octafluoropentoxide gave only about 37 wt% yield of carbohydrate, due to its low viscosity.

5.3.3 Wood Fractionation

The need for advancement towards innovative, economically viable fractionation systems that can yield value-added products upon conversion of lignocellulosic materials has been studied. As a result, several promising methods and solvent systems have been developed.[48] Unfortunately, the likelihood of the use of some of these methods is limited due to the cost of the whole processing cycle. Ionic liquids (ILs) have been widely studied as effective fractionation solvents for lignocellulosic materials, since their dissolution properties are unique and they can represent an even more environmentally friendly option with the choice of appropriate cations and/or anions. ILs are salts composed mainly of organic cations and organic or inorganic anions, having melting points usually below 100 °C. ILs exhibit also high thermal and chemical stability and a wide liquidus range.[49] Many ILs are able to dissolve lignocellulosic material or one of its major components, cellulose, hemicelluloses, and lignin.[50–54] Some of the concerns about the use of ILs as pretreatment/fractionation solvents for lignocellulosic materials are amongst others, the cost of the ILs, the system complexity related to IL recycling, toxicity, biomass solute separation and downstream processing. Switchable ionic liquids (SILs) have been studied as another alternative to ILs as fractionation solvents for lignocellulosic materials and the results of some of the promising studies are discussed herein.

5.3.3.1 Switchable Ionic Liquids for Selective Extraction of Hemicelluloses

The extraction of components from woody-biomass was studied using SILs, and several parameters for the fractionation of the biomass were studied. Furthermore, the efficiency of two difference SILs was tested. It was demonstrated that SILs, prepared by bubbling CO_2 through a mixture of DBU and alcohol, can be applied for the selective extraction of hemicelluloses from Norway spruce softwood.[55] The total weight reduction of the wood depended on the wood size and the alcohol used for the preparation of the SIL. With the hexanol SIL, the weight of the milled spruce was reduced by 12 wt%, while with spruce chips the weight reduction was 4 wt%. The corresponding values of weight reduction were 10 wt% for milled spruce, and 2 wt% for spruce chips with the butanol SIL. The hemicellulose content was reduced by 38 wt% for spruce treated with butanol SIL and by 29 wt% for spruce treated with the hexanol SIL. In reality, the SIL treatment was not very efficient in lignin removal. The residual lignin fraction (extractive-free wood) left in the undissolved wood after SIL treatment was 16 and 20 wt% of the total wood mass after the butanol or hexanol SIL treatment, respectively. There is clear evidence that mainly the monomeric sugar components from the sugar content of the wood hemicelluloses were extracted (see Figure 5.15). What is more noteworthy is that the extraction of hemicelluloses from wood with an SIL was carried out at 55 °C and at normal pressures.[55] Therefore, less energy was needed compared to many existing wood treatment technologies. Furthermore, the reversibility of the ionic liquid back to its molecular components improves their recyclability and reusability.

The morphology of the wood samples was investigated by SEM (Figure 5.16). From the pictures, it can be seen that after pre-treatment of the wood (*Picea abies* spruce) with the DBU–hexanol–CO_2 SIL (Figure 5.16, image 2), the middle lamella become clear. This result indicates that the dissolution of pectin (also seen as galacturonic acid) occurs (Figure 5.15), together with a significant reduction of the sugar content in the wood sample.

5.3.3.2 SO_2 vs. CO_2 Switched Ionic Liquids for Wood Treatment

The efficiency of SO_2 or CO_2 switched ionic liquids as solvents for the dissolution of lignocellulosic material was compared by measuring the weight loss of birch wood before and after SIL treatment. The study was performed with fresh birch chips containing a considerable amount of water. It should be mentioned that drying of the wood requires additional energy input and is undesirable from a process economy point of view. Industrial size chips (3 cm × 3 cm × 0.5 cm) were treated in a SIL for 5 days at 100 °C and after 5 days of treatment with a SIL the chip weight was reduced by 25% or over 50% for the chips treated with the DBU–glycerol–CO_2 SIL or DBU–glycerol–SO_2 SIL, respectively. The compositions of both the native and undissolved wood samples that had been treated with either the DBU–glycerol–CO_2 SIL or the DBU–glycerol–SO_2 SIL were determined. The relative weight composition of the native wood was: cellulose 41%, hemicelluloses 33% and lignin 21%. In addition, the remaining 5% was composed of mainly ash and extractives.[56]

After the treatment with either the DBU–glycerol–CO_2 SIL or the DBU–glycerol–SO_2 SIL, the undissolved wood residue contained 53% and 76% cellulose, respectively. These results indicate that relatively pure cellulose fibers were produced when using the DBU–glycerol–SO_2 SIL as a solvent for birch treatment. Furthermore, only small amounts of cellulose were dissolved during the SIL treatment. The majority of xylan (21%) ended up in the liquid phase upon DBU–glycerol–CO_2 SIL treatment, which corresponds to 76% of the original amount in the untreated wood. Furthermore, the undissolved solid, after SIL-treatment and washing, contained only 2 wt% lignin. In comparison, the treatment of birch chips with the DBU–glycerol–CO_2 SIL removed 10% of the xylan and 8% of the lignin. 95% of the hemicelluloses in the wood were dissolved with the DBU–glycerol–SO_2 SIL. Also, the weight of the treated wood sample was reduced by 50%, whereas only 25% weight loss occurred when using the DBU–glycerol–CO_2 SIL as a solvent. The analysis of the chemical compositions of the birch wood before and after SIL treatment confirmed that the main dissolved components were hemicelluloses, pectin and extractives.[55]

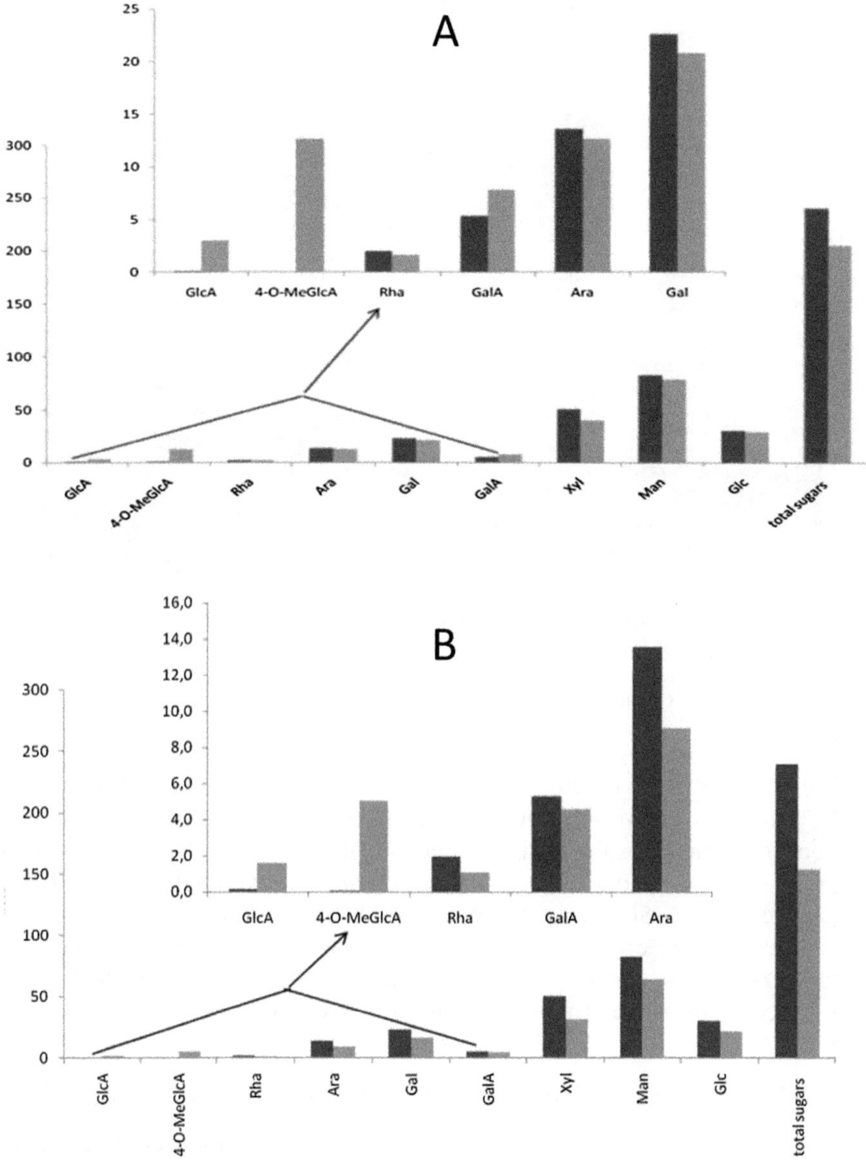

Figure 5.15 Hemicellulose content of wood samples before (darker shade) and after treatment (lighter shade) at 55 °C for 5 days, with no stirring. A: treated with the DBU–butanol–CO_2 SIL; B: treated with the DBU–hexanol–CO_2 SIL. Amounts are presented as mg per g of wood.

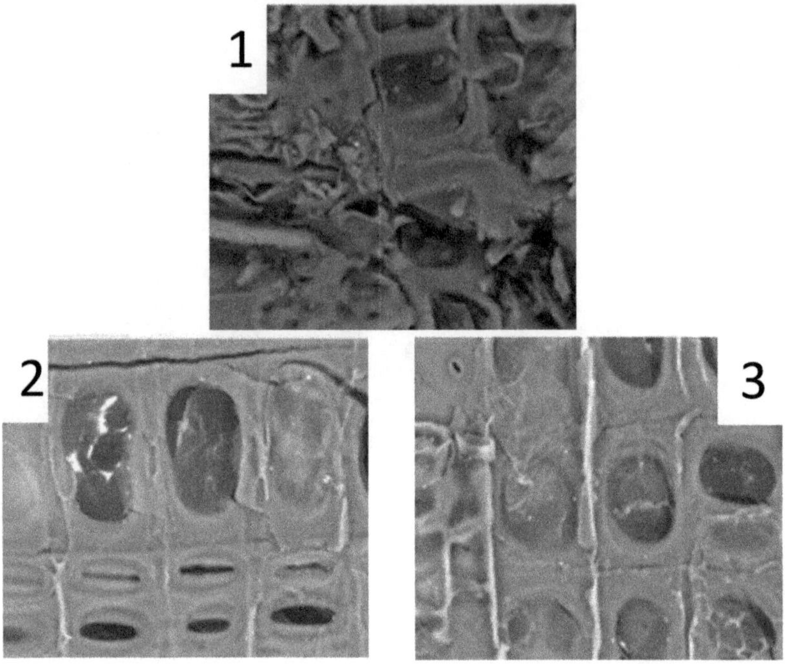

Figure 5.16 SEM images of the morphology of (1) native spruce wood, (2) spruce treated with the DBU–hexanol–CO_2 SIL for 1 day, and (3) spruce treated with the DBU–hexanol–CO_2 SIL for 5 days. The treatment occurred at 55 °C under ambient atmosphere and pressure.

The sugar content of the undissolved wood samples reduced to approx. 15 wt% and 75 wt% with the DBU–glycerol–CO_2 SIL and DBU–glycerol–SO_2 SIL, respectively. These results indicate that the DBU–glycerol–SO_2 SIL was much more efficient in the dissolution of birch (*Betula Pendula*) hardwood than the DBU–glycerol–CO_2 SIL. The efficiency was attributed to the acidity of the SILs since, the DBU–glycerol–SO_2 SIL could be correlated as having a similar acidity (pKa ~ −7) value to hydrochloric acid, while the acidity of the DBU–glycerol–CO_2 SIL could be related to that of trichloroacetic acid (pKa ~0.77).[57]

5.3.3.3 *Influence of Treatment Time for Wood Treated with a SIL*

The influence of treatment time in terms of the dissolution capacity of the DBU–glycerol–CO_2 SIL on birch chips was studied. As expected, a better dissolution capacity was achieved with increased time. The weight reduction difference between a one-day treatment and a five-day treatment was about 15 wt%. Based on the chemical composition of the treated material, it was concluded that hemicellulose dissolution occurs quite steadily during the whole treatment, while the dissolution of lignin occurs more towards the latter part of the treatment cycle. The sugar content decreased by 8% within the first 24 hours when using the DBU–glycerol–CO_2 SIL, while the corresponding decrease in lignin amount was 10 wt% after 24 hours and 21 wt% after 120 hours for the DBU–glycerol–CO_2 SIL.[56]

5.3.3.4 Selective Extraction of Components by SILs from Wood

Studies have shown that SILs can be selective towards one wood component. In a study by Anugwom *et al.* it was seen that SO_2-triggered SILs were evidently more selective towards lignin dissolution compared to CO_2-triggered SILs.[56] The reason for this was probably the presence of SO_2, since a comparable affinity toward lignin has been achieved using a sulfur-containing ionic liquid.[58,59] On the other hand, the removal of pectins and uronic acids was reported to occur in a quite similar way for both SILs.[56] Then again, the glucose yield was found to be much higher, even when a completely different SIL (DBU–monoethanol amine and SO_2 or CO_2) was used for the treatment of wood. With the DBU–MEA–SO_2 SIL (680 mg g^{-1}), the glucose yield was higher compared to both native wood and the DBU–MEA–CO_2 SIL-treated wood, whose yields were 420 mg g^{-1} and 660 mg g^{-1}, respectively.[9,43] With the SIL-treatment it was possible to remove over 90 wt% of the pectins and uronic acids from the wood. Furthermore, the removal of lignin was almost complete when the DBU–MEA–SO_2 SIL was applied, whereas about 50 wt% was removed upon treatment with the DBU–MEA–CO_2 SIL. Thus, it is evident that the DBU–MEA–SO_2 SIL was more selective for the removal of lignin compared to the DBU–MEA–CO_2 SIL. The selectivity towards lignin removal was further shown by the NMR results, which reflect the nearly complete removal of lignin from the wood. This was observed by the reduction of the peak in the region between 125 and 160 ppm, which was assigned mainly to the aromatic carbons of lignin, when compared of the spectra in Figure 5.17a (untreated birch) and Figure 5.17b and c (treated birch wood).[43] The ^{13}C NMR spectrum of native birch is compared with the spectra of the undissolved material after SIL treatment in Figure 5.17. A signal in the region between 50 and 105 ppm is assigned to cellulose, while the signal at 89 ppm relates to C-4 of the highly ordered cellulose of the crystallite interiors. However, the broader upfield signal at 84 ppm is also assigned to the C-4 of disordered cellulose.[43,60,61] The signals resonating at 21 and 173 ppm, respectively, denote the methyl and carboxylic carbon of acetyl groups attached to hemicelluloses and their intensities were reduced after both SO_2- and CO_2-based SIL treatment (Figure 5.17), thus indicating the removal of hemicelluloses from the wood.

5.3.3.5 SILs for Fractionation of Fast Growing Biomass: Grass, Agricultural Residues and Eucalyptus Bark

Herein, the power of SILs as a fractionation solvent for biomass was demonstrated on non-wood/fast growing biomass as a sustainable, environmentally friendly and cost-efficient approach. The primary fraction obtained upon the hydrated SIL fractionation process contains hemicelluloses as well as

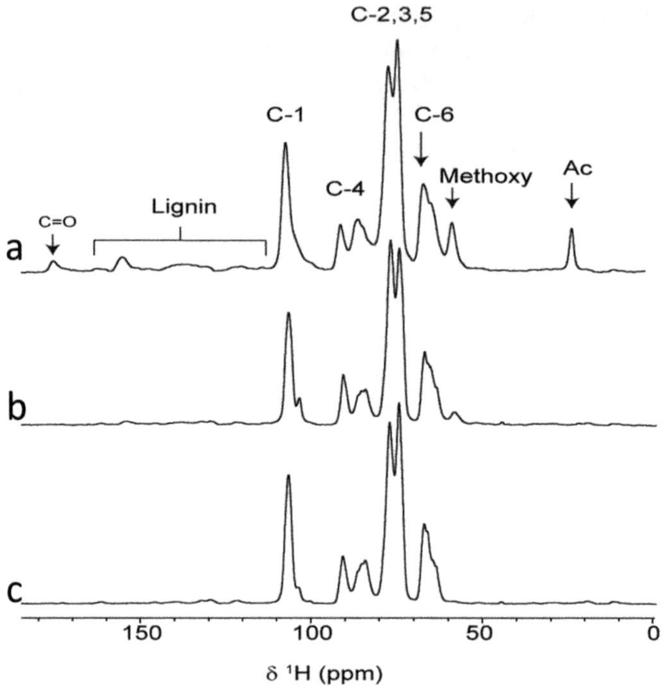

Figure 5.17 ^{13}C NMR spectra for (a) native birch (~1 mm), (b) DBU–MEA–CO_2 SIL-treated birch, and (c) DBU–MEA–SO_2 SIL-treated birch. Treatment was performed at 120 °C for 24 hours with stirring.

cellulose-rich pulp with a very low lignin content, which is analogous to what is obtained in the treatment of woody biomass. The chemical analysis results revealed that substantial removal of lignin occurred, which is consistent with the results of SIL treatment of wood. The aim was to assess the potential of this type of poorly explored biomass as a source of potentially valuable raw materials.

SILs have the ability to dissolve biomass by efficiently disrupting the complex network of non-covalent interactions between carbohydrates and lignin. It was demonstrated that switchable ionic liquids, specifically the SO_2-switched DBU–MEA SIL, can also be used to successfully fractionate non-conversional biomass types, such as grasses, agricultural residues and eucalyptus bark. The material size was not an issue since the SIL process uses big chunks for material (chips), thus reducing the energy demands of this process step. The SIL treatment for eucalyptus bark resulted in 48 wt% remaining as the non-dissolved fraction (of which 89.2 wt% was glucan, 7.3 wt% was hemicelluloses and 1 wt% was lignin), while also in the case of bamboo 49 wt% of the biomass remained in the non-dissolved fraction (of which 73.7 wt% was glucan, 9 wt% was hemicelluloses and 9 wt% was lignin). Still, in the case of wheat straw, again 50 wt% remained in the non-dissolved fraction (of which 67 wt% was glucan, 11 wt% was hemicelluloses and 13 wt% was lignin).[62]

5.3.3.6 Toward Industrial Utilization of SILs as Fractionation Solvents for Biomass

Steps towards actualizing a functional industrial process for the realization of a real biorefinery involved the study of various parameters affecting the process efficiency and the design of an optimized approach for the deconstruction of woody biomass into its main fractions. The main products from the SIL fractionation are a pulp containing hemicelluloses and a very low lignin content, as well as lignin in its very nearly native and non-degraded form. The approach studied by the authors was a short-time-high-temperature (STHT) process whereby a highly diluted, aqueous 'SO_2-switched' SIL based on an alkanol amine (monoethanol amine, MEA) and an organic superbase (1,8-diazabicyclo-[5.4.0]-undec-7-ene, DBU) was applied.[43,63,64] The ultimate objective was to develop a more sustainable, environmentally friendly and cost-efficient system for the efficient separation of the lignocellulosic fractions. Analogously, the NMR results reveal that substantial removal of lignin occurs even when a reasonably low amount of SIL is used. By using a very simple mathematical correlation, the authors described the dissolution of the lignocellulose components (hemicellulose and lignin) and the weight loss of wood as a function of time. The most efficient process involved the use of a SpinChem® rotating bed reactor, but promising results were also obtained using a flow through (loop) reactor at a treatment time of 4 hours.[63–65] All reactor systems studied gave rise to a rather low removal of hemicelluloses, which means that the solvent system is primary selective towards lignin dissolution. Furthermore, it was possible to fractionate the components of hard and softwood wood chips using the aqueous SO_2-switched DBU–MEA SIL. It was also demonstrated that different reaction conditions have a significant effect in terms of the final results, from the total weight reduction to the appearance of the wood.[63] Of the three types of reactors used, the most efficient was the SpinChem® rotating bed reactor (SRBR) that allows for 10-fold dilution in the SIL concentration with water (wt ratio 1 : 1 : 10 for wood : SIL : water) and still gives rise to reasonably good results. Also, the fractionations carried out in the SRBR gave similar results using either dry chips or soaked chips. The lignin analysis confirmed that the lignin in the wood was dissolved by the SIL, as expected, since the acidic effect of the SIL promotes the dissolution of the lignin matrix. Furthermore, it was observed that prolonging the fractionation time not only improved the dissolution of lignin, but also contributed to the dissolution of hemicelluloses. Thus, by carefully tuning the treatment conditions and the relative amounts of the SIL, water and wood, this process can give rise to good results in terms of fractionation of the wood components. The loop reactor showed consistency in terms of weight loss time in the modelling of the reactor and the same was true for the hemicellulose removal, which could be explained by the reactor type. However, all reactor types were modelled rather satisfactorily.[9]

5.3.3.7 Deconstruction of Hardwood in SILs and Acylation of the Dissolved Cellulose

The successful fractionation of hardwood (*Betula pendula*) was achieved using a batch autoclave equipped with a custom-made SpinChem® rotating bed reactor running at 120 °C using either CO_2- or CS_2-based SILs as solvents. The results of the chemical analysis of the undissolved residual woody material after treatment showed that 64 wt% of hemicelluloses and 70 wt% of lignin were removed from the native wood. Either by prolonging the processing time or by having sequential short-time treatments using new SILs would further decrease the amount of hemicelluloses and lignin in the residual woody material by 12 and 15 wt%, respectively. The derivatization process was carried out by partially dissolving the cellulose-rich fraction in an organic superbase and an ionic liquid system. Acylation of the partially dissolved cellulose was achieved both in the presence and in the absence of a catalyst, resulting in cellulose acetates with diverse degrees of substitution (DS), depending on the treatment conditions.[66] Comparing the cellulose content in the undissolved residual woody material after treatment against the native wood revealed that the amount of glucans remained basically unaffected by the SIL treatment, demonstrating that no degradation of cellulose has occurred. Higher delignification and hemicellulose solubility efficiency was achieved by the use of a CS_2-based SIL compared to the SIL based on CO_2. This was attributed to the strong nucleophilic characteristics and higher affinity of the sulphur-based anions towards lignin ether bonds. These observations are in agreement with literature data showing that under analogous conditions, delignification is the foremost occurrence upon the use of SILs based on SO_2 due to the presence of nucleophilic sulphur-containing anions.

5.3.4 Switchable Ionic Liquids as Reaction and Separation Media

One of the major parts requiring a lot of energy in chemical processes is the separation of the products from the solvent, since in many cases distillation is the only feasible way of achieving it. Liquid–liquid extraction gives one possibility to do this. However, if the polarity or hydrophobicity of the solvent could be changed, the process would be more efficient. Jessop *et al.* introduced these kinds of solvents when they studied the extraction of soybean oil from soy flakes and the separation of the oil from the extraction solvent.[67] The idea was to identify a solvent that is not miscible with water but could extract oil from soybean flakes. Then the solvent would react with CO_2 and water to form a water-soluble ionic liquid and the soybean oil would separate into a different phase. They tested several amidines for solubility in water and found two that were not soluble and could be utilized in the process. Blasucci *et al.* studied the extraction of oil from tar sand/oil shale with different reversible one- or two-component ionic liquids.[68,69] In this process, the oil could be extracted in the molecular form of the switchable ionic liquid, and then the oil could be separated from the solvent by switching it to an ionic form after the two solution phases separate and the oil can be easily decanted. The oil is used for further processing and the ionic liquid is recycled back for re-use in extraction. A different kind of example is separation of aliphatic and aromatic amines with CO_2-switchable ionic liquid aqueous two-phase systems, which was studied by Xiong *et al.*[70] In their system, both amines were dissolved in a traditional ionic liquid, such as 1-butylpyridinium trifluoromethane sulphonate, in water. After the solution was bubbled with CO_2, the aromatic amine formed an ionic liquid with water and was retained in the same phase as the traditional IL. However, the aliphatic amine forms a salt which affiliates with the water phase. These two phases can be separated by decantation. The aromatic amine can be separated from the IL by steam distillation and the IL can be recycled. The aliphatic amine could be separated from its salt by bubbling the water solution with nitrogen under elevated temperature, which breaks the salt and releases CO_2.

The ability to change the polarity of the solvent can also be utilized in chemical reactions when the solvent is used as reaction and separation media. The main benefit can be achieved when *e.g.* the reactants and catalyst are soluble in both the polar and non-polar form of the switchable solvent and the product is not soluble in the other form. Then the product can be easily separated by switching the solvent and filtrating it. Phan *et al.* demonstrated this phenomenon in the polymerization of styrene.[5] Styrene and an initiator were dissolved in a mixture of DBU and propanol. The reaction took place at 50 °C. After that the solution was bubbled with CO_2, which reacted with the DBU/propanol mixture and formed an ionic liquid. Polystyrene precipitated and was filtered out from the solution. The solution was then switched back to the molecular form by removing the CO_2 and recycled to be used in the next batch. Hart *et al.* successfully studied several coupling reactions, namely Claisen–Schmidt condensation of butanone and benzaldehyde, Heck reaction of bromobenzene and styrene, Cyano silylation of cyclohexanone and Michael addition of aniline to 1,3-diphenylpropenone, in different reversible ionic liquids.[71] In all of these processes, switchability was utilized in the separation of the products to a separate phase that could be easily removed, and the solvent as well as the catalyst could be recycled. Yadav *et al.* utilized a switchable ionic liquid and its capability to capture CO_2 in the hydrogenation of CO_2.[72] In this process, a DBU and methanol mixture was used to capture CO_2 and form an ionic liquid. A ruthenium catalyst was added to the ionic liquid and was allowed to react with hydrogen at elevated pressure and temperature. In the reaction, methyl carbonate was hydrogenated to methyl formate, which can further be hydrogenated to methanol (Figure 5.18).

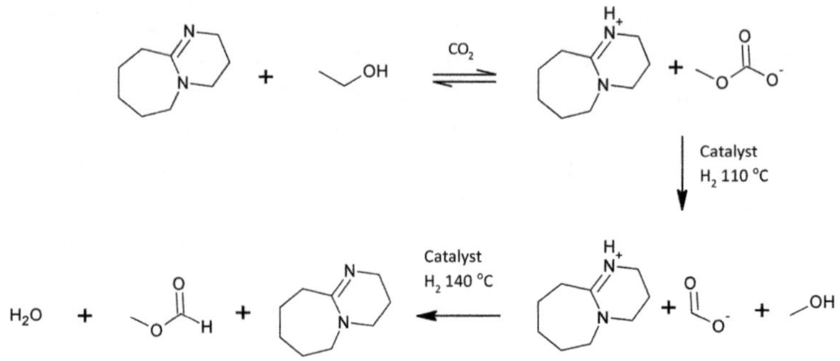

Figure 5.18 Proposed reaction scheme for the preparation of methyl formate from CO_2 and H_2 DBU/MeOH-based SILs. Adapted from ref. 72.

5.4 Conclusions

As demonstrated by these numerous examples, the development of switchable polarity solvents (SPSs), one-component switchable polarity systems, reversible ionic liquids (RevILs), switchable ionic liquids (SILs) and similar systems has only just begun. In the coming years, we will likely see many new examples of alternative 'switching' strategies and new solvent systems, as well as the demonstration of new application areas.

The biggest benefits of the 'switchable' systems arise from (i) the ease of synthesis and deconstruction of the ionic milieu, (ii) the facilitation of various separations initiated by the polarity changes, and (iii) the possibility to fine-tune the polarity by gradual change of the solvent system polarity. Thus far, nobody has demonstrated the concepts of *e.g.* a change from one polarity 'trigger' to another or the introduction of two separate 'triggers' to yield even more complex 'switchable' systems. Among other lines of research, we are currently investigating this type of system and expect to publish the results soon.

Acknowledgements

The Wallenberg Wood Science Center, the Bio4Energy programme and the Kempe Foundations (Kempestiftelserna) are gratefully acknowledged for their financial support. This work is also a part of the Åbo Akademi Process Chemistry Centre.

References

1. D. Koch and W. Leitner, *J. Am. Chem. Soc.*, 1998, **120**, 13398.
2. T. S. Chamblee, R. R. Weikel, S. A. Nolen, C. L. Liotta and C. A. Eckert, *Green Chem.*, 2004, **6**, 382.
3. P. G. Jessop, D. J. Heldebrant, X. Li, C. A. Eckert and C. L. Liotta, *Green Chem.*, 2005, **436**, 1102.
4. D. J. Heldebrant, P. G. Jessop, C. A. Thomas, C. A. Eckert and C. L. Liotta, *J. Org. Chem.*, 2005, **70**, 5335.

5. L. Phan, D. Chiu, D. J. Heldebrant, H. Huttenhower, E. John, X. Li, P. Pollet, R. Wang, C. A. Eckert, C. L. Liotta and P. G. Jessop, *Ind. Eng. Chem. Res.*, 2008, **47**, 539.
6. Y. Xie, R. Parnas, B. Liang, Y. Liu, C. Tao and H. Lu, *Chin. J. Chem. Eng.*, 2015, **23**, 1728.
7. D. J. Heldebrant, C. R. Yonker, P. G. Jessop and L. Phan, *Energy Environ. Sci.*, 2008, **1**, 487.
8. S. G. Khokarale, T. Le-That and J.-P. Mikkola, *ACS Sustainable Chem. Eng.*, 2016, **4**, 7032.
9. I. Anugwom, V. Eta, P. Virtanen, P. Mäki-Arvela, M. Hedenström, M. Hummel, H. Sixta and J.-P. Mikkola, *ChemSusChem*, 2014, **7**, 1170.
10. P. Rogne, T. Sparrman, I. Anugwom, J.-P. Mikkola and M. Wolf-Watz, *ChemSusChem*, 2015, **8**, 3764.
11. I. Anugwom, P. Mäki-Arvela, P. Virtanen, P. Damlin, R. Sjoholm and J.-P. Mikkola, *RSC Adv.*, 2011, **1**, 452.
12. E. Privalova, M. Nurmi, M. S. Marañón, E. V. Murzina, P. Mäki-Arvela, K. Eränen, D. Yu. Murzin and J.-P. Mikkola, *Sep. Purif. Technol.*, 2012, **97**, 42.
13. P. Pollet, C. A. Eckert and C. L. Liotta, *Chem. Sci.*, 2011, **2**, 609.
14. I. Kaljurand, T. Rodima, I. Leito, I. A. Koppel and R. Schwesinger, *J. Org. Chem.*, 2000, **65**, 6202.
15. K. Izutsu, *Acid–Base Dissociation Constants in Dipolar Aprotic Solvents*, Blackwell Science, Oxford, UK, 1990.
16. S. R. F. Zheng, D. J. Heldebrant, D. D. Caldwell, C. L. Aardahl, T. S. Autrey and J. C. Linehan, *Rev. Sci. Instrum.*, 2008, **79**, 084103.
17. D. R. Lide, *CRC Handbook of Chemistry and Physics*, CRC Press, Boca Raton, Florida, 84th edn, 2003.
18. A. Kraft, L. Peters, S. Johann, A. Reichert, F. Osterod and R. Frohlich, *Mater. Sci. Eng.*, 2001, **18**, 9.
19. A. N. M. Peeters, A. P. C. Faaij and W. C. Turkenburg, *Int. J. Greenhouse Gas Control*, 2007, **1**, 396.
20. D. R. Lide, *CRC Handbook of Chemistry and Physics*, CRC Press, Boca Raton, Florida, 84th edn, 2003.
21. J. S. Wilkes, *J. Mol. Catal. A: Chem.*, 2004, **214**, 11.
22. D. Xue, Y. Mu, Y. Mao, T. Yang and Z. Xiu, *Green Chem.*, 2014, **16**, 3218.
23. J. Bao, Y. Liu, R. Parnas, B. Lianga and H. Lu, *RSC Adv.*, 2015, **5**, 8311.
24. X. Cao, H. Xie, Z. Wu, H. Shen and B. Jing, *ChemCatChem*, 2012, **4**, 1272.
25. V. Blasucci, C. Dilek, H. Huttenhower, E. John, P. Pollet, C. A. Eckert and C. L. Liotta, *Chem. Commun.*, 2009, 116.
26. A. L. Rohan, J. R. Switzer, K. M. Flack, R. J. Hart, S. Sivaswamy, E. J. Biddinger, M. Talreja, M. Verma, S. Faltermeier, P. T. Nielsen, P. Pollet, G. F. Schuette, C. A. Eckert and C. L. Liotta, *ChemSusChem*, 2012, **5**, 2181.
27. J. R. Switzer, A. L. Ethier, K. M. Flack, E. J. Biddinger, L. Gelbaum, P. Pollet, C. A. Eckert and C. L. Liotta, *Ind. Eng. Chem. Res.*, 2013, **52**, 13159.
28. A. Bondi, *J. Phys. Chem.*, 1964, **68**, 441.
29. J. R. Switzer, A. L. Ethier, E. C. Hart, K. M. Flack, A. C. Rumple, J. C. Donaldson, A. T. Bembry, O. M. Scott, E. J. Biddinger, M. Talreja, M.-G. Song, P. Pollet, C. A. Eckert and C. L. Liotta, *ChemSusChem*, 2014, **7**, 299.

30. J. T. Yeh, H. W. Pennline and K. P. Resnik, *Energy Fuels*, 2001, **15**, 274.
31. H. Lepaumeier, D. Picq and P. L. Carrette, *Energy Procedia*, 2009, **1**, 893.
32. J. A. Mason, K. Sumida, Z. R. Herm, R. Krishna and J. R. Long, *Energy Environ. Sci.*, 2011, **4**, 3030.
33. R. Banerjee, A. Phan, B. Wang, C. Knobler, H. Furukawa, M. O'Keeffe and O. M. Yaghi, *Science*, 2008, **319**, 939.
34. R. P. Lively, R. R. Chance, B. T. Kelley, H. W. Deckman, J. H. Drese, C. W. Jones and W. J. Koros, *Ind. Eng. Chem. Res.*, 2009, **48**, 7314.
35. B. Kronberger, E. Johansson, G. Löffler, T. Mattisson, A. Lyngfelt and H. Hofbauer, *Chem. Eng. Technol.*, 2004, **27**, 1318.
36. G. T. Rochelle, *Science*, 2009, **325**, 1652.
37. J. L. Anthony, J. L. Anderson, E. J. Maginn and J. F. Brennecke, *J. Phys. Chem. B*, 2005, **109**, 6366.
38. E. Privalova, *Towards Novel Biogas Upgrading Processes*, Åbo Akademi University, Painosalama Oy, Turku, Finland, 2013, ISBN 978-952-12-2862-9.
39. J. L. Anthony, E. J. Maginan and J. F. Brennecke, *J. Phys. Chem. B*, 2002, **106**, 7315.
40. P. Walden, *Bull. Acad. Imp. Sci. St.-Petersbourg*, 1914, 405.
41. M. J. Muldoon, S. N. V. K. Aki, J. L. Anderson, J. K. Dixon and J. F. Brennecke, *J. Phys. Chem. B*, 2007, **111**, 9001.
42. R. Pezoa-Conte, A. Leyton, I. A. Anugwom, S. von Schoultz, J. Paranko, P. Mäki-Arvela, W. R. P. Lively, R. R. Chance, B. T. Kelley, H. W. Deckman, J. H. Drese, C. W. Jones and W. J. Koros, *Ind. Eng. Chem. Res.*, 2009, **48**, 7314.
43. I. Anugwom, V. Eta, P. Virtanen, P. Mäki-Arvela, M. Hedenström, M. Yibo, M. Hummel, H. Sixta and J.-P. Mikkola, *Biomass Bioenergy*, 2014, **70**, 373.
44. N. Zhou, Y. Zhang, X. Gong, Q. Wang and Y. Ma, *Bioresour. Technol.*, 2012, **118**, 512.
45. S. A. Choi, J. S. Lee, Y. K. Oh, M. J. Jeong, S. W. Kim and J. Y. Park, *Algal Res.*, 2014, **3**, 44.
46. K. S. Lovejoy, L. E. Davis, L. M. McClellan, A. M. Lillo, J. D. Welch, E. N. Schmidt, C. K. Sanders, A. J. Lou, D. T. Fox, A. P. Koppisch and R. E. Del Sesto, *J. Appl. Phycol.*, 2013, **25**(4), 973.
47. T. Yu, T. Yamada, G. C. Gaviola and R. G. Weiss, *Chem. Mater.*, 2008, **20**, 5337.
48. L. P. Ramos, *Quim. Nova*, 2003, **26**, 863.
49. P. Wasserscheid and T. Welton, *Ionic Liquids in Synthesis*, Wiley-VCH, Weinheim, 2006, vol. 1 & 2.
50. D. Fengel and E. Wegener, *Wood-chemistry Ultrastructure Reaction*, Walter de Gruyter, Berlin New York, 1984.
51. P. Mäki-Arvela, I. Anugwom, P. Virtanen, R. Sjöholm and J.-P. Mikkola, *Ind. Crops Prod.*, 2010, **32**, 175.
52. N. S. Mosier, C. Wyman, B. Dale, R. Elander, Y. Y. Lee, M. Holtzapple and M. R. Ladisch, *Bioresour. Technol.*, 2005, **96**, 673.
53. F. Teymouri, L. Laureano-Perez, H. Alizadeh and B. E. Dale, *Bioresour. Technol.*, 2005, **18**, 2014–2018.

54. R. P. Swatloski, S. K. Spear, J. D. Holbrey and R. D. Rogers, *J. Am. Chem. Soc.*, 2002, **124**, 4974.
55. I. Anugwom, P. Mäki-Arvela, P. Virtanen, S. Willför, R. Sjöholm and J.-P. Mikkola, *Carbohydr. Polym.*, 2011, **87**, 2005.
56. I. Anugwom, P. Mäki-Arvela, P. Virtanen, S. Willför, P. Damlin, M. Hedenström and J.-P. Mikkola, *Holzforschung*, 2012, **66**, 809.
57. D. Heldebrant, C. Yonker, P. Jessop and L. Phan, *Chem.-Eur. J.*, 2009, 7619.
58. Y. Pu, N. Jiang and A. Ragauskas, *J. Wood Chem. Technol.*, 2007, 23.
59. S. S. Y. Tan, D. R. MacFarlane, J. Upfar, L. A. Edye, W. O. S. Doherty, A. F. Patti, J. M. Pringle and J. L. Scott, *Green Chem.*, 2009, **11**, 339.
60. S. L. Maunu, *Prog. Nucl. Magn. Reson. Spectrosc.*, 2002, 151.
61. H. Wikberg and S. L. Maunu, *Carbohydr. Polym.*, 2004, 461.
62. I. Anugwom and J.-P. Mikkola, *Biomass Bioenergy*, (under review).
63. I. Anugwom, L. Rujana, J. Wärnå, M. Hedenström and J.-P. Mikkola, *Chem. Eng. J.*, 2016, **297**, 256.
64. V. Eta, I. Anugwom, P. Virtanen, K. Eränen, P. Mäki-Arvela and J.-P. Mikkola, *Chem. Eng. J.*, 2014, **238**, 242.
65. V. Eta, I. Anugwom, P. Virtanen, P. Mäki-Arvela and J.-P. Mikkola, *Ind. Crops Prod.*, 2014, **55**, 109.
66. V. Eta and J.-P. Mikkola, *Carbohydr. Polym.*, 2016, **136**, 459.
67. P. G. Jessop, L. Phan, A. Carrier, S. Robinson, C. J. Dürr and J. R. Harjani, *Green Chem.*, 2010, **12**, 809.
68. V. Blasucci, R. Hart, V. Llopis-Mestre, D. Hahne, M. Burlager, H. Huttenhower, B. Thio, C. L. Liotta and C. A. Eckert, *Fuel*, 2010, **89**, 1315.
69. V. M. Blasucci, R. Hart, P. Pollet, C. L. Liotta and C. A. Eckert, *Fluid Phase, Equilib.*, 2010, **294**, 1.
70. D. Xiong, Z. Li, H. Wang and J. Wang, *Green Chem.*, 2013, **15**, 1941.
71. R. Hart, P. Pollet, D. J. Hahne, E. John, V. Llopis-Mestre, V. Blasucci, H. Huttenhower, W. Leitner, C. A. Eckert and C. L. Liotta, *Tetrahedron*, 2010, **66**, 1082.
72. M. Yadav, J. C. Linehan, A. J. Karkamar, E. van der Eide and D. J. Heldebrant, *Inorg. Chem.*, 2014, **53**, 9849.

CHAPTER 6

Stimuli Responsive Smart Fluids Based on Ionic Liquids and Poly(ionic liquid)s

JIANBO YIN*, QI LEI, YUEZHEN DONG AND XIAOPENG ZHAO

Smart Materials Laboratory, Department of Applied Physics, Northwestern Polytechnical University, Xi'an 710129, P. R. China
*E-mail: jbyin@nwpu.edu.cn

6.1 Introduction

Smart materials can adaptively respond to an external stimulus and exhibit a useful physical or chemical change such as a volume change, mechanical stress, oxidization–deoxidization, and so on.[1] The stimulus may be mechanical stress, temperature, light, moisture, pH, or an electric or magnetic field. An important characteristic of the response of smart materials is reversibility or tunability by controlling the strength of the stimulus. To obtain available smart materials, a key aspect is to develop substances with high physical and chemical stability and strongly stimulus-responsive characteristics. In this regard, ionic liquids (ILs) might be very suitable compounds for such purposes owing to their robust and tunable properties compared to molecular compounds.

Ionic liquids are organic molten salts at temperatures below 100 °C. Because of the presence of strong electrostatic attraction between the cation

Smart Materials No. 29
Polymerized Ionic Liquids
Edited by Ali Eftekhari

Published by the Royal Society of Chemistry, www.rsc.org

part and the anion part, ionic liquids are usually very stable in terms of their physical and chemical properties, including negligible vapor pressure, high thermal stability, and even non-flammability. On the other hand, depending on the ionic constituents, alkyl chain length, and functional groups, the properties of ILs, such as the melting point, polarity, ionic conductivity, viscosity, density, hydrophilicity/hydrophobicity, solubility, and coordination ability, can be tailored.[2] Based on these characteristics, ionic liquids are not only being used as substitutes of volatile organic solvents in green chemistry and as safe electrolytes in electrochemical technology,[3] but have also been finding an increasing number of applications as novel functional materials, such as energetic materials, optical materials, electrochromic materials, ionic materials with switchable polarity, ionic liquid/carbon hybrid materials, and so on.[4] In particular, by functionalizing ILs themselves or incorporating ILs into traditional materials, researchers have explored smart IL materials with stimuli-responsive characteristics. For example, introducing photo-sensitive functional groups into ILs has yielded photo-responsive IL materials. Under light irradiation, these photo-responsive IL materials can actively change their physicochemical properties, such as melting points and conductivities.[5,6] Modifying indicator dyes with ILs has been demonstrated to give a more sensitive pH response in a wide range of solvents.[7] Because of their non-toxicity, wide temperature range of the liquid state, and volume change with temperature, ILs themselves can be used as more robust thermometric fluids in liquid-in-glass thermometers compared to classic volatile liquids such as ethanol and toxic mercury.[8] Because the negligible volatility of ILs can eliminate the possibility of weight loss due to liquid vaporization, ILs can also be directly used as gas sensing materials for the detection of polar and nonpolar organic vapors in quartz crystal microbalance techniques.[9] ILs themselves can be used as ion-responsive materials because their physicochemical properties are easy to vary by ion exchange.[10] In addition, introducing magnetic counter cations into ILs, such as Fe^{3+}, Co^{2+}, Gd^{3+}, Dy^{3+}, and so on, has produced novel ILs showing a strongly magneto-responsive characteristic.[11–14]

Besides small molecular ILs, recent research on polymeric ionic liquids or poly(ionic liquid)s (PILs) has stimulated new interest in developing new functional and smart materials, because PILs not only inherent the unique properties of ionic liquids mentioned above, but also possess intrinsic polymer properties (*e.g.* well-defined solid morphologies, processability and good mechanical properties) that are inaccessible by their low molecular weight counterparts.[15,16] For example, solid gas sensing materials based on tetraalkylammonium-based PILs have been demonstrated to have more sensitive CO_2 sensing behavior compared to liquid gas sensing materials based on ILs.[17] The combination of PILs and photonic structures has afforded a new class of self-reporting humidity sensing materials with excellent reversibility.[18] PIL brushes with tunable wettability have been successfully produced on the surface of silicon wafers by surface-initiated ATRP polymerization.[19] By combining temperature-sensitive poly(*N*-isopropylacrylamide) with pH-responsive PILs, thermal and pH dual-responsive PIL membranes have also been reported. These fabricated PIL membranes exhibit excellent mechanical properties and are able to act as pH-responsive soft actuators.[20] Porous PIL actuators that bend in response to acetone vapor have been prepared based on poly(3-cyanomethyl-1-vinylimidazolium bis(trifluoromethanesulfonyl)imide) and a carboxylic acid-substituted pillar[5]arene. Due to the unique combination of a porous morphology, a gradient structure and the interaction between solvent molecules and actuator materials, these actuators exhibit a speed that is an order of magnitude faster than the state-of-the-art, coupled with large-scale locomotion.[21] In addition, a porous PIL membrane actuator exhibiting exceptional sensitivity to low organic solvent concentrations has also been reported based on poly[3-cyanomethyl-1-vinylimidazolium bis(trifluoromethanesulfonyl)imide] and poly(acrylic acid). This actuator readily bent into an arc (curvature 0.076 mm^{-1}) upon adding as little as 0.25 mol% acetone molecules (1 acetone per 400 water molecules).[22] A novel imidazolium sulfonate-containing pentablock copolymer–ionic liquid composite has also been prepared and fabricated into a thermally stable electroactive actuator, which exhibits an effective actuation response under a low applied electrical potential of 4 V.[23]

Therefore, based on the broad range of properties of ILs, it is believed that in the near future the development of smart materials can significantly be improved by incorporating ILs into their designs. In particular, recent research on PILs has further provided new candidates for the development of new smart materials with well-defined solid morphologies, processability and good mechanical properties. Besides many research papers, some review papers have focused on topics related to functional materials based on ILs and PILs.[16,24,25]

6.2 Electro/magneto-responsive Smart Fluids

Using an external electric or magnetic field stimulus to control the flow or deformation of a fluid is very interesting for fundamental research and technical application. Electrorheological (ER) fluids and magnetorheological (MR) fluids are two kinds of very important stimulus-responsive smart fluids, whose viscosity and viscoelastic properties can be controlled by an external electric field and a magnetic field, respectively.[26]

ER fluids are composed of micro/nano-size dielectric particles with a relatively high dielectric constant in an insulating liquid with a relatively low dielectric constant, while MR fluids always consist of micrometer-sized soft magnetic particles in a liquid carrier. In the absence of an electric field or a magnetic field, the particles are freely dispersed in the carrier liquid, as shown in Figure 6.1 (left), and the fluid exhibits a low viscosity Newtonian fluid characteristic, as shown in Figure 6.2(b). In the presence of an electric field or a magnetic field, the dispersed particles are polarized and attracted to each other, forming a chain or column structure between two electrodes, as shown in Figure 6.1 (right). As a result, the viscosity of the fluid will be

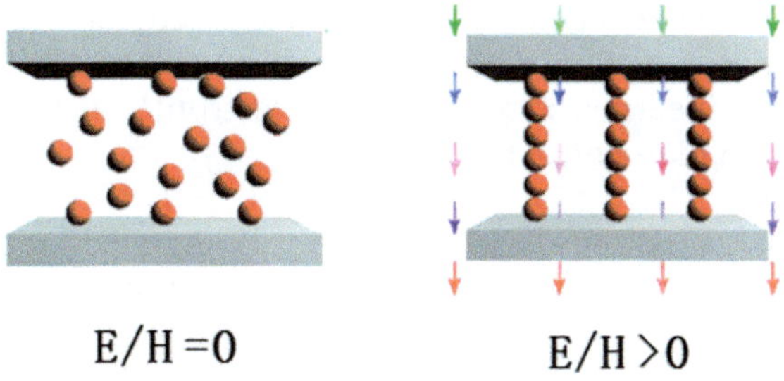

Figure 6.1 Schematic representation of the interactions between ER or MR particles dispersed in a liquid carrier in the absence (left) and in the presence (right) of an electric field or a magnetic field.

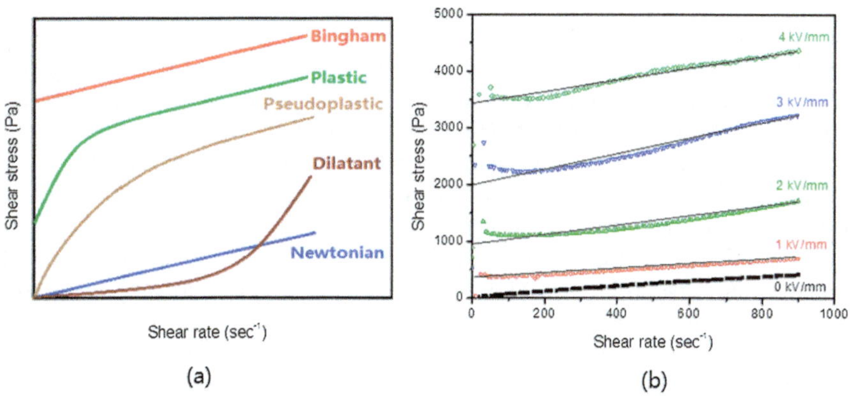

Figure 6.2 (a) Schematic flow curves of various fluids. (b) Flow curves of a typical ER fluid of mesoporous rare earth ion-doped titania particles under various electric fields (the symbol points represent the experimental data and the dotted lines are fitted by the Bingham mode).

largely increased within milliseconds, as shown in Figure 6.2(b). When the electric field or magnetic field strength is sufficiently strong, the fluid even transforms from a liquid-like state into a gel-like solid showing a significant yield stress to resist deformation. Meanwhile, the flow of the fluid can be approximately modeled as a Bingham fluid following the relation below:

$$\tau = \tau_y + \eta_{pl}\dot{\gamma} \tag{6.1}$$

Where $\dot{\gamma}$ is the shear rate, τ is the shear stress, τ_y is the dynamic yield stress, and η_{pl} is the plastic viscosity. This field-induced thickening is often referred to as the "ER effect" or "MR effect". The former was first discovered by W. M. Winslow in 1939,[27] and the latter was first reported by J. Rabinow in 1948.[28] When the external electric field or the magnetic field is removed, the fluid can rapidly return to its original liquid state. Due to the advantages of fast response, reversibility, low energy consumption, and so on, ER fluids and

MR fluids have attracted significant attention as electrical–mechanical interfaces for the active-control of various mechanical devices in the automotive industry, civil engineering, robotics, microelectronic devices and even in optical, sound and biomedical fields.[29]

Under a magnetic field of sufficient strength, an MR fluid can produce a yield stress over several tens of kPa. Due to this large field-responsive characteristic, MR fluids have been used in various technological applications, including semi-active shock absorbers in the automotive industry, dampers for seismic damage control in civil engineering, seals, valves, robotics, microelectronic devices, and so on.[30] MR fluids have been proposed for drug delivery and cancer therapeutic methods in medicine.[31] Currently the main problem with MR fluids is the settling of the dispersed magnetic particles, which may limit their further applications.[32] However, the use of ILs as new carriers for the preparation of stable MR fluids has been recently proposed, and it has been demonstrated that the use of ILs for this specific purpose may lead to MR fluids with a higher durability and a better stability against sedimentation due to the unique features of ILs such as high density, negligible vapor pressure and property tunability.[33] In a recent book chapter, Laura Rodríguez-Arco *et al.* have given a good overview of the latest advances and research highlights in the field of MR fluids concerning the use of room-temperature ILs as liquid carriers.[34]

Compared to MR fluids, ER fluids have some unique advantages including faster response, relatively high dispersion stability and lower energy consumption. In particular, ER fluids are also proving to be especially useful for small-size actuators in MEMs because an ER fluid only requires two thin electrodes connected by wires to activate it.[35] However, the large-scale utilization of ER fluids has not been achieved up to now because of some shortcomings, such as low electro-responsive efficiency, narrow working temperature range, and so on.[36] Therefore, current fundamental research on ER fluids still focuses mainly on the development of novel and high-performance ER materials in order to promote the real applications of ER technology. However, recent research into using ILs as new activators of first-generation ER fluids and using PILs as new anhydrous polyelectrolyte-based ER active components has provided new ways to develop ER fluids with high performance.[37,38] In the next sections, we will give a detailed description of these.

6.3 Electro-responsive Electrorheological Fluids

ER fluids are a typical two-phase system that consists of micrometer-size leaking dielectric particles in an insulating carrier liquid. The carrier liquid is always insulating oil with a low dielectric constant, low conductivity, low viscosity, high breakdown strength, high boiling point (over 200 °C), and a relatively high density. The most popular carrier is silicone oil. Other oils, such as vegetable oil, mineral oil, paraffin, kerosene, chlorinated hydrocarbons, transformer oil, and so on, have also been used.[36] Some high-density oils, such as fluoro- or phenyl-silicone oil, can also be used to decrease the sedimentation of particles. But these oils are usually high cost. In some cases, the ER effect is strongly related to the carrier liquid if the dielectric constant or conductivity of the particle phase is comparable to that of the carrier liquid. In most cases, however, the particle phase is very important to obtain an available ER effect. The particle phase must possess some required physical and chemical properties, including appreciable dielectric and conduction properties, low density, suitable size and shape, mechanical and chemical stability, and so on.[39] So the particle phase must be carefully selected and designed on the basis of these required physical and chemical properties.

Since Winslow first discovered the ER effect in a suspension of starch particles,[27] various particle materials ranging from inorganic to organic to composite have been developed to obtain versatile ER fluids for technological applications.[40–43] The first-generation ER fluids are based on water activated particles, such as silica gel, poly(lithium methacrylate), cellulose, starch, and so on.[44] These systems need to adsorb water to produce an ER effect. The adsorbed water has been considered to promote mobile charge carriers and induce particle polarization under electric fields and, thus, activate the ER effect.[45] However, adsorbed water also increases the current density and limits the working temperature stability. Other drawbacks including device corrosion and dispersion instability also need to be overcome. To overcome the drawbacks of water-activated ER fluids, researchers have used some other polar liquids with a high boiling point (*e.g.* glycerol, DMF, *etc.*) to replace water as an activator.[46] For example, we have developed an ER system based on glycerol-activated titania gel particles.[47] The gel particles can be prepared by an *in situ* sol–gel reaction of tetrabutyl titanate in ethanol containing glycerol. During the formation of the titania gel, glycerol can be confined in the gel network. It is found that confining glycerol can largely increase the dielectric constant of the titania gel and enhance the ER effect. The high boiling point and low volatility of glycerol can also improve the temperature stability of the titania gel ER fluid.

Compared to these polar liquids, ILs have significantly higher chemical and physical stability. In particular, ILs are entirely composed of ions, so they can largely improve the ion concentration in particles when they are confined into particles. This characteristic endows ILs with great potential as new alternative activators to help overcome the drawbacks of water and other polar liquids as activators in extrinsic ER systems. However, the use of ILs as activators has recently been proposed for the preparation of ER fluids with improved properties.[37] In Section 6.3.1, we give an overview of ER fluids with ILs as activators.

More current research studies are mainly concentrated on the development of second generation ER fluids based on dry particles. These systems are also called "intrinsic ER systems", and have a broad operating temperature range and a low leaking current density. Some inherently polarizable inorganic materials with high dielectric constant, such as rutile, perovskite, and so on, are ER active in the dry state when they are exposed to an AC electric field.[40] Some doped inorganic materials with improved ionic conductivity (*e.g.* zirconium-doped aluminosilicates, rare earth ion-doped titania, *etc.*) have been demonstrated to be ER active in the dry state under a DC electric field stimulus.[48] Due to their high density and stiffness, however, these inorganic ER particles exhibit problems involving particle sedimentation and abrasiveness. Different from inorganic materials, organic polymer particles have low density and low abrasiveness. These characteristics have attracted significant attention from researchers in the development of high-performance ER fluids.[42,49] Among various polymer particles, polyelectrolyte particles are the most promising candidates as the particle phase of ER fluids due to their low-cost and high ER effect. The classic polyelectrolyte ER particles include poly(lithium methacrylate), poly(sodium styrene sulfonate), and some ion-exchange resins.[50] The ER effect of polyelectrolyte particles is considered to arise from the interfacial polarization induced by the local migration of metal ions under electric fields. As a dry powder, however, the metal ions in the particles are strongly bound by the carboxylate or sulfonate groups. Therefore, the classic polyelectrolyte particles need to adsorb a small amount of water to promote the dissociation and migration of metal ions and thus activate the ER effect. But the adsorbed water inevitably causes the electrical and thermal problems mentioned before. To overcome these problems, the scientists at Bayer AG have developed, for the first time, anhydrous polyelectrolyte-based ER fluids based on poly(ethylene oxide)–salt (PEO–salt) particles by reacting polyethylene glycol ethers with toluene di-isocyanate to form cross-linked polyurethane elastomers, which are dissolving salts (*e.g.* lithium chloride or zinc dichloride).[50,51] Compared to classic polyelectrolytes, the metal cations in the PEO–salt polyelectrolytes are weakly bound by the electron pairs on the oxygen of the PEO segments and thus the particles contain mobile ions themselves in the absence of water or other small molecule activators. This new polyelectrolyte ER fluid exhibits light weight, low off-field viscosity and high field-induced yield stress in the dry state. However, there are some problems with this PEO–salt ER fluid. For example, the leaking current density of the PEO–salt ER fluid is still too high, in particular at high temperature, due to the fact that the small-size mobile metal counter ions can easily leach into the polymer matrix. In addition, the PEO–salt polyelectrolyte is still moisture-sensitive because of the hydrophilic nature of the PEO units.[52] Therefore, to promote real applications, a new anhydrous polyelectrolyte substance manifesting large intrinsic bulk or surface polarization under an electric field, a relatively low level of conductivity without affinity for a moisture-rich environment, and a stable ER effect in a wide working temperature range still needs to be found.

Different from classic polyelectrolytes, most PILs are water insoluble polyelectrolytes due to the presence of fluoric counter-ions, such as hexafluorophosphates (PF_6^-) and fluorinated imides ($((C_nF_{2n+1}SO_2)_2N^-)$).[53] Although the polymerization limits to some extent the long-range motion of the ions in PILs, the short-range motion of ions can still be promoted by external electric fields. Importantly, different from classic polyelectrolytes, making the ions locally move in PILs does not need the presence of water or other solvents.

At the same time, PILs contain a high density of cation/anion parts and therefore have strong dipoles. In light of this, we have recently developed a new ER system based on hydrophobic PIL particles and demonstrated that this new system seems very promising for the preparation of new anhydrous polyelectrolyte-based ER systems to help to overcome the disadvantages of the water-activated classic polyelectrolyte-based ER systems. In Section 6.3.2, we introduce microwave-assisted dispersion polymerization preparation, the anhydrous ER effect and its dependence on the molecular structure of ER systems of PILs.

6.3.1 Electrorheological Fluids Based on Ionic Liquids

In the first-generation extrinsic ER systems, the presence of adsorbed water or other polar liquids is the key to the ER effect. The adsorbed water or other polar liquid is considered to dissolve the ions in the particles and thus promote mobile charge carriers to induce particle polarization under electric fields. However, the adsorbed water or other polar liquid also increases the current density and limits the working temperature range. Having considered the unique properties of ionic liquids, such as excellent thermal and chemical stability, good ionic conductivity, and negligible vapor pressure, recent studies using ionic liquids as new alternative activators have proposed a new approach to prepare an extrinsic ER fluid with improved properties.[37]

J. A. Marins *et al.* developed the first ER fluid with ionic liquids as an alternative activator.[37] The particle phase in this fluid is common silica spheres. The ILs are not adsorbed on the surface of the silica spheres like a conventional water-activated ER system, but confined in the silica amorphous network through a sol–gel process like previously reported glycerol-activated titania gel particles.[47] Two types of ILs whose structures are shown in Figure 6.3 were used in the preparation. They are typical water soluble ILs. Although no acidic or basic catalyst is added, it is found that the pH of the sol–gel medium and the velocity of the sol–gel process are significantly decreased in the presence of the ILs, in particular IL1. This is considered to probably be because of the presence of a carboxyl group in the structure of used ILs that can also catalyze the initial hydrolysis step. The particles prepared in the

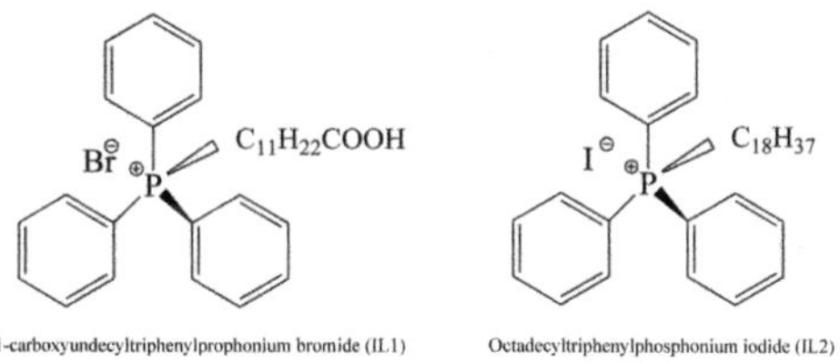

Figure 6.3 Chemical structure of phosphonium-based ILs. Reprinted from Journal of Colloid and Interface Science, 405, J. A. Marins, B. G. Soares, A. A. Silva, M. G. Hurtado, S. Livi, Electrorheological and dielectric behavior of new ionic liquid/silica systems, 64–70, Copyright (2013) with permission from Elsevier.[37]

presence of ionic liquids presented higher density and improved dielectric properties. The relative permittivity of the silica/11-carboxyundecyltriphenylphosphonium bromide sample was 1100 at 10 Hz, which is much higher than that of pure silica. Although it is comparable to that of barium titanate particles, the electrode polarization seems to be not removed from the dielectric constant of silica/ionic liquid hybrid particles.

An ER fluid based on silica/IL hybrid particles in silicone oil presents an enhanced ER effect, with values of shear stress at 1 kV mm^{-1} of electrical field that are comparable or even higher than those of several other systems reported in the literature, such as porous silica particles,[54] silica nanoparticles coated with polyaniline,[55] and polypyrrole confined in mesoporous silica.[56] Figure 6.4 shows the typical flow curves of the ER fluids of silica/IL hybrid particles. Without an electric field, these fluids behave as Newtonian fluids. In the presence of an electric field, the shear stress values significantly increase and the behavior changes from that of Newtonian fluids to that of Bingham plastic fluids. The ER effect seems to depend on the type of IL. For example, the increment in the shear stress at 10^{-3} s^{-1} for the silica/11-carboxyundecyltriphenylphosphonium bromide sample was around 140 times that of pure silica at 1 kV mm^{-1}, and that for the silica/octadecyltriphenylphosphonium iodide sample was around 12 times that of pure silica.

Further work involving other kinds of ILs confined in silica particles has also been reported in order to investigate the effect of ILs on the particle morphology and ER effect.[57] In this work, a similar sol–gel method was employed to prepare ILs confined in silica particles, using three different commercial ILs, including tri-isobutyl(methyl)phosphoniumtosylate (IL106),

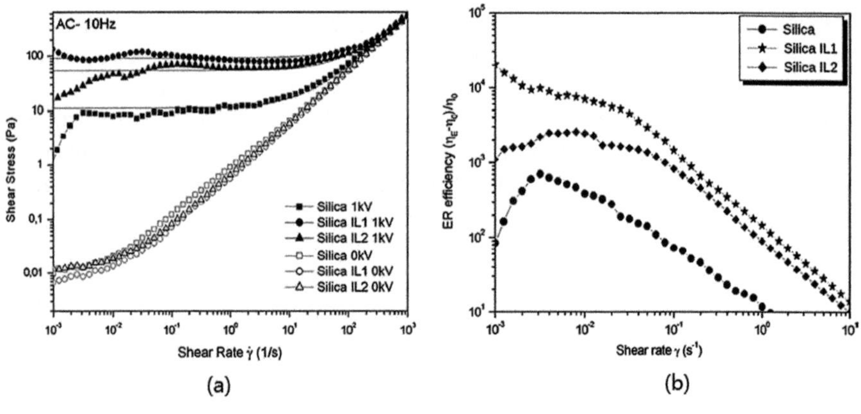

Figure 6.4 Shear stress (a) and ER efficiency (b) *vs.* shear rate measured in CSR mode of suspensions of pure silica, SiO_2/IL1, and SiO_2/IL2 particles. The lines in (a) represent the fit curve according to the Bingham mode. Reprinted from Journal of Colloid and Interface Science, 405, J. A. Marins, B. G. Soares, A. A. Silva, M. G. Hurtado, S. Livi, Electrorheological and dielectric behavior of new ionic liquid/silica systems, 64–70, Copyright (2013) with permission from Elsevier.[37]

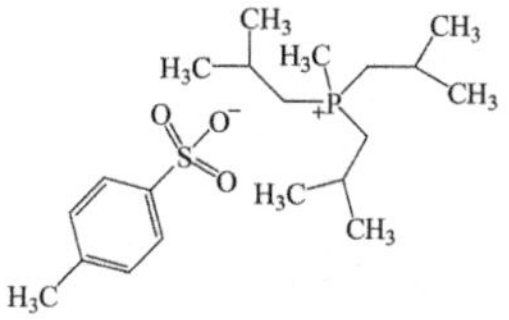

IL106

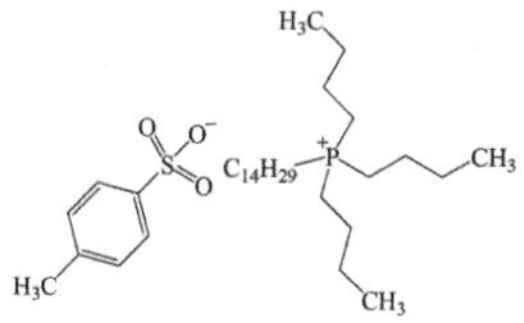

IL201

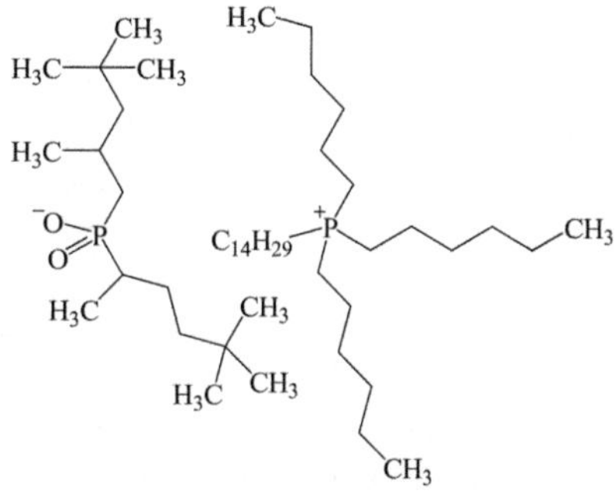

IL104

Figure 6.5 Chemical structures of ionic liquids.

tri(*n*-butyl)(tetradecyl)phosphonium-dodecylbenzene-sulfonate (IL201) and trihexyl-(tetradecyl)-phosphonium-bis-2,4,4-(trimethylpentyl)-phosphinate (IL104), whose structures are shown in Figure 6.5. It is found that the morphology of pure silica (SiO_2) is a spherical morphology with an average size of 170–600 nm, while the silica particles prepared in the presence of IL106 (SiO_2/IL106), IL104 (SiO_2/IL104) or IL201 (SiO_2/IL201) during the sol–gel synthesis are larger particle size aggregates, indicating a strong influence of the ionic liquids on the morphology of the silica particles (Figure 6.6).

Silica prepared in the presence of phosphonium-based ILs also displays a higher relative permittivity and improved interfacial polarization due to the enhanced accumulation of charge carriers from the confined ILs. Among the various silica/IL hybrid particles, SiO_2/IL104 has a higher permittivity value due to the higher mobility of the corresponding IL104 ions, but the SiO_2/IL104 ER fluid also presents a very large current density under an electric field and a decreased ER effect. In contrast, the SiO_2/IL106 fluid presents the highest ER effect and a relatively low current density, as indicated in Table 6.1. At the same time, the low current density of this system allows experiments to be performed in a higher electric field. Further comparing the ER effects of SiO_2/IL201 and SiO_2/IL106, it was found that the former displays the worst ER effect, despite the similar nature of the anion (sulfonate). Thus, the different ER effects may be due to the difference between them in terms of the size of the alkyl groups located on both the cation and anion moieties, which has an effect on the compatibility of the ionic liquids with the silicone oil. IL201 has long alkyl groups on both the cation and the anion, which imparts some

Figure 6.6 SEM images of silica, SiO_2/IL106, SiO_2/IL201 and SiO_2/IL104 particles. (Reproduced from ref. 57 with permission from the Royal Society of Chemistry).

Table 6.1 ER characteristics of ER fluids containing silica and silica/IL particles. (Reproduced from ref. 57 with permission from the Royal Society of Chemistry).

Sample	Electric field (kV mm^{-1})	Current density (A m^{-2})	Shear stress at 0.5 s^{-1}(Pa)
SiO_2	2	0.041	373–568
SiO_2/IL106	2	0.021	708
SiO_2/IL106	3	0.048	1215
SiO_2/IL201	2	0.099	72
SiO_2/IL104	2	0.117	104–149

affinity for the silicone oil and, thus, some molecules of IL201 previously confined inside the silica particles may be extracted by the oil, increasing the ionic conductivity of the oil. This results in an increase in the current density and a decrease in the ER effect of the suspension under an electrical field.

Therefore, due to their significant chemical and physical stability, ILs may be very promising alternative activators for extrinsic ER systems. However, IL-activated ER fluids still exhibit a high conductivity and large leaking current. In addition, the presently used ILs are still water-soluble systems, which cannot eliminate the influence of moisture. Thus, the real contribution of using ILs as an alternative activator still needs to be clarified by further investigations.

6.3.2 Electrorheological Fluids Based on Poly(ionic liquid)s

Polyelectrolyte particles are promising candidates due to their low-cost and relatively high ER effect.[50] However, the classic polyelectrolyte ER particles belong to extrinsic water-activated systems because they need to absorb a small amount of water to promote the dissociation and migration of ions and thus induce interfacial polarization for the ER effect. Although PEO–salt particles can show anhydrous ER effects, they are still moisture-sensitive because of the hydrophilic nature of the PEO units.[52] This largely limits the durability of PEO–salt-based ER fluids in practical applications.

As a new kind of polyelectrolyte, PILs have attracted significant attention recently, because PILs not only inherit some of the unique properties of ILs mentioned before, but also possess well-defined solid morphologies and mechanical properties that are inaccessible by ILs. In particular, different from classic polyelectrolytes, most PILs are hydrophobic and insoluble in water due to the presence of fluoric counter-ions, such as hexafluorophosphates (PF_6^-), and fluorinated imides ($(C_nF_{2n+1}SO_2)_2N^-$). Thus, the electrical conductivity of PILs almost completely lacks affinity to moisture or water content. Furthermore, the mobile counter ions in PILs are bulky polyatomic inorganic (*e.g.* PF_6^-) or organic (*e.g.* $(CF_3SO_2)_2N^-$) ion species and their transfer dynamics are also different from that of the small-size metal ions in the classic polyelectrolytes. After polymerization, the long-range transfer of counter ions can be restricted and the ionic conductivity of the PILs can be reduced to a moderate level. Thus, the local short-range motion of counter ions in addition to the high-density of cation/anion parts can endow PILs with a large intrinsic bulk or surface polarization response to an external electric field in the absence of any activators. Furthermore, most PILs are easy to synthesize by chain radical polymerization of monomeric ILs (*e.g.* vinyl, styrenic, (meth)acrylic, and (meth)acrylamide ionic liquids, *etc.*).[53] Therefore, it is easier to obtain the products in a powder or particle form compared to PEO–salt polyelectrolytes.

In light of these characteristics, our recent work has, for the first time, reported a new kind of anhydrous ER system based on hydrophobic poly[2-(methacryloyloxy)ethyltrimethylammonium bis(trifluoromethanesulfonylimide)] (poly($MTMA^+(CF_3SO_2)_2N^-$)) PIL particles.[38] The poly($MTMA^+(CF_3SO_2)_2N^-$) particles are easy to synthesize using microwave-assisted dispersion polymerization as shown in Figure 6.7. Compared to conventional heating, using microwave irradiation as a heat source can induce fast and effectively homogeneous heating of the monomers, which favors the homogeneous formation of growing centers of polymer chains in the initial stage and results in the reduction of side reactions. As a result, the product consists of clean and monodisperse microspheres with a narrow size distribution of ~1.8 μm.

The TGA trace shows that the poly($MTMA^+(CF_3SO_2)_2N^-$) particles have good thermal stability. The onset of degradation is at about 320 °C. The conductivity (σ_{DC}) of the poly($MTMA^+(CF_3SO_2)_2N^-$) particles is ~2.2×10^{-9} S cm^{-1}. Because the trimethylammonium cation parts are attached to the poly(meth)acrylic backbone, the conductivity mainly originates from the diffusion of mobile $(CF_3SO_2)_2N^-$. But it is noted that the conductivity of PILs is significantly lower than that of their low molecular weight monomer counterparts because of the fact that the high viscosity of the solid-like polymeric matrix restrains the long-range mobility of the ions in the particles. Thus, the conductivity range of the poly($MTMA^+(CF_3SO_2)_2N^-$) particles is very suitable for ER fluid applications according to the conduction model.

Due to their hydrophobic nature, the poly($MTMA^+(CF_3SO_2)_2N^-$) particles are easy to disperse in silicone oil and the resulting fluid shows good dispersion stability. Under an electric field, the poly($MTMA^+(CF_3SO_2)_2N^-$) particles can rapidly contact each other and form gap-spanning dense fibrous structures between the electrodes, as shown by optical microscopy in Figure 6.8(a). This aligned fibrous structure provides a large resistance to shear flow

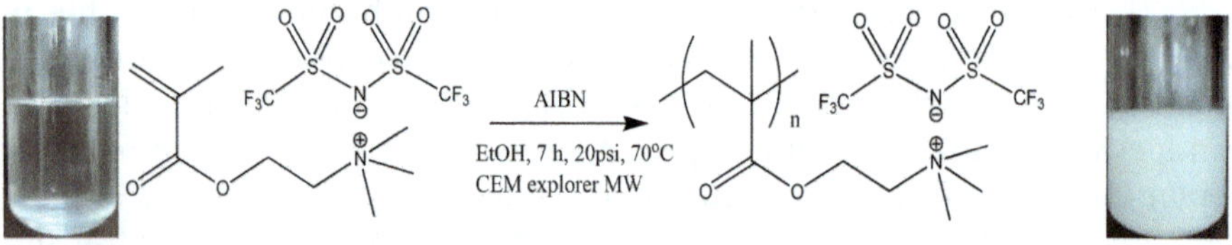

Figure 6.7 Preparation of poly($MTMA^+(CF_3SO_2)_2N^-$) particles by microwave-assisted dispersion polymerization. (Reproduced from ref. 38 with permission from the Royal Society of Chemistry.)

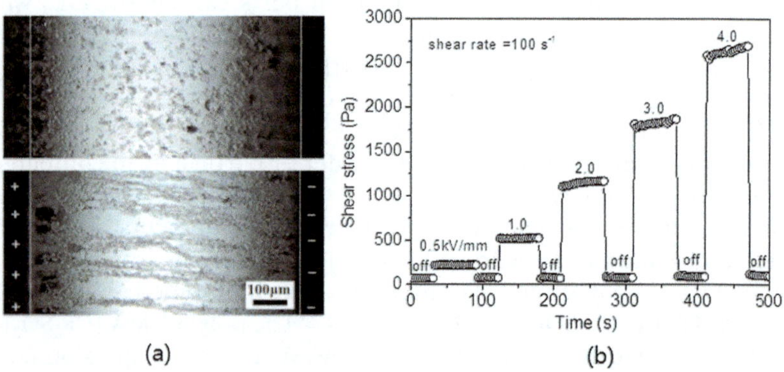

Figure 6.8 (a) Optical photo of the ER fluid of poly($MTMA^+(CF_3SO_2)_2N^-$) particles without and with an electric field (ϕ = 3 vol%, T = 23 °C). (b) Effect of switching the applied electric field on/off on the shear stress of the ER fluid of poly($MTMA^+(CF_3SO_2)_2N^-$) particles (ϕ = 28 vol%, T = 23 °C). (Reproduced from ref. 38 with permission from the Royal Society of Chemistry.)

perpendicular to the electric field and thus enhances the shear stress. Figure 6.8(b) shows the real-time response of the shear stress of the PIL-based ER fluid to an applied electric field when it is alternately turned on and off. It is found that when an electric field is applied, the shear stress of the fluid increases instantaneously. The value of the shear stress at 4.0 kV mm^{-1} is about 30 times as high as the off-field shear stress. When the electric field is removed, the shear stress drops rapidly back to the original level. As the electric field strength increases, the shear stress increases and the real-time response of the shear stress is still reversible and reproducible, revealing the good electro-responsive characteristic of the PIL-based ER fluid in the absence of any activators. According to the dielectric spectra analysis, the high ER activity of PILs can be attributed to their strong dielectric polarizability and adequate polarization response induced by the intrinsic high density of cation/anion parts.

It has been demonstrated that the ER effect depends on the type and size of the cation/anion parts. Using the same microwave-assisted dispersion polymerization technique, we further synthesized monodisperse PIL particles with a styrenic backbone and different sizes of cation/anion parts in order to understand the effect of molecular structure on the ER effect.[58] Figure 6.9 shows the chemical structures and corresponding morphology of the PIL particles. It is found that the PIL particles with a polystyrenic backbone and

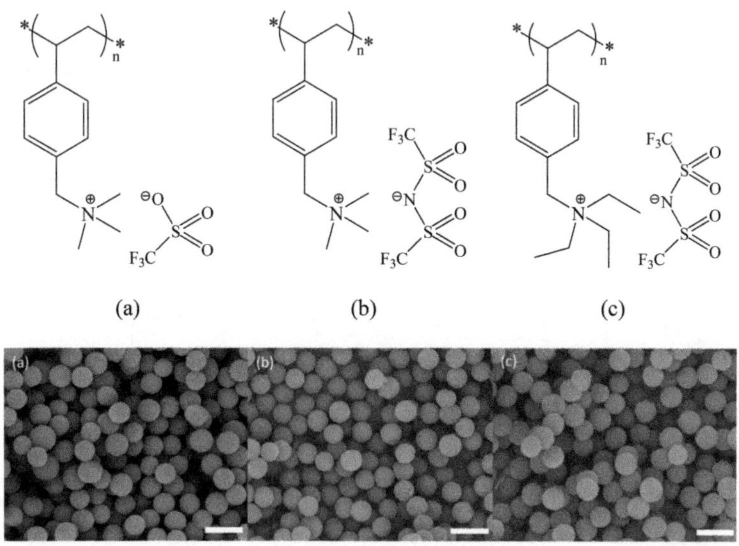

Figure 6.9 The chemical structure and corresponding SEM images of PIL particles: (a) poly($VBTMA^+CF_3SO_3^-$), (b) poly($VBTMA^+(CF_3SO_2)_2N^-$), and (c) poly ($VBTEA^+(CF_3SO_2)_2N^-$). The scale bar is 5.0 μm. Reprinted from Polymer, 97, Y. Dong, J. Yin, J. Yuan, X. Zhao, Microwave-assisted synthesis and high-performance anhydrous electrorheological characteristic of monodisperse poly(ionic liquid) particles with different size of cation/anion parts, 408–417, Copyright (2016) with permission from Elsevier.[58]

different sizes of quaternary ammonium/fluorinated imide ion parts are all monodisperse microspheres with similar particle sizes, which provides justification for studying the influence of the molecular structure on the properties of PIL particles by excluding the influence of the external particle shape.

The three PIL ER fluids all show a strong ER effect, but there is a significant difference in the magnitude of the yield stress and field-induced shear stress among them. At the same electric field strength, the fluid containing poly($VBTMA^+CF_3SO_3^-$) particles exhibits the highest yield stress, while the fluid containing poly($VBTEA^+(CF_3SO_2)_2N^-$) particles exhibits the lowest yield stress (see Figure 6.10). The different yield stresses indicate that the order of the ER effect of the PILs is poly($VBTMA^+CF_3SO_3^-$) > poly($VBTMA^+(CF_3SO_2)_2N^-$) > poly($VBTEA^+(CF_3SO_2)_2N^-$). Because the polymer backbone is the same for these three kinds of PIL particles, this order indicates that the ER effect strongly depends on the size of the cation/anion parts and that PIL particles with smaller cation/anion parts exhibit a larger ER effect.

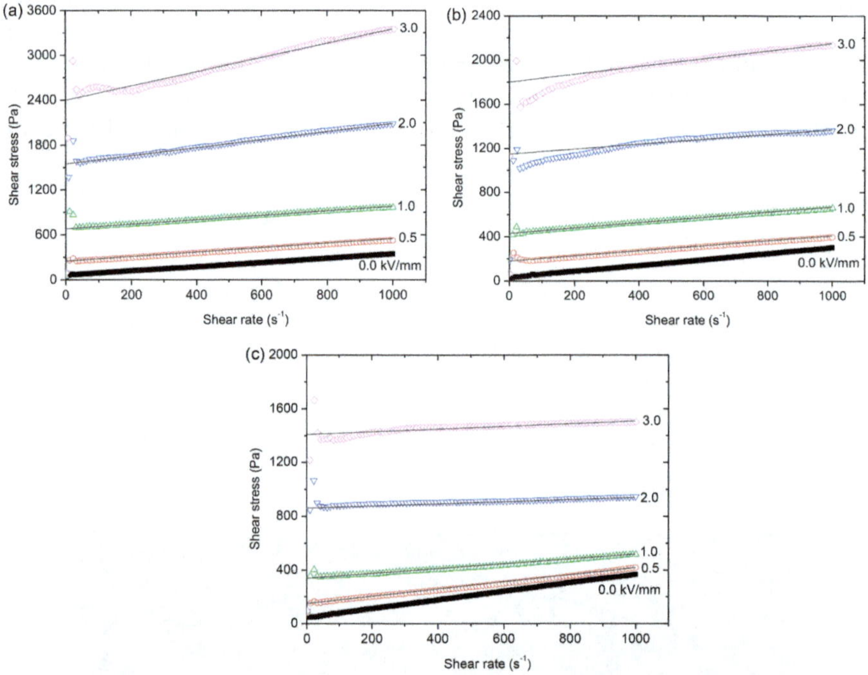

Figure 6.10 The flow curves of shear stress *vs.* shear rate for the suspensions of PIL particles under an electric field (ϕ = ~30 vol%, T = 23 °C): (a) poly ($VBTMA^+CF_3SO_3^-$), (b) poly($VBTMA^+(CF_3SO_2)_2N^-$), and (c) poly($VBTEA^+(CF_3SO_2)_2N^-$). The solid lines show the fit of the flow curves using eqn (1). Reprinted from Polymer, 97, Y. Dong, J. Yin, J. Yuan, X. Zhao, Microwave-assisted synthesis and high-performance anhydrous electrorheological characteristic of monodisperse poly(ionic liquid) particles with different size of cation/anion parts, 408–417, Copyright (2016) with permission from Elsevier.[58]

When comparing the dielectric properties of the three PIL ER fluids, it is noted that the order of the strength and rate of polarization is poly(VBTMA$^+$$CF_3SO_3^-$) > poly(VBTMA$^+$$(CF_3SO_2)_2N^-$) > poly(VBTEA$^+$$(CF_3SO_2)_2N^-$). This indicates that the PIL particles with smaller cation/anion parts possess the fastest dielectric relaxation time and the strongest interfacial polarization strength. The order agrees with the order of the ER effect. Therefore, the different ER effects of the PIL particles with different quaternary ammonium/fluorinated imide ion parts may be mainly attributed to their different dielectric polarizabilities. Two reasons may be responsible for this. The first reason is the difference in the number density of the mobile fluorinated imide anions. In terms of the monomer molecular weight and measured particle density, the number density ratio of mobile fluorinated imide anions can be approximately estimated to be poly(VBTMA$^+$$CF_3SO_3^-$) : poly(VBTMA$^+$$(CF_3SO_2)_2N^-$) : poly (VBTEA$^+$$(CF_3SO_2)_2N^-$) = 1.17 : 1 : 0.84 at the same particle volume. Thus, the poly(VBTMA$^+$$CF_3SO_3^-$) particles have the highest number of mobile anions, while the poly(VBTEA$^+$$(CF_3SO_2)_2N^-$) particles have the lowest number of mobile anions. The second reason is the difference in the mobility of the mobile counter anions. Comparing poly(VBTMA$^+$$CF_3SO_3^-$) with poly(VBTMA$^+$$(CF_3SO_2)_2N^-$), one can note that their polymer backbones and immobile quaternary ammonium cations are the same. Thus, the motion of $CF_3SO_3^-$ should be easier and faster than that of $(CF_3SO_2)_2N^-$ in the glassy state due to its smaller size.[59] Comparing poly(VBTMA$^+$$(CF_3SO_2)_2N^-$) with poly(VBTEA$^+$ $(CF_3SO_2)_2N^-$), one can note that their polymer backbones and mobile fluorinated imide anions are the same. Thus, the larger size of the immobile triethylammonium cations in the poly(VBTEA$^+$$(CF_3SO_2)_2N^-$) particles might have a higher potential barrier or steric effect for the motion of $(CF_3SO_2)_2N^-$ compared to the trimethylammonium cations in the poly(VBTMA$^+$$(CF_3SO_2)_2N^-$) particles.[60] Therefore, the poly(VBTMA$^+$$CF_3SO_3^-$) particles show the highest conductivity, which induces the fastest rate of polarization and the largest ER effect.

PIL particles have provided a new way to develop high-performance anhydrous polyelectrolyte-based ER fluids, but the working temperature range of the ER fluids based on PIL particles is still narrow due to the inherently low glass transfer temperature caused by the plasticization of the polyatomic organic counter ions in the PILs. To overcome the narrow working temperature range of the ER fluid based on poly(VBTMA$^+$$(CF_3SO_2)_2N^-$) PIL particles, we recently developed a new ER system based on cross-linked poly(VBTMA$^+$$(CF_3SO_2)_2N^-$) particles and demonstrated that cross-linking with a suitable degree does not significantly degrade the ER properties, but largely decreases the leaking current density and improves the working temperature range of the ER fluid of PIL particles.[61] Cross-linked poly(VBTMA$^+$$(CF_3SO_2)_2N^-$) particles with various weight ratios of cross-linker (divinylbenzene) can be synthesized by solution polymerization. After cross-linking with a cross-linker weight ratio of 5 wt%, the glass transfer temperature (T_g) of poly(VBTMA$^+$$(CF_3SO_2)_2N^-$) increases from 100 °C to 120 °C. When the weight ratio of cross-linker is higher than 12.5 wt%, no obvious glass transition occurs because the segment motion is significantly restrained and the polymer matrix becomes stiffer.

The rheological measurement results as shown in Figure 6.11 show that the off-field shear stress or off-field viscosity of the ER fluid of poly(VBTMA$^+$($(CF_3SO_2)_2N^-$) particles in the high shear rate region (>100 s^{-1}) gradually decreases with elevated temperature, but that in the low shear rate region it gradually increases with temperature and the fluid even exhibits a significant yield stress of ~15 Pa at 100 °C. This is attributed to the fact that the poly(VBTMA$^+$$(CF_3SO_2)_2N^-$) particles easily swell and become soft at high temperature due to the low inherent T_g and thus the relative particle volume fraction in the fluid increases and the particles easily cohere with each other, particularly at temperatures around T_g (~100 °C for poly(VBTMA$^+$$(CF_3SO_2)_2N^-$) particles). After cross-linking, T_g increases and the progressive uncoiling of the polymer chains and the enlargement of the free volume of the polymer matrix with temperature are restrained. Thus, the cross-linked poly(VBTMA$^+$$(CF_3SO_2)_2N^-$) particles do not swell and become soft as easily with temperature. In particular, the cross-linked poly(VBTMA$^+$$(CF_3SO_2)_2N^-$) particles with a high weight ratio of cross-linker are more mechanically stable due to the formation of a high-density cross-linking network. As a result, the colloid interaction among the cross-linked poly(VBTMA$^+$$(CF_3SO_2)_2N^-$) particles does not change significantly with temperature and the fluids of cross-linked poly(VBTMA$^+$$(CF_3SO_2)_2N^-$) particles with a high weight ratio of cross-linker do not present a significant yield stress or viscosity increase in the low shear rate region, as shown in Figure 6.11. Under an electric field, the ER effect of the ER fluid of poly(VBTMA$^+$$(CF_3SO_2)_2N^-$) particles increases with temperature, but the maximum leaking current also rapidly increases and exceeds the limit of the instrument at 60 °C. After cross-linking, the leaking current of the fluids is significantly decreased and the working temperature range is broadened. The typical fluid of cross-linked poly(VBTMA$^+$$(CF_3SO_2)_2N^-$) particles with 25 wt% cross-linker can maintain a stable rheological behavior in a wide shear rate region at temperatures above 100 °C.

By temperature-modulated dielectric relaxation studies, it was found that cross-linking can effectively suppress the thermally promoted long-range drift of mobile counter ions in the polymer matrix. The activation energy of the relaxation process and conductivity for the fluid of cross-linked poly(VBTMA$^+$$(CF_3SO_2)_2N^-$) particles are much higher than those of the fluid of poly(VBTMA$^+$$(CF_3SO_2)_2N^-$) particles, and the activation energy increases with an increase in the weight ratio of cross-linker. This means that the potential energy necessary to activate ion motion or ionic conductivity increases with increasing cross-linking degree. This can be attributed to the fact that, compared to the pure one, the cross-linked polymer matrix not only provides smaller cavities but also restrains the progressive uncoiling and the enlargement of the free volume of the polymer chains with temperature. Thereby, a higher activation energy is required to promote the dissociation and diffusion of $(CF_3SO_2)_2N^-$ in the cross-linked polymer matrix. As a result, as the temperature increases and comes near to T_g, the ionic conductivity of

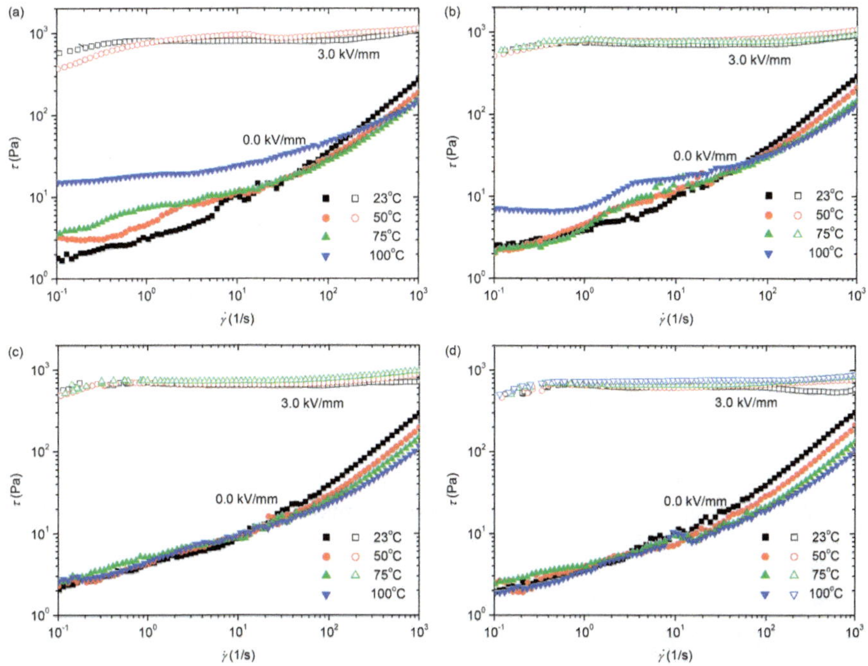

Figure 6.11 Flow curves of shear stress *versus* shear rate at several typical operating temperatures for the fluids of poly(VBTMA$^+$(CF_3SO_2)$_2$N$^-$) and cross-linked poly(VBTMA$^+$(CF_3SO_2)$_2$N$^-$) particles with different weight ratios of cross-linker: (a) 0 wt%; (b) 5 wt%; (c) 12.5 wt%; (d) 25 wt%. (Reproduced from ref. 61 with permission from the Royal Society of Chemistry.)

the fluid of poly(VBTMA$^+$(CF_3SO_2)$_2$N$^-$) particles suddenly increases and the values become larger than those predicted by the Arrhenius lines (see Figure 6.12) due to the accelerated uncoiling of the polymer chains. Thus, the formation of a large current leakage will come at the expense of the polarizability and cause the deterioration of the ER effects and even the failure of ER measurements. Therefore, the fluid of poly(VBTMA$^+$(CF_3SO_2)$_2$N$^-$) particles shows a narrow working temperature region due to its low T_g nature. After cross-linking, T_g is increased and the uncoiling of the polymer structure with temperature is retarded. As a result, the particles are not easy to soften at high temperature and the thermally promoted long-range drift of (CF_3SO_2)$_2$N$^-$ is suppressed and, thus, the working temperature range of the fluid of cross-linked poly(VBTMA$^+$(CF_3SO_2)$_2$N$^-$) particles is broadened.

Based on the research above, it can be demonstrated that in the absence of any activators, PILs can produce a large intrinsic bulk or surface polarization response to an electric field due to their anhydrous character and unique physical and chemical properties. This characteristic endows PILs with significant potential as a novel particle phase to develop high-performance anhydrous polyelectrolyte-based ER fluids.

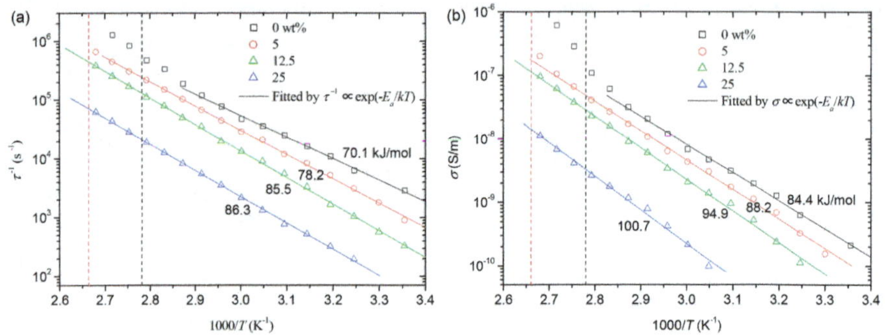

Figure 6.12 Temperature dependence of the relaxation time (a) and the direct current conductivity (b) of the fluids of poly(VBTMA$^+$(CF$_3$SO$_2$)$_2$N$^-$) and cross-linked poly(VBTMA$^+$(CF$_3$SO$_2$)$_2$N$^-$) particles with different weight ratios of cross-linker (ϕ = ~20 vol%, T = 23 °C). Solid lines represent the best fit curves of eqn (3) to the data at temperatures below T_g. (Reproduced from ref. 61 with permission from the Royal Society of Chemistry).

6.4 Summary

In this book chapter, we have given an overview of the latest advances and research highlights in the field of electro-responsive ER smart fluids concerning the use of ILs as possible alternative activators and the use of PILs as new anhydrous polyelectrolyte-based ER active particle phases. Due to their significant chemical and physical stability, ILs seem to be promising alternative activators to effectively solve the problems of extrinsic ER systems. The present experiments have given a preliminary study on this, but the real contribution of using ILs as alternative activators is not still clear. Hence, this research field is still in its infancy and more investigations need to be conducted in order to achieve a clear understanding of the contribution and advantages of ILs as alternative activators in extrinsic ER systems. The future work in this field should be concentrated on the study of the ER effect of particles containing various hydrophobic ILs. Meanwhile, the influence of moisture or other polar molecules should be eliminated and the ion leakage problem should also be specifically noted during material preparation.

On the other hand, our recent research into using hydrophobic PILs as the particle phase has provided a new strategy to develop a high-performance anhydrous polyelectrolyte-based ER fluid. PIL ER particles are easy to obtain in a powder or particle form by chain radical polymerization of monomeric ILs. The present experiments have demonstrated that in the absence of any activators, PILs can produce a strong ER effect under an electric field due to their large dielectric polarizability induced by the local short-range motion of the counter ions in addition to the high density of cation/anion parts. The ER effect has been demonstrated to depend on the structure of the PILs. By adjusting the ion pair and cross-linking degree of the polymer backbone, the physical and chemical properties of the PILs can be controlled and, as a result, the ER effect can be tuned. Based on these facts, it is expected that the use of PILs as the particle phase of ER fluids will open up new possibilities for the application of polyelectrolyte-based ER systems. However, there are still problems that have to be overcome. One of the most important problems of PILs is their cost, which can be much higher than that of conventional polyelectrolytes. In addition, ion migration in the existing PIL ER particles is still easily thermally promoted at a relative low temperature, and this often results in large leaking current and dielectric breakdown with elevated working temperature. Therefore, in the future, more work may be needed to specifically focus on the development of new low-cost PIL ER systems and improve our understanding of the effect of microscopic structure on the ion transfer dynamics and resulting ER effect. One interesting strategy to overcome these problems is to develop a PIL-based composite by combining PILs with low-lost inorganic or other polymer systems, because the contribution from the composite or nanocomposite effect may strongly influence the microscopic ion transfer dynamics and macroscopic ER response.

Because of their stable and fine-tuned physical and chemical properties, it is believed that, in the future, ILs and PILs will find more and more opportunities as suitable compounds in the development of stimuli-responsive smart materials, in particular as active compounds in novel magneto- and electro-responsive smart fluids.

Acknowledgements

We are thankful for the financial support from the National Natural Science Foundation of China (no. 51572225).

References

1. C. A. Rogers, An introduction to intelligent materials system and structures, in *Proc. Int. Workshop on Intelligent Structures*, ed. K. P. Chong, et al., Taipei, Taiwan, 1990, p. 42.
2. N. V. Plechkova and K. R. Seddon, *Chem. Soc. Rev.*, 2008, **37**, 123.
3. *Ionic Liquids as Green Solvents: Progress and Prospects*, ed. J. D. Holbrey, M. B. Turner and R. D. Rogers, Oxford University Press and The American Chemical Society, Washington, DC, 2003.
4. S. Zhang, Q. Zhang, Y. Zhang, Z. Chen, M. Watanabe and Y. Deng, *Prog. Mater. Sci.*, 2016, **77**, 80.
5. H. Tamura, Y. Shinohara and T. Arai, *Chem. Lett.*, 2010, **39**, 240.
6. S. Zhang, S. Liu, Q. Zhang and Y. Deng, *Chem. Commun.*, 2011, **47**, 6641.
7. K. Y. Yung, A. J. Schadock-Hewitt, N. P. Hunter, F. V. Bright and G. A. Baker, *Chem. Commun.*, 2011, **47**, 4775.
8. H. Rodríguez, M. Williams, J. S. Wilkes and R. D. Rogers, *Green Chem.*, 2008, **10**, 501.

9. C. Liang, C.-Y. Yuan, R. J. Warmack, C. E. Barnes and S. Dai, *Anal. Chem.*, 2002, **74**, 2172.
10. Y. F. Shen, Y. J. Zhang, D. Kuehner, G. F. Yang, F. Y. Yuan and L. Niu, *ChemPhysChem*, 2008, **9**, 2198.
11. S. Hayashi and H. O. Hamaguchi, *Chem. Lett.*, 2004, **33**, 1590.
12. T. Peppel, M. K. ckerling, M. Geppert-Rybczyńska, R. V. Ralys, J. K. Lehmann and S. P. Verevkin, *et al.*, *Angew. Chem., Int. Ed.*, 2010, **49**, 7116.
13. R. E. Del Sesto, T. M. McCleskey, A. K. Burrell, G. A. Baker, J. D. Thompson and B. L. Scott, *et al.*, *Chem. Commun.*, 2008, 447.
14. B. Mallick, B. Balke, C. Felser and A. V. Mudring, *Angew. Chem., Int. Ed.*, 2008, **47**, 7635.
15. N. Nishimura and H. Ohno, *Polymer*, 2014, **55**, 3289.
16. D. Mecerreyes, *Prog. Polym. Sci.*, 2011, **36**, 1629.
17. C. Willa, J. Yuan, M. Niederberger and D. Koziej, *Adv. Funct. Mater.*, 2015, **25**, 2537.
18. J. Huang, C. Tao, Q. An, C. Lin, X. Li, D. Xu, Y. Wu, X. Li, D. Shen and G. Li, *Chem. Commun.*, 2010, **46**, 4103.
19. X. He, W. Yang and X. Pei, *Macromolecules*, 2008, **41**, 4615.
20. F. Chen, J. Guo, D. Xu and F. Yan, *Polym. Chem.*, 2016, **7**, 1330.
21. Q. Zhao, J. W. C. Dunlop, X. Qiu, F. Huang, Z. Zhang, J. Heyda, J. Dzubiella, M. Antonietti and J. Yuan, *Nat. Commun.*, 2014, **5**, 4293.
22. Q. Zhao, J. Heyda, J. Dzubiella, K. Täuber, J. W. C. Dunlop and J. Yuan, *Adv. Mater.*, 2015, **27**, 2913.
23. R. Gao, D. Wang, J. R. Heflin and T. E. Long, *J. Mater. Chem.*, 2012, **22**, 13473.
24. M. Smiglak, J. Pringle, X. Lu, L. Han, S. Zhang and H. Gao, *et al.*, *Chem. Commun.*, 2014, **50**, 9228.
25. T. Torimoto, T. Tsuda, K. Okazaki and S. Kuwabata, *Adv. Mater.*, 2010, **22**, 1196.
26. T. C. Halsey and J. E. Martin, *Science*, 1992, **258**, 761.
27. W. M. Winslow, U.S. Patent 2,417,850, 1947.
28. J. Rabinow, *Tech. News Bull.*, 1948, **32**, 54.
29. S. P. Couter, K. D. Weiss and J. D. Carlson, *J. Intell. Mater. Syst. Struct.*, 1993, **4**, 248.
30. J. D. Carlson, D. M. Catanzarite and K. A. St. Clair, *Int. J. Mod. Phys. B*, 1996, **10**, 2857.
31. J. D. G. Durán, J. L. Arias, V. Gallardo and A. V. Delgado, *J. Pharm. Sci.*, 2007, **97**, 2948.
32. L. Rodríguez-Arco, M. T. López-López, J. D. G. Durán, A. Zubarev and D. Chirikov, *J. Phys.: Condens. Matter*, 2011, **23**, 455101.
33. C. Guerrero-Sanchez, T. Lara-Ceniceros and E. Jimenez-Regalado, *et al.*, *Adv. Mater.*, 2007, **19**, 1740.
34. C. Guerrero-Sanchez, T. Erdmenger and T. Lara-Ceniceros, *et al.* Smart Materials Based on Ionic Liquids: the Magnetorheological Fluid Case, *ACS Symp.*, 2009, **1030**, 147.
35. P. Sheng and W. Wen, *Annu. Rev. Fluid Mech.*, 2012, **44**, 143.

36. T. Hao, *Adv. Mater.*, 2001, **13**, 1847.
37. J. A. Marins, B. G. Soares, A. A. Silva, M. G. Hurtado and S. Livi, *J. Colloid Interface Sci.*, 2013, **405**, 6.
38. Y. Z. Dong, J. B. Yin and X. P. Zhao, *J. Mater. Chem. A*, 2014, **2**, 9812.
39. H. Block and P. Rattray, Recent developments in ER fluids, in *Progress in Electrorheology*, ed. K. O. Havelka, F. E. Filisko, Plenum Press, New York, 1995, p. 19.
40. K. D. Weiss, J. D. Carlson and J. P. Coulter, *J. Intell. Mater. Syst. Struct.*, 1993, **4**, 1177.
41. X. P. Zhao and J. B. Yin, *J. Ind. Eng. Chem.*, 2006, **12**, 184.
42. H. J. Choi and M. S. Jhon, *Soft Matter*, 2009, **5**, 1562.
43. J. B. Yin and X. P. Zhao, Chapter 14: Polyaniline Nanocomposites for Smart Electrorheological Fluid Applications, in *Fundamentals of Conjugated Polymer Blends, Copolymers and Composites*, ed. P. Saini, Scrivener Publishing LLC, 2015, p. 731.
44. H. Block and J. P. Kelly, *J. Phys. D: Appl. Phys.*, 1988, **21**, 1661.
45. R. Bloodworth and E. Wendt, ER fluids based on polyurethane dispersions: structure and properties, in *Progress in Electrorheology*, ed. O. Havelkako and F. E. Filisko, Plenum Press, New York, 1995, p. 185.
46. W. J. Wen, X. X. Huang, S. H. Yang, K. Q. Lu and P. Sheng, *Nat. Mater.*, 2003, **2**, 727.
47. J. B. Yin and X. P. Zhao, *J. Colloid Interface Sci.*, 2003, **257**, 228.
48. J. B. Yin and X. P. Zhao, *Chem. Mater.*, 2002, **14**, 4633.
49. J. B. Yin, X. P. Zhao, X. Xia, L. Q. Xiang and Y. P. Qiao, *Polymer*, 2008, **49**, 4413.
50. R. Bloodworth, Electrorheological Fluids Based on Polyurethane Dispersions, *4th International Conference on ER Fluids*, Feldkirch, Austria, 1993.
51. R. Bloodworth and E. Wendt, Materials for ER Fluids, *Actuator 96-5th Internatinal Conference on New Actuators*, Bremen, Germany, 1996.
52. S. Schneider and S. Eibl, *Appl. Rheol.*, 2008, **18**, 23956.
53. J. Yuan, D. Mecerreyes and M. Antonietti, *Prog. Polym. Sci.*, 2013, **38**, 1009.
54. V. Pavlínek, O. Quadrat, B. Porsch and P. Sáha, *Colloids Surf., A*, 1999, **155**, 241.
55. Y. D. Liu, F. F. Fang and H. J. Choi, *Soft Matter*, 2011, 7, 2782.
56. Q. Cheng, V. Pavlinek, A. Lengalova, C. Li, Y. He and P. Saha, *Microporous Mesoporous Mater.*, 2006, **93**, 263.
57. J. A. Marins, B. G. Soares, A. A. Silva and S. Livi, *RSC Adv.*, 2014, **4**, 50925.
58. Y. Z. Dong, J. B. Yin, J. H. Yuan and X. P. Zhao, *Polymer*, 2016, **97**, 4087.
59. K. Nakamura, K. Fukao and T. Inoue, *Macromolecules*, 2012, **45**, 3850.
60. U. H. Choi, Y. Ye, D. S. de la Cruz, W. Liu, K. I. Winey, Y. A. Elabd, J. Runt and R. H. Colby, *Macromolecules*, 2014, **47**, 777.
61. Y. Liu, J. H. Yuan, Y. Z. Dong, X. P. Zhao and J. B. Yin, *Soft Matter*, 2017, **13**, 1027.

CHAPTER 7

Thermo-responsive Poly(ionic liquid) Nanogels Prepared via One-step Cross-linking Copolymerization

J. R. YU, Y. ZUO AND Y. B. XIONG*

Key Laboratory of Eco-Environment-Related Polymer Materials, Ministry of Education, College of Chemistry and Chemical Engineering, Northwest Normal University, Lanzhou 730070, China
*E-mail: ybxiong@fudan.edu.cn

7.1 Introduction

The emergence of ionic liquids (ILs) has shed some new light on green chemistry in the past several decades. In view of their excellent properties, ILs have also attracted increasing interest in electrochemistry, catalysis, and materials science.[1–3] As far as polymer science is concerned, ILs are still in their infancy because some of their properties disappear when they are polymerized through covalent bonds.[4,5] Polymeric/polymerized ionic liquids, also called poly(ionic liquid)s (PILs), represent a special class of polyelectrolytes that feature an IL species in each of the monomer repeating units.[6] Thus, the cationic or anionic centers are connected through a polymeric backbone to form a macromolecular architecture. It is worth

Smart Materials No. 29
Polymerized Ionic Liquids
Edited by Ali Eftekhari

Published by the Royal Society of Chemistry, www.rsc.org

noting that although ILs are in a liquid state near room temperature, PILs are in fact solid in most cases, except for a couple of exceptions.[7] Nevertheless, as opposed to solid polyelectrolytes, PILs have a reportable glass transition temperature that in most cases is well below that of the usual ionic glasses.[8]

To date, a variety of IL monomers have been deliberately prepared for the fabrication of a variety of functional polymeric materials (see Scheme 7.1). Tang and co-workers prepared a series of IL monomers based on styrene and acryl compounds (Scheme 7.1, 1–4), and their corresponding PILs were also obtained through conventional radical polymerization and living/controlled polymerization.[9–11] They also found that PILs had a higher CO_2 absorption capacity than room temperature ILs, as well as faster CO_2 absorption and desorption rates. Recently, some advances have been achieved in PILs. Yuan and co-workers developed a strategy to fabricate porous membranes using PILs and carboxyl-containing compounds.[12–18] More interestingly, these porous materials are responsive to acetone. Additionally, PILs were also demonstrated to have potential applications in electrochemistry, CO_2 adsorption and self-assembly. For the first time, our group designed a novel PIL prepared through condensation polymerization using IL monomers with symmetrical hydroxyl or carboxyl groups.[19] This technique could also be utilized to modify polyethylene terephthalate (PET) through copolymerization.[20] The as-prepared PET materials presented defined wettabilities and good antibacterial performances due to the presence of IL units in the PET backbone. In consideration of the versatile performance of ILs, the exploration of PILs incorporating novel functionalities and properties is still a great challenge.[5–8]

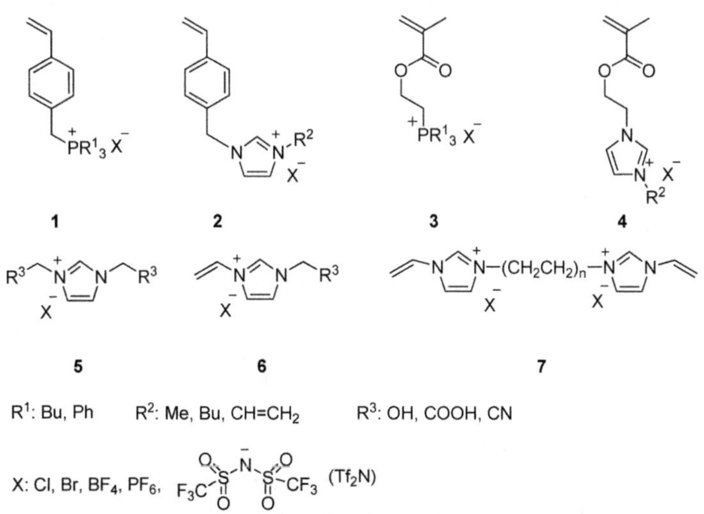

Scheme 7.1 Components of ionic liquid monomers in this study.

7.2 Thermo-responsive Systems Comprising ILs

Over the past several decades, the fabrication of stimuli-responsive polymers (SRPs) has attracted considerable interest due to their wide range of potential applications in fields such as smart materials, biomedicine, catalysis, and sensors.[21,22] These smart polymers can undergo reversible changes in their physical–chemical or colloidal properties in response to external stimuli, including temperature, pH, ionic strength, light, magnetic fields, electric fields, ultrasound, and redox/thiols.[23–25] Among the SRPs, thermo-responsive polymers have garnered significant attention due to their technical feasibility and practical advantages.[26–28] To date, many polymers have been found to exhibit changes in certain properties in response to temperature. The most widely studied thermo-responsive polymers are *N*-alkyl-substituted acrylamide-based polymers,[29,30] including poly(*N*-isopropylacrylamide), poly(*N*,*N*′-diethylacrylamide) (PDEAAm), and poly(2-carboxyisopropylacryl amide). These thermo-responsive polymers possess a lower critical solution temperature (LCST) in aqueous solutions. Further examples of thermo-responsive polymers include poly(ethylene oxide) (PEG)/oligo(ethylene glycol) and polypeptide-based polymers.[31–33] A number of exceptional thermo-responsive polymers have also been reported, such as poly(*N*-vinylcaprolactam), poly(2-oxazoline), amphiphilic polyethylenes, and polymers derived from L-proline.[34–37] Due to their versatile properties and tunable polarity, it is convenient to devise thermo-sensitive IL systems with a lower critical solution temperature (LCST) and an upper critical solution temperature (UCST). Therefore, it is much more desirable to fabricate PIL- and IL-containing polymers with thermo-responsive behavior.[38]

Normally, there are four techniques to fabricate a thermo-sensitive polymer system comprising an IL: (1) using the IL as the solvent for the polymer; (2) introducing ILs into thermo-sensitive polymers; (3) devising functional ILs with suitable hydrophobicity; (4) polymerizing and copolymerizing functional IL monomers.[39,40] The first two techniques are more feasible and convenient methods to build up thermo-responsive polymer systems.[41] Endo and co-workers prepared three *N*-vinylimidazolium salts, 1-(3-phenylpropyl)-3-vinylimidazolium bromide (PVI-Br), 1-(6-ethoxycarbonylhexyl)-3-vinylimidazolium bromide (EHVI-Br), and 1-(2-ethoxyethyl)- 3-vinylimidazolium bromide (EtOEVI-Br). Then, the polymerization of theses ILs was performed by a reversible addition–fragmentation chain transfer (RAFT)/macromolecular design *via* interchange of xanthate (MADIX) process. The polymerization of the *N*-vinylimidazolium salts using the dithiocarbonate-terminated poly(*N*-isopropylacrylamide) as a macro-chain-transfer agent provided well-defined thermo-responsive ionic liquid block copolymers. Thermally-induced phase separation behavior and assembled structures of the block copolymers were also studied in aqueous solution.[42]

Schubert and co-workers[43] found that the aqueous gelation of a quaternary ammonium oligo(propylene oxide)-based IL (see Scheme 7.2a) yielded ion gels with a reverse thermo-responsive behavior (*i.e.*, the mechanical moduli and viscosity increase with temperature) and enhanced ionic conductivities. Wang and co-workers[44] also reported a class of functionalized ILs (see Scheme 7.2b). By tuning the hydrophilicity and hydrophobicity of the cation and anion, respectively, these ILs could reversibly transfer between water and organics, triggered upon undergoing a temperature change. From a combination of multiple spectroscopic techniques, it was shown that the driving force behind the transfer originated from a change in the conformation of the PEG chain of the IL upon temperature variation. By utilizing the novel properties of this class of ILs, a highly efficient and controllable CuI-catalyzed cycloaddition reaction was achieved, wherein the IL was used to entrain, activate, and recycle the catalyst, as well as to control the reaction.

ILs derived from amino acids (see Scheme 7.3) also exhibit phase separation with a LCST after mixing with water. The phase-separation temperature of these mixtures depends on the ion structure and water content, and is lowered by an increase in the hydrophobicity of the IL. Although the mechanism of the LCST-type phase behavior of the ILs is not clear, these ILs could have a great impact on reaction and separation processes.[45]

Additionally, polymerizing functional IL monomers can afford novel polymers with thermo-responsive behavior. Ohno and co-workers have contributed a lot in this area. They polymerized tributylhexylphosphonium 3-sulfopropylmethacrylate with suitable hydrophobicity. The as-obtained PIL underwent a strongly temperature-sensitive LCST-type phase transition with pure water and also with aqueous salt solution, and the resulting hydrated PIL further undergoes a sharp and reversible liquid-to-gel transition at ambient temperatures.[46] A new series of LCST ILs has been copolymerized with cross-linkers of varying length to afford the first ever thermo-responsive PIL-based hydrogels. These hydrogels exhibit surprisingly broad LCST and volume transition behavior compared to standard thermo-responsive gels and linear ILs.[47] A new thermo-reversible ion gel was developed through the self-assembly of a triblock copolymer, by choosing appropriate associating end-blocks for the IL.[48]

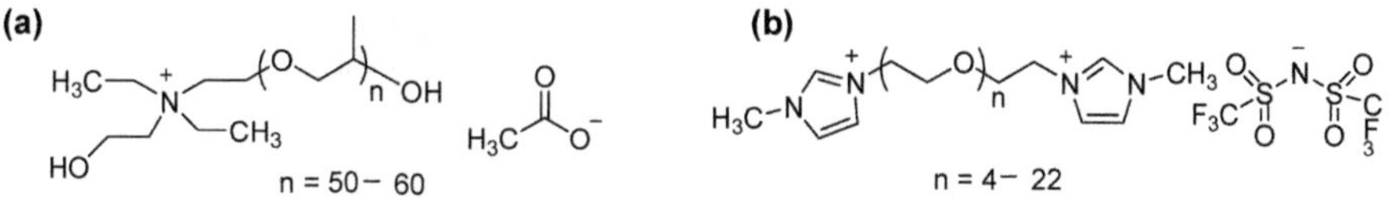

Scheme 7.2 Structures of the thermo-responsive ILs based on quaternary ammonium oligo(propylene oxide) (a) and bisimidazolium salt (b).

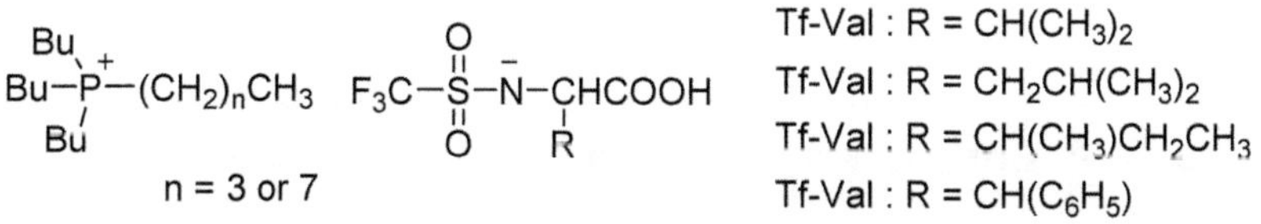

Scheme 7.3 Structure of amino acid-derived ILs.

Yuan and co-workers prepared a cationic polyelectrolyte based on the styrenic ionic liquid tributyl-4-vinylbenzylphosphonium pentanesulfonate (see Scheme 7.4). The cationic polyelectrolyte was found to undergo a LCST-type phase transition in aqueous solution. This phase transition occurs in a wide temperature range depending on the polymer concentration as well as the type and concentration of externally added salts. Anion exchange and salting out effects are responsible for the flexible phase transition temperature.[49]

7.3 Thermo-responsive PIL Nanogels Prepared *via* One-step Cross-linking Copolymerization

Recently, our group demonstrated a facile one-step synthesis strategy to prepare polymeric nanogels *via* the conventional radical copolymerization of IL monomers and the cross-linkers ethylene glycol dimethacrylate (EGDMA) and divinylbenzene (DVB) in selective solvents.[50,51] The as-prepared nanogels presented high activity for CO_2 cycloaddition reactions with epoxides (see Scheme 7.5). Nevertheless, when the imidazolium-based IL (ImIL) 1-vinyl-3-(ethoxycarbonyl)methyl imidazolium chloride was copolymerized

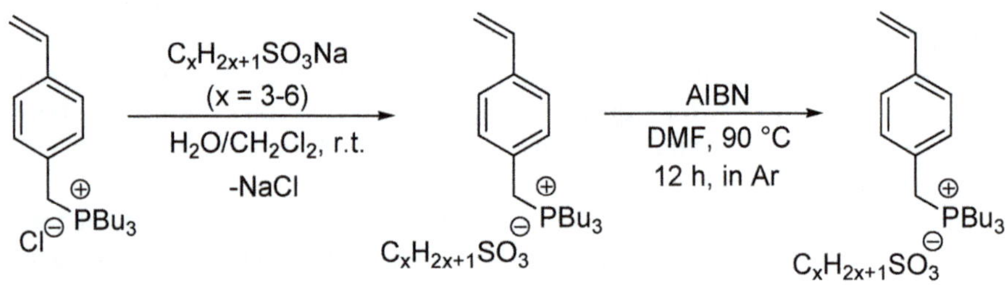

Scheme 7.4 Synthesis route of styrenic ionic liquids.

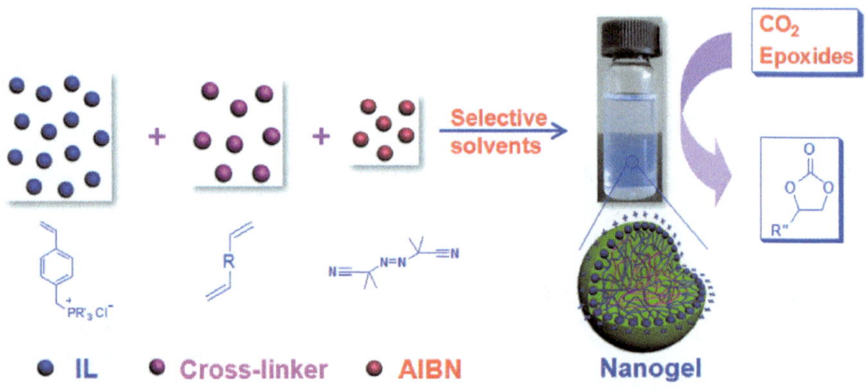

Scheme 7.5 Schematic illustration of the one-step synthesis of PIL nanogels for CO_2 cycloaddition reactions with epoxides.

with these cross-linkers under the same conditions, the copolymers precipitated from the solvent, resulting in particles in the submicrometer range.[52] Similar results were also reported by Han *et al.*[53,54] The precipitation of ImIL-based cross-linked copolymers from the solvent during the polymerization process was probably due to the lower content of poly(ImIL) chains in the copolymer.[51] This result was presumably caused by the lower polymerization reactivity of the ImIL monomers as compared with the cross-linkers.

To achieve stable nanogels in solution, geminal dicationic ImILs, 1,4-butanediyl-3,3′-bis-1-vinyl imidazolium halides ([BVIm]X = 1·X; X = Cl or Br), were devised and synthesized.[55] The copolymerization of 1·X with the cross-linkers in selective solvents was conducted under identical conditions (see Scheme 7.6). For comparison, 3-butyl-1-vinylimidazolium bromide ([C4VIm]Br) was also prepared. The results revealed that a stable ImIL-based nanogel could be easily obtained *via* a one-step synthesis. The as-prepared solutions with blue opalescence and without any precipitation are very stable. However, the copolymers precipitated from the solvent when [C4VIm]Br was copolymerized with the cross-linkers under the same conditions. The results confirmed the assumption that the polymer particles were not stable because of the low reactivity of the ImIL monomers as compared with the cross-linkers. In contrast, the polymer particles could be stabilized by introducing geminal ImIL monomers. More interestingly, the as-prepared nanogels could respond to the temperature. When the temperature decreased, the nanogels precipitated from the solution. This performance is reversible with a temperature change (see Scheme 7.6).

Dynamic light scattering (DLS) measurements demonstrated that the sizes of the ImIL-based nanogels could be controlled by altering the feed ratio of the ImIL monomer and the cross-linker. Moreover, the nanogel with a feed ratio of 5:1 (1·Br to EGDMA) precipitated from methanol

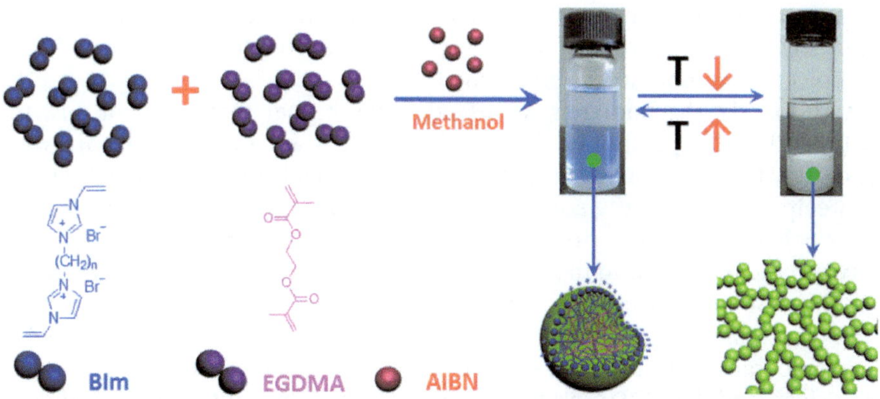

Scheme 7.6 Schematic illustration of the one-step synthesis of ImIL-based nanogels, and the thermo-sensitive behavior of the nanogel solutions with temperature change.

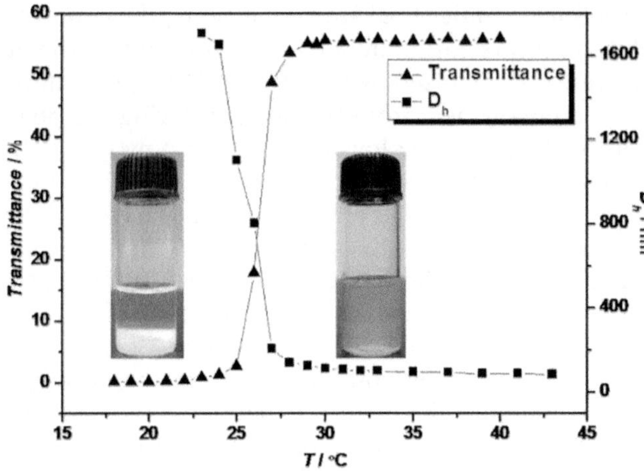

Figure 7.1 Temperature dependence of the transmittance (at 500 nm) and hydrodynamic diameter (D_h) for the 4.5 wt% nanogel (1·Br/EGDMA = 5/1) in methanol.

when the temperature decreased below 25 °C, but a stable nanogel solution was recovered immediately at temperatures above 25 °C. This behavior indicates that the 1·X-based nanogel was thermo-sensitive. Figure 7.1 shows the temperature dependence of the turbidity and diameter of the nanogel (4.5 wt%) solution in methanol. The blue opalescent solution was translucent at temperatures above 28 °C, and suddenly turned cloudy at temperatures below 25 °C. The discrete transition that occurred within a temperature difference of 3 °C indicates that the 1·X-based nanogels are thermo-sensitive particles with an upper critical solution temperature (UCST). The thermo-sensitive behavior was also testified by DLS. The hydrodynamic diameter (D_h) of the nanogel increased rapidly from less than 100 nm to micrometer levels in the same temperature range. These results suggest that the free nanogel was suddenly desolvated to form larger aggregates, leading to precipitation from the solvent at temperatures below the UCST.

Additionally, the nanogels with feed ratios higher than 10:1 of 1·Br to EGDMA did not precipitate from the solvent when the temperature was below 25 °C, even at 4 °C. Surprisingly, the nanogel solution gelled instead of precipitating from the solvent when the solutions were kept at temperatures below −10 °C. The preliminary results showed that some interactions between the nanogel and methanol exist, which can be enhanced by increasing the 1·Br content in the copolymers. Dynamic rheological analysis was employed to investigate the thermo-responsive performance of the nanogels. The temperature at which the storage modulus (G') curve intersects the curve of the loss modulus (G'') indicates the sol–gel transition point. Figure 7.2 shows the oscillatory temperature sweep profiles of the 4.5 wt% nanogel solution in methanol. When the temperature was above −7 °C, both nanogel solutions exhibited a viscoelastic response because the nanogel was in liquid state.

Both G' and G'' increased when the temperature was lower than −7 °C. However, G' increased more rapidly than G''. The sol–gel transition occurred when G' was higher than G'' at about −7 °C for both nanogel solutions. The results obtained from the temperature-dependent oscillatory shear rheological measurements further confirmed the sol–gel transition behavior of the nanogel solutions in methanol with a 1·Br feed ratio above 10.

To gain a deeper insight into the mechanism of the thermo-responsive behavior of the nanogels, ^{1}H NMR spectra of 1·Br in different protonated solvents were recorded. As shown in Figure 7.3, all of the signals confirmed the desired structure of 1·Br. A prominent characteristic found for 1·Br was the

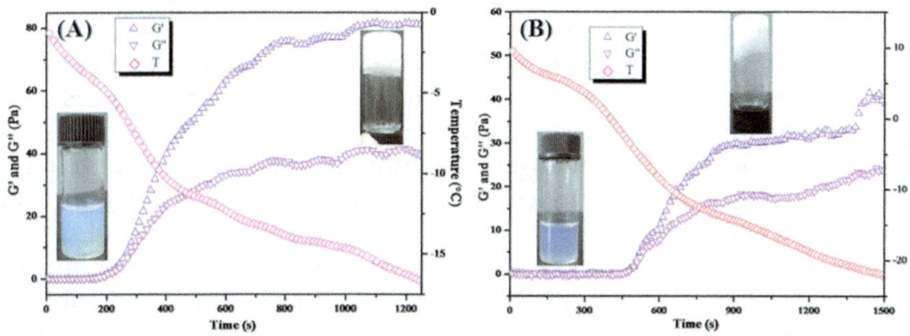

Figure 7.2 Shear storage modulus (G') and shear loss modulus (G'') for the nanogels (at a concentration of 4.5 wt% in methanol) as a function of temperature. (A) 1·Br/EGDMA = 10/1; (B) 1·Br/EGDMA = 15/1 (mol mol^{-1}). Photographs show the corresponding appearance.

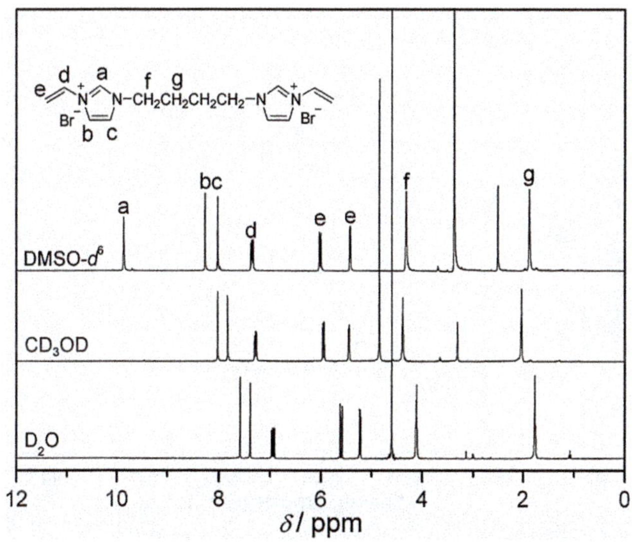

Figure 7.3 ^{1}H NMR spectra of 1·Br in different deuterium solvents.

signal associated with the C-2 proton of the imidazolium ring, which could not be observed when the measurements were conducted in deuterated water and methanol. However, the signal emerged when the measurements were conducted in deuterated DMSO. Moreover, the addition of methanol into the 1·Br solution in DMSO resulted in a significant decrease in the chemical shift and intensity of the C-2 proton of the imidazolium. The results indicate that the C-2 proton of the imidazolium can exchange with the protons of polar solvents such as water and methanol. Therefore, it is inferred that a strong H-bond can form between 1·Br and methanol.

In addition, the IR spectra of dried imidazolium-based ILs present a prominent absorption band at 3058 cm^{-1}, which is attributed to the C–H stretching vibration for C–H···X^-. Upon uptake of water, the peaks associated with the aromatic C–H stretching vibration and the C–H stretching vibration of the C–H···X^- shift to higher wavenumbers.[56] According to Pimentel and McClellan,[57] the stretching mode of an A–H moiety shifts to lower frequencies upon H-bond formation. To elucidate the interactions between the nanogel and methanol, FTIR spectra of nanogel-containing methanol were continuously collected during a temperature-controlled process. As shown in Figure 7.4a, two prominent bands centered at around 3125 and 3060 cm^{-1} were observed, corresponding to the vibration region of the aromatic C–H stretching and the C–H stretching of C–H···Br^-, respectively. However, when methanol was added, both of these bands shifted significantly to higher wavenumbers (3135 and 3087 cm^{-1}, respectively) (see Figure 7.4b). Compared with these shifts, most of the other bands of the nanogel remained unchanged upon the addition of methanol. The upward shift in the frequency for the C–H stretching was consistent with H-bond disruption/diminution between the

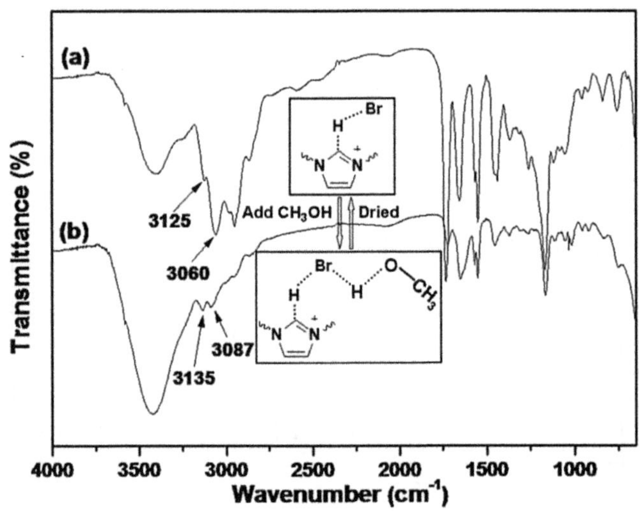

Figure 7.4 FTIR spectra of ImIL-based CLPN. (a) poly(EGDMA-*co*-1·Br); (b) poly (EGDMA-*co*-1·Br) containing methanol.

imidazolium ring and Br^- in the presence of methanol. In addition, the peaks downshifted to lower wavenumbers again when methanol was evaporated at higher temperature. Similarly, the downshift indicated the H-bond formation between the imidazolium ring and Br^- due to the evaporation of methanol. At the same time, the position of the O–H stretching band of methanol shifted from 3425 cm^{-1} to 3409 cm^{-1}, and the intensity decreased greatly. Therefore, the nanogel solution in methanol involves the partial replacement of C–H···Br^- bonds with H-bonds between the imidazolium ring and methanol. Considering these results, the formation of a thermo-reversible nanogel macrogel in methanol involved significant enhancement of H-bonds between the imidazolium ring, Br^-, and methanol at lower temperatures. The formation of a similar H-bonded network was also observed in water.[58] Such a network could serve to enhance the enclosure of methanol, and ultimately formed a physical macrogel (as illustrated in Scheme 7.7). When the temperature increased, the H-bonded network was destroyed, leading to the formation of a nanogel solution in methanol.

The morphology of the nanogel was examined using electron microscopy. From the transmission electron microscopy (TEM) images (see Figure 7.5A–C), spherical nanoparticles with sizes of less than 50 nm can be clearly observed. The sizes observed are smaller than those from the DLS measurements because the particles were swollen in the solution. Figure 7.5D and E show SEM images of the nanogel obtained after precipitating the reaction solution using diethyl ether. The nanogel aggregated and formed

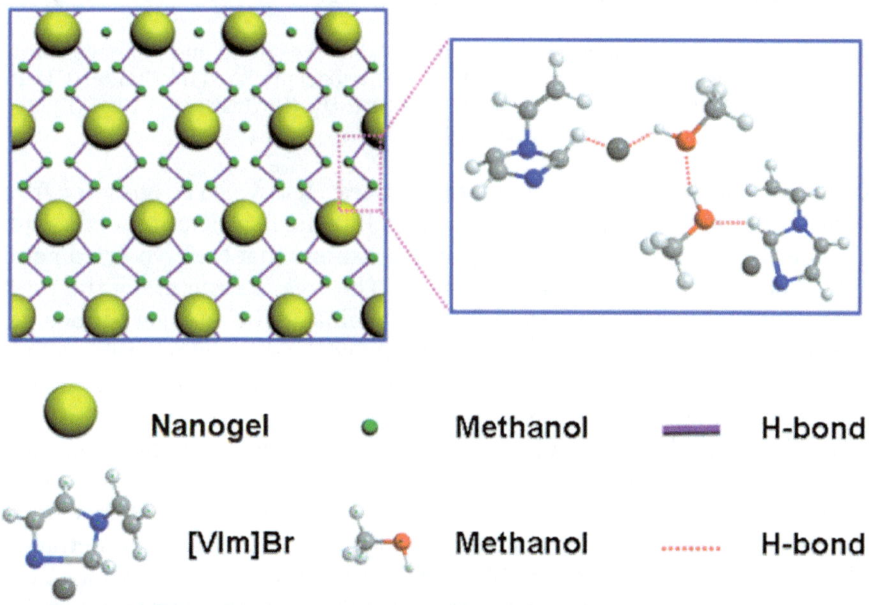

Scheme 7.7 Plausible H-bonded network among the imidazolium ring, Br^-, and methanol at lower temperatures.

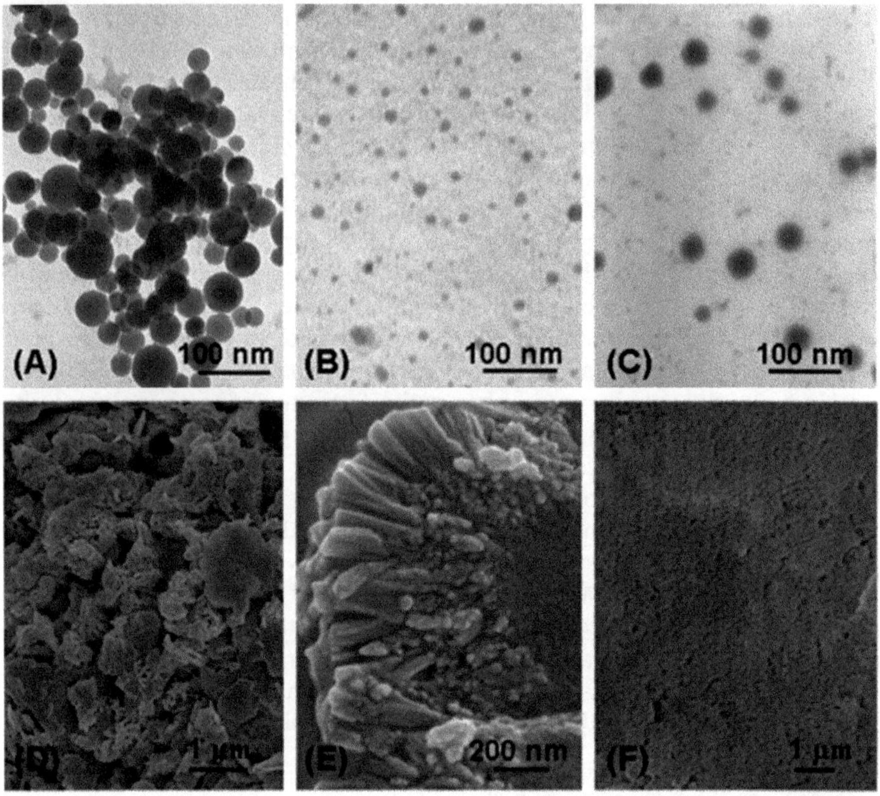

Figure 7.5 Representative TEM (A, B, and C) and SEM (D, E, and F) images of the nanogels. (A) 1·Br/DVB = 10/1; (B) nanogel prepared in ethanol; (C) 1·Br/EGDMA = 10/1; (D) 1·Br/DVB = 10/1; (E) 1·Br/EGDMA = 10/1; (F) cryo-dried gel (1·Br/EGDMA = 10/1).

larger particles during the precipitation process. It can also be observed from Figure 7.5E that the aggregates were crystalline-like particles that consisted of spherical nanogels. However, Figure 7.5F shows that the cryo-dried nanogel gel has a regularly arranged structure. The crystal structure of the ImIL-based nanogel may result from poly(1·Br) chains on the nanogel surface.

Moreover, we also prepared 1,*n*-butanediyl-3,3′-bis-1-vinyl imidazolium halides (BIm, n = 6, 8, or 12).[59] Stable BIm-based nanogels could also be obtained through one-step cross-linking copolymerization in a selective solvent. These BIm-based nanogels were found to be thermo-responsive. Their thermo-induced phase transformation behaviors were investigated using turbidimetry and DLS. Figure 7.6 shows the temperature dependence of the turbidity and particle diameter for 5.0 wt% nanogel solutions in methanol. All these nanogel solutions are blue opalescent and translucent above a certain temperature, and suddenly turned cloudy below that temperature. For example, the transmittance of one nanogel solution ([C_6VIm]Br : EGDMA = 5 : 1, Figure 7.6a) is above 80% at a temperature of 18 °C. Nonetheless, the

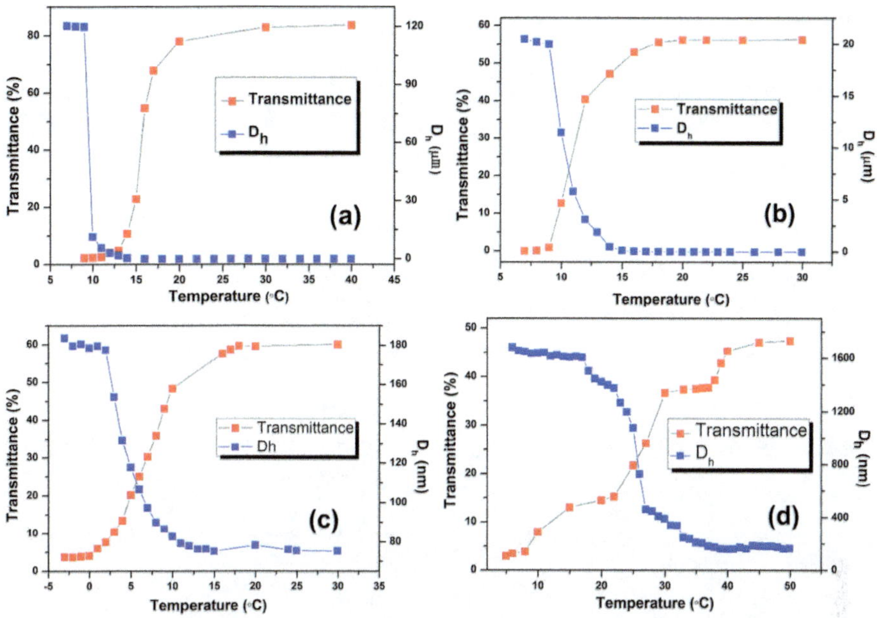

Figure 7.6 Temperature dependence of the transmittance at 500 nm and hydrodynamic diameter (D_h) of 5 wt% nanogels in methanol. (a) [C_6VIm]Br : EGDMA = 5 : 1; (b) [C_6VIm]Br : EGDMA = 10 : 1; (c) [C_6VIm]Br : DVB = 3 : 1; (d) [C_6VIm]Br : DVB = 10 : 1.

transmittance dropped promptly to below 10% when the temperature decreased to 13 °C. The discrete transition that occurred within a narrow temperature change indicates that the nanogels are thermo-sensitive particles with a UCST. The thermo-sensitive behavior of the nanogels was further confirmed by DLS measurements. The hydrodynamic diameter (D_h) of the nanogels increased rapidly from below 100 nm to the micrometer scale in the same temperature range. These results reveal that the free nanogels were suddenly desolvated to form large aggregates, and precipitated from the solvent below the UCST. The morphology of the as-prepared nanogels was also examined using SEM. As shown in Figure 7.7, spherical particles with sizes of less than 200 nm can be clearly observed. Most of the particles are smaller than the corresponding hydrodynamic diameter determined by DLS.

The results above revealed that the thermo-response of the PIL-based nanogels was attributed to hydrogen bonding interactions between the nanogels. Unfortunately, this strategy is infeasible when it comes to ILs that cannot form hydrogen bonding interactions.[51,52] As a result, it can be inferred that introducing H-bonding interactions is an efficient technique to fabricate novel thermo-responsive polymeric nanomaterials based on PILs. Hydroxyl, amino and carboxyl groups are the most commonly used groups to construct H-bonding systems. Therefore, hydroxyl- and carboxyl-containing ILs, including 1-vinyl-3-hydroxyethylimidazolium bromide ([VHEIm]Br) and

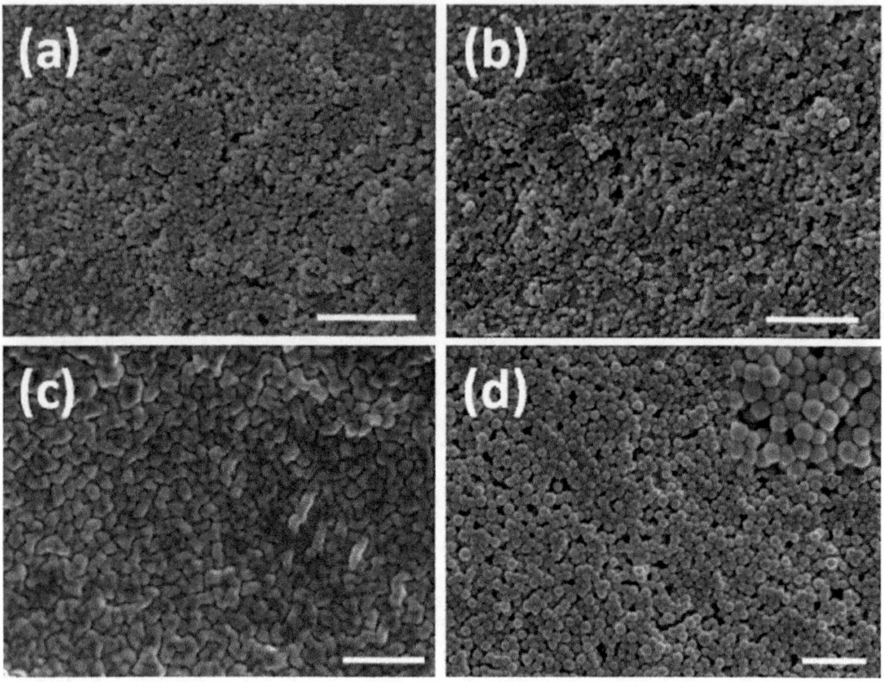

Figure 7.7 SEM images of BIm-based nanogels. (a) [C_6VIm]Br : EGDMA = 10 : 1; (b) [C_8VIm]Br : EGDMA = 10 : 1; (c) [C_6VIm]Br : EGDMA = 5 : 1; (d) [C_6VIm] Br : EGDMA = 3 : 1. The scale bar is 500 nm.

1-vinyl-3-carboxyl methylimidazolium chloride ([VCMIm]Cl) (as shown in Scheme 7.8a), were prepared and copolymerized with cross-linkers in methanol.[60] Owing to the H-bonding interactions between the hydroxyl (carboxyl) groups, the as-prepared nanogels were capable of reversible thermo-responsive performance. Thereby, our findings prove that introducing H-bonding interactions is an efficient strategy to fabricate thermo-responsive polymeric nanogels based on PILs through one-step synthesis (see Scheme 7.8b).

[VHEIm]Br and [VCMIm]Cl were synthesized according to the reported methods, and characterized using MS, ^{1}H and ^{13}C NMR measurements. However, it was found that the particles in the micrometer range precipitated from the solvent when the above monomers were copolymerized directly with the cross-linkers under the same conditions reported. This is possibly due to the lower reaction ratio of these as-prepared monomers compared with the cross-linkers.[51] To obtain stable nanogel solutions, two common IL-based monomers, 4-vinylbenzyl-triphenylphosphorous chloride (VBP-Ph) and 3-(4-vinylbenzene)-1-vinylimidazolium chloride ([VBVIm]Cl), were synthesized, and copolymerized with [VHEIm]Br and ([VCMIm]Cl) under identical conditions. The results revealed that stable and blue opalescent nanogel solutions could be obtained when VBP-Ph and [VBVIm]Cl were used as the co-monomers. In previous studies, it has been demonstrated that VBP-Ph

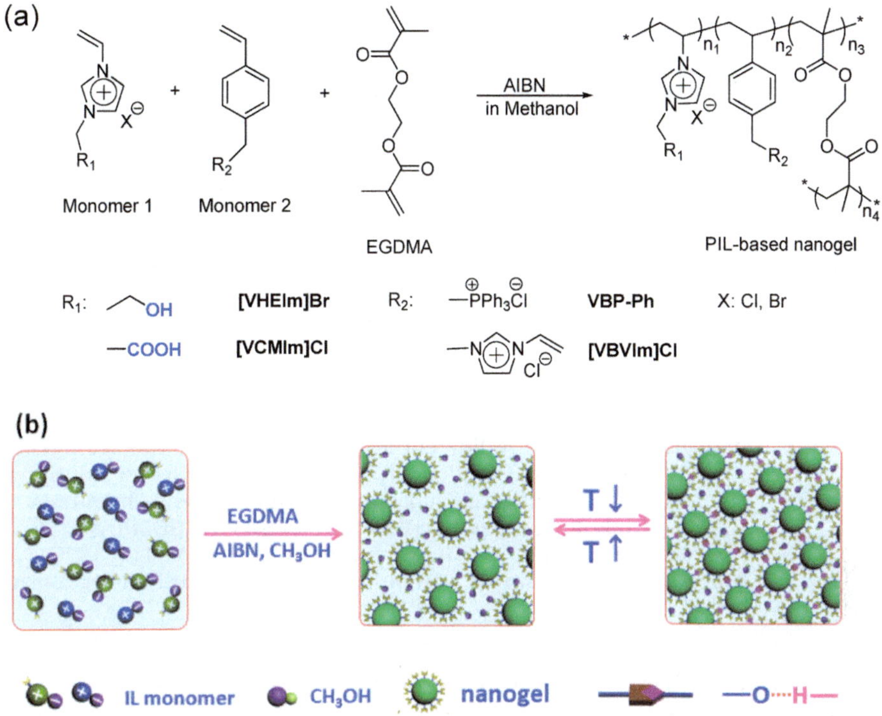

Scheme 7.8 Chemical structures of IL-based monomers and the as-prepared nanogels (a) and schematic illustration of the preparation of PIL-based nanogels and their thermo-responsive behavior (b).

and [VBVIm]Cl are more active monomers and can stabilize nanogels in methanol. Therefore, ternary copolymerization is an efficient approach to prepare stable nanogels comprising hydroxyl and carboxyl groups.

The as-prepared nanogel solutions are blue opalescent and translucent at higher temperatures. Owing to the presence of hydrogen-bonding interactions resulting from hydroxyl or carboxyl groups, the as-prepared nanogels are promising as thermo-responsive materials. It was found that they suddenly turned cloudy at lower temperatures. To specify the transition behaviors with the temperature change, the behavior of these nanogel solutions was monitored using the turbidimetric method. As summarized in Figure 7.8a, the transmittance of the PIL-based nanogel solutions is above 70% except for NG18, which is of lower transmittance due to its higher concentration and larger size. With a decrease in temperature, a sharp decrease in the transmittance could be observed. The discrete transition induced by the temperature indicates that the PIL-based nanogels are thermo-responsive and exhibit UCST transition performances. NG10 and NG12 presented a sharper transition compared with the others. This is possibly because the concentrations of the nanogels were different and the temperature was not automatically controlled. Additionally, the UCSTs of the as-prepared nanogels were

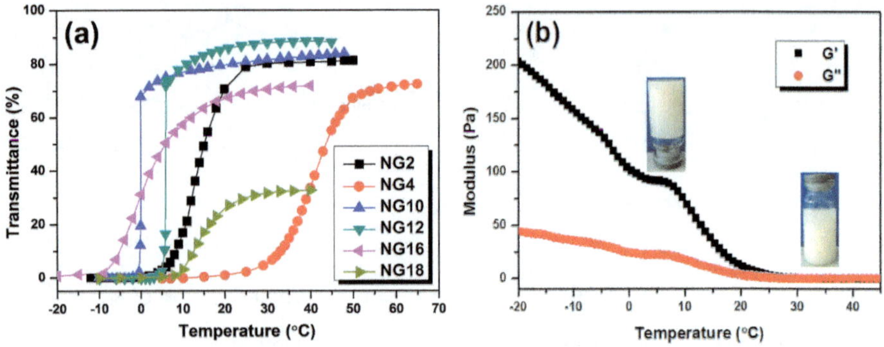

Figure 7.8 Thermo-responsive behavior of PIL-based nanogels. (a) Temperature dependence of the transmittance (at 500 nm) curves of the as-prepared nanogel solutions with different feed ratios. (b) Shear storage modulus (G') and shear loss modulus (G'') for NG9 in methanol solutions (12.5 wt%) as a function of temperature. ([VHEIm]Br : VBP-Ph : EGDMA = 3 : 1 : 2 (NG2); 8 : 1 : 2 (NG4); [VCMIm]Cl : VBP-Ph : EGDMA = 2 : 1 : 2 (NG10); 8 : 1 : 2 (NG12); [VCMIm]Cl : [VBVIm]Cl : EGDMA = 1 : 1 : 3 (NG16); 4 : 1 : 3 (NG18); [VHEIm]Br : [VBVIm]Cl : EGDMA = 10 : 1 : 2 (NG9), molar ratio).

further measured by DSC. The results indicated that the UCST of these nanogels is highly dependent on the composition. For example, the UCST of NG4 (32.5 °C) is much higher than that of NG2 (13.3 °C), and the UCST of NG16 (18 °C) is lower than that of NG18 (23.2 °C). Combining the DLS data,[60] it can be found that the more functional IL monomers ([VHEIm]Br and [VCMIm]Cl) in the feed ratio will enhance the UCST of the as-prepared nanogels. This result can be reasonably attributed to the hydrogen bonding interaction coming from [VHEIm]Br and [VCMIm]Cl. The more [VHEIm]Br and [VCMIm]Cl are fed in during ternary copolymerization, the more hydrogen bonding interactions will be produced. Therefore, a higher UCST could be achieved. When the feed ratio of [VCMIm]Cl, VBP-Ph and EGDMA is 10 : 1 : 2 (NG13), the nanogel solution can transform into macrogel at temperatures below 20 °C. As demonstrated in Figure 7.8b, both G' and G'' are very low and in the same range when the temperature is above 20 °C, which demonstrates that the nanogel solution exhibited a viscoelastic response. When the temperature decreases, both G' and G'' increased; however, G' increased much more rapidly than G''. A sol–gel transition occurred when G' was higher than G'' at around 20 °C. The results obtained from the temperature-dependent oscillatory shear rheological measurements confirmed the sol–gel transition behavior of the [VCMIm]Cl-based nanogel solution in methanol. Moreover, all of these temperature-induced transitions are reversible. DSC measurement revealed that the UCST of the PIL-based nanogels could be tuned from −1 °C to 40 °C *via* changing the feed ratio. Therefore, our findings further testify that introducing H-bonding interactions into IL monomers is an efficient approach to achieve thermo-responsive PIL-based nanogels.

Until recently, most thermo-responsive polymers were obtained through the polymerization of specific monomers whose structures or properties change reversibly in response to temperature after the reaction. Even in our cases of thermo-responsive PIL-based nanogels, it is prerequisite that the IL monomers are capable of forming hydrogen bonding interactions. Therefore, it will be a challenge to develop a versatile strategy suitable for monomers without the aforementioned features for the preparation of thermo-responsive polymers.

Reversible addition–fragmentation chain transfer (RAFT) polymerization is a versatile and efficient technique for the preparation of polymers with controlled structures and molecular weights. However, there have been few reports regarding the utilization of the functional groups of the chain transfer agent (CTA) to fabricate stimuli-responsive polymers through RAFT polymerization. To confirm our hypothesis, we prepared three CTAs (see Scheme 7.9). CTA3 without a carboxyl group was used as a control study. Then, the as-prepared IL monomers were copolymerized with cross-linkers *via* a RAFT polymerization process in a selective solvent. Due to the introduction of carboxyl groups from the CTAs, these PIL-based nanogels can transform reversibly from a nanogel solution into a macrogel with a change in temperature, and display reversible thermo-responsive behaviors. SEM, DLS, dynamic rheology, and variable temperature FTIR measurements were used to characterize the transformations, and a plausible mechanism for the thermo-reversible transition was suggested. PIL-based nanogels are also pH-sensitive and can be further decorated by surface grafting polymerization. Consequently, a simple and robust strategy for the preparation of stimuli-responsive polymer nanogels is demonstrated in this study, as well as a novel application of CTAs.[61]

As shown in Scheme 7.10, it can be seen that PIL-based nanogels were prepared successfully through RAFT cross-linking copolymerization. The nanogel solutions were very stable and could be stored without any precipitation

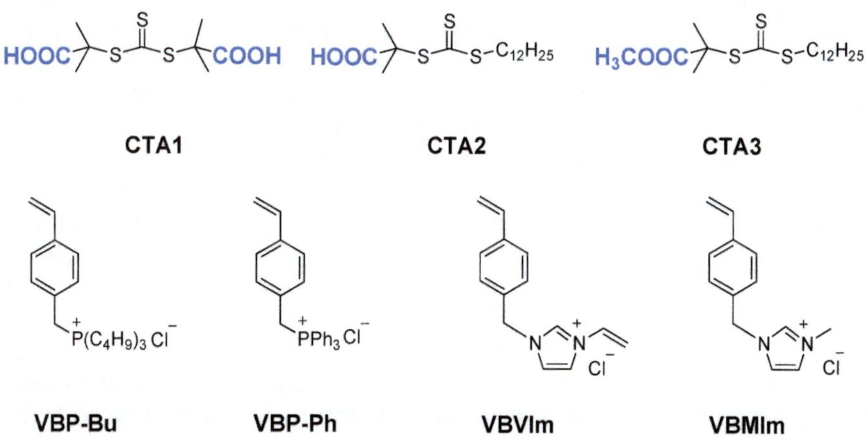

Scheme 7.9 Chemical structures of CTAs and IL monomers prepared in this study.

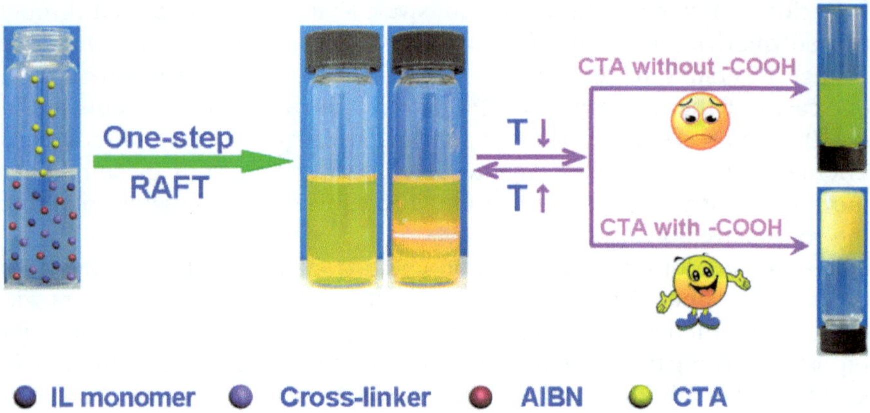

Scheme 7.10 Schematic representation of the preparation of PIL-based nanogels *via* RAFT cross-linking copolymerization and their thermo-responsive behaviors.

for more than several months. The PIL-based nanogels were observed to be thermo-responsive when trithiocarbonates containing carboxyl groups (CTA1 and CTA2) were used as the CTA. Phosphorous-based nanogel solutions in methanol, as well as imidazolium-based nanogel solutions, transformed reversibly into macrogels when the temperature changed in the range of −10 °C to 20 °C. The formation of free-standing physical macrogels was confirmed by a tube inversion test at lower temperatures. In contrast, the nanogel solutions remained in the form of free-flowing, low-viscosity fluids when CTA3 was used in the synthesis (see Scheme 7.10). Presumably, the gelation transition results from the formation of a hydrogen bonding network due to the presence of carboxyl groups. Hence, the nanogels prepared through CTA3 could not form gels due to the lack of carboxyl groups. As described in previous studies,[50,51] nanogels prepared using phosphorous-based monomers were not thermo-responsive due to the lack of groups capable of forming hydrogen bonding. However, through introducing CTAs containing carboxyl groups, thermo-responsive nanogels based on phosphorous IL monomers could also be achieved through RAFT cross-linking copolymerization.

The oscillatory temperature sweep profiles of the phosphorous- and imidazolium-based nanogel solutions in methanol are shown in Figure 7.9. A sharp increase in G' and G'' of both the phosphorous- and imidazolium-based nanogel solutions was clearly observed as the temperature decreased, which was associated with the gelation process. However, the rate of increase of G' was significantly faster, as elastic properties became dominant. Thus, the results indicate that the sol–gel phase transition of the PIL-based nanogel solutions can be achieved by decreasing the temperature. The T_{gel} of the phosphorous-based nanogels was observed to increase with the feed ratio of CTA1 in the RAFT copolymerization. For instance, the T_{gel} of the nanogel is above 10 °C when the molar feed ratio of CTA1 to AIBN is 10 (1P10E10C1), while

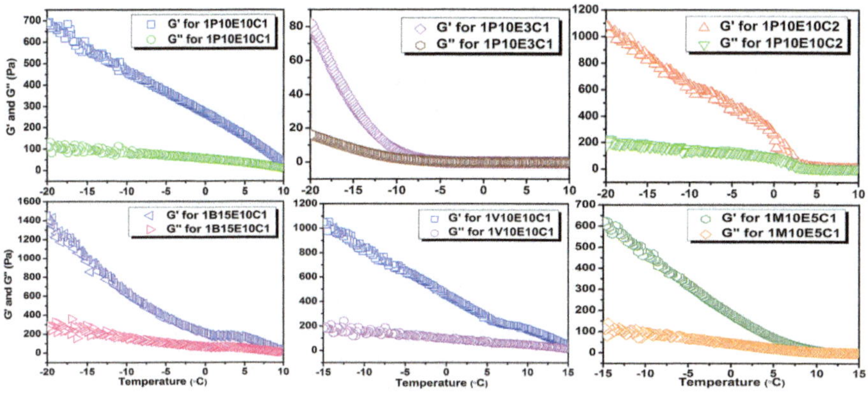

Figure 7.9 Shear storage modulus (G') and shear loss modulus (G'') for phosphorous- and imidazolium-based nanogels in methanol solutions (10.0 wt%) as a function of temperature.

T_{gel} is below zero when the molar feed ratio of CTA1 to AIBN is 3 (1P10E3C1). The same trend can be seen for the T_{gel} of the imidazolium-based nanogel solutions. Additionally, the T_{gel} of the nanogel prepared by using CTA1 in the RAFT copolymerization was found to be higher than that measured using CTA2 under identical conditions. Therefore, it can be inferred that a higher T_{gel} can be achieved when there are more carboxyl groups in PIL-based nanogels. It is also important to note that the T_{gel} of the VBVIm-based nanogel (above 15 °C) is the highest among all of the as-prepared nanogels. According to our previous study,[53] the C-2 protons in the imidazolium rings of the VIm-based ILs can exchange with the protons in protic solvents, such as water and alcohols. Similar behavior was also observed in VBIm-based IL monomers. Therefore, the fact that the VBVIm-based nanogels exhibit the highest T_{gel} can be attributed to the enhanced hydrogen bonding interactions resulting from the carboxyl groups of CTA and the C-2 proton in the imidazolium ring.

As demonstrated in previous studies, hydrogen bonding interactions are responsible for the thermo-responsive behaviors of PIL-based nanogels. To clearly establish the role of hydrogen bonding interactions in nanogels prepared *via* RAFT cross-linking copolymerization, control experiments using a CTA without a carboxyl group were conducted. As illustrated in Scheme 7.11, the nanogel prepared *via* CTA3 presented no response to a change in temperature. Additionally, the macrogel could be destroyed by the addition of Et_3N. These results demonstrate that the PIL-based nanogels are pH-responsive, further confirm that the thermo-responsive behavior of PIL-based nanogels is due to the presence of carboxyl groups in CTA. The destroyed macrogel could not be recovered when trifluoroacetic acid (TFA) was added, likely due to the formation of a TFA–Et_3N salt, which can interfere with hydrogen bonding. However, when the TFA–Et_3N salt was removed by dialysis, a macrogel can achieved again. Therefore, a plausible mechanism for the reversible thermo-responsive behaviors of the PIL-based nanogels is illustrated in Scheme 7.11.

PIL-based nanogels are solvated by methanol at higher temperatures (above T_{gel}), and PIL-based nanogels are well dispersed in the solution. As the temperature decreases below T_{gel}, the hydrogen bonds between the nanogel and methanol are replaced by those between the nanogels, which results in the formation of a nanogel network.

RAFT polymerization has been prevailingly employed to control the structure and molecular weight of polymers. However, the advantages of RAFT polymerization were not utilized during the preparation of PIL-based nanogels due to their highly cross-linked structure. To explore further advantages of RAFT polymerization using the current strategy, surface graft polymerization on PIL-based nanogels was investigated using a traditional thermo-responsive compound, *N*-isopropylacrylamide (NIPAAm), as the monomer. As illustrated in Scheme 7.12, graft polymerization on the PIL-based nanogels occurred readily at 70 °C in methanol in the presence of NIPAAm and AIBN.

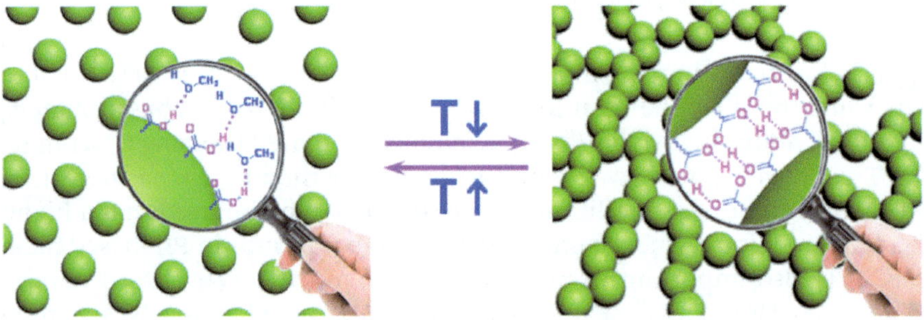

Scheme 7.11 Plausible mechanism for the reversible thermo-responsive behaviors of PIL-based nanogels.

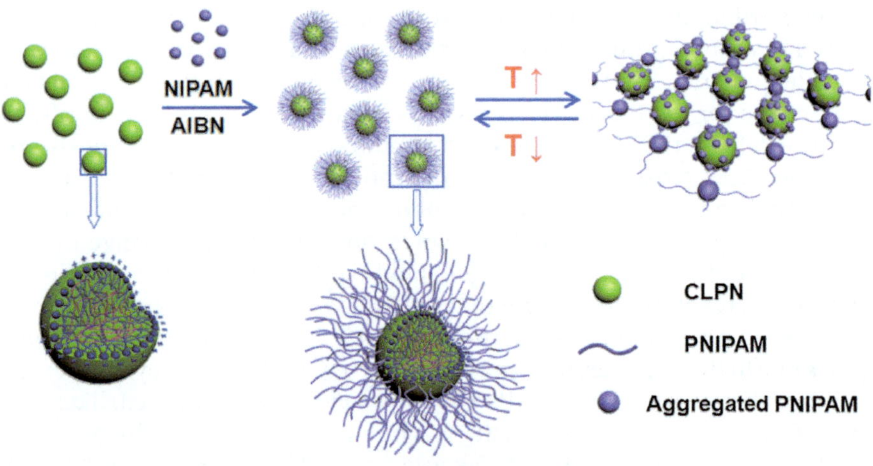

Scheme 7.12 Schematic illustration of surface graft polymerization on PIL-based nanogels and their thermo-responsive performances.

Following the polymerization, the possible free monomers and PNIPAAm homopolymers in the as-prepared solutions were thoroughly removed by dialysis against deionized water. The nanogel aqueous solutions were then lyophilized. In order to confirm the successful formation of PNIPAAm-*g*-nanogels, DLS, dynamic rheological analysis, FTIR, SEM, element analysis, and XPS measurements were conducted.

Firstly, the retention of the trithiocarbonate group derived from CTA in the PIL-based nanogels was confirmed by FTIR spectroscopy. The stretching vibration absorptions of S=C and C–S–C in the trithiocarbonate can be found at 1169 and 1065 cm^{-1}, respectively.[61] In addition, XPS results (see Figure 7.10) show the presence of S elements on the surface of the as-prepared nanogels. Therefore, the PIL-based nanogels could then be used as the macroCTA for graft copolymerization with NIPAAm. After the graft polymerization, the PIL-based nanogel solutions displayed uniform opalescence with a pale blue color in comparison to the bright yellow solution before graft polymerization. Additionally, strong scattering of light occurred when using a laser pointer on the samples after PNIPAAm grafting. XPS results show that the content of nitrogen increases a lot due to the grafting of PNIPAAm. Most importantly, these solutions were able to transform into free-standing physical macrogels when the temperature increased, as demonstrated by a tube inversion test. The sizes of the PNIPAAm-*g*-nanogels were also measured, and were observed to increase more rapidly at higher feed ratios of NIPAAm to AIBN. It is reasonable that PNIPAAm of higher molecular weight can be obtained when the NIPAAm feed is increased in the polymerization process. Grafting of PNIPAAm had a minimal influence on the PDI of the nanogels, and the nanogels were still positively charged, as determined by DLS measurements. These results provide evidence that PNIPAAm was successfully grafted onto the PIL-based nanogels.

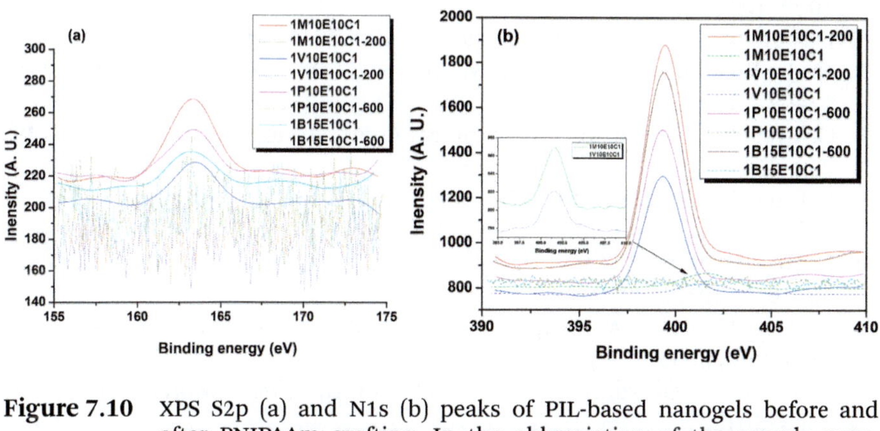

Figure 7.10 XPS S2p (a) and N1s (b) peaks of PIL-based nanogels before and after PNIPAAm grafting. In the abbreviation of the sample number "1P10E5C1", "1P10E" denotes that the feed ratio of VBP-Ph and EGDMA is 1 : 10; "5" denotes the feed ratio of CTA to AIBN; "C1" denotes CTA1. All other abbreviations are denoted similarly.

7.4 Summary and Outlook

Polymer nanoparticles have found extensive applications in versatile fields. However, the preparation of stable and functionalized polymer nanoparticles usually requires complicated manipulations. Incorporating the unique properties of ILs into the polymer nanoparticles gives a new class of polymeric materials. As an emerging interdisciplinary topic among polymer chemistry and physics, materials science, catalysis, separation, analytical chemistry, and electrochemistry, PILs are attracting increasing interest. Our one-step synthesis provides another optional technique to fabricate these kinds of polymer nanoparticles. Additionally, to endow nanoparticles with thermo-responsive behavior, we found that introducing hydrogen bonding interactions is a powerful strategy. Considering the designability of the cations and anions of ILs, polymer nanoparticles with multistimuli-responsive behavior could be devised and prepared. Our present studies demonstrated that photo- or redox-responsive nanogels based on PILs could also be achieved through introducing azobenzene units and disulfide bonds.[62] Our studies thus provide a new platform for the fabrication of stimuli-responsive nanomaterials, as well as to explore the potential applications of ILs in smart polymers.

References

1. N. V. Plechkova and K. R. Seddon, *Chem. Soc. Rev.*, 2008, **37**, 123–150.
2. T. Welton, *Chem. Rev.*, 1999, **99**, 2071–2083.
3. J. S. Wilkes, *Green Chem.*, 2002, **4**, 73–80.
4. R. Marcilla, J. A. Blazquez, J. Rodriguez, J. A. Pomposo and D. Mecerreyes, *J. Polym. Sci., Part A: Polym. Chem.*, 2004, **42**, 208–212.
5. J. Yuan, D. Mecerreyes and M. Antonietti, *Prog. Polym. Sci.*, 2013, **38**, 1009–1036.
6. J. Yuan and M. Antonietti, *Macromolecules*, 2011, **44**, 744–750.
7. H. L. Ricks-Laskoski and A. W. Snow, *J. Am. Chem. Soc.*, 2006, **128**, 12402–12403.
8. O. Green, S. Grubjesic, S. Lee and M. A. Firestone, *Polym. Rev.*, 2009, **49**, 339–360.
9. J. Tang, H. Tang, W. Sun, M. Radosz and Y. Shen, *J. Polym. Sci., Part A: Polym. Chem.*, 2005, **43**, 5477–5489.
10. J. Tang, W. Sun, H. Tang, M. Radosz and Y. Shen, *Macromolecules*, 2005, **38**, 2037–2039.
11. J. Tang, H. Tang, W. Sun, H. Plancher, M. Radosz and Y. Shen, *Chem. Commun.*, 2005, 3325–3327.
12. J. Yuan, C. Giordano and M. Antonietti, *Chem. Mater.*, 2010, **22**, 5003–5012.
13. J. Yuan, A. G. Márquez, J. Reinacher, C. Giordano, J. Janek and M. Antonietti, *Polym. Chem.*, 2011, **2**, 1654–1657.
14. Q. Zhao, M. Yin, A. P. Zhang, S. Prescher, M. Antonietti and J. Yuan, *J. Am. Chem. Soc.*, 2013, **135**, 5549–5552.

15. Q. Zhao, J. W. C. Dunlop, X. Qiu, F. Huang, Z. Zhang, J. Heyda, J. Dzubiella, M. Antonietti and J. Yuan, *Nat. Commun.*, 2014, **5**, 4293.
16. Q. Zhao, P. Zhang, M. Antonietti and J. Yuan, *J. Am. Chem. Soc.*, 2012, **134**, 11852–11855.
17. Q. Zhao, J. Heyda, J. Dzubiella, K. Täuber, J. W. C. Dunlop and J. Yuan, *Adv. Mater.*, 2015, **27**, 2913–2917.
18. J. Steinkoenig, F. R. Bloesser, B. Huber, A. Welle, V. Trouillet, S. M. Weidner, L. Barner, P. W. Rroesky, J. Y. Yuan, A. S. Goldmann and C. Barner-Kowollik, *Polym. Chem.*, 2016, **7**, 451–461.
19. Y. Xiong, H. Wang, Y. Wang and R. Wang, *Polym. Adv. Technol.*, 2012, **23**, 835–840.
20. N. Guo, D. Hu, H. Wang, R. Wang and Y. Xiong, *Polym. Bull.*, 2013, **70**, 3031–3040.
21. D. Roy, J. N. Cambre and B. S. Sumerlin, *Prog. Polym. Sci.*, 2010, **35**, 278–301.
22. M. A. C. Stuart, W. T. S. Huck, J. Genzer, M. Muller, C. Ober, M. Stamm, G. B. Sukhorukov, I. Szleifer, V. V. Tsukruk, M. Urban, F. Winnik, S. Zauscher, I. Luzinov and S. Minko, *Nat. Mater.*, 2010, **9**, 101–113.
23. F. Liu and M. W. Urban, *Prog. Polym. Sci.*, 2010, **35**, 3–23.
24. M. Motornov, Y. Roiter, I. Tokarev and S. Minko, *Prog. Polym. Sci.*, 2010, **35**, 174–211.
25. P. Theato, B. S. Sumerlin, R. K. O'Reilly and T. H. Epps, *Chem. Soc. Rev.*, 2013, **42**, 7055–7056.
26. D. Roy, W. L. A. Brooks and B. S. Sumerlin, *Chem. Soc. Rev.*, 2013, **42**, 7214–7243.
27. Y. Lu and M. Ballauff, *Prog. Polym. Sci.*, 2011, **36**, 767–792.
28. F. D. Jochum and P. Theato, *Chem. Soc. Rev.*, 2013, **42**, 7468–7483.
29. I. Idziak, D. Avoce, D. Lessard, D. Gravel and X. X. Zhu, *Macromolecules*, 1999, **32**, 1260–1263.
30. Y. Yang, A. J. Mijalis, H. C. Agosto, K. J. Tan, J. D. Batteas and D. E. Bergbreiter, *J. Am. Chem. Soc.*, 2012, **134**, 7378–7393.
31. W. Wu, W. Driessen and X. Q. Jiang, *J. Am. Chem. Soc.*, 2014, **136**, 3145–3155.
32. C. Chen, Z. Wang and Z. Li, *Biomacromolecules*, 2011, **12**, 2859–2863.
33. J. Lutz, *Adv. Mater.*, 2011, **23**, 2237–2243.
34. Y. Maeda, T. Nakamura and I. keda, *Macromolecules*, 2002, **35**, 217–222.
35. V. A. Kryuchkov, J. Daigle, K. M. Skupov, J. P. Claverie and F. M. Winnik, *J. Am. Chem. Soc.*, 2010, **132**, 15573–15579.
36. H. Mori, H. Iwaya, A. Nagai and T. Endo, *Chem. Commun.*, 2005, 4872–4874.
37. R. Hoogenboom, *Angew. Chem., Int. Ed.*, 2009, **48**, 7978–7994.
38. Y. Kohno, S. Saita, Y. Men, J. Yuan and H. Ohno, *Polym. Chem.*, 2015, **6**, 2163–2178.
39. D. Mecerreyes, *Prog. Polym. Sci.*, 2011, **36**, 1629–1648.
40. H. Yoshimitsu, A. Kanazawa, S. Kanaoka and S. Aoshima, *Macromolecules*, 2012, **45**, 9427–9434.

41. E. Karjalainen, N. Chenna, P. Laurinmaki, S. J. Butcher and H. Tenhu, *Polym. Chem.*, 2012, **4**, 1014–1024.
42. H. Mori, M. Yahagi and T. Endo, *Macromolecules*, 2009, **42**, 8082–8092.
43. J. Ribot, C. Guerrero-Sanchez, R. Hoogenboom and U. Schubert, *Chem. Commun.*, 2010, **46**, 6971–6973.
44. W. Yao, H. Wang, G. Cui, Z. Li, A. Zhu, S. Zhang and J. Wang, *Angew. Chem., Int. Ed.*, 2016, **55**, 7934–7938.
45. K. Fukumoto and H. Ohno, *Angew. Chem., Int. Ed.*, 2007, **46**, 1852–1855.
46. Y. Kohno, Y. Deguchi and H. Ohno, *Chem. Commun.*, 2012, **48**, 11883–11885.
47. B. Ziółkowski and D. Diamond, *Chem. Commun.*, 2013, **49**, 10308–10310.
48. Y. He and T. P. Lodge, *Chem. Commun.*, 2007, 2732–2734.
49. Y. Men, H. Schlaad and J. Yuan, *ACS Macro Lett.*, 2013, **2**, 456–459.
50. Y. Xiong, H. Wang, R. Wang, Y. Yan, B. Zheng and Y. Wang, *Chem. Commun.*, 2010, **46**, 3399–3401.
51. Y. Xiong, Y. Wang, H. Wang and R. Wang, *Polym. Chem.*, 2011, **2**, 2306–2315.
52. Y. Xiong, Y. Wang, H. Wang, R. Wang and Z. Cui, *J. Appl. Polym. Sci.*, 2012, **123**, 1486–1493.
53. Y. Xie, Z. Zhang, T. Jiang, J. He, B. Han, T. Wu and K. Ding, *Angew. Chem., Int. Ed.*, 2007, **46**, 7255–7258.
54. B. Hu, T. Wu, K. Ding, X. Zhou, T. Jiang and B. Han, *J. Phys. Chem. C*, 2010, **114**, 3396–3400.
55. Y. Xiong, J. Liu, Y. Wang, H. Wang and R. Wang, *Angew. Chem., Int. Ed.*, 2012, **51**, 9114–9118.
56. B. Sun, Q. Jin, L. Tan, P. Wu and F. Yan, *J. Phys. Chem. B*, 2008, **112**, 14251–14259.
57. G. C. Pimentel and A. L. McClellan in *The Hydrogen Bond*, ed. W. H. Freeman, San Francisco, 1960, pp. 348–363.
58. M. A. Firestone, P. G. Rickert, S. Seifert and M. L. Dietz, *Inorg. Chim. Acta*, 2004, **357**, 3991–3998.
59. J. Zhang, J. Liu, Y. Zuo, R. Wang and Y. Xiong, *Molecules*, 2015, **20**, 17378–17392.
60. Y. Zuo, Z. Jiao, L. Ma, P. Song, R. Wang and Y. Xiong, *Polymer*, 2016, **98**, 287–293.
61. Y. Zuo, N. Guo, Z. Jiao, P. Song, X. Liu, R. Wang and Y. Xiong, *J. Polym. Sci., Part A: Polym. Chem.*, 2016, **54**, 169–178.
62. C. Miao, F. Li, Y. Zuo, R. Wang and Y. Xiong, *RSC Adv.*, 2016, **6**, 3013–3019.

CHAPTER 8

Redox-active Immobilized Ionic Liquids and Polymer Ionic Liquids

THUAN-NGUYEN PHAM-TRUONG, JALAL GHILANE AND HYACINTHE RANDRIAMAHAZAKA*

Université Paris Diderot, Sorbonne Paris Cité, ITODYS, UMR 7086 CNRS, Surface- Ionic liquid - Electrochemistry and Energy (SIELE group), 15 rue Jean-Antoine de Baïf, 75205 Paris, France
*E-mail: hyacinthe.randria@univ-paris-diderot.fr

8.1 Introduction

> *"I am enough of the artist to draw freely upon my imagination. Imagination is more important than knowledge. Knowledge is limited. Imagination encircles the world."* Albert Einstein.[1]

Although the first synthesized ionic liquid was reported at the beginning of the last century, we had to wait until the 1990s to see the great innovation of this family of compounds. Since then, ionic liquids and their derivatives have attracted a lot of interest from the scientific world and have become a real pluridisciplinary field. So, what makes ionic liquids different from other compounds? According to the common definition, an ionic liquid is a molten salt with a melting temperature below 100 °C. The characteristics that

Smart Materials No. 29
Polymerized Ionic Liquids
Edited by Ali Eftekhari

Published by the Royal Society of Chemistry, www.rsc.org

can be attributed to ionic liquids are a wide liquidus range, very low vapor pressure, and thermal stability with unusual solvent properties. These properties make ionic liquids one of the best choices as solvents for green chemistry and for sustainability science. Overcome with knowledge about ionic liquids at this time, researchers' and scientists' imaginations make ionic liquids rather more useful than only being used in organic synthesis. Thus, the limits are being further pushed to tailor the physico-chemical properties of ionic liquids by generating new structures for use in the field of electrochemistry, in which the development of this type of compound becomes unlimited. Within this chapter, we will focus on describing the use of ionic liquids and their derivatives in electrochemistry and their potential applications in electrochemical devices.

8.2 Electrochemistry in Ionic Liquids

In 1982, dialkylimidazolium chloroaluminate-based ionic liquids were the very first room-temperature ionic liquids (RTILs) to enter into the electrochemistry field for use in the electrodeposition of aluminum, according to the reported work by Wilkes and co-workers.[2] Since then, different studies have been performed in order to investigate the electrochemical behaviors not only of ionic liquids themselves, but also of other materials in ionic liquids. From an electrochemical point of view, ionic liquids simultaneously play the role of a solvent and a supporting electrolyte. A good electrolyte for electrochemical measurements and for devices should exhibit a strong resistance to electro-oxidation and reduction. This behavior is indicated by a large electrochemical window (EW). For different families of ionic liquids (imidazolium, pyrrolidium, ammonium, *etc.*), a broad EW was observed (3.5 to 6 V), which depends upon the composition of the ionic liquid as listed in Table 8.1.

Compared with the usual organic solvents in use nowadays, whose EWs result partially from the interactions between the electrolyte and the solvent, the EW of ionic liquids is as an intrinsic value that is characteristic for each ionic liquid structure. In addition, the vapor pressure of RTILs is much lower than that of other organic solvents providing strong stability over time, which is extremely beneficial for long-term, non-volatile electrochemical devices. By possessing advantageous electrochemical properties, ionic liquids are being used in numerous applications in the electrochemistry field, especially for electrodeposition processes. Indeed, a large variety of metals, metal oxides and semiconductors have been deposited in ionic liquid media.

8.2.1 Electrodeposition of Metals, Metal Alloys and Semiconductors

As mentioned in the previous subsection, the first metal to be electrodeposited in an ionic liquid was aluminum. The main problem with this process is that the precursor, $AlCl_3$, which is a Lewis acid, is highly sensitive to the

Table 8.1 Typical electrochemical windows of different types of ionic liquid.

Ionic liquid[Ref]	Cathodic potential limit (V)	Anodic potential limit (V)	Electrochemical window (V)	Reference electrode	Working electrode
[1-Et-3-MeIm$^+$][Tf_2N^-][3]	−1.8	2.3	4.1	I^-/I_3^-	Pt
[1-Et-3-MeIm$^+$][OTf^-][3]	−1.7	2	3.7	I^-/I_3^-	Pt
[1-Bu-3-MeIm$^+$][BF_4^-][4]	−1.6	2.5	4.1	Pt QRE	Pt
[1-Bu-3-MeIm$^+$] [PF_6^-][4]	−1.6	2.55	4.15	Pt QRE	Pt
[1,2-Me_2-3-PrIm$^+$][Tf_3C^-][5]	0.28	5.65	5.37	Li^+/Li	GC
[1,2-Me_2-3-PrIm$^+$][Tf_2N^-][5]	0.2	5.4	5.2	Li^+/Li	GC
[1-Me-3-PrPyrro$^+$][Tf_2N^-][6]	−2.5	2.8	5.3	Ag Wire	Pt
[1-Bu-3-MePyrro$^+$][Tf_2N^-][7]	−3	2.5	5.5	Ag/Ag^+	GC
[Et_3HexAmm$^+$][Tf_2N^-][8]	−2.5	2	4.5	Ag/Ag^+	GC
[ButPyri$^+$][BF_4^-][9]	−1	2.4	3.4	Ag wire	Pt
[Et_3S^+][Tf_2N^-][10]	−2	2.7	4.7	Pt	GC
[But_3S^+][Tf_2N^-][10]	−2.1	2.8	4.8	Pt	GC

presence of water, which induces the formation of basic $AlCl_3$ and as a consequence inhibits the electrodeposition process. However, this deposition process becomes much more feasible in an ionic liquid[11] and the morphology of the deposits can also be controlled.[12] From this successful process, a large number of metals and metal alloys (Pd, Cu, Fe, Al–Mn, *etc.*) have been electrodeposited onto different substrates in ionic liquid media.[13–15]

Besides the electrodeposition of different metals, the electrodeposition of semiconductors in ionic liquids has also been investigated. Only a few published studies mention the use of ionic liquids for the electrodeposition of semiconductors.[15–17] Despite the feasibility of generating different SCs on a surface (IV, III–V, and p-type SCs) from an IL, the main problem with this process is the homogeneity of the generated surface.

8.2.2 Electropolymerization of Conducting Polymers

In this context, it would be remiss not to discuss the possible electropolymerization of conducting polymers in ionic liquid media. High quality, homogeneous and smooth-surface poly(3-alkylthiophene)s were electrochemically prepared in [1-But-3-MeIm$^+$][PF_6^-] as promising materials for electrochromic devices.[18] Another typical example is the electrochemical polymerization of poly(aniline), which is well-known for its use in energy applications. In addition, CF_3COOH-doped poly(aniline) nanotubule structures, with a diameter of around 120 nm, were obtained after electrochemical polymerization in an IL solution containing aniline monomers and trifluoroacetic acid[19] (Figure 8.1).

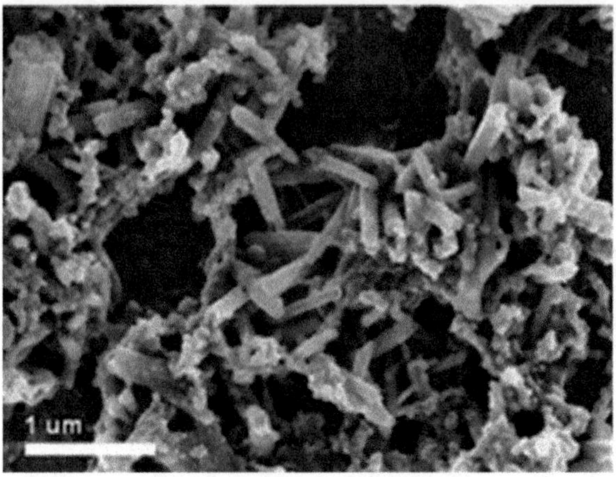

Figure 8.1 SEM images of PANI deposited onto ITO synthesized from $BMIPF_6$ containing 1.0 M CF_3COOH. Reprinted from Electrochemistry Communications, 8 (10), D. Wei, C. Kvarnström, T. Lindfors, A. Ivaska, Polyaniline nanotubules obtained in room-temperature ionic liquids, 1563–1566, Copyright (2006) with permission from Elsevier.[19]

Others conducting polymers have also been electrochemically synthesized in ionic liquids, such as poly(pyrrole),[20] poly(*p*-phenylene),[21] and poly(3,4-ethylenedioxythiophene).[22] Compared to a classical solvent, the use of ionic liquids as an electrolytic medium provides different morphological structures and offers new interface properties.

8.3 Redox-active Ionic Liquid

In the previous subsection, the different aspects of electrochemistry in ionic liquids were briefly described. Nevertheless, the potential of ionic liquids in electrochemistry is much more considerable than just being limited to their use as a solvent–electrolyte. Recently, new families of ionic liquids known as "task specific ionic liquids" have been introduced that are based on ionic liquids bearing functional groups. As an example, the introduction of a redox group to an ionic liquid framework led to the formation of a new class of ionic liquids labeled as redox-active ionic liquids. The first redox-active ionic liquids were prepared by Grätzel's group in 2004; they contained a redox active counter-anion ([I^-] and [$SeCN^-$]) and were successfully used in a dye sensitized solar cell system.[23,24] R. Balasubramanian *et al.* in 2006 reported the electrochemical behaviors of ferrocenated imidazolium compounds[25] whose ferrocenyl group is connected to the imidazolium ring by an organic linker (Figure 8.2, left).

This work shows that the viscosity of the redox ionic liquid depends on the length of the linker and the side chains. In Figure 8.2, right, the electrochemical response of the undiluted ferrocenated ionic liquid is strongly influenced by the temperature and, as a consequence, the viscosity of the

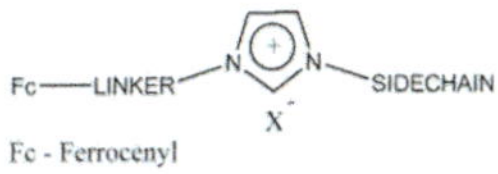

	linker	side chain	η, cP (25 °C)[b]
1	C_4H_8	CH_3 or C_4H_9	na[c]
2	C_6H_{12}	CH_3 or C_4H_9	8.5×10^3
3	$COO-C_3H_6$	CH_3 or C_4H_9	1.4×10^7
4	$COO-C_6H_{12}$	CH_3 or C_4H_9	5.8×10^5
5	$COO-C_{11}H_{22}$	CH_3 or C_4H_9	1.6×10^5
6	$COO-C_2H_4OC_2H_4$	CH_3 or C_4H_9	5.4×10^5
7	$COO-C_3H_6$	$C_2H_4OCH_3$	na[c]
8	$COO-C_6H_{12}$	$C_2H_4OCH_3$	na[c]

[a] $X = Br^-$, PF_6^-, or BF_4^-. All ILs listed are either fluid or glassy materials except for **1**·CH_3·Br, **2**·CH_3·Br, and **2**·CH_3·PF_6, which are solids. [b] All viscosities are for $X = PF_6$ and side chain = C_4H_9. [c] Not measured.

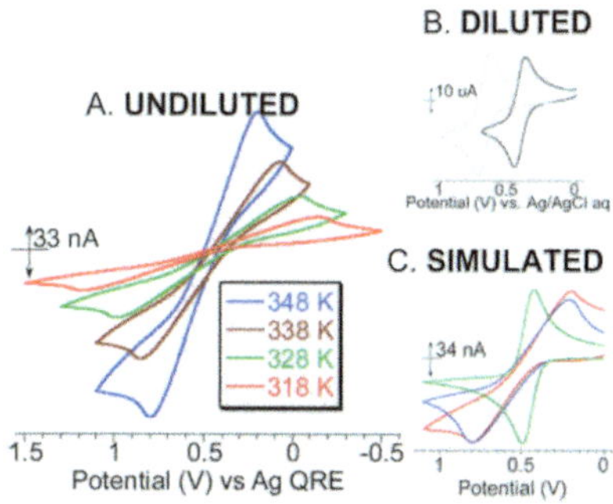

Figure 8.2 (Left) Intrinsically electroactive ionic liquids. (Right) (A) CV (10 mV s^{-1}) of undiluted 8.PF_6 with a 12.5 μm radius Pt microdisk electrode, under vacuum at the indicated temperatures. (B) CV of **1**,CH_3,PF_6 (right curve) and carboxylated **3**,CH_3,PF_6 (left curve) in dilute CH_2Cl_2 solution with 0.2 M Bu_4NClO_4 electrolyte. (C) Digitally simulated (blue curve) CV (10 mV s^{-1}) of undiluted **8**.PF_6 at 75 °C, compared to the experimental results (red curve). The green curve is the ideal CV curve with no iRUNC effect. Adapted with permission from ref. 25. Copyright (2006) American Chemical Society.

medium following the Williams–Landel–Ferry (WLF) model.[26] Compared to the electrochemical behavior of ferrocene in solution, a positive shift in terms of the formal potential is observed for the ferrocenated ionic liquid. Indeed, linking an electron attractor group (imidazolium) to the ferrocene induces a withdrawing effect due to the positive charge. The electron density of ferrocene decreases, resulting in a lower energy level of the highest occupied molecular orbital (HOMO) and an increase in terms of the oxidation potential.[27]

Recently, a hybrid redox system based on imidazolium ionic liquids was investigated by X. Chen *et al.*[28] A mixture of 2 redox ionic liquids containing imidazolium functionalized with 2,2,6,6-tetramethyl-piperidinyl-1-oxyl (TEMPO)[$TFSI^-$] and [1-Pro-3-$MeIm^+$][I^-] was used to produce a Gratzel-type cell. The reported bi-redox couple electrolyte showed strong absorption only in the UV region and a higher redox potential compared with the I^-/I_3^- redox couple.

Later, a bi-redox ionic liquid molecule was reported by O. Fontaine *et al.*[29,30] In these studies, a series of ionic liquids containing anions and cations that are functionalized with anthraquinone (AQ) and TEMPO moieties, respectively, was electrochemically studied (Figure 8.3).

Tethering redox-active moieties onto both the cation and the anion allows the density of the redox-active groups in the liquid state to be increased up to the bulk density of a redox-active solid. In addition, the bi-redox IL provided double the specific energy in a supercapacitor based on activated carbon and graphene oxide electrodes compared to the same approach without redox-active groups. At the same time, the power capability and cycling stability

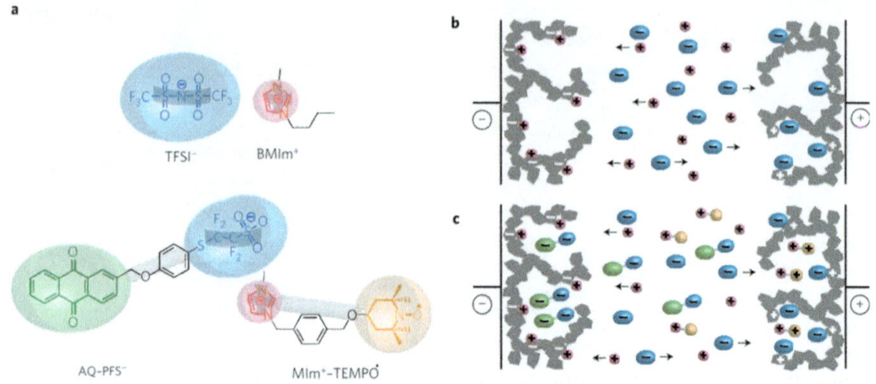

Figure 8.3 Comparison of charge storage in an EDLC with an IL electrolyte and a bi-redox IL-enhanced pseudocapacitor. (a) Structure of the herein used BMImTFSI IL and the bi-redox IL comprising a perfluorosulfonate anion bearing an anthraquinone moiety ($AQ–PFS^-$) and a methyl imidazolium cation bearing a TEMPO moiety ($MIm^+–TEMPO^{\cdot}$). (b) and (c) Charge storage in a purely capacitive EDLC comprising porous carbon electrodes and an IL electrolyte. Reprinted with permission from Macmillan Publishers Ltd: *Nature Materials* (ref. 30). Copyright (2016).

were maintained. The enhanced capacitance is sustained for 2000 cycles without degradation. Moreover, the bi-redox IL enabled high-capacity/high-rate charge storage, overcoming a major hurdle for high-energy supercapacitors, opening up a wide new field for redox materials and their applications.

8.4 Immobilization of Ionic Liquids and Redox-active Ionic Liquids

8.4.1 Introduction

From laboratory research to real-life applications, the role of surface modification with organic layers is always crucial. Indeed, the atoms at the surface of a material behave completely differently compared to the bulk atoms because of the leakage in terms of the coordination number. As a consequence, they have a different free energy, density of states, structure and reactivity. Thus, the attachment of organic molecules onto the surface is considered to be a facile approach to provide novel physico-chemical properties to materials.[31,32] The immobilization of organic molecules has been successfully integrated in several applications, including smart surfaces, molecular electronics, photovoltaics, catalysis and surface wettability, and several reviews have been devoted to this topic and related applications.[33–40] It is well-known that the surface atoms are more reactive than the other atoms in the bulk of the material, which implies the enhancement of the physisorption and chemisorption of molecules from the ambient environment at the surface.

Introducing an organic layer onto the surface could act as an organic barrier for protection and/or to generate new interfaces involving novel properties, which can be helpful in different fields such as anti-corrosion, super-hydrophobic/super-hydrophilic surfaces, inks and paints, *etc.* Ionic liquids could also be deposited onto a surface, providing new properties to the interface.[41–45]

8.4.2 Approaches to form Thin Layers of Ionic Liquids

Self-assembled monolayers (SAMs) of organic molecules consist of organic assemblies formed by the well-arranged adsorption of organic molecules that have a specific functionalized head group from the solution or the gas phase onto the surface of a solid or onto the surface of a liquid (in the case of liquid metals and its alloys). The most widely studied branch of SAMs is the adsorption of alkanthiols on gold, silver and copper. The strong affinity of thiols (–SH) for the surface of noble metals is the key factor leading to the generation of well-defined organic layers that have other functionalities at the exposed interface.

Thiol-terminated imidazolium-based ionic liquids have been immobilized onto Au substrates by dipping the Au substrate into a millimolar ethanolic solution of the thiol-terminated IL for a few hours.[41] This procedure is widely used for the formation of alkanethiols on gold substrates (Figure 8.4).

Actually, a well-defined structure of the SAM can only be achieved by immersing the substrate for a few hours in order to overcome the slow rearrangement rate of the adsorbates. Different parameters such as temperature, purity of the adsorbate, concentration of oxygen in the solution, solvent, immersion time, *etc.*, should be considered to achieve a good SAM. A generalized scheme of the surface modification of a noble metal by a thiol-terminated IL SAM is presented in Figure 8.4.

B. Lee *et al.*[41] also demonstrated direct counter-anion exchange from the 1.9 nm IL-SAM layer by simple immersion of the ionic liquid-modified substrate into different solutions containing the target counter-anion.

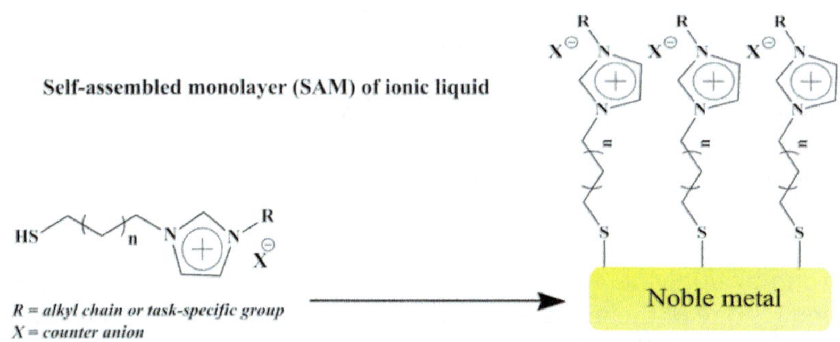

Figure 8.4 Self-assembled monolayer of a thiol-functionalized imidazolium-based ionic liquid on a noble metal (Au, Ag, Cu, *etc.*).

The influence of the counter-anion on the water wettability of the substrate was also demonstrated. Different counter-anions were found to influence the water contact angle of the substrate in the following order: $NTf_2^- > PF_6^- > CF_3SO_3^- > ClO_4^- > BF_4^- > Br^-$. Shortly after this work, an IL-SAM on a gold substrate was reported to be able to exhibit selective redox-switchable behaviors toward Ru^{3+}/Ru^{2+} in the presence of $[Fe(CN)_6]^{3-}$ as a counter-anion.[42] The electrons pass through the IL-SAM layer in the presence of $[Fe(CN)_6]^{3-}$ showing the redox signal of the Ru^{3+}/Ru^{2+} couple. However, the detected signal is cut-off by adding SCN^- or OCN^- into the solution. A self-assembled monolayer of ionic liquid was also applied to gold nanoparticles in order to selectively detect counter-anions[46] or to make a new class of enzymatic biosensors[47] by studying the displacement of the plasmonic band of the IL-functionalized Au nanoparticles.

Besides the formation of a self-assembled monolayer using thiol-terminated ILs on noble metal surfaces, IL-SAMs can also be formed on hydroxyl-modified surfaces by using alkoxysilanes bearing imidazolium moieties. As reported by B. Gadenne *et al.*,[48] well-ordered silica particles were synthesized *via* template-directed polycondensation of triethoxysilylated imidazolium. The materials prepared *via* this technique depend strongly on the length of the alkyl side chain. Whereas a short alkyl chain (methyl, allyl) gave a lamellar structure, longer chains lead to the formation of nanostructured silica with hexagonal symmetry, which can be attributed to the micellar arrangement of the surfactant molecules. In the case of short chains, the ionic liquid molecules interact only with the polar head of the surfactant, while in the case of longer chains, in addition to the interaction described for the previous case, the alkyl chain of the ionic liquid can also interact with the hydrophobic tail of the surfactant by being incorporated into a micellar arrangement. More recently, the surface of modified silica gels functionalized with a monolayer of covalently attached 1-(1-trimethoxysilane)propyl-2,3-dimethyl-imidazolium chloride, tetrafluoroborate or hexafluorophosphate ionic moieties was used to catalyze the aldol reaction. Later, M. Zhang *et al.*[49] proposed the use of silica spheres modified with an ionic liquid to make stationary phases for HPLC, thanks to their facile preparation, high bonding amount and excellent design ability as illustrated in Figure 8.5. Besides the alkoxysilane group, other functional groups can be used for the formation of a self-assembled monolayer *via* a condensation process, such as alkylphosphonium-bearing ionic liquids.[50]

In parallel with the self-assembled monolayers of ionic liquids with highly oriented molecules, the electrochemical grafting of a thin layer of ionic liquid with an electrochemically assisted method is also considered to be an efficient way to immobilize ionic liquid moieties onto a conductive substrate. This method involves the generation of radicals from oxidizable or reducible group by applying a suitable potential to the substrate. Although this technique is one of the most powerful, there are very few reports on surface functionalization with molecules based on ionic liquids.

Electrochemically assisted oxidation grafting of ionic liquid moieties can be carried out in one or multiple steps. To the best of our knowledge, there

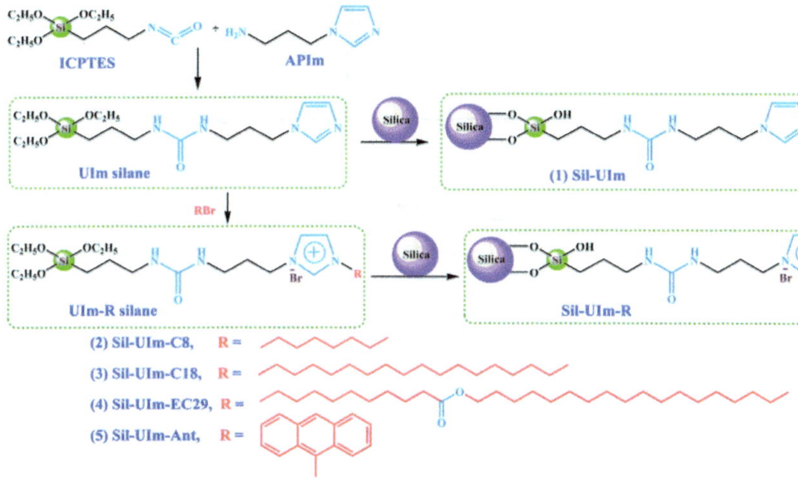

Figure 8.5 Synthetic strategy for SCIL stationary phases. Reprinted with permission from ref. 49. Copyright (2004) The Royal Society of Chemistry.

are very few reports that mention the oxidative grafting of an ionic liquid. The oxidative grafting of a redox-molecule-based ionic liquid onto an electrode surface was first reported by M. Gomez-Mingot *et al.*[51] In this report, *N*-(2-aminoethyl)-*N*-(4-nitrophenylethylimidazolium) bromide was grafted onto a glassy carbon electrode *via* electrochemical oxidation of the amine group at around 1.2 V *vs.* SCE. During the grafting process, aminyl radicals that were electrochemically generated near the electrode surface reacted rapidly with the surface leading to the formation of a thin ionic liquid layer (Figure 8.6).

The immobilization of the ionic liquid on the surface was evidenced by XPS. Figure 8.6 shows the electrochemical characterization of the modified electrode recorded in 0.1 M sulfuric acid aqueous solution, and exhibits the presence of a reduction wave at −0.4 V *vs.* SCE corresponding to the irreversible reduction of the nitro group. Next, during the second cycle, a new redox reversible signal of the generated NHOH/NO group is observed at 0.3 V. These results give rise to a promising material functionalization technique, which combines the attachment of ionic liquids to the surface, the presence of a redox active interface, and the ability to perform anion exchange.

Later on, the immobilization of a ferrocene-based ionic liquid onto a carbon electrode using a stepwise procedure was investigated.[52] The general procedure is based on the electrochemical grafting of a first layer bearing terminal groups that can be engaged in a chemical reaction, followed by a second step to attach the desired molecules. Figure 8.7a presents an example of this two-step procedure leading to the immobilization of ferrocene-based ionic liquid moieties through aryl–aryl coupling. First, oxidative grafting of 2-(4-aminophenyl)acetic acid was performed onto a glassy carbon electrode. Next, in the presence of sodium nitrite and perchloric acid, a diazotization

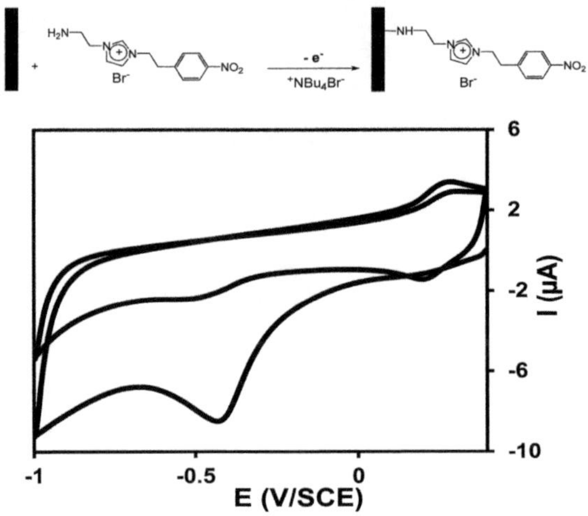

Figure 8.6 Schematic illustrating the grafting of a redox-based ionic liquid through the electrochemical oxidation of a primary amine onto a carbon electrode and the cyclic voltammetry curve of the modified carbon electrode in 0.1 M H_2SO_4. Scan rate: 0.1 V s^{-1}. Adapted with permission from M. Gomez-Mingot, V. Bui-Thi-Tuyet, G. Trippe-Allard, J. Ghilane and H. Randriamahazaka, *ChemElectroChem*,[51] John Wiley and Sons. Copyright © 2014 WILEY-VCH Verlag GmbH & Co. KGaA, Weinheim.

reaction occurs at the surface leading to the conversion of the attached NH_2 head group into a diazonium group. The latter is accessible for further transformation. Indeed, the Gomberg–Bachmann reaction through aryl–aryl coupling allows the attachment of an aryl radical to one of the cyclopentadienyl moieties of the ferrocene-based ionic liquid.[53]

In this work, the immobilization of the ferrocene-based ionic liquid was evidenced by electrochemical characterization, showing the reversible redox signal of the ferrocene group. In addition, a two-step functionalization process was also investigated that involved the electrochemical grafting of 4-azidoaniline, followed by a copper(I) catalyzed click reaction with 1-ferrocenylethyl-3-propargylimidazolium bromide (Figure 8.7b).[45] The as-prepared electrode contained a compact monolayer of redox-active ionic liquid, which was evidenced by using scanning electron microscopy, resulting in 2-D electron transport within the ionic liquid layer.

In parallel with oxidative grafting, electrochemical reduction grafting of ionic liquids has also been a subject of scientific investigation. Among the different functional groups, diazonium derivatives are the most investigated molecules for electrochemical grafting. S. Bouden *et al.*[44] immobilized a bi-redox ionic liquid, 1-nitrophenylethyl-3-methylferrocenylimidazolium Tf_2N, onto an electrode surface by the electrochemical reduction of *in situ*-generated diazonium functional groups (Figure 8.8). By sweeping the potential to the cathodic side (from −0.2 V to −1.1 V *vs.* SCE) in an MeCN solution containing a bi-redox ionic liquid, nitro-phenyl groups in the vicinity of the

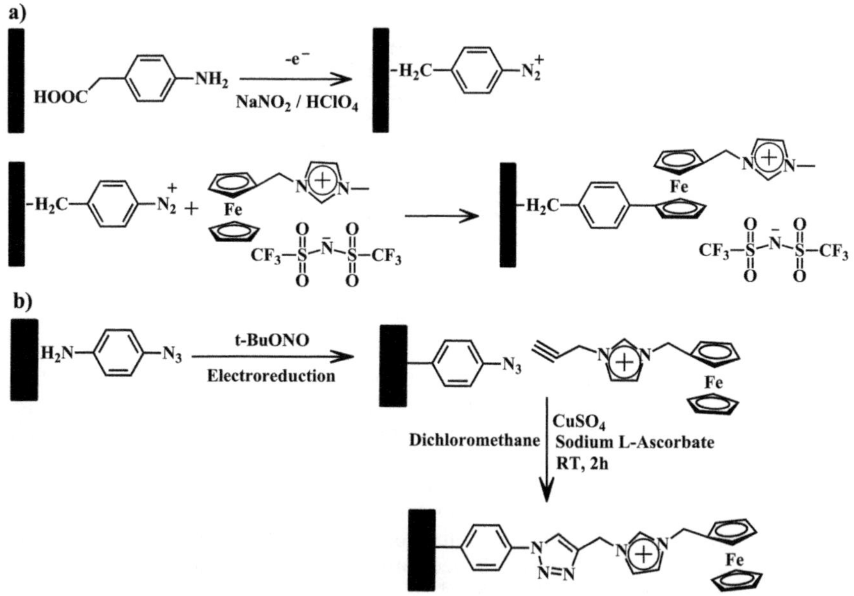

Figure 8.7 Two strategies for stepwise processes for the immobilization of ferrocene-based ionic liquids. (a) Reprinted from Electrochemistry Communications, 70, T. N. Pham-Truong, F. Lafolet, J. Ghilane, H. Randriamahazaka, Surface functionalization with redox active molecule-based imidazolium *via* click chemistry, 13–17, Copyright (2016) with permission from Elsevier[45] and (b) reprinted from *Journal of Electroanalytical Chemistry*, 713, V. Bui-Thi-Tuyet, G. Trippé-Allard, J. Ghilane, H. Randriamahazaka, Surface functionalization of ferrocene based ionic liquid onto carbon surface using stepwise grafting, 28–31, Copyright (2014) with permission from Elsevier.[52]

electrode surface were reduced *via* 6 electron pathways to give amines, which were transformed into diazonium groups in the presence of sodium nitrite under acidic conditions. Under negative potential, the *in situ*-generated diazonium groups within the diffusion layer were reduced into radicals and then attached rapidly to the electrode surface.

The immobilized ionic liquid layer was evidenced by XPS and electrochemistry by observing the signature of the ferrocene moieties (Figure 8.8). This approach leads to the generation of a multifunctional material bearing redox-active molecules and ionic liquid components. This new interface shows versatile applications toward electrochemically switchable wettability.

Recently, a simple and general route for the immobilization of molecules containing an ionic liquid framework was proposed as shown in Figure 8.9.

Inspired by the classical synthesis of ionic liquids, surface-initiated synthesis was investigated. This process is based on building the ionic liquid structure on the surface. Thus, a bromide-terminated layer was immobilized

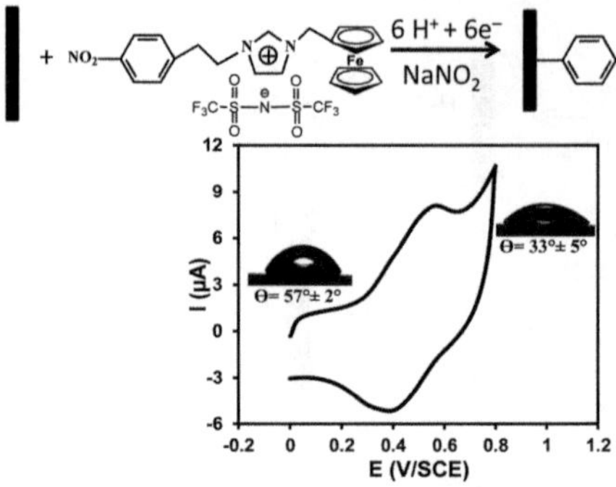

Figure 8.8 Proposed pathway for the electrochemical grafting of ferrocene-based ionic liquid molecules. Cyclic voltammogram of the grafted electrode in acetonitrile solution containing 0.1 M LiTFSI as the supporting electrolyte; scan rate: 0.1 V s^{-1}. Inset: contact angles obtained on the modified electrode before and after electrochemical oxidation. Reprinted from *Electrochemistry Communications*, 58, S. Bouden, G. Trippé-Allard, J. Ghilane, H. Randriamahazaka, Electrochemical immobilization of redox active molecule based ionic liquid, 65–68, Copyright (2015) with permission from Elsevier.[44]

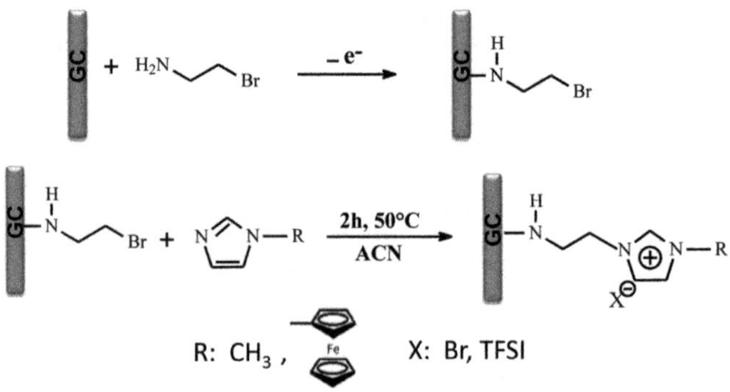

Figure 8.9 Proposed approach for the surface-initiated attachment of molecules based on ionic liquids. Reprinted with permission from ref. 43. Copyright (2016) American Chemical Society.

on the electrode surface, and the modified surface was then immersed in a solution containing imidazole derivatives. The generated structures provide multifunctional systems containing ions, immobilized cations and mobile anions, and redox species.[43]

Originally, J. Ghilane *et al.*[54] proposed a novel route for the surface modification of a Pt electrode using a bi-redox system containing an imidazolium-based ionic liquid. The Pt electrode was polarized under cathodic polarization at −2.2 V *vs.* SCE in the presence of 1-ferrocenylethyl-3-methylimidazolium NTf_2 as a supporting electrolyte (0.1 M) in dry acetonitrile solution.

As a result, Pt reduction occurs concomitant with the insertion of the supporting electrolyte cation, generating a new metal–organic phase, $[Pt_n^{\delta-}, A^{\delta+}, AX]$ (where A^+ is the cation and X^- is the anion electrolyte), as illustrated in Figure 8.10. This surface functionalization procedure changes greatly the nature of the starting material by forming ferrocene imidazolium platinide-like phases.

Furthermore, a second redox couple was introduced by spontaneous reaction between a 4-nitrophenyldiazonium tetrafluoroborate salt and the reduced platinum phase. Indeed, thanks to the reducing power of the cathodically activated platinum, the spontaneous reduction of diazonium occurs leading to the attachment of nitrophenyl groups onto the reduced Pt, as shown in Figure 8.10. As a result, bi-redox molecules were attached onto the Pt, with the ferrocenyl-based ionic liquid inside the Pt and nitro-phenyl at the surface. The electrochemical and XPS characterizations confirm the presence of two redox systems on the Pt surface (Figure 8.10). In summary, the immobilization of a thin layer of redox ionic liquids will bring new insight in the understanding of the properties of an electrical double layer at ionic liquid/electrode interfaces and will open new applications in various electrochemical devices.

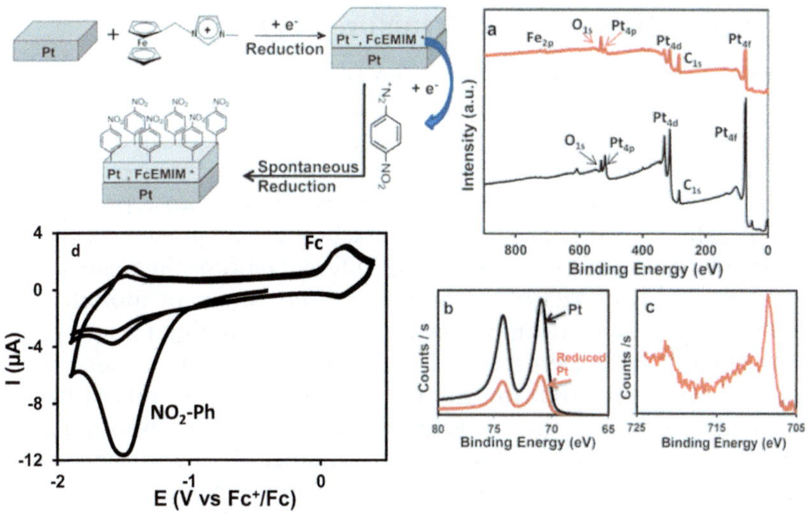

Figure 8.10 Schematic illustrating the spontaneous grafting of a nitrophenyl layer onto activated platinum. (a) XPS survey scans of the platinum substrate (black line) without any polarization and (red line) after 200 s of cathodic polarization at −2.2 V in a ferrocene-based ionic liquid. High resolution XPS spectra for the (b) Pt(4f) and (c) Fe(2p) core levels. (d) Electrochemical characterization of the modified Pt electrode in a solution containing 0.1 M Bu_4NBF_4 in ACN. The scan rate was 0.1 V s^{-1}. Reproduced with permission from ref. 54. Copyright (2013) American Chemical Society.

8.5 Approaches for Polymer Ionic Liquids

Recent developments in polymer science have allowed the synthesis of various polyelectrolytes that are continuously being introduced with controlled morphology, such as linear polymers, nano/micro particles, and dendritic and star structures. The term poly(ionic liquid) (PIL) refers to a subclass of polyelectrolytes that feature an ionic liquid (IL) species in each monomer repeating unit, connected through a polymeric backbone to form a macromolecular structure.[55] In the last decade, poly(ionic liquid)s have been widely investigated as new polyelectrolytes, leading to rapid growth in terms of publications and applications in diverse fields.[56–61] Ohno *et al.* were the pioneers of the use of PILs as solid electrolytes in electrochemical applications.[62–65] These studies inspired researchers to use PILs with different structures and properties in several applications. There are some reviews provided by Firestone, Long, Texter, Mecerreyes and Antonietti that summarize the synthesis, characterization, and applications of PILs.[55,66–71] In this subsection, the state-of-the-art in this area will be described, presenting a systematic view of the preparation and typical properties of PILs that have already been reported in the literature.

8.5.1 Synthetic Route and Structure of PILs

Monomer-based ionic liquids have been studied leading to the emergence of novel solid materials bearing ionic liquid frameworks. Several reviews were devoted to this new class of ionic polymer, highlighting the formation of new structures, new interfacial properties and potential applications.[66–68] Depending on the nature of the polymeric backbone, two types of PIL are classified, called polymerized cationic ionic liquids and polymerized anionic ionic liquids. For each series of PILs, different approaches can be used, such as direct polymerization of ionic liquid monomers or modification of polymeric polymers by forming repeating units containing ionic liquids. In 1973, the first PILs were synthesized by Salamone *et al.*[72] However, this work did not receive much attention from the scientific community at the time, and we had to wait until the 1990s for the next reports by Ohno *et al.* on the free radical polymerization of ionic liquids. Since the importance of polymer-based ionic liquids has been recognized, numerous publications have appeared reporting different structures of ionic liquid monomers bearing commonly vinyl or acryloyl groups, as shown in Figure 8.11.

Vinyl-functionalized ionic liquid monomers can be prepared in a one-step synthesis *via* a quaternization reaction by mixing halo-alkane molecules bearing vinyl groups and N–R imidazolium molecules (R = alkyl, redox-active or other functional groups). Further anion exchange might offer interesting properties to the monomer. Acryloyl-functionalized ionic liquid monomers can be synthesized in a 3-step process. Firstly, (meth)acryloyl chloride is treated with a hydroxyl-containing halo-alkane resulting in the corresponding ester. In the second step, the as-prepared halo-alkane bearing a

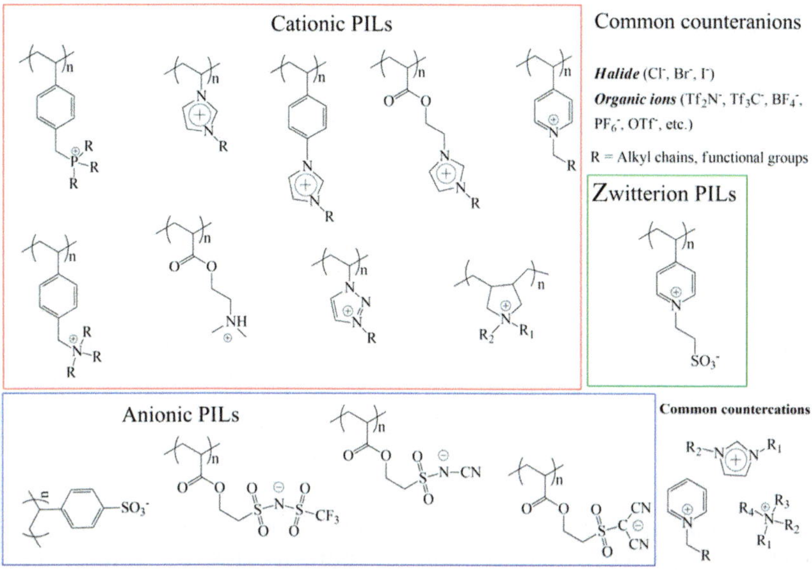

Figure 8.11 Examples of poly(ionic liquid) chemical structures.

(meth)acryloyl group is added leading to a quaternization reaction where the imidazolium is connected to the (meth)acryloyl group. Finally, most of the PILs can be commonly synthesized by conventional radical polymerization. In addition, other polymerization approaches can also be applied for PIL systems, such as atom transfer radical polymerization (ATRP)[73,74] and reversible addition–fragmentation transfer polymerization (RAFT).[75,76] The only part that determines the structure and properties of the PIL is the chemical structure of the ionic liquid monomer. Most PILs were successfully synthesized in bulk solution, in micro-emulsions or in dispersed media. However, it has been shown to be feasible to polymerize ionic liquids onto material surfaces by surface-initiated atom transfer radical polymerization (SI-ATRP).[61,77] In addition to linear polymers, cross-linked PIL networks have been synthesized by classical radical polymerization. To achieve a PIL network, an ionic liquid monomer bearing bi-, tri-, or multi-vinyl/acryloyl groups must first be prepared. The cross-linking reaction is initiated under thermal- or photo-activation leading to the formation of a PIL network.[78]

Beside these polymeric structures, block polymers have received some attention thanks to their ability to provide new and well-defined structures. Indeed, hyperbranched,[79–84] dendrimeric[85–88] and star-like[89–94] poly- and oligo(ionic liquid)s have been successfully synthesized. A hyperbranched PIL with an onion-like topography, containing a poly(1,3 diether) core, a polar imidazolium inner shell and a non-polar outer alkyl shell, was synthesized.[79] This system was proven to be highly stable under harsh conditions resulting in it being a promising material for use in nano-reactors and as a host for various functional materials. K. Pu *et al.*[95] reported a novel route to prepare

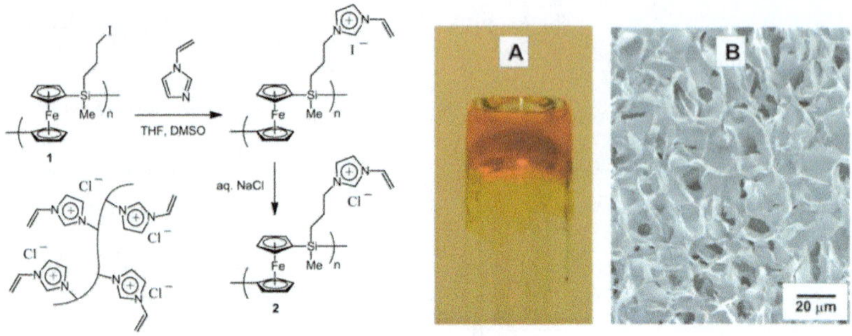

Figure 8.12 (Left) Synthesis of the cross-linkable PFS–PIL. (Right) (A) Photograph of a PFS–PIL macroscopic hydrogel. (B) SEM image of a PFS–PIL hydrogel after freeze-drying. Reproduced with permission from ref. 98. Copyright (2012) American Chemical Society.

fluorescent core–shell nanoparticles based on a hyper-branched PIL (HCPE) for cell imaging. Hyper-branched cationic poly-fluorine (PF) constitutes the core component of the HCPE and serves as a stable light-emitting center. This PF is covered by linear poly(ethylene glycol) (PEG), which acts as a passive layer on the macromolecular surface and, moreover, adds good cytocompatibility properties to the whole structure. Because of the shape persistence of the rigid core, the HCPE molecules form single-molecular nano-spheres, facilitating their uptake for cell imaging. In a number of publications, star-like PIL structures were synthesized by using mainly polyhedral oligomeric silsesquioxane (POSS)-bearing ionic liquids and tetraalkylammonium compounds.[91,96,97] For smart material applications, redox-active PILs have great potential. This kind of PIL is made from an ionic liquid monomer bearing redox-active sites.[77,98] Like others PILs, redox-active PILs can be synthesized either in bulk solution or at a surface. X. Sui *et al.*[98] reported the preparation of cross-linkable PILs bearing ferrocene moieties by using a PES–PIL hydrogel as described in Figure 8.12.

Poly(ferrocenyl(3-iodopropyl)-methylsilane) was mixed with 1-vinylimidazolium leading to quaternarization. A PES–PIL hydrogel (Figure 8.12) was then prepared by performing photo-initiated polymerization under UV-irradiation in the presence of a photoinitiator (Irgacure 2959). A relatively low glass transition and high decomposition temperature was observed for the PES–PIL (T_g = 48 °C, T_d = 350 °C). Reversible oxidation and the reverse reaction were carried out by adding an oxidant and a reducer to the hydrogel, leading to a color change of the gel. This work highlighted the feasibility of this new class of redox-responsive organometallic polymers. Later, Randriamahazaka *et al.* prepared surfaces functionalized with poly(3-(2-methacryloyloxy ethyl)-1-(*N*-(ferrocenylmethyl) imidazolium Tf_2N)) by using SI-ATRP.[77,99] The initiator was first immobilized onto the surface by electrochemical oxidation of the primary amine in aqueous solution (chronoamperometry at a potential of 1.2 V *vs.* SCE for 200 s). The polymerization process was performed by

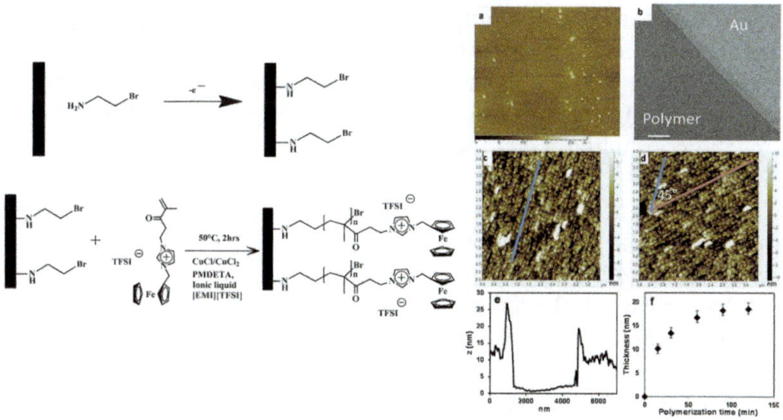

Figure 8.13 (Left) Attachment of the initiator layer through electrochemical oxidative grafting of bromo-ethylamine followed by polymerization using ATRP in an ionic liquid medium. (Right) (a) AFM topography image of Au modified with the initiator layer. (b) SEM image of poly(FcIL) growth on the Au. Scale bar: 1 μm. (c) AFM topography image of poly(FcIL) on the Au substrate; image size: 4 × 4 μm^2. (d) AFM topography image of the same area as in (c) using a scanning angle of 45°. (e) Cross-section after a scratch AFM experiment. (f) Evolution of the poly(FcIL) film thickness as a function of polymerization time for the SI-ATRP. Reproduced with permission from ref. 77. Copyright (2016) American Chemical Society.

dipping the surface-initiated electrode into an ionic liquid solution containing a millimolar concentration of the monomer, a catalytic amount of activator, a small amount of deactivator and a complexing agent under an inert atmosphere (Figure 8.13).

The thickness of the polymer film was followed over time by both electrochemistry and surface characterization techniques (XPS, AFM, and SEM) as demonstrated in Figure 8.13. The Fc-PIL-modified electrodes were electrochemically characterized in usual organic solvents and ionic liquid media, showing a high stability with a small decrease in the ferrocene signal. Interestingly, it was proven that the Fc-PIL could act as a self-double layer. Indeed, the ferrocene electrochemical signal of the attached poly(FcIL) was observed in free-supporting electrolyte acetonitrile solution. The authors also demonstrated a possible application of the poly(FcIL) as an electrochemically reversible surface wettability system and as an electrochemical sensor for catalytic activity toward the oxidation of tyrosine.

8.5.2 Physicochemical Properties

Poly(ionic liquid)s are used nowadays as important ionic polyelectrolytes for solid state devices. Recent studies have been performed in order to gain insight into the intrinsic properties of PILs (ionic conductivity, glass transition temperature, decomposition, *etc.*). Different parameters should be taken into consideration for influencing the ionic conductivity of PILs. The nature of the cation can strongly affect the ionic conductivity of the PIL. Thus, when changing the cation from ammonium, to imidazolium or pyrrolium, the conductivity changes considerably (10^{-7} S cm^{-1}, 3×10^{-7} S cm^{-1} and 3×10^{-6} S cm^{-1}, respectively).[100] In addition, the length of the spacer also plays an important role. It was proven that shortening the spacer between the cation and the polymeric main chain results in a decrease in the conductivity. As an example, the conductivity changes from 2.4×10^{-10} S cm^{-1} to 3.2×10^{-6} S cm^{-1} when the spacer is changed from ethyl to undecyl in poly(imidazolium)-based ILs.[100] When introducing a flexible spacer, such as poly- or oligo(ethyleneoxide), a drastic increase in the conductivity was recorded with a 2 orders of magnitude enhancement (10^{-4} S cm^{-1}) compared to the equivalent alkyl spacer.[101] Unlike for cationic PILs, in the case of anionic PILs, the spacer has to be as short as possible to get a high conductivity.[62] When keeping the same cation ($EtIm^+$), the conductivity increases from poly(*p*-styrenesulfonate) ($\sigma = 1.1 \times 10^{-8}$ S cm^{-1}) to poly(sulfonate) ($\sigma = 1.1 \times 10^{-4}$ S cm^{-1}). The explanation for this change might be attributed to the size effect of the anion on the mobility of the imidazolium cation. In summary, the ionic conductivity of a PIL is partially due to the degree of freedom of the positively charged ion, either in the main chain or free in the anionic polymer matrix. This is evidenced by the augmentation of the conductivity with an increase in the length of the spacer for cationic polymers and a decrease in the linker length for anionic PILs. Moreover, for cationic PILs, the anions also have an impact on the conductivity of the corresponding PIL. The influence of different anions on the conductivity of poly(1-Et-3-MeIm$^+$) was investigated and found to be in the following order (50 Hz, 20 °C): $CF_3SO_3^-$ (6.52×10^{-11} S cm^{-1}) < $(CF_3SO_2)_2N^-$ (1.27×10^{-10} S cm^{-1}) < $(CN)_2N^-$ (2.09×10^{-6} S cm^{-1}).[102] The influence of the anions was also investigated for 1-[2-(methacryloyloxy)ethyl]-3-butylimidazolium-based PILs, where PF_6^- (2.5×10^{-6} S cm^{-1}) < BF_4^- (5×10^{-6} S cm^{-1}) < $CF_3SO_3^-$ (1.25×10^{-5} S cm^{-1}) < $(CF_3SO_2)_2N^-$ (3.2×10^{-4} S cm^{-1}) at 110 °C.[103] This effect is attributed to several parameters, including the anion size, the delocalization of the negative charge and the ability of the anion to interact with the polymer backbone. Besides the impact of the PIL structure on its conductivity, the environment also plays an important role. Typically, the addition of an ionic liquid solution to the polymer matrix results in an enhancement in the ionic conductivity of PIL electrolytes. The ionic conductivities were 6.76×10^{-3} S cm^{-1}, 2.92×10^{-3} S cm^{-1} and 5.83×10^{-3} S cm^{-1} for the IL-, poly[BVIm][Tf_2N]/IL mixture- and poly[BVIm][HIm][Tf_2N]/IL mixture-based electrolytes, respectively.[104] Other parameters should also be taken into consideration, such as the water content in the PIL matrix or the impact of the co-block polymer on the final conductivity, *etc.*

In parallel with the ionic conductivity of the PIL, the thermal properties of the material are a key factor that makes PILs interesting materials for various applications. Generally, ionic liquids and poly(ionic liquid)s have rather low T_g values in comparison to other salts. As there is a relationship between conductivity and T_g, the parameters mentioned above also have an impact on T_g. At the same temperature, a PIL that has a lower T_g is more flexible and more dynamic than the others, resulting in a high mobility of the residual ions. Poly[VIm][TMEN][Tf_2N] bearing 2 cations (imidazolium and ammonium) exhibits a relatively high glass transition temperature of around 106.6 °C[105] compared to monocationic PILs (T_g = 80 °C for poly[2-(Methacryloyloxy) ethyl]-*N*,*N*,*N*-trimethylammonium] and 28 °C for poly[1-[2-(Methacryloyloxy)ethyl]-3-methylimidazolium][100]). The introduction of boron into the IL monomer structure causes the corresponding polymer to have a very low glass transition temperature (T_g = −59 °C to −45 °C).[106] It has been empirically accepted that there is a linear relationship between the ionic conductivity and the glass transition temperature (when a PIL has a lower glass transition temperature, its conductivity is higher).

8.6 Applications of Poly(ionic liquid)s

8.6.1 Nanostructuration

Since ionic liquids and poly(ionic liquid)s were considered as advanced materials for green chemistry, new efforts to make micro/nano-structures from and/or with poly(ionic liquid)s have been launched. Among the different forms of nanostructuration, nanoparticles, which are the bridge between the bulk material and molecules, have attracted a great deal of scientific interest. Recently, poly(ionic liquid) nanoparticles with controllable and tunable shapes and well-organized inner structural domains below 5 nm were reported for the first time.[81,107] It was also proven that a small change in the chemical structure of the ionic liquid monomers or the pH of the medium changes the morphology of the synthesized particles. As a consequence, by increasing the length of the alkyl chain of 1-vinyl-1,2,4-triazolium-type ionic liquids, self-assembled wasp-like ellipsoids and onion-like particles were obtained. This results from the energy balance between 3 components, the hydrophobic side chain, the positively charged triazolium and the surface energy of the particle. Indeed, with a longer hydrophobic tail (TILM-C16Br), the exposure of the surface to the aqueous phase is energetically unfavorable, which causes a morphological change from lamellar ellipsoid to vesicular particles. Understanding the physico-chemical properties and the self-ordering in a complex polymer system can be utilized to make functional polymer nanoparticles.

Benefitting from the extremely low vapor pressure of RTILs, poly(ionic liquid)s were prepared under high vacuum using focused ion beam (FIB)[108] and electron beam (EB) irradiation[109] (Figure 8.14). These techniques use high energy particles to generate radicals inside the ionic liquid monomer, which makes the polymerization process possible. The difference between the two techniques is related to the penetration distance of the irradiated beam. The polymerization using FIB irradiation is carried out at the ionic liquid/vacuum interface, while electron beams penetrate deeper inside the

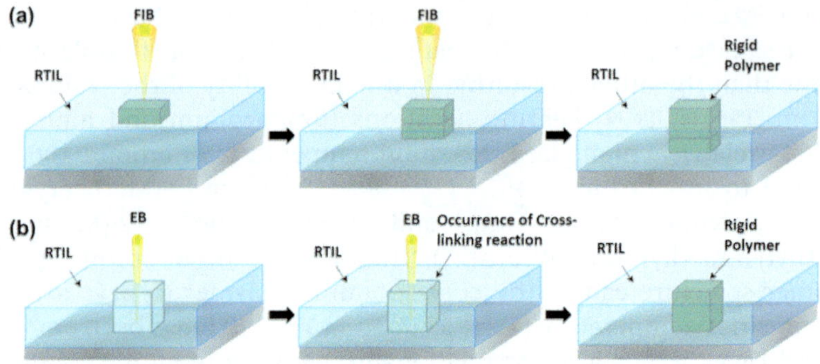

Figure 8.14 Illustrations of the polymer formation mechanism for (a) the FIB-RTIL method and (b) the EB-RTIL method. Reproduced with permission from ref. 109. Copyright (2015) American Chemical Society.

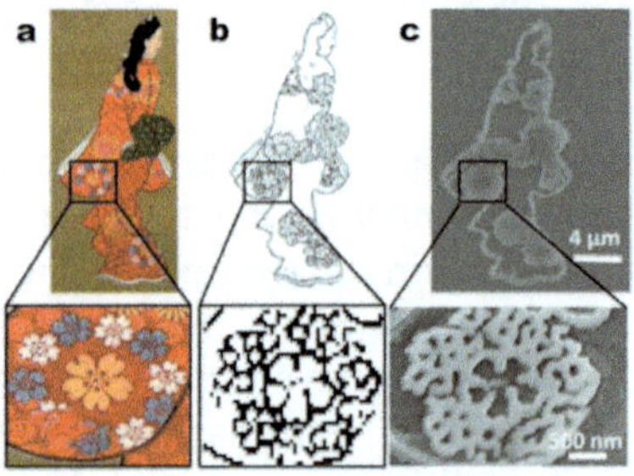

Figure 8.15 (a) A famous Japanese picture entitled, "Beauty Looking Back." This picture was painted by a Japanese artist, Moronobu Hishikawa, in the Edo period. (b) Bitmap image of "Beauty Looking Back" created for FIB drawing. (c) SEM image of a 2D polymer structure fabricated from the bitmap image. The enlarged view of the flower painted on the "Kimono" is shown under each picture. The irradiation conditions for fabrication of this structure were an ion dose of 100×10^{15} ions cm^{-2} and a number of scans of 33. Reprinted with permission from Macmillan Publishers Ltd: *Scientific Reports*, (ref. 108) Copyright (2014).

liquid resulting in polymerization throughout the whole irradiated volume (Figure 8.15). The resolution of the FIB and EB approaches can reach sub-hundred nanometer scales (Figure 8.16). The development of these techniques for making poly(ionic liquid)s with well-defined patterns can be helpful to fabricate micro-devices and micro-machines, or even for ionic liquid-based electronic devices. However, the biggest drawback of the ion/electron beam techniques is that they are time consuming, high cost, and only suitable for small-scale fabrication, so they are not suitable for industrial production. Large-area preparation of ionic liquid patterns is continuously improving through ongoing research. A mesoporous ionic liquid network was successfully prepared by using a hard-template of silica nanoparticles, in which the ionic liquid was grown by polymerization at the interstitial voids, resulting

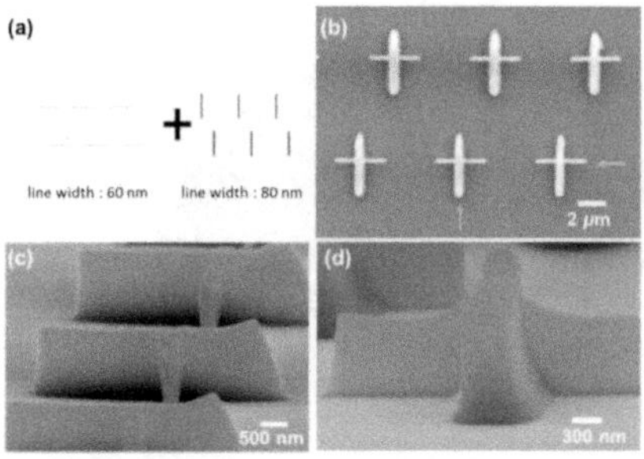

Figure 8.16 Cross-shaped structures prepared by two-step irradiation: (a) irradiation design indicating the two-step irradiation, (b) top-view SEM image of the deposited structures, and (c and d) side-view SEM images observed from the direction represented in (b) by (c) orange and (d) green arrows. The dose condition was 200 mC cm^{-2}. Reproduced with permission from ref. 109. Copyright (2015) American Chemical Society.

in inorganic–organic networks. By etching the silica template in 1 M NaOH, a mesoporous ionic liquid structure is produced with a specific surface area of around 150–220 $m^2\ g^{-1}$ (by applying Brunauer–Emmett–Teller theory) and a determined pore volume in the range of 0.45 $cm^3\ g^{-1}$.

On the other hand, based on poly(ionic liquid)s, different nanostructured materials have already been made. Briefly, Imidazolium-based ionic liquid polymers were used as precursors for the preparation of mesoporous graphitic carbon materials.[110,111] Typically, 1-vinyl-3-ethylimidazolium bromide (IL-a), 1-vinyl-3-ethylimidazolium dicyanamide (IL-b), 1-vinyl-3-cyanomethylimidazolium bromide (IL-c), and 1-vinyl-3-cyanopropylimidazolium bromide (IL-d) were pyrolyzed at 900 °C–1000 °C in the presence of $FeCl_2$, and the polymerization process occurred simultaneously with the carbonization process resulting highly conductive, mesoporous, N-doped graphitic carbon materials. These recent results opened a new and facile route to prepare interesting metal-free materials for energy applications.

8.6.2 Switchable Devices

Like some chameleon species, PILs can adapt their behaviors upon external stimuli or environmental change. Researchers in the field of polymer materials have been trying to mimic this interesting behavior for the last three decades. Their efforts have given birth to a novel concept, smart polymers. By following this rational idea, poly(ionic liquid)s with their unique properties were used for the construction of smart polymers for switchable devices. Among different aspects, the wettability of the material surface plays an important role for self-cleaning materials and anticorrosion. As ionic liquids and poly(ionic liquid)s are sensitive to water, PILs can be used for the efficient switchable wetting of surfaces based on counter-anion exchange. X. He *et al.*[61] reported the preparation of poly[1-(4-vinylbenzyl)-3-butylimidazoliumPF_6] by

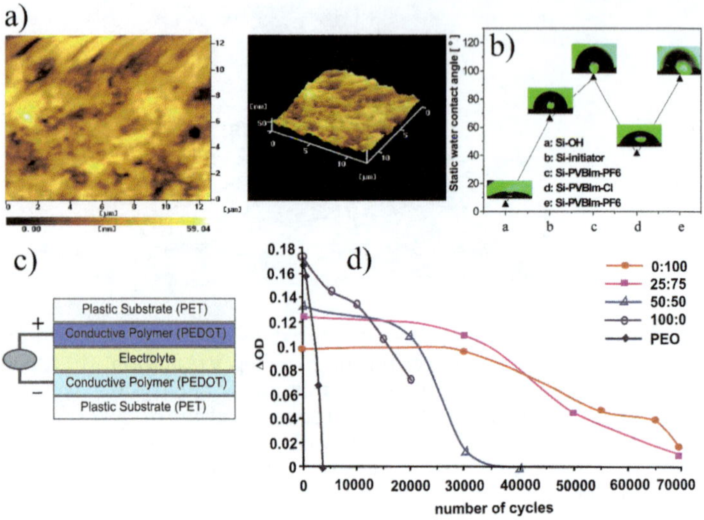

Figure 8.17 (a) AFM topography and 3D image of the PVBIm-PF6 brush-modified silicon surface and (b) changes in static water contact angles corresponding to various surfaces of silicon substrates. The insets show five photographs of the shapes of water droplets on different surfaces. Reproduced with permission from ref. 61. Copyright (2008) American Chemical Society. (c) All-polymeric five-layer electrochromic device configuration, and (d) variation of ΔOD as a function of the number of cycles for ECDs using polymer electrolytes with different [EMIm][Br]:poly[ViEtIm][Br] ratios and using a PEO/lithium triflate electrolyte. Reprinted from Electrochemistry Communications, 8(3), R. Marcilla, F. Alcaide, H. Sardon, J. A. Pomposo, C. Pozo-Gonzalo, D. Mecerreyes, Tailor-made polymer electrolytes based upon ionic liquids and their application in all-plastic electrochromic devices, 482–488, Copyright (2006) with permission from Elsevier.[113]

SI-ATRP on a silicon wafer and its application to static tunable-wettability surfaces. The as-prepared PIL with PF_6^- as the anion showed a rather hydrophobic response (95°). Then ion exchange was carried out by dipping the substrate into NaCl solution, which promoted the formation of a hydrophilic surface due to the formation of PIL–Cl^- (contact angle = 41°). By switching again from Cl^- to PF_6^- anions, the hydrophobic behavior was reversed leading to reversible and tunable wetting of the surface (Figure 8.17a and b). Based on the same ion exchange concept, switchable permeability poly(ionic liquid) hollow particles were prepared by using suspension polymerization utilizing the SaPSeP method.[60] Tf_2N^- and Br^- were chosen as exchangeable anions for the surface wetting study. As expected, the particles containing Tf_2N^- exhibited hydrophobic behavior, whereas the particles containing Br^- exhibited a hydrophilic response. The concept of ion exchange is interesting, but it is not suitable for practical use. Electroactive PILs have entered the field as novel materials for electro-stimuli-switchable devices. The redox-active PILs are based on concerted electron transfer coupled with anion transport. Randriamahazaka *et al.*[77] introduced poly(3-(2-methacryloyloxyethyl)-1-(*N*-(ferrocenylmethyl)imidazolium Tf_2N)) prepared by SI-ATRP for this concept. The contact angle in the neutral state of the PIL-modified Au substrate was determined to be around 90°, while after oxidizing the ferrocene moieties, the contact angle decreased to rather hydrophilic (60°) and *vice versa*. Originally, PEDOT-bearing ionic liquid moieties were demonstrated to be eligible not only for tunable wettability surfaces but also for electrochromic devices.[112]

A reversible hydrophobic/hydrophilic switch was observed depending on the nature of the counter-anion. Higher fluorinated anions yield higher contact angles. By flushing $LiTf_2N$ solution onto a rough surface, the contact angle was determined to be around 107 ± 3°, while when flushing $LiBF_4$ solution onto the same surface, the contact angle decreased dramatically to 24 ± 3°. This process was fully reversible. By heating/cooling oxidized PEDOT-ImB-F_4Cl between 20 °C and 70 °C, the color of the membrane changes reversibly from blue to purple. Previously, this group also proposed the assembly of a symmetric device in which the PIL layer is sandwiched between 2 PEDOT-electrochromic layers.[113] Different IL/PIL mixtures were tested for their ECD performance. It was found that the poly[1-vinyl-3-ethylIm][Br^-]/[1-vinyl-3-ethylIm][Br^-] mixture (75:25) offered the best performance, exceeding 70 000 pulsed cycles (0 and 3 V) (Figure 8.17c and d). However, no long-life requirements are needed, giving preference to using mixtures with a higher content of IL in order to get a higher optical contrast (ΔOD). This is a promising result for the development of higher performance ECDs using PILs rather than traditional polyelectrolytes.

8.6.3 Energy Applications

8.6.3.1 Supercapacitors

Electrochemical capacitors or supercapacitors are referred to as "electric double-layer capacitors" or "pseudo-capacitors". The term double-layer refers to capacitors that physically store electrical charge at the surface–electrolyte interface, which has a high surface-area, while pseudo-capacitors are related to a fast surface redox reaction.[114] Supercapacitors can be used as a complement to batteries or as an alternative solution for energy storage. Since the development of ionic liquids and poly(ionic liquid)s, they have been considered to be interesting materials for supercapacitors thanks to the ionic species contained inside their structures. In the past 5 years, poly(ionic liquid) s have started to be used in high performance supercapacitors. T. Kim and co-workers[115] investigated the energy storage possibility of a poly(1-ethyl-3-vinylimidazolium) Tf_2N-modified graphene electrode (Figure 8.18).

The poly(ionic liquid) is linked to reduced-graphene oxide (RGO) through non-covalent interactions (cation–π interactions) resulting in a hybrid structure with advantageous performance for energy storage. A synergetic effect due to the hybrid RGO and poly(ionic liquid) was revealed by using cyclic

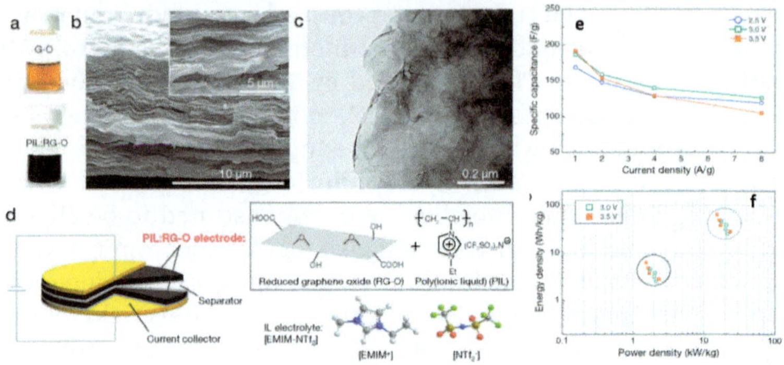

Figure 8.18 (a) Optical images of a suspension of graphene oxide (GO) in propylene carbonate (PC) and poly(ionic liquid)-modified reduced graphene oxide (PIL:RGO) in PC. (b) Scanning electron microscopy (SEM) and (c) transmission electron microscopy (TEM) images of PIL:RGO platelets. (d) Schematic diagram of the supercapacitor based on the PIL:RGO electrodes and an ionic liquid electrolyte (EMIM–NTf$_2$). (e) Plot of the specific capacitance *versus* the current density. (f) Plot of the energy density *versus* the power density at operating voltages of 3.0 and 3.5 V. The energy and power density were normalized to the total mass of the two electrodes employed including the electrolyte and current collector (solid circle), and the mass of two PIL:RGO electrodes (dashed circle). Reproduced with permission from ref. 115. Copyright (2011) American Chemical Society.

voltammetry and galvanostatic charge–discharge with a specific capacitance of around 187 F g^{-1}, a power density of 2.4 kW h^{-1} and an energy density of 6.5 W h kg^{-1} as shown in Figure 8.18e and f. Recently, a zwitterionic poly(ionic liquid) gel electrolyte, poly(propylsulfonate dimethylammonium propylmethacrylamide), deposited on a flexible graphene electrode was reported to have a high stability (10 000 charge–discharge cycles) and a high volume capacitance (300.8 F cm^{-3} at 0.8 A cm^{-3}).[57] A bi-redox ionic liquid bearing anthraquinone in the IL anion and tetramethylpiperidinyl-1-oxyl (TEMPO) in the IL cation has been proposed to enhance the performance of supercapacitors.[29] In this work, the specific energy of the supercapacitors based on activated carbon and graphene oxide electrodes could be doubled with the use of a bi-redox ionic liquid electrolyte (Figure 8.19). An immobilized redox-PIL based on imidazolium bearing ferrocene and anthraquinone was also proven to provide not only electrochromic properties but also electrochemical energy storage.[99]

Beside these highlighted works, other structures have also been investigated that exhibit an electrochemical responses toward energy storage.[59,116,117]

8.6.3.2 Li-batteries

Li-ion batteries, which were first reported by M. S. Whittingham,[118] were commercialized by Sony in 1991. Since their initial introduction, Li-ion batteries have become the most produced and consumed batteries in the world. Due to the nativity of the technique, polymer electrolytes have already been used[119]

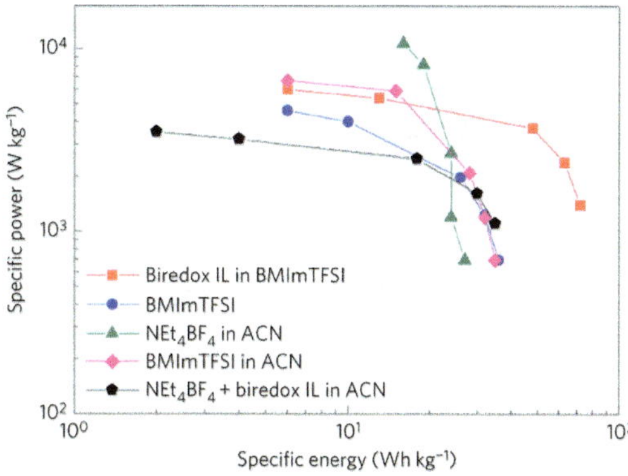

Figure 8.19 Ragone plot of supercapacitors using various electrolytes. 0.5 M bi-redox IL in BMImTFSI is compared to other electrolytes in symmetric PICA-based supercapacitors: 0.5 M NEt_4BF_4 in acetonitrile (ACN), 0.5 M BMImTFSI in acetonitrile and 0.5 M NEt_4BF_4 + 0.5 M bi-redox IL in acetonitrile. The energy and power densities are relative to the weight of PICA carbon in both electrodes. Reprinted with permission from Macmillan Publishers Ltd: *Nature Materials* (ref. 30), copyright (2016).

in order to prevent the decomposition of the organic electrolyte, and in the worst case, explosions or fires caused by using a liquid electrolyte. However, we had to wait until 2010 for the first report on the use of poly(dyallildimethylammonium) NTf_2 as a polyelectrolyte for Li-ion batteries.[120] This ternary polymer electrolyte was proven to have good mechanical and chemical stability with an ionic conductivity above 10^{-4} S cm^{-1}. This material was tested as both an electrolyte separator and a binder for the cathode material in Li/LFP batteries. The resulting battery was able to deliver 140 mA h g^{-1} with good capacity retention up to the medium current rate (C/5) at 40 °C. Later, M. Li *et al.*[121,122] reported the use of an ion gel based on guanidinium-based polymeric ionic liquids for rechargeable Li-batteries. It was found that their system showed good activity for Li/LFP batteries (discharge capacity around 140 mA h g^{-1} at C/10). However, the working temperature is relatively high for industrial applications ($T = 80$ °C). To fix this problem, K. Yin *et al.*[105] proposed the use of a dicationic polymeric ionic liquid (Figure 8.20), poly(*N*,*N*,*N*-trimethyl-*N*-(1-vinlyimidazolium-3-ethyl)-ammonium Tf_2N). The Li/LFP cell assembled with this polyelectrolyte can deliver discharge capacities of about 160 mA h g^{-1}, 140 mA h g^{-1} and 120 mA h g^{-1} at 40 °C, 30 °C and 25 °C, respectively. It should be noted that the batteries exhibited good capacity retention at each temperature.

Toward effective ion-conducting channels with high ionic conductivity, surface-initiated atom transfer radical polymerization (ATRP) of poly(*N*,*N*-diethyl-*N*-(2-methacryloylethyl)-*N*-methylammonium Tf_2N) on monodisperse silica particles (SiO_2 NPs with d = 130 nm) was performed and reported by Sato *et al.*[123] Living radical polymerization of the ionic liquid monomer lead to the formation of a SiO_2/poly(ionic liquid) core–shell structure in which the polymer layer had M_n = 25 000 and M_w/M_n = 1.13 at a monomer conversion of around 53%. By casting the hybrid particles onto an ionic liquid solution,

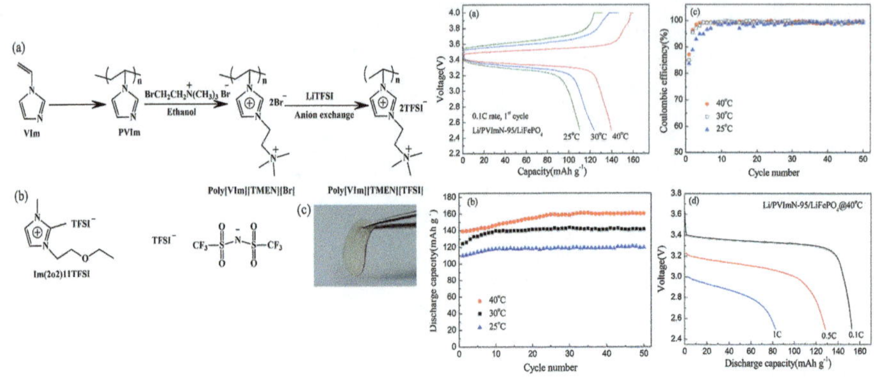

Figure 8.20 (Left) (a) Synthetic route for the imidazolium–tetraalkylammonium-based PIL. (b) Structure of IM(2o2)11TFSI. (c) Picture of a PIL–LiTFSI–IM(2o2)11TFSI electrolyte (PVImN-95 sample). (Right) Cell performance of Li/PVImN-95/$LiFePO_4$ cells. (a) Charge–discharge curves for the initial cycles at three temperatures, (b) cycling performance of the cells at 0.1 C as a function of temperature, (c) cycle-number dependence of the coulombic efficiency and (d) discharge curves of selected half cycles at various current rates at 40 °C, charging at a 0.1 C rate and discharging at various current rates. Reproduced from ref. 105 with permission from The Royal Society of Chemistry.

N,*N*-diethyl-*N*-(2-methoxyethyl)-*N*-methylammonium NTf_2, colloidal crystals were formed with an adequate composition of PSiP/DEME-NTf_2 (75 : 25 wt%), and a solid film was formed with sufficient physical strength to maintain the film shape. The fabricated solid electrolyte was tested in bipolar Li batteries where two unit-cells were connected with a bipolar electrode of $Li_4Ti_5O_{12}$ and $LiMn_2O_4$ layers. The charge and discharge operation were observed between 3.0 and 6.0 V, twice as high as the unit-cell voltage (Figure 8.20). In addition, the discharge specific capacity was found to be at the mA h scale (2.3 mA h) with 95% Coulombic efficiency, which was the best performance for a leak- and vapor-free (solid) bipolar LIB of prismatic aluminum-laminated, thin film-type. Very recently, Y. Yuan *et al.*[124,125] demonstrated that poly(ionic liquid)s act as efficient binders for strong improvement of the charge transfer between the solvent/salt, binder/salt and active material. Different choices of polymer moieties can give more stable binders in the high voltage region in order for them to be applied to cathode materials such as $LiCoPO_4$.

8.6.3.3 *Fuel Cells*

Polymer electrolytes that can conduct protons or hydroxyl groups have been investigated to develop electrolyte membranes for fuel cells. A variety of PILs have already been used for ion-exchange membranes in fuel cells.[126–131] Recent reports from Y. Feng *et al.*[132–134] have opened up a promising route to prepare anion-exchange membranes in alkaline media. C2-substituted imidazolium PILs exhibit a high ionic conductivity above 10^{-2} S cm^{-1} with excellent stability over time for up to 60 h in 2 M KOH without any deterioration (poly[DMVIm][Br]). However, serious care has to be taken when preparing fuel cells with PILs bearing imidazolium rings because the imidazolium can undergo a ring opening reaction at high temperature in concentrated alkaline solution leading to the degradation of the membrane. Very recently, a membrane based on different quaternary ammonium triblock copolymers was reported to have advantageous properties for use as an anion-exchange membrane.[135] A small amount of degradation of the membrane was observed after exposure to 1 M NaOH at 80 °C for 4 weeks. During a full life test for 110 h at 60 °C and 80 °C, even though the cell voltage decreased over time, there was no sign of degradation of the membrane. High fuel cell performance with a peak power density of 223 mW cm^{-2} was observed at 60 °C. When the temperature is raised up to 80 °C, the maximum power density was obtained (240 mW cm^{-2}). Membranes based on PILs represent an important innovation in the development and application of fuel cells.

8.6.3.4 Solid State Dye-sensitized Solar Cells (SS-DSSCs)

The use of poly(ionic liquid)s in solid state DSSCs was first investigated by studying the behaviors of poly(ionic liquid)s containing iodide as a counter-anion.[136,137] PILs can prevent problems caused by ionic liquid electrolytes, such as leakage of the liquid electrolyte and the evaporation of iodine in solution. A poly(pyridinium)-based IL containing iodide as a counter-anion/SWNTs was found to exhibit high performance (η = 3.7% under 100 mW cm^{-2} illumination) and good stability for more than 1000 h at room temperature without sealing.[137] Poly(1-vinyl-3-propylimidazolium) iodide also gave a similar light-to-electricity conversion efficiency of 3.73% under 10 mW cm^{-2} illumination.[136] Later, poly(1-[(4-ethenylphenyl)methyl]-3-butyl-imidazolium iodide) (PEBII)[138] was used as a solid electrolyte in a DSSC. The polymer was cast directly onto the prepared photoelectrode and then covered with a counter-electrode. The as-prepared solid state DSSC showed a higher performance of up to 5.93% at 100 mW cm^{-2}, which is one of the highest energy conversion efficiencies reported for N719 dye-based, I_2-free solid state DSSCs. A bis-imidazolium-based poly(IL), poly(1-butyl-3-(1-vinylimidazolium-3-hexyl) imidazolium NTf_2), which was mixed with a RTIL solution containing I_2, gave a quasi-solid-state gel electrolyte.[104] The resultant DSSCs exhibit a higher efficiency than those based on the monocationic PIL poly(1-butyl-3-vinylimidazolium NTf_2) (η = 5.92%). These results provide a novel method to synthesize (quasi) solid-state DSSCs for future applications.

8.6.3.5 Actuators

The concept of an electromechanical actuator involves the migration of charged species and the solvation shell under an electric field to the electrode surface. When these species pile up near the electrode surface, they create an over pressure at the electrode, which leads to the deformation of the material.

A flexible thin membrane, which was originally flat, tends to exhibit spontaneous bending depending on the applied electric field. From this notion, Long *et al.* developed a triblock and pentablock copolymer in which one of the components is a poly(ionic liquid).[139–141] The polymer materials were swelled in different ionic liquids in order to achieve an effective actuation response at low applied potential (typically <4 V). Mechanical property characterization using dynamic mechanical analysis displayed well-defined rubbery plateaus with high modulus storage (around 700 MPa). It should be noted that the fabrication of the ionic liquid-containing *n*-block copolymer afforded tunable polarity toward polar guest molecules, such as ionic liquids. The incorporation of an ionic liquid into the polymer matrix increased the ionic conductivity, which can decrease the response time and increase the curvature amplitude to 0.26 mm^{-1} within 20 s. The device then decayed back to the original position during 46 s and bent to the other side (cathode) with a final curvature of 0.67 mm^{-1} after 120 s.

Very recently, J. Yuan and co-workers[142–144] reported a series of materials based on poly(ionic liquid)s for multi-responsive behavior driven by solvent molecule adsorption. This novel approach can turn any inert object into an actuator by coating it with a cationic poly(ionic liquid). The $PILTf_2N$/C-pillar[5]arene membrane exhibits an unprecedented fast actuation speed in response to acetone stimulus (Figure 8.21). Under acetone vapor (24 kPa, 20 °C), the flat membrane bent rapidly into a closed loop within 0.1 s, with the top surface inwards (Figure 8.21a), and the twisting continued until the maximum curvature was reached (after 0.4 s). After removing the membrane from the acetone vapor chamber, the ribbon switched back to the initial form after 2 s. The actuators were tested under harsh conditions. After storage in liquid nitrogen, the membrane was cooked for 24 h at 150 °C under high pressure (100 MPa). The treated membrane remained active but with a slower response rate (0.5 s for bending into a semicircle). This novel concept opens a new route for making soft actuators that can work with a huge number of materials, like human hair, or complex systems by capture of perfume or tea tree oil vapor.[143]

8.6.4 Sensors

The worldwide increase in human health troubles and environmental pollution has been a driving force for the development of analytical techniques that can help people to quantify target substances. Following this direction, poly(ionic liquid)s have also been used for making sensor devices. In this subsection, we will highlight the most recent important developments in this field using poly(ionic liquid)s across different approaches. K. Cui *et al.* reported[148] the use of fluorescent nanoparticles (FLNPs) based on poly(ionic liquid)s for the selective detection of copper ions. Their FLNPs were synthesized by mixing poly(pyridinium)-based ionic liquids with fluorescent organic molecules (2-(4-amino-2-hydroxyphenyl) benzothiazole derivatives) and additional secondary ions (azo-compounds bearing different groups).

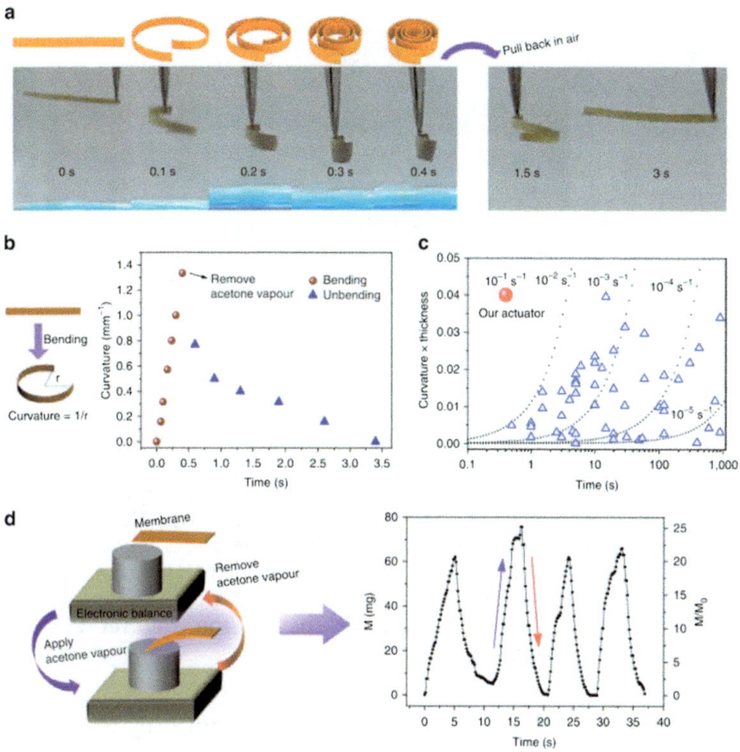

Figure 8.21 Actuation of the membrane actuator. (a) Adaptive movement of a PILT-f_2N/C-pillar[5]arene membrane (1 mm × 20 mm × 30 mm) placed in acetone vapor (24 kPa, 20 °C, left) and then back in air (right). Note: acetone solvent is stained in blue. (b) Plot of curvature against time for the membrane actuator in (a); on the left: a schematic of how the curvature is calculated. (c) Plot of 'curvature × thickness' against time for the membrane actuator in (a) and from literature results (blue triangle) compared with lines of constant actuation speed. (d) Plot of the force generated by the membrane actuator in (a) against time when the membrane was exposed to acetone vapor (24 kPa, 20 °C) and air alternately; *M* is the force that the actuator exerted on the balance; M_0 is the weight of the actuator membrane. On the left: schematic illustration of the experimental force measurement set-up. Reprinted with permission from Macmillan Publishers Ltd: *Nature Communications* (ref. 142), Copyright (2014) Nature Publishing Group.

The inter-connection and the hydrophobic balance between these 3 components resulted in their assembly into nanoparticles, in which the fluorescent organic molecules quench the poly(ionic liquid) and secondary ion in the inner part of the particles, while free poly(ionic liquid) chains cover the outside part of the particles. The FL molecule–poly(ionic liquid) complex structure is crucial for enhancing the fluorescence properties of the organic molecules. Indeed, a common problem with FL organic molecules is photobleaching in the presence of oxygen in the solution. By using a poly(ionic

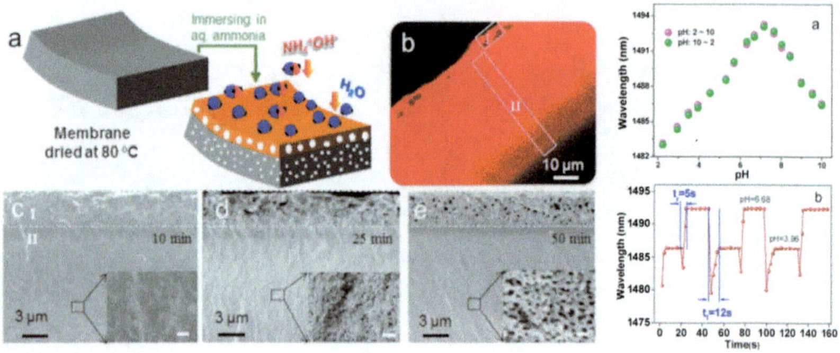

Figure 8.22 (Left) (a) Cartoon illustrating pore generation controlled by the diffusion of aqueous NH_3 into the membrane. (b) Fluorescent confocal laser scanning microscopy image of the PPM cross-section after the film was soaked for 2 h in 0.2 wt% aqueous NH_3 doped with 15 ppm rhodamine B. (c–e) Time-dependent cross-section structures of the membrane soaked in 0.2 wt% aqueous NH_3 for 10, 25, and 50 min, respectively. The scale bars in the insets are 100 nm. (Right) (a) Response of the TCFMI pH sensor *versus* pH. (b) Dynamic responses of the TCFMI pH sensor in solution with the pH alternating between 6.68 and 3.96. Reproduced with permission from ref. 145. Copyright (2013) American Chemical Society.

liquid) as the backbone, the final complex leads not only to enhanced FL intensity, with an increase in thermal stability and overall stability, but also retardation of the photo-bleaching problem. In addition, these nanoparticles exhibit highly selective recognition for copper ions due to the formation of a AHBTA–Cu^{2+} complex, which dramatically decreases the FL intensity (down to 10%). For pH sensing applications, Q. Zhao *et al.* demonstrated that poly(ionic liquid)-coated optical fibers can provide a high pH sensing performance with a fast response time as shown in Figure 8.22 (right).[145] The charged membrane using poly[1-cyanomethyl-3-vinylimidazolium Tf_2N] and poly(acrylic acid) was activated by immersion in aqueous ammonia resulting in the formation of a thin layer of macropores, followed by a thick layer of nanopores (Figure 8.22 (left)). The activated membrane is stable in both aqueous and non-aqueous solution with high sensibility (2.04 and −2.48 nm per pH unit corresponding to acidic and basic medium, respectively).

Poly(ionic liquid)s were also used for detecting bio-molecules such as dopamine, a neurotransmitter that is released by neurons to send a signal to other cells in order to control our feelings. 1-Vinyl-3-ethylimidazole bromide that was polymerized onto poly(pyrrole)-modified graphene oxide (PIL/PPy/GO) was used for dopamine detection in the presence of ascorbic acid.[56] For the PPy/GO and GO electrodes, the oxidation signals of 3 mM of L-ascorbic acid (AA) and 2 μM of dopamine (DA) cannot be distinguished, resulting in 1 oxidation peak at around 0.4 V *vs.* Ag/AgCl. However, in the case of poly(IL)/PPy/GO, two clearly separated peaks were observed at 0.18 V and 0.4 V for the oxidation of AA and DA, respectively. This result might be explained by a double catalysis system where the hybrid material catalyzed the oxidation of AA by shifting the oxidation potential to a lower value (from 0.4 V to 0.18 V *vs.* Ag/AgCl) and the oxidation of DA by enhancing the current intensity. The double catalysis system is applicable in real-life because in biological media, AA and DA always coexist with a concentration ratio of around 100 to 1000, and the poly(IL)/PPy/GO-modified GC electrode can identify the presence of DA at the same concentration ratio as reported. On the other hand, the detection limit of this system is much better than that in previously reported work (down to 73.3 nM) and the linear range is determined to be between 4 and 18 μM. Beside these cited applications of poly(ionic liquid)s, these materials can also be used for the determination of humidity[146] and for the detection of hydrogen peroxide.[147]

8.7 Concluding Remarks

During the past two decades, ionic liquids and their derivatives have attracted increasing research interest as solvent–electrolytes for electrochemical processes. However, recent studies have indicated the feasibility of using ionic liquids and ionic liquid-based polymers, called task-specific ionic liquids, as the main component for a large spectrum of applications. Despite the development of these materials, the number of reports is relatively limited and a fundamental understanding of IL and PIL properties is still open for further investigation. In this chapter, we have attempted to provide recent advances from fundamental research into practical uses in the field of ionic liquids. Even when being used as a green solvent, ionic liquids and their derivatives are quite harmful to the environment and human health. To resolve this problem, different approaches have been proposed in order to immobilize the ionic liquid onto various substrates, which can reduce the quantity of IL required and generate new interesting interfaces from thin layer ionic liquids to polymeric ionic liquids. In this context, self-assembled monolayers and electrochemical tools are among the most powerful techniques to graft ILs and PILs onto material surfaces from the nanoscale to a large area. These IL-modified substrates exhibit unusual behaviors compared to other organic materials, providing the possibility to use them for sensor, catalysis and energy applications, and so on. In parallel with the traditional ionic liquids, redox-active ionic liquids (RILs) and redox-active polymer ionic liquids have appeared as new candidates for integration into electrochemical processes. By introducing ionic liquid moieties onto redox active compounds, the intrinsic properties of both the ionic liquid and the redox-active group change, resulting in multifunctional materials. These materials have been successfully used as electro-responsive materials for reversible electrochemically switchable wettability and electrochemical energy storage. We expect that redox-active ILs and polymers will boost the whole field of ionic liquids toward multifunctional materials and will open up possibilities for their integration into various applications.

References

1. G. S. Viereck, *What Life Means to Einstein : An Interview by George Sylvester Viereck*, The Saturday Evening Post, 26 October 1929, p. 117.
2. J. S. Wilkes, J. A. Levisky, R. A. Wilson and C. L. Hussey, *Inorg. Chem.*, 1982, **237**, 1263–1264.
3. P. Bonhôte, A.-P. Dias, M. Armand, N. Papageorgiou, K. Kalyanasundaram and M. Grätzel, *Inorg. Chem.*, 1996, **35**, 1168–1178.
4. U. Schroder, J. D. Wadhawan, R. G. Compton, F. Marken, P. A. Z. Suarez, C. S. Consorti and F. De Souza, *New J. Chem.*, 2000, **24**, 1009–1015.
5. V. R. Koch, L. A. Dominey, C. Nanjundiah and M. Ondrechen, *J. Electrochem. Soc.*, 1996, **143**, 798–803.
6. J. N. Barisci, G. G. Wallace, D. R. MacFarlane and R. H. Baughman, *Electrochem. Commun.*, 2004, **6**, 22–27.
7. D. MacFarlane and P. Meakin, *J. Phys. Chem. B*, 1999, **103**, 4164–4170.
8. J. Sun, M. Forsyth and D. R. Macfarlane, *J. Phys. Chem. B*, 1998, **102**, 8858–8864.
9. A. Noda and M. Watanabe, *Electrochim. Acta*, 2000, **45**, 1265–1270.
10. H. Matsumoto, T. Matsuda and Y. Miyazaki, *Chem. Lett.*, 2000, **29**, 1430–1431.
11. S. Z. El Abedin, E. M. Moustafa, R. Hempelmann, H. Natter and F. Endres, *Electrochem. Commun.*, 2005, **7**, 1111–1116.
12. T. Jiang, M. J. Chollier Brym, G. Dubé, A. Lasia and G. M. Brisard, *Surf. Coat. Technol.*, 2006, **201**, 1–9.
13. F. Endres, M. Bukowski, R. Hempelmann and H. Natter, *Angew. Chem., Int. Ed.*, 2003, **42**, 3428–3430.
14. S. Z. El Abedin, M. Pölleth, S. A. Meiss, J. Janek and F. Endres, *Green Chem.*, 2007, **9**, 549–553.
15. F. Endres, *ChemPhysChem*, 2002, **3**, 144–154.
16. S. Zein El Abedin, A. Y. Saad, H. K. Farag, N. Borisenko, Q. X. Liu and F. Endres, *Electrochim. Acta*, 2007, **52**, 2746–2754.
17. O. Mann, G. B. Pan and W. Freyland, *Electrochim. Acta*, 2009, **54**, 2487–2490.
18. Y. Pang, X. Li, H. Ding, G. Shi and L. Jin, *Electrochim. Acta*, 2007, **52**, 6172–6177.
19. D. Wei, C. Kvarnström, T. Lindfors and A. Ivaska, *Electrochem. Commun.*, 2006, **8**, 1563–1566.
20. J. M. Pringle, M. Forsyth, G. G. Wallace and D. R. Macfarlane, *Macromolecules*, 2006, **39**, 7193–7195.
21. S. Zein El Abedin, N. Borissenko and F. Endres, *Electrochem. Commun.*, 2004, **6**, 422–426.
22. K. Liu, Z. Hu, R. Xue, J. Zhang and J. Zhu, *J. Power Sources*, 2008, **179**, 858–862.
23. P. Wang, S. M. Zakeeruddin, J. E. Moser, R. Humphry-Baker and M. Grätzel, *J. Am. Chem. Soc.*, 2004, **126**, 7164–7165.
24. P. Wang, S. M. Zakeeruddin, R. Humphry-Baker and M. Grätzel, *Chem. Mater.*, 2004, **16**, 2694–2696.

25. R. Balasubramanian, W. Wang and R. W. Murray, *J. Am. Chem. Soc.*, 2006, **128**, 9994–9995.
26. S. A. Mirkhani, F. Gharagheizi, P. Ilani-Kashkouli and N. Farahani, *Thermochim. Acta*, 2012, **543**, 88–95.
27. O. Fontaine, C. Lagrost, J. Ghilane, P. Martin, G. Trippé, C. Fave, J. C. Lacroix, P. Hapiot and H. N. Randriamahazaka, *J. Electroanal. Chem.*, 2009, **632**, 88–96.
28. X. J. Chen, D. Xu, L. H. Qiu, S. C. Li, W. Zhang and F. Yan, *J. Mater. Chem. A*, 2013, **1**, 8759–8765.
29. E. Mourad, L. Coustan, S. A. Freunberger, A. Mehdi, A. Vioux, F. Favier and O. Fontaine, *Electrochim. Acta*, 2015, **206**, 513–523.
30. E. Mourad, L. Coustan, P. Lannelongue, D. Zigah, A. Mehdi, A. Vioux, S. A. Freunberger, F. Favier and O. Fontaine, *Nat. Mater.*, 2016, **16**, 449–453.
31. J. C. Love, L. A. Estroff, J. K. Kriebel, R. G. Nuzzo and G. M. Whitesides, *Chem. Rev.*, 2005, **105**, 1103–1169.
32. D. Bélanger and J. Pinson, *Chem. Soc. Rev.*, 2011, **40**, 3995–4048.
33. P. Uhlmann, L. Ionov, N. Houbenov, M. Nitschke, K. Grundke, M. Motornov, S. Minko and M. Stamm, *Prog. Org. Coat.*, 2006, **55**, 168–174.
34. J. E. Anthony, *Chem. Rev.*, 2006, **106**, 5028–5048.
35. C. Querner, P. Reiss, J. Bleuse and A. Pron, *J. Am. Chem. Soc.*, 2004, **126**, 11574–11582.
36. A. Dong, X. Ye, J. Chen, Y. Kang, T. Gordon, J. M. Kikkawa and C. B. Murray, *J. Am. Chem. Soc.*, 2011, **133**, 998–1006.
37. B. A. Kakade and V. K. Pillai, *J. Phys. Chem. C*, 2008, **112**, 3183–3186.
38. S. R. Forrest, *Nature*, 2004, **428**, 911–918.
39. Q. H. Wang and M. C. Hersam, *Nat. Chem.*, 2009, **1**, 206–211.
40. C. Lao, Y. Li, C. P. Wong and Z. L. Wang, *Nano Lett.*, 2007, **7**, 1323–1328.
41. B. S. Lee, Y. S. Chi, J. K. Lee, I. S. Choi, C. E. Song, S. K. Namgoong and S. G. Lee, *J. Am. Chem. Soc.*, 2004, **126**, 480–481.
42. Y. S. Chi, S. Hwang, B. S. Lee, J. Kwak, I. S. Choi and S. G. Lee, *Langmuir*, 2005, **21**, 4268–4271.
43. S. Bouden, M. Gómez-Mingot, H. Randriamahazaka and J. Ghilane, *Anal. Chem.*, 2016, **88**, 1017–1021.
44. S. Bouden, G. Trippé-Allard, J. Ghilane and H. Randriamahazaka, *Electrochem. Commun.*, 2015, **58**, 65–68.
45. T. N. Pham-Truong, F. Lafolet, J. Ghilane and H. Randriamahazaka, *Electrochem. Commun.*, 2016, **70**, 13–17.
46. H. Itoh, K. Naka and Y. Chujo, *J. Am. Chem. Soc.*, 2004, **126**, 3026–3027.
47. M. Ratel, A. Provencher-Girard, S. Shuo Zhao, J. Breault-Turcot, J. Labrecque-Carbonneau, M. Branca, J. N. Pelletier, A. R. Schmitzer and J.-F. Masson, *Anal. Chem.*, 2013, **85**, 5770–5777.
48. P. Hesemann and J. J. E. Moreau, *Chem. Commun.*, 2004, 1768–1769.
49. M. Zhang, J. Chen, H. Qiu, A. K. Mallik, M. Takafuji and H. Ihara, *RSC Adv.*, 2014, **4**, 34654–34658.
50. J. A. Marins, B. G. Soares, A. A. Silva and S. Livi, *RSC Adv.*, 2014, **4**, 50925–50931.

51. M. Gomez-Mingot, V. Bui-Thi-Tuyet, G. Trippe-Allard, J. Ghilane and H. Randriamahazaka, *ChemElectroChem*, 2014, **1**, 1467–1470.
52. V. Bui-Thi-Tuyet, G. Trippe-Allard, J. Ghilane and H. Randriamahazaka, *J. Electroanal. Chem.*, 2014, **713**, 28–31.
53. H. Hazimeh, S. Pioge, C. Combellas and F. I. Podvorica, *Chem. Mater.*, 2013, **25**, 605–612.
54. J. Ghilane and J. C. Lacroix, *J. Am. Chem. Soc.*, 2013, **135**, 4722–4728.
55. J. Yuan, D. Mecerreyes and M. Antonietti, *Prog. Polym. Sci.*, 2013, **38**, 1009–1036.
56. H. Mao, J. Liang, H. Zhang, Q. Pei, D. Liu, S. Wu, Y. Zhang and X. M. Song, *Biosens. Bioelectron.*, 2015, **70**, 289–298.
57. X. Peng, H. Liu, Q. Yin, J. Wu, P. Chen, G. Zhang, G. Liu, C. Wu and Y. Xie, *Nat. Commun.*, 2016, **7**, 11782.
58. B. Qiu, B. Lin and F. Yan, *Polym. Int.*, 2013, **62**, 335–337.
59. T. Zhou, X. Gao, B. Dong, N. Sun and L. Zheng, *J. Mater. Chem. A*, 2015, **4**, 1112–1118.
60. R. Nakamura, M. Tokuda, T. Suzuki and H. Minami, *Langmuir*, 2016, **32**, 2331–2337.
61. X. He, W. Yang and X. Pei, *Macromolecules*, 2008, **41**, 4615–4621.
62. H. Ohno, M. Yoshizawa and W. Ogihara, *Electrochim. Acta*, 2004, **50**, 255–261.
63. W. Ogihara, S. Washiro, H. Nakajima and H. Ohno, *Electrochim. Acta*, 2006, **51**, 2614–2619.
64. S. Washiro, M. Yoshizawa, H. Nakajima and H. Ohno, *Polymer (Guildf.)*, 2004, **45**, 1577–1582.
65. H. Nakajima and H. Ohno, *Polymer (Guildf.)*, 2005, **46**, 11499–11504.
66. J. Yuan and M. Antonietti, *Polymer (Guildf.)*, 2011, **52**, 1469–1482.
67. D. Mecerreyes, *Prog. Polym. Sci.*, 2011, **36**, 1629–1648.
68. A. S. Shaplov, R. Marcilla and D. Mecerreyes, *Electrochim. Acta*, 2015, **175**, 18–34.
69. J. Lu, F. Yan and J. Texter, *Prog. Polym. Sci.*, 2009, **34**, 431–448.
70. M. D. Green and T. E. Long, *Polym. Rev.*, 2009, **49**, 291–314.
71. O. Green, S. Grubjesic, S. Lee and M. A. Firestone, *Polym. Rev.*, 2009, **49**, 339–360.
72. J. C. Salamone, S. C. Israel, P. Taylor and B. Snider, *Polymer (Guildf.)*, 1973, **14**, 639–644.
73. H. He, M. Zhong, D. Luebke, H. Nulwala and K. Matyjaszewski, *J. Polym. Sci., Part A: Polym. Chem.*, 2014, **52**, 2175–2184.
74. H. He, D. Luebke, H. Nulwala and K. Matyjaszewski, *Macromolecules*, 2014, **47**, 6601–6609.
75. H. Mori, M. Yanagi and T. Endo, *Macromolecules*, 2009, **42**, 8082–8092.
76. K. Vijayakrishna, S. K. Jewrajka, A. Ruiz, R. Marcilla, J. A. Pomposo, D. Mecerreyes, D. Taton and Y. Gnanou, *Macromolecules*, 2008, **41**, 6299–6308.
77. V. Bui-Thi-Tuyet, G. Trippe-Allard, J. Ghilane and H. Randriamahazaka, *ACS Appl. Mater. Interfaces*, 2016, **8**, 28316–28324.

78. A. Wilke, J. Yuan, M. Antonietti and J. Weber, *ACS Macro Lett.*, 2012, **1**, 1028–1031.
79. F. Schüler, B. Kerscher, F. Beckert, R. Thomann and R. Mülhaupt, *Angew. Chem., Int. Ed.*, 2013, **52**, 455–458.
80. V. V. Shevchenko, A. V. Stryutsky, N. S. Klymenko, M. A. Gumenna, A. A. Fomenko, V. N. Bliznyuk, V. V. Trachevsky, V. V. Davydenko and V. V. Tsukruk, *Polymer (Guildf.)*, 2014, **55**, 3349–3359.
81. J. Yuan, S. Soll, M. Drechsler, A. H. E. Mueller and M. Antonietti, *J. Am. Chem. Soc.*, 2011, **133**, 17556–17559.
82. N. Canilho, M. Scholl, H. Klok, R. Mezzenga, C. Muse, C. Fribourg, L. Polyme, I. Mate and Ä. P. Fe, *Macromolecules*, 2007, **40**, 8374–8383.
83. K. Schadt, B. Kerscher, R. Thomann and R. Mu, *Macromolecules*, 2013, **46**, 4799–4804.
84. B. Kerscher, A. Appel, R. Thomann and R. Mu, *Macromolecules*, 2013, **46**, 4395–4402.
85. D. Porphyrins, N. Tomioka and D. Takasu, *Angew. Chem., Int. Ed.*, 1998, **37**, 1531–1534.
86. J. Huang, H. Luo, C. Liang, I. W. Sun, G. A. Baker and S. Dai, *J. Am. Chem. Soc.*, 2005, **127**, 12784–12785.
87. M. Marcos, R. Alcala and L. Serrano, *Chem. Mater.*, 2008, **20**, 5209–5217.
88. J. Vergara, M. Cano, J. Barber, P. Romero, L. Serrano and M. B. Ros, *Chem. Mater.*, 2011, **23**, 4931–4940.
89. H. Mori, Y. Ebina, R. Kambara and K. Nakabayashi, *Polym. J.*, 2012, **44**, 550–560.
90. E. Stathatos, V. Jovanovski, B. Orel, I. Jerman and P. Lianos, *J. Phys. Chem. C*, 2007, **111**, 6528–6532.
91. V. Jovanovski, B. Orel, R. Jes and S. Angela, *J. Phys. Chem. B*, 2005, **109**, 14387–14395.
92. J.-H. Jeon, K. Tanaka and Y. Chujo, *RSC Adv.*, 2013, **3**, 2422–2427.
93. J. Tan, X. Sun, S. Feng and C. Zhang, *Dalton Trans.*, 2013, **42**, 4337–4339.
94. P. Majumdar, E. Lee, N. Gubbins, S. J. Stafslien, J. Daniels, C. J. Thorson and B. J. Chisholm, *Polymer (Guildf.)*, 2009, **50**, 1124–1133.
95. K. Pu, K. Li, J. Shi and B. Liu, *Chem. Mater.*, 2009, **21**, 3816–3822.
96. P. Cardiano, G. Lazzara, S. Manickam, P. Mineo, S. Milioto and S. Lo, *Eur. J. Inorg. Chem.*, 2012, 5668–5676.
97. S. Manickam, P. Cardiano, P. G. Mineo and S. Lo, *Eur. J. Inorg. Chem.*, 2014, 2704–2710.
98. X. Sui, M. A. Hempenius and G. J. Vancso, *J. Am. Chem. Soc.*, 2012, **9**, 4023–4025.
99. H. Randriamahazaka, J. Ghilane, G. Trippe-Allard, V. Bui-Thi-Tuyet and T. N. Pham-Truong, Fr. Pat. 370121D36047, PCT/FR2016/051069, 2016.
100. A. S. Shaplov, E. I. Lozinskaya, D. O. Ponkratov, I. A. Malyshkina, F. Vidal, P. Aubert, V. Okatova, G. M. Pavlov, L. I. Komarova, C. Wandrey and Y. S. Vygodskii, *Electrochim. Acta*, 2011, **57**, 74–90.
101. M. Yoshizawa and H. Ohno, *Chem. Lett.*, 1999, 889–890.

102. Y. S. Vygodskii, O. A. Mel, E. I. Lozinskaya, A. S. Shaplov, I. A. Malyshkina, N. D. Gavrilova, K. A. Lyssenko, M. Y. Antipin, D. G. Golovanov, A. A. Korlyukov, N. Ignat and U. Welz-biermann, *Polym. Adv. Technol.*, 2007, **18**, 50–63.
103. Y. Ye and Y. A. Elabd, *Polymer (Guildf.)*, 2011, **52**, 1309–1317.
104. X. Chen, J. Zhao, J. Zhang, L. Qiu, D. Xu, H. Zhang, X. Han, B. Sun, G. Fu, Y. Zhang and F. Yan, *J. Mater. Chem.*, 2012, **22**, 18018.
105. K. Yin, Z. Zhang, X. Li, L. Yang, K. Tachibana and S. Hirano, *J. Mater. Chem. A*, 2015, **3**, 170–178.
106. N. Matsumi, K. Sugai, M. Miyake and H. Ohno, *Macromolecules*, 2006, **39**, 6924–6927.
107. W. Zhang, Z. Kochovski, Y. Lu, B. V. K. J. Schmidt, M. Antonietti and J. Yuan, *ACS Nano*, 2016, **10**, 7731–7737.
108. S. Kuwabata, H. Minamimoto, K. Inoue, A. Imanishi, K. Hosoya, H. Uyama, T. Torimoto, T. Tsuda and S. Seki, *Sci. Rep.*, 2014, **4**, 3722.
109. H. Minamimoto, H. Irie, T. Uematsu, T. Tsuda, A. Imanishi, S. Seki and S. Kuwabata, *Langmuir*, 2015, **31**, 4281–4289.
110. J. Yuan, C. Giordano and M. Antonietti, *Chem. Mater.*, 2010, **22**, 5003–5012.
111. P. Zhang, J. Yuan, T. P. Fellinger, M. Antonietti, H. Li and Y. Wang, *Angew. Chem., Int. Ed.*, 2013, **52**, 6028–6032.
112. M. Döbbelin, R. Tena-Zaera, R. Morcilla, J. Iturri, S. Moya, J. A. Pomposo and D. Mecerreyes, *Adv. Funct. Mater.*, 2009, **19**, 3326–3333.
113. R. Marcilla, F. Alcaide, H. Sardon, J. A. Pomposo, C. Pozo-Gonzalo and D. Mecerreyes, *Electrochem. Commun.*, 2006, **8**, 482–488.
114. P. Simon and Y. Gogotsi, *Nat. Mater.*, 2008, **7**, 845–854.
115. T. Y. Kim, H. W. Lee, M. Stoller, D. R. Dreyer, C. W. Bielawski, R. S. Ruoff and K. S. Suh, *ACS Nano*, 2011, **5**, 436–442.
116. J. P. C. Trigueiro, R. L. Lavall and G. G. Silva, *J. Power Sources*, 2014, **256**, 264–273.
117. Y. J. Kang, Y. Yoo and W. Kim, *ACS Appl. Mater. Interfaces*, 2016, **8**, 13909–13917.
118. M. S. Whittingham, *Science*, 1976, **192**, 1126–1127.
119. W. H. Meyer, *Adv. Mater.*, 1998, **10**, 439–448.
120. G. B. Appetecchi, G. T. Kim, M. Montanino, M. Carewska, R. Marcilla, D. Mecerreyes and I. De Meatza, *J. Power Sources*, 2010, **195**, 3668–3675.
121. M. Li, L. Yang, S. Fang, S. Dong, S. I. Hirano and K. Tachibana, *J. Power Sources*, 2011, **196**, 8662–8668.
122. M. Li, L. Yang, S. Fang, S. Dong, S. I. Hirano and K. Tachibana, *Polym. Int.*, 2012, **61**, 259–264.
123. T. Sato, T. Morinaga, S. Marukane, T. Narutomi, T. Igarashi, Y. Kawano, K. Ohno, T. Fukuda and Y. Tsujii, *Adv. Mater.*, 2011, **23**, 4868–4872.
124. K. Grygiel, J. S. Lee, K. Sakaushi, M. Antonietti and J. Yuan, *ACS Macro Lett.*, 2015, **4**, 1312–1316.
125. J.-S. Lee, K. Sakaushi, M. Antonietti and J. Yuan, *RSC Adv.*, 2015, **5**, 85517–85522.

126. M. Díaz, A. Ortiz, M. Vilas, E. Tojo and I. Ortiz, *Int. J. Hydrogen Energy*, 2014, **39**, 3970–3977.
127. Z. Wojnarowska, J. Knapik, M. Díaz, A. Ortiz, I. Ortiz and M. Paluch, *Macromolecules*, 2014, **47**, 4056–4065.
128. O. I. Deavin, S. Murphy, A. L. Ong, S. D. Poynton, R. Zeng, H. Herman and J. R. Varcoe, *Energy Environ. Sci.*, 2012, **5**, 8584–8597.
129. J. Wang, S. Li and S. Zhang, *Macromolecules*, 2010, **43**, 3890–3896.
130. Y. Ye and Y. A. Elabd, *Macromolecules*, 2011, **44**, 8494–8503.
131. B. Qiu, B. Lin, Z. Si, L. Qiu, F. Chu, J. Zhao and F. Yan, *J. Power Sources*, 2012, **217**, 329–335.
132. B. Lin, L. Qiu, B. Qiu, Y. Peng and F. Yan, *Macromolecules*, 2011, **44**, 9642–9649.
133. B. Lin, L. Qiu, J. Lu and F. Yan, *Chem. Mater.*, 2010, **22**, 6718–6725.
134. B. Lin, H. Dong, Y. Li, Z. Si, F. Gu and F. Yan, *Chem. Mater.*, 2013, **25**, 1858–1867.
135. A. D. Mohanty, C. Y. Ryu, Y. S. Kim and C. Bae, *Macromolecules*, 2015, **48**, 7085–7095.
136. E. Azaceta, R. Marcilla, A. Sanchez-Diaz, E. Palomares and D. Mecerreyes, *Electrochim. Acta*, 2010, **56**, 42–46.
137. R. Kawano, T. Katakabe, H. Shimosawa, M. K. Nazeeruddin, M. Graetzel, H. Matsui, T. Kitamura, N. Tanabe and M. Watanabe, *Phys. Chem. Chem. Phys.*, 2010, **12**, 1916–1921.
138. W. S. Chi, J. K. Koh, S. H. Ahn, J.-S. Shin, H. Ahn, D. Y. Ryu and J. H. Kim, *Electrochem. Commun.*, 2011, **13**, 1349–1352.
139. M. D. Green, D. Wang, S. T. Hemp, J. H. Choi, K. I. Winey, J. R. Heflin and T. E. Long, *Polymer*, 2012, **53**, 3677–3686.
140. R. Gao, D. Wang, J. R. Heflin and T. E. Long, *J. Mater. Chem.*, 2012, **22**, 13473.
141. T. Wu, D. Wang, M. Zhang, J. R. Heflin, R. B. Moore and T. E. Long, *ACS Appl. Mater. Interfaces*, 2012, **4**, 6552–6559.
142. Q. Zhao, J. W. C. Dunlop, X. Qiu, F. Huang, Z. Zhang, J. Heyda, J. Dzubiella, M. Antonietti and J. Yuan, *Nat. Commun.*, 2014, **5**, 4293.
143. Q. Zhao, J. Heyda, J. Dzubiella, K. Täuber, J. W. C. Dunlop and J. Yuan, *Adv. Mater.*, 2015, **27**, 2913–2917.
144. H. Song, H. Lin, M. Antonietti and J. Yuan, *Adv. Mater. Interfaces*, 2016, **3**, 1–7.
145. Q. Zhao, M. Yin, A. P. Zhang, S. Prescher, M. Antonietti and J. Yuan, *J. Am. Chem. Soc.*, 2013, **135**, 5549–5552.
146. L. L. Wang, X. Duan, W. Xie, Q. Li and T. Wang, *Chem. Commun.*, 2016, **52**, 8417–8419.
147. Q. Wang and Y. Yun, *Microchim. Acta*, 2013, **180**, 261–268.
148. K. Cui, L. Xuemin, C. Wei, W. Jun, C. Xumeng and L. Qinghua, *Chem. Commun.*, 2011, **47**, 920–922.

CHAPTER 9

Doping Polymers with Ionic Liquids to Manipulate Their Morphology and Membrane Gas Separation Properties

XIANDA HOU[†], JUNYI LIU[†], HIEN NGUYEN AND HAIQING LIN*

Department of Chemical and Biological Engineering, University at Buffalo, The State University of New York, Buffalo, NY 14260, USA
*E-mail: haiqingl@buffalo.edu

9.1 Introduction

Polymeric membranes have emerged as an important technology for industrial gas separations, due to their inherent advantages such as high energy-efficiency, simplicity, small footprint, and easy scalability.[1–3] The core of membrane technology is polymers with superior gas separation properties, such as high gas permeability and gas selectivity. However, there exists an intrinsic trade-off, *i.e.*, polymers with higher gas permeability

[†]These authors contributed equally.

Smart Materials No. 29
Polymerized Ionic Liquids
Edited by Ali Eftekhari

Published by the Royal Society of Chemistry, www.rsc.org

tend to exhibit lower gas selectivity.[4–7] Therefore, there is great interest in developing polymer-based hybrid materials to overcome this limitation, while retaining good processability for membrane formation derived from polymers.

Polymers doped with ionic liquids (ILs) have been widely explored for membrane gas separation. ILs exhibit essentially no vapor pressure, giving them good stability and low viscosity, providing high gas diffusivity and thus high permeability.[8–10] More importantly, ILs containing imidazolium exhibit high CO_2 solubility and high CO_2/N_2 solubility selectivity, due to specific interactions between amine groups in imidazolium and CO_2.[11–15] While these ILs can be fabricated into liquid membranes supported on porous supports to achieve good separation properties for CO_2/N_2 and CO_2/CH_4, such membranes are not stable under high pressures across the membranes.[16–20] Therefore, ILs have often been blended with polymers to achieve both good separation properties and good stability.[12,21,22]

ILs doped in polymers behave like plasticizers and they can significantly change the polymer morphology, as shown in Figure 9.1.[13] IL doping increases the polymer chain flexibility or decreases the glass transition temperature (T_g) of the blends; for semi-crystalline polymers, IL doping only occurs in the amorphous phase, and it can decrease the polymer melting temperature (T_m) and even the crystallinity,[12,13,23–26] which can inevitably influence the gas separation properties.

This chapter critically reviews the effect of IL doping on the physical properties of polymers, such as T_g and crystallinity, and their gas transport properties. Without exhaustively reviewing the literature on polymer/IL blends for membrane gas separation, we aim to elucidate a quantitative relationship between changes in T_g, crystallinity, and gas permeation properties derived from IL doping, providing guidelines to select suitable ILs for polymer doping to enhance gas separation properties.

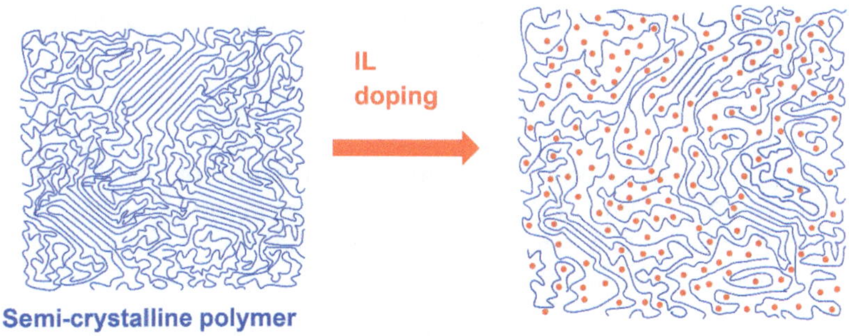

Figure 9.1 Schematic showing the effect of IL doping on chain flexibility and crystallinity in a semi-crystalline polymer. Data from ref. 13.

9.2 Background

The permeability of gas component A in a polymeric thin film, P_A, can be written as:[27]

$$P_A = \frac{N_A \bullet l}{A_m(p_{2A} - p_{1A})} \tag{9.1}$$

where N_A is the steady state gas flux (cm^3 (STP) s^{-1}), l is the film thickness (cm), A_m is the film area (cm^2), and p_{2A} and p_{1A} are the feed and permeate pressure (cmHg) of component A, respectively. The permeability coefficient has a unit of Barrer (1 Barrer = 10^{-10} cm^3 (STP) cm cm^{-2} s^{-1} cmHg^{-1}). Within a framework of the solution–diffusion model for gas permeation in polymers, P_A can also be expressed as:[27]

$$P_A = S_A \times D_A \tag{9.2}$$

where S_A is the solubility coefficient of the gas in the polymer (cm^3 (STP) cm^{-3} atm^{-1}), and D_A is the gas diffusion coefficient (cm^2 s^{-1}). The gas selectivity ($\alpha_{A/B}$) is the ratio of the gas permeabilities, and it is also a combination of the solubility selectivity (S_A/S_B) and the diffusivity selectivity (D_A/D_B):

$$\alpha_{A/B} = \frac{P_A}{P_B} = \left(\frac{S_A}{S_B}\right) \times \left(\frac{D_A}{D_B}\right) \tag{9.3}$$

Gas solubility in a polymer is determined by the gas condensability and its interaction with the polymer.[6] Therefore, by doping a polymer with an IL exhibiting a stronger interaction with gas A than the polymer, S_A can be increased, as well as S_A/S_B. Gas diffusivity is determined by gas molecular size, polymer chain rigidity and free volume.[6,28] The IL doping of a polymer typically decreases the polymer chain rigidity, and thus increases D_A and decreases the polymer size sieving ability.

For semi-crystalline polymers, crystallinity can significantly influence gas permeability, since crystals are not accessible for gas sorption and diffusion. The following simple model is often used to describe the apparent gas solubility ($S_{A,SC}$) in a semi-crystalline polymer:[13,29–31]

$$S_{A,SC} = S_{A,am} \times \phi_{am} \tag{9.4}$$

where $S_{A,am}$ is the estimated solubility in the wholly amorphous polymer, and ϕ_{am} is the volume fraction of the amorphous phase in the polymer.

The gas diffusivity in a semi-crystalline polymer ($D_{A,SC}$) can be described using the following simplified model:[13,29–31]

$$D_{A,SC} = D_{A,am} \times \phi_{am} \tag{9.5}$$

where $D_{A,am}$ is the estimated gas diffusivity in the wholly amorphous polymer. Combining eqn (4) and eqn (5) gives the following equation:[13,29,30]

$$P_{A,SC} = P_{A,am} \times \phi_{am}^{\ 2} \tag{9.6}$$

where $P_{A,SC}$ is the measured permeability in the semi-crystalline polymer, and $P_{A,am}$ is the estimated permeability in the wholly amorphous polymer. Eqn (6) indicates that the gas permeability in semi-crystalline polymers depends sensitively on the crystallinity.

9.3 Effect of IL Doping on the T_g of Blends

9.3.1 T_g Depression and Modeling Using the Gordon–Taylor Equation

Figure 9.2 shows the effect of IL doping on the T_g of two representative blends, CTA/[emim][BF_4][13] and PMMA/[emim][Tf_2N].[32] The detailed names are shown in the caption, and the chemical structure of the ILs is also shown in Figure 9.3. While PMMA is amorphous, CTA is semi-crystalline. Therefore,

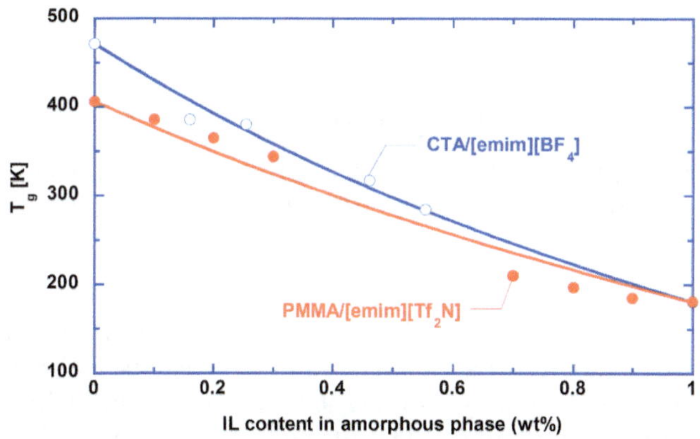

Figure 9.2 Effect of IL content in the amorphous phase (wt%) on the T_g of two blends, cellulose triacetate/1-ethyl-3-methylimidazolium tetrafluoroborate (CTA/[emim][BF_4]),[13] and poly(methyl methacrylate)/1-ethyl-3-methylimidazolium bis(trifluoromethylsulfonyl)imide (PMMA/[emim][Tf_2N]).[32] The curves are best fits based on eqn (7).

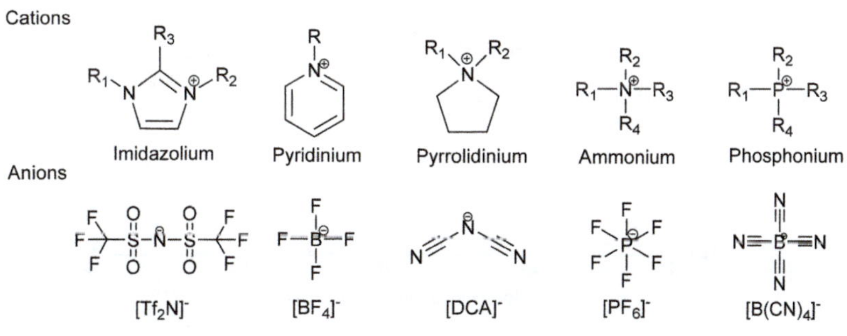

Figure 9.3 Chemical structures of representative cations and anions for ILs.

the IL content in the amorphous phase is used for correlation, assuming that the ILs do not co-crystallize with the polymers and only exist in the amorphous phase. In general, increasing the IL content decreases the T_g, since the ILs have a much lower T_g than the polymers.

The effect of IL doping on the T_g of the blends can be described using the Gordon–Taylor equation:[13,33]

$$T_g = \frac{\omega_{P,am} T_{g,P} + K\omega_{IL,am} T_{g,IL}}{\omega_{P,am} + K\omega_{IL,am}} \tag{9.7}$$

where ω_{am} is the weight content in the amorphous phase, K is an adjustable parameter, and the subscripts P and IL represent the pure polymer and IL, respectively. As shown in Figure 9.2, the Gordon–Taylor equation can satisfactorily model the data, with K values of 1.4 and 2.0 for CTA/[emim][BF_4] and PMMA/[emim][Tf_2N], respectively. Similar behavior has also been observed for poly(vinyl alcohol) (PVA),[34] poly(L-lactide) (PLA),[35] poly(vinyl chloride) (PVC),[36] and poly(styrene-*b*-2-vinylpyridine)[37] doped with ILs.

9.3.2 Estimation of the T_g for ILs

To predict the T_g for a polymer/IL blend, the $T_{g,IL}$ values need to be determined, which can also be estimated using the following equation:[38]

$$T_{g,IL} = K_1 \delta_{IL}{}^{0.25} \tag{9.8}$$

where δ_{IL} is the Hildebrand solubility parameter for ILs ($MPa^{0.5}$), and K_1 is an adjustable parameter (K $MPa^{-0.125}$). The value of δ_{IL} can be estimated from the ionic cohesive energy density.[39] Figure 9.4 presents the modeling of $T_{g,IL}$ and δ_{IL} for several ILs using eqn (8). The fitting is satisfactory with a K_1 value of 84 K $MPa^{-0.25}$. This is expected since T_g is an indication of the flexibility of liquid molecules, while δ_{IL} indicates the strength of ionic interactions in ILs, which would influence the molecule flexibility.

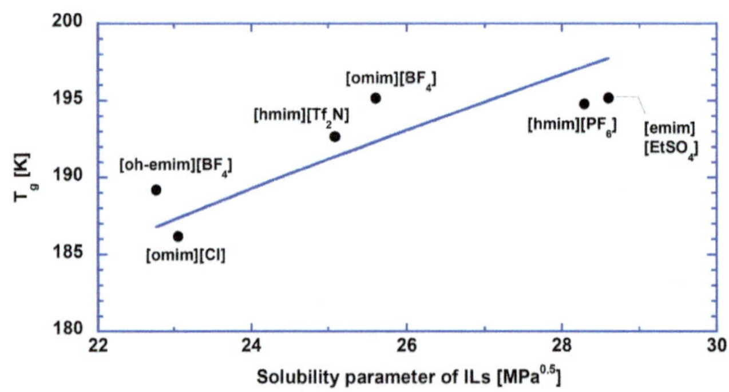

Figure 9.4 Relationship between the solubility parameter (δ_{IL}) and T_g for ILs. Data from ref. 39–43. The line is the best fit based on eqn (8).

9.4 Effect of IL Doping on Polymer Crystallization

9.4.1 Effect of IL Doping on T_m Depression

The crystallization of polymer chains is driven by intermolecular forces such as hydrogen bonds and van der Waals forces. ILs have polarity and the anions (such as Cl^-[44] and Br^-[45]) can act as good hydrogen bond acceptors. Therefore, the IL doping of semi-crystalline polymers weakens the intermolecular forces among the polymer chains and thus disrupts crystallization, resulting in a decrease in both T_m and crystallinity,[46,47] and even a change in the crystalline structure.[48,49] Figure 9.5 shows the decrease of T_m in representative polymers such as P(VDF-HFP),[50] PVA,[34] PVDF,[46] and PA 6[49] caused by IL doping. Similar behaviors have also been observed in poly(vinyl pyrrolidone) (PVP)[51] and PVC.[52]

For semi-crystalline polymers, the depression of T_m in the polymers by IL doping can be estimated using the equation shown below:[46,53]

$$\frac{1}{T_m} - \frac{1}{T_m^0} = \frac{R}{\Delta H_m}\frac{V_P}{V_{IL}}\left(\phi_{IL} - \chi\phi_{IL}^2\right) \tag{9.9}$$

where χ is the interaction parameter between the polymer and IL at T_m, $T_m{}^0$ is the melting temperature of the pure polymer (K), R is the gas constant (8.314 J mol^{-1} K^{-1}), ΔH_m is the melting enthalpy of the crystalline phase in the polymer (J mol^{-1}), and V_P and V_{IL} are the molar volumes of the polymer repeat unit and IL (cm^3 mol^{-1}), respectively. The χ parameter may be estimated from the solubility parameter using the following equation:[54,55]

$$\chi = \frac{V_{IL}}{RT}\left(\delta_p - \delta_{ILs}\right)^2 \tag{9.10}$$

where δ_p is the solubility parameter of the polymer (MPa$^{0.5}$).

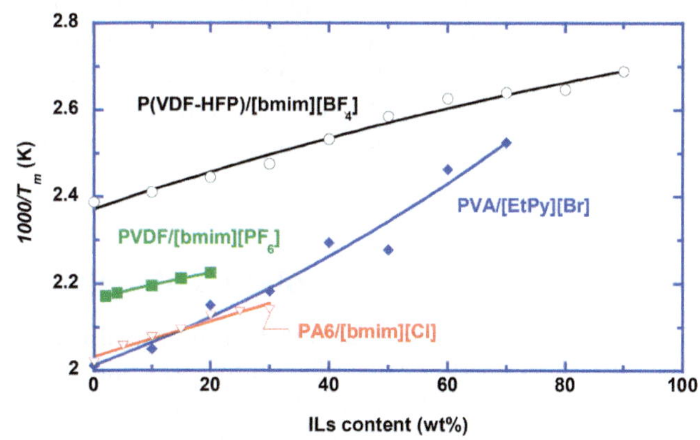

Figure 9.5 Effect of the IL content in the blends on the T_m of polymers, including poly(vinylidenefluoride-*co*-hexafluoropropylene)/1-butyl-3-methylimidazolium tetrafluoroborate) (P(VDF-HFP)/[bmim][BF_4]),[50] PVA/1-ethylpyridinium bromide (PVA/[EtPy][Br]),[34] polyamide 6/1-butyl-3-methylimidazolium chloride (PA 6/([bmim][Cl]),[49] and PVDF/[bmim][PF_6].[47] The curves are best fits using eqn (9).

As shown in Figure 9.5, the T_m data can be satisfactorily described using eqn (9) with ΔH_m and χ as adjustable parameters. More details on the modeling of PVDF/[EtPy]Br blends are also presented in the literature.[46] However, for other blends, a comparison of the adjustable parameters with the literature is not available, due to the lack of parameters (such as χ) in the literature. In general, with the right choice of ILs, the T_m of the blends could be reduced to the membrane operating temperature, and thus, the crystallinity is eliminated to achieve high gas permeability.

9.4.2 Effect of IL Doping on Polymer Crystallinity

The crystallization of polymers can be influenced by the chemical structure and processing conditions.[53] The ultimate crystallinity ($\phi_{c,max}$ or the attainable maximum crystallinity) in polymers has been related to the parameter of T_g/T_m, as shown below:[56,57]

$$\phi_{c,max} = -13475\left(\frac{T_g}{T_m}\right)^3 + 19.664\left(\frac{T_g}{T_m}\right)^2 - 10.898\left(\frac{T_g}{T_m}\right) + 2.987 \quad (9.11)$$

This empirical equation was generated from more than 10 polymers, and it is valid for polymers with T_g/T_m values ranging from 0.352 to 0.791. Generally, the equation predicts that $\phi_{c,max}$ increases with decreasing T_g/T_m, presumably because polymers with chain structures suitable for crystalline packing tend to have low T_g (or good chain flexibility) and high T_m (strong intermolecular interaction).[59]

This model can also shed some light on the effect of IL doping on polymer crystallinity. IL doping decreases T_g, as indicated by eqn (7), and decreases T_m, as indicated by eqn (9), though there may not be a single trend for the T_g/T_m values. As shown in Figure 9.6, increasing the IL content increases $\phi_{c,max}$ for PVDF and P(VDF-HFP), and decreases $\phi_{c,max}$ for CTA and CA. Nevertheless, the $\phi_{c,max}$ (which is defined for the crystallinity in the polymers, instead of the polymer/IL blends) can be estimated for the polymer/IL blends.

9.4.3 Effect of ILs Doping on Dissolution of Cellulose Acetate

ILs have been extensively explored to dissolve semi-crystalline cellulose and its derivatives, due to the wide applications of these materials.[60,61] On the other hand, cellulose acetate with different degrees of acetylation (ranging from 2.0 to 3.0) has been widely used for commercial membranes for industrial CO_2/CH_4 separation, due to its low price and availability on an industrial scale.[62–64] Therefore, it would be useful to review the effect of ILs on CA dissolution, which can elucidate the interactions between ILs and CA, and thus the change in CA morphology due to IL doping.

The dissolution of semi-crystalline cellulose follows two steps: (1) cellulose decrystallization to an amorphous polymer, and (2) disentanglement of the polymer chains for dissolution in the solvent.[65] In general, ILs with greater hydrogen bond basicity and polarity tend to have higher efficiency for cellulose dissolution.[66,67] For example, coordinating anions with good hydrogen bond acceptors (such as OAc^-, $HCOO^-$, Cl^-, and Br^-[68–71]) can dissolve cellulose much more rapidly than non-coordinating ones such as TF_2N^-, BF_4^-, and PF_6^-.[70] Figure 9.7 depicts the interactions between an example IL ([emim][Cl]) and CA, which lead to the dissolution of CA. The anion of Cl^- forms a new hydrogen bond with the OH groups in CA, and the [emim] cation forms a bond with the oxygen atom to replace the hydrogen atom. Therefore, the hydrogen bond leading to the crystallization is disrupted and CA is decrystallized.[45]

Figure 9.8 shows the effect of the hydrogen bond accepting ability of the anions (as indicated by the Kamlet–Taft parameter) on the cellulose solubility. The OAc^- anion shows the highest Kamlet–Taft parameter and also the highest cellulose solubility. The interactions between OAc^- and the OH groups in cellulose have also been confirmed using nuclear magnetic resonance (NMR).[72]

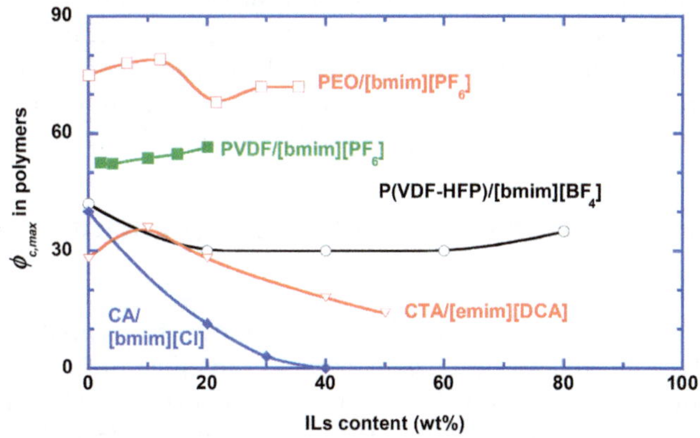

Figure 9.6 Effect of the IL weight percentage in the blends on the polymer crystallinity ($\phi_{c,max}$), including (P(VDF-HFP)/[bmim][BF_4]),[50] CA/[bmim][Cl],[58] CTA/[emim][DCA],[13] PVDF/[bmim][PF_6],[46] and PEO/[bmim][PF_6].[59] The lines are to guide the eye.

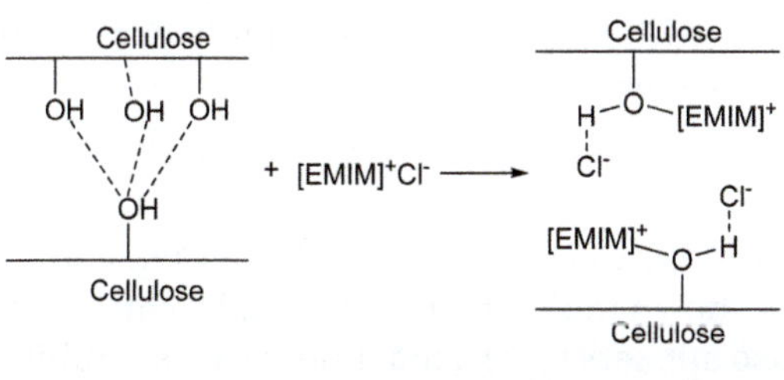

Figure 9.7 Schematic of the dissolution of cellulose in [emim][Cl] by disrupting the hydrogen bonding in cellulose. For further reading see ref. 45.

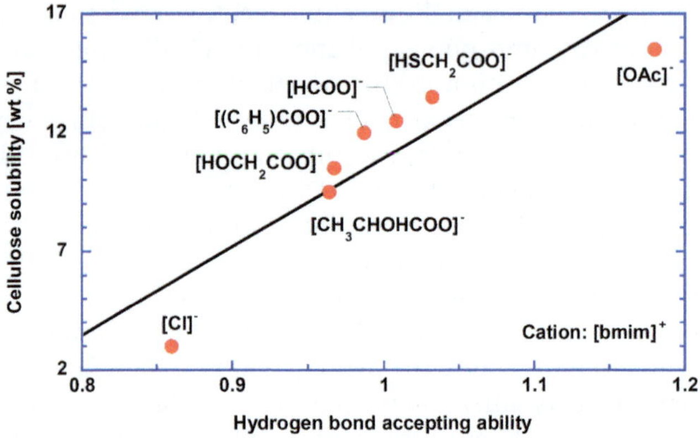

Figure 9.8 Effect of the hydrogen bond accepting ability (Kamlet–Taft parameter) of the anions in various ILs containing [bmim] on the cellulose solubility at 70 °C. Data from ref. 70, 72 and 73.

The type of cation also has an important effect on the cellulose solubility in ILs. Cations with longer alkyl chains tend to have greater hydrophobicity, resulting in lower cellulose solubility. For example, [C6mim][Cl] shows much higher cellulose solubility than [C8mim][Cl] at 100 °C.[70] Pyridinium and imidazolium cations contain nitrogen-based aromatic rings and thus high polarizability, which leads to a high capability to dissolve cellulose.[74,75]

9.5 Effect of IL Doping on Gas Permeation Properties

ILs have been extensively evaluated for CO_2 removal from mixtures containing N_2 or CH_4.[11,12] This section briefly reviews the effect of IL doping on CO_2 solubility and diffusivity, and the separation properties of CO_2/N_2 and CO_2/CH_4. Quantitative models to define such effects are also described.

9.5.1 Gas Solubility in ILs

Figure 9.9 shows CO_2 solubility at 35 °C in various ILs with different solubility parameter values (δ_{IL}). These ILs contain imidazolium with amine groups to interact favorably with CO_2. The CO_2 solubility can be correlated with δ_{IL} using the equation below:[11,76]

$$\ln S_{CO_2} = a + b(\delta_{IL} - \delta_{CO_2})^2 \tag{9.12}$$

where δ_{CO_2} is the solubility parameter ($MPa^{0.5}$) of CO_2, and a and b are adjustable constants. As shown in Figure 9.9, this model can satisfactorily describe the data. There also appears to be a maximum CO_2 solubility when the IL has a δ_{IL} value of 38 $MPa^{0.5}$, which would suggest that δ_{CO_2} has a value of 38 $MPa^{0.5}$. This result is different from an early study (showing a δ_{CO_2} value of 21.5 $MPa^{0.5}$)

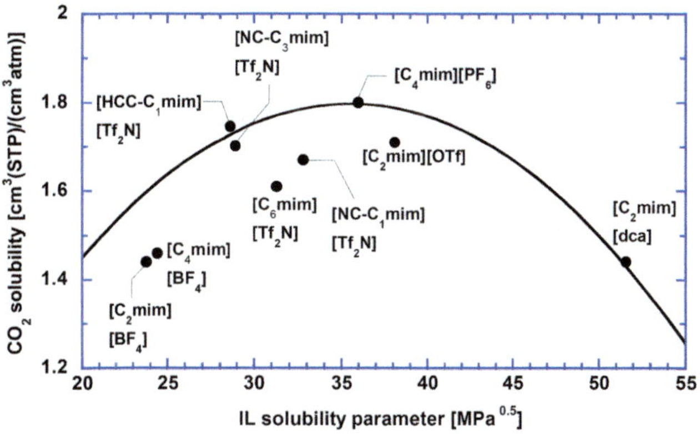

Figure 9.9 CO_2 solubility at 35 °C in ILs as a function of δ_{IL}.[11,78,79] The curve is the best fit based on eqn (12).

based on the CO_2 solubility in various hydrocarbon-based liquids,[76] partially because the δ_{IL} values are overestimated.[77] Nevertheless, eqn (12) provides a useful guideline in designing ILs to achieve high CO_2 sorption.

Light gases such as N_2 and CH_4 have very low solubility (S_A) in ILs, and the following equation has been used to correlate S_A with δ_{IL}:[11,76]

$$\ln S_A = \alpha + \beta \delta_{IL}{}^2 \tag{9.13}$$

where α and β are adjustable constants. Eqn (13) is mathematically consistent with eqn (12), since N_2 and CH_4 have solubility parameters of 5.3 and 11.6 $MPa^{0.5}$, respectively,[11] which are much lower than those of the ILs.

9.5.2 Effect of IL Doping on Gas Solubility in Polymer/IL Blends

Figure 9.10 shows the effect of IL doping on the CO_2 solubility in three polymer/IL blends, PVDF/[emim][B(CN)$_4$],[80] CTA/[emim][BF$_4$],[13] and Pebax 1657/[bmim][CF$_3$SO$_3$].[25] Except for the blend based on Pebax 1657 (which is a block copolymer containing amorphous PEO and semi-crystalline PA6 with low crystallinity), the estimated solubility in the amorphous phase is used for comparison to eliminate the effect of crystallinity. Two opposite trends can be observed. For glassy polymers such as CTA, increasing the IL content decreases the CO_2 solubility;[13] for rubbery polymers such as PVDF[80] and Pebax 1657,[25] increasing the IL content increases the CO_2 solubility. These behaviors can be explained by the unique gas sorption behavior in glassy polymers, which is often described using a dual mode model:

$$S_A = k_{D,A} + \frac{C'_H b}{1 + b p_A} \tag{9.14}$$

where $k_{D,A}$ (cm^3(STP) cm^{-3} atm^{-1}) is Henry's sorption coefficient, C'_H (cm^3(STP) cm^{-3} polymer) is the Langmuir capacity constant, and b (atm^{-1}) is the Langmuir affinity of the gas. Rubbery polymers do not have Langmuir sorption sites (*i.e.*, $C'_H = 0$). Therefore, glassy polymers often have a higher gas solubility than rubber polymers. The addition of an IL in glassy polymers decreases the free volume or Langmuir sites available (C'_H) and thus gas sorption. The decrease in gas sorption has been correlated with the decrease in T_g of the polymer/IL blends.[81] On the other hand, when the ILs have a higher gas sorption than the rubber polymers, increasing the IL content is expected to increase the gas sorption in the blends.

The gas solubility in the blends can be related to that in the individual IL and polymer components using the following equation:[82,83]

$$\ln S_{am} = \phi_{am,p} \ln S_{am,p} + \phi_{am,IL} \ln S_{IL} \quad (9.15)$$

where $S_{am,p}$ and S_{IL} are the gas solubilities in the amorphous phase polymer and IL, respectively. As shown in Figure 9.10a, the gas solubility can be satisfactorily described by the equation, despite the different trends for glassy and rubbery polymers.

Figure 9.10b shows that increasing the IL content also increases the CO_2/N_2 and CO_2/CH_4 solubility selectivity, since the ILs (such as [emim][DCA] and [emim][BF_4]) exhibit much higher CO_2/light gas selectivity than the polymer matrix (*i.e.*, CTA),[13] though CTA has been widely used for commercial membrane applications for CO_2/CH_4 separation.

9.5.3 Effect of IL Doping on Gas Diffusivity in Polymer/IL Blends

Increasing the IL content in the blends often increases the gas diffusivity, mainly because of the plasticization by ILs, increasing the flexibility of the polymer chains. For example, in CTA/[emim][BF_4] blends, increasing the IL content from 0 to 50% decreases T_g from 198 °C to 11 °C, and increases the CO_2 diffusivity from 1.4×10^{-8} to 7.4×10^{-8} cm^2 s^{-1}.[13] The increase in the diffusivity cannot be described using the free volume model, presumably because of the high uncertainty in estimating the free volume in ILs.[13]

Figure 9.11 shows the effect of IL content on the estimated CO_2 diffusivity in the following amorphous polymer/IL blends: CTA/[emim][BF_4],[13] P(VDF-HFP)/[emim][TFSI],[26] PVDF/[emim][B(CN)$_4$],[80] and polyimide (PI)/[bmim][TF_2N][81] at 35 °C. In the first three blends, the amorphous phase CO_2 diffusivity increases continuously with increasing IL content. However, for the last blend, increasing the IL content initially decreases the gas diffusivity (such as CO_2) before increasing it, presumably due to the anti-plasticization effect, *i.e.*, the addition of a low content of ILs decreases the free volume in the polymers, leading to lower gas diffusivity.[84] At higher contents, the IL can plasticize the polymer chains and increase the gas diffusivity. The minimal CO_2 diffusivity reflects the balance of these two counteracting factors in PI/[bmim][TF_2N].[81]

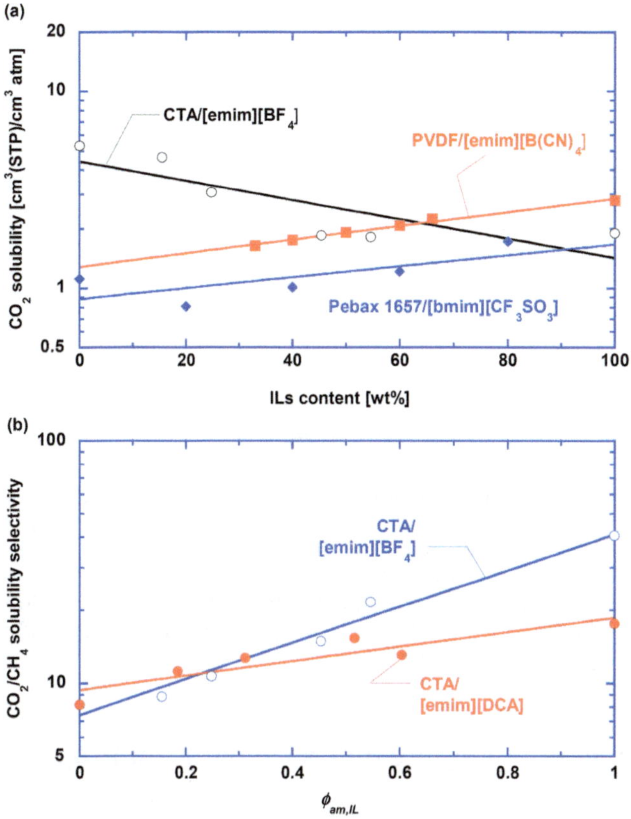

Figure 9.10 Effect of IL content on (a) the estimated CO_2 solubility in amorphous polymer/IL blends, PVDF/[emim][B(CN)$_4$],[80] CTA/[emim][BF$_4$],[13] and Pebax 1657/[bmim][CF$_3$SO$_3$],[25] and (b) CO_2/CH_4 solubility selectivity in CTA/[emim][BF$_4$], and CTA/[emim][DCA] at 35 °C.[13] The lines are best fits using eqn (15).

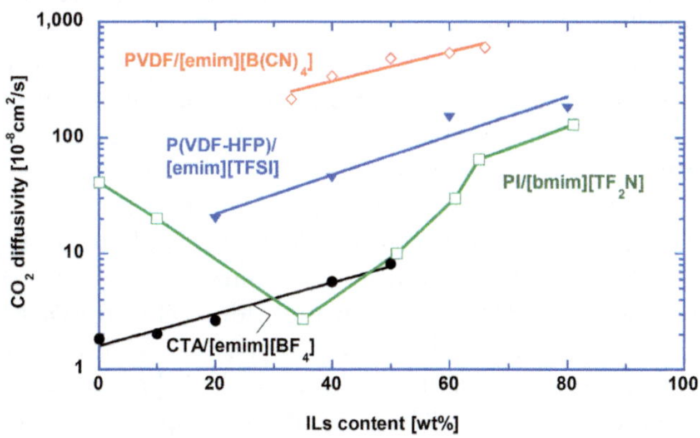

Figure 9.11 Effect of the IL content on the CO_2 diffusivity at 35 °C in the following amorphous polymer/IL blends: CTA/[emim][BF$_4$],[13] P(VDF-HFP)/[emim][TFSI],[26] PVDF/[emim][B(CN)$_4$],[80] and PI/[bmim][TF$_2$N].[81] The curve and lines are to guide the eye.

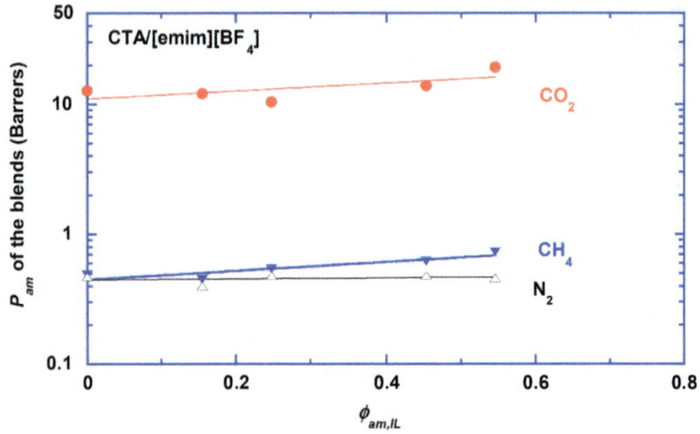

Figure 9.12 Effect of the IL loading ($\phi_{am,IL}$) in the amorphous phase of the CTA/[emim][BF_4] blends on the amorphous phase gas permeability at 35 °C. The lines are the best fits based on eqn (16).[13] Reprinted with permission Polymer, 89, B. Lam, M. Wei, L. Zhu, S. Luo, R. Guo, A. Morisato, P. Alexandridis, H. Lin, Cellulose triacetate doped with ionic liquids for membrane gas separation, 1–11, Copyright (2016) with permission from Elsevier.

9.5.4 Effect of IL Doping on Gas Separation Properties

The following equation has often been used to model gas permeability in homogeneous blends using the properties of the individual components:[13,82,83]

$$\ln P_{am} = \phi_{am,CTA} \ln P_{am,CTA} + \phi_{am,IL} \ln P_{IL} \quad (9.16)$$

where $P_{am,CTA}$ and P_{IL} are the gas permeability in the amorphous phase CTA and IL, respectively. As shown in Figure 9.12, the fittings are generally good for CTA/[emim][BF_4] blends.[13]

Figure 9.13 shows examples of IL doping to improve the CO_2/N_2 separation properties in a Robeson's plot. The upper bound line shows the trade-off between CO_2 permeability and CO_2/N_2 selectivity.[5–7] Pure ILs such as [emim][DCA] and [emim][BF_4] exhibit superior CO_2/N_2 separation properties, as their performance is above the upper bound. As shown in Figure 9.13a, the addition of [emim][DCA] in poly(ionic liquid)s such as poly(1-vinyl-3-butylimidazolium dicyanamide) (P[vbim][DCA]) and poly(1-vinyl-3-heptylimidazolium dicyanamide) (P[vhim][DCA]) increases both the CO_2 permeability and CO_2/N_2 selectivity.[85–89] Similar behavior has been observed when [emim][DCA] and [emim][BF_4] are used to dope CTA, as shown in Figure 9.13b.[13,22] These results demonstrate the effectiveness of this strategy in improving polymer-based materials for membrane gas separation.

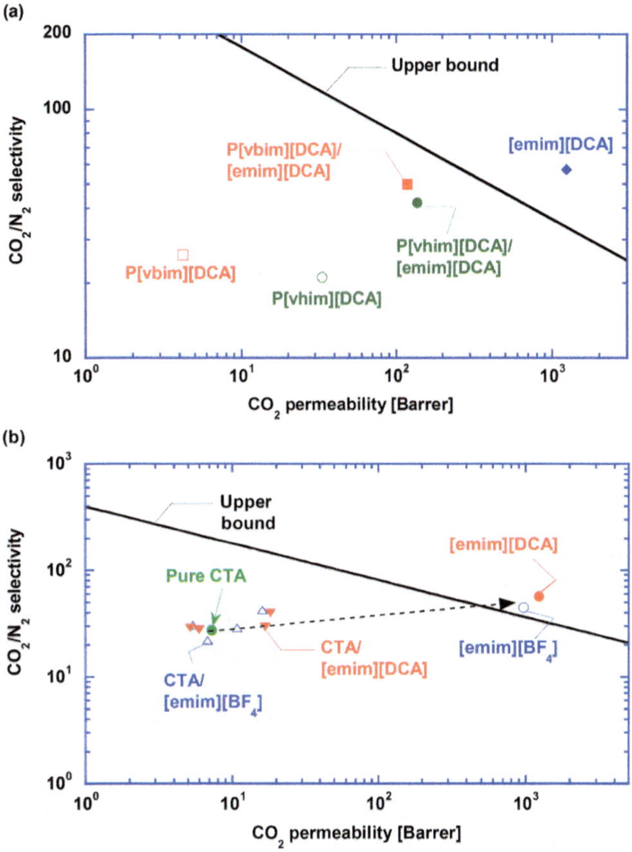

Figure 9.13 Enhancement of CO_2/N_2 separation properties in polymers by IL doping. (a) Poly(ionic liquid)s doped with [emim][DCA],[22,85–89] and (b) CTA doped with [emim][BF_4] and [emim][DCA] at 35 °C.[13,22] Data for Figure 9.13a is taken from ref. 22, and Figure 9.13b is reprinted from Polymer, 89, B. Lam, M. Wei, L. Zhu, S. Luo, R. Guo, A. Morisato, P. Alexandridis, H. Lin, Cellulose triacetate doped with ionic liquids for membrane gas separation, 1–11, Copyright (2016) with permission from Elsevier.[13]

9.6 Conclusion

This review aims to elucidate the effect of the interactions between ILs and polymers on physical properties, such as morphology and gas transport properties, and thus provides guidelines to design and identify suitable ILs for doping to enhance the membrane gas separation properties in polymers. IL doping decreases T_g, and depresses T_m and the crystallinity for semi-crystalline polymers, which, in turn, increases the gas permeability. ILs based on [emim] exhibit favorable interactions with CO_2, and thus, their doping can significantly increase the CO_2 permeability and CO_2/N_2 selectivity.

This review also uses examples of cellulose derivatives (including CTA and CA, commercial membrane materials for CO_2/CH_4 separation) to elucidate the interaction between ILs and semi-crystalline polymers. The effect of IL structures on cellulose decrystallization is presented, where doping using two promising ILs ([emim][DCA] and [emim][BF_4]) results in enhancement in the CO_2/N_2 separation properties according to the Robeson's upper bound.

Acknowledgements and Disclaimer

We gratefully acknowledge the financial support of this work by the Korean Carbon Capture and Sequestration R&D Center (KCRC), and the partial support from the U.S. National Science Foundation (NSF) under the CAREER award number 1554236.

References

1. D. E. Sanders, Z. P. Smith, R. Guo, L. M. Robeson, J. E. McGrath, D. R. Paul and B. D. Freeman, *Polymer*, 2013, **54**, 4729–4761.
2. R. W. Baker and B. Low, *Macromolecules*, 2014, **47**, 6999–7013.
3. H. Lin, *Curr. Opin. Chem. Eng.*, 2014, **4**, 54–61.
4. L. M. Robeson, *J. Membr. Sci.*, 1991, **62**, 165–185.
5. L. M. Robeson, *J. Membr. Sci.*, 2008, **320**, 390–400.
6. B. D. Freeman, *Macromolecules*, 1999, **32**, 375–380.
7. H. Lin and M. Yavari, *J. Membr. Sci.*, 2015, **475**, 101–109.
8. N. V. Plechkova and K. R. Seddon, *Chem. Soc. Rev.*, 2008, **37**, 123–150.
9. J. E. Bara, D. E. Camper, D. L. Gin and R. D. Noble, *Acc. Chem. Res.*, 2010, **43**, 152–159.
10. P. Scovazzo, D. Camper, J. Kieft, J. Poshusta, C. Koval and R. Noble, *Ind. Eng. Chem. Res.*, 2004, **43**, 6855–6860.
11. J. E. Bara, T. K. Carlisle, C. J. Gabriel, D. Camper, A. Finotello, D. L. Gin and R. D. Noble, *Ind. Eng. Chem. Res.*, 2009, **48**, 2739–2751.
12. Z. Dai, R. D. Noble, D. L. Gin, X. Zhang and L. Deng, *J. Membr. Sci.*, 2016, **497**, 1–20.
13. B. Lam, M. Wei, L. Zhu, S. Luo, R. Guo, A. Morisato, P. Alexandridis and H. Lin, *Polymer*, 2016, **89**, 1–11.
14. J. L. Anthony, J. L. Anderson, E. J. Maginn and J. F. Brennecke, *J. Phys. Chem. B*, 2005, **109**, 6366–6374.
15. C. Cadena, J. L. Anthony, J. K. Shah, T. I. Morrow, J. F. Brennecke and E. J. Maginn, *J. Am. Chem. Soc.*, 2004, **126**, 5300–5308.
16. Y.-Y. Jiang, Z. Zhou, Z. Jiao, L. Li, Y.-T. Wu and Z.-B. Zhang, *J. Phys. Chem. B*, 2007, **111**, 5058–5061.
17. J. H. Lee, I. S. Chae, D. Song, Y. S. Kang and S. W. Kang, *Sep. Purif. Technol.*, 2013, **112**, 49–53.
18. S. Kasahara, E. Kamio, T. Ishigami and H. Matsuyama, *Chem. Commun.*, 2012, **48**, 6903–6905.

19. J. Albo, E. Santos, L. Neves, S. Simeonov, C. Afonso, J. Crespo and A. Irabien, *Sep. Purif. Technol.*, 2012, **97**, 26–33.
20. P. Scovazzo, D. Havard, M. McShea, S. Mixon and D. Morgan, *J. Membr. Sci.*, 2009, **327**, 41–48.
21. A. Car, C. Stropnik, W. Yave and K.-V. Peinemann, *J. Membr. Sci.*, 2008, **307**, 88–95.
22. J. Y. Liu, X. D. Hou, H. B. Park and H. Q. Lin, *Chem.–Eur. J.*, 2016, **22**, 15980–15990.
23. L. Liang, Q. Gan and P. Nancarrow, *J. Membr. Sci.*, 2014, **450**, 407–417.
24. H. Rabiee, A. Ghadimi and T. Mohammadi, *J. Membr. Sci.*, 2015, **476**, 286–302.
25. P. Bernardo, J. C. Jansen, F. Bazzarelli, F. Tasselli, A. Fuoco, K. Friess, P. Izák, V. Jarmarová, M. Kačírková and G. Clarizia, *Sep. Purif. Technol.*, 2012, **97**, 73–82.
26. K. Friess, J. C. Jansen, F. Bazzarelli, P. Izák, V. Jarmarová, M. Kačírková, J. Schauer, G. Clarizia and P. Bernardo, *J. Membr. Sci.*, 2012, **415–416**, 801–809.
27. J. Wijmans and R. Baker, *J. Membr. Sci.*, 1995, **107**, 1–21.
28. R. S. Prabhakar, R. Raharjo, L. G. Toy, H. Lin and B. D. Freeman, *Ind. Eng. Chem. Res.*, 2005, **44**, 1547–1556.
29. A. S. Michaels and H. J. Bixler, *J. Polym. Sci.*, 1961, **50**, 393–412.
30. H. Lin and B. D. Freeman, *J. Membr. Sci.*, 2004, **239**, 105–117.
31. Z. Mogri and D. R. Paul, *Polymer*, 2001, **42**, 7765–7780.
32. M. M. Mok, X. Liu, Z. Bai, Y. Lei and T. P. Lodge, *Macromolecules*, 2011, **44**, 1016–1025.
33. L. An, D. He, J. Jing, Z. Wang, D. Yu, B. Jiang, Z. Jiang and R. Ma, *Eur. Polym. J.*, 1997, **33**, 1523–1528.
34. J. Yoon, H.-J. Lee and C. M. Stafford, *Macromolecules*, 2011, **44**, 2170–2178.
35. P. Zhang, L. Peng and W. Li, *e-Polym.*, 2008, **8**, 1970–1975.
36. S. Y. Choi, H. Rodríguez, A. Mirjafari, D. F. Gilpin, S. McGrath, K. R. Malcolm, M. M. Tunney, R. D. Rogers and T. McNally, *Green Chem.*, 2011, **13**, 1527–1535.
37. J. M. Virgili, A. Hexemer, J. A. Pople, N. P. Balsara and R. A. Segalman, *Macromolecules*, 2009, **42**, 4604–4613.
38. T. Tsutsui and T. Tanaka, *Polymer*, 1977, **18**, 817–821.
39. S. H. Lee and S. B. Lee, *Chem. Commun.*, 2005, 3469–3471.
40. J. G. Huddleston, A. E. Visser, W. M. Reichert, H. D. Willauer, G. A. Broker and R. D. Rogers, *Green Chem.*, 2001, **3**, 156–164.
41. L. C. Branco, J. N. Rosa, J. J. Moura Ramos and C. A. M. Afonso, *Chem.–Eur. J.*, 2002, **8**, 3671–3677.
42. M. Alavianmehr, S. Hosseini, A. Mohsenipour and J. Moghadasi, *J. Mol. Liq.*, 2016, **218**, 332–341.
43. A. Fernández, J. S. Torrecilla, J. García and F. Rodríguez, *J. Chem. Eng. Data*, 2007, **52**, 1979–1983.
44. G. Bentivoglio, T. Röder, M. Fasching, M. Buchberger, H. Schottenberger and H. Sixta, *Lenzinger Ber.*, 2006, **86**, 154–161.

45. L. Feng and Z.-l. Chen, *J. Mol. Liq.*, 2008, **142**, 1–5.
46. C. Y. Xing, M. M. Zhao, L. P. Zhao, J. C. You, X. J. Cao and Y. J. Li, *Polym. Chem.*, 2013, **4**, 5726–5734.
47. J. C. Jansen, G. Clarizia, P. Bernardo, F. Bazzarelli, K. Friess, A. Randova, J. Schauer, D. Kubicka, M. Kacirkova and P. Izak, *Sep. Purif. Technol.*, 2013, **109**, 87–97.
48. L. Guo, Y. Liu, C. Zhang and J. Chen, *J. Membr. Sci.*, 2011, **372**, 314–321.
49. Y. Tian, H. Qin, X. Yang, C. Chi and S. Liu, *Mater. Lett.*, 2016, **180**, 200–202.
50. S. Chaurasia, R. Singh and S. Chandra, *J. Phys. Chem. B*, 2013, **117**, 897–906.
51. A. Saroj, R. Singh and S. Chandra, *Mater. Sci. Eng., B*, 2013, **178**, 231–238.
52. M. Rahman and C. S. Brazel, *Polym. Degrad. Stab.*, 2006, **91**, 3371–3382.
53. L. H. Sperling, *Introduction to Physical Polymer Science*, Wiley-Interscience, New York, 3rd edn, 2001.
54. T. Nishi and T. Wang, *Macromolecules*, 1975, **8**, 909–915.
55. J.-L. Lin and R.-J. Roe, *Macromolecules*, 1987, **20**, 2168–2173.
56. D. W. Van Krevelen and K. Te Nijenhuis, *Properties of Polymers: Their Correlation with Chemical Structure; Their Numerical Estimation and Prediction from Additive Group Contributions*, Elsevier, New York, 2009.
57. J. Bicerano, *J. Macromol. Sci., Rev. Macromol. Chem. Phys.*, 1998, **C38**, 391–479.
58. A. Bendaoud and Y. Chalamet, *Carbohydr. Polym.*, 2014, **108**, 75–82.
59. L. T. Costa, R. L. Lavall, R. S. Borges, J. Rieumont, G. G. Silva and M. C. Ribeiro, *Electrochim. Acta*, 2007, **53**, 1568–1574.
60. A. Pinkert, K. N. Marsh, S. Pang and M. P. Staiger, *Chem. Rev.*, 2009, **109**, 6712–6728.
61. H. Wang, G. Gurau and R. D. Rogers, *Chem. Soc. Rev.*, 2012, **41**, 1519–1537.
62. R. W. Baker and K. Lokhandwala, *Ind. Eng. Chem. Res.*, 2008, **47**, 2109–2121.
63. H. Lin, L. S. White, K. A. Lokhandwala and R. W. Baker, in *Encyclopedia of Membrane Science and Technology*, ed. E. M. V. Hoek and V. V. Tarabara, Wiley, New Jersey, 2013, pp. 1644–1667.
64. M. Tang, United States Pat., 2008/0143014 A1, 2008.
65. M. Ghasemi, P. Alexandridis and M. Tsianou, *Cellulose*, 2016, 1–20.
66. S. Park and R. J. Kazlauskas, *Curr. Opin. Biotechnol.*, 2003, **14**, 432–437.
67. J. L. Anderson, J. Ding, T. Welton and D. W. Armstrong, *J. Am. Chem. Soc.*, 2002, **124**, 14247–14254.
68. G. Cheng, P. Varanasi, R. Arora, V. Stavila, B. A. Simmons, M. S. Kent and S. Singh, *J. Phys. Chem. B*, 2012, **116**, 10049–10054.
69. Y. Fukaya, K. Hayashi, M. Wada and H. Ohno, *Green Chem.*, 2008, **10**, 44–46.
70. R. P. Swatloski, S. K. Spear, J. D. Holbrey and R. D. Rogers, *J. Am. Chem. Soc.*, 2002, **124**, 4974–4975.
71. S. Barthel and T. Heinze, *Green Chem.*, 2006, **8**, 301–306.
72. A. Xu, J. Wang and H. Wang, *Green Chem.*, 2010, **12**, 268–275.

73. Y. Fukaya, Y. Iizuka, K. Sekikawa and H. Ohno, *Green Chem.*, 2007, **9**, 1155–1157.
74. A. R. Ferreira, M. G. Freire, J. C. Ribeiro, F. M. Lopes, J. G. Crespo and J. A. P. Coutinho, *Ind. Eng. Chem. Res.*, 2011, **50**, 5279–5294.
75. T. Heinze, K. Schwikal and S. Barthel, *Macromol. Biosci.*, 2005, **5**, 520–525.
76. H. Lin and B. D. Freeman, *J. Mol. Struct.*, 2005, **739**, 57–74.
77. M. S. Shannon, J. M. Tedstone, S. P. Danielsen, M. S. Hindman, A. C. Irvin and J. E. Bara, *Ind. Eng. Chem. Res.*, 2012, **51**, 5565–5576.
78. D. Camper, J. Bara, C. Koval and R. Noble, *Ind. Eng. Chem. Res.*, 2006, **45**, 6279–6283.
79. L. S. Wang, X. X. Wang, Y. Li, K. Jiang, X. Z. Shao and C. J. Du, *AIChE J.*, 2013, **59**, 3034–3041.
80. H. Z. Chen, P. Li and T.-S. Chung, *Int. J. Hydrogen Energy*, 2012, **37**, 11796–11804.
81. S. Kanehashi, M. Kishida, T. Kidesaki, R. Shindo, S. Sato, T. Miyakoshi and K. Nagai, *J. Membr. Sci.*, 2013, **430**, 211–222.
82. Y. Maeda and D. R. Paul, *Polymer*, 1985, **26**, 2055–2063.
83. D. R. Paul, *J. Membr. Sci.*, 1984, **18**, 75–86.
84. Y. Maeda and D. R. Paul, *J. Polym. Sci., Part B: Polym. Phys.*, 1987, **25**, 1005–1016.
85. J. E. Bara, S. Lessmann, C. J. Gabriel, E. S. Hatakeyama, R. D. Noble and D. L. Gin, *Ind. Eng. Chem. Res.*, 2007, **46**, 5397–5404.
86. T. K. Carlisle, J. E. Bara, A. L. Lafrate, D. L. Gin and R. D. Noble, *J. Membr. Sci.*, 2010, **359**, 37–43.
87. T. K. Carlisle, E. F. Wiesenauer, G. D. Nicodemus, D. L. Gin and R. D. Noble, *Ind. Eng. Chem. Res.*, 2012, **52**, 1023–1032.
88. K. Simons, K. Nijmeijer, J. E. Bara, R. D. Noble and M. Wessling, *J. Membr. Sci.*, 2010, **360**, 202–209.
89. M. G. Cowan, D. L. Gin and R. D. Noble, *Acc. Chem. Res.*, 2016, **49**, 724–732.

CHAPTER 10

Ionic Liquid-modified Poly(Vinylidene Fluoride): from High Performance Anti-static Materials to Flexible Dielectric Materials

C. Y. XING AND Y. J. LI*

Hangzhou Normal University, College of Material, Chemistry and Chemical Engineering, No. 16 Xuelin Rd., Hangzhou, 310036, People's Republic of China
*E-mail: yongjin-li@hznu.edu.cn

10.1 Introduction

Poly(vinylidene fluoride) (PVDF) has stimulated intensive interest due to its intriguing features in applications in the field of electronic devices.[1–5] The inherent properties of PVDF, such as high polarity, high dielectric constant, good strength and favourable flexibility, conferred by the CF_2–CH_2 structure units that alternately form the PVDF backbone, make it a promising candidate as a matrix or a precursor for the design of electronic devices among common polymers. However, neat PVDF cannot meet all of the requirements for practical application due to its insulating characteristics, non-polar phase, and relatively low dielectric constant compared with inorganic ceramics.[6–8]

Smart Materials No. 29
Polymerized Ionic Liquids
Edited by Ali Eftekhari

Published by the Royal Society of Chemistry, www.rsc.org

Therefore, considerable effort has been devoted to modifying PVDF for specific purposes, such as improvement of the polar phase content,[9–24] thermal conductivity,[25,26] dielectric performance,[27–46] and so on.

Room temperature ionic liquids (RTILs) have recently been utilized as modifiers in common polymers due to their unique physical properties, including low melting temperature (commonly lower than 100 °C), non-volatility, non-flammability, high ionic conductivity, unique capacity against bacteria, *etc.*[47–49] The entirely ion-based chemical structures of ionic liquids (ILs) are responsible for their intriguing physical properties. The reported roles that ILs have played in the modification of polymers include as plasticizers for PVDF,[50] poly(methyl methacrylate) (PMMA),[51,52] poly(vinyl chloride) (PVC)[53] and polylactic acid (PLA),[54] compatibilizers for polymer matrices with nanofillers including carbon nanotubes (CNTs)[55–58] and montmorillonite (MMT),[59] anti-static agents for PVDF,[50] polycarbonate (PC),[60] thermoplastic polyurethanes (TPU)[61] and polypropylene (PP),[62] a flame retardant for PP[63] and an anti-bacterial agent for TPU.[61] It is suggested that the successful modification of polymers with ILs is based on the good thermodynamic miscibility of ILs with polymers.[50,60,61,67] On the other hand, research into the modification of PVDF or its fluoride copolymers by using ILs has been investigated intensively.[64,65] Overall, several preparation strategies to form PVDF/IL composites can include spin-coating,[66] melt-blending,[50] electrospinning,[67] dip-coating,[68] and so on. Most studies have demonstrated good physical interactions between PVDF and the IL, namely static interactions of the positively charged cations of the IL with the negatively charged $>CF_2$ dipoles[50] and/or the negatively charged anions of the IL with the positively charged $>CH_2$ dipoles.[66] However, the former is more reasonable and meaningful.[69] To the best of our knowledge, there were no studies regarding the chemical modification of PVDF by an IL until the very recent reports by our group, which may open up a new avenue to create PVDF/IL-based conductive nanostructures.[70–73]

In this chapter, IL-integrated PVDF composites (PVDF/ILs) are introduced. The structure–property relationships of these PVDF/IL composites are also discussed in detail. In terms of the different potential applications of such PVDF/IL composites, two distinguished cases based on PVDF and ILs are presented, as shown in Figure 10.1: (1) high-performance anti-static composites (Figure 10.1a–c); (2) dielectric composites (Figure 10.1d–f).

10.2 Anti-static PVDF/IL Composites

10.2.1 Anti-static Miscible PVDF/IL Films

Like most common plastics, semi-crystalline PVDF is insulating; on the other hand, ILs are highly ion-conductive and are built entirely from cations and anions. It is therefore reasonable to integrate PVDF with ILs to improve the electrical properties of PVDF. In this case, the thermodynamic compatibility of PVDF with the IL is of great importance, and plays a crucial role in

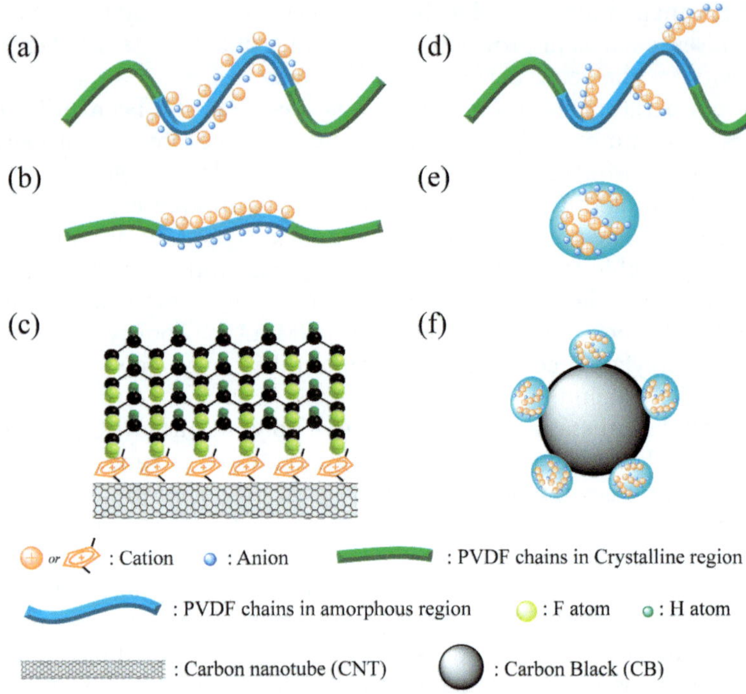

Figure 10.1 Various kinds of PVDF/IL-based system: (a) a miscible PVDF/IL system, in which the IL molecules are only located in the amorphous regions of PVDF and are fully miscible with the amorphous PVDF chains;[50] (b) electrospun PVDF/IL composite nanofibers;[67] (c) IL-coated carbon nanotube (CNT)-integrated PVDF (PVDF/IL–CNT) nanocomposites with all-trans conformations of the PVDF chains;[74] (d) IL-grafted PVDF amorphous chains (PVDF-*g*-IL), forming block-like copolymers with structures of crystalline PVDF-*b*-amorphous PVDF-*g*-IL-*b*-crystalline PVDF chains;[70,72] (e) nanostructured PVDF/IL composites, formed by microphase separation of PVDF-*g*-IL chains from ungrafted PVDF chains;[73] (f) PVDF nanocomposites with simultaneous organic PVDF-*g*-IL nanodomains and inorganic carbon black (CB) nanoparticles with nanodomain@CB nanoparticle structures.[71]

determining the physical properties of the PVDF/IL system. It was suggested in our previous study[50] that PVDF is thermodynamically miscible with the imidazolium-based IL 1-butyl-3-methylimidazolium hexafluorophosphate ([BMIM][PF_6]), and a negative value of the Huggins–Flory interaction parameter was found (−2.84) in the melt (*i.e.* at 190 °C), obtained by using the revised Nishi–Wang equation.[75] The absolute value is much larger than that of the classically miscible systems of PVDF/PMMA[75] and PVDF/PBSU.[76] Moreover, such an evaluation of the compatibility between PVDF and the IL can be supported by the depression of the melting point (T_m), equivalent melting point (T_m^0), melt-crystallization temperature (T_c) and glass transition temperature (T_g) of PVDF as the loading level of IL increases.[50] As a consequence, miscible binary PVDF/IL blend films were obtained by melt-blending with uniform morphologies, as shown in Figure 10.2a (with 10% IL) and Figure 10.2b (with 20% IL). It is noteworthy that the IL molecules are only located in the amorphous regions of PVDF and that they are fully miscible with PVDF's amorphous chains, strongly supported by both the depression of T_g measured by dynamic mechanical analysis (DMA) and the increase in crystal long period (L) obtained by small angle X-ray scattering (SAXS) with an increase in IL content.[50]

A good PVDF/IL miscibility results in excellent physical properties for PVDF/IL blend films, especially electrical, mechanical and optical properties. Table 10.1 shows the electrical properties of PVDF/IL blend films with 300 μm thickness. The surface resistivity (R_s) of the PVDF/IL samples is significantly decreased by the addition of IL, and the best electrical properties of the PVDF/ILs are found in the case of PVDF/IL 100/20. Moreover, this sample

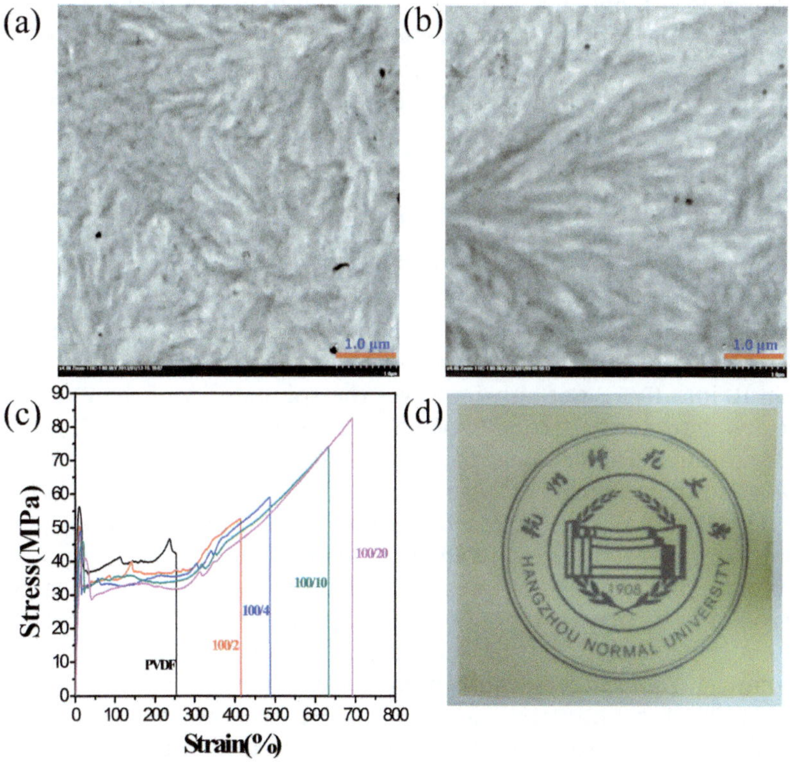

Figure 10.2 Morphologies and physical properties of miscible PVDF/IL blend films prepared by melt-blending: transmission electron microscopy (TEM) images of PVDF/IL blend films with 10% (a) and 20% (b) IL loadings; strain–stress curves of PVDF/IL blend films (c) and a photo of a PVDF/IL blend film (with 20% IL) with a surface resistivity value of 1.27×10^8 ohm per square (d). Reprinted from ref. 50 with permission from the Royal Society of Chemistry.

Table 10.1 Anti-static properties of PVDF/IL blend films with 300 μm thickness. The evaluation of the anti-static effect of the PVDF/IL blend films was performed as described in the literature.[77,78] Reprinted from ref. 50 with permission from the Royal Society of Chemistry.

PVDF/IL blend films	Surface resistivity (ohm/square)	Volume resistivity (ohm cm)	Anti-electrostatic effect
Neat PVDF	Out of range	Out of range	None
100/2	1.44×10^{11}	2.06×10^{10}	Good
100/4	3.01×10^{10}	2.4×10^{9}	Very good
100/10	3.13×10^{9}	3.92×10^{8}	Excellent
100/20	1.27×10^{8}	1.52×10^{7}	Excellent

simultaneously exhibits considerable improvements in tensile strength (82.7 MPa *vs.* 42.5 MPa for neat PVDF) and elongation at break (693% *vs.* 252% of neat PVDF), as shown in Figure 10.2c. It is noteworthy that the 20% IL-filled PVDF film is still optically transparent (higher than 85% transparency) and no bleeding phenomena are observed in this sample, as shown in Figure 10.2d. Their desirable anti-static properties, together with these excellent mechanical and optical properties, enable the PVDF/IL blend films to be promising in electronics design.

10.2.2 Anti-static PVDF/IL Nanofibrous Films

Owing to the good miscibility of PVDF with ILs, in addition to compact PVDF/IL blend films, PVDF/IL composite nanofibrous films/mats can also be readily fabricated through an electrospinning technique. Figure 10.3 shows the morphologies of PVDF/IL-based nanofibers.[67] It has been shown that adding an IL into PVDF not only improves the spinnability of the PVDF nanofibers by increasing the conductivity of the PVDF solution, which results in the disappearance of beads on the fibers (Figure 10.3a–f), but also makes the PVDF nanofibers rougher (Figure 10.3D and F) relative to neat PVDF nanofibers (Figure 10.3A). The increased roughness of the PVDF nanofiber surface improves the hydrophobicity of PVDF, as shown below.

Similar to neat PVDF films, neat PVDF nanofibrous mats have a large surface resistivity (R_s), which is out of the instrument range as shown in Figure 10.4a. Adding an IL into the PVDF precursor solution can lead to PVDF/IL composite nanofibers with enhanced electrical properties compared with the PVDF counterpart without IL. Moreover, the PVDF/IL composite nanofibrous mats also possess enhanced stretchability, as shown in Figure 10.4b, and exhibit good flexibility, as shown in the inset of Figure 10.4b. Compared with compact PVDF-based films, PVDF-based nanofibrous mats are known to have a large specific surface area and porous structure, as well as being lightweight. Therefore, neat PVDF in the form of nanofibers has a water contact angle (WCA) value of 140° (Figure 10.4c2), while a neat

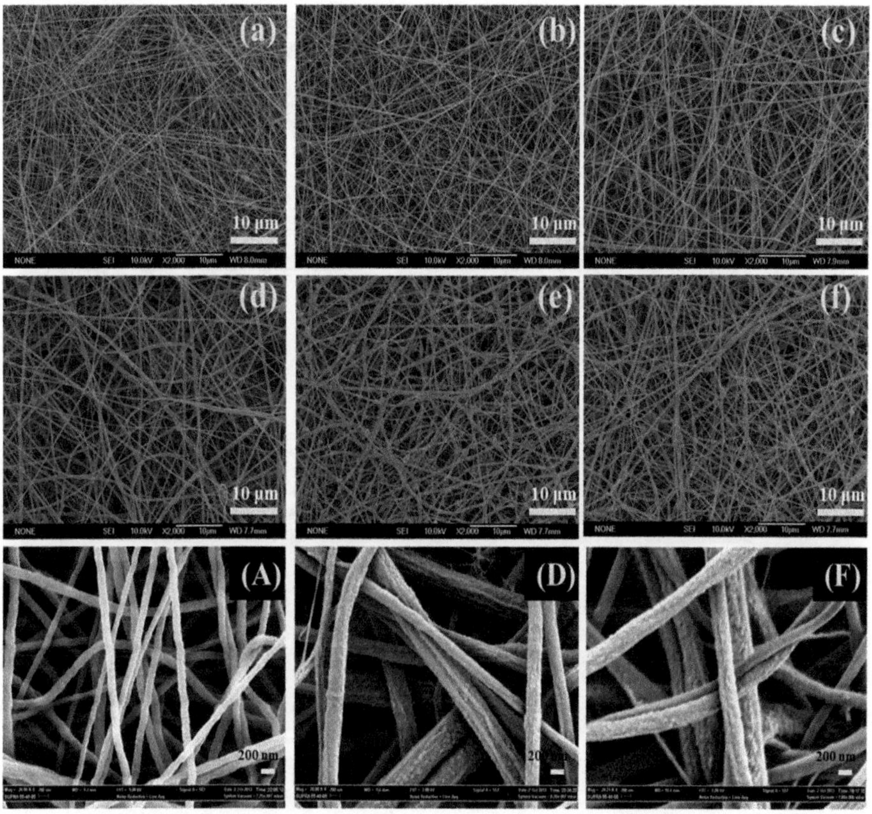

Figure 10.3 Field emission scanning electron microscopy (FESEM) micrographs of electrospun PVDF and PVDF/IL composite nanofibers with different IL contents: (a and A) 0 wt% (*i.e.* neat PVDF); (b) 5 wt%; (c) 10 wt%; (d and D) 20 wt%; (e) 30 wt%; (f and F) 40 wt%. Note that the micrographs in A, D, and F are high-magnification SEM images with a scale bar of 200 nm. Reprinted with permission from ref. 67. Copyright (2014) American Chemical Society.

PVDF film only exhibits a WCA value of 90° (Figure 10.4c1). The WCA value is increased to about 147° in the case of a PVDF/IL composite nanofibrous mat with 20% IL, as shown in Figure 10.4c3. The significantly increased WCA value can be attributed to the nano-size effect of the nanofibers and the good hydrophobicity of the IL. Due to the favourable hydrophobicity of the PVDF/IL-based nanofibrous mats, a mixture of CCl_4 and water can be effectively separated using these films. In Figure 10.4d1, the size of the water droplets in the CCl_4/water mixture is about 2–10 μm; after separation, as shown in Figure 10.4d2, the water content in the CCl_4 is as low as 45.6 ppm, which is slightly lower than the 48.2 ppm obtained by using a neat PVDF nanofibrous mat.

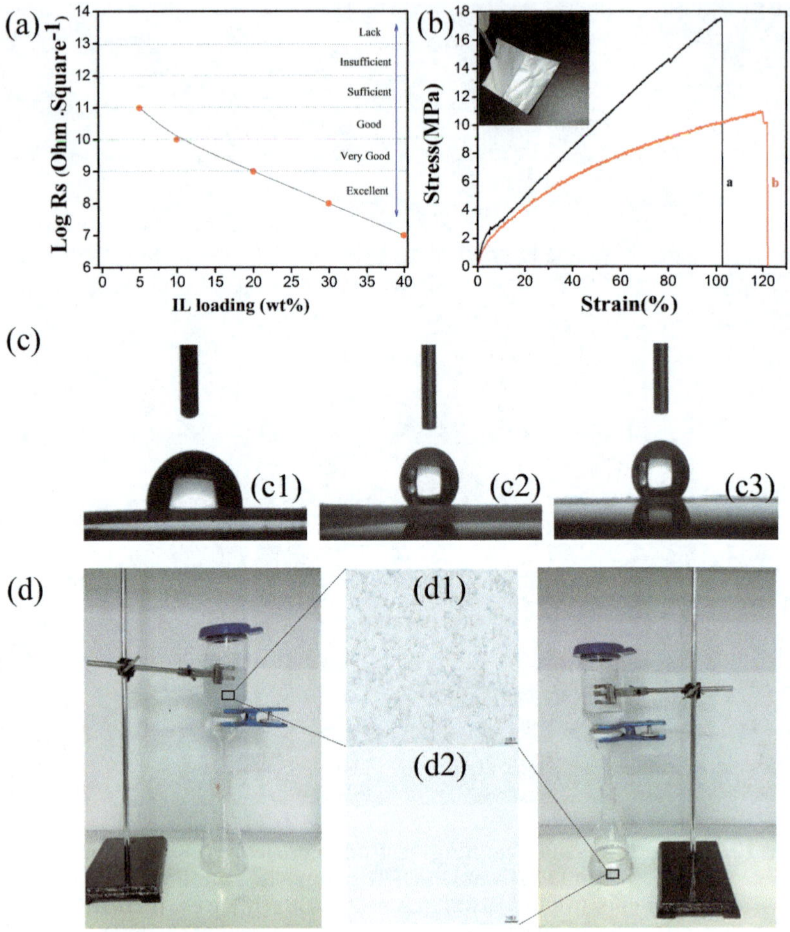

Figure 10.4 Physical properties of electrospun PVDF/IL nanofibrous mats: (a) surface resistivity (R_s) as a function of IL loading in the PVDF/IL nanofibrous mat; (b) strain–stress curves of neat PVDF (black) and the PVDF/IL nanofibrous mat with 20% IL (red); (c) water contact angle (WCA) of a neat PVDF hot-pressed film (c1), neat PVDF (c2) and the PVDF/IL nanofibrous mat with 20% IL loading (c3); (d) CCl_4–water separation experiment for the PVDF/IL nanofibrous mat with 20% IL loading, with optical microscope images of the CCl_4–water mixture before (d1) and after (d2) separation. Reprinted with permission from ref. 67. Copyright (2014) American Chemical Society.

10.2.3 Anti-static PVDF/IL–CNT Nanocomposites

Carbon nanotubes (CNTs) have gained considerable attention in recent years in the polymer modification research field due to their extraordinary unique properties, including outstanding electrical, thermal, transport, and mechanical properties. However, the poor compatibility of inorganic CNTs with organic polymers makes the strategy of integrating CNTs with polymers to fabricate conductive composites challenging. It has been reported that ILs can enable CNTs to be disentangled *via* a green physical interaction, that is, cation–π and/or π–π interactions.[55–58] Together with the good miscibility of PVDF with ILs, fabricating conductive PVDF–CNT-based composites with the aid of ILs is promising in this context. We have reported that IL-coated multi-walled carbon nanotubes (MWCNTs) (*i.e.* IL–CNT hybrids) can not only be dispersed homogeneously within a PVDF matrix, but also effectively induce 100% polar-PVDF crystals (mainly β phase) in the final PVDF/IL–CNT nanocomposites.[74] As shown in Figure 10.5, compared with raw CNTs (Figure 10.5a), the IL can not only disentangle the CNTs (Figure 10.5b) but also coat the surface of the CNTs forming IL layers (Figure 10.5c). As a result, the IL-modified CNTs can be well dispersed in the PVDF matrix (Figure 10.5e and f) relative to the PVDF/CNT sample without using an IL (Figure 10.5d), leading to the improvement of the electrical properties of PVDF. As for the PVDF crystal forms, it is believed that the IL-coated CNTs with the IL cations aligned on the CNT surface create long-range templates for PVDF all-trans conformations (*i.e.* the β phase) through interactions between the cations of the IL and the CF_2 groups of PVDF, as shown in Figure 10.1c.

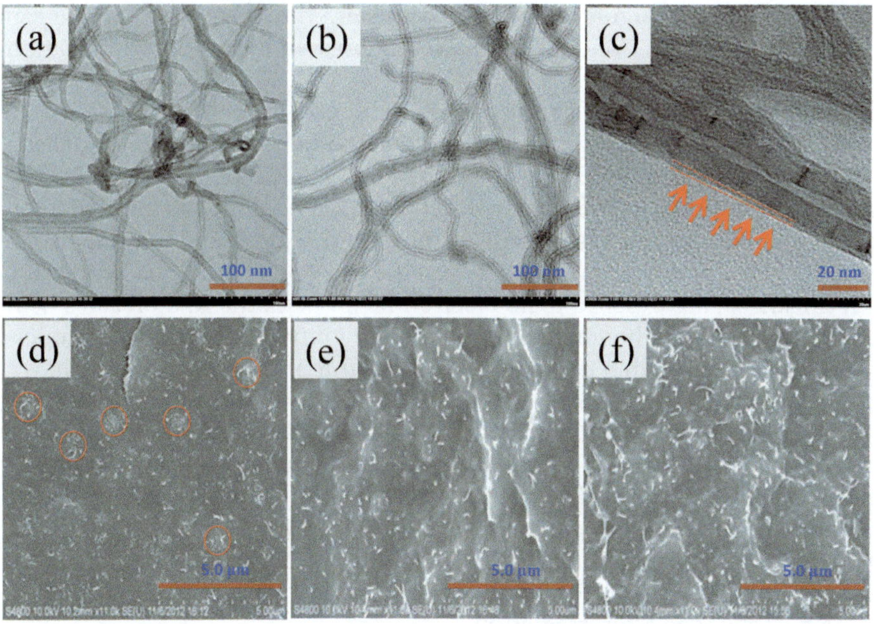

Figure 10.5 TEM images of raw multi-walled carbon nanotubes (MWCNTs) (a), IL-modified CNTs (IL–CNT: 1 : 5) (b) and the corresponding magnified image (c). SEM images of PVDF sample cross-sections: (d) PVDF/CNTs (100/2) (d), PVDF/IL–CNTs (100/10–2) (e) and PVDF/IL–CNTs (100/20–2) (f).

10.3 Dielectric PVDF/IL Composites

10.3.1 Formation of PVDF-*g*-IL Films

PVDF is of great interest in the fabrication of flexible dielectric devices as an important precursor due to its large dielectric constant, good flexibility and excellent processability. Generally, PVDF's dielectric performance can be further improved by increasing its dielectric constant, which can be realized by blending PVDF with inorganic nanofillers with a very high dielectric constant, such as dielectric ceramic, carbon nanotubes (CNTs), noble metal nanoparticles, *etc.* Like these rigid inorganic nanofillers, ILs have also been reported to exhibit great dielectric constants.[79,80] More importantly, ILs are soft and thermodynamically miscible with the PVDF matrix and therefore may be used as fillers to improve the permittivity of PVDF.[50] However, the small molecule characteristics of ILs, as well as their ionic structure, make this idea challenging due to the fact that the IL ions can generate significant loss in the PVDF matrix due to the movement of the ions under an external electric field.[72,73,81] The IL-induced loss in the PVDF/IL system is detrimental to the overall dielectric performance of PVDF. As shown in Figure 10.6, adding a small amount of IL into PVDF can largely increase the dielectric constant of PVDF. For example, in the case of the PVDF sample with 8% IL, the permittivity of PVDF can reach up to 52, which is ~8 fold that of neat PVDF with a value of 7.5 (Figure 10.6a). Unfortunately, the loss of this sample can reach up to 2.1, which should be avoided/depressed in practical application (Figure 10.6b).[72]

In order to reduce the ion-movement-induced dielectric loss of the IL in the PVDF matrix, our group employed a solid electron beam irradiation (EBI) method to immobilize the IL on the PVDF chains for the first time.[72] As shown in Figure 10.6c (left), in a typical PVDF/IL crystalline structure, IL molecules are only located in the PVDF's amorphous regions and they are miscible with these amorphous chains. When the PVDF/IL blend films (with a 300 μm thickness) are exposed to an electron beam field, the IL molecules with unsaturated double bonds in the cations are grafted onto the PVDF amorphous chains through a possible radical grafting polymerization, as shown in Figure 10.6c (right). PVDF-based macromolecular radicals are probably formed by breaking C–F, C–H and even C–C bonds with the bombardment of accelerated electrons during the irradiation procedure. It is noted that the electrons with an applied accelerated voltage can readily, homogeneously and completely penetrate the PVDF/IL films with 300 μm thickness. In other words, the IL grafting onto the PVDF chains is uniform and homogeneous. Moreover, the crystalline form, morphology, and crystal long period of PVDF remain unchanged after irradiation.[72]

After irradiation, the formed PVDF-*g*-IL films exhibited effectively reduced dielectric loss, as shown in Figure 10.6e. At the same time, due to the immobilization of the cations of the IL in the PVDF-*g*-IL films, the dielectric constant, along with the electrical properties, are also reduced, but still higher

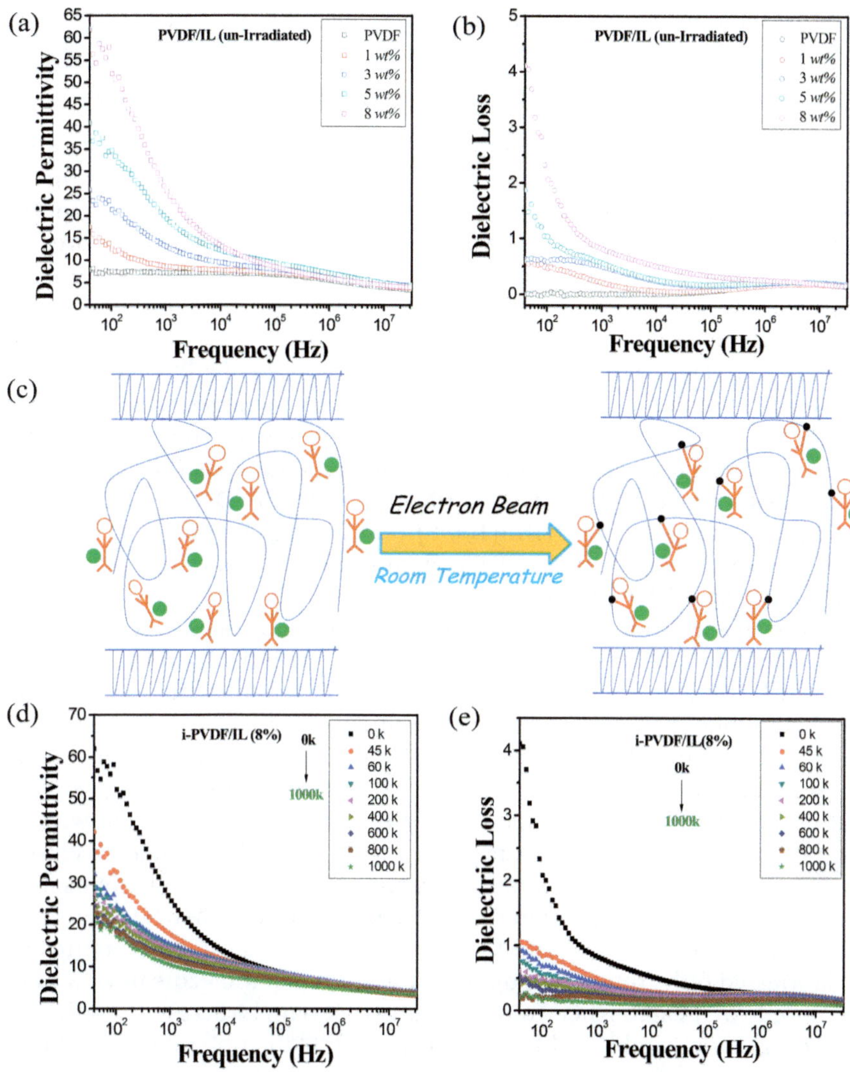

Figure 10.6 Dielectric properties and mechanism schematic of a PVDF-*g*-IL system. (a) and (b) Frequency dependence of the dielectric permittivity and loss tangent of miscible PVDF/IL blend films with 1, 3, 5 and 8% IL loading (*i.e.* un-irradiated samples), respectively; (c) the formation of PVDF-*g*-IL with the aid of solid electron beam irradiation at room temperature; (d) and (e) frequency dependence of the dielectric permittivity and loss tangent of PVDF-*g*-IL blend films with 8% IL loading at various absorbed doses, respectively. Reprinted with permission from ref. 72. Copyright (2015) American Chemical Society.

than those of neat PVDF, as shown in Figure 10.6d. Optimal IL loading in PVDF, as well as the optimal applied absorbed dose, can lead to PVDF-based dielectric composites with flexibility, high dielectric constant and low loss.[72]

10.3.2 Dielectric PVDF/IL Nanostructured Composites

The reported PVDF/IL-based composites in the literature are usually miscible and it is the miscibility between PVDF and the IL that can lead to intriguing physical properties, including good electrical conductivity, nice optical transparency and good mechanical properties, *etc.*[50,66–68] Furthermore, uniform structures can be observed in this kind of miscible PVDF/IL system.[50] In sharp contrast, our group has reported an interesting type of PVDF/IL composite with nanostructures, as shown in Figure 10.7.[73] Typically, these nanostructured PVDF/IL (100/8) composites are composed of a PVDF matrix with PVDF-*g*-IL nanodomains that are 20–30 nm in size homogeneously dispersed within the matrix (Figure 10.7a–c). Further elemental mapping analysis confirmed the chemical composition of the nanodomains (Figure 10.7d–f). Compared with neat PVDF (Figure 10.7g), miscible PVDF/IL (100/8) (Figure 10.7h) and irradiated PVDF/IL (100/8) (*i.e.* PVDF-*g*-IL composites) (Figure 10.7i), these nanostructured PVDF/IL composites can show properties different from those of the other three.

Figure 10.8 illustrates the formation of nanostructures based on PVDF and an IL. First, neat PVDF (state I) is melt-blended with the IL to form miscible IL-filled PVDF composites (state II). The PVDF/IL films with 300 μm thickness were then exposed to electron beam irradiation at room temperature in air. In this irradiation process, the IL's cations in the amorphous regions are grafted onto the PVDF amorphous chains *in situ* by grafting polymerization (state III).[72] The formed PVDF-*g*-IL chains are immiscible with the ungrafted PVDF chains and subsequent melting of the irradiated samples can result in microphase separation, forming PVDF-*g*-IL nanodomains in the melt (state IV). Nanostructured PVDF/IL composites can then be obtained by a crystallization process (state V). It is interesting that nanostructures can be fabricated only from a homopolymer and a small molecule in view of the commonly well-established routes to construct nanostructures in polymers, including self-assembly of block copolymers,[82–84] melt-blending[85] and high-shear processing.[86,87]

When it comes to dielectric properties, the nanostructured PVDF/IL (100/8) composite shows higher performance relative to other PVDF-based composites. As shown in Figure 10.9a, an increased dielectric constant is observed in the case of the nanostructured PVDF/IL composites compared with neat PVDF. Moreover, the nanostructured sample also has a lower dielectric loss than that of the PVDF/IL (100/8) and irradiated PVDF/IL (100/8) samples, as shown in Figure 10.9b. This is due to the further confinement of the IL ions in the PVDF-*g*-IL nanodomains. What is more, the obtained nanostructured PVDF/IL sample exhibited excellent mechanical properties and flexibility, as shown in Figure 10.9c and d, respectively.

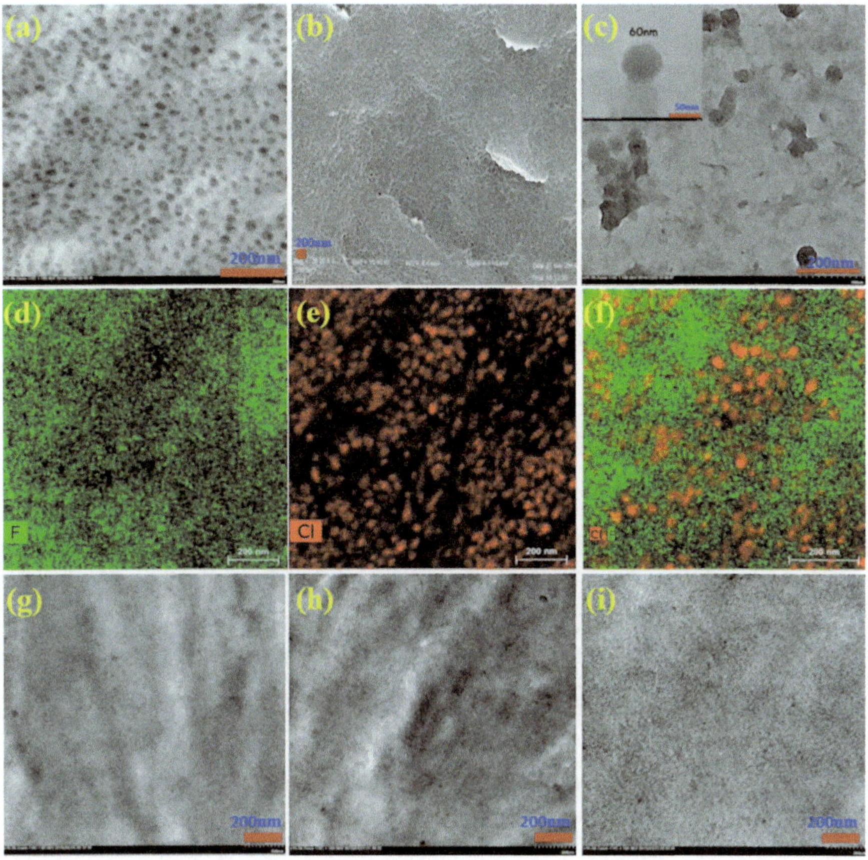

Figure 10.7 Morphologies of nanostructured PVDF/IL (100/8) samples (a–f) and their counterparts, including neat PVDF (g), a PVDF/IL (100/8) blend (h) and irradiated PVDF/IL (100/8) (*i.e.* a PVDF-*g*-IL composite) at an absorbed dose of 45 kGy (i). (a), (c) and (g–i) Transmission electron microscopy (TEM) images; (b) high-resolution scanning electron microscopy (HRSEM) image; (d–f) elemental mapping images (EMI). Reprinted with permission from ref. 73. Copyright (2015) American Chemical Society.

10.3.3 Block-like Copolymers of PVDF-*g*-IL Chains and Their Microphase Separation Behaviours

From the chain structure point of view, neat PVDF, PVDF/IL blends and irradiated PVDF/IL blends were compared with each other, as shown in Figure 10.10. Just like other semi-crystalline polymers, neat PVDF is comprised of a crystalline region, amorphous regions and the interfaces between them (Figure 10.10a). Due to the presence of tie molecules that bridge the crystalline region and amorphous regions, neat PVDF can be regarded as a block-like copolymer with structures of crystalline PVDF-*b*-amorphous PVDF-*b*-crystalline

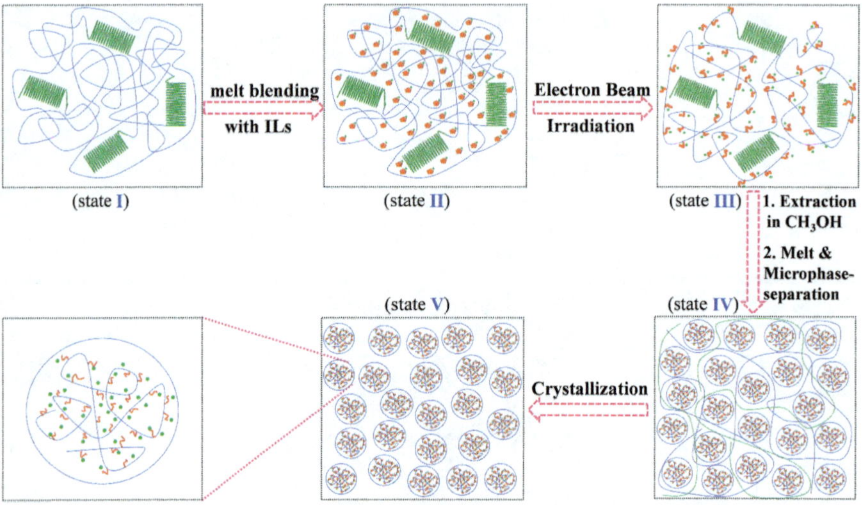

Figure 10.8 Formation of nanostructures originating from homopolymer PVDF and a small molecule ionic liquid. The typical PVDF/IL-based states in the preparation procedure include: neat PVDF (state I), miscible PVDF/IL blends (state II), irradiated PVDF/IL blends, namely, PVDF-*g*-IL composites (state III), and nanostructured PVDF/IL composites in the melted (state IV) and crystallized states (state V). Reprinted with permission from ref. 73. Copyright (2015) American Chemical Society.

PVDF chains (Figure 10.10b). When blended with an IL, the PVDF block-like copolymer chain structures remain unchangeable, as shown in Figure 10.10c and d, due to the fact that PVDF interacts with the IL through static physical interactions.[50] However, in the case of irradiated PVDF/IL samples (Figure 10.10e and f), a block-like copolymer with the structure of crystalline PVDF-*b*-amorphous PVDF-*g*-IL-*b*-crystalline PVDF chains is formed. Other possible chains of PVDF-*g*-IL are also shown in Figure 10.10g and f. More importantly, the amorphous PVDF-*g*-IL chains are immiscible with the crystalline PVDF (*i.e.* ungrafted PVDF chains, Figure 10.10i) and microphase separation of the PVDF-*g*-IL chains can then occur in the PVDF matrix in the melt, thus creating nanostructures.[70] Such a simple method of producing a block-like copolymer is greatly different from the traditional radical polymerization route. Therefore, it may be safe to draw the conclusion that a common crystalline polymer and its miscible unsaturated ionic liquid can be readily fabricated into conductive nanostructures using our strategy.[70]

To deeply understand the microphase separation of these block-like copolymers in the melt, *in-situ* SAXS measurements were performed, as shown in Figure 10.11.[70] Neat PVDF (Figure 10.11A-a1), irradiated PVDF/IL (100/10) (Figure 10.11A-b1) and irradiated PVDF/IL (100/20) (Figure 10.11A-c1) exhibited crystal scattering signals at 80 °C. In the melt, neat PVDF only shows signals of the beam stopper and air (Figure 10.11A-a2), just like air (Figure

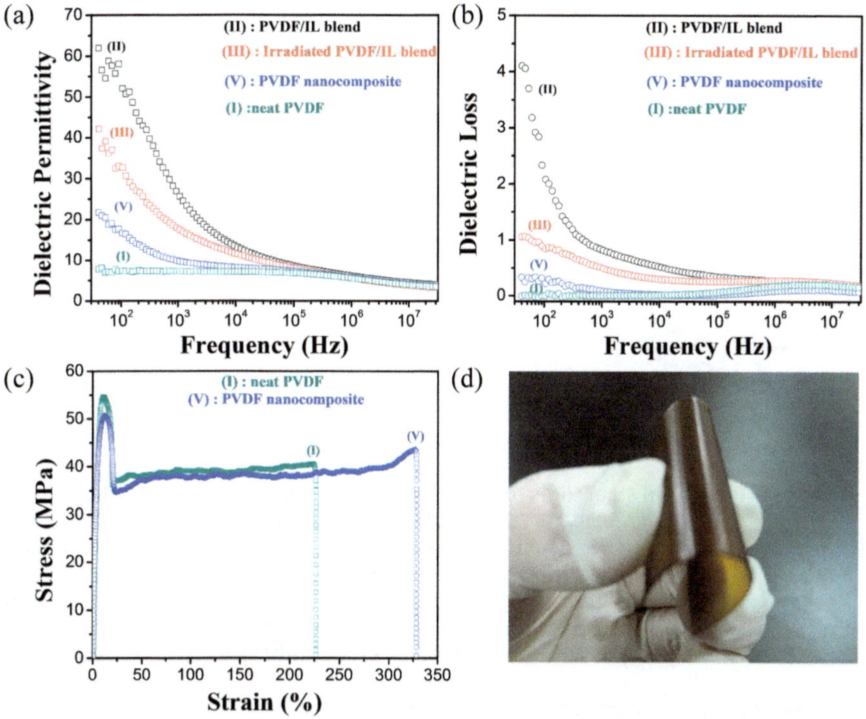

Figure 10.9 Physical properties of the nanostructured PVDF/IL (100/8) composite and its counterparts including neat PVDF, a miscible PVDF/IL (100/8) blend and an irradiated PVDF/IL (100/8) blend. (a) Frequency dependency of the dielectric permittivity; (b) frequency dependency of the loss tangent; (c) strain–stress curves and (d) a photo of the nanostructured PVDF/IL (100/8) composite. Reprinted with permission from ref. 73. Copyright (2015) American Chemical Society.

10.11A-a3), suggesting the homogeneous phase of neat PVDF. However, in the case of the irradiated PVDF/IL samples, small and clear scattering signals can be observed, as shown in Figure 10.11A-b2 and Figure 10.11A-c2, respectively. Such scattering signals originate from the PVDF-*g*-IL nanodomains in the melt (200 °C), and they can also be reflected in both the $I \sim q$ and $Iq^2 \sim q$ curves, as shown in Figure 10.11B and C, respectively.

Figure 10.12a and b show that the occurrence of PVDF-*g*-IL nanodomains in the melt is accompanied by a gradual increase in the scattering intensity with time. Moreover, the q values of the PVDF-*g*-IL nanodomains are gradually decreased with time, and then reached a steady state (Figure 10.12c). This phenomenon means that the size of the PVDF-*g*-IL nanodomains increased with time, which leads to a possible mechanism regarding the evolution of the PVDF-*g*-IL nanodomains, as shown in Figure 10.12d–g. Typically, irradiated PVDF/IL blends (*i.e.* PVDF-*g*-IL composites) are composed of PVDF

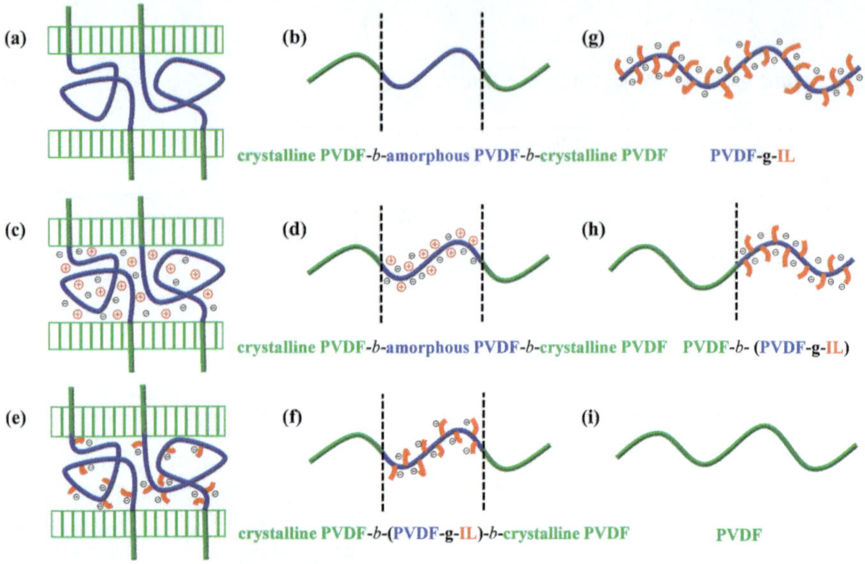

Figure 10.10 Chain structures of neat PVDF (a and b), miscible PVDF/IL (c and d), irradiated PVDF/IL, namely, PVDF-*g*-IL, (e–h), and ungrafted PVDF chains (i). Reproduced with permission from C. Xing, J. Li, C. Yang, Y. Li, Local Grafting of Ionic Liquid in Poly(vinylidene fluoride) Amorphous Region and the Subsequent Microphase Separation Behaviour in Melt, *Macromolecular Rapid Communications*, John Wiley and Sons, Copyright © 2016 WILEY-VCH Verlag GmbH & Co. KGaA, Weinheim.[70]

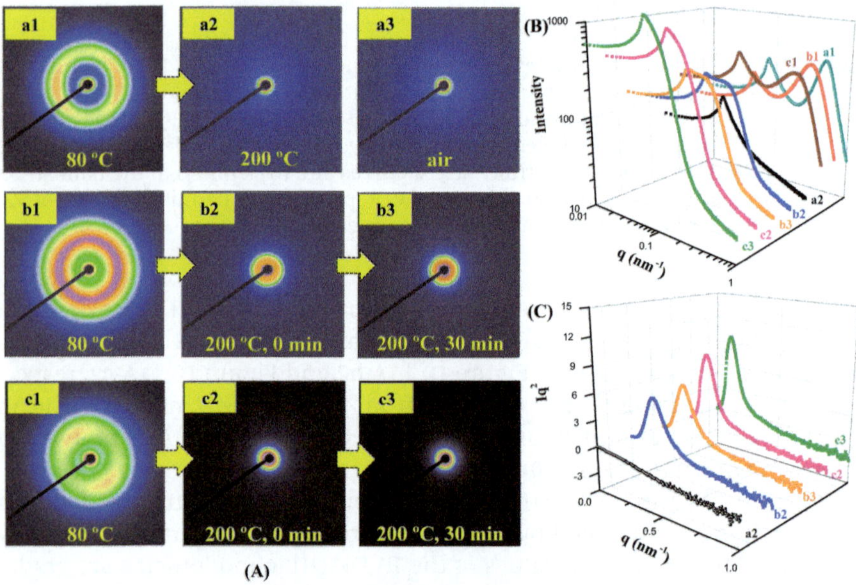

Figure 10.11 (A) SAXS patterns of neat PVDF (a1 and a2), air (a3), irradiated PVDF/IL (100/10) (b1–b3), and irradiated PVDF/IL (100/20) (c1–c3) at certain temperatures for different annealing times. (B) Intensity (I) ~ q curves; (C) Iq^2 ~ q curves. Reproduced with permission from C. Xing, J. Li, C. Yang, Y. Li, Local Grafting of Ionic Liquid in Poly(vinylidene fluoride) Amorphous Region and the Subsequent Microphase Separation Behaviour in Melt, *Macromolecular Rapid Communications*, John Wiley and Sons, Copyright © 2016 WILEY-VCH Verlag GmbH & Co. KGaA, Weinheim.[70]

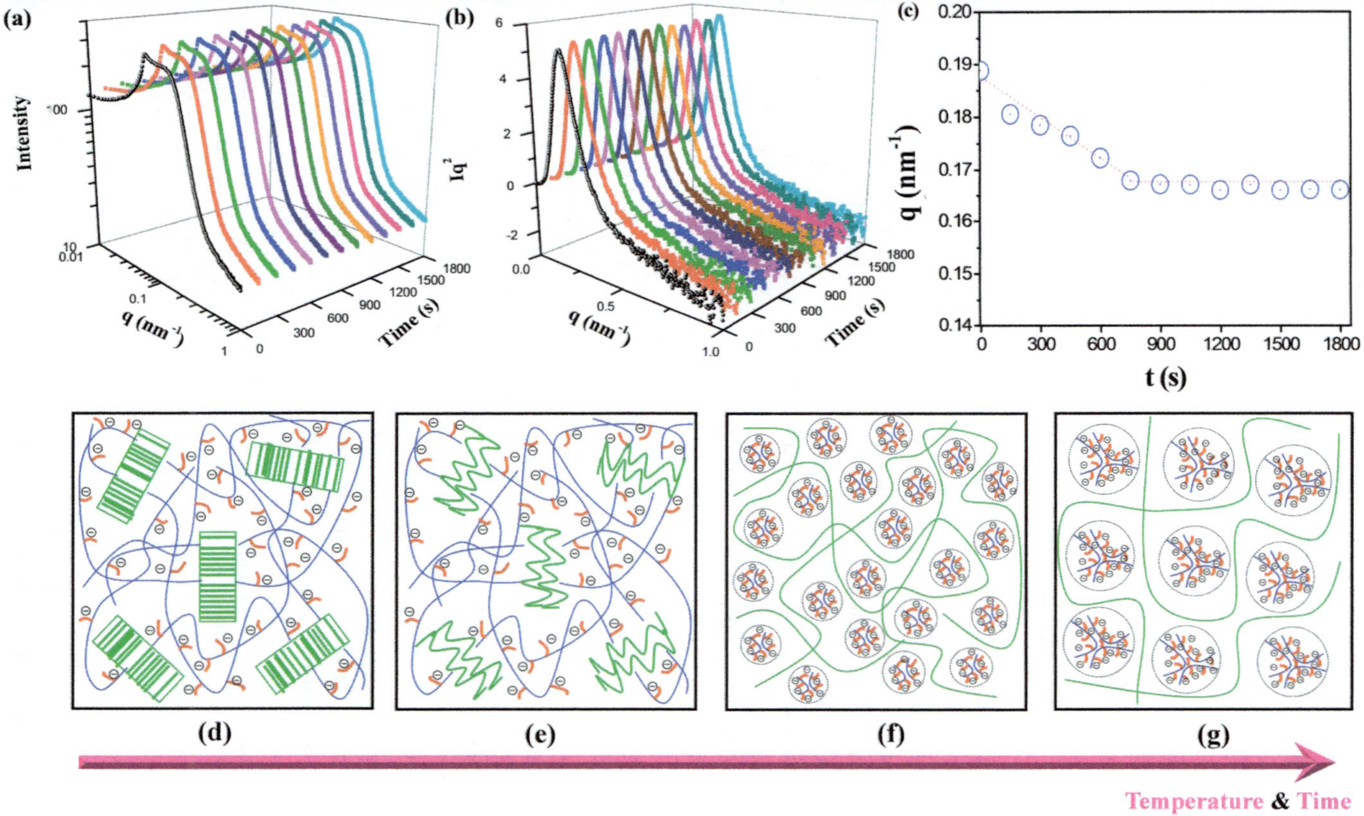

Figure 10.12 The evolution of PVDF-*g*-IL nanodomains in the melt (200 °C) with time: (a) $I \sim q$ curves; (b) $Iq^2 \sim q$ curves; (c) $q_{max} \sim$ time (t) curves, and the corresponding mechanism (d–g): (d) the crystalline structure of the PVDF-*g*-IL sample; (e) the melt of PVDF crystals with unchangeable PVDF-*g*-IL chains; (f) the nucleation of PVDF-*g*-IL nanodomains when the PVDF crystals are totally melted; (g) the growth of PVDF-*g*-IL nanodomains with time in the melt. Reproduced with permission from C. Xing, J. Li, C. Yang, Y. Li, Local Grafting of Ionic Liquid in Poly(vinylidene fluoride) Amorphous Region and the Subsequent Microphase Separation Behaviour in Melt, *Macromolecular Rapid Communications*, John Wiley and Sons, Copyright © 2016 WILEY-VCH Verlag GmbH & Co. KGaA, Weinheim.[70]

crystals, PVDF amorphous chains and PVDF-*g*-IL chains, especially tie molecules, in the amorphous region (Figure 10.12d). The microphase separation of the PVDF-*g*-IL chains only occurred when the PVDF crystals were totally fused because the PVDF crystal chains can strongly restrict the movement of the PVDF-*g*-IL chains (Figure 10.12e), and PVDF-*g*-IL nanodomains were thus formed (Figure 10.12f). The size of these nanodomains increased with time before reaching a steady state (Figure 10.12g). It is therefore concluded that the formation of PVDF-*g*-IL nanodomains probably occurs *via* nucleation and growth procedures, which can help in understanding the creation of nanostructures based on PVDF and ILs.[70]

10.3.4 Dielectric PVDF/IL–CB Nanocomposites

It is well-established that integrating inorganic nanofillers possessing great dielectric constants with polymers can give rise to flexible dielectric nanocomposites with excellent dielectric performances. Neither of them can ever meet the requirements for practical applications due to the relatively low dielectric constant for raw polymers and the lack of flexibility of inorganic nanofillers. At this point, dielectric ceramics (for instance, $BaTiO_3$) and conductive nanoparticles, such as carbon nanotubes (CNTs) and graphene, are of great effectiveness for enhancing the dielectric performance of polymers.[27–46] Moreover, given their very small loading levels within the polymer matrix, conductive nanoparticles have recently gained considerable attention.[41] However, the poor compatibility of inorganic nanofillers with polymers is very common in these systems, resulting in poor dispersion of the nanofillers with large domain sizes, considerable dielectric loss, and low breakdown strength. Therefore, the dispersion issue of nanoparticles in polymer matrices should be first addressed when it comes to integrating inorganic nanofillers into polymers to fabricate high dielectric constant nanocomposites.

On the other hand, when it comes to improving the dispersion of conductive nanofillers in polymers, especially CNTs, ionic liquids (ILs) can be regarded as promising candidates for two reasons: (1) ILs can interact with CNTs *via* physical cation–π and/or π–π interactions between the cations of the IL and the electron-rich structures of the CNTs;[55–58] (2) the good miscibility of ILs with certain polymers is found in some systems, such as PVDF/IL,[50,67,74] PMMA/IL,[51,52] TPU/IL,[61] *etc.* It is therefore feasible to disperse conductive nanofillers by using ILs. Despite these interesting findings, PVDF/IL systems suffer from considerable dielectric loss that is caused by the movement of the ions of the IL in an electric field, as also discussed in Section 10.2.[72,73]

Very recently, we have demonstrated interesting PVDF nanocomposites with simultaneous carbon black (CB) nanoparticles and PVDF-*g*-IL nanodomains (termed as nano-PVDF/IL–CB composites).[71] For the nano-PVDF/IL–CB composites, the use of an IL can help to disperse the CB dispersion within the PVDF matrix, and the cations of the IL are grafted onto the PVDF chains, and subsequently, together with the anions of the IL, are confined

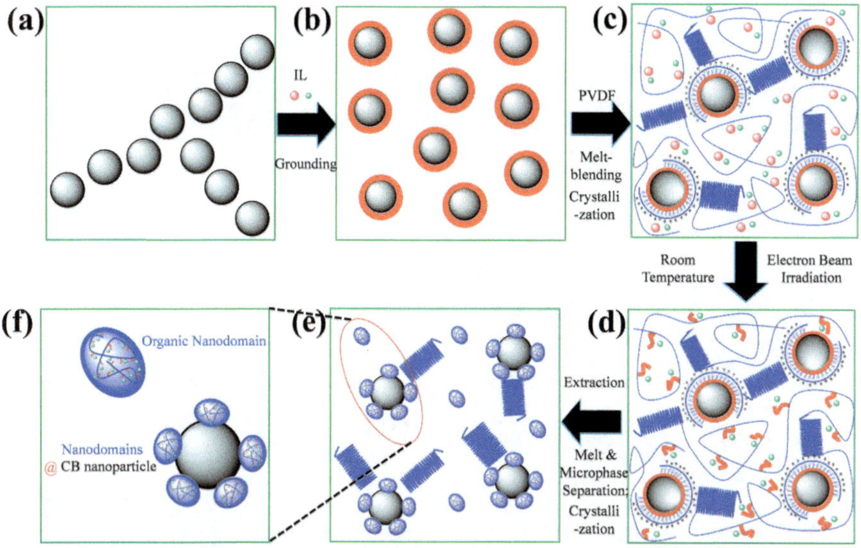

Figure 10.13 Fabrication of PVDF nanocomposites with simultaneous carbon black (CB) nanoparticles and PVDF-*g*-IL nanodomains (termed as nano-PVDF/IL–CB composites). (a) Raw CB nanoparticle; (b) IL-coated CB nanoparticles (termed as IL–CB) prepared by grounding CB with the IL; (c) the PVDF/IL–CB (100/16–2) nanocomposite; (d) the irradiated PVDF/IL–CB (100/16–2) nanocomposite at 45 kGy; (e) the final nano-PVDF/IL–CB (100/16–2) composites with morphologies of nanodomains@CB nanoparticles (f). Reprinted with permission from ref. 71. Copyright (2016) American Chemical Society.

within PVDF-*g*-IL nanodomains, which depresses the loss of IL considerably. Figures 10.13 and 10.14 depict the fabrication of the nano-PVDF/IL–CB composites and their corresponding microstructures, respectively. CB is incompatible with PVDF and large domains of CB can be observed in the PVDF matrix. In order to disperse CB homogeneously within the PVDF matrix, an IL was first ground with CB (Figure 10.13a) to form an IL–CB bulky gel (Figure 10.13b) and the as-prepared IL–CB was melt-blended with the PVDF matrix (Figure 10.13c). In physically mixed PVDF/IL–CB nanocomposites (Figure 10.13c), the IL molecules are free, partially coating the surface of the CB with others being miscible with the PVDF molecules. To suppress the dielectric loss of the IL, the IL was chemically grafted onto the amorphous PVDF chains by electron beam irradiation (Figure 10.13d), generating block-like chains of crystalline PVDF-*b*-(amorphous PVDF-*g*-IL)-*b*-crystalline PVDF, just like the binary PVDF/IL system under irradiation treatment. Microphase separation of the PVDF-*g*-IL chains from the ungrafted PVDF chains occurred in the presence of homogeneously dispersed CB nanoparticles in a subsequent melting procedure for the irradiated PVDF/IL–CB nanocomposites. Nano-PVDF/IL–CB composites can then be obtained *via* a crystallization process. Compared with binary PVDF/CB (Figure 10.14a) and PVDF/IL

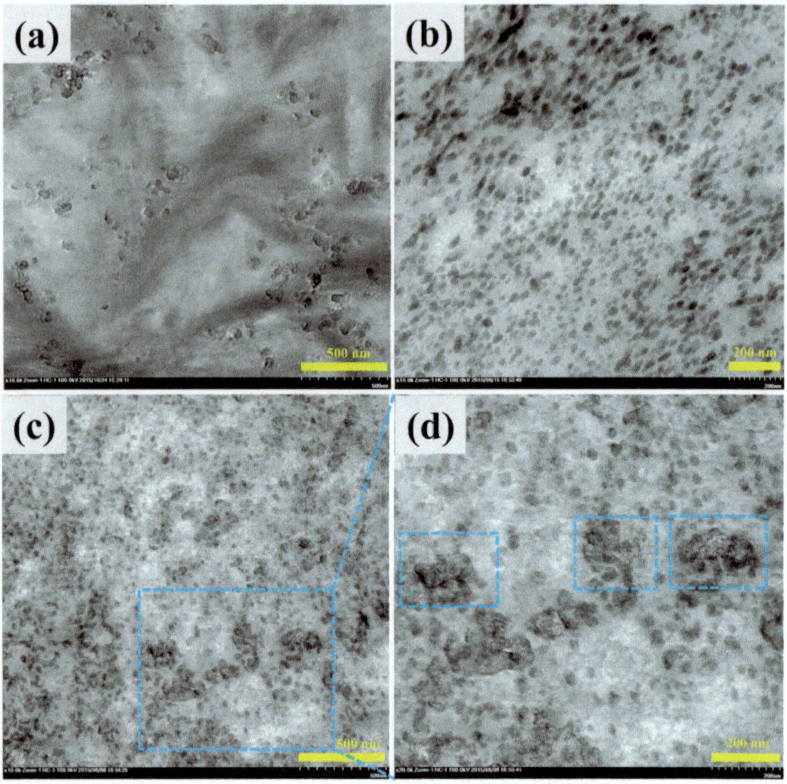

Figure 10.14 TEM images of (a) PVDF/CB (100/2), (b) nanostructured PVDF/IL (100/16), and (c) nano-PVDF/IL–CB (100/16–2) composites with nanodomain@CB nanoparticle morphologies (d). Reprinted with permission from ref. 71. Copyright (2016) American Chemical Society.

(Figure 10.14b), the nano-PVDF/IL–CB composites simultaneously contain inorganic CB nanoparticles with good dispersion and organic PVDF-*g*-IL nanodomains (Figure 10.14c). Interestingly, the CB nanoparticles adhered to the PVDF-*g*-IL nanodomains, forming nanodomain@CB nanoparticle structures (Figure 10.14d).

Figure 10.15 shows the dielectric properties of neat PVDF, binary PVDF/CB, binary PVDF/IL, nanostructured PVDF/IL and the nano-PVDF/IL–CB composites, respectively. Neat PVDF exhibits both a low dielectric constant and a low loss tangent, although it has excellent flexibility. Both CB and an IL can largely improve the dielectric constant of PVDF by enhancing the average electric field intensity and through dipole movements of the IL, respectively, as shown in Figure 10.15a. However, the poor dispersion of CB in the PVDF matrix results in a considerable leakage current in the binary PVDF/CB system, leading to severe dielectric loss (Figure 10.14b). Owing to the movement of the ions in the IL, binary PVDF/IL blend samples also exhibit serious dielectric loss (Figure 10.14b). These large loss tangents are detrimental for energy storage in dielectric materials. In our

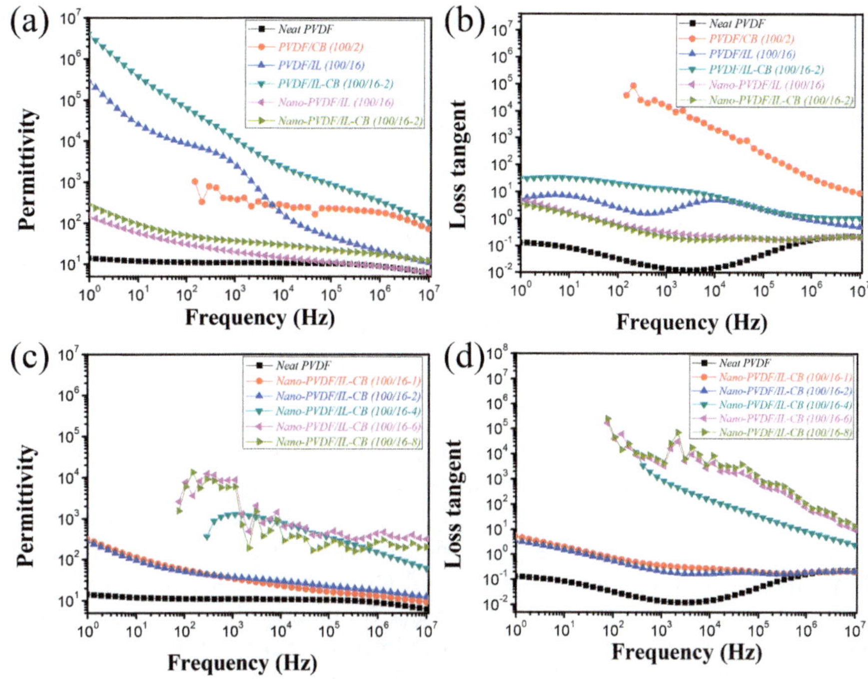

Figure 10.15 Dielectric properties of PVDF-based samples. (a) and (c) Frequency dependency of the dielectric permittivity; (b) and (d) frequency dependency of the loss tangent. Reprinted with permission from ref. 71. Copyright (2016) American Chemical Society.

work,[71] the CB agglomeration-induced loss can be suppressed by using an IL to help disperse CB in the PVDF matrix, and the IL ion movement-induced loss can be effectively decreased by confining the IL ions within PVDF-*g*-IL nanodomains. Therefore, the nano-PVDF/IL–CB composites with the suppression of these two kinds of losses exhibited the minimal dielectric loss relative to binary PVDF/CB, miscible PVDF/IL blends, and ternary PVDF/IL–CB nanocomposites, as shown in Figure 10.15b. More interestingly, it was found that nano-PVDF/IL–CB (100/16–2) and nanostructured PVDF/IL (100/16) show almost identical dielectric loss values, suggesting the complete suppression of the dielectric loss of CB, which is rare in the reported literature on PVDF/conductive nanofiller systems (Figure 10.15b). Additionally, further increases in the CB content in the nano-PVDF/IL–CB samples not only increase the dielectric constant of PVDF (Figure 10.15c), but also increase the dielectric loss of the system (Figure 10.15d) by forming conductive CB networks in the PVDF matrix. Therefore, the appropriate CB content in the PVDF matrix should be considered. On the other hand, the unique nanodomain@CB nanoparticle structures not only hinder the direct connection of the CB nanoparticles, but also provide a good PVDF–CB interface, which is responsible for the good dielectric performance of the nano-PVDF/IL–CB composites.

10.4 Conclusion and Outlook

PVDF-based materials modified by ILs were introduced in this chapter. It was demonstrated that PVDF is miscible with ILs through a physical static interaction between the cations of the IL and the negatively charged CF_2 groups of PVDF. Furthermore, IL molecules are only located in the amorphous regions of PVDF and are miscible with these amorphous chains. As a result, miscible PVDF/IL blend films, PVDF/IL fibrous nanofibers and PVDF/IL–CNT nanocomposites can be fabricated using a physical mixing method. Using this good miscibility of PVDF and ILs, along with a radiation grafting chemical strategy, IL-grafted PVDF composites (PVDF-*g*-IL) with block-like structures of crystalline PVDF-*b*-(amorphous PVDF-*g*-IL)-*b*-crystalline PVDF chains can be created, the synthesis route of which is totally different from the traditional radical polymerization used to synthesize block copolymers. Microphase separation of the PVDF-*g*-IL composites resulted in nanostructured PVDF composites with conductive PVDF-*g*-IL nanodomains. Based on the nanostructures in the PVDF/IL system, when CB nanoparticles were integrated, PVDF nanocomposites with simultaneous organic PVDF-*g*-IL nanodomains and homogeneously dispersed CB nanoparticles were produced. The microstructures of nanodomains@CB nanoparticles in these PVDF nanocomposites are responsible for the improvement in the dielectric performance of PVDF. The modification of PVDF by using an IL, either through a physical or chemical method, described in this chapter may offer opportunities to develop other common polymers in the future, and a large amount of work to gain a deep understanding regarding polymer–IL systems, especially PVDF–IL systems, should be done. Further investigations are needed to enable the use of these materials in real applications in industry.

References

1. A. J. Lovinger, *Science*, 1983, **220**(4602), 1115–1121.
2. P. Martins, A. C. Lopes and S. Lanceros-Mendez, *Prog. Polym. Sci.*, 2014, **39**(4), 683–706.
3. Z. Cui, N. T. Hassankiadeh, Y. Zhuang, E. Drioli and Y. M. Lee, *Prog. Polym. Sci.*, 2015, **51**, 94–126.
4. Z. Cui, E. Drioli and Y. M. Lee, *Prog. Polym. Sci.*, 2014, **39**(1), 164–198.
5. L. Persano, A. Camposeo and D. Pisignano, *Prog. Polym. Sci.*, 2015, **43**, 48–95.
6. R. Gregorio, *J. Appl. Polym. Sci.*, 2006, **100**(4), 3272–3279.
7. R. Hasegawa, Y. Takahashi, Y. Chatani and H. Tadokoro, *Polym. J.*, 1972, **3**(5), 600–610.
8. H. Guo, Y. Zhang, F. Xue, Z. Cai, Y. Shang, J. Li, Y. Chen, Z. Wu and S. Jiang, *CrystEngComm*, 2013, **15**(8), 1597–1606.
9. B. S. Ince-Gunduz, R. Alpern, D. Amare, J. Crawford, B. Dolan, S. Jones, R. Kobylarz, M. Reveley and P. Cebe, *Polymer*, 2010, **51**(6), 1485–1493.
10. D. Mandal, K. J. Kim and J. S. Lee, *Langmuir*, 2012, **28**(28), 10310–10317.

11. A. C. Lopes, C. M. Costa, C. J. Tavares, I. C. Neves and S. Lanceros-Mendez, *J. Phys. Chem. C*, 2011, **115**(37), 18076–18082.
12. J. Yang, J. Wang, Q. Zhang, F. Chen, H. Deng, K. Wang and Q. Fu, *Polymer*, 2011, **52**(21), 4970–4978.
13. J. Yang, Q. Chen, F. Chen, Q. Zhang, K. Wang and Q. Fu, *Nanotechnology*, 2011, **22**(35), 355707.
14. A. C. Lopes, C. Caparros, S. Ferdov and S. Lanceros-Mendez, *J. Mater. Sci.*, 2013, **48**(5), 2199–2206.
15. S. P. Bao, G. D. Liang and S. C. Tjong, *Carbon*, 2011, **49**(5), 1758–1768.
16. T. Miyazaki and Y. Takeda, *Eur. Polym. J.*, 2014, **61**, 1–12.
17. S. Garain, S. Jana, T. K. Sinha and D. Mandal, *ACS Appl. Mater. Interfaces*, 2016, **8**(7), 4532–4540.
18. J. Chang, M. Dommer, C. Chang and L. Li, *Nano Energy*, 2012, **1**(3), 356–371.
19. N. Levi, R. Czerw, S. Xing, P. Iyer and D. L. Carroll, *Nano Lett.*, 2004, **4**(7), 1267–1271.
20. G. H. Kim, S. M. Hong and Y. Seo, *Phys. Chem. Chem. Phys.*, 2009, **11**(44), 10506–10512.
21. S. Manna and A. K. Nandi, *J. Phys. Chem. C*, 2007, **111**(40), 14670–14680.
22. Q. Liu, J. Tu, X. Wang, W. Yu, W. Zheng and Z. Zhao, *Carbon*, 2012, **50**(1), 339–341.
23. G. X. Chen, Y. Li and H. Shimizu, *Carbon*, 2007, **45**(12), 2334–2340.
24. P. Costa, J. Silva, V. Sencadas, C. M. Costa, F. W. J. Van Hattum, J. G. Rocha and S. Lanceros-Méndez, *Carbon*, 2009, **47**(11), 2590–2599.
25. Y. Li, X. Huang, Z. Hu, P. Jiang, S. Li and T. Tanaka, *ACS Appl. Mater. Interfaces*, 2011, **3**(11), 4396–4403.
26. W. Zhang, Z. Zhang, J. Yang, T. Huang, N. Zhang, X. Zheng, Y. Wang and Z. Zhou, *Carbon*, 2015, **90**, 242–254.
27. F. He, S. Lau, H. L. Chan and J. Fan, *Adv. Mater.*, 2009, **21**(6), 710–715.
28. Z. Wang, M. Fang, H. Li, Y. Wen, C. Wang and Y. Pu, *Compos. Sci. Technol.*, 2015, **117**, 410–416.
29. Y. N. Hao, X. H. Wang, S. Obrien, J. Lombardi and L. T. Li, *J. Mater. Chem. C*, 2015, **3**(37), 9740–9747.
30. Y. Hou, Y. Deng, Y. Wang and H. Gao, *RSC Adv.*, 2015, **5**(88), 72090–72098.
31. S. Liu, S. Xiao, S. Xiu, B. Shen, J. Zhai and Z. An, *RSC Adv.*, 2015, **5**(51), 40692–40699.
32. S. Liu, S. Xiu, B. Shen, J. Zhai and L. Kong, *Polymers*, 2016, **8**(2), 45.
33. M. Zhu, X. Huang, K. Yang, X. Zhang, J. Zhang, J. He and P. Jiang, *ACS Appl. Mater. Interfaces*, 2014, **6**(22), 19644–19654.
34. X. Huo, W. Li, J. Zhu, L. Li, Y. Li, L. Luo and Y. Zhu, *J. Phys. Chem. C*, 2015, **119**(46), 25786–25791.
35. S. Luo, S. Yu, R. Sun and C. Wong, *ACS Appl. Mater. Interfaces*, 2014, **6**(1), 176–182.
36. Y. Li, J. Yuan, J. Xue, F. Cai, F. Chen and Q. Fu, *Compos. Sci. Technol.*, 2015, **118**, 198–206.

37. Y. Song, Y. Shen, P. Hu, Y. Lin, M. Li and C. W. Nan, *Appl. Phys. Lett.*, 2012, **101**(15), 152904.
38. S. Liu, J. Zhai, J. Wang, S. Xue and W. Zhang, *ACS Appl. Mater. Interfaces*, 2014, **6**(3), 1533–1540.
39. S. Liu, S. Xue, B. Shen and J. Zhai, *Appl. Phys. Lett.*, 2015, **107**(3), 032907.
40. D. Wang, Y. Bao, J. Zha, J. Zhao, Z. Dang and G. Hu, *ACS Appl. Mater. Interfaces*, 2012, **4**(11), 6273–6279.
41. Z. M. Dang, M. S. Zheng and J. W. Zha, *Small*, 2016, **12**(13), 1688–1701.
42. Z. M. Dang, H. Y. Wang, Y. H. Zhang and J. Q. Qi, *Macromol. Rapid Commun.*, 2005, **26**(14), 1185–1189.
43. B. Satish, K. Sridevi and M. S. Vijaya, *J. Phys. D: Appl. Phys.*, 2002, **35**, 2048–2050.
44. X. Huang, P. Jiang and L. Xie, *Appl. Phys. Lett.*, 2009, **95**(24), 242901.
45. L. Zhang, S. Yuan, S. Chen, D. Wang, B. Han and Z. Dang, *Compos. Sci. Technol.*, 2015, **110**, 126–131.
46. Z. M. Dang, Y. H. Lin and C. W. Nan, *Adv. Mater.*, 2003, **15**(19), 1625–1629.
47. D. Mecerreyes, *Applications of Ionic Liquids in Polymer Science and Technology [M]*, Springer, 2015.
48. J. Lu, F. Yan and J. Texter, *Prog. Polym. Sci.*, 2009, **34**(5), 431–448.
49. A. Taubert, *Eur. J. Inorg. Chem.*, 2015, **2015**(7), 1148–1159.
50. C. Y. Xing, M. Zhao, L. Zhao, J. You, X. Cao and Y. J. Li, *Polym. Chem.*, 2013, **4**(24), 5726–5734.
51. M. P. Scott, C. S. Brazel, M. G. Benton, J. W. Mays, J. D. Holbrey and R. D. Rogers, *Chem. Commun.*, 2002, **13**, 1370–1371.
52. M. P. Scott, M. Rahman and C. S. Brazel, *Eur. Polym. J.*, 2003, **39**(10), 1947–1953.
53. M. Rahman and C. S. Brazel, *Polym. Degrad. Stab.*, 2006, **91**(12), 3371–3382.
54. K. I. Park and M. Xanthos, *Polym. Degrad. Stab.*, 2009, **94**(5), 834–844.
55. T. Fukushima, A. Kosaka, Y. Ishimura, T. Yamamoto, T. Takigawa, N. Ishii and T. Aida, *Science*, 2003, **300**(5628), 2072–2074.
56. T. Fukushima, A. Kosaka, Y. Yamamoto, T. Aimiya, S. Notazawa, T. Takigawa, T. Inabe and T. Aida, *Small*, 2006, **2**(4), 554–560.
57. L. Zhao, Y. Li, X. Cao, J. You and W. Dong, *Nanotechnology*, 2012, **23**(25), 255702.
58. S. Bellayer, J. W. Gilman, N. Eidelman, S. Bourbigot, X. Flambard, D. M. Fox, H. C. De Long and P. C. Trulove, *Adv. Mater.*, 2005, **15**(6), 910–916.
59. H. Xu, F. Tong, J. Yu, L. Wen, J. Zhang and J. He, *Polym. Int.*, 2012, **61**(9), 1382–1388.
60. C. Y. Xing, X. Zheng, L. Xu, J. Jia, J. Ren and Y. J. Li, *Ind. Eng. Chem. Res.*, 2014, **53**(11), 4304–4311.
61. C. Y. Xing, J. Guan, Z. Chen, Y. Zhu, B. Zhang, Y. J. Li and Y. Li, *Nanotechnology*, 2015, **26**(10), 105704.
62. Y. Ding, H. Tang, X. Zhang, S. Wu and R. Xiong, *Eur. Polym. J.*, 2008, **44**(4), 1247–1251.
63. X. Yang, N. Ge, L. Hu, H. Gui, Z. Wang and Y. Ding, *Polym. Adv. Technol.*, 2013, **24**(6), 568–575.

64. S. Livi, J. Duchet-Rumeau and J. F. Gérard, *Polymer*, 2011, **52**(7), 1523–1531.
65. J. Yang, S. Pruvost, S. Livi and J. Duchet-Rumeau, *Macromolecules*, 2015, **48**(13), 4581–4590.
66. F. Wang, A. Lack, Z. Xie, P. Frübing, A. Taubert and R. Gerhard, *Appl. Phys. Lett.*, 2012, **100**(6), 062903.
67. C. Y. Xing, J. Guan, Y. Li and J. Li, *ACS Appl. Mater. Interfaces*, 2014, **6**(6), 4447–4457.
68. D. Okada, H. Kaneko, K. Kato, S. Furumi, M. Takeguchi and Y. Yamamoto, *Macromolecules*, 2015, **48**(8), 2570–2575.
69. C. L. Liang, Z. H. Mai, Q. Xie, R. Y. Bao, W. Yang, B. H. Xie and M. B. Yang, *J. Phys. Chem. B*, 2014, **118**(30), 9104–9111.
70. C. Y. Xing, J. Y. Li, C. M. Yang and Y. J. Li, *Macromol. Rapid Commun.*, 2016, **37**(19), 1559–1565.
71. C. Y. Xing, Y. Wang, X. Huang, Y. J. Li and J. Y. Li, *Macromolecules*, 2016, **49**(3), 1026–1035.
72. C. Y. Xing, Y. Wang, C. Zhang, L. Li, Y. J. Li and J. Y. Li, *Ind. Eng. Chem. Res.*, 2015, **54**(38), 9351–9359.
73. C. Y. Xing, J. You, Y. J. Li and J. Y. Li, *J. Phys. Chem. C*, 2015, **119**(36), 21155–21164.
74. C. Y. Xing, L. Zhao, J. You, W. Dong, X. Cao and Y. Li, *J. Phys. Chem. B*, 2012, **116**(28), 8312–8320.
75. T. Nishi and T. T. Wang, *Macromolecules*, 1975, **8**(6), 909–915.
76. J. P. Penning and R. St. J. Manley, *Macromolecules*, 1996, **29**(1), 77–83.
77. J. Pernak and A. Czepukowicz, *Ind. Eng. Chem. Res.*, 2001, **40**(11), 2379–2383.
78. J. Pernak, K. Sobaszkiewicz and J. Foksowicz-Flaczyk, *Chem.–Eur. J.*, 2004, **10**(14), 3479–3485.
79. C. Wakai, A. Oleinikova, M. Ott and H. Weingärtner, *J. Phys. Chem. B*, 2005, **109**(36), 17028–17030.
80. J. Wu and J. P. W. Stark, *Meas. Sci. Technol.*, 2006, **17**(4), 781–788.
81. G. X. Chen, S. Zhang, Z. Zhou and Q. Li, *Polym. Compos.*, 2015, **36**(1), 94–101.
82. A. J. Ryan, *Nat. Mater.*, 2002, **1**(1), 8–10.
83. A. V. Ruzette and L. Leibler, *Nat. Mater.*, 2005, **4**(1), 19–31.
84. K. V. Peinemann, V. Abetz and P. F. W. Simon, *Nat. Mater.*, 2007, **6**(12), 992–996.
85. Y. Li, Y. Iwakura, L. Zhao and H. Shimizu, *Macromolecules*, 2008, **41**(9), 3120–3124.
86. H. Shimizu, Y. Li, A. Kaito and H. Sano, *Macromolecules*, 2005, **38**(19), 7880–7883.
87. C. Rotella, S. Tencé-Girault, M. Cloitre and L. Leibler, *Macromolecules*, 2014, **47**(14), 4805–4812.

CHAPTER 11

Ionic Liquids as Tools in the Production of Smart Polymeric Hydrogels

S. S. SILVA*[a,b] AND R. L. REIS[a,b]

[a]3B's Research Group - Biomaterials, Biodegradables, Biomimetics, Univ. Minho, Headquarters of the European Institute of Excellence on Tissue Engineering and Regenerative Medicine, AvePark-Parque de Ciencia e Tecnologia, 4805-017 Barco, Guimarães, Portugal; [b]ICVS/3B's - PT Government Associate Laboratory, Braga/Guimarães, Portugal
*E-mail: simonesilva@dep.uminho.pt

11.1 Introduction

The application of ionic liquids (ILs) in the development of functional tailor-made materials has grown over the years. This interest has been motivated by the unique physico-chemical properties of ILs, enabling the utilization of IL-based materials in a broad range of fields, namely the chemical, pharmaceutical, energy, and biotechnology fields, among others. ILs are organic salts that contain organic cations such as imidazolium, pyridinium, pyrrolidinium or ammonium derivatives.[1,2] These can be associated with organic anions such as CH_3COO^- or inorganic anions such as Cl^-, Br^-, I^- or BF^{4-} to design a particular IL. ILs are stable in the liquid form at a temperature below 100 °C. They have high ionic conductivity (10^{-4} to 10^{-2} S cm^{-1}), are inflammable and exhibit negligible vapor pressure.[1] Furthermore, they also exhibit high thermal (up to 300 °C)

Smart Materials No. 29
Polymerized Ionic Liquids
Edited by Ali Eftekhari

Published by the Royal Society of Chemistry, www.rsc.org

and electrochemical stability (4–5.7 V).[3] Since ILs have low volatility, they can be recycled through their separation from liquid solutions by evaporation of water or another volatile solvent (*e.g.* ethanol) and re-used.[4]

The role of ILs not only as a solvent but also as a reaction medium for difficult biomacromolecules has been demonstrated in a number of different publications in the literature. In fact, renewable materials like natural polymers have many properties, such as availability in nature, biocompatibility, and biodegradability, that are useful for the development of renewable smart materials, *e.g.* hydrogels, membranes, and scaffolds. Despite these benefits, some natural biomacromolecules, such as chitin, chitosan and cellulose, exhibit a lack of solubility in water and organic solvents due to their high crystallinity and rigid inter/intramolecular hydrogen bonds. Therefore, the dissolution, functionalization, and blending of natural biomacromolecules using certain ILs have opened up new possibilities for the potential use of the aforementioned polymers. The dissolution mechanism of these macromolecules with certain ILs has been associated with their ability to disrupt the hydrogen bonds, creating quasi-solid materials termed as ion gels, which offer good mechanical strength and conductivity. Other authors have also reported the application of ILs to extract biopolymers such as agarose[5] or chitin[6] from biomass, based on simple, easy and efficient methodologies.

In the literature, it is also possible to find references to the ability of ILs to create hydrogels.[7–10] Hydrogels are three-dimensional (3D) polymeric networks that are highly swollen and possess hydrophilic character. They are able to absorb large amounts of water or biological fluids.[11] Polymeric hydrogels have been made using natural (*e.g.* cellulose, chitin, chitosan, xanthan gum, agarose, or silk fibroin)[7,8,10] and synthetic polymers, like poly(*N*-isopropyl acrylamide) (PNIPAm) and poly(benzyl methacrylate) (PBzMA),[9] or a combination of the two. These approaches have also been used to produce stimuli responsive (SR) hydrogels. SR hydrogels can change their physicochemical properties in response to variations in external stimuli, *e.g.* pH, temperature and magnetic or electric fields.[9,11]

Another possible approach to prepare polymeric hydrogels is the production of poly(ionic liquid)s, PILs, which result from the combination of ILs with polymers.[12–14] PILs are a subclass of polyelectrolytes that feature IL species in each monomer repeating unit, connected to a polymeric backbone to form a macromolecular architecture.[13,15,16]

In this chapter, we will review the most important and interesting features of polymeric materials, mainly hydrogels resulting from the combination of natural and synthetic polymers prepared in ILs (Figure 11.1), highlighting their potential application in different fields.

11.2 Polymeric Hydrogels Using Ionic Liquids

Many reports on ILs and polymers have dealt with the use of ILs as solvents for poorly soluble macromolecules. For instance, different studies have reported that ILs containing strong hydrogen-bond-accepting anions such as

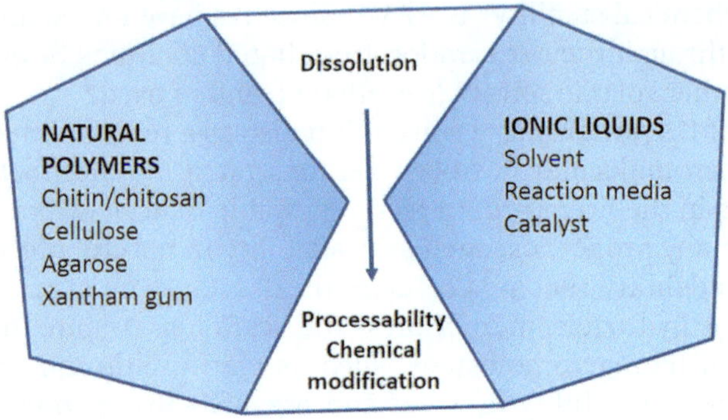

Figure 11.1 Overview of the strategies applied for using natural polymers and ionic liquids for the development of materials.

chloride and acetate anions can dissolve natural polymers. Most of the work in the literature related to biopolymer solubility in ILs is mainly focused on polysaccharides (*e.g.* agarose, cellulose, chitin/chitosan, xanthan gum and starch) and proteins (gelatin, and silk fibroin).[7,17–19] Their dissolution is a useful tool as it allows them to be processed into composites, fibres, ion gels, microspheres, and hydrogels.

The preparation of IL-based polymer gels can be based on diverse methodologies, which can be classified into three major types: gelation of ILs within polymers/biopolymers, *in situ* polymerization of vinyl monomers in ILs, and polymerization of ILs containing polymerizable groups.[20]

In this chapter, the physicochemical properties and characteristics of the natural and synthetic polymers most used in the preparation of polymeric ionogels and hydrogels in ILs are also reviewed.

11.2.1 Agarose

Agarose is a linear polysaccharide present in red algae. It is composed of (1→3)-β-D-galactopyranose-(1→4)-3,6-anhydro-β-L-galactopyranose units.[21] Recent studies have explored the use of ILs as a medium for the efficient extraction of agarose *via* the dissolution of red algae, under various conditions of heating or microwave irradiation.[5] This method is easy, simple and highly efficient when compared to the conventional methods of agarose extraction. When solubilized in water, this neutral polysaccharide forms thermoreversible gels, which solidify upon cooling below 35–40 °C.[8] To take advantage of its thermoresponsive character, agarose has been associated with chitosan, gelatin, and fibrin.[9,22–24] These approaches have been utilized in the formation of hydrogels useful for biomedical applications.

The preparation of composites of agarose/chitosan in 1-butyl-3-methylimidazolium chloride (BMIMCl) has been demonstrated.[25] Besides composite preparation, it is also possible to form ionogels by cooling biopolymer/IL solutions at room temperature (RT). Ionogels can act as smart polymeric conducting materials that combine the chemical versatility of an IL with the morphological versatility of a biopolymer. In such ionogels, the IL is immobilized in a way that involves the formation of a three dimensional network. As an example, agarose ionogels have been produced from agarose in ammonium-based ILs and further mixed with imidazolium- or pyridinium-based ILs.[26] The obtained agarose/IL ionogels present good strength and conductivity when the agarose/IL solutions are cooled to RT. Such ionogels can be suitable for application in electrochemical devices. In another approach, agarose (CAG) ionogels were prepared from a protic–aprotic mixed-IL system (BMIMCl and *N*-(2-hydroxyethyl)ammonium formate).[27] These ionogels demonstrated superior self-healing properties. In addition, these ionogels were suitable as a flexible solid electrolyte for activated-carbon-based supercapacitors. This type of smart polymeric-material ionogel will also have broad applications in the biomedical and electronic fields.

11.2.2 Cellulose

Cellulose is the most abundant polysaccharide in nature. It is constituted by β(1→4)-linked D-glucopyranose units. This polymer is recognized for its wide availability, low cost and biocompatibility. Nevertheless, its application has been limited due to its high crystallinity degree and rigid intra/intermolecular hydrogen bonds, resulting in its insolubility in water as well as organic solvents, thus reducing its applications. A series of studies has been initiated on dissolving cellulose in ILs. In 2002, Swatloski and co-workers reported, for the first time, the dissolution of cellulose using an IL, BMIMCl, in relatively high concentrations.[28] The success of this work stimulated other studies on the dissolution mechanism, processability, and functionalization of cellulose in ILs. Besides BMIMCl, other ILs, namely 1-ethyl-3-methylimidazolium acetate (BMIMAc), 1-allyl-3-methylimidazolium chloride (AMIMCl), and 1-allyl-2,3-dimethylimidazolium bromide (AMIMBr), have been investigated either to dissolve cellulose or as a reaction media for the functionalization of cellulose.[28–33] Therefore, many aspects related to the dissolution mechanism and interactions of cellulose in ILs have been reviewed by several authors.[32,34,35] These studies provide molecular insights into the interactions of cellulose with ILs and suggest that hydrogen-bonding is critical to governing/managing cellulose dissolution. In addition, the chemical versatility of both cellulose and ILs has been investigated for the next generation of cellulosic materials for application in different fields. For instance, shape-persistent and tough cellulose hydrogels can be moulded by step-wise solvent exchange from a homogeneous cellulose/IL solution to methanol vapour exposure (Figure 11.2). These tough cellulose hydrogels may find applications as scaffolds for tissue engineering or micro-reactors using microfluidics.

Cellulose and its derivatives are also excellent building blocks for hydrogel fabrication. Kadokawa and co-workers reported a facile approach for the preparation of a flexible cellulose hydrogel material from cellulose/BMIMCl (15 wt%) solution by keeping it at room temperature for 7 days.[29] In addition,

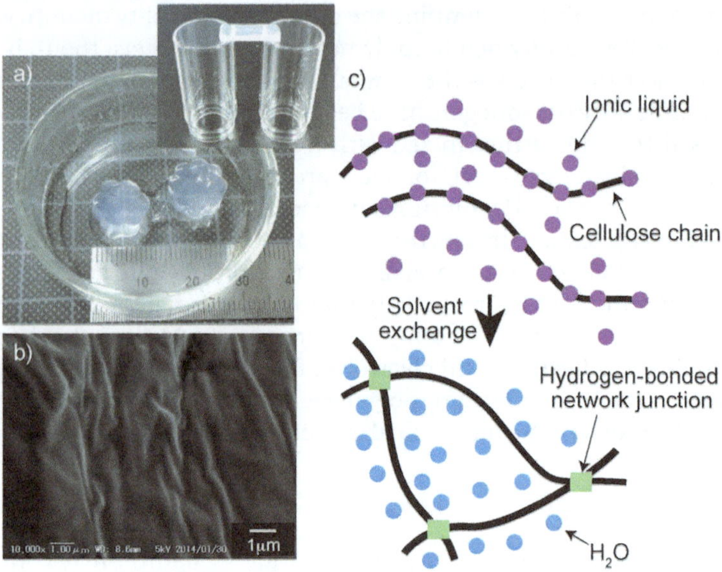

Figure 11.2 (a) Optical images of flower-shaped cellulose hydrogels prepared from a 5 wt% IL solution of wood pulp. The inset is an optical image of the bridged cellulose hydrogel. (b) SEM image of the surface of the dried cellulose hydrogel. (c) Schematic gelation process of the cellulose solution. Reprinted from Macmillan Publishers Ltd: *Scientific Reports*, ref. 36. Copyright 2015.

it has been found that the preparation of cellulose/biopolymer composites, *e.g.* cellulose/collagen hydrogel beads,[37] cellulose/chitosan hydrogel beads and microspheres,[38] and cellulose/agarose beads,[39] relies on the ability of the IL to dissolve cellulose and its regeneration in ethanol or water. These hydrogels have been used for heavy metal adsorption and enzyme immobilization, and as an adsorbent for dye contaminants. Another interesting approach is one described by Xu and co-workers where cellulose/graphene hydrogels (CGH) were manufactured by regenerating a mixture of wood pulp and reduced graphene oxide (rGO) from a BMIMCl solution using water as a coagulant.[40] It was found that incorporating rGO into the cellulose matrix improved its mechanical properties, forming highly tough composite hydrogels. These materials hold great potential for applications in wide fields, such as biomedicine and the environment.

11.2.3 Chitin and Chitosan

Chitin is the second most abundant polymer after cellulose.[21,41] This structure is composed of *N*-acetyl-D-glucosamine units linked through β-(1→4)-glycosidic linkages. It is found in the shells of crustaceans such as crabs, shrimps, and squid.[42] Chitin itself has attractive propertiesfor biomedical applications, namely the acceleration of woundhealing, tumor cell growth suppression, nontoxicity, and biocompatibility.[21,41] Despite these properties, and the huge annual production and easy accessibility of biomass resources, chitin has some limitations due to its lack of solubility in water and some organic solvents. The solubility behaviour of chitin is affected by its crystalline structure. α-chitin (*e.g.* chitin isolated from crab and shrimp cells) is formed by strong intermolecular hydrogen bonding, which implies a lack of solubility, while β-chitin (*e.g.* chitin extracted from squid pens) has a parallel arrangement, resulting in weaker intermolecular forces, higher reactivity and higher affinity for solvents when compared to α-chitin.[42] Therefore, much attention has been paid to applying ILs as a solvent for chitin as a way to expand its processability and potential uses. In fact, many ILs have been reported in the literature with the ability to dissolve chitin, such as AMIMBr, BMIMAc, 1-ethyl-3-methylimidazolium acetate (EMIMAc), and BMIMCl.[43–45] It seems that chitin requires a more basic anion such as acetate due to the increased number of hydrogen bond donors and acceptors.[46] It has also been shown that not only the structure of the anion or the cation is important, but also the features of chitin such as its origin, polymorphic form, molecular weight and degree of acetylation.[47] The complete dissolution of chitin was achieved using AMIMBr to obtain a flowing solution or an ionic gel depending on the chitin concentration.[48] It was suggested that the reason behind chitin dissolution in AMIMBr could be related to the IL composition, which is a combination of an allyl substituent and a bromide counter anion on the imidazolium. Despite the promising findings revealed by these studies, it can be difficult to predict the behavior of chitin in a given IL, probably due to the large number of structural variations of ILs.

An IL platform could also lead to economically and environmentally sustainable isolation methodologies and processing of chitin directly from marine sources, *e.g.* crab shells, replacing the use of hazardous solvents and reducing the energy and time involved in the process.[6] For instance, Qin *et al.* have shown that EMIMAc can dissolve raw crustacean shells, leading to a high molecular weight chitin powder, but it can produce fibers directly from the extract solution.[49] In another study, chitin fibers could be electrospun in a one-pot process directly from a chitin/EMIMAc solution from dried shrimp shells.[46] Recent studies also demonstrated that ILs, such as hydroxylammonium acetate ([NH_3OH][OAc]), can reactively demineralize and remove proteins from shrimp shells in an efficient one-pot pulping process, thus allowing the isolation of native chitin with >80% purity and a high degree of acetylation and crystallinity.[6]

To date, there are few reports involving the dissolution of chitin to form hydrogels. Shen *et al.* prepared hydrogels from IL-extracted biopolymers (chitin and cellulose).[10] These hydrogels exhibited properties substantially different from those made from the commercially available biopolymers. The hydrogels were supercritically dried producing amorphous, lightweight, and porous aerogels. The findings obtained from dye (indigo camine) release suggest the potential use of both chitin and cellulose hydrogels for drug delivery applications.

Chitosan is a polysaccharide obtained by the alkaline deacetylation of chitin.[41] This polymer has many properties such as hemostatic action, film-forming ability, chelating activities, adsorption abilities, bacteriostatic action, cationic nature, biodegradability, and biocompatibility.[41] These particular properties arise from the presence of primary amines along the chitosan backbone. Chitosan is recognized as a key component in materials applied in several fields, namely the food processing, packaging, cosmetic, biomedical, and pharmaceutical fields, and the removal of metal ions from waste water.[41,50–54]

Similar to chitin, research on the use of ILs has also been extended to chitosan.[44] The processability of chitosan has some limitations due to its insolubility in pure water and common organic solvents. This feature is associated with the strong hydrogen bonding in chitosan and the presence of amorphous and crystalline regions in its structure. A series of ILs, namely 1-EMIMCl, BMIMCl, EMIMAc, and BMIMAc, have been studied for the dissolution and processability of chitosan.[44] Therefore, the ability of certain ILs, *e.g.* BMIMCl, to act as a solvent and reaction medium for chitosan occurs through disrupting its hydrogen bonds, allowing its amino group to be exposed completely to react with other chemical reagents easily.

A simple one-step method based on the ability of ILs to co-dissolve cellulose and chitosan has been reported for preparing magnetic biopolymer hybrid hydrogels.[55] These hydrogels had high adsorption abilities for some heavy metals ions, such as Cu^{+2}, Fe^{+2} and Pb^{+2}, and they could be efficiently recycled and reused.

11.2.4 Silk Fibroin

Silk fibroin (SF) is a natural protein. It is composed mainly of glycine, alanine, and serine in different percentages.[21,56] Silks produced by silkworms are biopolymers, and they can be classified into mulberry (*Bombyx mori*) and non-mulberry (*e.g. muga, eri, tasar*) silks.[57]

Silk-based materials have been produced into versatile formats such as membranes, nanofibers, scaffolds and hydrogels.[58,59] The processing of silk proteins into hydrogels, for instance, can be achieved using different methods based on chemical and physical principles.[56] The methodologies demonstrated that SF hydrogel formation could be dependent on temperature, pH and protein concentration, among other parameters.

Many reports emphasized the use of certain ILs to solubilize mulberry silkworm *Bombyx* protein fibroin alone[60] or in combination with chitosan[8] and cellulose.[61] Other studies used ILs to dissolve and process fibroin obtained from the cocoons of mulberry silkworms *Antheraea mylitta*,[7] *Antheraea assamensis*,[19] and *Samia/Philosamia ricini (Eri)*.[18] However, only a few reports suggested the production of SF hydrogels through the dissolution of SF in certain ILs. SF hydrogel-based constructs were prepared to solubilize both materials in BMIMAc as a common solvent of chitosan/silk (CSF)/BMIMAc blended systems.[8] The CSF solutions have extended stability and easy processability, allowing the moulding of the CSF hydrogels into a range of shapes and sizes. Moreover, *in vitro* assays showed that these CSF hydrogels provide a good environment for the adhesion and growth of primary human dermal fibroblasts.

Chen and co-workers report the preparation of solutions of regenerated silk fibroin (RSF), obtained from degummed silks of *Bombyx mori* and *Antherae pernyi* in EMIMAc.[62] The results provided an insight into the sol–gel transition of both RSF/EMIMAc/water from the viewpoint of thermodynamics, which is helpful to interpret the conformational changes of RSF, as well as to create a route for the production of biomedical SF hydrogels from SF/IL solutions.

11.2.5 Xanthan Gum

Xanthan gum is an anionic polysaccharide produced by *Xanthomonas campestris* and a useful food hydrocolloid. It is composed of cellulose-type main chains (β(1→4)-glucan) with trisaccharide side chains (mannose-β(1→4)-glucuronic acid-β(1→2)-mannose-α(1→3)-) attached to alternate main chain glucose units.[63,64] Xanthan gum does not readily form hydrogels *via* the usual gelation processes. In fact, an aqueous dispersion of xanthan gum exhibits only a weak gel-like behaviour in the presence of a sufficient amount of inorganic salt. Since ILs have specific properties to dissolve polysaccharides, they have been considered as an efficient solvent to produce xanthan gum hydrogels.[65,66] Xanthan gum/BMIMCl gels were prepared by a heating–cooling process.[65] These gels showed good mechanical properties and a thermally induced shape-memory effect. Later, the gels were converted into xanthan gum hydrogels by soaking them in water, and the resulting hydrogels showed good elasticity (Figure 11.3). Further soaking the obtained hydrogels in $CaCl_2$ aqueous solution gave rise to an ionically cross-linked hydrogel with Ca^{+2}, which exhibited much better mechanical properties and salt concentration-induced swelling/shrinking behaviour. The authors supposed that the shape memory behaviour of the ion gel was created by the effect of ion-exchange of the carboxylate metal salts in xanthan gum with BMIMCl during the gel formation.

Setoyama and co-workers prepared cellulose/xanthan gum composite films and hydrogels through gelation with BMIMCl.[67] In this work, the solutions of cellulose and xanthan gum in BMIMCl were heated and left to gelate at room temperature, followed by Soxhlet extraction with ethanol. Films and hydrogel materials were obtained by the methods of regeneration and exchange of the dispersing media from the ion gels, respectively.

11.3 Smart Polymeric Hydrogels

Recent advances in the creation of smart materials involve ILs and PILs that exhibit stimuli responsiveness (*e.g.* critical solution temperature, pH, magnetic field and thermoreversibility). When such ILs are used to compose new

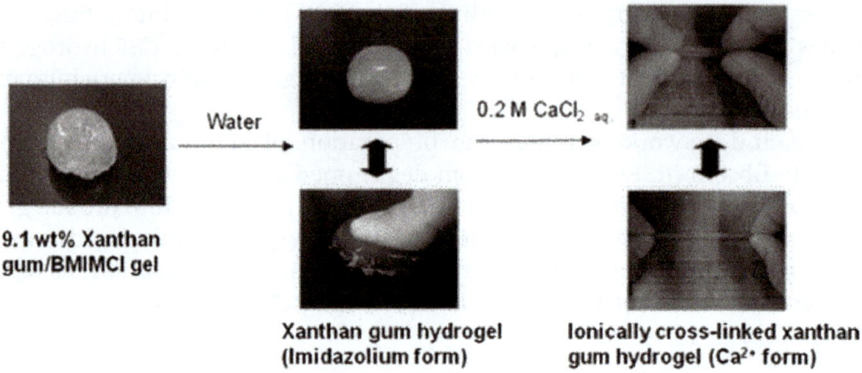

Figure 11.3 Conversion of a 9.1 wt% xanthan gum/BMIMCl gel into a xanthan gum hydrogel (imidazolium form), which was further converted into an ionically cross-linked hydrogel (Ca^{+2} form) by treatment with a 0.2 M $CaCl_2$ aqueous solution. Reproduced from ref. 65 with permission from The Royal Society of Chemistry.

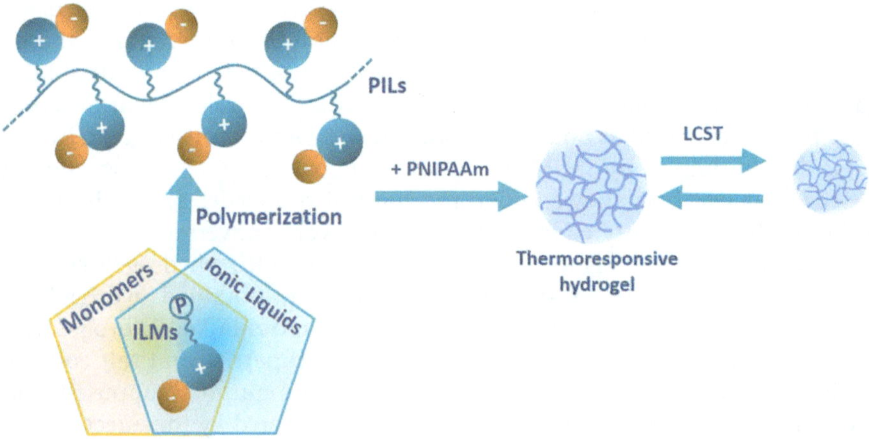

Figure 11.4 Schematic illustration of the relationship between ionic liquids and poly(ionic liquid)s, and the formation of a thermoresponsive hydrogel. ILMs: ionic liquid monomers. "P": polymerizable group. Adapted from *Progress in Polymer Science*, 38 (7), J. Yuan, D. Mecerreyes, M. Antonietti, Poly(ionic liquid)s: An update, 1009–1036, Copyright (2013) with permission from Elsevier.[15]

polymers, the resulting polymers will often exhibit responsive phenomena. Therefore, SR hydrogels have also been constructed by incorporation of thermosensitive polymers like PNIPAAm and PBzMA into PILs (Figure 11.4). PNIPAAm is the most widely studied thermosensitive polymer in temperature-responsive hydrogels. It shows a lower critical solution temperature (LCST)-type phase separation in aqueous solutions at approximately body temperature.[11]

Such materials have potential use in highly diverse areas ranging from controlled release to chemomechanical actuators to supports of catalysis, among other applications.

The combination of ILs with polymers permits the creation of smart soft materials. T. Ueki has reviewed the thermodynamic aspects and recent developments in stimuli-responsive polymers in ILs, as a new class of smart soft materials.[68] It is suggested that polymers in ILs require prolonged periods to reach thermodynamic equilibrium as a result of the high viscosity of ILs. Therefore, the viscosity of polymer/IL composites must be supressed to achieve a fast response time for the resulting soft materials.

Sui *et al.* produced a new class of cross-linkable redox-responsive poly(ferrocenylsilane)-based PILs (PFS-PILs) *via* a post-polymerization modification approach.[69] PFS-PILs were obtained by reacting poly[ferrocenyl(3-iodopropyl)-methylsilane] with 1-vinylimidazole at 60 °C for 24 h. PFS-PILs self-cross-link at low concentrations into nanogels or form macroscopic hydrogel networks at higher concentrations. TEM analysis revealed nanoparticles that were 20–40 nm in size, corresponding to the cross-linked PFS-PIL nanogels (Figure 11.5). These nanogels were anion-sensitive and redox-responsive. They also proved to be efficient dispersants in the microemulsion polymerization of methyl methacrylate, producing stable PFS–poly(methyl methacrylate) latex suspensions.

PILs have also been used as a route to obtain dual-responsive hydrogels.[9] These systems are prepared by incorporating two functional monomers that respond to different stimuli, offering additional functional control. Feng and co-workers reported the preparation of dual-responsive hydrogels, which are highly swellable, consisting of thermo-responsive PNIPAM and redox-responsive poly(ferrocenylsilane) (PFS)-based PILs, which are produced by photo-polymerization.[9] The swelling ability and

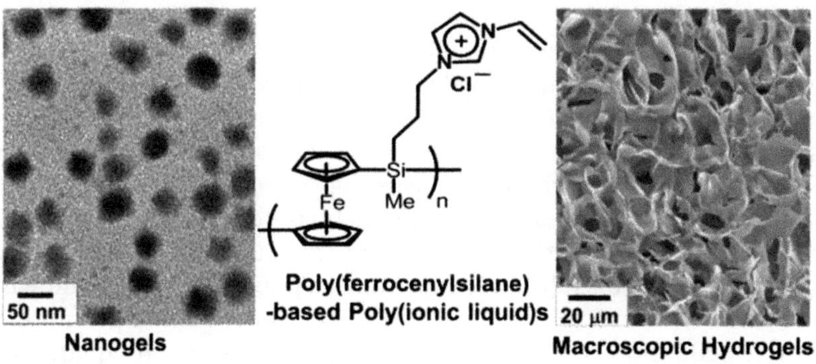

Figure 11.5 Nanogels and macroscopic hydrogels prepared from poly(ferrocenylsilane)-based PILs (chemical structure shown in the middle), which could be cross-linked through the vinylimidazolium functional groups. Reprinted with permission from ref. 69. Copyright (2016) American Chemical Society.

thermo-responsivity of the hydrogels could betuned by changing the counterion type due to the presence of thepoly(ionic liquid) part. This hydrogel has been employed as a reducing environment for the *in situ* fabrication of gold nanoparticles (AuNPs), forming AuNP–hydrogel composites.[9] Another strategy employed for the preparation of responsive hydrogels based on PILs has been described by Tudor *et al.*[13] In this work, semi-interpenetrating (sIPN) hydrogels were synthesized by adding increasing amounts of poly(*N*-isopropylacrylamide-*co*-spiropyran-*co*-acrylic acid) to a cross-linked PIL matrix. All of the hydrogels, polymerized for approximately 120 s, have different mechanical properties that can be modulated through the content of the copolymer. Gallagher *et al.* described the sIPN hydrogels prepared using varying concentrations of linear pNIPAAM incorporated into a thermo-responsive PIL hydrogel, namely tributyl-hexyl phosphonium 3-sulfopropylacrylate (P-SPA).[70] The swelling and shrinking responses of the obtained hydrogels were modulated by the presence of linear pNIPAAM in the polymer matrix.

Other authors indicated that PNIPAAM-based ionic hydrogels could be synthesized by free-radical polymerization with *N*-isopropylacrylamide as a monomer and imidazolium-based dicationic IL as a cross-linker.[12] The obtained hydrogels had good swelling properties and exhibited strongly interfacial interaction with anionic dyes such as methyl orange, methyl blue, Congo red, orange G, thymol blue and bromothymol blue in aqueous solution. This behaviour was dependent on the chemical structure of the anionic dyes in solution.

11.4 Conclusions

In this chapter, we review the recent advances in the use of ILs in the process of synthesizing polymeric gels (ionogels) and hydrogels. The tunability and versatility of ILs aligned with the features of natural (*e.g.* chitin or cellulose) and/or synthetic polymers have opened up new routes to process them not only as hydrogels, but also as matrices with different shapes and sizes. In particular, polymeric hydrogels processed in ILs can be useful for a broad range of applications from the removal of anionic dyes to biomaterials. However, concerning biomedical applications, despite the promising findings, little has been reported regarding the *in vitro* and *in vivo* biocompatibility of the developed matrices, which could limit their potential use as biomaterials. Most of the research reported suggests the use of ILs as a tool not only in the dissolution/processing of polymers in high value matrices but also to extract biopolymers directly from biomass promoting the economic reduction of the process and an increase in the quality of the obtained biopolymer. The synthesis of PILs has also contributed to certain developments in the creation of responsive hydrogels. Despite this, only a few reports can be found in the literature suggesting that it is at an early stage. However, due to its potential, more developments are expected in the near future.

List of Abbreviations

AMIMBr	1-Allyl-3-methylimidazolium bromide
AMIMCl	1-Allyl-3-metylimidazolium chloride
BMIMCl	1-Butyl-3-methylimidazolium chloride
BMIMAc	1-Ethyl-3-methylimidazolium acetate
CGH	Cellulose/graphene hydrogels
CSF	Chitosan/silk
3D	Three dimensional
rGO	Graphene oxide
AuNPs	Gold nanoparticles
ILs	Ionic liquids
ILMs	Ionic liquid monomers
LCST	Lower critical solution temperature
PILs	Poly(ionic liquid)s
PFS	Poly(ferrocenylsilane)
PFS-PILs	Poly(ferrocenylsilane)-based PILs
PBzMA	Poly(benzyl methacrylate)
PNIPAm	Poly(N-isopropyl acrylamide)
P-SPA	Tributyl-hexyl phosphonium 3-sulfopropylacrylate
RSF	Regenerated silk fibroin
RT	Room temperature
SF	Silk fibroin
SR	Responsive hydrogels
sIPN	Semi-interpenetrating

Acknowledgements

The authors are thankful for the financial support from the Portuguese Foundation for Science and Technology (FCT) for the fellowship grant of Simone S. Silva (SFRH/BPD/112140/2015) "Fundo Social Europeu"- FSE and "Programa Diferencial de Potencial Humano POPH".

References

1. J. Anthony, J. Brennecke, J. Holbrey, E. Maginn, P. Mantz, A. V. Trulove and T. Welton, in *Ionic Liquids in Synthesis*, ed. P. Wasserscheid and T. Welton, Wiley-VCH, 2002, pp. 41–126.
2. M. Hirao, H. Sugimoto and H. Ohno, *J. Electrochem. Soc.*, 2000, **147**, 4168–4172.
3. N. Lourenço, A. Nunes, C. Duarte and P. Vidinha, in *Applications of Ionic Liquids in Science and Technology*, ed. P. S. Handy, 2011, p. 155.
4. P. S. Barber, J. L. Shamshina and R. D. Rogers, *Pure Appl. Chem.*, 2013, **85**, 1693–1701.
5. T. Trivedi and A. Kumar, *Green Sustainable Chem.*, 2014, **4**, 190–201.

6. J. L. Shamshina, P. S. Barber, G. Gurau, C. S. Griggs and R. D. Rogers, *ACS Sustainable Chem. Eng.*, 2016, **4**, 6072–6081.
7. S. S. Silva, E. G. Popa, M. E. Gomes, M. B. Oliveira, S. Nayak, B. Subia, J. F. Mano, S. C. Kundu and R. L. Reis, *Acta Biomater.*, 2013, **9**, 8972–8982.
8. S. S. Silva, T. C. Santos, M. T. Cerqueira, A. P. Marques, L. L. Reys, T. H. Silva, S. G. Caridade, J. F. Mano and R. L. Reis, *Green Chem.*, 2012, **14**, 1463–1470.
9. X. L. Feng, K. H. Zhang, P. Chen, X. F. Sui, M. A. Hempenius, B. Liedberg and G. J. Vancso, *Macromol. Rapid Commun.*, 2016, **37**, 1939–1944.
10. X. P. Shen, J. L. Shamshina, P. Berton, J. Bandomir, H. Wang, G. Gurau and R. D. Rogers, *ACS Sustainable Chem. Eng.*, 2016, **4**, 471–480.
11. J. F. Mano, *Adv. Eng. Mater.*, 2008, **10**, 515–527.
12. X. J. Zhou, J. Wang, J. J. Nie and B. Y. Du, *Polym. J.*, 2016, **48**, 431–438.
13. A. Tudor, L. Florea, S. Gallagher, J. Burns and D. Diamond, *Sensors*, 2016, **16**, DOI: 10.3390/s16020219.
14. B. Ziolkowski and D. Diamond, *Chem. Commun.*, 2013, **49**, 10308–10310.
15. J. Yuan, D. Mecerreyes and M. Antonietti, *Prog. Polym. Sci.*, 2013, **38**, 1009–1036.
16. J. Yuan and M. Antonietti, *Polymer*, 2011, **52**, 1469–1482.
17. R. N. L. de Carvalho, N. M. T. Lourenco, P. M. V. Gomes and L. J. P. da Fonseca, *J. Polym. Sci., Polym. Phys.*, 2013, **51**, 817–825.
18. S. S. Silva, N. M. Oliveira, M. B. Oliveira, D. P. S. da Costa, D. Naskar, J. F. Mano, S. C. Kundu and R. L. Reis, *Acta Biomater.*, 2016, **32**, 178–189.
19. N. Goujon, R. Rajkhowa, X. Wang and N. Byrne, *J. Appl. Polym. Sci.*, 2013, **128**, 4411–4416.
20. J. Lu, F. Yan and J. Texter, *Prog. Polym. Sci.*, 2009, **34**, 431–448.
21. S. S. Silva, E. M. Fernandes, S. Pina, J. Silva-Correia, S. Vieira, J. M. Oliveira and R. L. Reis, in *Reference Module in Materials Science and Materials Engineering*, Elsevier, 2017, pp. 228-252. DOI: 10.1016/B978-0-12-803581-8.10134-1.
22. S. P. Miguel, M. P. Ribeiro, H. Brancal, P. Coutinho and I. J. Correia, *Carbohydr. Polym.*, 2014, **111**, 366–373.
23. R. Imani, S. H. Emami, P. R. Moshtagh, N. Baheiraei and A. M. Sharifi, *J. Macromol. Sci., Part B*, 2012, **51**, 1606–1616.
24. A.-M. Ionescu, M. Alaminos, J. d. l. C. Cardona, J. d. D. García-López Durán, M. González-Andrades, R. Ghinea, A. Campos, E. Hita and M. d. M. Pérez, *J. Mech. Behav. Biomed. Mater.*, 2011, **4**, 1963–1973.
25. T. J. Trivedi, K. S. Rao and A. Kumar, *Green Chem.*, 2014, **16**, 320–330.
26. T. J. Trivedi, D. N. Srivastava, R. D. Rogers and A. Kumar, *Green Chem.*, 2012, **14**, 2831–2839.
27. T. J. Trivedi, D. Bhattacharjya, J.-S. Yu and A. Kumar, *Chemsuschem*, 2015, **8**, 3294–3303.
28. R. P. Swatloski, S. K. Spear, J. D. Holbrey and R. D. Rogers, *J. Am. Chem. Soc.*, 2002, **124**, 4974–4975.

29. J.-i. Kadokawa, M.-a. Murakami and Y. Kaneko, *Carbohydr. Res.*, 2008, **343**, 769–772.
30. X. Y. Hu, K. Hu, L. L. Zeng, M. M. Zhao and H. H. Huang, *Carbohydr. Polym.*, 2010, **82**, 62–68.
31. A. Kimura, N. Nagasawa and M. Taguchi, *Radiat. Phys. Chem.*, 2014, **103**, 216–221.
32. M. Isik, H. Sardon and D. Mecerreyes, *Int. J. Mol. Sci.*, 2014, **15**, 11922.
33. M. Işık, H. Sardon and D. Mecerreyes, in *Applications of Ionic Liquids in Polymer Science and Technology*, ed. D. Mecerreyes, Springer Berlin Heidelberg, Berlin, Heidelberg, 2015, pp. 135–152.
34. H. Wang, G. Gurau and R. D. Rogers, *Chem. Soc. Rev.*, 2012, **41**, 1519–1537.
35. K. M. Gupta and J. Jiang, *Chem. Eng. Sci.*, 2015, **121**, 180–189.
36. M. Kimura, Y. Shinohara, J. Takizawa, S. Ren, K. Sagisaka, Y. Lin, Y. Hattori and J. P. Hinestroza, *Sci. Rep.*, 2015, **5**, 16266.
37. J. L. Wang, L. G. Wei, Y. C. Ma, K. L. Li, M. H. Li, Y. C. Yu, L. Wang and H. H. Qiu, *Carbohydr. Polym.*, 2013, **98**, 736–743.
38. M. F. Li, Z. M. Wang and B. J. Li, *Desalin. Water Treat.*, 2016, **57**, 16970–16980.
39. M. H. Kim, S. An, K. Won, H. J. Kim and S. H. Lee, *J. Mol. Catal. B: Enzym.*, 2012, **75**, 68–72.
40. M. Xu, Q. Huang, X. Wang and R. Sun, *Ind. Crops Prod.*, 2015, **70**, 56–63.
41. S. S. Silva, J. F. Mano and R. L. Reis, *Crit. Rev. Biotechnol.*, 2010, **30**, 200–221.
42. L. L. Reys, S. S. Silva, J. M. Oliveira, S. G. Caridade, J. F. Mano, T. H. Silva and R. L. Reis, *Biomed. Mater.*, 2013, **8**, 045002.
43. J. Kadokawa, *Pure Appl. Chem.*, 2016, **88**, 621–629.
44. S. S. Silva, J. F. Mano and R. L. Reis, *Green Chem.*, 2017, **19**, 1208–1220.
45. S. S. Silva, A. R. C. Duarte, A. P. Carvalho, J. F. Mano and R. L. Reis, *Acta Biomater.*, 2011, **7**, 1166–1172.
46. P. S. Barber, C. S. Griggs, J. R. Bonner and R. D. Rogers, *Green Chem.*, 2013, **15**, 601–607.
47. W. T. Wang, J. Zhu, X. L. Wang, Y. Huang and Y. Z. Wang, *J. Macromol. Sci., Part B*, 2010, **49**, 528–541.
48. K. Prasad, M. Murakami, Y. Kaneko, A. Takada, Y. Nakamura and J. Kadokawa, *Int. J. Biol. Macromol.*, 2009, **45**, 221–225.
49. Y. Qin, X. M. Lu, N. Sun and R. D. Rogers, *Green Chem.*, 2010, **12**, 968–971.
50. I. Younes and M. Rinaudo, *Mar. Drugs*, 2015, **13**, 1133–1174.
51. M. Rinaudo, *Prog. Polym. Sci.*, 2006, **31**, 603–632.
52. R. A. A. Muzzarelli, *Carbohydr. Polym.*, 2009, **76**, 167–182.
53. M. Cardoso, R. Costa and J. Mano, *Mar. Drugs*, 2016, **14**, 34.
54. M. Prabaharan and J. F. Mano, *Macromol. Biosci.*, 2006, **6**, 991–1008.
55. Z. Liu, H. Wang, C. Liu, Y. Jiang, G. Yu, X. Mu and X. Wang, *Chem. Commun.*, 2012, **48**, 7350–7352.
56. M. Floren, C. Migliaresi and A. Motta, *J. Funct. Biomater.*, 2016, **7**, 26.

57. S. C. Kundu, B. Kundu, S. Talukdar, S. Bano, S. Nayak, J. Kundu, B. B. Mandal, N. Bhardwaj, M. Botlagunta, B. C. Dash, C. Acharya and A. K. Ghosh, *Biopolymers*, 2012, **97**, 455–467.
58. S. S. Silva, D. Maniglio, A. Motta, J. F. Mano, R. L. Reis and C. Migliaresi, *Macromol. Biosci.*, 2008, **8**, 766–774.
59. S. S. Silva, A. Motta, M. T. Rodrigues, A. F. M. Pinheiro, M. E. Gomes, J. F. Mano, R. L. Reis and C. Migliaresi, *Biomacromolecules*, 2008, **9**, 2764–2774.
60. D. M. Phillips, L. F. Drummy, D. G. Conrady, D. M. Fox, R. R. Naik, M. O. Stone, P. C. Trulove, H. C. De Long and R. A. Mantz, *J. Am. Chem. Soc.*, 2004, **126**, 14350–14351.
61. Y. B. Yao, E. J. Zhang, X. L. Xia, J. C. Yu, K. J. Wu, Y. M. Zhang and H. P. Wang, *Cellulose*, 2015, **22**, 625–635.
62. C. Zhang, X. Chen and Z. Shao, *ACS Biomater. Sci. Eng.*, 2016, **2**, 12–18.
63. K. Limori, K. Yamamoto and J.-I. Kadokawa, *Plast. Polym. Technol.*, 2016, **4**, 63–72.
64. L. D. Melton, L. Mindt and D. A. Rees, *Carbohydr. Res.*, 1976, **46**, 245–257.
65. H. Izawa, Y. Kaneko and J. Kadokawa, *J. Mater. Chem.*, 2009, **19**, 6969–6972.
66. H. Izawa and J.-i. Kadokawa, *J. Mater. Chem.*, 2010, **20**, 5235–5241.
67. M. Setoyama, K. Yamamoto and J. Kadokawa, *J. Polym. Environ.*, 2014, **22**, 298–303.
68. T. Ueki, *Polym. J.*, 2014, **46**, 646–655.
69. X. Sui, M. A. Hempenius and G. J. Vancso, *J. Am. Chem. Soc.*, 2012, **134**, 4023–4025.
70. S. Gallagher, L. Florea, K. Fraser and D. Diamond, *Int. J. Mol. Sci.*, 2014, **15**, 5337.

CHAPTER 12

Preparation of Functional Polysaccharides and Related Materials Combined with Ionic Liquids

J. KADOKAWA

Graduate School of Science and Engineering, Kagoshima University, 1-21-40 Korimoto, Kagoshima 890-0065, Japan
*E-mail: kadokawa@eng.kagoshima-u.ac.jp

12.1 Introduction

Ionic liquids have been identified to dissolve various organic and inorganic materials with poor solubility, such as polysaccharides.[1–6] Polysaccharides are widely distributed in nature and exhibit important biological functions.[7,8] Abundant polysaccharides such as cellulose and chitin (Figure 12.1) are the most well-known organic resources because of their huge annual productions. Besides being abundantly present in nature, natural polysaccharides show specific functions mainly related to biological systems, such as biocompatibility, biodegradability, and non-toxicity. Accordingly, the efficient use of natural polysaccharides as components can be expected to provide new biomedical and environmentally benign materials.[9–11] However, structural polysaccharides such as cellulose and chitin show solubility problems with common solvents owing to the formation of numerous hydrogen bonds

Smart Materials No. 29
Polymerized Ionic Liquids
Edited by Ali Eftekhari

Published by the Royal Society of Chemistry, www.rsc.org

among the molecules and stiff chain packing, leading to poor processability. Even besides cellulose and chitin, most of the natural polysaccharides exhibit poor solubility for the same reasons, and accordingly, are mostly soluble only in aqueous media.

Over the past decade, ionic liquids have been noted to exhibit the ability to dissolve natural polysaccharides[1–6] since Rogers *et al.* reported that an ionic liquid, 1-butyl-3-methylimidazolium chloride (BMIMCl), dissolved cellulose in relatively high concentrations (Figure 12.2).[12] In 1934, it had already been found that molten *N*-ethylpyridinium chloride, in the presence of nitrogen-containing bases, could dissolve cellulose.[13] This was probably the first report on the dissolution of cellulose with ionic liquid-type solvents. However, this study was considered to be of little practical value at the time because the concept of ionic liquids had not been put forward. Then, the aforementioned dissolution study of cellulose with BMIMCl in 2002 opened

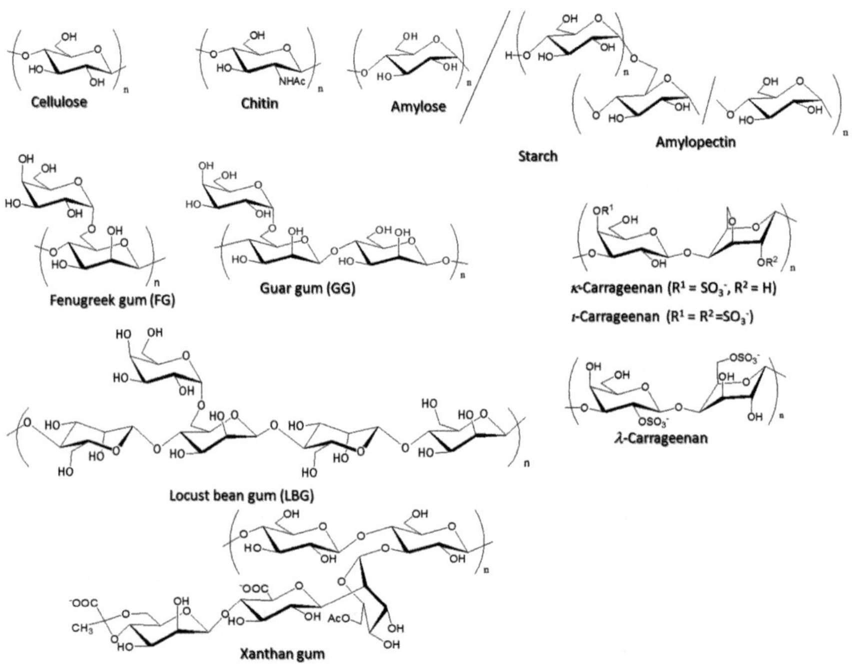

Figure 12.1 Typical polysaccharides used for functional materials combined with ionic liquids.

Figure 12.2 Ionic liquids used for the dissolution and gelation of polysaccharides.

up a new route for studies concerning a class of cellulose solvent systems. Since this study, ionic liquids have been identified as good solvents for cellulose and other polysaccharides, and thus used as media for their material processing, such as derivatization, modification, and regeneration.[5,6]

Ionic liquids have been extensively used as functional components in polysaccharide-based materials, besides their above ordinary uses as solvents and media.[14,15] The concept in these research studies is based on the property that ionic liquids exhibit good affinity for polysaccharides, which subsequently leads to good composites, of course, depending on their chemical structures. For example, ionic liquids have been used as dispersion media in polysaccharide-based gels, so-called 'ion gels', which include ionic liquids in the polysaccharide network matrix. Furthermore, polysaccharide ion gels have been converted into value-added sustainable materials *via* suitable treatments, such as exchange of the dispersion medium and regeneration. In this chapter, on the basis of this background information and these viewpoints, the preparation of polysaccharide-based materials by combining them with ionic liquids is described. The former part deals with polysaccharide ion gels and further conversion into value-added materials. In relation to ion gels, the latter part discloses the other type of polysaccharide–ionic liquid composite material, that is, composites with polymeric ionic liquids that have mainly been prepared by means of *in situ* polymerization of polymerizable ionic liquids.

12.2 Polysaccharide Ion Gels

12.2.1 Ion Gels of Abundant Polysaccharides with Ionic Liquids

It was found that solutions of cellulose in BMIMCl (9.1–13.0 wt%), which were prepared by heating mixtures of cellulose and BMIMCl at 100 °C for 24 h, would form ion gels with the exclusion of excess BMIMCl from the gel matrix when left standing at room temperature for 7 days (Figure 12.3(a)).[16] The resulting gel was purified by washing with ethanol and dried under reduced pressure, and exhibited flexible and manipulatable properties. Thermogravimetric analysis (TGA) of the ion gel showed two weight loss steps, one at temperatures below 100 °C attributed to the evaporation of water and another at higher than 250 °C due to degradation of cellulose, strongly indicating that the ion gel was composed not only of cellulose and BMIMCl, but also of water (*ca.* 15 wt%). This result suggested that the absorption of water into the solution, due to the strong hygroscopic nature of BMIMCl, probably caused the gelation of cellulose with BMIMCl, and the following gelation process was proposed. The ionic liquid, BMIMCl, does not dissolve cellulose perfectly from a molecular point of view as the crystalline structure of cellulose is slightly retained, as suggested by the powder X-ray diffraction (XRD) data of the ion gel. Accordingly, aggregates of the crystalline parts, partially present among the cellulose molecules, were formed through the gradual

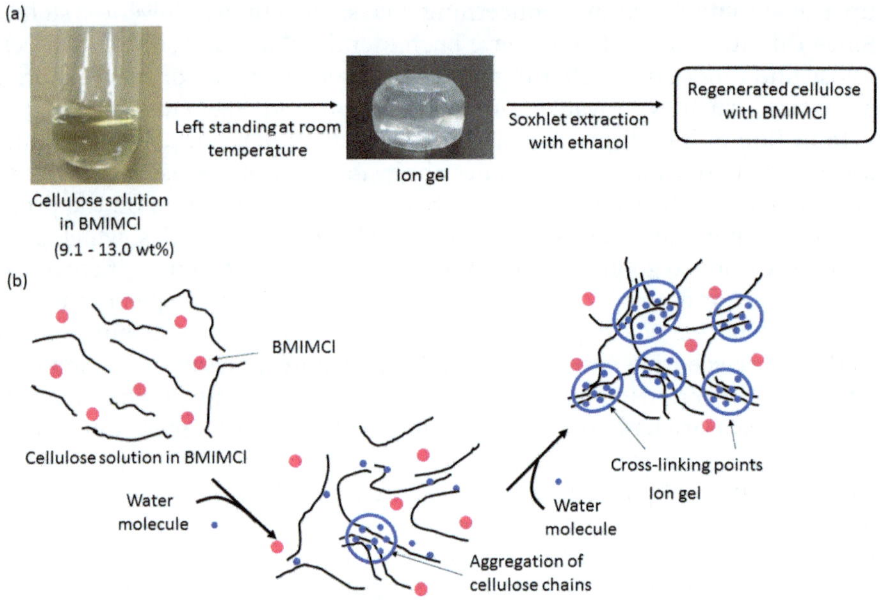

Figure 12.3 Formation of a cellulose ion gel with BMIMCl and conversion into regenerated cellulose with BMIMCl (a) and plausible gelation mechanism (b).

absorption of water into the solution, owing to the insolubility of crystalline cellulose in water, and these aggregates acted as cross-linking points for gelation (Figure 12.3(b)).

Soxhlet extraction of the ion gels with ethanol for the desired times gave regenerated cellulose materials with different BMIMCl contents (Figure 12.3(a)). With prolonged extraction times, the content of BMIMCl in the product decreased, and consequently, BMIMCl was mostly removed following extraction for 4 h. The mechanical properties of the regenerated cellulose materials were evaluated by tensile testing, and depended on the BMIMCl content.

The preparation procedure for the cellulose ion gel with BMIMCl was extended to provide a binary gelling system with starch. When a solution of cellulose and starch (9.1 and 4.8 wt%, respectively) in BMIMCl, which was prepared by heating the corresponding mixture at 100 °C for 24 h, was left standing at room temperature for several days, a binary ion gel was formed that excluded excess BMIMCl.[17] The XRD profile of the resulting gel suggested that the crystalline structures of the polysaccharides in the gel were mostly disrupted. A viscous mixture of cellulose and starch (9.1 wt% each) in BMIMCl could be drawn up from the solution surface. The resulting fine linear material from the surface was soaked in acetone for regeneration to fabricate a cellulose/starch binary polysaccharide fiber. The SEM image of the material exhibited a fibrous morphology with the fibers *ca.* 100–200 μm in width. The XRD result indicated no reconstruction of the crystalline structures of cellulose and starch during the regeneration process, suggesting the good composition of the two polysaccharides in the fibrous material.

Different from cellulose, only a few examples of the dissolution of chitin in ionic liquids have been reported.[18–23] 1-Allyl-3-methylimidazolium acetate, 1-butyl-3-methylimidazolium acetate (BMIMOAc), and 1-ethyl-3-methylimidazolium acetate have been found to show the ability to dissolve chitin at certain concentrations (Figure 12.2).[19,24,25] Cooling chitin solutions with BMIMOAc to ambient temperature resulted in gelation, and the gel could be further converted into sponge and film materials by regeneration with water or a methanol coagulant. 1-Allyl-3-methylimidazolium bromide (AMIMBr) was also found to dissolve chitin and the maximum concentration for dissolution was 4.8 wt% by heating at 100 °C for 48 h (Figure 12.4(a)).[26] The dissolution ability of AMIMBr is considered to be due to the combination of an ion pair, imidazolium and bromide. This type of ionic liquid has also been used as a solvent for the synthesis of polyamides and polyimides,[27] which are well-known poorly soluble polymers due to the formation of strong hydrogen bonds by the –N–C=O groups in the main chains, similar to the acetamido

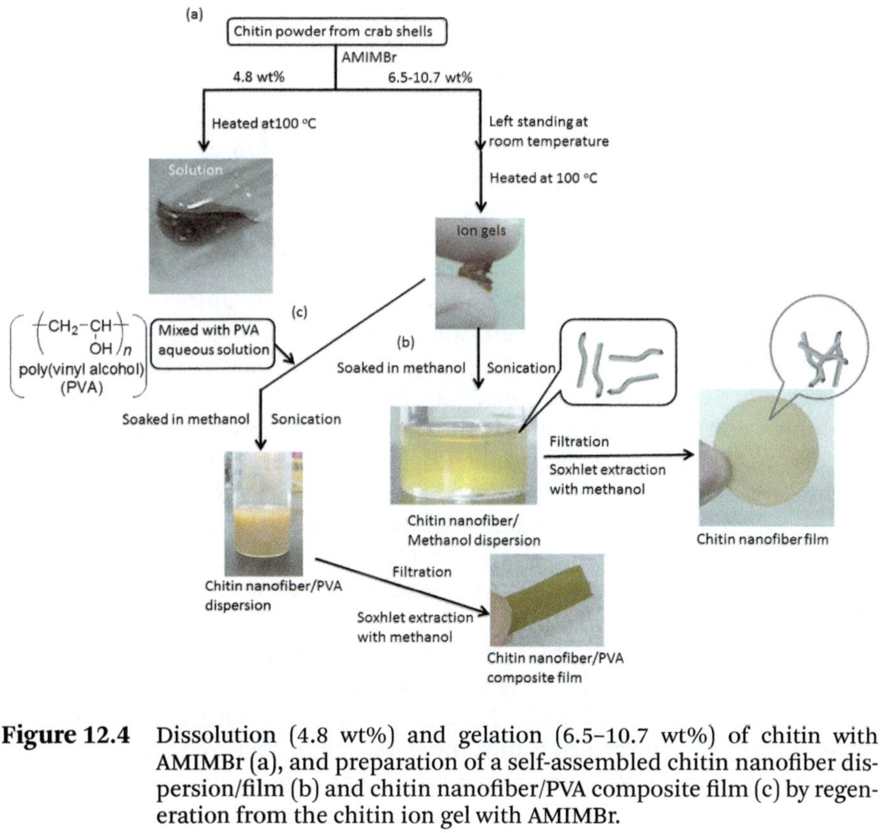

Figure 12.4 Dissolution (4.8 wt%) and gelation (6.5–10.7 wt%) of chitin with AMIMBr (a), and preparation of a self-assembled chitin nanofiber dispersion/film (b) and chitin nanofiber/PVA composite film (c) by regeneration from the chitin ion gel with AMIMBr.

groups in chitin. Accordingly, the chemical species of imidazolium bromide probably show the ability to break hydrogen bonds formed by –N–C=O groups for the dissolution of chitin. However, other imidazolium bromides, such as 1-methyl-3-propylimidazolium and 1-butyl-3-methylimidazolium bromides do not dissolve chitin at all, suggesting that the allyl substituent in AMIMBr also affects the dissolution ability, although the reason why AMIMBr specifically dissolves chitin is not yet clear. The analytical results of the regenerated chitin from the solution with AMIMBr indicated that the degradation and depolymerization of chitin did not frequently occur during the dissolution process.

When larger amounts of chitin (6.5–10.7 wt%) were successively immersed in AMIMBr at room temperature, heated at 100 °C for 24 h, and cooled to room temperature, gel-like materials (ion gels) with higher viscosity were formed (Figure 12.4(a)). Dynamic rheological measurements showed that both the 4.8 wt% and 6.5 wt% liquids of chitin with AMIMBr behaved as weak gels.

When the regeneration of chitin from the 9.1–10.7 wt% ion gels with AMIMBr was carried out by soaking in methanol at room temperature for 24 h, followed by sonication, dispersions of the regenerated chitin were obtained (Figure 12.4(b)).[28,29] The SEM image of the resulting dispersion indicated a nanofiber morphology with the fibers *ca.* 20–60 nm in width and several hundred nm in length, indicating the self-assembly of chitin nanofibers during the regeneration process from the ion gels using methanol. The regenerated chitin formed a film by isolation through filtration of the dispersion. The SEM image of the resulting film showed a highly entangled nanofiber morphology, which probably contributed to the formation of the film. Self-assembled chitin nanofiber/poly(vinyl alcohol) composite films were also fabricated using a co-regeneration approach (Figure 12.4(c)).[28]

As chitin is regarded to be a basic polysaccharide because of the presence of several% of free amino groups in the total repeating units formed by partial deacetylation of acetamido groups, the self-assembled chitin nanofibers were used as a reinforcing agent for acidic carboxymethyl cellulose (CMC) through electrostatic interactions.[30] A CMC film was immersed in the self-assembled chitin nanofiber methanol dispersion, followed by centrifugation and drying, which gave a composite film (Figure 12.5(a)). The self-assembled chitin nanofibers were also employed as a reinforcing agent for a cellulose film obtained from an ion gel with BMIMCl.[31] When the thinly prepared cellulose ion gel with BMIMCl was soaked in the self-assembled chitin nanofiber methanol dispersion, the two polysaccharides were made compatible through regeneration of cellulose. The mixture was then subjected to successive centrifugation, Soxhlet extraction with methanol for the complete removal of BMIMCl, and drying to fabricate a chitin nanofiber-reinforced cellulose film (Figure 12.5(b)). The SEM images of both the surface and cross-section of the resulting film indicated nanofiber morphologies, suggesting that chitin nanofibers were present not only on the surface but also inside the film.

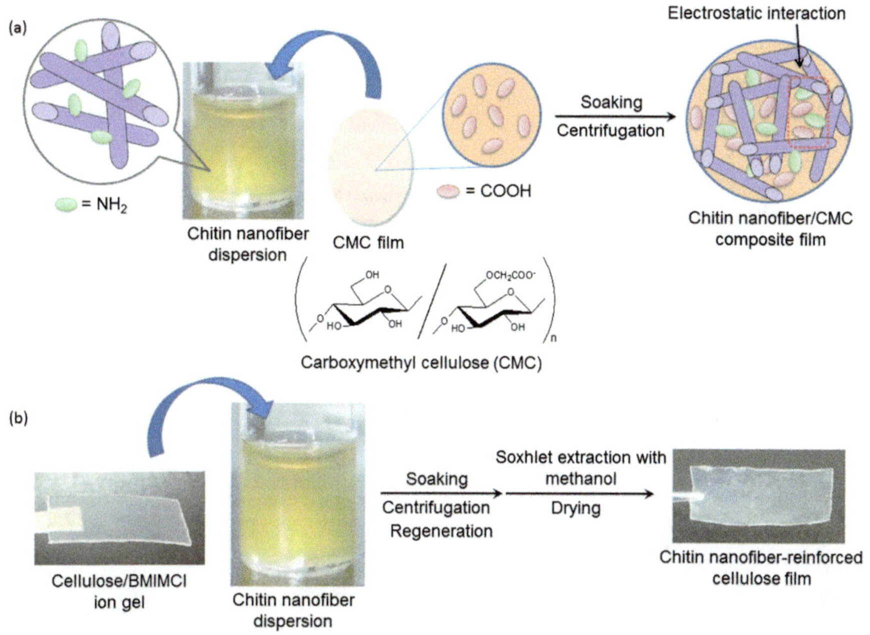

Figure 12.5 Preparation of a self-assembled chitin nanofiber/CMC composite film (a) and self-assembled chitin nanofiber/cellulose composite film (b).

Self-assembled chitin nanofiber-*graft*-synthetic polymer composite films were prepared by surface-initiated graft polymerization (Figure 12.6).[32] For example, chitin nanofiber-*graft*-biodegradable polyester and polypeptide composite films were fabricated by surface-initiated graft ring-opening polymerization of the corresponding cyclic monomers, *i.e.*, L-lactide (LA)/ε-caprolactone (CL) and γ-benzyl L-glutamate-NCA (BLG-NCA), from the chitin nanofiber film.[33,34] The surface-initiated graft atom transfer radical polymerization of vinyl monomers such as 2-hydroxyethyl acrylate (HEA) was also conducted from a self-assembled chitin macroinitiator incorporating initiating sites on the fiber surfaces.[35,36]

A cellulose/chitin binary ion gel using two ionic liquids, BMIMCl and AMIMBr, was also fabricated.[37] A 9.1 wt% cellulose solution with BMIMCl and a 4.8 wt% chitin solution with AMIMBr were first mixed at 100 °C to obtain a homogeneous solution. The resulting solution was left standing at room temperature for 4 days, followed by washing with ethanol to form a cellulose/chitin binary ion gel that excluded excess ionic liquids (Figure 12.7(a)). The analytical results suggested the formation of a relatively good composite between cellulose, chitin, and the ionic liquids. A cellulose/chitin binary film was fabricated by regeneration from the thinly prepared binary ion gel (Figure 12.7(b)).[38]

The cellulose/chitin binary ion gel was used as a novel electrolyte for an electric double layer capacitor (EDLC). The binary ion gel was first treated

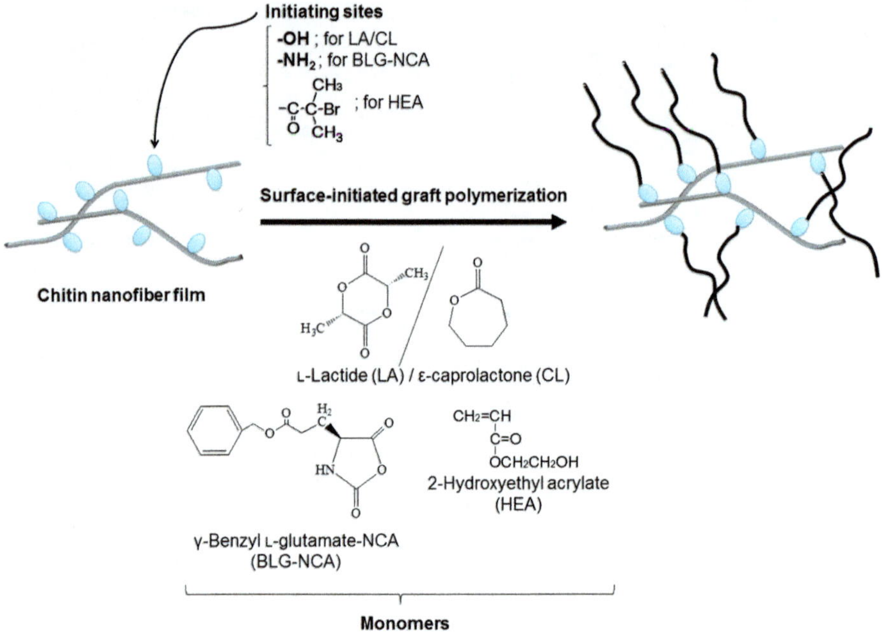

Figure 12.6 Surface-initiated graft (*co*)polymerization of several cyclic and vinyl monomers from chitin nanofiber films with appropriate initiating sites.

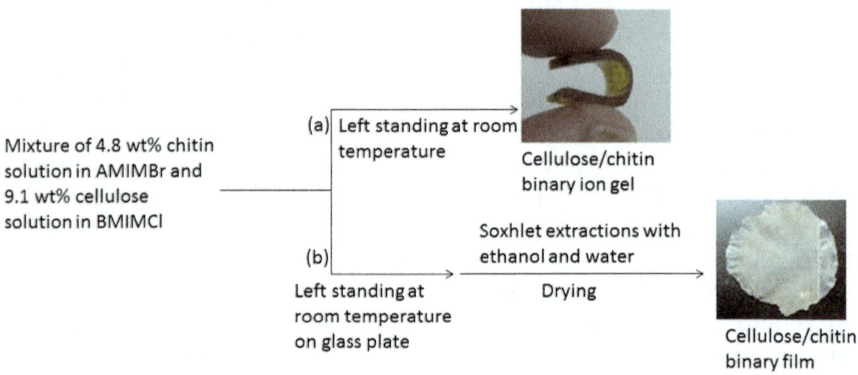

Figure 12.7 Preparation of a cellulose/chitin binary ion gel (a) and film (b) using ionic liquids.

with a 2.0 mol L^{-1} H_2SO_4 aqueous solution for 3 h to give an acidic cellulose/chitin binary gel.[39–41] The electrochemical characteristics of the resulting acidic gel electrolyte were evaluated by galvanostatic charge–discharge measurements. The test cell with the acidic binary gel electrolyte exhibited a specific capacitance of 162 F g^{-1} at room temperature, which was higher than that for a cell with an ordinary H_2SO_4 electrolyte (155 F g^{-1}). The acidic binary gel electrolyte showed an excellent high-rate discharge capability in a wide range of current densities, as well as H_2SO_4 aqueous solution. In addition, the discharge capacitance of the test cell retained over 80% of its initial value after 10^5 cycles, even at a high current density of 5000 mA g^{-1}. The above results indicated that the acidic cellulose/chitin binary gel electrolyte had practical applicability for an advanced EDLC.

12.2.2 Ion Gels of Hydrocolloid Polysaccharides with Ionic Liquids

Besides abundant polysaccharides, such as cellulose and chitin, there are a number of polysaccharides from various sources such as plants, animals, seaweeds, and bacteria in nature, which often have complicated structures composed of one or more kinds of monosaccharide residues and have sometimes exhibited branched structures.[42] Because such polysaccharides form highly viscous aqueous solutions, they have been used as hydrocolloids for stabilizers, viscous agents, and structure providers in the food industry. However, there have not been many investigations into the use of hydrocolloid polysaccharides as components in practical materials. To provide value-added materials from such polysaccharides, therefore, the fabrication of ion gels has also been attempted.[15]

Galactomannans are representative food hydrocolloids, consisting of a β(1→4)-linked mannopyranose main chain with branched α-galactopyranose units at the 6-position.[43] The proportion of galactose and mannose residues depends on the source, and for example, the major galactomannans fenugreek gum (FG), guar gum (GG), and locust bean gum (LBG) have average galactose/mannose ratios of *ca.* 1 : 1, 1 : 1.8, and 1 : 3.5, respectively (Figure 12.1). Solutions of GG with BMIMCl (9.1–28.6 wt%), which were prepared by heating the corresponding mixture at 100 °C for 5 h, completely turned into an ion gel after being left standing at room temperature for 30 min.[44,45] The mechanical properties of the ion gels under both compressive and tensile stress depended on the content of GG in the ion gels. From FG and LBG, the corresponding ion gels were also produced by the same procedures.[46]

It has been reported that imidazolium-type ionic liquids exhibit excitation-wavelength-dependent fluorescent behavior due to the presence of energetically different associated species. For example, BMIMCl typically exhibits emissions at around 450–600 nm depending on the excitation wavelength. The imidazolium-type ionic liquids, which form many different species, have potential as components to contribute to developing new fluorescent functional materials. Accordingly, a unique fluorescent ion gel was obtained by exploiting the above gelling system of GG with BMIMCl, because the fluorescence spectra of the GG ion gel exhibited similar excitation-wavelength-dependent fluorescence behavior following excitation at 260–600 nm.[47]

The unique fluorescent function in the GG ion gel was achieved by means of fluorescence resonance energy transfer (FRET). FRET is an interaction between the electronic excited states of two fluorescent substrates, a donor and an acceptor, in which excitation energy is transferred from the former to the latter without emission of a photon. Besides exhibiting emission following excitation at the characteristic wavelength of each fluorescent dye, fluorescent materials in practical applications are often required to exhibit fluorescence emissions following excitation at different wavelengths. For the purpose of developing such dye materials, the FRET technique has been used in designed fluorescent materials to obtain a large shift of the excitation wavelength from that natively shown by the dyes. The occurrence of FRET in the GG ion gel with BMIMCl containing rhodamine 6G (1.5 mmol L^{-1}), which is a representative red fluorescent dye and exhibits emission maxima at *ca.* 540–610 nm following excitation at around 520 nm, has been found as follows, where BMIMCl and rhodamine 6G act as a donor and an acceptor, respectively. When the fluorescence spectra of the ion gel were measured with excitation at 260–600 nm, emissions due to the dye were observed in all the spectra, whereas no emissions due to BMIMCl were observed (Figure 12.8(a)). These results indicated the occurrence of FRET from BMIMCl

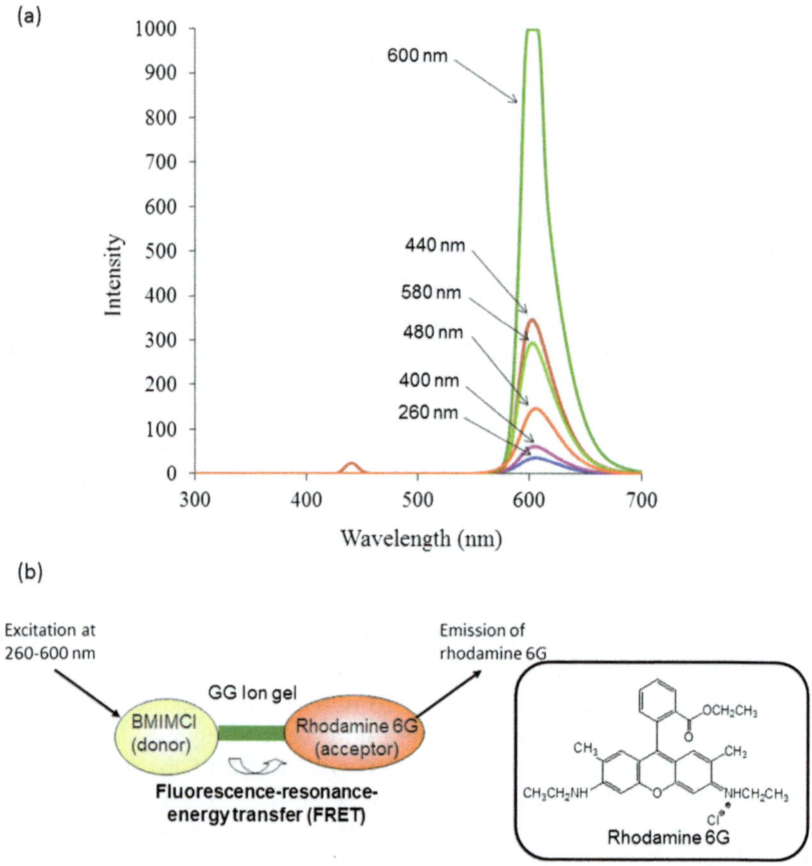

Figure 12.8 Fluorescence spectra of a guar gum/rhodamine 6G/BMIMCl ion gel with excitation at 260–600 nm (a) and image of FRET from BMIMCl (donor) to rhodamine 6G (acceptor) in the GG ion gel (b).

to rhodamine 6G in the ion gel (Figure 12.8(b)). Indeed, the gel showed red emission with photoirradiation at various wavelengths. It was further confirmed that the emission peaks in the fluorescence spectra of the GG ion gel with BMIMCl excited at various wavelengths overlapped with the absorption peak of rhodamine 6G at 545 nm in its solution with BMIMCl.

The GG ion gels were converted into films using a compression technique (60 °C, 4 MPa).[44,45] The film showed temperature-induced shaping ability, in which it became hard upon heating and returned back to being soft again upon cooling. The TGA and XRD analyses of the film indicated that this property was induced by the construction and disruption of crystalline structures of GG in the film by evaporation and absorption of water during the heating–cooling process. Moreover, the ion gels were converted into porous materials by soaking them in ethanol and water, followed by lyophilization.

Anionic hydrocolloid polysaccharides with sulfate and carboxylate groups, such as carrageenans and xanthan gum, have also been used for the fabrication of ion gels with BMIMCl. During the gelation procedure using such anionic polysaccharides, ion-exchange of anion metal salts with BMIMCl takes place to form sulfate- and carboxylate-BMIM ion pairs on the polysaccharide main chains (Figure 12.9), which contribute to enhancement of the mechanical properties and the appearance of unique function. Carrageenans are water-soluble phycocolloids extracted from red algae and are used mostly as a stabilizer and a structural provider in the food and ice cream industries.[48] Three major varieties of carrageenans, *i.e.*, κ-, λ-, and ι-carrageenans, are known, all of which show anionic nature owing to the presence of sulfate groups (Figure 12.1). For the formation of ion gels, carrageenans were first dissolved with BMIMCl (13 wt%) at 100 °C for 10 h with stirring. When the solutions were cooled to room temperature, ion gels were obtained, which could suitably be manipulated after standing for 12 h upon addition of acetone and excess BMIMCl excluded from the gels was washed out with ethanol.[49] κ-Carrageenan formed a hard gel, whereas the other carrageenans formed soft gels.

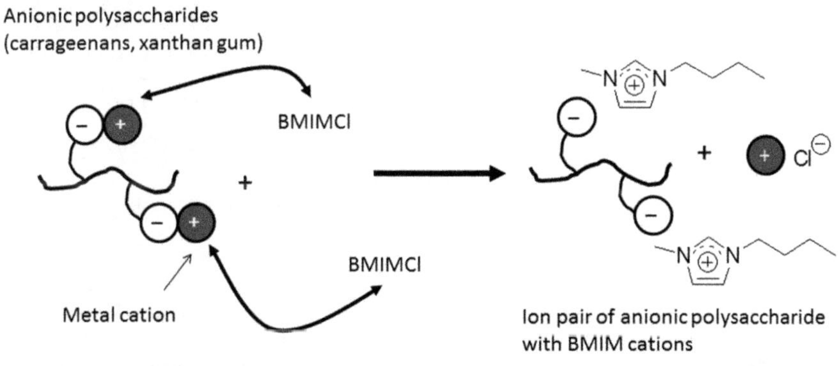

Figure 12.9 Ion exchange between anionic polysaccharides and BMIMCl during gelation.

Cellulose/carrageenan binary ion gels with BMIMCl were obtained by the same procedure as above.[49] When a homogeneous solution of cellulose and carrageenan (15% w/w each) in BMIMCl was prepared by heating at 100 °C for 10 h and then cooled to room temperature, the mixture could suitably be handled after 12 h of standing, followed by treatment with acetone and ethanol, affording a binary ion gel. The mechanical properties of the binary ion gel were superior to those of the cellulose ion gel with BMIMCl. The presence of sulfate–BMIM ion pairs on the carrageenan main chain, which were produced during the gelation procedure, probably contributed to the enhancement of the mechanical properties of the resulting ion gel.

Xanthan gum, produced by *Xanthomonas canpestris*, is composed of a cellulose-type main chain (β(1→4)-glucan) with trisaccharide side chains attached to alternating glucose units in the main chain. The side chains include carboxylate groups, resulting in the anionic nature of xanthan gum (Figure 12.1).[50] Solutions of xanthan gum (9.1–50 wt%), which were prepared by heating the corresponding mixtures at 100 °C for 12 h, completely turned into a gel form after standing at room temperature for 20 min (Figure 12.10(a)).[51–53] The 9.1 wt% ion gel showed a very elastic nature and the gel gradually became harder with an increasing content of xanthan gum in the gel. Furthermore, the following experiment suggested the appearance of thermally induced shape-memory behavior in the xanthan gum ion gel. The designed permanent shape of the gel was first fabricated by a heating–cooling process using an appropriate mold. The gel was then softened and deformed by heating at around 50 °C, and the intended temporary shape was subsequently fixed by cooling at room temperature. When the resulting gel was left standing at around 50 °C, the temporary shape gradually returned back to the permanent shape. The shape memory behavior of the ion gel was induced by the

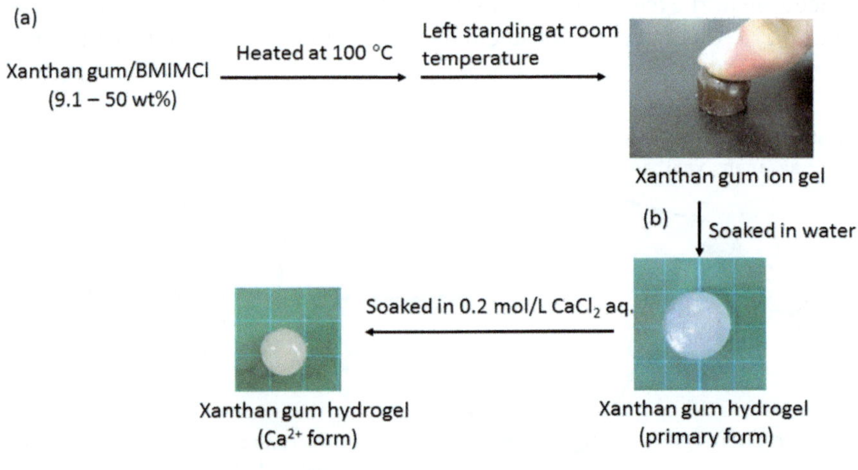

Figure 12.10 Preparation of a xanthan gum ion gel with BMIMCl (a) and conversion into a xanthan gum hydrogel from the ion gel (b); primary form and Ca^{2+} form.

specific association between xanthan gum and BMIMCl owing to the occurrence of ion-exchange of the carboxylate metal salts in the xanthan gum with BMIMCl during the gelation procedure and also to the rigid nature of the xanthan gum main chain (cellulose-type). The analytical data for the ion gel suggested the specific association of BMIMCl in the ion gel, which was constructed on the basis of the regularly ordered intermolecular interaction of BMIM counter-cations of xanthan gum with other BMIMCl molecules present in the gel because of the rigidity of the main chain, contributing to the shape-memory behavior of the ion gel. For example, the UV-vis spectrum of the 23.1 wt% ion gel showed strong absorptions at 200–450 nm due to this specific association, which were not observed in the UV-vis spectrum of liquid BMIMCl.

On the basis of the UV-vis result, the fluorescence behaviors of the ion gels with different xanthan gum contents were investigated.[54] The fluorescence spectra of the ion gels following excitation at 360–480 nm exhibited emission maxima that were shifted to longer wavelengths (from 508 nm to 539 nm) depending on the xanthan gum content, and the color changed from yellow to red–brown with increasing xanthan gum content (Figure 12.11). This red-shift was potentially due to the extended conjugation of BMIMCl, because of the presence of the specific association states of BMIMCl.

Exchange of the dispersion medium from BMIMCl to water was carried out by soaking the 9.1 wt% ion gel in water (1.15 g per 100 mL), affording a xanthan gum hydrogel with 94.1 wt% water content (Figure 12.10(b)).[51,52] This hydrogel was then soaked in 0.2 mol L^{-1} $CaCl_2$ aqueous solution to produce an ionically cross-linked hydrogel with divalent Ca^{2+} (Figure 12.10(b)). The water content of the product was reduced to 85.2 wt% by this procedure, indicating shrinkage by ionic cross-linking. This material showed reversible swelling–shrinking behavior. When the ionically cross-linked hydrogel was soaked in water for 1 day, it was gradually swollen. Soaking the swollen hydrogel in 0.2 mol L^{-1} $CaCl_2$ aqueous solution for 1 day resulted in shrinkage again. The swelling–shrinkage cycles of the hydrogel were repeated further twice. This behavior was explained by the conformation change of xanthan gum, depending on the ionic strength of the aqueous solution. Under compression, the hydrogel with Ca^{2+} showed much better mechanical properties than the primary formed hydrogel, suggesting that cross-linking with Ca^{2+} in the hydrogel strongly contributed to the enhancement of the mechanical properties.

Figure 12.11 Photograph of ion gels with various xanthan gum contents.

Mineralization of hydroxyapatite upon the xanthan gum hydrogel was investigated by an alternate soaking process to produce a composite hydrogel, where the xanthan gum matrix acted as a scaffold owing to the presence of carboxylate groups.[55] The content of hydroxyapatite in the resulting composite hydrogels depended on the soaking cycles. The mineralization of hydroxyapatite induced a microstructure change from a layered structure to a porous structure.

Solutions of cellulose/xanthan gum with BMIMCl, which were prepared by heating the corresponding mixtures at 100 °C for 9 h, were left standing at room temperature for 1 day, leading to the formation of binary ion gels.[56] Furthermore, the thinly prepared binary ion gels on a Petri dish were subjected to Soxhlet extraction with ethanol for regeneration, followed by drying under ambient conditions, affording cellulose/xanthan gum binary films. The XRD profiles of the resulting films indicated the relatively good composition of the two polysaccharides. The binary ion gels were soaked in water to convert them into binary hydrogels. The water contents of the binary hydrogels were always higher than 90% regardless of the cellulose/xanthan gum weight ratio. When increasing the weight ratio of cellulose to xanthan gum, the mechanical properties of both the films under tensile stress and the hydrogels under compressive stress were hardened.

12.3 Polysaccharide–Polymeric Ionic Liquid Composite Materials

12.3.1 Polymeric Ionic Liquids

Polymeric ionic liquids are defined as the polymers obtained by the polymerization of ionic liquids with polymerizable groups (polymerizable ionic liquids).[57,58] Thus, polymeric ionic liquids are not necessarily liquids at room temperature or even at some ambient temperatures. Polymeric ionic liquids, therefore, are often called 'polymerized ionic liquids' or 'poly(ionic liquid)s'. The major advantages of employing polymeric ionic liquids are improved processability and feasibility, and enhanced stability in application as practical materials. Polymerizable ionic liquid monomers can be produced by incorporating polymerizable groups at anionic or cationic sites. In the former case, the counter-anions in general ionic liquids are exchanged with polymerizable anions, affording polymerizable ionic liquids. In the latter case, vinyl, meth(acryloyl), and vinyl benzyl substituents have typically been incorporated as polymerizable groups at the cationic site. The quaternization reaction of vinylbenzyl halides or haloalkyl (meth)acrylates with 1-alkylimidazoles gives the corresponding imidazolium-type polymerizable ionic liquids with the vinyl benzyl or (meth)acryloyl polymerizable group. As 1-vinylimidazole is commercially available, vinylimidazolium-type polymerizable ionic liquids are provided by quaternization of 1-vinylimidazole with a variety of alkyl halides. When quaternization with vinylbenzyl halides or haloalkyl (meth)acrylates is conducted, polymerizable ionic liquids with two polymerizable groups are produced. Because the resulting polymerizable ionic liquids can be converted into insoluble and stable polymeric ionic liquids with a cross-linked structure by radical polymerization, they have great potential to be used as a cross-linker for the production of practical polymeric materials.

For example, a transparent cross-linked polymeric ionic liquid film with the chemical structure of poly(**1**-*co*-**2**) was obtained by radical copolymerization of two polymerizable ionic liquids, 1-methyl-3-(4-vinylbenzyl)imidazolium chloride (**1**) and 1-(3-methacryloyloxypropyl)-3-vinylimidazolium bromide (**2**) (10 : 1) in the presence of AIBN as an initiator at 65 °C for 30 min and subsequently at 75 °C for 2 h (Figure 12.12).[59] The UV-vis spectrum of the film showed small absorptions at wide wavelengths of 280–550 nm, which were probably related to the fluorescence emissions of the imidazolium structure, besides large absorptions at wavelengths below 280 nm. The fluorescence spectra of the film showed excitation-wavelength-dependent fluorescence emission maxima at around 430–470 nm following excitation at 260–400 nm. Indeed, the film exhibited blue emission with UV light irradiation at 365 nm.

Based on the principle of three primary colors, polymeric ionic liquid films that exhibit multicolor emissions depending on combinations of the primary colors have been designed.[60] Three fluorescent dye monomers, namely, rhodamine (red emission), 7-(diethylamino)coumarin-3-carboxylic acid (DEAC, green emission), and pyranine (blue emission) derivatives (**3**, **4**, and **5**), with a methacrylate group were synthesized. Radical copolymerization of **1** and **2**, with **3**, **4**, or **5** was conducted in the presence of AIBN to obtain the polymeric ionic liquid films **6**, **7**, and **8** carrying the respective dye moieties (Figure 12.13). The fluorescence spectra of the resulting films with excitation at 260–400 nm suggested the occurrence of FRET from the units **1** and **2** to the respective dye group in the films.

Three combinations of polymerizable dyes, that is, **3** + **4**, **3** + **5**, and **4** + **5**, were then copolymerized with **1** and **2** in the presence of AIBN to produce polymeric ionic liquid films exhibiting tunable color emissions. The fluorescence spectra of all the resulting films exhibited two kinds of emission due to the incorporated dye moieties with excitation at 260–400 nm, suggesting that the respective dye groups in the films individually emitted *via* direct excitation or FRET. A polymeric ionic liquid film carrying three dye moieties

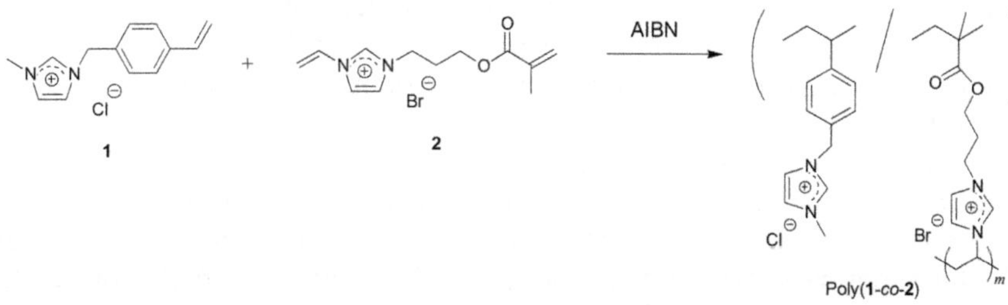

Figure 12.12 Radical copolymerization of polymerizable ionic liquids **1** and **2**.

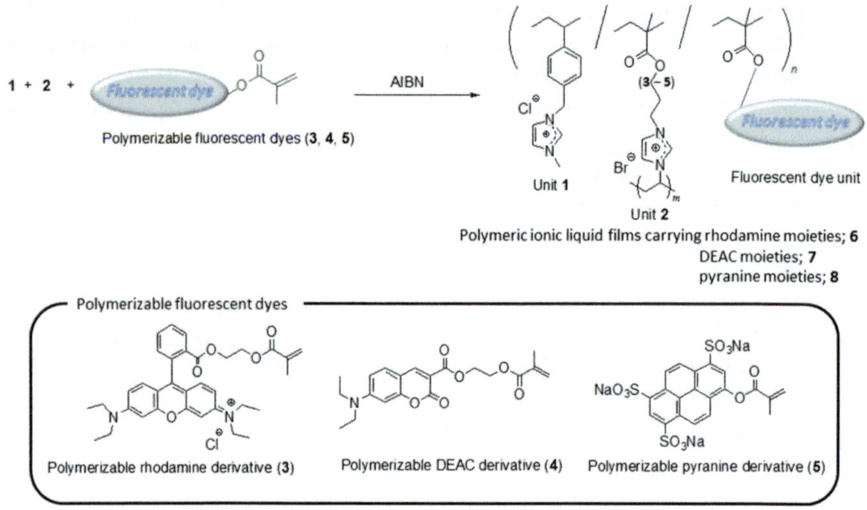

Figure 12.13 Radical copolymerization of **1** and **2**, with **3**, **4**, or **5** to give polymeric ionic liquid films (**6**, **7**, and **8**) carrying primary color fluorescent dye moieties.

was also prepared by copolymerization of **1** and **2**, with the three polymerizable dyes. The fluorescence spectra of the resulting film also showed three kinds of emission excited at 260–400 nm. Accordingly, the resulting films with multiple dye moieties exhibited yellow, magenta, cyan, and white fluorescence emissions, respectively, with UV light irradiation at 365 nm. These results indicated that the polymeric ionic liquid films carrying suitable fluorescent dye moieties emitted tunable multicolors following excitation at a single wavelength.

12.3.2 Preparation of Polysaccharide Films Reinforced by Polymeric Ionic Liquids

Poly(**1**-*co*-**2**) has been used as a reinforcing agent for cellulose/chitin composite films, which were prepared from ionic liquid solutions.[61] For the preparation of the polymeric ionic liquid-reinforced composite film by *in situ* polymerization of **1** and **2**, however, the aforementioned AMIMBr was not suitable as the chitin solvent, because it suppresses the progress of radical polymerization, probably due to the presence of an allyl carbon–carbon double bond. Therefore, the ionic liquid used as the solvent of chitin was BMIMOAc, because it has also been found to dissolve chitin as aforementioned. A solution of **1** + **2** and AIBN in a small amount of methanol was added to a 4.8 wt% chitin solution in BMIMOAc and the resulting solution was mixed with the separately prepared 9.1 wt% cellulose solution in BMIMCl. The homogeneous mixture was then thinly cast onto a Petri dish and heated at 100 °C for 10 h to allow radical polymerization. The resulting material was left standing at room temperature for the formation of an ion gel. The gelled product was subjected to Soxhlet extraction with ethanol, followed by drying to give a cellulose/chitin film composited with poly(**1**-*co*-**2**). The presence of the polymeric ionic liquid in the product was confirmed by IR measurement. The XRD profile of the product suggested that the crystalline structures of cellulose and chitin were largely disrupted, indicating the good composition of the two polysaccharides in the film. The stress–strain curve of the film from tensile testing supported the reinforcing effect of poly(**1**-*co*-**2**) present in the cellulose/chitin composite film.

An *in situ* polymerization approach has been employed to fabricate a FG film reinforced by polymeric ionic liquids.[46] A suspension of FG, **1**, and **2** with AIBN in BMIMCl was thinly cast onto a Petri dish. The suspension was then heated at 100 °C for 9 h for the simultaneous dissolution of FG with BMIMCl and the progress of radical polymerization. The resulting homogeneous material was left standing at room temperature for 3 h for the formation of an ion gel. The gel-like product was subjected to Soxhlet extraction with ethanol, followed by drying under reduced pressure to give a FG film composited with poly(**1**-*co*-**2**). Several composite films were prepared with varying feed ratios of **1** to **2** and their mechanical properties were evaluated by tensile testing. The elongation values at break increased with an increasing unit ratio of **1** in the polymeric ionic liquids, whereas the tensile strength at break decreased. These data indicated that the incorporation of unit **1** in the polymeric ionic liquids affected the higher elastic modulus of the composite films.

12.3.3 Preparation of Polysaccharide–Polymeric Ionic Liquid Composites

As imidazolium-type ionic liquids such as BMIMCl are good solvents for polysaccharides, imidazolium-type polymeric ionic liquids have been considered to show good affinity for them, and accordingly employed as components in composites with polysaccharides, such as cellulose.[62,63] To fabricate a composite with cellulose by *in situ* polymerization, a polymerizable ionic liquid, 1-(4-acryloyloxybutyl)-3-methylimidazolium bromide (**9a**) was first used. Cellulose was dissolved with BMIMCl (9.1 wt%) by heating the corresponding mixture at 100 °C for 24 h. An equimolar amount of **9a** with a repeating glucose unit was then added to the solution and the mixture was heated at 80 °C for 5 h in the presence of AIBN to allow radical polymerization (Figure 12.14).[64] The polymerization mixture was washed with acetone, treated further with refluxed acetone, and subsequently subjected to Soxhlet extraction with methanol to give the composite. The TGA and XRD results suggested the efficient conjugation of cellulose with poly**9a** in the composite.

Although the above approach has been considered to be efficient for the preparation of cellulose–polymeric ionic liquid composites, the procedure should be improved due to the following disadvantageous points: the necessity of two kinds of ionic liquids and the instability of the obtained composites toward solvents. To overcome the second problem, **10** with two polymerizable groups, vinyl and acryloyl, was employed to be copolymerized with **9b**, affording an insoluble and stable cross-linked polymeric ionic liquid, poly(**9b**-*co*-**10**). Furthermore, it was found that a mixture of the two polymerizable ionic liquids mostly disrupted the crystalline structure of cellulose by pre-treatment (immersing cellulose in the mixed solution), suggesting the possible preparation of composites without the use of a solvent ionic liquid such as BMIMCl, unlike the case in the above study. For the preparation of composites, accordingly, cellulose (30 wt%) was first pre-treated in mixtures of **9b** + **10** (feed weight ratio = 100 : 0–50 : 50) at 7 °C for 24 h. The XRD and TGA data of the pre-treated mixtures indicated that the crystalline structure of cellulose was largely disrupted by the pre-treatment. The fabrication of composites was then performed by *in situ* polymerization of the polymerizable ionic liquids in the pre-treated mixtures by heating at 80 °C for 24 h in the presence of AIBN. Consequently, the composites were completely fabricated without any further isolation and purification procedures (Figure 12.15).[65] The XRD and TGA results of the resulting composites indicated the good composition of cellulose and poly(**9b**-*co*-**10**) in all the composites. The

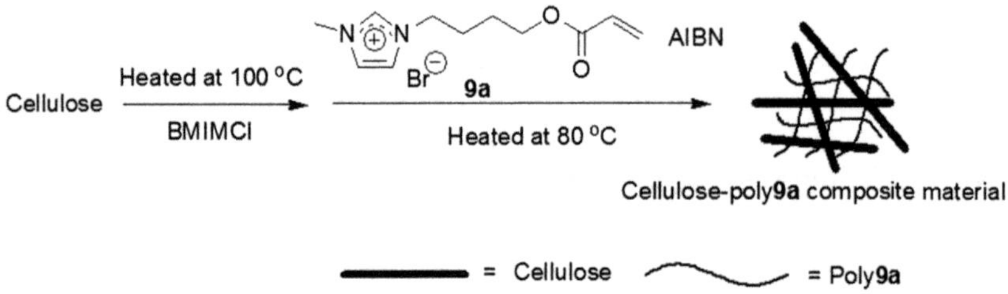

Figure 12.14 Preparation of a cellulose–poly**9a** composite material by *in situ* polymerization of **9a** in BMIMCl.

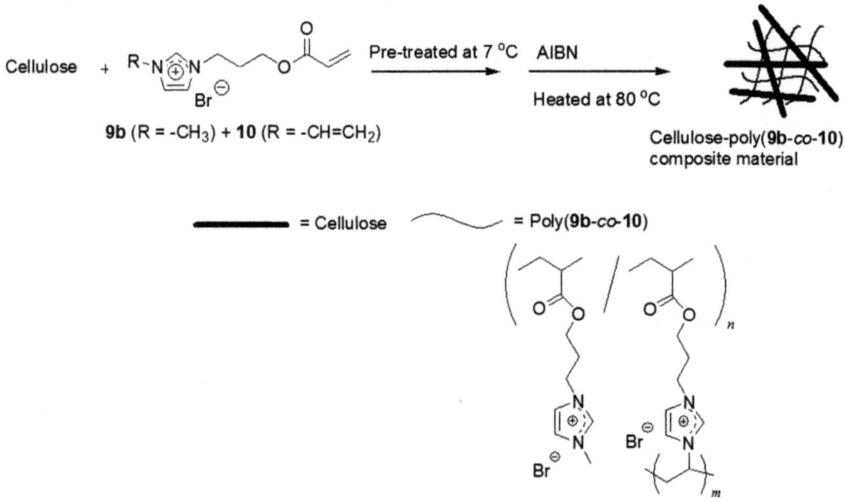

Figure 12.15 Preparation of a cellulose–poly(**9b**-*co*-**10**) composite material by *in situ* copolymerization of **9b** with **10**.

SEM images of the surface and cross-sectional area in the composite showed a completely different morphology from that of standard cellulose, also supporting the good composition of the two components in the composite. The mechanical properties of the composites under tensile stress were affected by the feed ratio of **10** to **9b**.

The above *in situ* polymerization approach was extended to the use of another polymerizable ionic liquid, that is, 1-vinyl-3-(4-vinylbenzyl)imidazolium chloride (**11**) with vinylbenzyl and vinyl polymerizable groups, which gave cross-linked poly**11** by radical polymerization (Figure 12.16).[66] After cellulose (9.1–50.0 wt%) was pre-treated with **11**, the mixtures were heated at 80 °C for 24 h in the presence of AIBN for radical polymerization to occur, affording composites without any further isolation and purification procedures. The XRD and TGA results of all the composites indicated the partial disruption of the crystalline structure of cellulose. The SEM images of the composite supported the good composition between cellulose and poly**11**.

The preparation of a cellulose-based ionic porous material with a polymeric ionic liquid was investigated by *in situ* polymerization of polymerizable ionic liquids. This was successfully conducted by combining the *in situ* polymerization method with a templating technique using an oil/ionic liquid emulsion system.[67] *In situ* polymerization of a mixture of **9b** and **10** was first performed in a solution of cellulose in BMIMCl. Sonication of the mixture in the presence of corn oil and sorbitan monooleate and subsequent treatment with the appropriate solvents gave a porous material. The pore sizes in the materials evaluated from the SEM images were found to be around 0.15–1.3 μm, accompanied by smaller sizes of 30–70 nm.

Following the preparation of the above cellulose composites with polymeric ionic liquids, composites using ι-carrageenan were also prepared by composition with polymeric ionic liquids by means of the *in situ* polymerization technique using **10** and **11**.[68] Because of the presence of sulfate anions, the formation of well-compatible materials with cationic polymeric ionic liquids through cross-linking *via* ion exchange was expected, which possibly contributed to improving the mechanical properties of the composites. For the preparation of a composite using **10**, BMIMCl was used for the dissolution of ι-carrageenan because its direct immersion in **10** was not possible due to its high viscosity. After the dissolution of ι-carrageenan with BMIMCl by

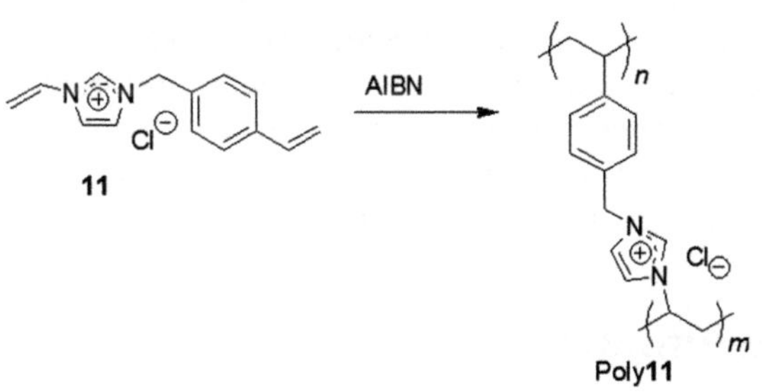

Figure 12.16 Radical polymerization of **11**.

heating at 100 °C for 7 h, radical polymerization of **10** with AIBN in the solution was performed at 90 °C for 5 h, followed by a work-up procedure to give the composite. Because the viscosity of **11** was lower than that of **10**, on the other hand, the preparation of a composite using **11** was carried out without the use of any solvent for the dissolution of ι-carrageenan. Incompatible components were not seen in the charge coupled device camera views (mm scale) of the surfaces of the composites, whereas the SEM images at the μm scale of the two composites exhibited different morphologies. The SEM image of the composite with poly**10** revealed loose structures with nanoscale distributions, while closer packed nanostructures were seen in the SEM image of the composite with poly**11**. The looser structure for the former material was probably constructed during the removal of BMIMCl. The mechanical properties of the ι-carrageenan composites under compressive stress were superior to those of the cellulose composites with the respective polymeric ionic liquids. The better mechanical properties of the ι-carrageenan composites were reasonably explained by the good composition with the polymeric ionic liquids, which was probably due to the cross-linking through ion exchange between ι-carrageenan and the polymeric ionic liquids. The electrical conductivities of the composites with poly**10** and poly**11** were measured to be 6.2×10^{-4} and 1.6×10^{-4} S cm^{-1}, respectively, which were comparable to the conductance of semiconductors and higher than those of the cellulose composites with the respective polymeric ionic liquids.

12.4 Conclusion

This chapter provided an overview of the research into the combined use of ionic liquids with polysaccharides to provide functional materials. Taking the good affinity of ionic liquids for polysaccharides into account, ion gels are efficiently produced from the corresponding solutions. Other types of materials are also obtained from the ion gels *via* the appropriate procedures. Polymeric ionic liquids are also combined with polysaccharides to produce composite materials using an *in situ* polymerization approach. The resulting polysaccharide materials combined with ionic liquids show unique and high-performance properties. Polysaccharides are generally known to show poor solubility and processability due to numerous hydrogen bonds; however, ionic liquids have opened up a new route for their efficient functionalization and materialization as representatively described in this chapter. As some properties of the polysaccharide materials combined with ionic liquids show potential for practical applications, research into approaches using ionic liquids for functionalization and materialization of polysaccharides will increasingly attract much attention in the future.

Acknowledgements

The author is indebted to his co-workers, whose names are found in the references from his papers, for their enthusiastic collaborations.

References

1. O. A. El Seoud, A. Koschella, L. C. Fidale, S. Dorn and T. Heinze, *Biomacromolecules*, 2007, **8**, 2629–2647.
2. T. Liebert and T. Heinze, *BioResources*, 2008, **3**, 576–601.
3. L. Feng and Z. I. Chen, *J. Mol. Liq.*, 2008, **142**, 1–5.
4. A. Pinkert, K. N. Marsh, S. S. Pang and M. P. Staiger, *Chem. Rev.*, 2009, **109**, 6712–6728.
5. M. Gericke, P. Fardim and T. Heinze, *Molecules*, 2012, **17**, 7458–7502.
6. M. Isik, H. Sardon and D. Mecerreyes, *Int. J. Mol. Sci.*, 2014, **15**, 11922–11940.
7. C. Schuerch, in *Encyclopedia of Polymer Science and Engineering*, ed. H. F. Mark, N. Bilkales and C. G. Overberger, John Wiley & Sons, New York, 2nd edn, 1986, vol. 13, pp. 87–162.
8. J. M. Berg, J. L. Tymoczko and L. Stryer, *Biochemistry*, W. H. Freeman, New York, 7th edn, 2012.
9. D. Klemm, B. Heublein, H. P. Fink and A. Bohn, *Angew. Chem., Int. Ed.*, 2005, **44**, 3358–3393.
10. K. Kurita, *Mar. Biotechnol.*, 2006, **8**, 203–226.
11. A. O. Ashogbon and E. T. Akintayo, *Starch-Starke*, 2014, **66**, 41–57.
12. R. P. Swatloski, S. K. Spear, J. D. Holbrey and R. D. Rogers, *J. Am. Chem. Soc.*, 2002, **124**, 4974–4975.
13. C. Graenacher, US patent, 1943176, 1934.
14. J. Kadokawa, in *Handbook of Ionic Liquids: Properties, Applications and Hazards*, ed. J. Mun and H. Sim, Nova Science Publishers, New York, 2012, pp. 455–466.
15. A. Takada and J. Kadokawa, *Biomolecules*, 2015, **5**, 244–262.
16. J. Kadokawa, M. A. Murakami and Y. Kaneko, *Carbohydr. Res.*, 2008, **343**, 769–772.
17. J. Kadokawa, M. Murakami, A. Takegawa and Y. Kaneko, *Carbohydr. Polym.*, 2009, **75**, 180–183.
18. M. E. Zakrzewska, E. Bogel-Łukasik and R. Bogel-Łukasik, *Energy Fuels*, 2010, **24**, 737–745.
19. W. T. Wang, J. Zhu, X. L. Wang, Y. Huang and Y. Z. Wang, *J. Macromol. Sci., Part B: Phys.*, 2010, **49**, 528–541.
20. M. M. Jaworska, T. Kozlecki and A. Gorak, *J. Polym. Eng.*, 2012, **32**, 67–69.
21. J. Kadokawa, *Green Sustainable Chem.*, 2013, **03**, 19–25.
22. J. Kadokawa, *RSC Adv.*, 2015, **5**, 12736–12746.
23. J. Kadokawa, *Pure Appl. Chem.*, 2016, **88**, 621–629.
24. Y. Wu, T. Sasaki, S. Irie and K. Sakurai, *Polymer*, 2008, **49**, 2321–2327.
25. Y. Qin, X. M. Lu, N. Sun and R. D. Rogers, *Green Chem.*, 2010, **12**, 968–971.
26. K. Prasad, M. Murakami, Y. Kaneko, A. Takada, Y. Nakamura and J. Kadokawa, *Int. J. Biol. Macromol.*, 2009, **45**, 221–225.
27. Y. S. Vygodskii, E. L. Lozinskaya and A. S. Shaplov, *Macromol. Rapid Commun.*, 2002, **23**, 676–680.
28. J. Kadokawa, A. Takegawa, S. Mine and K. Prasad, *Carbohydr. Polym.*, 2011, **84**, 1408–1412.

29. R. Tajiri, T. Setoguchi, S. Wakizono, K. Yamamoto and J. Kadokawa, *J. Biobased Mater. Bioenergy*, 2013, **7**, 655–659.
30. D. Hatanaka, K. Yamamoto and J. Kadokawa, *Int. J. Biol. Macromol.*, 2014, **69**, 35–38.
31. J. Kadokawa, R. Endo, D. Hatanaka and K. Yamamoto, *J. Polym. Environ.*, 2015, **23**, 348–355.
32. J. Kadokawa, *Coatings*, 2016, **6**, 27.
33. T. Setoguchi, K. Yamamoto and J. Kadokawa, *Polymer*, 2012, **53**, 4977–4982.
34. J. Kadokawa, T. Setoguchi and K. Yamamoto, *Polym. Bull.*, 2013, **70**, 3279–3289.
35. K. Yamamoto, S. Yoshida and J. Kadokawa, *Carbohydr. Polym.*, 2014, **112**, 119–124.
36. R. Endo, K. Yamamoto and J. Kadokawa, *Fibers*, 2015, **3**, 338–347.
37. A. Takegawa, M. Murakami, Y. Kaneko and J. Kadokawa, *Carbohydr. Polym.*, 2010, **79**, 85–90.
38. J. Kadokawa, K. Hirohama, S. Mine, T. Kato and K. Yamamoto, *J. Polym. Environ.*, 2012, **20**, 37–42.
39. S. Yamazaki, A. Takegawa, Y. Kaneko, J. Kadokawa, M. Yamagata and M. Ishikawa, *Electrochem. Commun.*, 2009, **11**, 68–70.
40. S. Yamazaki, A. Takegawa, Y. Kaneko, J. Kadokawa, M. Yamagata and M. Ishikawa, *J. Power Sources*, 2010, **195**, 6245–6249.
41. S. Yamazaki, A. Takegawa, Y. Kaneko, J. Kadokawa, M. Yamagata and M. Ishikawa, *J. Electrochem. Soc.*, 2010, **157**, A203–A208.
42. A. M. Stephen, G. O. Phillips and P. A. Williams, *Food Polysaccharides and Their Applications*, CRC/Taylor & Francis, Boca Raton, FL, 2nd edn, 2006.
43. I. C. M. Dea and A. Morrison, in *Advances in Carbohydrate Chemistry and Biochemistry*, ed. R. S. Tipson and H. Derek, Academic Press, 1975, vol. 31, pp. 241–312.
44. K. Prasad, H. Izawa, Y. Kaneko and J. Kadokawa, *J. Mater. Chem.*, 2009, **19**, 4088.
45. S. Mine, K. Prasad, H. Izawa, K. Sonoda and J. Kadokawa, *J. Mater. Chem.*, 2010, **20**, 9220.
46. J. Kadokawa, T. Kato, M. Setoyama and K. Yamamoto, *J. Polym. Environ.*, 2013, **21**, 512–519.
47. H. Izawa, S. Wakizono and J. Kadokawa, *Chem. Commun.*, 2010, **46**, 6359–6361.
48. V. D. Prajapati, P. M. Maheriya, G. K. Jani and H. K. Solanki, *Carbohydr. Polym.*, 2014, **105**, 97–112.
49. K. Prasad, Y. Kaneko and J. Kadokawa, *Macromol. Biosci.*, 2009, **9**, 376–382.
50. L. D. Melton, L. Mindt and D. A. Rees, *Carbohydr. Res.*, 1976, **46**, 245–257.
51. H. Izawa and J. Kadokawa, *J. Mater. Chem.*, 2010, **20**, 5235–5241.
52. H. Izawa, Y. Kaneko and J. Kadokawa, *J. Mater. Chem.*, 2009, **19**, 6969–6972.
53. K. Iimori, K. Yamamoto and J. Kadokawa, *Plast. Polym. Technol.*, 2016, **4**, 63.

54. J. Kadokawa, in *Ionic Liquids–New Aspects for the Future*, ed. J. Kadokawa, InTech, Rijeka, Croatia, 2013, pp. 653–672.
55. H. Izawa, S. Nishino, H. Maeda, K. Morita, S. Ifuku, M. Morimoto, H. Saimoto and J. Kadokawa, *Carbohydr. Polym.*, 2014, **102**, 846–851.
56. M. Setoyama, K. Yamamoto and J. Kadokawa, *J. Polym. Environ.*, 2014, **22**, 298–303.
57. O. Green, S. Grubjesic, S. W. Lee and M. A. Firestone, *J. Macromol. Sci., Part C: Polym. Rev.*, 2009, **49**, 339–360.
58. D. Mecerreyes, *Prog. Polym. Sci.*, 2011, **36**, 1629–1648.
59. S. Wakizono, K. Yamamoto and J. Kadokawa, *J. Photochem. Photobiol., A*, 2011, **222**, 283–287.
60. S. Wakizono, K. Yamamoto and J. Kadokawa, *J. Mater. Chem.*, 2012, **22**, 10619–10624.
61. M. Setoyama, T. Kato, K. Yamamoto and J. Kadokawa, *J. Polym. Environ.*, 2013, **21**, 795–801.
62. J. Kadokawa, in *Ionic Liquids: Application and Perspectives*, ed. A. Kokorin, InTech, Rijeka, Croatia, 2011, pp. 95–114.
63. J. Kadokawa, *J. Biobased Mater. Bioenergy*, 2013, **7**, 3–11.
64. M. Murakami, Y. Kaneko and J. Kadokawa, *Carbohydr. Polym.*, 2007, **69**, 378–381.
65. A. Takegawa, M. Murakami, Y. Kaneko and J. Kadokawa, *Polym. Compos.*, 2009, **30**, 1837–1841.
66. J. Kadokawa, M. Murakami and Y. Kaneko, *Compos. Sci. Technol.*, 2008, **68**, 493–498.
67. K. Prasad, S. Mine, Y. Kaneko and J. Kadokawa, *Polym. Bull.*, 2010, **64**, 341–349.
68. K. Prasad and J. Kadokawa, *Polym. Compos.*, 2010, **31**, 799–806.

CHAPTER 13

Tailoring Transport Properties Aiming for Versatile Ionic Liquids and Poly(Ionic Liquids) for Electrochromic and Gas Capture Applications

ROBERTO M. TORRESI*[a], CINTIA M. CORRÊA[a], TÂNIA M. BENEDETTI[b] AND VITOR L. MARTINS[a,c]

[a]Dpto de Química Fundamental, Instituto de Química, Univ. de São Paulo, Av. Prof. Lineu Prestes 748, 05508 São Paulo, Brazil; [b]School of Chemistry, University of New South Wales, Kensington NSW 2052, Sydney, Australia; [c]Chemical and Biological Engineering, University of Sheffield, Sir Robert Hadfield Building, Mappin Street, Sheffield S1 3JD, England, UK
*E-mail: rtorresi@iq.usp.br

13.1 Introduction

Ionic liquids (ILs) are unique materials since they are fluids composed solely of ions. This feature has a wide range of implications in terms of their physicochemical characteristics. For example, they possess intrinsic ionic conductivity due to the movement of their ions, and this is one of the most interesting properties, alongside electrochemical stability, when they are

Smart Materials No. 29
Polymerized Ionic Liquids
Edited by Ali Eftekhari

Published by the Royal Society of Chemistry, www.rsc.org

considered for use as electrolytes in electrochemical devices, such as batteries and electrochemical double-layer capacitors, among others.

Poly(ionic liquid)s (PILs) are polymers in which the repeating unit contains moieties of ILs. It is possible to polymerize units that contain the cation or the anion of the IL, and in this way, there are PILs derived from many different monomers (Scheme 13.1). If the cationic species is polymerized, then a polycation-type PIL is formed; if the anionic species is polymerized, then a polyanion-type PIL is formed. Moreover, it is possible to polymerize both the cation and the anion or different regions of the same cation building up a network.[1] In the same way, as with most polymers, it is also possible to produce co-polymers; thus, the structure possibilities are endless.

Most of the studied monomers are those that result in cationic PILs (Scheme 13.1, I), particularly the ones based on vinylimidazolium to produce imidazolium-based PILs, which are the most studied PILs.[1–5] Anionic-type PILs, however, are less studied due to the difficulty of synthesizing these materials.[6] Furthermore, PILs that contain spacer groups are common. Spacer groups are structures that connect the polymer chain and the bonded IL moiety, resulting in increased conductivity, for example. Figure 13.1 shows one example of a PIL with a spacer group.[7]

Shaplov *et al.* (2011)[22] investigated how the length of the alkyl spacer influences the properties of PILs (Figure 13.2). In this case, it was shown that increasing the spacer group from 2 to 10 carbon atoms resulted in an increase of the PIL conductivity by two orders of magnitude from 3.0×10^{-8} ($-(CH_2)_2-$, Figure 13.2A) to 3.2×10^{-6} S cm^{-1} ($-(CH_2)_{10}-$, Figure 13.2C). Moreover, the glass transition temperature (T_g) decreases as the spacer group length increases.[1,15]

Moreover, it is possible to modify the structures of the typical IL moieties to produce ILs or monomers with properties of interest. These modified/functionalized ILs or PILs are generally called task-specific ILs or PILs. He *et al.* (2016) prepared ILs and PILs functionalized with $-NH_2$ or ether to improve the capacity for capturing CO_2 (this will be discussed in Section 13.4).[23] Practically all functionalized ILs – chiral, acidic, basic, metal-containing, guanidinium, and –OH group ILs, among others[24] (some examples are shown in Scheme 13.1, II) – can contain a polymerizable group and become a functionalized PIL. As with ILs, the properties of PILs can be tuned by changing the type (anionic, cationic and others), the nature of the cation and anion, the quantity of ionic centers in each monomer repeating unit and the absence or presence of spacer groups and their nature.[22] After obtaining the monomer, the polymerization can occur in several ways: free radical polymerization using 2,2′-azobis(2-methylpropionitrile) (AIBN)[22,25] or 2,2′-azobis(2-methylpropianamidine)dihydrochloride (AAPH),[26,27] atom transfer radical polymerization,[28] photopolymerization,[5] ring-opening metathesis[28] and cyclopolymerization.[22,28]

PILs are generally solids; consequently, their properties are different from those of their liquid counterparts. An important example is the lower conductivity of PILs,[29] which is a result of the exponential increase in viscosity.

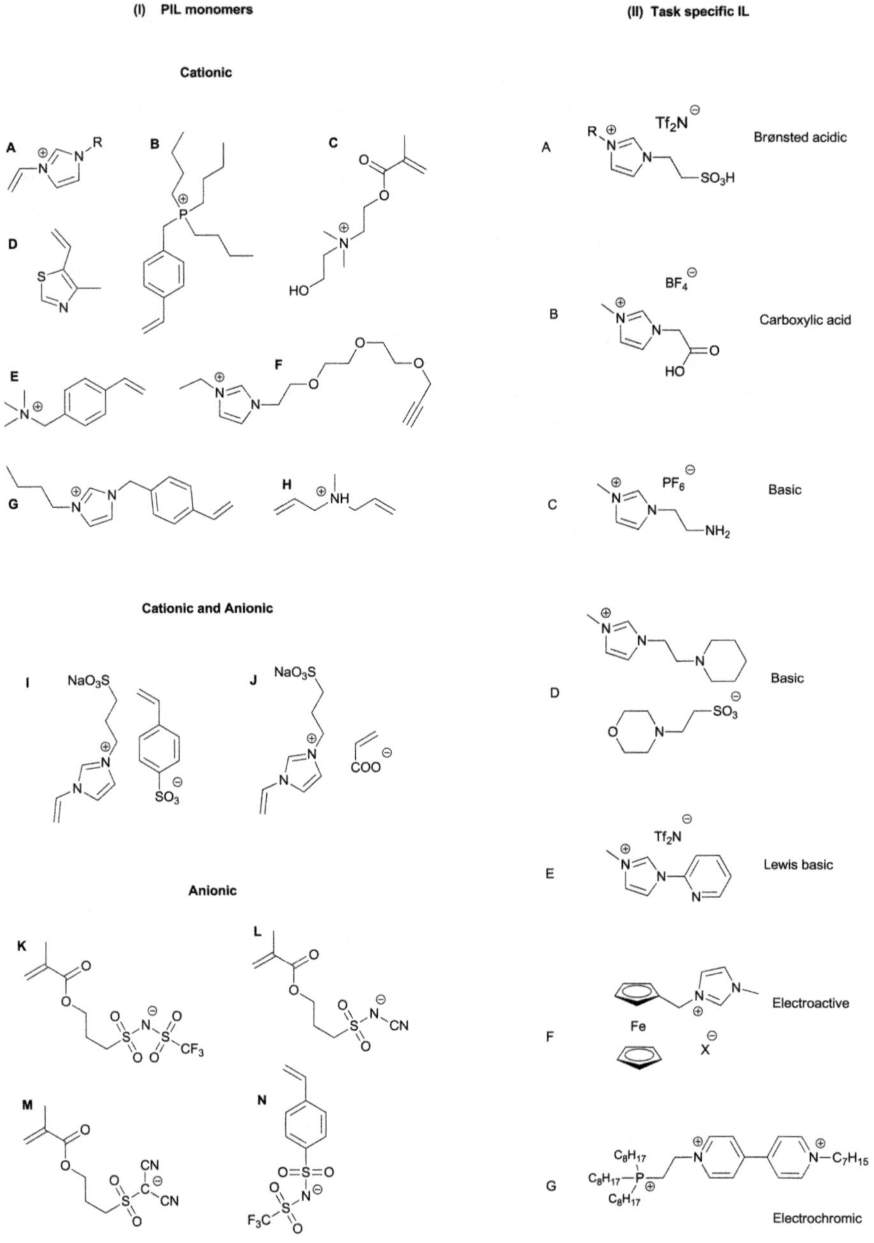

Scheme 13.1 (I) Structures of some PIL monomers. (II) Task-specific ILs.[2–4,8–21]

However, PILs are still very conductive compared with electrically neutral polymers and can be used as polyelectrolytes, for example. The ionic conductivities of some PILs are shown in Table 13.1.

Ionic conductivity affects the transport properties of both ILs and PILs and can be modulated with different combinations of cations and anions,

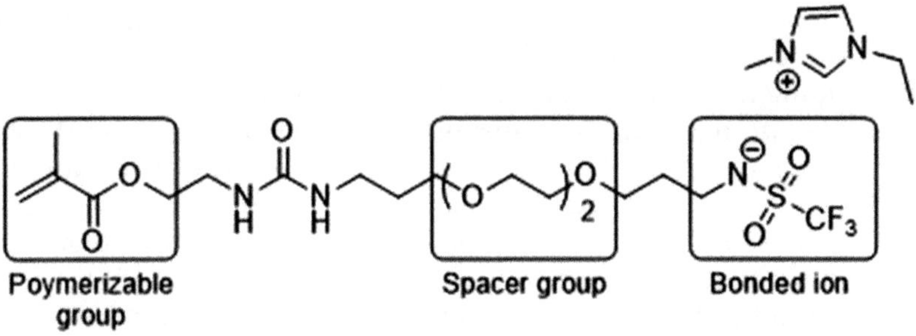

Figure 13.1 A type of spacer group used by Juger *et al.* (2009). Reprinted from *Tetrahedron Letters*, 50 (1), J. Juger, F. Meyer, F. Vidal, C. Chevrot, D. Teyssié, Synthesis, polymerization and conducting properties of an ionic liquid-type anionic monomer, 128–131, Copyright (2009) with permission from Elsevier.[7]

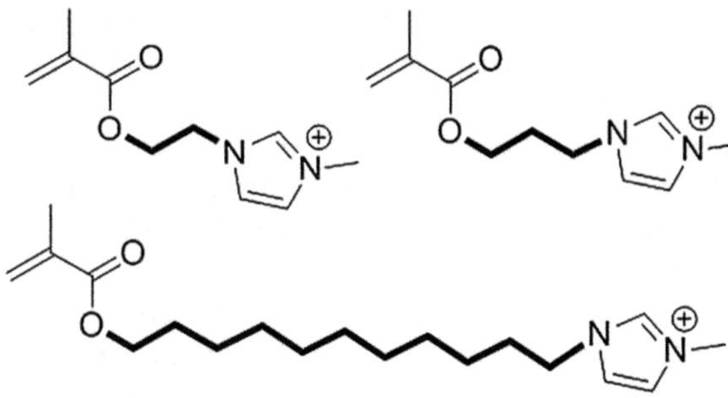

Figure 13.2 Different spacer group lengths used by Shavlov (2011). For further reading please see ref. 22.

Table 13.1 Ionic conductivities of some PILs. Data from ref. 30.

	PIL	Counter ion	Conductivity (S cm^{-1})
A	poly(1-ethyl-3-vinyl-imidazolium)	$[N(CN)_2]$	1.4×10^{-5}
B	poly(1-ethyl-3-vinyl-imidazolium)	$[N(CF_3SO_2)_2]$	2.5×10^{-11}
C	poly(1-ethyl-3-vinyl-imidazolium)	$[CF_3SO_3]$	4.9×10^{-12}
D	poly(1-[(2-methacryloyloxy) ethyl]-3-butylimidazolium	$[N(CF_3SO_2)_2]$	4.0×10^{-4}
E	poly(1-[(2-methacryloyloxy) ethyl]-3-butylimidazolium	$[CF_3SO_3]$	1.5×10^{-5}
F	poly(1-[(2-methacryloyloxy) ethyl]-3-butylimidazolium	$[BF_4]$	6.5×10^{-6}
G	poly(1-[(2-methacryloyloxy) ethyl]-3-butylimidazolium	$[PF_6]$	3.8×10^{-6}
H	Vinyl acids with carboxylic, phosphoric, *p*-benzenesulfonic and sulfuric groups		Highest conductivity 1.1×10^{-4}

and variations in the size and nature of the lateral chains, and type of PIL (anionic, cationic and others). Therefore, many efforts have been devoted to understanding the influence of these structural changes.

Moosavi *et al.* reported the transport properties of dicationic ILs with a short alkyl chain length. Dicationic ILs are composed of two cations that are covalently linked by a spacer, typically an alkyl chain. These types of ILs are generally used as surfactants, solvents for high-temperature organic reactions, high-temperature lubricants, and electrolytes for secondary batteries, electrospray ionization mass spectroscopy, dye-sensitized solar cells and others.[31] Compared with conventional monocationic ILs, dicationic ILs offer more possible combinations to generate different compounds. Therefore, it is possible to achieve new or enhanced properties with a specific combination of cations and anions.

Regarding PILs, Lee *et al.* (2013)[32] studied the transport properties of polyvinylimidazolium PILs and the influence of variations in the alkyl chains or other pendant groups (ethyl, mono or triethylene oxide) while maintaining the anionic species [NTf_2]. In this case, the glass transition temperature of the PILs decreased and the ionic conductivity increased as the ethylene oxide pendant groups were elgonated.[32,33]

Tuning the physicochemical properties of ILs and PILs by changing their structures is clearly a widely considered approach. Research groups have studied aspects ranging from small changes in the alkyl chains in the cations to completely new anion structures to obtain smart ILs and PILs that can be specifically used in a desired application. In this regard, combinations between different cations and anions, changes in the length of the alkyl side chain and functionalization of ionic structures greatly influence the transport properties of this class of material, primarily because of the physicochemical interactions among the ions. Therefore, with a specific application and possible challenge to overcome in mind, one may design an IL or PIL with the desired characteristics for a specific task by utilizing the options mentioned above. With this in mind, the following sections present considerations on how these modulations affect the physicochemical properties of ILs and PILs and consequently their transport properties, as well as the design of electrochromic devices based on ILs and PILs; additionally, gas capture, specifically CO_2 capture, is also discussed.

13.2 Physicochemical Properties of ILs and PILs and the Effect on Transport

Designing and preparing a smart IL or PIL for a specific task is a laborious and challenging task. It is necessary to take into account that changes in the structures of ILs and PILs for a desired application will affect the physicochemical properties of the materials. Therefore, having a good design and synthesis route is as important as performing reliable physicochemical characterization of the product. It is important to purify the product as much as possible and report any impurity, such as halide remaining from the synthesis and any trace of moisture, because it is well known that such impurities have a great impact on the physicochemical and electrochemical properties of ILs.[34,35]

13.2.1 Density of ILs

Although the material density, ρ (g cm^{-3}), is not directly related to the transport property characteristics, it is important to determine the density when analyzing the ionic conductivity of ILs since it is a key factor in calculating the molar ionic conductivity. If one has the material ionic conductivity σ (S cm^{-1}), then the molar ionic conductivity Λ (S cm^2 mol^{-1}) is easily calculated by multiplying the first by the IL molar mass M (g mol^{-1}) and dividing it by ρ, as shown in eqn (1):

$$\Lambda\left(\mathrm{S\ cm^2\ mol^{-1}}\right) = \frac{\sigma\left(\mathrm{S\ cm^{-1}}\right)\cdot M\left(\mathrm{g\ mol^{-1}}\right)}{\rho\left(\mathrm{g\ cm^{-3}}\right)} \tag{13.1}$$

With this in mind, the search for ILs that present a low density, such as those containing anions based on cyano groups, is understandable.[36–38] Ye and Shreeve proposed the equation $\rho_{\mathrm{calc.}} = M/V \times 0.6022$, where V is the theoretical volume of each ion,[39] and a few examples are presented in Table 13.2. Later, Gardas and Coutinho proposed an extended version of this equation (eqn (2)), which takes pressure and temperature into account:

$$\rho_{\mathrm{calc.}} = \frac{M\left(\mathrm{g\ mol^{-1}}\right)}{N\cdot V\left(\mathrm{\AA^3}\right)\left[a + b\cdot T(\mathrm{K}) + c\cdot p(\mathrm{MPa})\right]} 10^{24}\left(\mathrm{\AA^3 cm^{-3}}\right) \tag{13.2}$$

where a, b and c are the correction factors 0.8005, 6.652×10^{-4} K^{-1}, and -5.919×10^{-4} MPa^{-1}, respectively.[40] The extended equation has been shown to be

Table 13.2 Volumes of ion groups for predicting ionic liquid density. Data from ref. 39.

Ions or groups	Volume (Å^3)
[mmim]	154
[mmpyr]	169
[N_{1111}]	136
[P_{1111}]	163
$-CH_2-$	+28
$-CH_3$	+35
–H	+7
[PF_6]	107
[BF_4]	73
[NTf_2]	248
[$B(CN)_4$][41]	143
[$N(CN)_2$]	86

very accurate in predicting the densities of various ionic liquids in the temperature range suggested by the authors (273.15 to 393.15 K) and even in further temperatures outside this range.

The accuracy of the equation can be observed in Figure 13.3, where the calculated density and experimental measurements for different families of ILs in a broad range of temperatures are presented. Note that one can use the theoretical volume of a chemical group to be replaced by a cation or anion and estimate the density of the designed IL.

13.2.2 Transport Properties of ILs and PILs

The relationship between ionic conductivity and viscosity is straightforward to understand: as viscosity increases, the ionic conductivity must decrease because the charge carriers will have to overcome a higher frictional force to move, thus resulting in lower mobility. The following equations (eqn (3) and (4)) describe why this phenomenon occurs. From the Nernst–Einstein equation,

$$\Lambda_{\mathrm{m}}^{0} = \frac{F^2}{RT}\left(\upsilon_{+} z_{+}{}^{2} D_{+} + \upsilon_{-} z_{-}{}^{2} D_{-}\right) \tag{13.3}$$

we can affirm that the ionic conductivity is directly proportional to the diffusion coefficient (D) of the charged species, and from the Stokes–Einstein equation,

$$D_{\mathrm{i}} = \frac{k_{\mathrm{B}} T}{6\pi \eta r_{\mathrm{i}}} \tag{13.4}$$

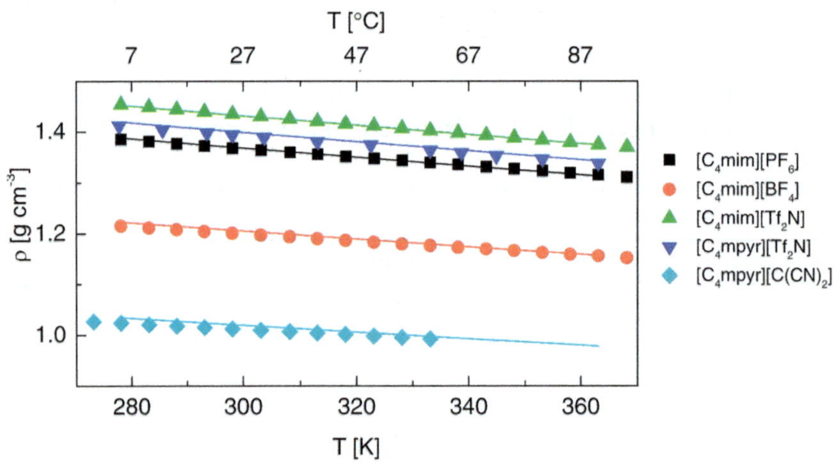

Figure 13.3 Experimental densities of ILs (symbols) and the densities calculated using eqn (2) of Gardas and Coutinho (lines). Experimental data from $[C_4mim][PF_6]$ (squares),[42] $[C_4mim][BF_4]$ (circles),[42] $[C_4mim][NTf_2]$ (upward triangles),[42] $[C_4mpyr][NTf_2]$ (downward triangles),[43] and $[C_4mpyr][C(CN)_2]$ (diamonds).[44]

we find that D is inversely proportional to viscosity. In addition, D is also inversely proportional to the charge carrier's radius and increases with temperature.

Based on the relationship between ionic conductivity and viscosity and the use of the Walden rule for dilute aqueous electrolytes,[45] and later for non-aqueous electrolytes[46] and even molten salts,[47] Angell's group[48] classified a range of ILs based on this relationship. The specific ionic conductivity or molar ionic conductivity is considered to be proportional to the liquid fluidity, which in turn is the reciprocal viscosity, as shown in eqn (5):

$$\Lambda = a \cdot \eta^{-1} \tag{13.5}$$

where a is a constant. There are two plots that rely on this relationship. The first one is the so-called Walden plot, which presents $\log(\Lambda/\mathrm{S\ cm^2\ mol^{-1}})$ *vs.* $\log(\eta^{-1}/\mathrm{Poise^{-1}})$. Interestingly, the plot of a dilute aqueous KCl solution is a line with a slope of 1.0. This works as a calibration curve, where ILs may be classified according to their position relative to the "ideal line", as shown in Figure 13.4a. ILs located in the upper right top corner are considered *good ionic liquids*, and as they drop down relative to the ideal line, the classification goes to *poor ionic liquids* until it reaches the right bottom corner, which

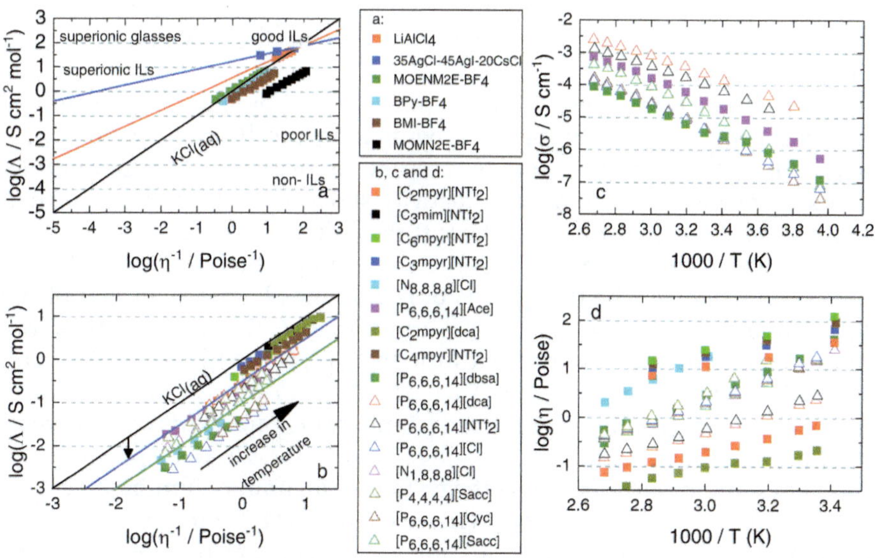

Figure 13.4 (a and b) Walden plot and Arrhenius-like plot for (c) the ionic conductivity and (d) the viscosity of several ionic liquids. (a) Reprinted (adapted) with permission from ref. 48. Copyright (2003) American Chemical Society, and (b–d) Reproduced from ref. 49 with permission from The Royal Society of Chemistry.

is classified as *non-ionic liquids*. Figure 13.4b shows a wide range of different ILs, with varying cations and anions, in the Walden plot.[48,49] MacFarlane *et al.*[49] considered the observations of Bonhôte and collaborators[50,51] and Abbott[52] that take the radii of the ions into account to explain and better understand the ionicity of ILs. They found that the Walden plot can be useful for analyzing the transport properties of ILs; however, to describe their ionicity – *i.e.*, the availability of ions to carry charge – it is necessary to have information on the thermodynamic activity of the ions in the ILs.[49]

The transport properties, such as ionic conductivity (Figure 13.4c), viscosity (Figure 13.4d) and self-diffusion, are often shown in an Arrhenius-like plot, *i.e.*, the transport property *vs.* $1/T$. Rather than the usual Arrhenius linear profile, ILs generally show a certain curvature and are fitted with the Vogel–Fulcher–Tammann (VFT) equation (eqn (6)):

$$A = A_0 \exp\left(\frac{B}{T - T_0}\right) \tag{13.6}$$

where A is the transport property and A_0, B and T_0 are adjustable parameters. Ionic conductivity and the self-diffusion coefficient present a negative B. Furthermore, the relationship between $|B|/T_0$ is inversely related to the liquid fragility; thus, if the relationship decreases, then the fragility of the liquid increases, which means that the transport property will experience more changes with temperature changes.[48,53] Although the IL fragility is often evaluated in the literature, this characteristic is also important when measured near the T_g, and changes in the curve profile may be observed at different temperatures.[54]

The transport properties of ILs are more affected by changing the anions than the cations. For example, ILs containing the $[C_4mim]$ cation, with $[PF_6]$[55], $[BF_4]$[55,56] or $[NTf_2]$[57] anions, present viscosities of 312, 180 and 50 mPa s, respectively. The ionic conductivity shows the expected trend when keeping the $[C_4mim]$ cation and these three anions are used: it increases as the viscosity decreases, but not in the same proportion. For example, the ionic conductivity of the $[NTf_2]$ IL is only 2.6 times higher than that of the $[PF_6]$ liquid, whereas the viscosity is more than 6 times lower. However, ILs containing $[C_4mpyr]$ paired with the $[NTf_2]$[58], $[B(CN)_4]$[41], $[C(CN)_3]$[59] and $[N(CN)_2]$[44,60] anions present viscosities of 85, 45, 21 and 34 mPa s, respectively; for instance, $[C_4mpyr][BF_4]$ and $[C_4mpyr][PF_6]$ are solids at room temperature, and their melting temperatures are 138 and 70 °C, respectively.[61,62] Another example worth noting is the change of an ammonium-based cation for a phosphonium-based cation. Whereas $[N_{2225}][NTf_2]$[63] presents a viscosity and ionic conductivity of 167 mPa s and 1.7 mS cm^{-1}, respectively, its phosphonium analogue $[P_{2225}][NTf_2]$[63,64] presents values of 85 mPa s and 2.1 mS cm^{-1}, respectively. As discussed previously, viscosity is not the only parameter that impacts the ionic conductivity, and one also needs to consider the considerably higher volume of $[NTf_2]$ (248 *vs.* 107 Å, see Table 13.2), among other parameters involved in the mobility of ions.

ILs that contain imidazolium cations present relatively higher ionic conductivities than other ILs. For example, [C_4mim][NTf_2] has an ionic conductivity that is 77% higher than that of its pyrrolidinium analogue, [C_4mpyr][NTf_2], as discussed above, due to the delocalized positive charge over the aromatic ring decreasing the coulombic interaction with the anion. However, the hydrogen at carbon-2 in the imidazolium ring presents a certain acidity and reacts with metallic lithium to produce carbene and H_2.[57] Replacing the hydrogen at carbon-2 with a methyl group eliminated the IL reactivity with metallic lithium; conversely, the ionic conductivity of [C_4mmim][NTf_2] drops almost 60%, being even lower than that of [C_4mpyr][NTf_2], and the viscosity is almost twice as high as that of [C_4mim][NTf_2].[57] This concern is important when using ILs as electrolytes for Li-ion batteries and must be considered. Kowsari and Fakhraee studied the influence of the side chain in imidazolium-based ILs containing the [NTf_2] anion using molecular dynamics simulations; although they noted that the nonpolarizable force field used in their study underestimates conductivity,[65] their analyses are still valuable for comparing the cations that they considered. They showed that among the ILs studied, [C_4mim][NTf_2] presents higher ionic conductivity and that the functionalization of the butyl chain with a polar –NH_2 group decreases the ionic conductivity. The IL [mim][NTf_2] presents even lower ionic conductivity, but the lowest ionic conductivity among the four ILs studied was the already mentioned [C_4mmim][NTf_2].[65]

MacFarlane and collaborators produced and characterized an extensive list of ILs.[58,66,67] In their search for low-viscosity ionic liquids, they proposed the use of the dicyanamide anion paired with cations based on imidazolium, ammonium and pyrrolidinium.[36] Specifically, the viscosity of [C_4mpyr][$N(CN)_2$] is 34 mPa s, and its ionic conductivity is 13 mS cm^{-1}, which is one of the highest conductivities for an ionic liquid at room temperature, as discussed above. Later, Wolff *et al.* characterized and used pyrrolidinium ILs containing [$N(CN)_2$] as electrolytes for supercapacitors.[68] The high ionic conductivity of liquids containing this anion is very beneficial for the amount of charge stored in this type of device and has resulted in supercapacitors that can store more energy than those containing [C_4mpyr][NTf_2] as the electrolyte. Nevertheless, the drawback of these liquids is the low electrochemical stability compared to those containing the [NTf_2] anion; whereas [C_4mpyr][NTf_2] may operate at 3.5 V, the supercapacitors with [$N(CN)_2$] operate at 2.6 V, which constitutes the energy density.[68]

Another class of ILs that present better transport properties than [C_4mpyr][NTf_2], but also a lower electrochemical stability, is those that contain sulfonium cations and the [NTf_2] anion. Rennie *et al.* characterized three sulfonium ILs, [S_{221}], [S_{222}] and [S_{223}], all containing [NTf_2].[43] These three liquids presented higher ionic conductivities than the [C_4mpyr] analogue throughout the entire range of temperatures studied. All liquids presented similar ionicity according to the Walden plot, but [S_{223}][NTf_2] presented a slightly lower slope, indicating the formation of more aggregates than for its counterparts. Interestingly, although this may result in hindered performance when the supercapacitor is operating at a high current rate, the capacitance stored by the [S_{223}] supercapacitor was the highest when operating at a low current. The authors suggested that this result occurs because of the higher number of aggregates in the [S_{223}] liquid and that at low currents there is sufficient time for those aggregates to be broken and play their role in the capacitance. However, at high currents, there is not sufficient time for the larger aggregates to participate in the charge storage.[43]

Belhocine *et al.* prepared a new class of cations containing a seven-membered alicyclic secondary amine, azepanium. These ILs present high electrochemical stability but poor transport properties.[69] Balducci and collaborators investigated these ILs for use as electrolytes for supercapacitors operating at 60 °C.[70] They showed that the devices containing these ILs store less energy than those containing [C_4mpyr][NTf_2].[70] It is clear that the search for ILs that present good transport properties and still have a wide electrochemical window is still a challenge. Promising ILs based on anions with low coordinating ability such as [$B(CN)_4$] were suggested by Bernhardt and collaborators.[71,72] Later, Kuang *et al.* reported the use of an IL containing this anion in solar cells,[73] and Koller *et al.* and Sanchez-Ramirez *et al.* characterized ILs containing the [$B(CN)_4$] anion.[41,74] These liquids present a low density (lower than 1.0 at 25 °C), low viscosity and high ionic conductivity, yielding a molar ionic conductivity of 1.8 S cm^2 mol^{-1} for [C_4mpyr][$B(CN)_4$], and still present high electrochemical stability.[41] The drawback of these ionic liquids, interestingly, is related to their high ionic conductivity: weak coordination of cations. Consequently, the solubility of the Li^+ salt is considerably lower than that when other anions are used.

Mixtures of metallic salts with ILs are important for different reasons, from metal electrodeposition to batteries.[75] Figure 13.5 shows the effect of Li salt

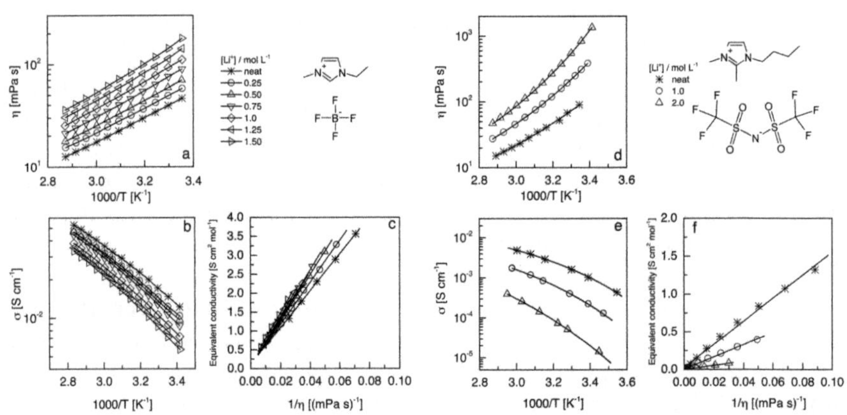

Figure 13.5 Transport properties of ionic liquids with Li salt addition. (a, b and c) [C3mim][BF_4] figures reprinted (adapted) with permission from ref. 76. Copyright (2004) American Chemical, and (d, e and f) [C_4mmim][NTf_2] figures reprinted with permission from ref. 77. Copyright (2004) American Chemical Society.

addition on the transport properties of two ILs: $[C_3mim][BF_4]$ and $[C_4mmim][NTf_2]$.[76,77] As the amount of Li salt increases, the viscosity also increases and the ionic conductivity consequently decreases for both mixtures. The presence of the small Li^+ cation completely changes the ionic structure of the liquid, resulting in different aggregates that will incur more friction from ionic movements. In the case of the $[NTf_2]$ IL, it is clear that there is the formation of larger and stiffer aggregates, thus lowering the molar ionic conductivity when compared at the same liquid fluidity. However, mixtures of the $[BF_4]$ IL oddly presented a higher molar ionic conductivity at the same fluidity when more Li salt was added, meaning that the new aggregates are more free to move. According to the authors, the presence of Li^+ in $[C_3mim][BF_4]$ releases the IL cations from their aggregates, thereby enabling them to play a more important role in the ionic conductivity than in less concentrated mixtures.

These studies also provided the experimental determination of the self-diffusion coefficient, *D*, by pulsed gradient spin echo nuclear magnetic resonance (PGSE-NMR). The different ions in the mixture can be followed by different probes; thus, the value of *D* for the IL cation and anion as well as the Li^+ can be determined using ^1H, ^{19}F and ^7Li PGSE-NMR measurements. In this technique, the movements of all species containing the probe are taken into account; therefore, if the ions are forming different aggregates and do not carry any charge, their movement will also be considered. Both studies found that Li^+ has a lower *D* than the other ions in the mixture, clearly indicating aggregate formation; moreover, the IL cation has a higher *D* than the anion, even with a greater radius in the case of $[C_3mim]$ and $[BF_4]$, indicating that the anion always participates in aggregation. With the diffusion of ions, it is possible to estimate the ionic conductivity of the IL + Li salt mixtures using the Nernst–Einstein equation, eqn (7):

$$\sigma_{\mathrm{NMR}} = \frac{e^2}{kT}\left(\sum n_{\mathrm{i}} D_{\mathrm{i}}\right) \tag{13.7}$$

The relationship between the experimental ionic conductivity and the estimated ionic conductivity based on NMR should be 1.0 if all ions in the mixtures can carry their unit charge. As discussed above, the ions tend to form aggregates, making this relationship normally much lower than 1.0. Mixtures of Li salts and ionic liquids containing an anion with a weak ability for coordination, such as the example above, $[B(CN)_4]$, can yield a high value of charge carriers, and the relationship $\sigma_{\mathrm{exp.}}/\sigma_{\mathrm{NMR}}$ will be close to 1.0.[41]

Another class of ILs, the protic ionic liquids, is worth mentioning. These ILs are composed of an equimolar mixture of a Brønsted acid and a Brønsted base, and good examples are those based on protonated ammonium cations. Anouti *et al.* studied the transport properties of several protic ILs and their mixtures with water.[78] The neat protic IL $[pyr][NO_3]$ presents an ionic conductivity as high as 50.1 mS cm^{-1}, which is considerably higher than that of the aprotic ILs presented here. They also showed that for mixtures with water, there is a maximum ionic conductivity with increasing water content, and after this maximum is reached, the ionic conductivity decreases with the further addition of water. The maximum ionic conductivity is a function of the number of carbons in the alkyl chain in the anion: the longer the alkyl chain is, the lower the maximum ionic conductivity and the higher the amount of water in the mixture required to reach this maximum.[78]

The search for an all-solid-state electrolyte is driven by the need for safer and/or flexible devices and has attracted attention for applications in energy storage systems. Currently, flammable organic solvents are used in electrolyte compositions; thus, the danger of a battery catching fire due to a malfunction must be considered. Replacing the flammable organic solvent with an IL or PIL is already a considered approach.[79] Although important and reported by many research groups, we are excluding from this chapter gel polymers of standard polymers (with no charge) with ionic liquids. Osada *et al.* published an extensive review on this matter, covering from binary mixtures (polymer and ionic liquid) to those with the addition of lithium salts (ternary mixtures).[80] The focus will be only when an IL is a part of the polymeric matrix.

As expected, PILs present a considerably lower ionic conductivity than an IL with a similar structure because one of the ions is attached to the polymeric matrix. For example, poly(1-ethyl-3-vinylimidazolium) [PEVim] has a structure similar to the [C_3mim] cation; whereas [C_3mim][BF_4] and [C_3mim][NTf_2] present ionic conductivities of 15 and 10 mS cm^{-1}, respectively,[81] [PEVim][BF_4] and [PEVim][NTf_2] present ionic conductivities of $<10^{-9}$ and 2.5×10^{-8} mS cm^{-1}, respectively.[82] These values are several orders of magnitude lower than those of their liquid analogues.

Many approaches have been described to overcome the low ionic conductivity of PILs and take advantage of their other properties. Lee *et al.* studied imidazolium-based PILs and changed the mobility of the polycation side chain to improve the membrane transport properties *via* changing the alkyl side chain *via* ethylene oxide groups. Increasing the chain mobility drastically decreased the T_g of the PILs, consequently increasing the ionic conductivity. They also showed that the addition of Li[NTf_2] salt to the membrane is beneficial to the ionic conductivity; however, the increment is more significant for the alkyl-chain-containing PIL than for the ethylene-oxide-containing PILs. They attributed this effect to the interaction between Li^+ and the oxygen in the functional groups.[32] Cowan *et al.* presented a series of phosphonium-based PILs and discussed how the alkyl chain length in the polymer affects the ionic conductivity. They found that the polycation with the shortest alkyl chain presented the lowest ionic conductivity at 25 °C, but at temperatures higher than 90 °C, the inverse is observed: a shorter alkyl chain presents a higher ionic conductivity, as observed in other studies.[33] The authors suggested that the inverse trend to that expected at low temperature is observed due to stronger cation–anion interactions, decreasing ion mobility, whereas once the temperature is higher, this interaction loses importance in the overall ion mobility.[33]

The transport properties in a PIL greatly depend on its structure. Shaplov, Marcilla and Mecerreyes recently summarized in a review the effects of the chemical structures of the cation and anion on the PIL conductivity.[30] The cation structure will impact the ionic conductivity of cationic or anionic PILs. This means that regardless of whether the positive charge is attached to the polymeric matrix, it will play a role in the ionic conductivity. Moreover, if the cation is attached near the polymeric matrix, it can decrease the ionic conductivity; consequently, the length of the spacer between the polymer backbone and the cation will have a great effect on the charge movement. In addition, interactions between the anion and polymeric matrix, such as hydrogen bonds, will also affect the ionic conductivity.[30]

Shaplov *et al.* highlighted in their review that comparisons among PILs produced by different research groups may be hindered due to several factors, such as moisture content, purification process and molecular weight of the polymeric matrix.[30] To overcome the irreproducibility among different groups, Obadia *et al.* described the effect of anions in cationic PILs, specifically triazolium-based polycations.[83] They showed that poly(3-methyl-1,2,3-triazolium) [NTf_2] presents the highest ionic conductivity among the PILs studied, as shown in Figure 13.6.

Marcilla and collaborators reported the use of PILs as electrolytes for all-solid-state supercapacitors.[84,85] Using the cationic PIL poly(diallyldimethylammonium) [NTf_2] ([PDDA][NTf_2]), they first showed the optimization of the electrolyte impregnation in activated carbon as the electrode material using a mixture of PIL:[C_4mpyr][NTf_2] in acetone. The symmetrical device showed a high specific capacitance operating at 3.5 V (*ca.* 100 F g^{-1}).[84] Later, they used the previous optimization process to prepare a solid electrolyte using the same PIL, but mixing with ILs containing different anions. Figure 13.6 shows the prepared mixtures and the ionic conductivity at different temperatures. At 25 °C, the mixture [PDDA][NTf_2]:[C_4mpyr][Tf] presented the highest ionic conductivity among the investigated systems. These membranes showed higher electrochemical stability than conventional organic solvent electrolytes, although these analyses were performed with stainless steel electrodes, and different stability is expected when activated carbon is used as the electrodes. Nonetheless, the symmetrical supercapacitor was shown to possess a high specific capacitance, and the PILs are promising materials to be used as electrolytes in many electrochemical devices.[85]

The previous discussion clearly shows that by changing the anions, cations and substituents, it is possible to gain accurate control over the transport properties. In addition, if we consider the melting temperature of ILs or PILs, an on/off control over conductivity is possible. The transport properties can also be controlled in magnetic ILs,[86,87] but the discussion of these types of ILs is not inside the scope of this chapter. Thus, it is clear that considering the architectures of ILs and PILs and modifying the transport properties through external perturbation make it possible to classify these materials as smart materials.

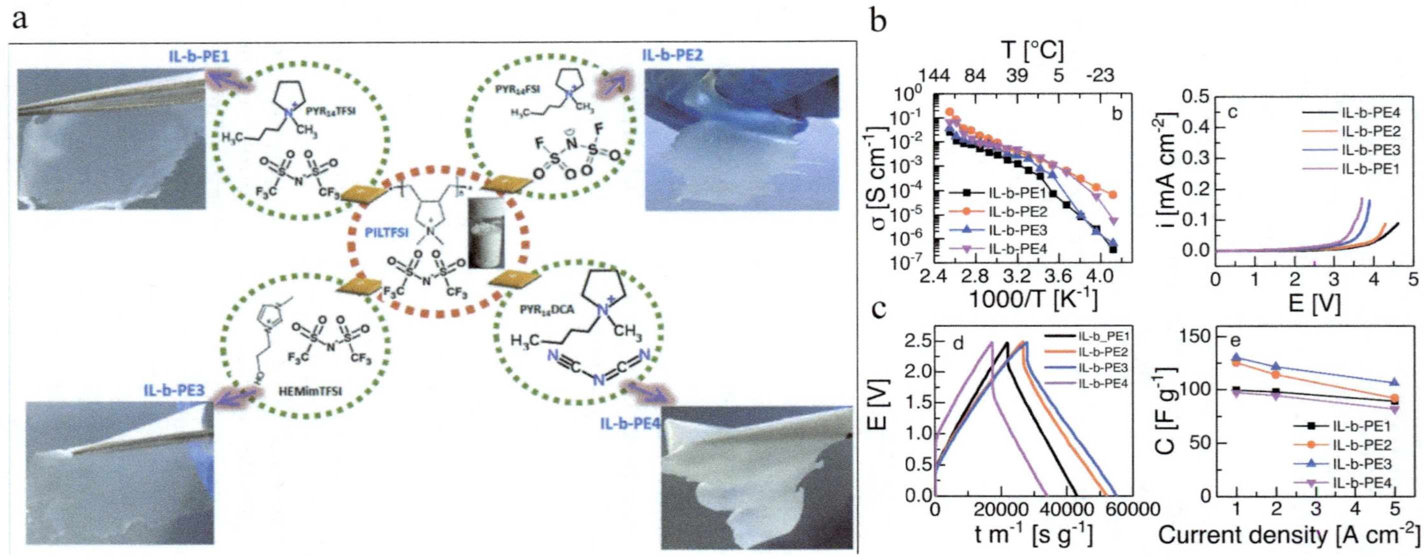

Figure 13.6 (a) Gel polymer electrolytes containing a cationic poly(ionic liquid) and mixtures with ionic liquids containing different anions, (b) their ionic conductivity and electrochemical stability, and (c) their performance as supercapacitor electrolytes. Reprinted from *Journal of Power Sources*, 326, G. A. Tiruye, D. Muñoz-Torrero, J. Palma, M. Anderson, R. Marcilla, Performance of solid state supercapacitors based on polymer electrolytes containing different ionic liquids, 560–568, Copyright (2016) with permission from Elsevier.[85]

13.3 Ionic Liquids, Polymeric Ionic Liquids and Electrochromism

Electrochromic materials are chemical compounds that can change their optical properties electrochemically through the application of a specific potential. There are many types of electrochromic materials, such as transition metal oxides, Prussian Blue, viologens, conducting polymers, metallopolymers, phthalocyanines, and buckminsterfullerene.[88–92] Thus, devices manufactured with such materials are called electrochromic devices (ECDs).

Due to the thermal stability of ILs,[93] it is possible to use ILs in devices more securely compared to conventional electrolytes, which can explode in closed devices, for example. Moreover, in general, aqueous and organic electrolytes are unfavorable for ECDs due to their narrow potential stability window; additionally, some compounds are unstable in aqueous medium, such as tungsten trioxide (WO_3), which is widely used in these types of devices. Non-aqueous and aprotic electrolytes could be better but present drawbacks in terms of long-term stability. As shown in the following, ILs could solve these problems and enhance some properties of electrochemical devices such as ECDs.[94]

In 2002, Lu *et al.* published a report in which they describe the beneficial aspects of using ILs in ECDs based on a conducting polymer (CP). First, ILs could be used as a chemical or electrochemical reaction medium for the preparation of several compounds. To this end, Lu *et al.* prepared thin films of some conducting polymers (PEDOT and PANI) electrochemically in an ionic liquid ([C_4mim][BF_4]). They found that the polymer prepared in the IL had different properties compared with the same polymer prepared in propylene carbonate. This finding opened a new line of research that studies ILs as a reaction medium for the preparation of many compounds and their influence on the properties of the generated materials. Thus, it is possible to optimize the properties of various materials for future application in ECDs. This again shows how broad the applications of ILs are.[95]

Furthermore, an electrochromic window based on PANI electrochemically polymerized in an IL cycled (−0.6 to 1.1 V) using [C_4mim][BF_4] as the electrolyte presented exceptional stability for 1 000 000 cycles. Similarly, PANI fibers and other examples containing a conducting polymer were tested with ILs as the electrolyte and compared with conventional electrolytes; the lifetime tests also showed an unprecedented level of performance, thus confirming the benefits of ILs for the stability of these devices. For PANI, a relevant aspect is the different redox mechanism in ILs compared with aqueous electrolytes, which prevents the known electrochemical degradation of PANI in aqueous medium; there is no gain or loss of solvated protons, and thus, the electroactivity has no pH dependence.[95]

Continuing with ECDs based on a CP, Lu *et al.* (2004) also prepared CPs (PPy, PANI and PEDOT) in IL media. Note that the IL is a pure medium for electrochemical polymerization, in contrast to aqueous or organic solvents that contain water or solvents and a soluble salt. Furthermore, the IL acts as an electrolyte and dopant. The oxidation potentials in [C_4mim][BF_4] were 0.5 V for pyrrole and 0.8 V for EDOT and aniline. The cyclic voltammetry of the obtained PANI showed characteristic peaks without the middle peak related to the degradation that occurs in aqueous media, indicating the stability of the film; subsequently, the material was cycled from leucoemeraldine to pernigraniline for 1 000 000 cycles and remained stable. More stable ILs such as [C_4mim][NTf_2] and [C_4mim][PF_6] further improved the stability. The final ECDs prepared using PANI, PEDOT and [C_4mim][BF_4] presented high performance with low operating potentials (approximately 1.5 V), high coloration contrast (greater than 50%), fast coloration speed (less than 100 ms) and high coulombic efficiency (greater than 98%).[96]

Brazier *et al.* (2007) worked with the IL *N*-butyl-*N*-methylpyrrolidinium bis(trifluoromethanesulfonyl)imide ([C_4mpyr][NTf_2]), which has the lowest melting point of 0 °C and is very well rated for ECD applications due to the good visible light transmission of the pure IL, and mixed it with [$P(EO)_{10}$][Li][NTf_2] to form a polymer electrolyte. Using a typical ECD system containing a WO_3 thin layer electrode and a V_2O_5 thin layer electrode interposed by a Li^+ conducting electrolyte, they compared the use of [Li][C_4mpyr][NTf_2] and polypropylene carbonate–Li^+ electrolytes. Electrochromism occurs because of the transport of Li^+ between the two electrodes reversibly by a specific applied potential. The ECD transmittance is very similar for both electrolytes: 57.6 and 61.0% for the PIL and PC, respectively. The stability is better in the IL electrolyte, and this is confirmed by the decrease of transmittance observed using only the PC electrolyte; moreover, the mixture of lithium and IL electrolyte has a very stable electrochemical window of approximately 5 V. Additionally, using the IL electrolyte, it is possible to incorporate it in a polymeric matrix as [$P(EO)_{10}$][Li][NTf_2], resulting in a solid polyelectrolyte, and to assemble an all-solid ECD that, in this work, showed a very promising durability of approximately 2000 cycles.[94]

Ouyang *et al.* (2015) enhanced electrochromic devices using a substrate of ITO functionalized with an IL ([C_4mim][BF_4]) and poly(4,4′,4″-tris[4-(2-bithienyl)-phenyl]amine) (PTBTPA) to form [PTBTPA][C_4mim][BF_4] films on the ITO surface by electrochemical polymerization. The first result showed that the surface morphology changes with the insertion of the IL, becoming smaller and more uniform compared with pure PTBTPA. This morphological change may lead to a larger contact area and more channels that could facilitate ionic exchange, thereby improving the switching speed. The pure PTBTPA is electrochromic, reversibly changing its color from orange to olive green to dark gray, and with the insertion of the IL, the film preserves this multicolor electrochromism. By comparing pure PTBTPA with PTBTPA–IL, it becomes clear how the IL positively influences the film properties. The bleaching time and coloring time were improved from 1.76 and 4.51 s to less than 0.87 and 2.91 s at 1100 nm (Figure 13.7a). Moreover, the switching speed of PTBTPA–IL was faster compared to the pure PTBYPA. The reasons for these behaviors are that the pure PTBTPA film has ClO_4 as a counter ion, whereas PTBTPA–IL has ClO_4 and BF_4 to compensate charge during the redox event, thereby accelerating the process. Furthermore, the IL is a linker between the polymer and the electrode, decreasing the charge transfer barrier (Figure 13.7b). The cyclic stabilities (Figure 13.7c) of the films containing IL were also significantly improved due to the [BF_4] size being smaller than ClO_4, thus requiring less expansion/contraction of the polymer film, resulting in improvements in stability.[97] Therefore, to corroborate the speculations, the electrochemical impedance measurements (Figure 13.7d) showed semicircle

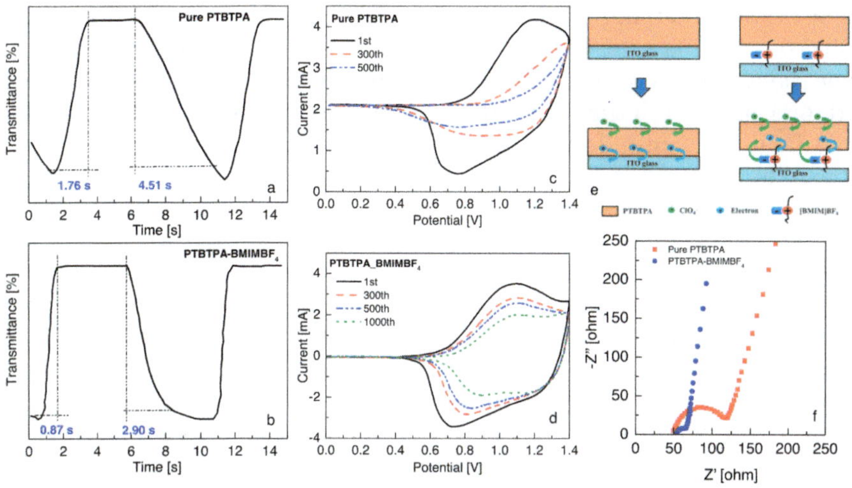

Figure 13.7 Switching response time of: (a) a pure PTBTPA film and (b) a [PTBTPA][C_4mim][BF_4] film monitored at 1100 nm. Electrochemical stability of (c) the pure PTBTPA film and (d) the [PTBTPA][C_4mim][BF_4] film at 300 mV s^{-1} for 1000 cycles. (e) Schematic plot of electron transfer and counter ion diffusion and (f) electrochemical impedance measurements of the pure PTBTPA film and the [PTBTPA][C_4mim][BF_4] film. Reproduced from ref. 97 with permission from The Royal Society of Chemistry.

diameters of 13.18 Ω and 65.5 Ω for PTBTPA–IL and pure PTBTPA, respectively, indicating that PTBTPA–IL presents a much lower charge transfer resistance.

Sydam *et al.* (2015) prepared a new type of viologen-quaternized electrolyte and set up an ECD using the electrochromic viologen, an IL ([C_3mim][NTf_2]), and DMF. The electrolyte presented a high ionic conductivity that was invariant with temperature, good thermal stability (above 150 °C) and stable electrochemical window (approximately 3.6 V). The focus in this work was the new viologen with high redox reversibility for the di-reduced viologen. The assembled ECD switched color from colorless to dark blue with a high contrast of 52%, coloration efficiency of 533 $cm^2\ C^{-1}$, good color uniformity, fast response time (2 seconds) and excellent write–erase efficiency. The IL was crucial for the high thermal stability and ionic conductivity.[98]

However, ILs remain liquids and may have leakage drawbacks in closed devices. Thus, the search for materials with good ionic conductivity in the solid state has been significantly increasing. In this context, the main application of PILs is as polymer electrolytes because of their ionic conductivity and ability to enhance the device properties when compared with usual polymer electrolytes. Their high ionic conductivity of up to 10^{-3} S cm^{-1} at 25 °C[30] and good thermal and electrochemical stabilities (above 350 °C and EW up to 5 V, respectively)[30] place PILs as an alternative to current solid polyelectrolytes. For these reasons, they are potential materials for many applications, such as batteries,[18,99] supercapacitors,[100,101] electrochromic devices,[102,104] dye-sensitized solar cells,[5,105,106] transistors[107] and others, generally as a polyelectrolyte.

For ECDs containing a PIL as the electrolyte, the first research was published by Marcilla *et al.* (2006).[102] They prepared three PILs based on imidazole (poly(1-vinyl-ethyl-imidazolium)) with 3 different anions, $[NTf_2]$, $[BF_4^-]$ and $[Br^-]$, and mixed them with the respective 1-ethyl-3-methylimidazolium ionic liquids ($[C_4mim][NTf_2]$, $[C_4mim][BF_4]$, and $[C_4mim][Br]$) in different ratios. Films prepared from these mixtures by casting presented different features depending on the ratio between the PIL and IL, being a transparent film for 0 : 100 PIL and a very viscous and sticky gel for 75 : 25 IL : PIL.

The mixtures of only the IL and PIL containing the same anion made it possible to achieve the best interaction and compatibility, resulting in homogeneous phases that were stable for weeks. The ionic conductivity of the materials depends on the anion and varies from 10^{-2} to 10^{-5} S cm^{-1} at room temperature; furthermore, the ionic conductivity decreases as the quantity of polymer in the mixture increases. This result is expected because the ILs have two free charged species while PILs present one of the charged species attached to the polymeric matrix. Moreover, ILs and PILs containing Br^- as the anion have lower conductivities due to the interaction between cationic and anionic species; thus, $[NTf_2]$ is larger than Br^- and presents a structure that leads to a large delocalization of the negative charge, thus decreasing the interaction with the cation.

These polyelectrolytes were employed in electrochromic cells or devices, generating an all-polymeric five-layer device configuration. The replacement of the IL with a PIL as the electrolyte has brought advantages, such as the ability to mount the cell with good adhesion and leak-free closing. Furthermore, the use of a PIL resulted in the longest life: the device containing only IL exceeded 20 000 cycles (square-wave 0–3 V), whereas for the PIL, it was 70 000 cycles. Therefore, the use of a PIL not only facilitated the assembly of the device but also improved its performance.[102]

As previously mentioned, leakage is a considerable problem related to conventional electrolytes and even ILs. PILs are solid electrolytes, which eliminates possible leaks. Shaplov *et al.* (2014) published a study called "A first truly all-solid state organic electrochromic device based on polymeric ionic liquids".[103] In this work, they prepared a PIL based on imidazole and using a spacer group, $(-CH_2CH_2O)_2$, with an ionic conductivity of 1×10^{-5} S cm^{-1}. The PEDOT electrodes were thin films of 90 nm thickness on an ITO surface, and the final device/cell presented a coloration efficiency of 390 cm^2 C^{-1}, which is a very high value for solid-state organic ECDs. Furthermore, after 1000 cycles of cyclic voltammetry (from 0 V to 2.5 V), there was no significant decrease of current, thereby demonstrating the stability of the cell. Finally, the switching time was 3 seconds, which is comparable to ECDs using liquid electrolytes.

In the same year, Shaplov *et al.* continued their research[104] and assembled a fully organic electrochromic cell (Figure 13.8) using the PIL (poly[1(2-2-(2-(methacryloyloxy)ethoxy)ethyl)-3-methylimidazolium] bis(trifluoromethylsulfonyl)imide or tetracyanoborate) as the polyelectrolyte and PEDOT as the electrode to produce an all-polymer-based thin-film symmetrical ECD. They highlighted the tunable properties of the PIL by the virtually unlimited

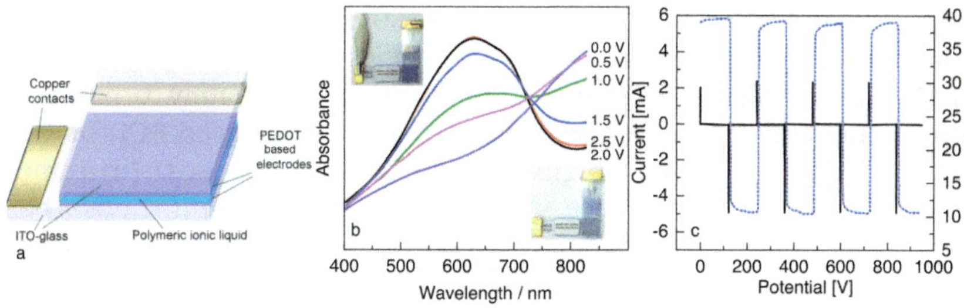

Figure 13.8 (a) Assembly of the fully solid cell. (b) Optical characterization at different potentials. (c) Evolution of the current and the optical contrast at 620 nm under application of subsequent double potential steps (between 0 and 2.5 V during 1000 seconds) of the PEDOT/PIL/PEDOT ECD. Reprinted from *Polymer*, 55 (16), A. S. Shaplov, D. O. Ponkratov, P.-H. Aubert, E. I. Lozinskaya, C. Plesse, A. Maziz, P. S. Vlasov, F. Vidal, Y. S. Vygodskii, Truly solid state electrochromic devices constructed from polymeric ionic liquids as solid electrolytes and electrodes formulated by vapor phase polymerization of 3,4-ethylenedioxythiophene, 3385–3396, Copyright (2014) with permission from Elsevier.[104]

combination of cations and anions to obtain desired characteristics, such as high conductivity, and proposed the preparation of a new PIL polycation in combination with two anions: $[NTf_2]$ and tetracyanoborate ([TCB]). These anions were chosen because the ILs that contain them present high conductivities of approximately 10^{-3}–10^{-2} S cm^{-1} at room temperature. After the radical polymerization of the monomers, the PILs presented conductivities of 1.0×10^{-5} and 1.5×10^{-6} S cm^{-1} for the PIL $[NTf_2]$ and the PIL [TCB] at room temperature, respectively, increasing to 10^{-4} S cm^{-1} as the temperature was increased to 55 (PIL $[NTf_2]$) and 60 °C (PIL [TCB]). In addition, the electrochemical stability was demonstrated by cyclic voltammetry with an electrochemical window from −2.0 V to +2.0 V (Figure 13.8a).

The PILs solubilized in acetonitrile were cast directly on a PEDOT electrode deposited on an ITO surface to produce the ECD. The UV-Vis spectrum of this device (Figure 13.8c) showed a strong absorption at λ_{max} = 620 nm, and this wavelength was used in the spectroelectrochemistry studies. Moreover, the type of PIL influences the switching time, which for the most conductive PIL $[NTf_2]$ was 3–5 s, and for the less conductive PIL [TCB], the switching time occurred less quickly. A similar behavior was observed for the coloration efficiency, being higher for the PIL $[NTf_2]$, and the highest value obtained was 430 cm^2 C^{-1}. Finally, the stability was studied by cyclic voltammetry along with spectrometry, and after 1000 cycles (0–2.5 V), the electroactivity remained practically unchanged.

Lu *et al.* (2016) worked with bipyridine to prepare an ECD using an electrochromic gel composed of a salt based on bipyridine quaternized with heptyl, octyl or nonyl groups,[108] a PIL as a polyelectrolyte (1-allyl-3-methylimidazolium bis(trifluoromethylsulfonyl)imide), and ferrocene as a redox mediator. This gel was sandwiched between two ITO glass plates. The ECD presented a large transmittance change (56.7% at 6050 nm), a coloration efficiency of approximately 60.4 $cm^2\ C^{-1}$, switching times of 2.1 s for bleaching and 2.13 for coloring and a long-term stability of approximately 10 000 cycles. Improvements related to the use of a PIL in a viologen-based ECD were proposed, including (i) the PIL bringing additional ions to the ECD, (ii) a smaller operating potential being necessary to reach the maximum absorbance change, (iii) the PIL providing a lower driving energy, and (iv) the coloration efficiency increasing significantly with increasing PIL concentration due to the decrease of the measured charge density.[108]

As previously mentioned, ILs and PILs can be functionalized and exhibit properties of interest. Branco *et al.*[109] prepared some ILs that are intrinsically electrochromic and magnetic. This was only possible due to the combination of appropriate electrochromic and magnetic anions based on ethylenediaminetetraacetic metal complexes with cations based on imidazolium, phosphonium and tetraalkylammonium (Figure 13.9). Thus, the IL showed an ionic conductivity higher than the usual ILs, electrochromic properties and, according to the complexed metal, it is possible to observe diamagnetic and paramagnetic states upon oxidation or reduction (Figure 13.8). The IL preparation was performed using a simple mixture of cobalt(III), chromium(III) or iron(III) ethylenediaminetetraacetic complex anions and cations such as 1-ethyl-3-methylimidazolium and others. Another similar study conducted by Branco *et al.* (2013) showed five intrinsically electrochromic ILs, which were based on vanadium oxide.[110] Again, it is the anion that provides the color change feature, in this case, vanadium oxide (VO_3^-), which presents three colored states: V(V), V(IV) and V(III) – yellow, blue and green, respectively.

Kavanagh and co-workers (2013) prepared a novel phosphonium-based ionic liquid with electrochromic properties based on a bipyridine covalently linked to the cation IL structure (Figure 13.9). This IL behaves as both an electrolyte and an electrochromic material, and a solid-state platform was prepared using a hybrid sol–gel. The electrochromic platform presented coloration kinetics of 221 seconds for 95% of absorbance and a coloration efficiency of 10.72 $cm^2\ C^{-1}$. The most interesting feature is that the viologen is part of the IL structure, and consequently, there is no leaking.

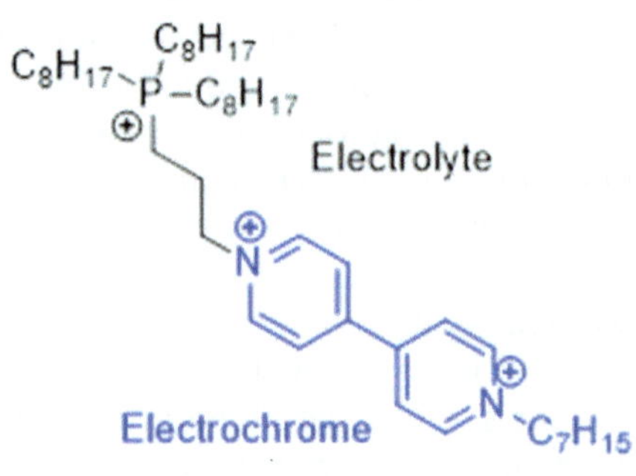

Figure 13.9 Ionic liquid functionalized with bipyridine. For further reading please see ref. 111.

The electrochromism is generated from the insertion of one electron into the quaternized bipyridine (V^{2+}) present in the IL cation, producing a cation radical ($V^{\bullet +}$) that causes the color to change from colorless to blue.[111]

If the mentioned ILs had a cation or anion-polymerizable fraction, for example, after the polymerization, a PIL containing a bipyridine moiety would be obtained. Therefore, a solid polyelectrolyte with electrochromic properties could be studied. In this context, ILs and PILs are considered smart materials because their properties can be modulated by different combinations of cations and anions, variations of the lateral chain size or nature and functionalization of the cation or anion structures by inserting moieties of interest. There are endless possibilities. Moreover, the use of a PIL in the assembly of ECDs – as the electrolyte – improves several properties compared to conventional electrolytes. Finally, the functionalization of these materials may generate a complete material with, for example, ionic conductivity and electrochromic properties, in other words, color variation by applying an external specific potential in a unique chemical structure.

13.4 Transport of Gases by Ionic Liquids and Poly(ionic liquid)s: CO_2 Separation

Carbon dioxide (CO_2) is the main gas emitted to the atmosphere from the combustion of fossil fuels. It causes the greenhouse effect, which contributes to global climate change. Despite the recent advances in the search for alternative sources of energy, the burning of fossil fuels is still the most economical and widely used energy source. Consequently, a short- to mid-term approach is to prevent the CO_2 from entering the atmosphere by separating it from its main emission sources –power plant flue (mainly N_2, with 3–20% of CO_2), syngas production (mainly H_2, with 20–40% CO_2), natural gas production (mainly CH_4, with 1–8% CO_2) and biogas production (mainly CH_4, with 35–40% CO_2)[112] – for further sequestration and/or recycling.[113,114] The separation of CO_2 during natural gas production is currently performed using aqueous amines, such as monoethanolamine (MEA), because they can chemically react with CO_2; however, it comes with high energy costs to remove the CO_2 due to the high heat capacity of water, thereby compromising the system efficiency.[115] In power plants, which represent approximately 40% of total emissions, there is no current method for separating CO_2.[116]

Herein, the development of new smart materials for CO_2 separation represents an important challenge to be explored in the coming years with potential to improve their energy efficiency and capacity.[114,117–121] Among the different possible options, the ones based on ILs have been highlighted due to the possibility of designing a task-specific system for a given emission stream composition. ILs are smart materials whose structures can be modified to potentially separate CO_2 from a specific gas mixture and at a given pressure and temperature. In addition to their significant CO_2 solubility and high selectivity, the negligible vapor pressure of ILs reduces contamination issues and the energy required to remove the captured gas.[122] In the next sections, the aspects driving the CO_2 separation process by ILs, as well as their incorporation in solid matrices and the use of PILs to create membranes for gas permeation, will be discussed.

13.4.1 Why Can Ionic Liquids Selectively Dissolve CO_2?

Perhaps one of the most important features of ILs for CO_2 separation is that they can *selectively* capture CO_2 from a mixture of gases. In the pioneering work of Anthony *et al.*,[123] the solubilities of different gases and water in the IL 1-*n*-butyl-3-methylimidazolium hexafluorophosphate ([C_4mim][PF_6]) were studied and correlated with parameters such as polarizability, quadrupole moment and Henry's constant. From Figure 13.10, one can observe that the solubility of CO_2 is significantly higher than that of the other gases, where the solubilities of H_2, CO and N_2 are under the detection limit of the apparatus used for the measurements.

The solubilities of the gases calculated in terms of Henry's constant follow a trend with their polarizability values, except for CO_2, H_2O, N_2 and CO, suggesting that other factors are also governing the solubility in these cases. For N_2 and CO, the predictable solubility based on the trend is higher than the detection limit of the equipment and should be measurable, whereas for CO_2 and water, the predictable solubility values are considerably lower than the actual ones. The reason for the high solubility of water is attributed to its strong dipole moment and the formation of hydrogen bonds, whereas the high solubility for CO_2 is attributed to its large quadrupole moment, which interacts with the electrical charges of the IL.

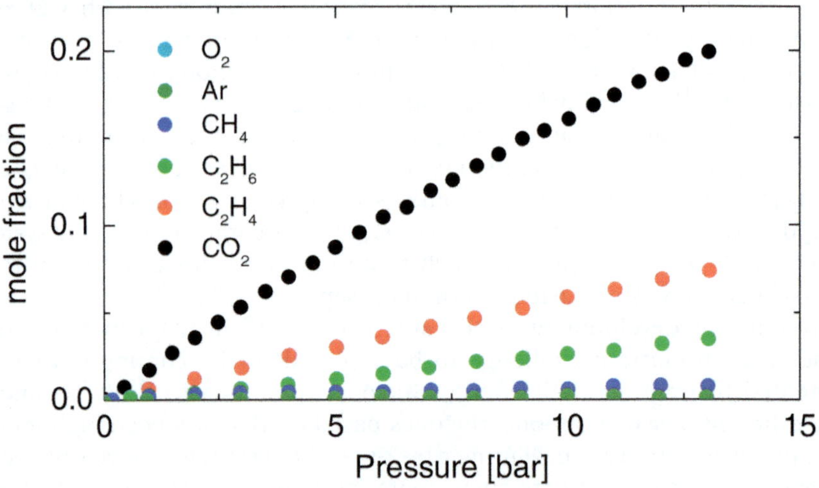

Figure 13.10 CO_2, C_2H_4, C_2H_6, CH_4, Ar, and O_2 solubilities in [C_4mim][PF_6] at 25 °C. Reprinted with permission from ref. 123. Copyright (2002) American Chemical Society.

Despite the high selectivity for CO_2 absorption over other gases commonly found in the gas flue from different processes, the absorption capacity by the most common ILs, whose absorption mechanism is physical and not chemical as is the case with amines, is not sufficiently high for power plant flue. However, they potentially have sufficient capacity and selectivity for other separation processes, such as from natural gas and syngas.[116] The absorption capacity depends on the IL structure, which can be modified to increase the physical absorption and to allow chemical absorption.

13.4.2 Factors Affecting CO_2 Solubility in Ionic Liquids

The physical absorption of CO_2 in ILs is dependent on their structure and on the transport properties, as discussed previously. For instance, both cation and anion structures affect the absorption capacity, with the influence of the anion being more significant. For imidazolium cations, it has been shown that increasing the lateral alkyl chain slightly increases the absorption capacity[124,125] due to an increased volume available for CO_2. In fact, both simulations and experimental results have shown that the density of ILs increases with gas absorption as there is no volume change,[126] which means that the free space available in the IL structure affects the absorption capacity. Changing the nature of the cation from imidazolium to pyridinium has very little influence over the CO_2 absorption capacity, whereas changing to a cation with a stronger interaction with the anion such as choline decreases the CO_2 solubility, which has been attributed to the lower availability of the ions for interaction with CO_2.[127]

The nature of the IL anion has a significantly higher influence on the CO_2 solubility than the cation. The gas molecule interacts strongly with the anions, as shown by molecular simulations.[126] For a given cation, the solubility of CO_2 does not follow a trend with the anion basicity but is dependent on their fluorine content, suggesting a stronger interaction of the gas molecule with fluorinated species than an acid/base interaction.[127] In fact, the use of fluorinated alkyl chains at the surface of substrates has increased the solubility in supercritical CO_2.[128] However, fluorination of the IL cation still has little effect on CO_2 compared with the influence of the anion, corroborating the fact that the CO_2 is mainly interacting with the negative species.[127] To illustrate these points, the Henry's law constants for some ILs with varying cation lateral chain lengths and degrees of fluorination, as well as different types of anion, are presented in Table 13.3.[127,129]

The lower the Henry's law constant, the higher the solubility. The data presented in Table 13.3 for the ILs containing the [NTf_2] anion show that changing the imidazolium cation lateral chain length from 4 to 6 carbon atoms, as well as fluorinating it, only slight affects the solubility. In contrast, a dramatic change in solubility can be observed for the phosphate-based anions, where the replacement of three of the F atoms by fluorinated alkyl chains results in a dramatic decrease in the Henry's law constant, which continues to decrease with increasing fluorinated chain length. However, because the

Table 13.3 Henry's law constants of ILs.

Ionic Liquid	Name	Cation	Anion	Henry's law constant (25 °C)
[C_4mim] [PF_6]	1-butyl-3-methylimidazolium hexafluorophosphate	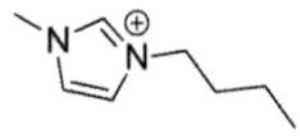	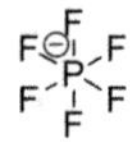	54.3 ± 0.3
[C_4mim] [BF_4]	1-butyl-3-methylimidazolium tetrafluoroborate	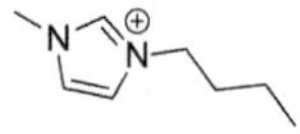		59.0 ± 2.6
[C_4mim] [NTf_2]	1-butyl-3-methylimidazolium bis(trifluoromethylsulfonyl) imide	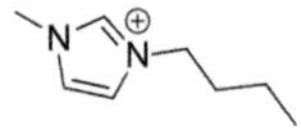	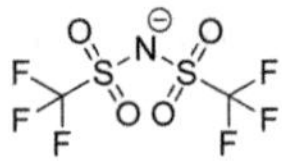	33.0 ± 0.3
[C_6mim] [NTf_2]	1-hexyl-3-methylimidazolium bis(trifluoromethylsulfonyl) imide	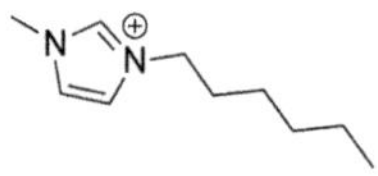	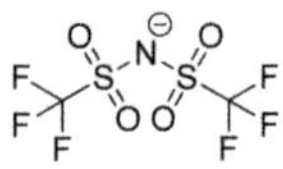	31.6 ± 0.2
[C_4mpyr] [NTf_2]	1-butyl-3-methylpyridinium bis(trifluoromethylsulfonyl) imide	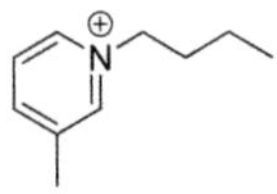	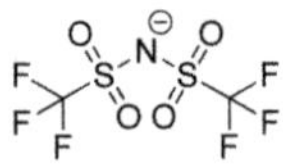	32.8 ± 0.2
[$C_6H_4F_9$im] [NTf_2]	1-methyl-3-(3,3,4,4,5,5,6,6,6-nonafluorohexyl)-imidazolium bis(trifluoromethylsulfonyl) imide	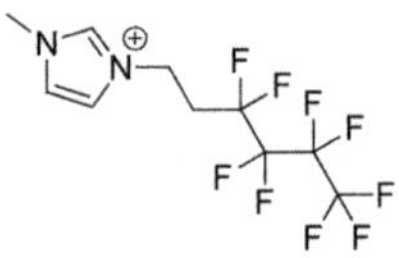	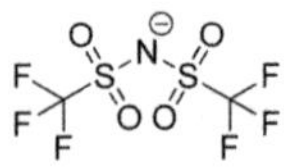	28.4 ± 0.1
[$C_8H_4F_{13}$im] [NTf_2]	1-methyl-3-(3,3,4,4,5,5,6,6,7,7,8,8,8-tridecafluorooctyl)-imidazolium bis(trifluoromethylsulfonyl) imide	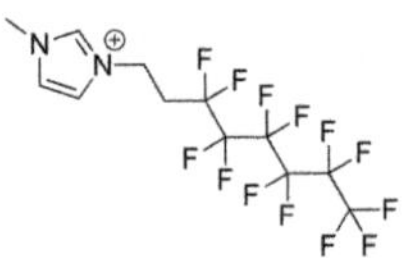	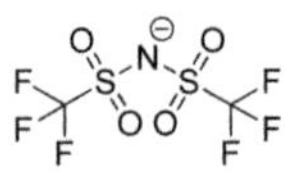	27.3 ± 0.2
[C_6mim] [FAP]	1-hexyl-3-methylimidazolium tris(pentafluoroethyl) trifluorophosphate	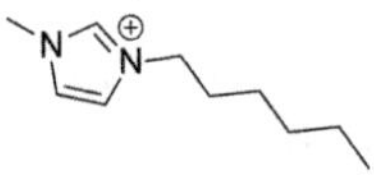	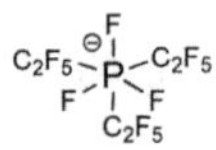	25.2 ± 0.1
[C_6mim] [pFAP]	1-hexyl-3-methylimidazolium tris-(heptafluoropropyl) trifluorophosphate	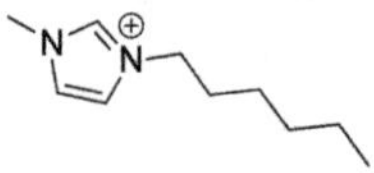	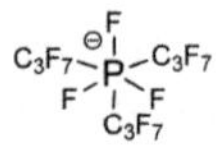	21.6 ± 0.1
[C_5mim] [bFAP]	1-pentyl-3-methylimidazolium tris(nonafluorobutyl) trifluorophosphate	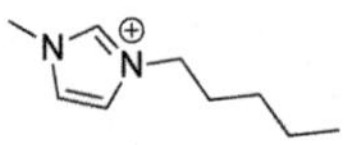	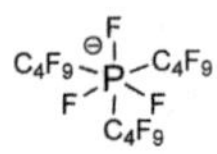	20.2 ± 0.1

solubility can also be dependent on the molar volume of the anion and both the F atom content and the molar volume are changing at the same time in this case, it is difficult to separate the effect of each one. If one considers the ILs containing the [C_4mim] cation, changing the anion from [BF_4] to [PF_6] (and increasing the F content) causes only a slight increase in CO_2 solubility; however, with the [NTf_2] anion, which has the same number of F atoms as [PF_6], the solubility is much higher. This fact has been attributed to the larger size of [NTf_2], and because the solubility is a function of the heat of vaporization and the molar volume, it is reasonable to assume that the energy of vaporization of both anions is similar.[130] Based on these data, it appears that increasing the molar volume of the anion plays a more significant role in the CO_2 solubility than fluorination.

Modifications such as fluorination and increasing the molar volume affect the CO_2 solubility, but the absorption mechanism is still physical. Despite the advantages of using ILs such as non-volatility and inflammability, as well as the high selectivity over other gases, their absorption capacities are still not competitive for practical applications. To further increase the CO_2 solubility, IL functionalization with groups that allow chemical absorption has been explored.

In the pioneering work of Bates *et al.*,[131] the inclusion of an $-NH_3$ group to the imidazolium cation lateral chain to form the IL 1-propylamide-3-butyl imidazolium tetrafluoroborate ([APC_4im][BF_4]) caused a 10-fold increase in CO_2 absorption capacity at ambient conditions compared with the best absorption results obtained with ILs that only allow physical absorption. The mechanism of CO_2 chemical absorption in both the MEA and the amino-functionalized IL is the formation of a carbamate, which is limited to 1 : 2 stoichiometry (Figure 13.11).

In this case, a close to stoichiometric 0.5 mol ratio was obtained, which is comparable to the standard materials such monoethanolamine (MEA) and corresponds to a 7.8% mass gain upon CO_2 absorption. The lower absorbed CO_2 mass percentage for the ILs (due to the higher molecular weight) when compared with aqueous MEA is compensated by the energetic gains to regenerate the absorbent and the high selectivity. Moreover, a close to 1 : 1 mol absorption ratio is possible when some IL anions are functionalized rather than the cation,[132,133] which can be attributed to the preferred stabilization of carbamic acid (eqn (8)) rather than subsequent proton transfer to another

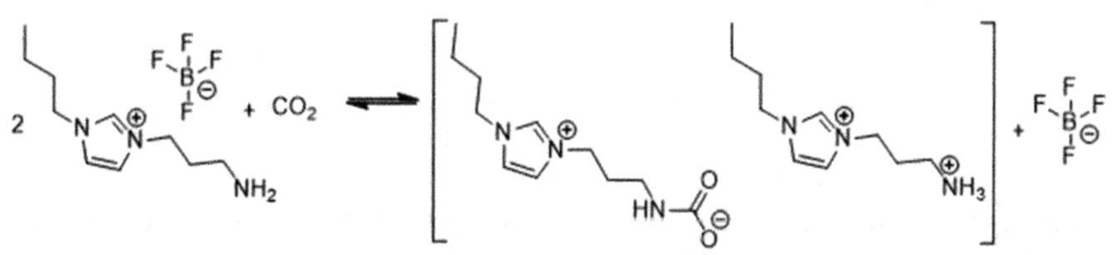

Figure 13.11 Reaction between APC_4imBF_4 and CO_2. Reprinted (adapted) with permission from ref. 131. Copyright (2002) American Chemical Society.

amine group (eqn (9) and (10)) to form the carbamate anion (eqn (11)).[134] In addition to stabilization of the carbamic acid when the amine is attached to a negative species, other factors, such as the stability of the derived carbamate dianion, must be considered, which can favor one or other stoichiometries.[135]

$$R-NH_2+CO_2 \rightarrow R-NHCOOH \tag{13.8}$$

$$R-NHCOOH \rightarrow R-NHCOO^- + H^+ \tag{13.9}$$

$$R-NH_3+H^+ \rightarrow R-NH_3^+ \tag{13.10}$$

$$2R-NH_2+CO_2 \rightarrow R-NHCOO^- + R-NH_3^+ \tag{13.11}$$

However, a 1:1 stoichiometry is not achieved at atmospheric pressures, where the partial pressure of CO_2 in air is approximately 0.4 mbar, or it takes a very long time under CO_2 gas flow to achieve this stoichiometry. This is due to the slow diffusion of CO_2 through the IL, favoring the formation of carbamic acid (1:1 IL:CO_2 stoichiometry) only at the surface of the liquid, with the formation of carbamate anions (2:1 IL:CO_2 stoichiometry) being predominant in the bulk, as previously demonstrated by XPS measurements.[136] Further improvements can be achieved by adding more amino groups to the IL cation or anion, increasing the CO_2 absorption capacity; however, the effects of slow CO_2 diffusion are more evident as the ILs become more viscous. In addition, strong hydrogen bonds formed between the carboxylic acid groups contribute to further reducing the diffusion of the gas molecule through the absorbing liquid.[137–139]

Thus, to improve the efficiency of CO_2 absorption, it is important to consider strategies to decrease the gas diffusion path, which can be performed by creating high-surface-area structures to support the absorbing IL.

13.4.3 Supported Ionic Liquids for CO_2 Separation

The separation of CO_2 using solid platforms for selectively capturing CO_2 can be performed using two different approaches: physical adsorption or membrane separation (Figure 13.12).[140]

In the physical adsorption approach, the solid material is under the flue gas stream. It selectively adsorbs the CO_2 and emits CO_2-free gas, and the solid adsorbent is further regenerated by increasing the temperature or pressure. ILs have been immobilized in materials such as mesoporous silica[141,142] and porous polymers,[143,144] resulting in faster adsorption at lower pressures compared with the liquid systems. The regeneration of the solid adsorbents can be performed by increasing the temperature or pressure, and fast desorption under mild conditions is crucial for reducing the energetic costs, resulting in a feasible system. Fast desorption at mild temperatures from 40 °C to 60 °C was possible with an "inverse" supported IL made of silica nanoparticles (Figure 13.13)[145] and at mild pressures of approximately 15 bar with solid ILs (ILs with melting points above room temperature).[146]

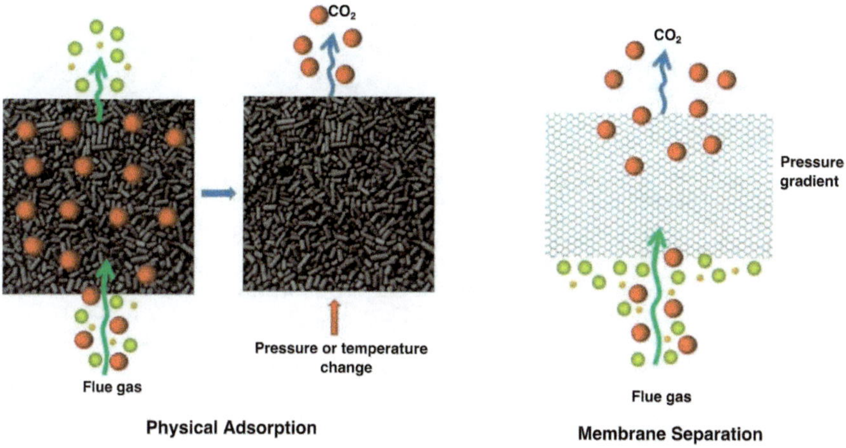

Figure 13.12 Physical adsorption and membrane separation approaches for CO_2 separation using solid adsorbents. Reproduced from ref. 140 with permission from The Royal Society of Chemistry.

Figure 13.13 IL encapsulated in silica nanoparticles as a solid adsorbent for CO_2 capture. Reprinted (adapted) with permission from ref. 145. Copyright (2014) American Chemical Society.

Membranes can be advantageous over physical adsorption due to their simpler configuration and continuous operation;[147] however, the involved energy costs for both systems must be considered. With membranes, the separation is based on a sieving or sorption–diffusion process for porous or non-porous materials, respectively, and in a supported ionic liquid membrane (SILM),

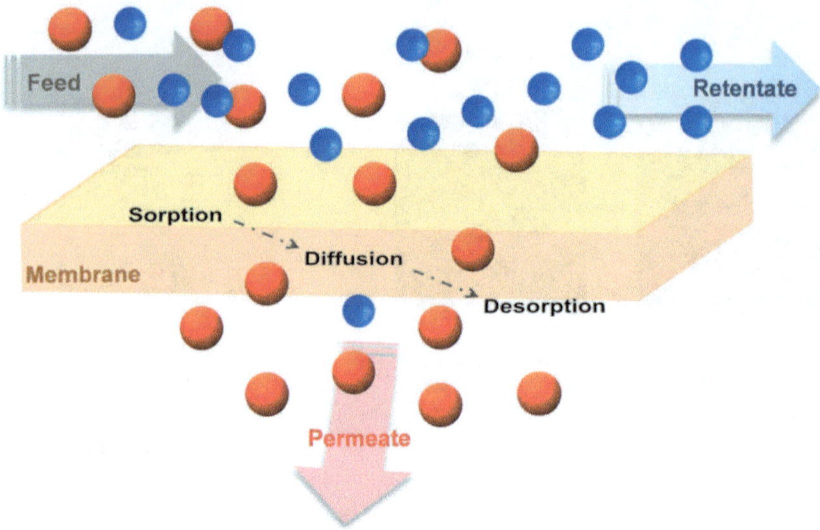

Figure 13.14 Gas transport through a non-porous dense membrane. Reproduced from ref. 112 with permission from The Royal Society of Chemistry.

CO_2 will preferentially pass through the membrane, being released at the other side due to a pressure gradient.[112] As gases generally have a small size difference among them, the use of non-porous membranes is preferable because it is based on gas solubility and diffusivity differences rather than size, as is the case for the porous ones. Thus, ideally, a difference in pressure will let only the CO_2 pass through the SILM, and the other gases will be blocked (Figure 13.14).

In the case of non-porous membranes, their performance is measured in terms of permeability and selectivity. Permeability is the rate of gas diffusion normalized by the thickness and is a function of gas diffusivity and solubility. Selectivity is the rate at which gas *a* moves across the membrane in comparison with gas *b*. These parameters are subjected to a *permeability/selectivity trade off*, where selectivity decreases when permeability increases, and this is visualized in the Robeson plot (Figure 13.15), where an empirical upper bound is represented by a line.[148,149] The performances of the different membranes can thus be compared in terms of permeability/selectivity proximity to this upper bound line.

A high-performance SILM must have high CO_2 permeability and selectivity, and many contributions report SILMs from ceramic materials and polymers impregnated with ILs.[150–157] However, the SILMs based on these solid materials are capable of maintaining the IL inside their structures at low pressures of less than 3 bar for microporous supports and approximately 7 bar for nanoporous supports,[158,159] compromising their stability. For application in natural gas and syngas processes, new materials with much smaller and uniform pore sizes that are capable of retaining the IL at transmembrane

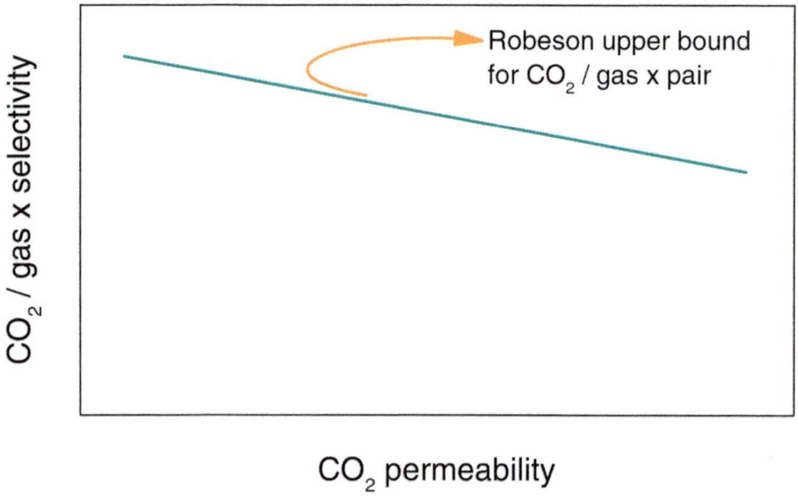

Figure 13.15 General Robeson plot with upper bound line for a given CO_2/gas pair.[148,149]

pressures as high as 15–60 bar are required. Simulation studies with metal–organic frameworks (MOFs),[160] as well as the first report of a material based on a high-surface-area mesoporous carbon with a pore size of less than 10 nm that was capable of operating at 10 bar,[161] show the feasibility of the development of membranes that could work at high-pressure conditions.

13.4.3.1 Poly(ionic liquid)s as Active Solid Supports

Another strategy for developing membranes for CO_2 separation based on ILs is to immobilize the IL cation or anion in a polymeric chain to create PILs.[162] This class of polyelectrolyte combines the properties of ILs in a macromolecular framework, being a promising alternative as a CO_2 adsorbent. In addition to the possibility of using it at high pressures, as the immobilized IL species will not leak from the solid matrix, CO_2 adsorption by PILs is in general significantly higher and faster than in their non-polymerized counterparts, as first shown for imidazolium-based PIL powders (Table 13.4).[163]

The reasons why the CO_2 sorption capacity and rate are higher in PILs with respect to their IL counterparts were explored in later studies, showing that in the case of the PIL membranes, the gas permeability dramatically increases with increasing cation lateral chain length.[33,164] This was attributed as being a consequence of more free volume due to the inefficient packing of the chains. In addition, because one of the ionic species is fixed in the polymer chain, the ions cannot continuously rearrange themselves to accommodate the ionic and lateral chain repulsions as they do when they are free (which is the case with the ILs), and this will also create free spaces in the polymer.

Table 13.4 Structures of the PIL powders studied as CO_2 adsorbents.

Monomer	Name	Cation	Anion
$VBBIBF_4$	1-(*p*-Vinylbenzyl)-3-butyl-imidazolium tetrafluoroborate	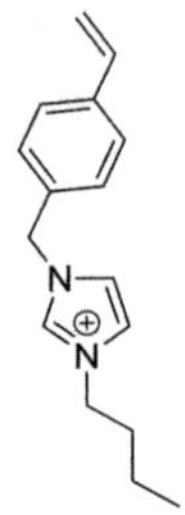	
VBBI N Tf_2	1-(*p*-Vinylbenzyl)-3-butyl-imidazolium bis(trifluoromethanesulfonil)imide		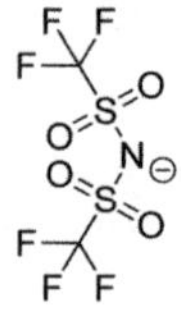
VBBISac	1-(*p*-Vinylbenzyl)-3-butyl-imidazoliu-mo-benzoicsulphimide		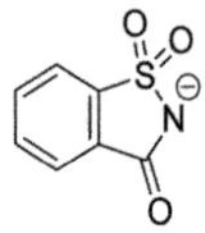
$MABIBF_4$	1-[2-(Methacryloyloxy)ethyl]-3-butyl-imidazolium tetrafluoroborate	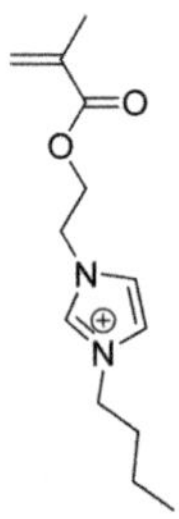	

Recent strategies to improve the performance of PIL membranes for CO_2 capture consist of creating nanoporous polymers,[165] immobilizing polymers on mesoporous silica to create core–shell structures,[166] using deep eutectic monomers,[167] co-polymerization[168] and the use of cross-linking agents,[169] combining ILs and PILs to create gels[113] and mixing the PIL with an amine-functionalized IL.[149] All these contributions show the feasibility of using PIL-based membranes for CO_2 separation, and many of them presented their performances in terms of the upper bound line of the Robeson plot, showing promising results. However, this field is still at its nascent stage, and more methodic studies to understand the relationship between the different varying chemical structures and CO_2 separation performance as well as measurements at higher pressures and for longer periods must be performed to achieve optimized and competitive performances.

13.5 Concluding Remarks

Tuning the properties of ILs and PILs by changing their structures is a highly useful approach for adapting physicochemical properties to a desired application, such as for chromogenic devices or solubility of a specific gas. In this regard, combinations between different cations and anions, changes in the length of the alkyl side chain and functionalization of ionic structures greatly influence the transport properties of this class of material, mainly because of the interactions among the ions. The possibilities are endless. Therefore, with a specific application in mind, it is possible to design an IL or PIL with the desired characteristics for this application. Additionally, some characteristics can be changed using external perturbation, as is the case of electrochromic reactions, where in addition to the redox chromophore redox couple, a good ionic conductivity is necessary. Moreover, using ILs and PILs as electrolytes for EDCs, it is possible to improve some characteristics, such as lifetime, coloration efficiency, switching times for beaching and coloring and electrochemical stability of the assembled devices. All these characteristics can be merged in an IL or PIL. Due to the wide range of applications where ILs and PILs can be employed, there is much room for improvement and innovation. For example, these materials are widely considered for use as electrolytes in supercapacitors and rechargeable batteries; however, the still high viscosity hinders the performance of these devices at room temperature. Designing and tailoring ILs and PILs may reveal the next state-of-the-art electrolyte.

Acknowledgements

The authors acknowledge the financial support from CAPES, CNPq and FAPESP (2015/263087). VLM and CMC thank FAPESP (14/14690-1 and 13/00322-8, respectively) for the fellowships granted.

References

1. J. Yuan and M. Antonietti, *Polymer (Guildf.)*, 2011, **52**, 1469–1482.
2. L. Wang, X. Duan, W. Xie, Q. Li and T. Wang, *Chem. Commun.*, 2016, **52**, 8417–8419.
3. W. Zhang, Z. Kochovski, Y. Lu, B. V. K. J. Schmidt, M. Antonietti and J. Yuan, *ACS Nano*, 2016, **10**, 7731–7737.
4. D. Cordella, A. Debuigne, C. Jérôme, Z. Kochovski, D. Taton and C. Detrembleur, *Macromol. Rapid Commun.*, 2016, **37**, 1181–1187.
5. S.-M. Chen, T.-L. Wang, P.-Y. Chang, C.-H. Yang and Y.-C. Lee, *React. Funct. Polym.*, 2016, **108**, 103–112.
6. M. M. Obadia, B. P. Mudraboyina, A. Serghei, T. N. T. Phan, D. Gigmes and E. Drockenmuller, *ACS Macro Lett.*, 2014, **3**, 658–662.
7. J. Juger, F. Meyer, F. Vidal, C. Chevrot and D. Teyssié, *Tetrahedron Lett.*, 2009, **50**, 128–131.
8. J. Zhang, D. Xu, J. Guo, Z. Sun, W. Qian, Y. Zhang and F. Yan, *Macromol. Rapid Commun.*, 2016, **37**, 1194–1199.
9. W. Zhang and J. Yuan, *Macromol. Rapid Commun.*, 2016, **37**, 1124–1129.
10. Z. Zheng, Q. Xu, J. Guo, J. Qin, H. Mao, B. Wang and F. Yan, *ACS Appl. Mater. Interfaces*, 2016, **8**, 12684–12692.
11. H. Wu, M. Kuang, L. Cui, D. Tian, M. Wang, G. Luan, J. Wang and L. Jiang, *Chem. Commun.*, 2016, **52**, 5924–5927.

12. X. Fan, H. Liu, Y. Gao, Z. Zou, V. S. J. Craig, G. Zhang and G. Liu, *Adv. Mater.*, 2016, **28**, 4156–4161.
13. A. Okafuji, Y. Kohno and H. Ohno, *Macromol. Rapid Commun.*, 2016, **37**, 1130–1134.
14. M. Isik, F. Ruiperez, H. Sardon, A. Gonzalez, S. Zulfiqar and D. Mecerreyes, *Macromol. Rapid Commun.*, 2016, **37**, 1135–1142.
15. T. Ikeda, S. Moriyama and J. Kim, *J. Polym. Sci., Part A: Polym. Chem.*, 2016, **54**, 2896–2906.
16. J. Steinkoenig, F. R. Bloesser, B. Huber, A. Welle, V. Trouillet, S. M. Weidner, L. Barner, P. W. Roesky, J. Yuan, A. S. Goldmann and C. Barner-Kowollik, *Polym. Chem.*, 2016, **7**, 451–461.
17. Y. Dong, J. Yin, J. Yuan and X. Zhao, *Polymer (Guildf.)*, 2016, **97**, 408–417.
18. K. Grygiel, J.-S. Lee, K. Sakaushi, M. Antonietti and J. Yuan, *ACS Macro Lett.*, 2015, **4**, 1312–1316.
19. T. Zhou, X. Gao, B. Dong, N. Sun and L. Zheng, *J. Mater. Chem. A*, 2016, **4**, 1112–1118.
20. A. S. Shaplov, P. S. Vlasov, M. Armand, E. I. Lozinskaya, D. O. Ponkratov, I. A. Malyshkina, F. Vidal, O. V. Okatova, G. M. Pavlov, C. Wandrey, I. A. Godovikov and Y. S. Vygodskii, *Polym. Chem.*, 2011, **2**, 2609.
21. P. Dimitrov-Raytchev, S. Beghdadi, A. Serghei and E. Drockenmuller, *J. Polym. Sci., Part A: Polym. Chem.*, 2013, **51**, 34–38.
22. A. S. Shaplov, E. I. Lozinskaya, D. O. Ponkratov, I. A. Malyshkina, F. Vidal, P.-H. Aubert, O. V. Okatova, G. M. Pavlov, L. I. Komarova, C. Wandrey and Y. S. Vygodskii, *Electrochim. Acta*, 2011, **57**, 74–90.
23. W. He, F. Zhang, Z. Wang, W. Sun, Z. Zhou and Z. Ren, *Ind. Eng. Chem. Res.*, 2016, **55**, 12616–12631.
24. C. Yue, D. Fang, L. Liu and T.-F. Yi, *J. Mol. Liq.*, 2011, **163**, 99–121.
25. J. Yuan, C. Giordano and M. Antonietti, *Chem. Mater.*, 2010, **22**, 5003–5012.
26. M. Döbbelin, I. Azcune, M. Bedu, A. Ruiz de Luzuriaga, A. Genua, V. Jovanovski, G. Cabañero and I. Odriozola, *Chem. Mater.*, 2012, **24**, 1583–1590.
27. W. Bi, B. Tang and K. H. Row, *Bioprocess Biosyst. Eng.*, 2013, **36**, 651–658.
28. A. S. Shaplov, D. O. Ponkratov and Y. S. Vygodskii, *Polym. Sci., Ser. B*, 2016, **58**, 73–142.
29. K. Nakamura, T. Saiwaki and K. Fukao, *Macromolecules*, 2010, **43**, 6092–6098.
30. A. S. Shaplov, R. Marcilla and D. Mecerreyes, *Electrochim. Acta*, 2015, **175**, 18–34.
31. M. Moosavi, F. Khashei, A. Sharifi and M. Mirzaei, *Ind. Eng. Chem. Res.*, 2016, **55**, 9087–9099.
32. J. H. Lee, J. S. Lee, J.-W. Lee, S. M. Hong and C. M. Koo, *Eur. Polym. J.*, 2013, **49**, 1017–1022.
33. M. G. Cowan, M. Masuda, W. M. McDanel, Y. Kohno, D. L. Gin and R. D. Noble, *J. Membr. Sci.*, 2016, **498**, 408–413.

34. V. L. Martins, B. G. Nicolau, S. M. Urahata, M. C. C. Ribeiro and R. M. Torresi, *J. Phys. Chem. B*, 2013, **117**, 8782–8792.
35. K. R. Seddon, A. Stark and M.-J. Torres, *Pure Appl. Chem.*, 2000, **72**, 2275–2287.
36. D. R. MacFarlane, S. A. Forsyth, J. Golding and G. B. Deacon, *Green Chem.*, 2002, **4**, 444–448.
37. J. X. Mao, A. S. Lee, J. R. Kitchin, H. B. Nulwala, D. R. Luebke and K. Damodaran, *J. Mol. Struct.*, 2013, **1038**, 12–18.
38. M. Marszalek, Z. Fei, D.-R. Zhu, R. Scopelliti, P. J. Dyson, S. M. Zakeeruddin and M. Grätzel, *Inorg. Chem.*, 2011, **50**, 11561–11567.
39. C. Ye and J. M. Shreeve, *J. Phys. Chem. A*, 2007, **111**, 1456–1461.
40. R. L. Gardas and J. A. P. Coutinho, *Fluid Phase Equilib.*, 2008, **263**, 26–32.
41. N. Sanchez-Ramirez, V. L. Martins, R. A. Ando, F. F. Camilo, S. M. Urahata, M. C. C. Ribeiro and R. M. Torresi, *J. Phys. Chem. B*, 2014, **118**, 8772–8781.
42. J. Salgado, T. Regueira, L. Lugo, J. Vijande, J. Fernández and J. García, *J. Chem. Thermodyn.*, 2014, **70**, 101–110.
43. A. J. R. Rennie, V. L. Martins, R. M. Torresi and P. J. Hall, *J. Phys. Chem. C*, 2015, **119**, 23865–23874.
44. N. Zec, M. Bešter-Rogač, M. Vraneš and S. Gadžurić, *J. Chem. Thermodyn.*, 2015, **91**, 327–335.
45. P. Walden, *Z. Phys. Chem.*, 1906, **55**, 683.
46. W. A. Adams and K. J. Laidler, *Can. J. Chem.*, 1968, **46**, 1978.
47. A. N. Campbell and E. T. van der Kouwe, *Can. J. Chem.*, 1968, **46**, 1293.
48. W. Xu, E. I. Cooper and C. A. Angell, *J. Phys. Chem. B*, 2003, **107**, 6170–6178.
49. D. R. MacFarlane, M. Forsyth, E. I. Izgorodina, A. P. Abbott, G. Annat and K. Fraser, *Phys. Chem. Chem. Phys.*, 2009, **11**, 4962–4967.
50. P. Bonhôte, A.-P. Dias, N. Papageorgiou, K. Kalyanasundaram and M. Grätzel, *Inorg. Chem.*, 1996, **35**, 1168–1178.
51. P. Bonhôte, A.-P. Dias, N. Papageorgiou, K. Kalyanasundaram, M. Grätzel, M. Armand, N. Papageorgiou, K. Kalyanasundaram and M. Grätzel, *Inorg. Chem.*, 1998, **37**, 166.
52. A. P. Abbott, *ChemPhysChem*, 2005, **6**, 2502–2505.
53. H. Weingärtner, *Angew. Chem., Int. Ed.*, 2008, **47**, 654–670.
54. M. C. C. Ribeiro, *J. Chem. Phys.*, 2010, **133**, 1–6.
55. P. A. Z. Suarez, S. Einloft, J. E. L. Dullius, R. F. de Souza and J. Dupont, *J. Chim. Phys. Phys.-Chim. Biol.*, 1998, **95**, 1626–1639.
56. T. Nishida, Y. Tashiro and M. Yamamoto, *J. Fluorine Chem.*, 2003, **120**, 135–141.
57. F. F. C. Bazito, Y. Kawano and R. M. Torresi, *Electrochim. Acta*, 2007, **52**, 6427–6437.
58. D. R. MacFarlane, P. Meakin, J. Sun, N. Amini and M. Forsyth, *J. Phys. Chem. B*, 1999, **103**, 4164–4170.
59. U. Domańska, M. Królikowska and K. Walczak, *J. Solution Chem.*, 2014, **43**, 1929–1946.

60. D.-X. Zhuang, M.-J. Deng, P.-Y. Chen and I.-W. Sun, *J. Electrochem. Soc.*, 2008, **155**, D575.
61. S. Forsyth, J. Golding, D. MacFarlane and M. Forsyth, *Electrochim. Acta*, 2001, **46**, 1753–1757.
62. U. L. Bernard, E. I. Izgorodina and D. R. MacFarlane, *J. Phys. Chem. C*, 2010, **114**, 20472–20478.
63. S. Seki, K. Hayamizu, S. Tsuzuki, K. Fujii, Y. Umebayashi, T. Mitsugi, T. Kobayashi, Y. Ohno, Y. Kobayashi, Y. Mita, H. Miyashiro and S. Ishiguro, *Phys. Chem. Chem. Phys.*, 2009, **11**, 3509–3514.
64. Y. Kunugi, H. Hayakawa, K. Tsunashima and M. Sugiya, *Bull. Chem. Soc. Jpn.*, 2007, **80**, 2473–2475.
65. M. H. Kowsari and M. Fakhraee, *J. Chem. Eng. Data*, 2015, **60**, 551–560.
66. D. R. Macfarlane, D. R. Macfarlane, J. Huang, J. Huang, M. Forsyth and M. Forsyth, *Nature*, 1999, **402**, 1998–2000.
67. D. R. MacFarlane, J. Sun, J. Golding, P. Meakin and M. Forsyth, *Electrochim. Acta*, 2000, **45**, 1271–1278.
68. C. Wolff, S. Jeong, E. Paillard, A. Balducci and S. Passerini, *J. Power Sources*, 2015, **293**, 65–70.
69. T. Belhocine, S. A. Forsyth, H. Q. N. Gunaratne, M. Nieuwenhuyzen, P. Nockemann, A. V. Puga, K. R. Seddon, G. Srinivasan and K. Whiston, *Green Chem.*, 2011, **13**, 3137.
70. S. Pohlmann, T. Olyschläger, P. Goodrich, J. Alvarez Vicente, J. Jacquemin and A. Balducci, *J. Power Sources*, 2015, **273**, 931–936.
71. E. Bernhardt, G. Henkel and H. Willner, *Z. Anorg. Allg. Chem.*, 2000, **626**, 560–568.
72. E. Bernhardt, M. Finze and H. Willner, *Z. Anorg. Allg. Chem.*, 2003, **629**, 1229–1234.
73. D. Kuang, P. Wang, S. Ito, S. M. Zakeeruddin and M. Grätzel, *J. Am. Chem. Soc.*, 2006, **128**, 7732–7733.
74. T. M. Koller, M. H. Rausch, J. Ramos, P. S. Schulz, P. Wasserscheid, I. G. Economou and A. P. Fröba, *J. Phys. Chem. B*, 2013, **117**, 8512–8523.
75. M. Armand, F. Endres, D. R. MacFarlane, H. Ohno and B. Scrosati, *Nat. Mater.*, 2009, **8**, 621–629.
76. K. Hayamizu, Y. Aihara, H. Nakagawa, T. Nukuda and W. S. Price, *J. Phys. Chem. B*, 2004, **108**, 19527–19532.
77. M. J. Monteiro, F. F. C. Bazito, L. J. A. Siqueira, M. C. C. Ribeiro and R. M. Torresi, *J. Phys. Chem. B*, 2008, **112**, 2102–2109.
78. M. Anouti, J. Jacquemin and D. Lemordant, *Fluid Phase Equilib.*, 2010, **297**, 13–22.
79. A. Eftekhari, Y. Liu and P. Chen, *J. Power Sources*, 2016, **334**, 221–239.
80. I. Osada, H. de Vries, B. Scrosati and S. Passerini, *Angew. Chem., Int. Ed.*, 2016, **55**, 500–513.
81. B. Garcia, S. Lavallée, G. Perron, C. Michot and M. Armand, *Electrochim. Acta*, 2004, **49**, 4583–4588.
82. Y. S. Vygodskii, O. A. Mel'nik, E. I. Lozinskaya, A. S. Shaplov, I. A. Malyshkina, N. D. Gavrilova, K. A. Lyssenko, M. Y. Antipin, D. G. Golovanov,

A. A. Korlyukov, N. Ignat'ev and U. Welz-Biermann, *Polym. Adv. Technol.*, 2007, **18**, 50–63.

83. M. M. Obadia, S. Fagour, Y. S. Vygodskii, F. Vidal, A. Serghei, A. S. Shaplov and E. Drockenmuller, *J. Polym. Sci., Part A: Polym. Chem.*, 2016, **54**, 2191–2199.
84. G. Ayalneh Tiruye, D. Muñoz-Torrero, J. Palma, M. Anderson and R. Marcilla, *J. Power Sources*, 2015, **279**, 472–480.
85. G. A. Tiruye, D. Muñoz-Torrero, J. Palma, M. Anderson and R. Marcilla, *J. Power Sources*, 2016, **326**, 560–568.
86. H. Zhang, S. Zhang and X. Zhang, *Chem. Mater.*, 2016, **28**, 8710–8714.
87. R. E. Del Sesto, T. M. McCleskey, A. K. Burrell, G. A. Baker, J. D. Thompson, B. L. Scott, J. S. Wilkes and P. Williams, *Chem. Commun.*, 2008, 447–449.
88. R. J. Mortimer, *Chem. Soc. Rev.*, 1997, **26**, 147.
89. P. R. Somani and S. Radhakrishnan, *Mater. Chem. Phys.*, 2003, **77**, 117–133.
90. A. A. Karyakin, *Electroanalysis*, 2001, **13**, 813–819.
91. V. D. Neff, *J. Electrochem. Soc.*, 1978, **125**, 886.
92. K. Itaya, T. Ataka and S. Toshima, *J. Am. Chem. Soc.*, 1982, **104**, 4767–4772.
93. M. Galiński, A. Lewandowski, I. Stepniak, M. Galinski, A. Lewandowski and I. Stepniak, *Electrochim. Acta*, 2006, **51**, 5567–5580.
94. A. Brazier, G. B. Appetecchi, S. Passerini, A. Surca Vuk, B. Orel, F. Donsanti and F. Decker, *Electrochim. Acta*, 2007, **52**, 4792–4797.
95. W. Lu, A. G. Fadeev, B. Qi, E. Smela, B. R. Mattes, J. Ding, G. M. Spinks, J. Mazurkiewicz, D. Zhou, G. G. Wallace, D. R. MacFarlane, S. A. Forsyth and M. Forsyth, *Science*, 2002, **297**, 983–987.
96. W. Lu, A. G. Fadeev, B. Qi and B. R. Mattes, *J. Electrochem. Soc.*, 2004, **151**, H33.
97. M. Ouyang, Y. Yang, X. Lv, Y. Han, S. Huang, Y. Dai, C. Su, Y. Lv, M. Sumita and C. Zhang, *New J. Chem.*, 2015, **39**, 5329–5335.
98. R. Sydam, A. Ghosh and M. Deepa, *Org. Electron.*, 2015, **17**, 33–43.
99. J.-S. Lee, K. Sakaushi, M. Antonietti and J. Yuan, *RSC Adv.*, 2015, **5**, 85517–85522.
100. J. P. C. Trigueiro, R. L. Lavall and G. G. Silva, *J. Power Sources*, 2014, **256**, 264–273.
101. T. Y. Kim, H. W. Lee, M. Stoller, D. R. Dreyer, C. W. Bielawski, R. S. Ruoff and K. S. Suh, *ACS Nano*, 2011, **5**, 436–442.
102. R. Marcilla, F. Alcaide, H. Sardon, J. A. Pomposo, C. Pozo-Gonzalo and D. Mecerreyes, *Electrochem. Commun.*, 2006, **8**, 482–488.
103. A. S. Shaplov, D. O. Ponkratov, P-H. Aubert, E. I. Lozinskaya, C. Plesse, F. Vidal and Y. S. Vygodskii, *Chem. Commun.*, 2014, **50**, 3191–3193.
104. A. S. Shaplov, D. O. Ponkratov, P.-H. Aubert, E. I. Lozinskaya, C. Plesse, A. Maziz, P. S. Vlasov, F. Vidal and Y. S. Vygodskii, *Polymer*, 2014, **55**, 3385–3396.
105. X. Chen, J. Zhao, J. Zhang, L. Qiu, D. Xu, H. Zhang, X. Han, B. Sun, G. Fu, Y. Zhang and F. Yan, *J. Mater. Chem.*, 2012, **22**, 18018.

106. J. Zhao, X. Shen, F. Yan, L. Qiu, S. Lee and B. Sun, *J. Mater. Chem.*, 2011, **21**, 7326.
107. M. M. Obadia and E. Drockenmuller, *Chem. Commun.*, 2016, **52**, 2433–2450.
108. H.-C. Lu, S.-Y. Kao, H.-F. Yu, T.-H. Chang, C.-W. Kung and K.-C. Ho, *ACS Appl. Mater. Interfaces*, 2016, **8**, 30351–30361.
109. A. Branco, L. C. Branco and F. Pina, *Chem. Commun.*, 2011, **47**, 2300–2302.
110. A. Branco, J. Belchior, L. C. Branco and F. Pina, *RSC Adv.*, 2013, **3**, 25627.
111. A. Kavanagh, K. J. Fraser, R. Byrne and D. Diamond, *ACS Appl. Mater. Interfaces*, 2013, **5**, 55–62.
112. L. C. Tomé and I. M. Marrucho, *Chem. Soc. Rev.*, 2016, **45**, 2785–2824.
113. M. G. Cowan, D. L. Gin and R. D. Noble, *Acc. Chem. Res.*, 2016, **49**, 724–732.
114. J. Liu, X. Hou, H. B. Park and H. Lin, *Chem.–Eur. J.*, 2016, **22**, 1–12.
115. K. Goto, K. Yogo and T. Higashii, *Appl. Energy*, 2013, **111**, 710–720.
116. J. F. Brennecke and B. E. Gurkan, *J. Phys. Chem. Lett.*, 2010, **1**, 3459–3464.
117. Y. Zeng, R. Zou and Y. Zhao, *Adv. Mater.*, 2016, 2855–2873.
118. Q. Zhuang, B. Clements, J. Dai and L. Carrigan, *Int. J. Greenhouse Gas Control*, 2016, **52**, 449–460.
119. X. Lu, D. Jin, S. Wei, Z. Wang, C. An and W. Guo, *J. Mater. Chem. A*, 2015, **3**, 12118–12132.
120. J. M. Huck, L.-C. Lin, A. H. Berger, M. N. Shahrak, R. L. Martin, A. S. Bhown, M. Haranczyk, K. Reuter and B. Smit, *Energy Environ. Sci.*, 2014, **7**, 4132–4146.
121. D. M. D'Alessandro, B. Smit and J. R. Long, *Angew. Chem., Int. Ed.*, 2010, **49**, 6058–6082.
122. X. X. Zhang, X. X. Zhang, H. Dong, Z. Zhao, S. Zhang and Y. Huang, *Energy Environ. Sci.*, 2012, **5**, 6668.
123. J. L. Anthony, E. J. Maginn and J. F. Brennecke, *J. Phys. Chem. B*, 2002, **106**, 7315–7320.
124. R. E. Baltus, B. H. Culbertson, S. Dai, H. Luo and D. W. DePaoli, *J. Phys. Chem. B*, 2004, **108**, 721–727.
125. Y. Hou and R. E. Baltus, *Ind. Eng. Chem. Res.*, 2007, **46**, 8166–8175.
126. C. Cadena, J. L. Anthony, J. K. Shah, T. I. Morrow, J. F. Brennecke and E. J. Maginn, *J. Am. Chem. Soc.*, 2004, **126**, 5300–5308.
127. J. L. Anderson, J. K. Dixon and J. F. Brennecke, *Acc. Chem. Res.*, 2007, **40**, 1208–1216.
128. E. J. Beckman, *Chem. Commun. (Camb.)*, 2004, 1885–1888.
129. M. J. Muldoon, S. N. V. K. Aki, J. L. Anderson, J. K. Dixon and J. F. Brennecke, *J. Phys. Chem. B*, 2007, **111**, 9001–9009.
130. J. L. Anthony, J. L. Anderson, E. J. Maginn and J. F. Brennecke, *J. Phys. Chem. B*, 2005, **109**, 6366–6374.
131. E. D. Bates, R. D. Mayton, I. Ntai and J. H. Davis, *J. Am. Chem. Soc.*, 2002, **124**, 926–927.

132. B. E. Gurkan, J. C. de la Fuente, E. M. Mindrup, L. E. Ficke, B. F. Goodrich, E. A. Price, W. F. Schneider and J. F. Brennecke, *J. Am. Chem. Soc.*, 2010, **132**, 2116–2117.
133. B. F. Goodrich, J. C. De La Fuente, B. E. Gurkan, D. J. Zadigian, E. A. Price, Y. Huang and J. F. Brennecke, *Ind. Eng. Chem. Res.*, 2011, **50**, 111–118.
134. C. Wu, T. P. Senftle and W. F. Schneider, *Phys. Chem. Chem. Phys.*, 2012, **14**, 13163.
135. Q. Yang, Z. Wang, Z. Bao, Z. Zhang, Y. Yang, Q. Ren, H. Xing and S. Dai, *ChemSusChem*, 2016, **9**, 806–812.
136. I. Niedermaier, M. Bahlmann, C. Papp, C. Kolbeck, W. Wei, S. Krick Calderón, M. Grabau, P. S. Schulz, P. Wasserscheid, H.-P. Steinrück and F. Maier, *J. Am. Chem. Soc.*, 2014, **136**, 436–441.
137. S. Saravanamurugan, A. J. Kunov-Kruse, R. Fehrmann and A. Riisager, *ChemSusChem*, 2014, **7**, 897–902.
138. Y. S. Sistla and A. Khanna, *Chem. Eng. J.*, 2015, **273**, 268–276.
139. B. F. Goodrich, J. C. de la Fuente, B. E. Gurkan, Z. K. Lopez, E. A. Price, Y. Huang and J. F. Brennecke, *J. Phys. Chem. B*, 2011, **115**, 9140–9150.
140. S. Zulfiqar, M. I. Sarwar and D. Mecerreyes, *Polym. Chem.*, 2015, **6**, 6435–6451.
141. V. Hiremath, A. H. Jadhav, H. Lee, S. Kwon and J. G. Seo, *Chem. Eng. J.*, 2016, **287**, 602–617.
142. K. N. Ruckart, R. A. O'Brien, S. M. Woodard, K. N. West and T. G. Glover, *J. Phys. Chem. C*, 2015, **119**, 20681–20697.
143. X. Wang, N. G. Akhmedov, Y. Duan, D. Luebke, D. Hopkinson and B. Li, *ACS Appl. Mater. Interfaces*, 2013, **5**, 8670–8677.
144. X. Wang, N. G. Akhmedov, Y. Duan, D. Luebke and B. Li, *J. Mater. Chem. A*, 2013, **1**, 2978.
145. G. E. Romanos, P. S. Schulz, M. Bahlmann, P. Wasserscheid, A. Sapalidis, F. K. Katsaros, C. P. Athanasekou, K. Beltsios and N. K. Kanellopoulos, *J. Phys. Chem. C*, 2014, **118**, 24437–24451.
146. G. R. M. Dowson, D. G. Reed, J.-M. Bellas, C. Charalambous and P. Styring, *Faraday Discuss.*, 2016, **192**, 511–527.
147. S. Wang, X. Li, H. Wu, Z. Tian, Q. Xin, G. He, D. Peng, S. Chen, Y. Yin, Z. Jiang and M. D. Guiver, *Energy Environ. Sci.*, 2016, **9**, 1863–1890.
148. L. M. Robeson, *J. Membr. Sci.*, 2008, **320**, 390–400.
149. B. D. Freeman, *Macromolecules*, 1999, **32**, 375–380.
150. S. D. Hojniak, A. L. Khan, O. Hollóczki, B. Kirchner, I. F. J. Vankelecom, W. Dehaen and K. Binnemans, *J. Phys. Chem. B*, 2013, **117**, 15131–15140.
151. J. E. Bara, R. D. Noble and D. L. Gin, *Ind. Eng. Chem. Res.*, 2009, **48**, 4607–4610.
152. J. E. Bara, D. L. Gin and R. D. Noble, *Ind. Eng. Chem. Res.*, 2008, **47**, 9919–9924.
153. J. E. Bara, C. J. Gabriel, T. K. Carlisle, D. E. Camper, A. Finotello, D. L. Gin and R. D. Noble, *Chem. Eng. J.*, 2009, **147**, 43–50.

154. K. M. Gupta, Y. Chen and J. Jiang, *J. Phys. Chem. C*, 2013, **117**, 5792–5799.
155. Y. Ban, Z. Li, Y. Li, Y. Peng, H. Jin, W. Jiao, A. Guo, P. Wang, Q. Yang, C. Zhong and W. Yang, *Angew. Chem., Int. Ed.*, 2015, **54**, 15483–15487.
156. P. Tamilarasan and S. Ramaprabhu, *J. Mater. Chem. A*, 2015, **3**, 797–804.
157. A. S. L. Gouveia, L. C. Tomé and I. M. Marrucho, *J. Membr. Sci.*, 2016, **510**, 174–181.
158. W. Zhao, G. He, F. Nie, L. Zhang, H. Feng and H. Liu, *J. Membr. Sci.*, 2012, **411–412**, 73–80.
159. Q. Gan, D. Rooney, M. Xue, G. Thompson and Y. Zou, *J. Membr. Sci.*, 2006, **280**, 948–956.
160. K. M. Gupta, Y. Chen, Z. Hu and J. Jiang, *Phys. Chem. Chem. Phys.*, 2012, **14**, 5785–5794.
161. S. H. Chai, P. F. Fulvio, P. C. Hillesheim, Z. A. Qiao, S. M. Mahurin and S. Dai, *J. Membr. Sci.*, 2014, **468**, 73–80.
162. J. Tang, H. Tang, W. Sun, H. Plancher, M. Radosz and Y. Shen, *Chem. Commun.*, 2005, 3325–3327.
163. J. Tang, H. Tang, W. Sun, M. Radosz and Y. Shen, *J. Polym. Sci., Part A: Polym. Chem.*, 2005, **43**, 5477–5489.
164. J. E. Bara, S. Lessmann, C. J. Gabriel, E. S. Hatakeyama, R. D. Noble and D. L. Gin, *Ind. Eng. Chem. Res.*, 2007, **46**, 5397–5404.
165. I. Azcune, I. García, P. M. Carrasco, A. Genua, M. Tanczyk, M. Jaschik, K. Warmuzinski, G. Cabañero and I. Odriozola, *ChemSusChem*, 2014, **7**, 3407–3412.
166. H. Cheng, P. Wang, J. Luo, J. Fransaer, D. E. De Vos and Z.-H. Luo, *Ind. Eng. Chem. Res.*, 2015, **54**, 3107–3115.
167. M. Isik, S. Zulfiqar, F. Edhaim, F. Ruiperez, A. Rothenberger and D. Mecerreyes, *ACS Sustain. Chem. Eng.*, 2016, **4**, 7200–7208.
168. S. Soll, Q. Zhao, J. Weber and J. Yuan, *Chem. Mater.*, 2013, **25**, 3003–3010.
169. A. Wilke, J. Yuan, M. Antonietti and J. Weber, *ACS Macro Lett.*, 2012, **1**, 1028–1031.

CHAPTER 14

Wearable Energy Storage Based on Ionic Liquid Gels

STEPHANIE F. ZOPF*[b], ANTHONY J. D'ANGELO[a], HUAN QIN[a] AND MATTHEW J. PANZER[a]

[a]HP Inc, San Diego, CA, USA; [b]Tufts University, Department of Chemical and Biological Engineering, 4 Colby St., Medford, MA 02155, USA
*E-mail: stephanie.flores@gmail.com

14.1 Introduction

14.1.1 Wearable Technology

The field of wearable technology has been around since the early 1990s when high-tech 'wearables' were being researched and developed at the Massachusetts Institute of Technology Media Lab in an effort to make people smarter.[1] This class of technology refers to electronic devices worn on the body. Companies and research institutions have been developing wearables since that time,[1] and technological acceleration has largely been due to the miniaturization and ubiquity of hardware. The popularity of wearables today is evident in the adoption of such technologies in fashion, healthcare, and military applications. Its surge in the marketplace creates consumer excitement, but is also often met with skepticism[2] that impacts user acceptance of future wearable technologies.

According to a 2014 survey conducted by Morgan Stanley on wearable devices, 30–40% of respondents stated that features such as: (1) more

Smart Materials No. 29
Polymerized Ionic Liquids
Edited by Ali Eftekhari

Published by the Royal Society of Chemistry, www.rsc.org

sensors, (2) more comfort, (3) greater computing power, and (4) more accurate and reliable data would increase wearable device usage. Increasing the number of sensors would mean more data collected and processed; improving computing power and generating more accurate and reliable data would require developing better hardware or algorithms to ensure that data can be processed more efficiently. Although addressing these features will have a direct impact on the computing side of wearables, they will also have an impact on its energy consumption and usable lifetime. On the other hand, providing more comfort to a wearable device means enabling users to have normal movements without any physical constraint.[3] This has typically been addressed by utilizing materials with flexibility and small form factors.[3] However, since wearables essentially evolved from portable electronics,[1] wearable electronics still incorporate the rigid hardware components found in portable devices, making these devices uncomfortable to wear. To improve wearable technology, incorporating materials that enable portable device performance while mimicking properties conducive to comfort in clothing would help.

14.1.2 Energy Storage for Wearable Applications

When choosing to power portable electronics, rigid batteries are selected due to the reliability and availability of these energy storage devices.[4] Thus, commercial wearable devices such as Fitbit, Google Glass and Apple Watch can be seen as redesigned portable devices made to adhere to the body as an accessory. Wearable sensors have found market success since people have become more interested in improving their health.[2] These wearable devices, however, are minimal in their computational ability in that they only collect simple vital signs like heart rate in order to preserve battery life. As more sophisticated body sensors develop, more body-conformal wearables are desired for more accurate vital sign data collection.[5,6] Moreover, wearable robotic devices will move towards being as flexible and compliant as possible in order to avoid causing potential body damage by rigid wearable devices.[7] For example, the Soft Exosuit developed by Harvard University's Wyss Institute is a lower limb assistive device that utilizes soft, flexible materials that interface with the body.[8] While this wearable device has demonstrated comfort over its rigid exoskeleton predecessors, the control system is powered by two sets of 4-cell bulky lithium polymer battery packs that require specific body placement for proper weight distribution, which still acts as a bodily obstruction.[8] A guideline for wearability requires that wearable objects have a form that is "within the wearer's intimate space, so that perceptually they become a part of the body."[9] An example of wearable energy storage that follows this guideline is the US Army's Conformal Wearable Battery (CWB). It is meant to be worn on the front of the torso so that the Warfighter's CWB can power portable electronics while on a dismounted mission.[10] Rated at 17 V, the CWB achieves conformability by being constructed as an array of discrete, rigid lithium ion-based cells.[11] While lithium-based batteries find use in wearables due to their unparalleled performance, they are potentially unsafe and have low cyclability. In addition to batteries, supercapacitors should be considered when developing energy storage for wearable electronics. As improvement in power and energy density is usually the focus when developing energy storage devices, ensuring that such technologies are safe and comfortable for the user should also be kept in mind when directing them towards wearable applications.

Batteries and supercapacitors are energy storage systems that are constructed similarly but generate and store energy differently.[12] Both systems have two electrodes, an electrolyte, and a separator. Batteries produce electrical energy through the conversion of chemical energy generated by electrochemical oxidation–reduction reactions.[13] When a battery is connected to an electrical load (discharging process), electrons flow from the negatively charged electrode (anode) through the load to the positively charged electrode (cathode), thus oxidizing and reducing the anode and cathode, respectively.[13] Negative ions (anions) and positive ions (cations) in the electrolyte flow to the anode and cathode, respectively, to balance the overall charge.[13] In the charging process for secondary (rechargeable) batteries, electrons flow in the opposite direction, making what was once the anode and cathode in the discharging process the cathode and anode, respectively.[13] As in the discharging process, electrolyte ions flow to the electrodes for overall charge balance. Supercapacitors, on the other hand, store energy electrostatically – *i.e.*, at the electrolyte/electrode interface between ions and electrode surface charges.[12] Recently, supercapacitors have become a focal point of numerous investigations due to their advantages of long cycle lifetime (>100 000 cycles), high power and energy densities, high safety, and relatively low series resistance.[14,15] In comparison to batteries, supercapacitors generally possess lower energy density but higher power density, due to the different internal mechanisms occurring within each device. While batteries have been the primary energy source for portable electronics, supercapacitors are finding their place in micropower systems as powering sensor nodes in wireless sensor systems, or as intermediary storage in conjunction with energy harvesting systems.[16]

Guidelines for developing wearable energy storage can emerge from the flexible electronics community by adhering to recommended qualities of being "thin, bendable, and mechanically compliant."[17] Additionally, the device should be safe and comfortable to wear. Attaining comfort for wearable devices can be achieved by following the guidelines presented by Gemperle *et al.*[9] in working closely with the human form. These guidelines include placement, form language, human movement, proxemics, sizing, and attachment; such considerations should be kept in mind when designing energy storage for wearables. On the other hand, achieving safety for wearable energy storage devices will involve considering new materials and device constructions that pose low danger – such as low flammability and no skin irritation – to the wearer.

14.2 Ionic Liquid Gels for Energy Storage

14.2.1 Ionic Liquid Gels Overview

With their unique combination of physical properties, namely ultralow volatility (which confers nonflammability), and high electrochemical and thermal stability, as well as moderate room temperature ionic conductivity, ionic liquids are an emerging class of materials with tremendous potential to enable safe, wearable electrochemical energy storage devices. In spite of their many benefits, however, ionic liquids remain subject to one limitation that all liquid electrolytes face: the possibility of leaking out from a damaged or physically compromised device. By incorporating a solid, volume-spanning internal scaffold (such as a cross-linked polymer network) inside the ionic liquid for structural support, an *ionic liquid gel* can be created to eliminate this concern. Ionic liquid gels, also referred to as ionogels or ion gels, not only alleviate the possibility of electrolyte leakage, but they can also be designed to achieve a high degree of mechanical flexibility while retaining many of the desirable properties of the pure ionic liquid, including high ionic conductivity and electrochemical/thermal stability.

In general, an ionic liquid gel contains a minimum of approximately 50% ionic liquid, either on a volume or mass basis, integrated with a supporting scaffold material that is generally nonconductive. When the ionic liquid content is much lower than 50%, the material properties of the composite may more closely mirror those of the scaffold, rather than those of the ionic liquid itself. For example, a polymer/ionic liquid mixture that contains only a small amount of ionic liquid can yield a solid polymer electrolyte, in which ionic motion may be strongly coupled to the polymer segmental dynamics (*i.e.* a salt-in-polymer system).[18] Thus, ionic conductivity, in particular, can be reduced to levels that are too low for practical electrochemical energy storage devices ($\ll 1$ mS cm^{-1}) if the ionic liquid is not the majority component within an ionic liquid gel.

Several approaches have been successfully employed to create ionic liquid gels in recent years. Here, they have been grouped into three major categories: (1) macromolecular, (2) colloidal, and (3) molecular, following the nomenclature used by Marr and Marr,[19] who recently classified ionic liquid gels according to the nature of their supporting scaffolds. The following sections provide a general overview of typical ionic liquid gel formulations/synthesis methods and material properties reported to date. It is intended to be illustrative, rather than comprehensive. Several recent review articles on the topic of ionic liquid gels are recommended for those seeking additional information and perspectives.[19–22] It should also be noted that the focus here is primarily on ionic liquid gels that feature hydrophobic or water-immiscible ionic liquids (typically those having polyfluorinated anions), as these generally provide higher electrochemical stability, and thus may be better suited for wearable energy storage devices that can provide large energy densities. For the sake of brevity, the ionic liquid ion abbreviations shown in Table 14.1 are used:

Table 14.1 Ion abbreviations for some common ionic liquids.

Cations		Anions	
EMI	1-Ethyl-3-methylimidazolium	Tf	Trifluoromethanesulfonate
BMI	1-Butyl-3-methylimidazolium	TFSI	Bis(trifluoromethanesulfonyl)imide
HMI	1-Hexyl-3-methylimidazolium	BF_4	Tetrafluoroborate
OMI	1-Octyl-3-methylimidazolium	PF_6	Hexafluorophosphate
PYR_{13}	*N*-Propyl-*N*-methylpyrrolidinium	FAP	Tris(pentafluoroethyl) trifluorophosphate
PYR_{14}	*N*-Butyl-*N*-methylpyrrolidinium	TCB	Tetracyanoborate

14.2.1.1 Formulations and Synthesis Methods

14.2.1.1.1 Macromolecular Scaffolds (Polymers)

Linear or Chemically Cross-linked Polymers. Polymers represent a highly versatile set of scaffold materials from which to create ionic liquid gels. They can be synthesized with an endless variety of chemical functionalities and can be used to create both physically and chemically cross-linked networks to support ionic liquids. Polymer scaffolds are integrated into ionic liquid gels mainly using one of two different approaches. The first is to blend a desired amount of polymer with an ionic liquid, typically using an organic co-solvent to promote miscibility, and then allow the polymer to self-assemble or entangle with itself to form a supporting network after the co-solvent has been evaporated *via* a heating and/or vacuum treatment step.[23,24] The second is to synthesize a polymer scaffold *in situ* within the ionic liquid, for example by thermally- or UV-initiated free radical polymerization.[25] An advantage of the *in situ* approach is that the solid ionic liquid gel electrolyte can be formed from a liquid phase precursor solution that contains the scaffold-forming monomers/cross-linkers, an initiator, and the ionic liquid after the solution has made intimate contact with other material layers inside a device structure, such as nanostructured battery or supercapacitor electrodes. UV-initiated polymerization can generally provide shorter gelation times (under 10 minutes)[26] compared to thermally-initiated polymerization, which is typically performed at 80 °C for up to 24 hours.[27] An exception to this rule was reported by Visentin *et al.*, who demonstrated that *in situ* polymerization/cross-linking *via* rapid microwave heating could generate ionic liquid gels in as little as 10 seconds.[28]

One of the earliest reports of an ionic liquid gel was made by Fuller, Breda, and Carlin in 1997;[23] this team at Covalent Associates, Inc. created a gel with EMI Tf supported by the copolymer poly(vinylidene fluoride)-*co*-hexafluoropropylene (PVDF-HFP) that contained a little more than 50 wt% ionic liquid. A few years later, Noda and Watanabe synthesized EMI BF_4 gels supported by a chemically cross-linked poly(2-hydroxyethyl methacrylate) scaffold created *in situ* within the ionic liquid at varying molar ratios, up to a maximum of 50 mol% ionic liquid.[25] The formation of cross-linked polymer scaffolds *in situ* was also reported by the groups of Winterton[29] and Rogers[30] using OMI

TFSI and HMI TFSI, respectively. In 2005, a landmark paper in the field was published by the Watanabe group, wherein they described the *in situ* fabrication of a series of cross-linked poly(methyl methacrylate) (PMMA)-supported EMI TFSI gels that contained up to 70 mol% ionic liquid.[27] Jiang *et al.* created linear PMMA-supported BMI PF_6 gels with up to 50 wt% ionic liquid *via in situ* polymerization.[31] Nitrile rubber was combined with *N*-ethylimidazolium TFSI to generate gels with ionic liquid contents as high as 50 wt%, as reported by the Ohno group.[32]

More recently, polymer-supported ionic liquid gels have typically contained in the range of 70–95 wt% ionic liquid, since a higher ionic liquid fraction leads to a greater ionic conductivity of the composite gel. Pandey *et al.* created EMI Tf gels with PVDF-HFP scaffolds that included 80 wt% ionic liquid;[33] analogous gels featuring the ionic liquid EMI TCB were also reported.[34] *In situ* polymerization of the cross-linkable macromer poly(ethylene glycol) diacrylate (PEGDA) successfully enabled the synthesis of EMI TFSI gels containing up to 95 wt% ionic liquid by Visentin *et al.*[26] Cross-linked PEGDA was also employed to generate EMI TCB and EMI FAP gels that included as much as 93 wt% or 96 wt% ionic liquid, respectively.[35] Various methacrylate monomers bearing different chemical functional groups were used to synthesize a series of cross-linked poly(methacrylate) scaffolds in EMI TFSI by D'Angelo *et al.* that yielded gels with at least 80 mol% ionic liquid.[36] In late 2016, the Panzer group reported the first zwitterionic copolymer-supported ionic liquid gels, which contained 76 vol% EMI TFSI.[37]

Triblock Copolymers. ABA triblock copolymer scaffolds have also been employed to created ionic liquid gels; this approach has largely been spearheaded by the Lodge group.[24] Triblock copolymers offer the advantage of spontaneous scaffold self-assembly due to the presence of both ionic liquid compatible and incompatible blocks; however, their synthesis is generally more complex. Polystyrene-*b*-poly(ethylene oxide)-*b*-polystyrene (PS-PEO-PS) triblock copolymers were used to achieve BMI PF_6 gels with as high as 95 wt% ionic liquid.[24] Replacing the middle block with PMMA has also been successful in realizing EMI TFSI gels using PS-PMMA-PS scaffolds.[38] In 2011, Gu and Lodge reported triblock copolymer ionic liquid gels having a poly(ionic liquid) middle block to support 85 wt% EMI TFSI.[39] The addition of azide groups to the polystyrene blocks of PS-PEO-PS triblock copolymers yielded tough, chemically cross-linked scaffolds for EMI TFSI gels featuring 90 wt% ionic liquid.[40] Photoreversible gelation of up to 90 wt% BMI PF_6 using azobenzene-functionalized triblock copolymer scaffolds has also recently been described.[41]

Other Polymer Scaffolds and Lithium Ion-containing Electrolytes. Fujii *et al.* utilized a mixture of two different tetrafunctional macromers to create a highly cross-linked polymer scaffold *in situ* within EMI TFSI, yielding gels with up to 97 wt% ionic liquid.[42] Festin *et al.* synthesized nitrile butadiene rubber/PEO interpenetrating network scaffolds to support EMI TFSI gels designed for mechanical actuators.[43] Poly(ionic liquid)s, such as poly(diallyldimethylammonium) TFSI (PDADMA TFSI), have also been successfully used to create ionic liquid gels in combination with a neat ionic liquid (*e.g.* 70 mol% PYR_{14} TFSI).[44]

Polymer scaffolds have been employed to support lithium ion-containing ionic liquid gels to serve as nonvolatile battery electrolytes. Poly(acrylonitrile) (PAN) was synthesized *in situ* within a solution of 0.5 M lithium bis(trifluoromethanesulfonyl)imide (LiTFSI) in PYR_{14} TFSI by Patel *et al.* to create gels with up to 85 wt% ionic liquid.[45] PVDF-HFP is another common scaffold for creating ionic liquid gels for batteries; this polymer has been used to create gelled solutions of lithium hexafluorophosphate in EMI FAP,[46] as well as LiTFSI in PYR_{14} TFSI.[47] The Schönhoff group described the formation of LiTFSI/PYR_{14} TFSI gel electrolytes using either PMMA or a poly(ionic liquid), PDADMA TFSI, as the supporting scaffold.[48] Horowitz and Panzer employed the acid-catalyzed hydrolysis/condensation reaction of triethoxysilylethyl-terminated poly(dimethylsiloxane) (PDMS) macromers in EMI TCB or EMI TFSI to create PDMS-supported gels containing up to 80 wt% ionic liquid; LiTFSI was also included in some gel formulations.[49]

14.2.1.1.2 Colloidal Scaffolds. In contrast to polymer supporting scaffolds, colloidal (nanoparticle)-based scaffolds are typically inorganic in nature (*e.g.* silicon dioxide, SiO_2) and yield ionic liquid gels that display a less elastic rheological response. On the other hand, the use of an oxide-based scaffold typically provides greater thermal stability to the gel composite in comparison to some polymer networks. Ionic liquid gels have been realized by mechanically mixing SiO_2 nanoparticles into EMI TFSI, using as little as 3 wt% SiO_2, as reported by Ueno *et al.*[50] Polymer-coated SiO_2 nanoparticles synthesized *via* chemical grafting of PMMA have also been added to EMI TFSI to create gels that exhibit structural color tunability.[51] Lithium cation-modified SiO_2 nanoparticles were used to fabricate PYR_{13} TFSI gels by Sun *et al.*[52]

While the examples cited above describe gels formed by the simple addition of nanoparticles into an ionic liquid, several manifestations of gels have also been created by the *in situ* synthesis of an oxide nanoparticle-based scaffold (*e.g.* the acid-catalyzed sol–gel reaction of a methoxysilane precursor to form a nanostructured silica support). One of the earliest reports of the creation of a silica scaffold through the reaction of tetramethoxysilane (TMOS) and formic acid (FA) in EMI TFSI was made by Dai *et al.*,[53] although the focus at the time was on the silica structure itself (the ionic liquid was washed out by solvent reflux). The TMOS/FA approach was utilized by the Vioux group in BMI TFSI, where a 2:1 TMOS:ionic liquid molar ratio led to the creation of silica scaffold-supported ionic liquid gels.[54] Addition of methyltrimethoxylsilane to the TMOS/FA mixture and up to 50 mol% ionic liquid was also shown to yield BMI TFSI gels.[55] Horowitz and Panzer further tuned the TMOS/FA/ionic liquid ratio to generate soft ionic liquid gels containing up to 94 wt% EMI TFSI.[56] The generality of the TMOS/FA approach was also demonstrated using several additional ionic liquids of various cations and anions.[57] An important limitation of the *in situ* sol–gel oxide scaffold approach, however, is the possibility of significant gel shrinkage and/or cracking that can occur as volatile reaction byproducts evaporate during the course of gel ageing times that can last more than one week at room temperature.[56] Nanocomposite hybrid scaffolds formed using a mixture of tetraethoxysilane and trimethoxysilyl-modified PMMA for sol–gel reaction in BMI TFSI have been described by Gayet *et al.*[58,59]

14.2.1.1.3 Molecular Scaffolds. The addition and self-assembly of low molecular weight gelators (LMWGs) to ionic liquids has been used to create molecular framework gel scaffolds. After adding a small amount of LMWG to an ionic liquid, the mixture is heated to promote dissolution and then cooled back down to room temperature. In some cases, less than 1 wt% of the LMWG is necessary to induce gelation.[60] However, these gels are typically extremely soft and can disassemble upon heating. An interesting hybrid scaffold was recently described by Kataoka *et al.*, which employed a LMWG/cross-linked PMMA double network to gel EMI TFSI with 90 wt% ionic liquid.[61]

14.2.1.2 Gel Properties

14.2.1.2.1 Ionic Liquid Gel Deposition Methods. In order to measure their material properties, ionic liquid gel samples or thin gel films can be prepared using various deposition methods. The most common approach is to simply cast the precursor solution into a mold with the desired shape, or as a thin film on a smooth substrate, and then initiate gelation (*e.g.* evaporate the co-solvent,[34] or irradiate with UV light[26]). Spin-coating has been used to generate triblock copolymer-supported ionic liquid gel thin films.[38] If the ionic liquid gel film is first formed *via* spin-coating on an elastomeric stamp, it can subsequently be transfer printed onto another substrate.[62] Aerosol ink-jet printing has been successfully employed by the Frisbie group to create patterned ionic liquid gels for flexible electronic circuits.[63] Spray-coating of ionic liquid gel films has also been demonstrated.[64]

14.2.1.2.2 Ionic Conductivity. The ionic conductivity of ionic liquid gels is typically measured *via* electrical impedance spectroscopy, with the ionic liquid electrolyte contacted by two blocking electrodes in a well-defined geometry. The real component of impedance measured at high frequencies (~100 kHz) is used to determine the gel ionic conductivity, accounting for the series resistance of the test setup.[26] The temperature dependence of the ionic conductivity for both ionic liquids and ionic liquid gels generally follows Vogel–Tamman–Fulcher (VTF) behavior,[65] which can be approximated by simple Arrhenius behavior over small temperature ranges and for values significantly above the glass transition temperature of the neat ionic liquid.[36] When a scaffold is introduced into an ionic liquid to create an ionic liquid gel, the ionic conductivity of the composite is reduced due mainly to the lower ionic liquid concentration and higher tortuosity with respect to ion motion, but ionic liquid–scaffold interactions can also play a role. At room temperature (RT), some ionic liquids have ionic conductivities near 10 mS cm^{-1} or higher (*e.g.* EMI TFSI, EMI TCB). Several manifestations of EMI TFSI gels have displayed RT ionic conductivity values of approximately 9–10 mS cm^{-1}.[26,42,50,56] A 93 wt% EMI TCB gel exhibited a RT ionic conductivity of 13 mS cm^{-1}, which is one of the highest values reported for an ionic liquid gel to date.[35]

It should be noted that many ionic liquid gels display excellent thermal stability, and can exhibit essentially no mass loss for temperatures up to ~300 °C.[34,55] A high thermal stability of the scaffold component is essential to realizing the outstanding thermal stability of the ionic liquid in the gel composite. When the temperature dependence of ionic conductivity is fit to the Arrhenius model, the activation energy parameter that one extracts is generally larger for ionic liquid gels compared to that of the neat ionic liquid.[23] Moreover, the activation energy of ionic conductivity has been observed to increase with increasing scaffold content, for the case of highly compatible polymer scaffolds.[36] In contrast, a PDMS-supported EMI TCB gel (80 wt% ionic liquid), in which the PDMS scaffold was not miscible with the ionic liquid, exhibited an activation energy of ionic conductivity nearly equal to that of the neat ionic liquid itself.[49]

14.2.1.2.3 Mechanical Properties/Gel Rheology. Ionic liquid gel mechanical responses are generally probed using either tensile[42,66] or compressive[37,42,56,67] stress–strain testing, or *via* shear rheometry.[24,39,45] The typical elastic modulus (or shear modulus) values for the softest gels, those having the highest ionic liquid contents, are on the order of 1–10 kPa. Elastic modulus values of 15–20 MPa have been observed for ionic liquid gels containing high polymer contents and/or scaffold cross-link densities.[26,66] Typically, however, there is a trade-off between increasing mechanical stiffness *via* polymer addition and the resulting decrease in gel ionic conductivity.[36]

14.2.1.2.4 Electrochemical Stability Window. For both supercapacitor and battery applications, a wide potential window of electrochemical stability is desirable. Generally speaking, the electrochemical stability of a well-designed ionic liquid gel will mimic that of its neat ionic liquid component. This can be achieved by selecting a scaffold material that is very electrochemically stable. Representative ionic liquid gel device operating voltage stability windows range from approximately 3–5 V for several hydrophobic ionic liquids.[34,45,68]

14.2.2 Ionic Liquid Gel Electrolytes for Battery Applications

Lithium-ion (Li-ion, Li^+) batteries are the most dominant energy storage device to meet the current demands due to their excellent cyclability, robustness, and high energy density.[69] Rechargeable batteries typically consist of a cathode, an anode, and an ion-containing electrolyte. Conventional battery electrolytes are composed of a lithium salt dissolved in an organic solvent (*e.g.* ethylene carbonate or propylene carbonate) that possesses a high dielectric constant. However, organic solvent-based electrolytes exhibit a non-negligible vapor pressure, making them flammable components in the design of an energy storage device; therefore, in the event of a shorting of the device or if the device is operated at elevated temperatures, fire or an explosion is a likely scenario.[20]

To circumvent this safety hazard, ionic liquid-based electrolytes possessing a negligible vapor pressure are of great interest for battery applications. Furthermore, solid-state electrolytes (ionic liquid gels) confining an ionic liquid offer the advantages of a leakproof design, are lightweight materials, and are applicable for flexible electronics.[20] Typically, to form a lithium-containing electrolyte, a lithium salt (*e.g.* LiTFSI) is dissolved into an ionic liquid in which an optimum concentration with respect to ionic conductivity is experimentally determined (~1 M).[70] The increase in viscosity due to the addition of a salt to an ionic liquid, which effectively reduces the ionic conductivity of the electrolyte, must be considered. Furthermore, an important property for lithium-based devices is the Li^+ *transference* number, which is approximately equal to the fraction of current carried by Li^+ cations through the device, calculated by direct current polarization techniques;[71] a related property is the Li^+ *transport* number, which is calculated from the fraction of the Li^+ diffusivity relative to the diffusivities of all mobile ionic species measured *via* nuclear magnetic spectroscopy techniques.[72,73] A higher Li^+ transference number promotes lower ion concentration gradients at the surfaces of both electrodes, which can reduce the internal resistance of the device and result in longer cell lifetime and faster discharge rates.[74] Due to the presence of the ionic liquid cations/anions, such electrolytes contain three mobile ionic species (*i.e.* Li^+, ionic liquid cation, and ionic liquid anion), which limits the Li^+ transference number to a typical range of ~0.05–0.5.[70] Progress is being made for ionic liquid-based systems to increase the Li^+ transference number by various means of anchoring the anions to the scaffold in order to develop a more stable electrolyte for lithium-based batteries.[75–77]

14.2.2.1 Physically Cross-linked Polymer Scaffolds

The most prevalent polymer-based scaffold confining a lithium-containing ionic liquid has been PVDF-HFP. This scaffold contains strong electron withdrawing groups that generate a high dielectric constant polymer with suitable anodic stability (*i.e.* interfacial stability against a lithium metal anode) that also promotes lithium salt dissociation. Many investigations have focused on the modification of the liquid electrolyte by means of functionalization of the ionic liquid in order to meet the demands of lithium-based battery applications. An early investigation of this system was reported by Fernicola and co-workers, who confined an ionic liquid solution of *N*-butyl-*N*-ethylpiperidinium TFSI (PP_{24} TFSI) containing 0.2 mol kg^{-1} LiTFSI; various gel compositions were synthesized, featuring 20 wt% to 40 wt% PVDF-HFP.[75] An important consideration for lithium batteries is the stability of the lithium metal anode, which is very reactive and forms dendrites capable of shorting the device and causing a fire hazard. Due to the shear modulus of most gels being below the GPa range (*i.e.* too soft to form a mechanical barrier for lithium dendrite formation), a mechanism to form a stable lithium metal/electrolyte interface is the formation of a passivating solid electrolyte interphase (SEI) layer. The SEI layer is monitored by measuring the resistance of the lithium/electrolyte interface, where one seeks to verify that a steady state value is achieved in a short amount of time. Fernicola *et al.* determined that SEI layer stability was achieved for a 20 wt% PVDF-HFP-supported ionic liquid gel after 3 days during open-circuit conditions, reaching a steady state resistance of ~2400 Ω cm^{-2}.

Furthermore, Ferrari *et al.* attempted to increase the thermal and interfacial stability of a PVDF-HFP-based ionic liquid gel *via* the incorporation of silica nanoparticles as an additive.[78] The gel confined an ionic liquid solution of *N*-ethyl(methylether)-*N*-methylpyrrolidinium TFSI (PYR_{12O1} TFSI) with a LiTFSI concentration of 0.41 mol LiTFSI kg^{-1}. The authors also added nano-sized silica (10 wt%) possessing a well-ordered hexagonal mesoporous structure with a narrow pore size distribution of 6 nm. The synthesized composites containing 30 wt% PVDF-HFP and 10 wt% silica filler achieved a stable interfacial stability after 10 days, possessing a steady state resistance of 150 Ω cm^{-2} compared to a resistance of ~800 Ω cm^{-2} for the gel without silica filler loading. Cyclability and discharge capacity at specific discharge rates is also critical for the practical operation of lithium-based batteries. Ferrari and co-workers fabricated composite coin cells with a configuration of lithium metal/ionic liquid gel (containing silica filler)/lithium iron phosphate ($LiFePO_4$). At a C-rate of C/10 (*i.e.* a discharge time of 10 hours) and C/3 (*i.e.* a discharge time of 3 hours), the discharge capacity using the 30 wt% PVDF-HFP ionic liquid gel with a silica loading of 10 wt% was measured to be ~140 mA h g^{-1} (on a mass of cathode basis) and ~70 mA h g^{-1}, respectively, while demonstrating a high cyclability with an average Coulombic efficiency of ~99% up to 180 cycles; it should be noted that the $LiFePO_4$ cathode possesses a theoretical capacity of ~170 mA h g^{-1}.

Moreover, an alternative electrolyte layer synthesis option for battery applications is to utilize electrospinning to synthesize polymer-based membranes. The electrospinning technique allows for the facile synthesis of a homogenous polymer membrane containing pores in the size range of 1 μm to 1 nm and exhibiting high mechanical strength due to the fibrous network.[78] A high porosity also allows for significant electrolyte uptake, leading to high ionic conductivities of gel membranes with electrospun polymer scaffolds. In 2007, Cheruvally and co-workers synthesized electrospun polymer membranes supporting a 0.5 M solution of LiTFSI in BMI TFSI with a PVDF-HFP polymer content of ~12 wt%.[79] A lithium/ionic liquid gel/$LiFePO_4$ cell was tested using a C/10 discharge rate that achieved a capacity of 152 mA h g^{-1} after 24 charge/discharge cycles. The discharge capacity of the identical electrolyte solution soaked in a commercial fiber-based separator was ~130 mA h g^{-1} after only 6 cycles. At a faster rate of C/2, the discharge performance was 132 mA h g^{-1} after 10 cycles. However, the interfacial resistance of the device was not measured. In 2012, Rao *et al.* demonstrated an electrospun polymer-supported ionic liquid gel electrolyte utilizing PAN/PMMA as the polymer scaffold.[80] PAN acts as a mechanical support for the membrane while also exhibiting an attribute for the improvement of lithium ion mobility *via* its polar –CN groups, while PMMA acts as a solubilizing agent for the lithium-containing ionic liquid electrolyte. The interfacial resistance of this ionic liquid gel against a lithium metal anode was measured to start at 280 Ω and reach a steady state value of ~580 Ω after 30 days, performing markedly better than commercial PAN-based membranes. Discharge measurements were performed at a C/10 rate to yield a capacity of 139 mA h g^{-1}. Cyclic performance at a C/5 rate was tested for 50 cycles and a 92% retention of initial capacity was achieved for the PAN/PMMA-supported ionic liquid gel compared to only 84% retention for the commercial fiber-supported electrolyte.

Polymerized ionic liquids (PILs) are a more recent class of polymer scaffolds that have attracted much attention for ionic transport applications. Appetecchi and co-workers cast a 1 M LiTFSI solution with the poly(ionic liquid) PDADMA TFSI to form an ionic liquid gel membrane, in which a minimum of 27.6 wt% of PIL was needed to obtain a self-standing gel.[81] Interfacial resistance measurements made after 92 days in a lithium/ionic liquid gel/lithium cell exhibited an increase in resistance from 900 Ω cm^{-2} to 2700 Ω cm^{-2}, while achieving a steady state value after 64 days. Discharge measurements were performed in a Li/ionic liquid gel/$LiFePO_4$ cell and exhibited a capacity of 141.1 mA h g^{-1} and 16.9 mA h g^{-1} at C/10 and 1C rates, respectively.

14.2.2.2 *Chemically Cross-linked Polymer Scaffolds*

Utilizing a chemically cross-linked polymer scaffold for ionic liquid gel electrolytes has not been extensively explored relative to physically cross-linked networks for lithium battery applications to date. In 2002, Nakagawa and co-workers polymerized PEGDA (M_W = 600 g mol^{-1}) inside a 1 M lithium tetrafluoroborate ($LiBF_4$)/EMI BF_4 solution to create a 15 wt% polymer scaffold.[82] The discharge capacity of a lithium metal cell containing this gel electrolyte was measured to be 103 mA h g^{-1} after the 5th cycle at a C/10 rate; a comparable cell containing only the ionic liquid solution exhibited a capacity of 132 mA h g^{-1}. Cyclability tests for the ionic liquid gel electrolyte demonstrated a capacity retention of 73.1% after 50 cycles. The authors attributed the decrease in discharge capacity to increased lithium interaction with the polymer scaffold (*i.e.* ethylene oxide groups) that induced a concentration polarization due to the diffusion of lithium being relatively slow compared to the other ionic species in the system. In addition, the instability of the SEI layer also contributed to the decrease in capacity of the battery system.

In 2011, a PAN-based ionic liquid gel system was investigated by Patel *et al.* that exhibited an improved interfacial stability of the gel electrolyte relative to the lithium salt/ionic liquid solution using a lithium metal anode.[45] Discharge measurements also verified an increased capacity using the ionic liquid gel electrolyte (by ~10%) at a C-rate of C/5; however, the discharge capacity decreased from 112 mA h g^{-1} to 60 mA h g^{-1} after 80 cycles, proving this electrolyte system to be unsuitable for practical applications. In 2012, Yun and co-workers fabricated an ionic liquid gel system with increased SEI stability by the addition of vinylene carbonate (10 wt%), an SEI-forming organic solvent.[83] The gel composites were synthesized *via in situ* polymerization of poly(ethylene glycol) dimethacrylate (PEGDMA, M_n = 550 g mol^{-1}) in the range of 4–10 wt% with a 1 M LiTFSI in PYR_{14} TFSI ionic liquid solution. For a 4 wt% PEGDMA ionic liquid gel doped with vinylene carbonate, the discharge capacity in a lithium metal cell at a C/5 rate was measured to be 149.1 mA h g^{-1}, which decreased to 141.5 mA h g^{-1} after 50 cycles, whereas the liquid electrolyte (no scaffold) cell dropped from 160 mA h g^{-1} to ~80 mA h g^{-1} during the same cycle period. The authors reported that a stable SEI layer was formed after only one charge–discharge cycle, while flammability tests proved that the addition of the organic solvent did not alter the advantageous properties of the ionic liquid-based electrolyte.

14.2.2.3 Inorganic Colloidal Scaffolds

Inorganic oxide-based compounds and nanoparticles are commonly used as gelling agents to synthesize solid-state ionic liquid gel composites. Inorganic-based networks can offer the advantages of facile synthesis and high thermal stability.[77] Silica-supported gels have been the most commonly investigated composites in the literature due to the commercial availability of silica nanoparticles and their easy fabrication. Wu and co-workers fabricated a silica-supported ionic liquid gel containing an epoxy-functionalized silane coupling agent with LiTFSI dissolved in PYR_{13} TFSI, where the scaffold was formed *via* the sol–gel method.[84] Lithium/ionic liquid gel/$LiFePO_4$ cells exhibited a discharge capacity after 30 cycles of 154.9 mA h g^{-1} and 168.6 mA h g^{-1} at 30 °C and 90 °C (Coulombic efficiency of ~99% at both temperatures), respectively, demonstrating enhanced capacity at elevated temperatures. A lithium titanium oxide ($Li_4Ti_5O_{12}$)/ionic liquid gel/$LiFePO_4$ cell tested for 200 cycles demonstrated a capacity of 165.7 mA h g^{-1} with close to 100% efficiency.

Another use of nanoparticles is to immobilize one of the ions of the ionic liquid on the surface of the nanoparticles in order to create a gel that is capable of selectively promoting lithium-ion transport. Lu *et al.* tethered the cation of the ionic liquid 1-methyl-3-propylimidazolium TFSI to the surface of 7 nm silica nanoparticles while doping the composite with LiTFSI.[77] The shear modulus values for these gel electrolytes were on the order of 0.1 MPa. Thermal stability up to 400 °C was measured, together with electrochemical stability up to 4.25 V (*vs.* Li/Li^+) and a Li^+ transference number of 0.54. However, no charge–discharge measurements or interfacial electrode stability tests were performed on this ionic liquid gel system.

14.2.3 Ionic Liquid Gel Electrolytes for Supercapacitors

The development of new solid-state electrolytes possessing high ionic conductivity and wide electrochemical stability window is necessary to further improve the performance of supercapacitors, as the energy density is proportional to the square of the maximum operating voltage, which is generally determined by the breakdown limit of the electrolyte. Therefore, considerable research has been performed on ionic liquid gel materials with wide electrochemical stability windows and relatively high ionic conductivity, which can offer a promising option for future electrolytes in solid-state supercapacitors.[19,20] Currently, the majority of these studies are focused on the ionic liquid gel electrolyte-based electrical double layer capacitors (EDLCs) with some common electrode materials, such as activated carbon (AC),[85] porous carbon,[86] and carbon nanotubes (CNTs).[34,87,88] Based on ionic liquid gel electrolytes, flexible and stretchable supercapacitors have been demonstrated, and are described in Section 14.3.1.1.

The properties of an ionic liquid gel are mainly dependent on the nature of the ionic liquid, the polymer scaffold, and the interaction between the ionic liquid and its polymer scaffold. Therefore, the selection of an appropriate polymer scaffold becomes a crucial factor for achieving good supercapacitor performance with ionic liquid gels. A variety of polymers have been studied as scaffolds for ionic liquid gels, such as poly(vinyl alcohol) (PVA),[89] poly(ethylene oxide) (PEO),[90] PMMA,[91] PEGDA,[26,92] and PVDF-HFP.[34,93,94] Visentin and Panzer[26] prepared UV-initiated cross-linked PEGDA ionic liquid gels based on EMI TFSI. These gels were tested in capacitive structures and showed potential for application in supercapacitor devices. Other than polymer scaffolds, silica-supported ionic liquid gels could also be a viable option for supercapacitors.[87,95,96]

The properties of ionic liquids can vary greatly, so it is crucial to select and design the ionic liquid in an ionic liquid gel electrolyte for the desired applications. Pandey and Hashmi[34] utilized a low viscosity ionic liquid, EMI TCB, to create an ionic liquid gel electrolyte for EDLCs in order to overcome the low conductivity that typically results from the high viscosity of many ionic liquids. They entrapped EMI TCB in PVDF-HFP copolymer to form the gel electrolyte, and fabricated solid-state EDLCs based on the prepared gel. The assembled EDLCs showed a specific energy and power density of 3.5 W h kg^{-1} and 4.2 kW kg^{-1}, respectively (in terms of the mass of the CNT electrodes). The same group also examined the effect of the addition of lithium salt into a PVDF-HFP-supported EMI FAP gel electrolyte,[46] where they demonstrated that the addition of lithium salt could significantly improve the specific capacitance of the EDLCs from 76 F g^{-1} to 127 F g^{-1}. However, the application of neat ionic liquids at sub-zero temperatures is often limited by their high viscosity and melting point near room temperature. Tsai *et al.*[97] reported a eutectic mixture of ionic liquids (1 : 1 molar ratio of *N*-methyl-*N*-propylpiperidinium bis(fluorosulfonyl)imide (PIP_{13} FSI) and PYR_{14} FSI) as an electrolyte for supercapacitor applications over a wide temperature range from −50 °C to 80 °C. The same group prepared an ionic liquid gel using a sol–gel method to create a SiO_2 scaffold supporting the eutectic mixture, and further utilized the ionic liquid gel in an all solid-state supercapacitor.[98] The ionic liquid gel exhibited high ionic conductivity over a wide temperature range (from 0.2 mS cm^{-1} at −40 °C up to 10 mS cm^{-1} at 60 °C) and the fabricated supercapacitor showed a capacitance of up to 90 F g^{-1} at room temperature.

14.3 Fabrication Techniques for Ionic Liquid Gel Integration into Wearable Systems

14.3.1 Device Assembly Techniques

Ionic liquid gel-based energy storage devices in the literature that could find application in wearable systems have been demonstrated as standalone devices, on-chip microelectronics components, and elements in electronic textiles. Standalone energy storage devices can be made wearable if they can adhere to clothing or be worn as an accessory. On-chip micro-energy storage can find a place in the hardware inside wearable sensor systems. Lastly, energy storage woven into the fabric of clothing for powering interactive textiles[99] is a logical path forward as wearable computing becomes more ubiquitous. In Section 14.2.1.2.1, different methods of depositing ionic liquid gels were described. However, it is worth understanding the sequential steps in which ionic liquid gels are synthesized and assembled with other energy storage device components to inform how these devices can be scaled-up for commercialization. The way in which components are assembled is important because it influences the overall resistance of the device, ultimately affecting final device performance. In general, ionic liquid gel electrolytes[100–105] for batteries have been demonstrated in coin cell devices to demonstrate the improved safety that ionic liquids can impart on Li-ion batteries. Coin cells use a parallel plate architecture, where the electrolyte is sandwiched between two parallel electrodes and rigid packaging helps electrical components achieve low device resistance. Since the coin cell battery architecture is an established configuration, its assembly will not be described here. In addition to the conventional parallel plate architecture, ionic liquid gel supercapacitor devices have also adopted the in-plane interdigital electrode architecture, where electrodes exist in the same plane on a single substrate. The following section will outline how different ionic liquid gel synthesis methods have lent themselves to different device architectures. Table 14.2 summarizes methods of ionic liquid gel incorporation with respect to each architecture, as well as considerations for device assembly for a particular method of incorporation.

14.3.1.1 Polymer-supported Ionic Liquid Gels Integrated into Wearable Energy Storage Devices

Polymer-supported ionic liquid gel electrolytes have mechanical characteristics that make these materials suitable for wearable energy storage devices. Demonstrated devices in the literature with 'wearable potential' typically have

Table 14.2 Methods of incorporating ionic liquid gel electrolytes for different device architectures.

Device Architecture	Ways to incorporate ionic liquid gel	Device assembly considerations
Parallel plate	Electrode coating	• Mold required for defining ionic liquid gel dimensions • Ionic liquid gel to have characteristics that prevent electrical shorting when device is flexed • Sealing components together (can be done by using ionic liquid gel as an adhesive or through use of external packaging)
	Freestanding film to be assembled with other freestanding components	• Pre-wetting of electrodes to decrease electrode/electrolyte interfacial resistance • Sealing components together (through use of external packaging or external adhesive)
	Foundational support for attaching other electrical components	• Electrical components need to be added subsequently • Pre-treatment of ionic liquid gel for proper adhesion of electrical components
Interdigital	Thin film deposition	• Electrical components usually deposited on substrate before ionic liquid gel deposition • Ionic liquid gel dimensions defined by mold or finite volume

ionic liquid gels that are fabricated in one of two ways: (1) polymer scaffold solubilization through the use of a co-solvent, or (2) scaffold polymerization through UV-initiated cross-linking of monomers. Removal of a co-solvent can take up to 24 hours,[28] while UV-initiated polymerization can be achieved in only a few minutes.[28] Polymer-supported ionic liquid gel supercapacitors typically utilize the conventional parallel plate architecture due to the higher structural integrity of these electrolytes over silica-supported ionic liquid gels. Researchers, however, are starting to incorporate polymer-supported electrolytes into flexible interdigital architectures as flexible microelectronics and low power computing are on the rise.

Early polymer-supported ionic liquid gels were based on a PVDF-HFP scaffold and tested in the parallel plate device configuration[106–108] in order to understand the best type of carbon electrode with which to pair these ionic liquid gels. Lu *et al.* demonstrated the earliest parallel plate supercapacitor device with PVDF-HFP ionic liquid gels by pairing them with high surface area carbon electrodes/aluminum foil current collectors.[106] They constructed a freestanding ionic liquid gel electrolyte that was then sandwiched between two AC electrode/aluminum foil current collectors epoxied to a polypropylene plate.[106] Epoxy was then used to seal the edges of the components together to yield a device. Although the device was not advertised to be flexible, the fabrication steps provide insight into two considerations to heed when constructing a parallel plate flexible device for wearability. The first is wetting the entirety of the electrode surface area with ionic liquid to "ensure good contact and low interfacial [electrode/electrolyte] resistance";[106] and the second is proper adhesion of all the electrical components – electrode, current collectors and electrolyte – to each other, so as to also keep the overall device resistance low.

Polymer-supported ionic liquid gels can serve as an "adhesive" between two electrodes (Figure 14.1). In their work using CNT/bacterial nanocellulose electrodes, Kang *et al.* utilized a PS-PEO-PS triblock copolymer scaffold to produce a gel *via* the co-solvent method, claiming that "the hydrophobic polystyrene block ensures good adhesion to the hydrophobic CNT electrode," while the PEO block retains the ionic liquid in the polymer scaffold.[109] On the other hand, Tamilarasan and Ramaprabhu used an ionic liquid gel based on a PAN scaffold and BMI TFSI. PAN was chosen due to its mechanical robustness.[110] A dispersion of graphene in isopropanol was spray-coated onto each side of the freestanding, PAN-supported ionic liquid gel, which adhered well to the gel. This electrode/ionic liquid gel/electrode assembly was subsequently sandwiched between copper foil current collectors and acrylic sheets so that electrical contact and low interfacial resistance, respectively, could be achieved.

A robust ionic liquid gel with durable mechanical characteristics is desirable for wear and was demonstrated in a series of parallel plate supercapacitor devices by Liu *et al.*[111–113] Two of this team's earlier ionic liquid gels involved formation of a polymer scaffold by self-initiation using the chlorine anion of the ionic liquid. This is atypical of UV-initiated polymerization of ionic liquid gel scaffolds, which usually requires a separate photoinitiator in addition to the monomer and ionic liquid. In another instance of their ionic liquid gel formulation, inorganic titanium oxide nanoparticles were added as a UV initiator. A cotton mask was also added for more robustness. For practical use, however, flexible current collectors needed to be integrated with the device (the authors of this work integrated it between rigid current collector plates for testing).[111–113] Furthermore, packaging needs to be incorporated to hold freestanding components together. Li and Ardebili demonstrated a flexible battery using an ionic liquid gel electrolyte based on PVDF-HFP (polymer scaffold), EMI dicyanamide (EMI DCA) (ionic liquid), and lithium perchlorate ($LiClO_4$) (lithium ion source), with PVDF-based electrodes and aluminum foil current collectors laminated together.[114] Before lamination, the electrodes were pre-wetted with $LiClO_4$/EMI DCA solution.[114] Researchers from the US Army Natick Soldier Research, Development and Engineering Center[115] were able to show that graphene conductive sheet current collectors made intimate contact with activated carbon fabric (ACF) electrodes in flexible supercapacitor devices incorporating UV-cured polymer scaffold-supported ionic liquid gels[26,36] formulated by the Panzer group. Device components were assembled and packaged by a silicone overmolding process that interfaced all of the components together and yielded device resistances as low as 7 Ω.[115] The silicone packaging helped to prevent the components

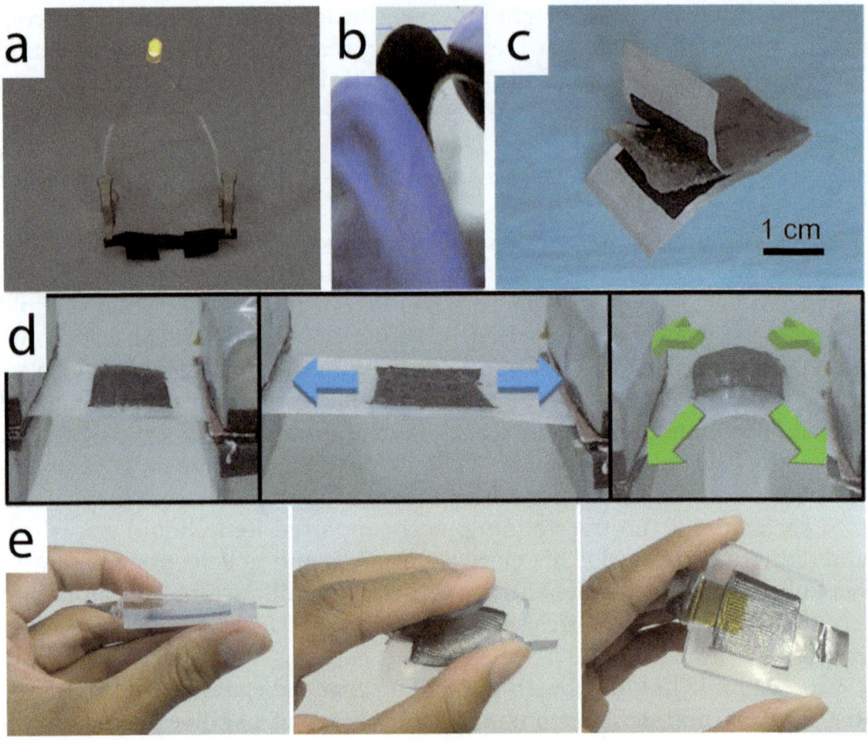

Figure 14.1 Flexible parallel plate supercapacitors based on polymer ionic liquid gel electrolytes. (a) PS-PEO-PS ionic liquid gel electrolyte paired with CNT bacterial nanocellulose electrodes. Reprinted with permission from ref. 109. Copyright (2012) American Chemical Society. (b) PAN–ionic liquid gel electrolytes with spray-coated graphene. Reprinted from Energy, 51, P. Tamilarasan, S. Ramaprabhu, Graphene based all-solid-state supercapacitors with ionic liquid incorporated polyacrylonitrile electrolyte, 374–381, Copyright (2013) with permission from Elsevier.[110] (c) Chitosan–hydroxyethyl methacrylate (HEMA) ionic liquid gel electrolyte paired with activated carbon (AC) electrodes. Reproduced from ref. 111 with permission from The Royal Society of Chemistry. (d) Stretchy PVDF-HFP ionic liquid gel combined with double-walled CNTs dispersed in an ionic liquid gel matrix. Reprinted with permission from Institute of Physics – Journals, W. Kim, W. Kim, 3V omni-directionally stretchable one-body supercapacitors based on a single ion-gel matrix and carbon nanotubes, Nanotechnology, 27 (22), 1–9, 22 April 2016 © IOP Publishing. Reproduced with permission. All rights reserved.[93] (e) Flexible supercapacitor encapsulated in silicone incorporating PEGDA/EMI TFSI ionic liquid gel electrolyte, AC fabric electrodes and graphene foil current collectors.[115]

from delaminating, which can be a major problem for flexible energy storage devices that undergo mechanical deformation.[17]

As researchers have focused on stronger component adhesion to prevent delamination, flexible energy storage devices with stretchable characteristics have also emerged. Stretchable ionic liquid gel supercapacitors[91,93,116] were demonstrated using PVDF-HFP. Lee *et al.* demonstrated stretchable ionic liquid gels based on EMI TFSI and PVDF-HFP by spreading the ionic liquid gel precursor onto stretchy electrodes.[116] The electrodes were made from CNT sprayed onto micro-roughened PDMS. The authors optimized adhesion between the CNT electrode and ionic liquid gel by drop casting ionic liquid gel precursor onto pre-stretched CNT electrodes. Using PMMA as the polymer scaffold, a highly stretchable BMI TFSI-based ionic liquid gel electrolyte was prepared by Tamilarasan and Ramaprabhu, which was also used as a supporting medium for a hydrogen exfoliated graphene electrode material.[91] The assembled supercapacitor showed a specific capacitance of 83 F g^{-1}, an energy density of 26.1 W h kg^{-1}, and a power density of 5 kW kg^{-1} at a specific current of 2.67 A g^{-1}, in terms of mass of total electrode material. Kim and Kim demonstrated stretchable ionic liquid gel supercapacitors with the same electrolyte formulation as Lee *et al.*; similar to the work of Tamilarasan and Ramaprabhu, the electrodes in these devices incorporated the ionic liquid gel to facilitate intimate contact between the electrode and electrolyte.[93] The energy and power density of these devices were 20.3 W h kg^{-1} and 3.7 kW kg^{-1}, respectively, and those values remained stable after 500 cycles of 50% strain.[93]

Interdigital architectures can be used over parallel plate architectures to minimize the risk of electrically shorting the device. This is especially advantageous through device handling. The Ha group at Korea University has demonstrated a few examples[92,117–119] of polymer-supported ionic liquid gel-based interdigital devices, which are all shown in Figure 14.2. Furthermore, their demonstrations provide design strategies for imparting characteristics such as stretchability and a waterproof nature, which is advantageous for wearability. One of their first devices (Figure 14.2a) used an ionic liquid gel composed of the triblock copolymer PS-PMMA-PS in EMI TFSI.[117] The PS-PMMA-PS-supported ionic liquid gel precursor solution was deposited on top of CNT electrodes spray-coated onto patterned titanium/gold (Ti/Au) current collectors. Individual devices were assembled into an array, and stretchability in the assembly was achieved by using electrical interconnects in a serpentine pattern, along with a PDMS packaging material. Subsequent microsupercapacitor device arrays[92,119] utilized an ionic liquid gel synthesized *via* UV-initiated polymerization of a PEGDA scaffold *in situ*. The PEGDA-supported ionic liquid gel enables patterning of the electrolyte in a desired area; in this case, it was printed atop multiwalled carbon nanotube (MWCNT) electrodes and Ti/Au current collectors on a flexible polyethylene terephthalate (PET) substrate. This construction constitutes the supercapacitor device cell seen in Figure 14.2b. This device cell was seen to be stable in air without packaging (Figure 14.2c).[118] Waterproof and stretchable arrays were demonstrated with this cell using liquid metal interconnects and Ecoflex rubber material (Figure 14.2d).[119] Lim *et al.* reported highly stretchable microsupercapacitors exhibiting high performance using a PEGDA-supported EMI TFSI ionic liquid gel electrolyte and CNT electrodes.[92] The microsupercapacitors were tested in an array, which consisted of a pair of parallel connected and four

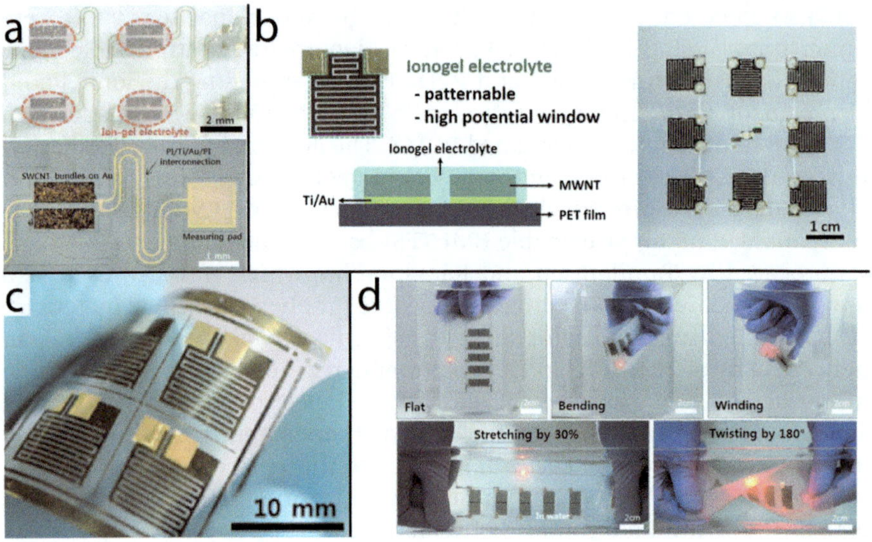

Figure 14.2 Flexible interdigital supercapacitors based on polymer ionic liquid gel electrolytes. (a) Ionic liquid gel deposited on top of printed coplanar single-walled CNT electrodes connected with serpentine interconnects Reprinted with permission from ref. 117. Copyright (2013) American Chemical Society. (b) Ionic liquid gel patterned on top of interdigital MWNT electrodes printed on flexible PET. A stretchable array is made by using liquid metal to connect single supercapacitor cells and Ecoflex as packaging material. Reprinted with permission from ref. 92. Copyright (2014) American Chemical Society. (c) Single supercapacitor device cell as made in (b), but with no packaging, and with demonstrated stability in air for 8 weeks. Reprinted with permission from ref. 118. Copyright (2015) American Chemical Society. (d) Stretchable and stacked array made of single supercapacitor cells as in (b), encapsulated in Ecoflex elastomer and demonstrated to work under different mechanical stresses and underwater. Reprinted with permission from ref. 119. Copyright (2016) American Chemical Society.

serially connected capacitors. The formed supercapacitor array exhibited a high energy and power density of 25 mW h cm^{-3} and 32 W cm^{-3} (in terms of the total volume of the electrodes) and remained electrochemically stable up to 100% uniaxial and 50% biaxial stretching strain.

14.3.1.2 Colloidal Ionic Liquid Gels Integrated into Potential Wearable Energy Storage Devices

As mentioned previously, colloidal ionic liquid gels are fabricated in one of two ways: (1) through the addition of nanoparticles to an ionic liquid, or (2) *in situ via* sol–gel reaction. Silica-based ionic liquid gels can take anywhere from fifteen minutes to several days[28] to render into solid gel form, depending on the formulation. A typical fabrication process consists of using the ionic liquid gel in a 'slurry' state to apply onto the electrodes. The low viscosity of silica-based ionic liquid gel precursors enables good wetting of the electrodes,[120] which is advantageous for achieving very thin films (~100 μm).[121] However, precursor slurry application requires careful control of its wetting behavior so that a clean electrical connection can be made to the electrode. Application of the ionic liquid gel is achieved by utilizing a finite amount of precursor volume or using a mask to define areas on porous electrodes. Integration of silica-based ionic liquid gel electrolytes with *in-plane interdigital* microsupercapacitor architectures has mostly been explored due to the established methods used by the semiconductor industry to fabricate thin films. For instance, Leung *et al.* incorporated ionic liquid gels synthesized *via* sol–gel reaction of vinyltriethoxysilane, TMOS, and FA in BMI TFSI. The authors used a mask to deposit ionic liquid gels on top of carbon electrodes and gold current collectors sequentially printed onto SiO_2.[121] On the other hand, Wang *et al.* also incorporated their fumed silica-supported EMI TFSI ionic liquid gel onto interdigitated carbon electrodes printed on a silicon wafer. Electrical connection was made to the device using silver epoxy to copper wire. No mention of a mask was stated as being used for the ionic liquid gel integration, but the volume was probably controlled to obtain a repeatable electrolyte/electrode interface.[86] Their technique was claimed to be easily scalable, which is promising for ionic liquid gel entry into commercialized micro-integrated circuits. Silica-supported ionic liquid gels have also been demonstrated to be stable at the high temperatures present in reflow soldering,[120] a technique used to connect electrical components, which is also advantageous for integration of these materials into microelectronics.

Silica-based ionic liquid gel microsupercapacitors have also been fabricated in the conventional parallel plate architecture using AC,[68] electrophoretic deposition-grown CNTs,[122] and silicon nanowires (SiNWs)[120] as electrode materials. For parallel plate architectures, a thin gasket between the electrodes is typically utilized for the precursor solution to control the geometry of the ionic liquid gel as it cures.[122] Alternatively, prepared electrodes can be coated with silica/ionic liquid gel slurries and subsequently assembled;[68,123] however, there needs to be something that physically holds the coated electrodes together during use. Kang *et al.* demonstrated how silica-supported ionic liquid gel alone could be used to adhere two CNT-coated office paper electrodes in a parallel plate architecture for a flexible supercapacitor device.[87] The authors coated a slurry of their ionic liquid gel – which was made by adding fumed silica to EMI TFSI – onto two CNT/office paper electrodes, and waited for the slurry to dry. Another layer of ionic liquid gel slurry was subsequently applied to 'gently' adhere the two dried ionic liquid gel-coated CNT/office paper electrodes together. The device had a wide electrochemical window of 3 V, retained almost 100% of its capacitance after 4000 cycles and was able to bend to a radius of curvature of 4.5 mm. The maximum power and energy density of the CNT-based supercapacitors were 164 kW kg^{-1} and 41 W h kg^{-1}, respectively. The device performed the same after being bent 100 times; however, the authors did not show how the device performs during bending. This would be helpful to know for dynamic structural integrity assessment of the ionic liquid gel electrolyte for wearable applications. Parallel plate architectures have a higher risk of short-circuiting[17] if the electrodes make contact with one another through the ionic liquid gel while being mechanically flexed.

As mentioned previously, to minimize the risk of device short-circuiting that can occur in flexible devices, it is best to use the in-plane interdigital electrode architecture.[16] Interdigital supercapacitors incorporating silica-supported ionic liquid gels *via* fumed silica addition[124] and sol–gel reaction within the ionic liquid[125] have been demonstrated. El Kady and Kaner incorporated their ionic liquid gel (fumed silica in BMI TFSI) onto interdigitated graphene electrodes that were laser scribed on a sheet of PET plastic.[124] The authors used polyimide (Kapton) tape to define and control the area where the ionic liquid gel electrolyte was to be coated. Electrochemical results showed that the device retained the same capacitive behavior as in its flat state while being twisted at different angles and bent at different radii of curvature (2.5–7 mm).[124] The electrochemical window achieved during cyclic voltammetry testing and galvanostatic charge/discharge testing for a single device was 2.5 V.[124] Hsia *et al.* integrated an ionic liquid gel based on the sol–gel silica formulation developed by Horowitz and Panzer[56] with chemical vapor deposition-grown, vertically-aligned carbon nanotubes transferred onto a thin polycarbonate sheet.[125] The device reached a higher electrochemical window of 3 V and could be bent to radii of curvature ranging from 1.5–6.5 cm.

14.3.1.3 Ionic Liquid Gels in Textile Energy Storage

The durable nature of polymer-supported ionic liquid gels suggests a vision of energy storage devices integrated into things we wear daily. The studies summarized above describe energy storage devices that, after being assembled, could be fabricated separately and then adhered to clothing or worn as part of an accessory to power wearable flexible electronics. Textile energy storage has recently been explored by researchers, and the concept of utilizing nontoxic ionic liquid gels as an alternative electrolyte to boost energy density has been suggested.[126] In this section, some thoughts on how ionic liquid gel electrolytes can realistically be incorporated into textiles by examining relevant work and how it pertains to different steps of textile manufacturing are outlined. Figure 14.3 shows the textile manufacturing process steps conducive to ionic liquid gel integration, based on recent developments of ionic liquid gel–textile composites found in the literature.

14.3.1.3.1 Fiber and Yarn Production. Examples of fiber and yarn energy storage devices have been reported in the literature (ref. 126 includes a comprehensive overview); however, very few devices incorporate ionic liquid gels. Utilizing ionic liquid gel electrolytes can enable easy synthesis of such devices. Kozlov *et al.* reported the earliest description of an ionic liquid gel

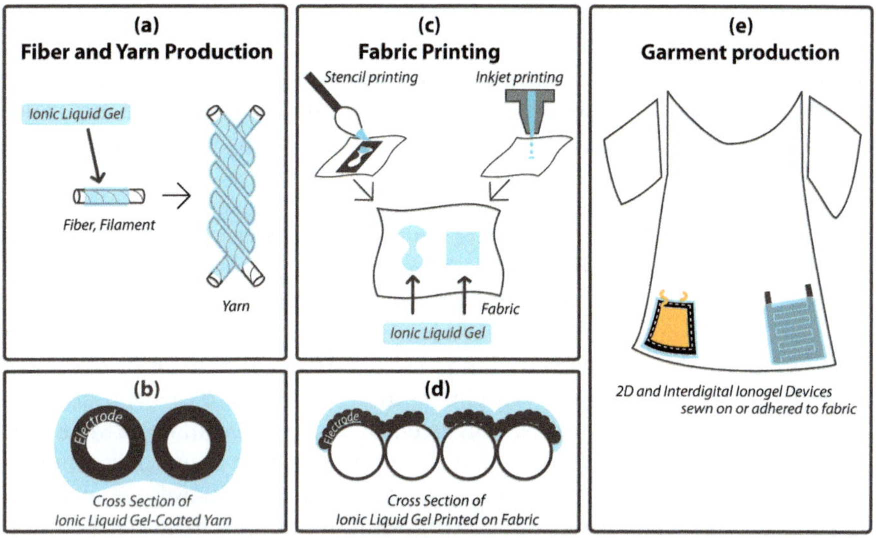

Figure 14.3 Steps in textile production where ionic liquid gels can be integrated. Ionic liquid gels can be coated, printed, or incorporated into textiles as a standalone device. (a) and (b) Ionic liquid gels can coat the surface of fibers and yarn to create "3D" devices (each fiber can act as an electrode). (c) and (d) Ionic liquid gels can be printed onto fabric. (e) Freestanding ionic liquid gel devices can be sewn or adhered onto fabric before assembling a wearable system.

fiber supercapacitor in 2005.[127] In their work, hollow single-walled carbon nanotube fibers were synthesized and separately coated with a "polymer ionic liquid–solid electrolyte system [comprised of] (methyl methacrylate polymerized in the presence of the ionic liquid electrolyte, ethylmethylimidazolium trifluoromethylsulfonyl imide)"[127] (details of the ionic liquid gel synthesis in this work were not reported). Two coated fibers were then twisted to form a yarn supercapacitor with a specific capacitance of 7 F g^{-1} (normalized by device weight), as measured using cyclic voltammetry. More recently, Lyu *et al.* reported a yarn supercapacitor with a more detailed description of the ionic liquid gel synthesis and device fabrication. The ionic liquid gel utilized PVDF-HFP as its scaffold and EMI BF_4 as the ionic liquid.[128] As for yarn device fabrication, the core of the yarn was constructed of stainless steel and carbon nanotubes, acting as current collector and electrode, respectively. As with the device reported by Kozlov *et al.*, each yarn acts as one part of the current collector/electrode pair. Each yarn is then coated in ionic liquid gel twice, resulting in a coating that is 20–25 μm thick. Each coated yarn is then twisted together to create a 2-ply yarn supercapacitor device as depicted in Figure 14.3a and 14.3b.

14.3.1.3.2 Textile Printing. Printing onto textiles involves incorporating inks physically adsorbed onto the surface and into the bulk of the fabric.

Ionic liquid gels have been printed on textiles for wearable sensors, but not too many examples exist for ionic liquid gels printed on textiles for wearable energy storage. A device architecture that leverages textile printing of ionic liquid gels is an avenue worth exploring since they can easily be integrated with textiles. Now that microsupercapacitors are finding a place within integrated circuits and conductive materials are being printed onto textiles,[129–131] an interdigitated architecture is a design that could be used in order to enable ionic liquid gel-based energy storage on clothing. The UV curing of polymer scaffolds to form ionic liquid gels offers a way to achieve this. For example, Takamatsu *et al.* selectively printed UV-cured polymer scaffold-supported ionic liquid gels onto knitted textiles inspired by a Japanese kimono dyeing process.[5] This process involved filling the bulk of the fabric with PDMS to create a stencil in order to selectively choose areas on which to incorporate ionic liquid gel. The ionic liquid gel consisted of EMI ethyl sulfate (EMI $EtSO_4$) as the ionic liquid, and PEGDA and the photoinitiator 2-hydroxy-2-methylpropiophenone as polymer scaffold precursor and UV initiator, respectively.[5] Another promising method of printing ionic liquid gels is inkjet printing. Delaney *et al.* used UV-curable ionic liquid gel inks to print ionic liquid gels onto paper. The researchers used a piezoelectric-driven Autodrop system to deposit ionic liquid gel precursor solution onto paper, and then placed the assembly into a UV chamber to cure.[132] Although printing was not performed on a textile material, this substrate could be easily exchanged for one.

14.3.1.3.3 Freestanding Device Incorporation. Lastly, as outlined in this section, researchers have developed freestanding ionic liquid gel-based energy storage devices. If one were to incorporate these freestanding devices into a wearable system, one could simply use adhesive or devise a way to package it in order to sew it onto clothing. The packaging method described by researchers from the US Army Natick Soldier Research, Development and Engineering Center not only allowed supercapacitor components (and freestanding ionic liquid gel electrolytes) to interface intimately with one another, but also afforded a material that allowed the device to be sewn onto a textile (Figure 14.1e).[115] The next section outlines factors to take into account when incorporating freestanding devices into wearable systems.

14.3.2 Considerations for Integrating Ionic Liquid Gel-based Energy Storage into Wearable Systems

Developing wearable electronics involves considering both the design of printed circuit boards and the design of suits with mobility. Currently, wearable electronics has mostly focused on achieving the desired application performance, while powering the application remains an afterthought. This results in bulky systems that are uncomfortable to wear. In order to design an all-flexible electronic application, the integration of a flexible power source should be taken into account from the start by simultaneous consideration of preserving electronics and wearability.

14.3.2.1 Device Selection Based on Electrical Properties

Electronic designers are advised to choose a battery to power their portable electronic applications based on "price, energy density, shelf life, constancy of voltage during discharge, peak current capability, temperature range, and availability."[4] Although guidance for choosing supercapacitors is similar, it is largely design-based, since the energy density of supercapacitors limits them from being a primary energy source. A supercapacitor's voltage continuously drops during discharge, and so an additional parameter by which supercapacitors can be compared against one another is the characteristic discharge time (or 'time constant'). Since most ionic liquid gel-based energy storage technologies still reside in non-commercial sectors such as university and government labs, price and availability are currently unavailable as bases for comparison. Voltage, current and resistance values of energy storage devices are needed when designing integrated circuits for electronic applications, in addition to knowing energy density and capacity/capacitance. Tables 14.3 and 14.4 show an overview of the flexible ionic liquid gel-based energy storage devices discussed in the previous section that can aid with incorporating ionic liquid gel devices into electronic circuit designs.

14.3.2.2 Device Development Towards Wearability

14.3.2.2.1 Durability and Comfort. When making electronics wearable, durability and comfort are two properties to be maintained.[133] Body worn systems will have to move with the wearer, which means that future wearable flexible energy devices will undergo the constant mechanical stresses and strains that textiles endure. While no standardized testing exists for these devices, flexible ionic liquid gel-based energy storage devices are reported to be resilient to various types of mechanical stress. Table 14.5 points to references that include devices demonstrating electrical performance stability after enduring the listed mechanical stress.

Comfort, on the other hand, is affected by size, weight, shape and placement of the electronic component.[133] As a start, developers of wearable electronics can find guidance for the placement of flexible and stretchable energy storage systems on body-worn systems from the use of "lines of non-extension," a concept developed by Arthur Iberall in the 1960s for providing more mobility to space suits.[134] The left portion of Figure 14.4 depicts this concept on a body in motion. These lines indicate the direction in which the skin does not elongate during body movement.[135] Areas between the drawn lines on the body are potential areas of placement for stretchable energy storage; however, in areas of limited stretch, ionic liquid gel-based devices that have some flexibility but not stretchability may be incorporated on those parts of the body. According to Watkins and Dunne, it is difficult to outline a list of desirable characteristics for mobile clothing designs due to the varied movement needs required for each activity, which depends on the environment and the wearer.[135] Thus, wearable energy storage device developers need to

Table 14.3 Flexible parallel plate ionic liquid gel supercapacitor devices.

Gel Components		Electrical performance							
Ionic Liquid	Scaffold[a]	Electrode[b]	Capacitance[c] ($F\ g_{electrode}^{-1}$)	Voltage window (V)	ESR (Ω)	Energy density[c] ($W\ h\ kg^{-1}$)	Power density[c] ($kW\ kg^{-1}$)	Cycles[d]	Ref.
EMI Tf[e]	PVDF-HFP	MWCNT	41.0	2.0	20.0	17.0	13.0	>50 000	107
EMI TFSI	Silica	CNT	134.6	3.0	30.5	41.0	164	4000	87
EMI TFSI	PS-PEO-PS	CNT	50.5	3.0	31.3	15.5	1.5	5000	109
EMI FAP	PVDF-HFP	MWCNT	76.0	2.0	28.7	16.2	20.3	10 000	108
BMI TFSI	PAN	Graphene	108.0	3.0	4.4	32.3	82	1000	110
BMI TFSI	PMMA	Graphene	83	3.0	16.0	26.1	5	1000	91
BMI Cl	Chitosan-HEMA	AC	4.0	1.0	–	–	–	2000	111
BMI Cl	Chitosan-HEMA/water	AC	43.5	1.0	–	–	–	–	112
EMI TFSI	PVDF-HFP	SWCNT	47.3	3.0	45.0	15.7	1.48	3000	116
BMI BF_4	DMAA, MBAA, TiO_2 and cotton mask	AC	26.0	3.0	–	–	–	–	113
EMI TFSI	PVDF-HFP	DWCNT	66.3	3.0	20.0	20.3	3.7	5000	93
EMI TFSI	PEGDA	ACF	36.4	1.8	9.1	16.4	0.9	500	115
EMI TFSI	TFEMA/PETA	ACF	47.1	1.8	6.9	21.2	1.2	500	115

[a]Abbreviations in this column not previously defined: DMAA = *N,N*-dimethylacrylamide; MBAA = methylene-bis-acrylamide; TiO_2 = titanium oxide; TFEMA/PETA = 2,2,2-trifluoroethyl methacrylate/pentaerythritol tetraacrylate.
[b]Abbreviations in this column not previously defined: SWCNT = single-walled carbon nanotube; DWCNT = double-walled carbon nanotube.
[c]Highest reported.
[d]Defined by constant current charge–discharge testing.
[e]Combined with magnesium trifluoromethanesulfonate and a co-solvent of ethylene carbonate and propylene carbonate.

Table 14.4 Flexible interdigital ionic liquid gel supercapacitor devices (single cell).

Gel Components		Electrical performance							
Ionic Liquid	Scaffold	Electrode	Capacitance[a] (mF cm^{-3})	Voltage window (V)	ESR (Ω)	Energy density[a] (mW h cm^{-3})	Power density[a] (W cm^{-3})	Cycles[b]	Ref.
BMI TFSI	Silica	Graphene	2400	2.5	9	2.0	15	30 000	124
EMI TFSI	Silica	VACNT[c]	86	3.0	450	0.1	2	–	125
EMI TFSI	PEGDA	MWCNT	13 200	3.0	66	20.0	60	10 000	92
EMI TFSI	PEGDA	MWCNT	1800	1.5	234	–	–	30 000	119

[a]Highest reported.
[b]Defined by constant current charge–discharge testing.
[c]VACNT = vertically aligned carbon nanotubes.

Table 14.5 Flexible ionic liquid gel energy storage devices demonstrating physical capabilities.

Physical Stress	References
Flexing	87, 109, 111, 119, 125
Compression	112 and 113
Twisting	119
Stretching	91, 92, 93, 116, 117, 119

work with clothing designers to come up with an optimal way to harness the mechanical benefits of their devices into wearable systems.

14.3.2.2.2 Safety. Protection for flexible energy storage is also crucial not only for electrical insulation, but also for preventing the wearer from making contact with chemical components. An advantage of using ionic liquid gel electrolytes is reduced flammability,[136] which has been demonstrated by researchers in an effort to improve the safety of lithium ion batteries.[137] As for skin toxicity of ionic liquid gels – the effect of prolonged exposure to the material is unknown, although Takamatsu *et al.* demonstrated on-skin electroencephalography sensors based on formulations similar to supercapacitor electrolytes for a period of four days without any skin irritation.[5] To be safe, some kind of inert material should probably be used for packaging. Additionally, packaging can help reduce the permeation of oxygen and water into ionic liquids, which can degrade performance. There also exists the potential to use ionic liquid gel electrolytes without packaging. The ionic liquid gel microsupercapacitors developed by Kim *et al.*[118] using EMI TFSI as the ionic liquid showed stability by retaining 80% of its original capacitance

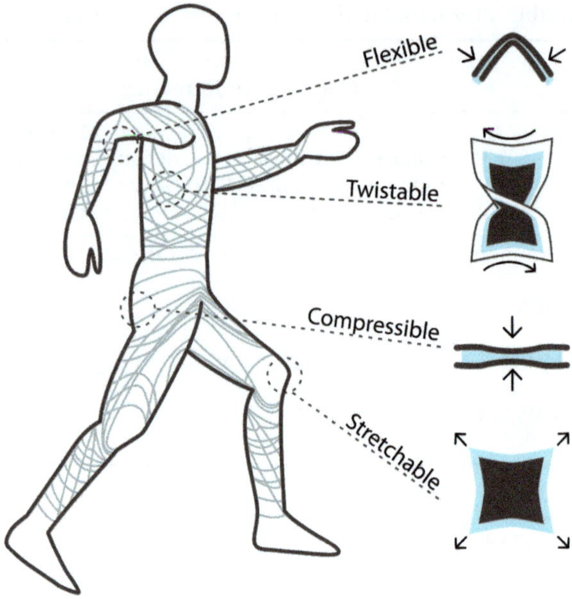

Figure 14.4 Lines of non-extension on a body (for further reading please see ref. 134). Potential locations of flexible energy storage for wearability.

for 8 weeks and 30 000 cycles when tested at room temperature (27 °C) and 50% relative humidity. Although these devices were geared towards on-chip microsupercapacitors, this demonstration showed the potential of the use of an easy-to-fabricate ionic liquid gel that could also be used for wearable applications. If ionic liquid gels were to be used as an outer coating, however, their mechanical characteristics would have to be optimized to withstand physical abrasion. However, the question of whether wearable electronics would be capable of being laundered[126] or rugged enough to weather the elements is always posed, implying that packaging of such devices is still desirable. Kim *et al.* demonstrated waterproof device packaging for ionic liquid gel supercapacitors by encapsulating their devices in Ecoflex, a silicone rubber material that offers stretchability.[119] When the encapsulated device was submerged underwater for four days, the authors found that 82% of the original capacitance was retained. Degradation of device performance was attributed to water permeation through the thin Ecoflex packaging film. Although four days is not a long time, a working, stretchable device that can be used underwater has nevertheless been demonstrated.

14.4 Conclusions

Ionic liquid gel electrolytes have characteristics that enable the realization of safer energy storage with increased energy density. On the material level, the choices of ionic liquid and supporting scaffold are parameters that can be adjusted in order to yield desirable ionic conductivities and electrochemical windows for the material itself. While many combinations of various polymeric or other scaffolds with different ionic liquids have been used to successfully create ionic liquid gels to date, these represent only the tip of the iceberg when it comes to exploring the nearly endless variations that may be achieved by varying the chemical identities of both components. Recognition of the important role of chemical interactions (*e.g.* ion–dipole, hydrogen bonding, Lewis acid–base) between the ionic liquid components and the scaffold on the properties of these composite materials is only just beginning to be considered for most ionic liquid gel systems. Gaining a deeper understanding of how these interactions can be leveraged to maximize the electrochemical characteristics of ionic liquid gels will naturally lead to better design rules for creating the next generation of nonvolatile gel electrolytes.

Ionic liquid gel electrolytes may prove to be imperative for the development and commercialization of safe, flexible, and environmentally friendly lithium metal-based energy storage devices. In terms of polymer-based scaffolds, the chemical nature of the polymer network needs to be more thoroughly investigated to gain a deeper understanding of the interactions between the polymer network and Li^+ as well as its counter anion. The most critical factor for the commercialization of lithium metal-based batteries is the interfacial stability of the lithium metal anode and the electrolyte as lithium metal is plated and stripped from the surface. Cyclic charge–discharge testing on lithium/ionic liquid gel/$LiFePO_4$ cells must be more thoroughly investigated up to at least ~500 cycles, which would match the typical lifetime of a commercially available lithium-ion battery; additionally, the interfacial resistance must be measured and remain relatively constant throughout the lifetime of the battery. Furthermore, the capacity must be further improved at higher C-rates (~1C) to prove the viability of ionic liquid gel electrolytes for lithium metal batteries, in which the lithium-ion transference number and its diffusivity must be further enhanced in order to increase the discharge capacity of the battery to meet the power and energy demands of contemporary applications.

In supercapacitors, ionic liquid gel materials with a wide electrochemical stability window and high ionic conductivity are promising as electrolytes. However, some problems remain that may limit the large-scale application of ionic liquid gels. Ionic liquids are often costly and potentially toxic, so it is necessary to find environmentally friendly substitutes and/or develop benign ways to synthesize ionic liquids. In addition, creating electrode materials with the optimal pore size and hierarchical structure to allow ionic liquid ions to effectively polarize within them, could also improve the performance of ionic liquid gel-based supercapacitors to some extent.

For wearable applications, a more thorough understanding of how best to incorporate ionic liquid gels into a final energy storage device is required. Several flexible devices for potential wearable application have been reported, although much room for improvement and innovation remains. Lastly, in order for ionic liquid gel devices to be viable, it is important that the devices meet the required demands from both the electrochemical and wearable viewpoints.

References

1. A. Pentland, *Scientific American Presents, Special Issue on Intelligence*, 1998, vol. 9, p. 90.
2. The wear, why and how, *The Economist*, http://www.economist.com/news/business/21646225-smartwatches-and-other-wearable-devices-become-mainstream-products-will-take-more, 2015, accessed January 2017.
3. V. G. Motti and K. Caine, *Proceedings of the Human Factors and Ergonomics Society Annual Meeting 2014 58*, SAGE Publishing, 2014.
4. P. Horowitz and W. Hill, *The Art of Electronics*, Cambridge University Press, New York, NY, 3rd edn. 2015.
5. S. Takamatsu, T. Lonjaret, D. Crisp, J.-M. Badier, G. G. Malliaras and E. Ismailova, *Sci. Rep.*, 2015, **5**, 15003.
6. A. J. Bandodkar, I. Jeerapan and J. Wang, *ACS Sens.*, 2016, **1**, 464.
7. D. Rus and M. T. Tolley, *Nature*, 2015, **521**, 467.
8. A. Asbeck, K. Schmidt, I. Galiana, D. Wagner and C. J. Walsh, *2015 IEEE International Conference on Robotics and Automation (ICRA)*, IEEE, Seattle, WA, 2015.
9. F. Gemperle, C. Kasabach, J. Stivoric, M. Bauer and R. Martin, *Wearable Computers, 1998. Digest of Papers. Second International Symposium on*, IEEE, Pittsburgh, PA, USA, 1998.
10. The U.S. Army, https://www.army.mil/article/107362, accessed January 2017.
11. I. Ayub, K. Habegger, W. M. Batts and A. Hopfer, US Pat., 2013/0295434 A1, 2013.
12. J. R. Miller and P. Simon, *Electrochem. Soc. Interface*, 2008, **17**, 31.
13. D. Linden in *Handbook of Batteries*, ed. D. Linden and T. B. Reddy, McGraw-Hill, New York, NY, 3rd edn, 2001.
14. P. Yang and W. Mai, *Nano Energy*, 2014, **8**, 274.
15. C. Zhong, Y. Deng, W. Hu, J. Qiao, L. Zhang and J. Zhang, *Chem. Soc. Rev.*, 2015, **44**, 7484.
16. M. Beidaghi and Y. Gogotsi, *Energy Environ. Sci.*, 2014, **7**, 867.
17. A. M. Gaikwad, A. C. Arias and D. A. Steingart, *Energy Technol.*, 2015, **3**, 305.
18. T. Ueki and M. Watanabe, *Macromolecules*, 2008, **41**, 3739.
19. P. C. Marr and A. C. Marr, *Green Chem.*, 2015, **18**, 105.
20. J. L. Bideau, L. Viau and A. Vioux, *Chem. Soc. Rev.*, 2011, **40**, 907.
21. Y.-S. Ye, J. Rick and B.-J. Hwang, *J. Mater. Chem. A*, 2013, **1**, 2719.
22. A. S. Shaplov, R. Marcilla and D. Mecerreyes, *Electrochim. Acta*, 2015, **175**, 18.
23. J. Fuller, C. Breda and R. T. Carlin, *J. Electrochem. Soc.*, 1997, **144**, L67.
24. Y. He, P. G. Boswell, P. Bühlmann and T. P. Lodge, *J. Phys. Chem. B*, 2007, **111**, 4645.
25. A. Noda and M. Watanabe, *Electrochim. Acta*, 2000, **45**, 1265.
26. A. F. Visentin and M. J. Panzer, *ACS Appl. Mater. Interfaces*, 2012, **4**, 2836.
27. M. A. B. H. Susan, T. Kaneko, A. Noda and M. Watanabe, *J. Am. Chem. Soc.*, 2005, **127**, 4976.

28. A. F. Visentin, T. Dong, J. Poli and M. J. Panzer, *J. Mater. Chem. A*, 2014, **2**, 7723.
29. P. Snedden, A. I. Cooper, K. Scott and N. Winterton, *Macromolecules*, 2003, **36**, 4549.
30. M. A. Klingshirn, S. K. Spear, R. Subramanian, J. D. Holbrey, J. G. Huddleston and R. D. Rogers, *Chem. Mater.*, 2004, **16**, 3091.
31. J. Jiang, D. Gao, Z. Li and G. Su, *React. Funct. Polym.*, 2006, **66**, 1141.
32. E. Marwanta, T. Mizumo, N. Nakamura and H. Ohno, *Polymer*, 2005, **46**, 3795.
33. G. P. Pandey, S. A. Hashmi and Y. Kumar, *Energy Fuels*, 2010, **24**, 6644.
34. G. P. Pandey and S. A. Hashmi, *J. Mater. Chem. A*, 2013, **1**, 3372.
35. A. F. Visentin, S. Alimena and M. J. Panzer, *ChemElectroChem*, 2014, **1**, 718.
36. A. J. D'Angelo, J. J. Grimes and M. J. Panzer, *J. Phys. Chem. B*, 2015, **119**, 14959.
37. F. Lind, L. Rebollar, P. Bengani-Lutz, A. Asatekin and M. J. Panzer, *Chem. Mater.*, 2016, **28**, 8480.
38. K. H. Lee, S. Zhang, T. P. Lodge and C. D. Frisbie, *J. Phys. Chem. B*, 2011, **115**, 3315.
39. Y. Gu and T. P. Lodge, *Macromolecules*, 2011, **44**, 1732.
40. Y. Gu, S. Zhang, L. Martinetti, K. H. Lee, L. D. McIntosh, C. D. Frisbie and T. P. Lodge, *J. Am. Chem. Soc.*, 2013, **135**, 9652.
41. T. Ueki, Y. Nakamura, R. Usui, Y. Kitazawa, S. So, T. P. Lodge and M. Watanabe, *Angew. Chem., Int. Ed.*, 2015, **54**, 3018.
42. K. Fujii, H. Asai, T. Ueki, T. Sakai, S. Imaizumi, U.-I. Chung, M. Watanabe and M. Shibayama, *Soft Matter*, 2012, **8**, 1756.
43. N. Festin, A. Maziz, C. Plesse, D. Teyssié, C. Chevrot and F. Vidal, *Smart Mater. Struct.*, 2013, **22**, 104005.
44. S. Jeremias, M. Kunze, S. Passerini and M. Schönhoff, *J. Phys. Chem. B*, 2013, **117**, 10596.
45. M. Patel, M. Gnanavel and A. J. Bhattacharyya, *J. Mater. Chem.*, 2011, **21**, 17419.
46. G. P. Pandey and S. A. Hashmi, *J. Power Sources*, 2013, **243**, 211.
47. K.-N. Jung, J.-I. Lee, J.-H. Jung, K.-H. Shin and J.-W. Lee, *Chem. Commun.*, 2014, **50**, 5458.
48. R. Bhandary and M. Schönhoff, *Electrochim. Acta*, 2015, **174**, 753.
49. A. I. Horowitz and M. J. Panzer, *Angew. Chem., Int. Ed.*, 2014, **53**, 9780.
50. K. Ueno, K. Hata, T. Katakabe, M. Kondoh and M. Watanabe, *J. Phys. Chem. B*, 2008, **112**, 9013.
51. K. Ueno, A. Inaba, T. Ueki, M. Kondoh and M. Watanabe, *Langmuir*, 2010, **26**, 18031.
52. J. Sun, P. Bayley, D. R. MacFarlane and M. Forsyth, *Electrochim. Acta*, 2007, **52**, 7083.
53. S. Dai, Y. H. Ju, H. J. Gao, J. S. Lin, S. J. Pennycook and C. E. Barnes, *Chem. Commun.*, 2000, 243.
54. M.-A. Néouze, J. L. Bideau, F. Leroux and A. Vioux, *Chem. Commun.*, 2005, 1082.

55. M.-A. Néouze, J. L. Bideau, P. Gaveau, S. Bellayer and A. Vioux, *Chem. Mater.*, 2006, **18**, 3931.
56. A. I. Horowitz and M. J. Panzer, *J. Mater. Chem.*, 2012, **22**, 16534.
57. A. I. Horowitz, Y. Wang and M. J. Panzer, *Green Chem.*, 2013, **15**, 3414.
58. F. Gayet, L. Viau, F. Leroux, F. Mabille, S. Monge, J.-J. Robin and A. Vioux, *Chem. Mater.*, 2009, **21**, 5575.
59. F. Gayet, L. Viau, F. Leroux, S. Monge, J.-J. Robin and A. Vioux, *J. Mater. Chem.*, 2010, **20**, 9456.
60. K. Hanabusa, H. Fukui, M. Suzuki and H. Shirai, *Langmuir*, 2005, **21**, 10383.
61. T. Kataoka, Y. Ishioka, M. Mizuhata, H. Minami and T. Maruyama, *ACS Appl. Mater. Interfaces*, 2015, **7**, 23346.
62. K. H. Lee, S. Zhang, Y. Gu, T. P. Lodge and C. D. Frisbie, *ACS Appl. Mater. Interfaces*, 2013, **5**, 9522.
63. J. H. Cho, J. Lee, Y. Xia, B. Kim, Y. He, M. J. Renn, T. P. Lodge and C. Daniel Frisbie, *Nat. Mater.*, 2008, **7**, 900.
64. S. Thiemann, S. J. Sachnov, M. Gruber, F. Gannott, S. Spallek, M. Schweiger, J. Krückel, J. Kaschta, E. Spiecker, P. Wasserscheid and J. Zaumseil, *J. Mater. Chem. C*, 2014, **2**, 2423.
65. S. A. M. Noor, P. M. Bayley, M. Forsyth and D. R. MacFarlane, *Electrochim. Acta*, 2013, **91**, 219.
66. K. Matsumoto, S. Sogabe and T. Endo, *J. Polym. Sci., Part A: Polym. Chem.*, 2010, **48**, 4515.
67. A. I. Horowitz, K. Westerman and M. J. Panzer, *J. Sol-Gel Sci. Technol.*, 2015, **78**, 34.
68. M. Brachet, T. Brousse and J. Le Bideau, *ECS Electrochem. Lett.*, 2014, **3**, A112.
69. X. Wang, X. Lu, B. Liu, D. Chen, Y. Tong and G. Shen, *Adv. Mater.*, 2014, **26**, 4763.
70. M. Galiński, A. Lewandowski and I. Stępniak, *Electrochim. Acta*, 2006, **51**, 5567.
71. P. G. Bruce, J. Evans and C. A. Vincent, *Solid State Ionics*, 1988, 918.
72. S. Zugmann, M. Fleischmann, M. Amereller, R. M. Gschwind, H. D. Wiemhöfer and H. J. Gores, *Electrochim. Acta*, 2011, **56**, 3926.
73. G. Annat, D. R. MacFarlane and M. Forsyth, *J. Phys. Chem. B*, 2007, **111**, 9018.
74. J. L. Schaefer, D. A. Yanga and L. A. Archer, *Chem. Mater.*, 2013, **25**, 834.
75. A. Fernicola, F. C. Weise, S. G. Greenbaum, J. Kagimoto, B. Scrosati and A. Soleto, *J. Electrochem. Soc.*, 2009, **156**, A514.
76. M. Safa, A. Chamaani, N. Chawla and B. El-Zahab, *Electrochim. Acta*, 2016, **213**, 587.
77. Y. Lu, S. S. Moganty, J. L. Schaefer and L. A. Archer, *J. Mater. Chem.*, 2012, **22**, 4066.
78. S. Ferrari, E. Quartarone, P. Mustarelli, A. Magistris, M. Fagnoni, S. Protti, C. Gerbaldi and A. Spinella, *J. Power Sources*, 2010, **195**, 559.

79. G. Cheruvally, J.-K. Kim, J.-W. Choi, J.-H. Ahn, Y.-J. Shin, J. Manuel, P. Raghavan, K.-W. Kim, H.-J. Ahn, D. S. Choi and C. E. Song, *J. Power Sources*, 2007, **172**, 863.
80. M. Rao, X. Geng, Y. Liao, S. Hu and W. Li, *J. Membr. Sci.*, 2012, **399–400**, 37.
81. G. B. Appetecchi, G. T. Kim, M. Montanino, M. Carewska, R. Marcilla, D. Mecerreyes and I. De Meatza, *J. Power Sources*, 2010, **195**, 3668.
82. H. Nakagawa, S. Izuchi, K. Kuwana, T. Nukuda and Y. Aihara, *J. Electrochem. Soc.*, 2003, **150**, A695.
83. Y. S. Yun, J.-A. Choi and D.-W. Kim, *Macromol. Res.*, 2012, **21**, 49.
84. F. Wu, N. Chen, R. Chen, L. Wang and L. Li, *Nano Energy*, 2017, **31**, 9.
85. G. A. Tiruye, D. Muñoz-Torrero, J. Palma, M. Anderson and R. Marcilla, *J. Power Sources*, 2015, **279**, 472.
86. S. Wang, B. Hsia, C. Carraro and R. Maboudian, *J. Mater. Chem. A*, 2014, **2**, 7997.
87. Y. J. Kang, H. Chung, C.-H. Han and W. Kim, *Nanotechnology*, 2012, **23**, 065401.
88. Y. J. Kang, Y. Yoo and W. Kim, *ACS Appl. Mater. Interfaces*, 2016, **8**, 13909.
89. C.-W. Liew, S. Ramesh and A. K. Arof, *Int. J. Hydrogen Energy*, 2014, **39**, 2953.
90. S. Ketabi and K. Lian, *Electrochim. Acta*, 2013, **103**, 174.
91. P. Tamilarasan and S. Ramaprabhu, *Mater. Chem. Phys.*, 2014, **148**, 48.
92. Y. Lim, J. Yoon, J. Yun, D. Kim, S. Y. Hong, S.-J. Lee, G. Zi and J. S. Ha, *ACS Nano*, 2014, **8**, 11639.
93. W. Kim and W. Kim, *Nanotechnology*, 2016, **27**, 225402.
94. M. Suleman, Y. Kumar and S. A. Hashmi, *J. Phys. Chem. B*, 2013, **117**, 7436.
95. L. Negre, B. Daffos, P.-L. Taberna and P. Simon, *J. Electrochem. Soc.*, 2015, **162**, A5037.
96. S. Wang, B. Hsia, J. P. Alper, C. Carraro, Z. Wang and R. Maboudian, *J. Power Sources*, 2016, **301**, 299.
97. W.-Y. Tsai, R. Lin, S. Murali, L. L. Zhang, J. K. McDonough, R. S. Ruoff, P.-L. Taberna, Y. Gogotsi and P. Simon, *Nano Energy*, 2013, **2**, 403.
98. L. Negre, B. Daffos, V. Turq, P.-L. Taberna and P. Simon, *Electrochim. Acta*, 2016, **206**, 490.
99. I. Poupyrev, N.-W. Gong, S. Fukuhara, M. E. Karagozler, C. Schwesig and K. E. Robinson, *Proceedings of the 2016 CHI Conference on Human Factors in Computing Systems*, ACM Press, New York, USA, 2016.
100. J. H. Lee, A. S. Lee, J.-C. Lee, S. M. Hong, S. S. Hwang and C. M. Koo, *J. Mater. Chem. A*, 2015, **3**, 2226.
101. R. Zhang, Y. Chen and R. Montazami, *Materials*, 2015, **8**, 2735.
102. M. Ravi, S. Song, J. Wang, T. Wang and R. Nadimicherla, *J. Mater. Sci.: Mater. Electron.*, 2015, **27**, 1370.
103. F. Wu, N. Chen, R. Chen, Q. Zhu, J. Qian and L. Li, *Chem. Mater.*, 2016, **28**, 848.

104. A. S. Lee, J. H. Lee, S. M. Hong, J.-C. Lee, S. S. Hwang and C. M. Koo, *Electrochim. Acta*, 2016, **215**, 36.
105. M. Baloch, A. Vizintin, R. K. Chellappan, J. Moskon, D. Shanmukaraj, R. Dedryvère, T. Rojo and R. Dominko, *J. Electrochem. Soc.*, 2016, **163**, A2390.
106. W. Lu, K. Henry, C. Turchi and J. Pellegrino, *J. Electrochem. Soc.*, 2008, **155**, A361.
107. G. P. Pandey, S. A. Hashmi and Y. Kumar, *J. Electrochem. Soc.*, 2010, **157**, A105.
108. G. P. Pandey and S. A. Hashmi, *Electrochim. Acta*, 2013, **105**, 333.
109. Y. J. Kang, S.-J. Chun, S.-S. Lee, B.-Y. Kim, J. H. Kim, H. Chung, S.-Y. Lee and W. Kim, *ACS Nano*, 2012, **6**, 6400.
110. P. Tamilarasan and S. Ramaprabhu, *Energy*, 2013, **51**, 374.
111. X. Liu, Z. Wen, D. Wu, H. Wang, J. Yang and Q. Wang, *J. Mater. Chem. A*, 2014, **2**, 11569.
112. X. Liu, D. Wu, H. Wang and Q. Wang, *Adv. Mater.*, 2014, **26**, 4370.
113. X. Liu, B. Wu, N. Brandon and Q. Wang, *Energy Technol.*, 2017, **5**, 220.
114. Q. Li and H. Ardebili, *J. Power Sources*, 2016, **303**, 17.
115. S. F. Zopf, *Presented at Final Project Review for Thin-film Supercapacitors for Integration with Uniforms and Equipment PAO# U16-611*, Natick, MA, 2016.
116. J. Lee, W. Kim and W. Kim, *ACS Appl. Mater. Interfaces*, 2014, **6**, 13578.
117. D. Kim, G. Shin, Y. J. Kang, W. Kim and J. S. Ha, *ACS Nano*, 2013, **7**, 7975.
118. D. Kim, G. Lee, D. Kim and J. S. Ha, *ACS Appl. Mater. Interfaces*, 2015, **7**, 4608.
119. H. Kim, J. Yoon, G. Lee, S.-H. Paik, G. Choi, D. Kim, B.-M. Kim, G. Zi and J. S. Ha, *ACS Appl. Mater. Interfaces*, 2016, **8**, 16016.
120. M. Brachet, D. Gaboriau, P. Gentile, S. Fantini, G. Bidan, S. Sadki, T. Brousse and J. L. Bideau, *J. Mater. Chem. A*, 2016, **4**, 11835.
121. G. Leung, L. Smith, J. Lau, B. Dunn and C. O. Chui, *Nanotechnology*, 2015, **27**, 035204.
122. D. Membreno, L. Smith, K.-S. Shin, C. O. Chui and B. Dunn, *Transl. Mater. Res.*, 2015, **2**, 015001.
123. L. Negre, B. Daffos, V. Turq, P.-L. Taberna and P. Simon, *Electrochim. Acta*, 2016, **206**, 490.
124. M. F. El-Kady and R. B. Kaner, *Nat. Commun.*, 2013, **4**, 1475.
125. B. Hsia, J. Marschewski, S. Wang, J. B. In, C. Carraro, D. Poulikakos, C. P. Grigoropoulos and R. Maboudian, *Nanotechnology*, 2014, **25**, 055401.
126. K. Jost, G. Dion and Y. Gogotsi, *J. Mater. Chem. A*, 2014, **2**, 10776.
127. M. E. Kozlov, R. C. Capps, W. M. Sampson, V. H. Ebron, J. P. Ferraris and R. H. Baughman, *Adv. Mater.*, 2005, **17**, 614.
128. X. Lyu, F. Su and M. Miao, *J. Power Sources*, 2016, **307**, 489.
129. K. Jost, C. R. Perez, J. K. McDonough, V. Presser, M. Heon, G. Dion and Y. Gogotsi, *Energy Environ. Sci.*, 2011, **4**, 5060.
130. K. Jost, D. Stenger, C. R. Perez, J. K. McDonough, K. Lian, Y. Gogotsi and G. Dion, *Energy Environ. Sci.*, 2013, **6**, 2698.

131. S. F. Zopf and M. Manser, *J. Eng. Fibers Fabr.*, 2016, **11**, 1.
132. J. T. Delaney Jr, A. R. Liberski, J. Perelaer and U. S. Schubert, *Macromol. Rapid Commun.*, 2010, **31**, 1970.
133. K. Hartman, *Make: Wearable Electronics*, O'Reilley Media, Sebastopol, CA, USA, 1st edn, 2014.
134. A. S. Iberall, 1964, Defense Technical Information Center (DTIC): get-tr-doc/pdf?AD= AD0610519.
135. S. M. Watkins and L. E. Dunne, *Functional Clothing Design*, Fairchild Books, an imprint of Bloomsbury Publishing, London, UK, 3rd edn, 2015.
136. A.-O. Diallo, A. B. Morgan, C. Len and G. Marlair, *Energy Environ. Sci.*, 2013, **6**, 699.
137. Q. Guo, Y. Han, H. Wang, X. Hong, C. Zheng, S. Liu and K. Xie, *RSC Adv.*, 2016, **6**, 101638.

CHAPTER 15

Ionic Liquids in Wearable Chemical Sensors†

S. A. GOODCHILD[a], M. R. SAMBROOK[a] AND A. J. S. MCINTOSH*[b]

[a]Defence Science and Technology Laboratory, Porton Down, Salisbury, SP4 0JQ, UK; [b]Imperial College London, South Kensington Campus, Exhibition Road, London, SW7 2AZ, UK
*E-mail: a.mcintosh@imperial.ac.uk

15.1 Introduction

The umbrella term "Wearable Technologies" has been coined to describe textiles, garments or devices with embedded sensors and data collection/transmission capabilities, which are worn by an individual. This allows the wearer to measure signatures generated by themselves (components of the Quantified Self) and/or sense environmental stimuli.[1,2] Such technologies may therefore include any one of a number of individual technologies, including smart/functional materials, electronic textiles (e-textiles), and flexible electronics, as well as an array of different technology platforms ranging from

Smart Materials No. 29
Polymerized Ionic Liquids
Edited by Ali Eftekhari

Published by the Royal Society of Chemistry, www.rsc.org

functional garments (clothing, footwear, gloves) through to miniaturised devices integrated into standard wearable items (watches, glasses, contact lenses *etc.*).[1–3] Access to these technology options has been supported by accelerating developments in materials science (smart fabrics, miniaturised sensing devices) in combination with continued drivers for increasing access to greater resolution of signals of importance to public health. A vast and disruptive commercial market for person-worn physiological monitoring technologies has therefore emerged, targeting markets in sport and healthcare, and catalysing a debate as to the validity of these types of technology in preventing injury/death and to enhance performance.[4–6]

15.2 Sensing with Wearable Technologies

As the technologies available in wearable form have become more sophisticated, the options to exploit wearable technologies beyond standard physiological measurements into wider sensing applications has grown.[7] Here, chemical sensing is discussed and defined as the transformation of chemical information, such as molecular identity, ion species, pH, and so forth, into a signal that can be interrogated analytically. Thus, chemical sensors are the technology being employed to interact with the target environment and analyte, provide a transduction mechanism, and furnish a readable response. To add further to this definition, the chemical targets considered herein are small molecules and these can be found in a range of environments and scenarios, ranging from small gas molecules, to metabolites, bio-markers, and compounds of environmental and/or health and safety concern. Markets for these classes of sensors would be available for those responsible for ensuring health and safety within work environments as well as medical practitioners responding to drivers for remote monitoring of patient health in addition to smaller communities such as emergency first responders and specialists within the defence and security sectors.[8–10] As such, development of wearable sensors for specific chemical materials present in environments and/or produced by an individual, as a result of exposure or acute/chronic disease, remains an area of high investment within the research community.

Not all sensor types require solvents to facilitate a transduction response; for example, spectroscopic and spectrometric approaches allow for direct detection of target analytes, and require no solvents or additional techniques to provide a signalling output. As a consequence, development of wearable sensors exploiting these technologies may be limited by fundamentals of miniaturisation, energy requirements, computation power, cost and so forth. However, many other sensing strategies require the use of sample collection from the environment and/or from the individual, processing of that sample into a form appropriate for analysis, and then generation of a signal output using multiple interconnected steps and/or techniques. All of these steps may impose power requirements and the need for solvent/liquid handling. Developing these classes of technology into a wearable solution requires solvents with particular physical and chemical properties to be compatible with integration and functionality in a wearable context.

15.3 The Benefits of Ionic Liquids for Use in Wearable Chemical Sensors

In a human-centric (worn close to the skin or on the skin of an individual) wearable concept, the sensing materials incorporate flexible electronics,[11–16] papers[17–19] and organic elastomers[20–22] that are soft, flexible and stretchable to conform to the non-linear shape of the body and maintain contact to the sensing surface throughout the constant multi-directional movements and mechanical stresses that the surface, *e.g.* skin, will undergo. These mechanical property requirements inherently lead to the consideration of liquid-based chemical sensors, as the liquid phase can undergo substantial deformations without detrimental impact on performance. Sensing components are also being embedded into functional yarns,[2,21,23–25] where the sensing component must be viable as a garment, displaying additional facets above a sensing capability, *i.e.* they must be flexible, durable, and ergonomic (breathable, comfortable) in order to adequately replace existing clothing.

As discussed in previous chapters, ILs are typically composed of large organic cations and small inorganic or organic anions. Through this combination of mismatched sizes and asymmetry the Coulombic interactions are reduced, crystal packing frustrated and entropic contributions enhanced leading to very low melting points resulting in them usually being liquids below 100 °C. As a class of solvents, ILs exhibit some unusual and highly desirable physical properties, and these will also have an impact on their selection as components in wearable sensors, including:

- Favourable solvation behaviour to enable uptake and concentration of analytes, potentially in a selective manner.
- High thermal stability and large liquid temperature range allowing for the use of the sensors in multiple environments and long-term operation.
- Negligible vapour pressure enabling long-term use through retention of sensor structure.
- Optical transparency facilitating interrogation of reporter groups *via* spectroscopic means.
- High electrochemical stability and conductivity window enhancing the options for electrochemical-based sensing of analytes.

It is noteworthy that many of these properties cannot be found, especially in conjunction with each other, in standard aqueous or organic solvents. Furthermore, each of these characteristics can be manipulated by judicious selection of the ions used to form the IL.

Although the application of ILs to novel wearable sensors is currently in its infancy, step changes in components of the sensing concept such as sampling and sensing capabilities have already been made. In addition, although out of the scope of this chapter, the impact of ILs on future power sources that could enable flexible batteries to be included in wearable devices should not be ignored. For example, Zhang *et al.*[26] have investigated the use of ionic liquid-doped gel polymers as electrolytes for batteries and have found that with 40–50 vol% ionic liquid, the resulting electrolyte has superior ionic properties and lower internal resistance compared with the undoped gel polymer (ethylene carbonate/propylene carbonate), combined with a high rest voltage and discharge capacity (3.81 V, 87.6 mA h g^{-1} and 3.94 V, 91.8 mA h g^{-1}, respectively). Research has also shown that IL hybrid systems have potential for use in lithium batteries, such as the use of IL-tethered nanoparticles, which have been shown to retard lithium dendrite growth at moderate current densities, which could improve cell operating times by several hundred hours.[27] Compared with pure propylene carbonate (PC), the use of IL nanoparticle hybrids dispersed within the PC retarded lithium dendrite growth, and increased the time before short circuit by one order of magnitude. While ILs have a lower thermal conductivity compared to solid-state materials, they have shown promise due to their tuneable properties and the possible coupling with polymers or the direct use of polymeric ionic liquids could allow for the manufacture of flexible generators, which would be ideal for wearable devices.[28]

In the context of this chapter, the term sensor is therefore considered to refer predominantly to material-based systems that present an interface for interaction (*e.g.* molecular recognition, chemical reactivity, solvation) with the environment and target analytes. These requirements, combined with other beneficial properties (*vide infra*), have resulted in the assessment of ILs as components in many sensor materials. This chapter will therefore consider case studies of where ILs have already had a beneficial impact on optimising sensing applications through (a) enhancement of the selectivity/specificity of analyte interactions, (b) enhancements of sensor performance, such as electrochemical response, and (c) the imparting of novel properties to material structure and assembly. Where ILs have already been incorporated into early wearable technologies, consideration of how these solutions might transition to real world requirements for man-worn sensors to provide a viable sensing concept is also discussed.

15.4 Exploiting the Selective Solvation of Ionic Liquids in Sensor Systems

With any sensing device, the sampling of the target material from the environment (vapour, gas, liquid or solid) and effective delivery of that sample to the transduction system is a fundamental requirement. Ideally, where low power *in situ* detection systems are desirable, this process would be entirely passive.

The tuneable solvent properties of ILs have long been understood to be an advantage in enhancing the selective (the ability to respond only to the target) uptake of materials and thus improve the sensitivity (the ability to discriminate between small differences in concentration) and accessibility of sampling strategies[29–31] in the ideal wearable system. The design potential of ILs is almost limitless with an estimated 10^8 different possible single anion and cation combinations and billions more combinations possible if individual ILs are combined.[32,33] Within these possible combinations, particular chemical functionalities are well understood. For example, potential issues experienced in atmospheric sampling due to water contamination may be alleviated through the use of long alkyl chains on the cation and appropriate selection of counter-anions (*e.g.* per-fluoro-based ions) to furnish hydrophobic ILs.[34–36]

Taking full advantage of the tuneable solvation behaviour of ILs, to facilitate selective analyte uptake, requires an understanding of the underlying solvation properties of ILs and a way to quantify these to support predictive solvent design. In an attempt to aid the design and application of ILs, several reviews have been written about their solvation properties based on different solvation models, including the Rohrschneider–McReynolds model,[37] the Abraham solvation parameter model,[38,39] the electrostatic interaction model[40] and the reference interaction site model.[41] Solvation behaviour is commonly described by three factors: solvent strength (synonymous with polarity), solvent selectivity (the ability of the solvent to interact with compounds through specific interactions), and solubility. Kamlet–Taft parameters can be utilised to understand solvent properties through the generation of well-understood data on the solvent polarisability, hydrogen bond donor ability and hydrogen bond acceptor ability.[29] Through both computational and experimental data analysis the *a priori* design of task-specific ILs may be feasible. Aspects of all of these have been used to select liquids for particular applications including in the selective uptake of specific analytes from complex backgrounds – a key technical challenge imposed on wearable technologies. In an environment where a wearable device is being used, many of the materials being targeted tend to be vapours or gases. The utility of ILs to solubilise a range of gases including CO_2,[42–54] N_2,[44,45,51–53,55–57] O_2,[45,51–54,56,58] and CO[45,51–54,56] has been discussed by Lei *et al.*[59] in their extensive review highlighting the applicability of ILs as solvents within this class of sensor.

15.4.1 Towards Selective Sampling Using Ionic Liquid Solvents

An area where differential solvation characteristics have found the greatest application is in solid phase sampling technologies where concerns of sensitivity, selectivity and specificity can also be found. Consideration of some recent developments may indicate the potential for exploiting the same understanding of IL behaviour in new wearable chemical sensors. Solid phase microextraction (SPME)[60] is a very common technique for the collection of environmental samples, leaving the analysis to be performed downstream using standard benchmarked analytical techniques. In this scenario,

ILs have shown compatibility with a range of analytical techniques including headspace gas chromatography (GC),[61–65] liquid phase microextraction (LPME),[66–69] and matrix-assisted laser desorption/ionization-mass spectrometry (MALDI-MS).[70–72] For further specific details the reader is directed to two reviews.[31,73] Additionally, it should be noted that ILs have been shown to allow for greater control of retention factors and separation selectivity when used as the GC stationary phase; this is out of the scope of the current chapter but interested readers are referred to a number of recent examples of ILs used in this manner.[74–78]

SPME is an advantageous technique to consider in the context of wearable sensing as it encompasses both the sample collection and sample extraction in a single step. Furthermore, it can be used in active (direct immersion) or passive (sampling of air) modes.[79,80] ILs have begun to find exploitation pathways in SPME methodologies improving the sensitivity, versatility and extraction efficiency of standard commercial SPME fibre substrates.[81] Recent examples have attempted to illustrate the application of IL-functionalised SPME substrates as a wearable passive sampling solution, for example, for the monitoring of air for quantities of illicit drugs, *e.g.* ketamine.[82] In this example, four polymeric IL formulations (PILs 1–4), with C_4 and C_{12} spacers, based on mono and di-cationic imidazolium IL components, were selected on the basis of parameters considered likely to be selective for ketamine and tested against a range of small molecules with differences in polarity to test their selectivity. In three of the tested PIL formulations (PIL4 was discarded due to poor adhesion to the fibres), selectivity to ketamine was observed above low level binding of toluene (selectivity factor, SF, of 1, 1 and 1 for PILs 1, 2, and 3, respectively), limonene (SF: 0.6, 1.3, 0.1) and ethanol (SF: 2.9, 1.1, 5.8). Cross-reactivity to methamphetamine, which has similar chemical structure and properties, indicated that hydrophobicity, H-bonding, dipole–dipole interactions and π–π stacking were all important in the selectivity towards ketamine. Figure 15.1 shows the chromatograms obtained *via* SPME-GC-MS after environmental sampling using a commercial polydimethylsiloxane/divinylbenzene (PDMS/DVB) fibre (C) and a PIL1-coated fibre (D). The PIL1-based fibre is shown to have a substantially lower sensitivity to possible environmental interferents such as coffee, alcoholic and non-alcoholic beverages, and cigarettes, compared to the commercial PDMS/DVB fibres.

Testing of the SPME fibres against controlled ketamine aerosols indicated that the best performing PIL formulation (polymerised 1-vinyl-3-butylimidazolium with a *bis*(trifluoromethylsulfonyl)imide ($[N(Tf)_2]^-$) counter ion) could achieve two-fold improvement compared to commercially available silicon-based SPME fibre substrates with extraction efficiencies of up to 84.2 ± 3.3% (at 0.1 mg m^{-3}) and 93.6 ± 2.6% (at 1.0 mg m^{-3}) ($n = 3$), showed excellent thermal stability (up to 350 °C) and could be reused up to 200 times without loss of function, demonstrating the potential for passive collection targeted to downstream analytical processes where appropriate.[82]

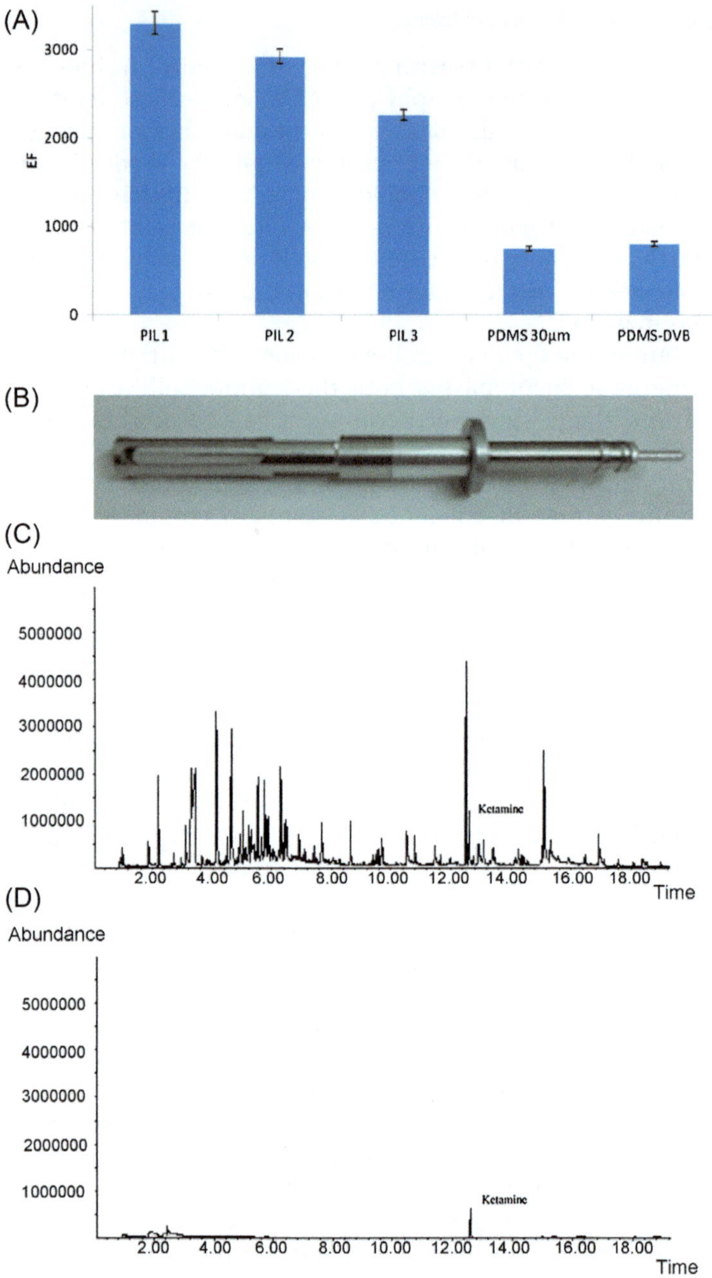

Figure 15.1 Performance of PIL formulations in wearable SPME devices for environmental monitoring of ketamine in recreational areas. (A) Enhancement factors (EFs) for PIL formulation-coated SPME *vs.* commercial coatings. (B) Example SPME fibre prior to insertion into a wearable device. SPME-GC-MS full scan using (C) PDMS/DVB fibres and (D) PIL1 fibres. (Reprinted from *Journal of Chromatography A*, 1331, F. Bianchi, F. Bisceglie, S. Dugheri, G. Arcangeli, V. Cupelli, E. del Borrello, L. Sidisky, M. Careri, Ionic liquid-based solid phase microextraction necklaces for the environmental monitoring of ketamine, 1–9, Copyright (2014) with permission from Elsevier).[82]

15.4.2 Improved Selectivity and Specificity of Sensing Strategies Achieved Using Ionic Liquids

Ambitions within the IL research community extend further than achieving selective sampling and are now reaching towards enabling selective sampling linked to transduction methods. The ability of ILs to support sensor selectivity was first examined on a quartz crystal microbalance (QCM) where differential binding of ILs containing methylimidazolium cations and tetrafluoroborate ($[BF_4]^-$) and $[N(Tf)_2]^-$ anions,[83] imidazolium cations with non-halogen anions, and phosphonium or ammonium cations with alkylsulfonate or alkylbenzenesulfonate anions[84] were demonstrated for a variety of organic vapours.

Arrays of small numbers of broadly selective ILs have since transitioned to platforms that are aiming at portable wearable systems that provide a signal output based on the chemical changes that occur within the solvent environment due to the uptake of particular analytes. Optical detection platforms based on porous silicon (PSi)[85] can be readily functionalised with a variety of IL formulations to generate surfaces with differential selectivity.[86,87] Sensing signals can be generated from these arrays by taking advantage of refractive index and volume changes that occur in the ILs as they take up volatile organic compounds (VOCs). These changes create microdroplets on the surface of the PSi resulting in light scattering and reflectance peak intensities of the IL, which can be used as a simple transduction mechanism.[88] This phenomenon has recently been demonstrated in an array format using 1-butyl-3-methylimidazolium ($[C_4C_1im]^+$) in combination with six different anions: hexafluorophosphate ($[PF_6]^-$), $[BF_4]^-$, trifluoromethanesulfonate ($[OTf]^-$), trifluoroacetate ($[CF_3CO_2]^-$), perchlorate ($[ClO_4]^-$), and $[N(Tf)_2]^-$ in addition to 1-octyl-3-methylimidazolium tetrafluoroborate ($[C_8C_1im][BF_4]$), tributylhexyl ammonium *bis*(trifluoromethylsulfonyl)imide ($[N_{4446}][N(Tf)_2]$), and trihexyl(tetradecyl)phosphonium ($[P_{66614}]^+$) with imidazole, triazole and tetrazole anions.[89] The differential responses of each IL were used to generate principal component plots that could clearly discriminate between seven analytes (Figure 15.2).

Electronically-driven approaches have also been established with a number of semi-selective IL formulations, namely $[C_4C_1im][OTf]$, $[N_{6222}][N(Tf)_2]$, $[C_4C_1im][BF_4]$, $[C_4C_1im][ClO_4]$, and $[C_4C_1im][PF_6]$, having been transitioned towards a filter paper-like graphene oxide (GO)-supported small array where differential responses measured by electrical conductance were observed for model gas and vapour molecules (O_2, NO_2, Cl_2, toluene, hexane and ethylacetate).[90] It is hypothesised that these measurements are based on changes in viscosity and/or volume due to dipole interactions of polar VOCs with IL anions and π–π interactions between imidazolium groups and aromatic VOCs (Figure 15.3).

Further work targeting the design of ILs with selectivity to particular volatile components, *e.g.* aldehydes,[91,92] and acyclic and cyclic ketone gases,[92] has also been performed. Development of silver(I)-based ILs of the general

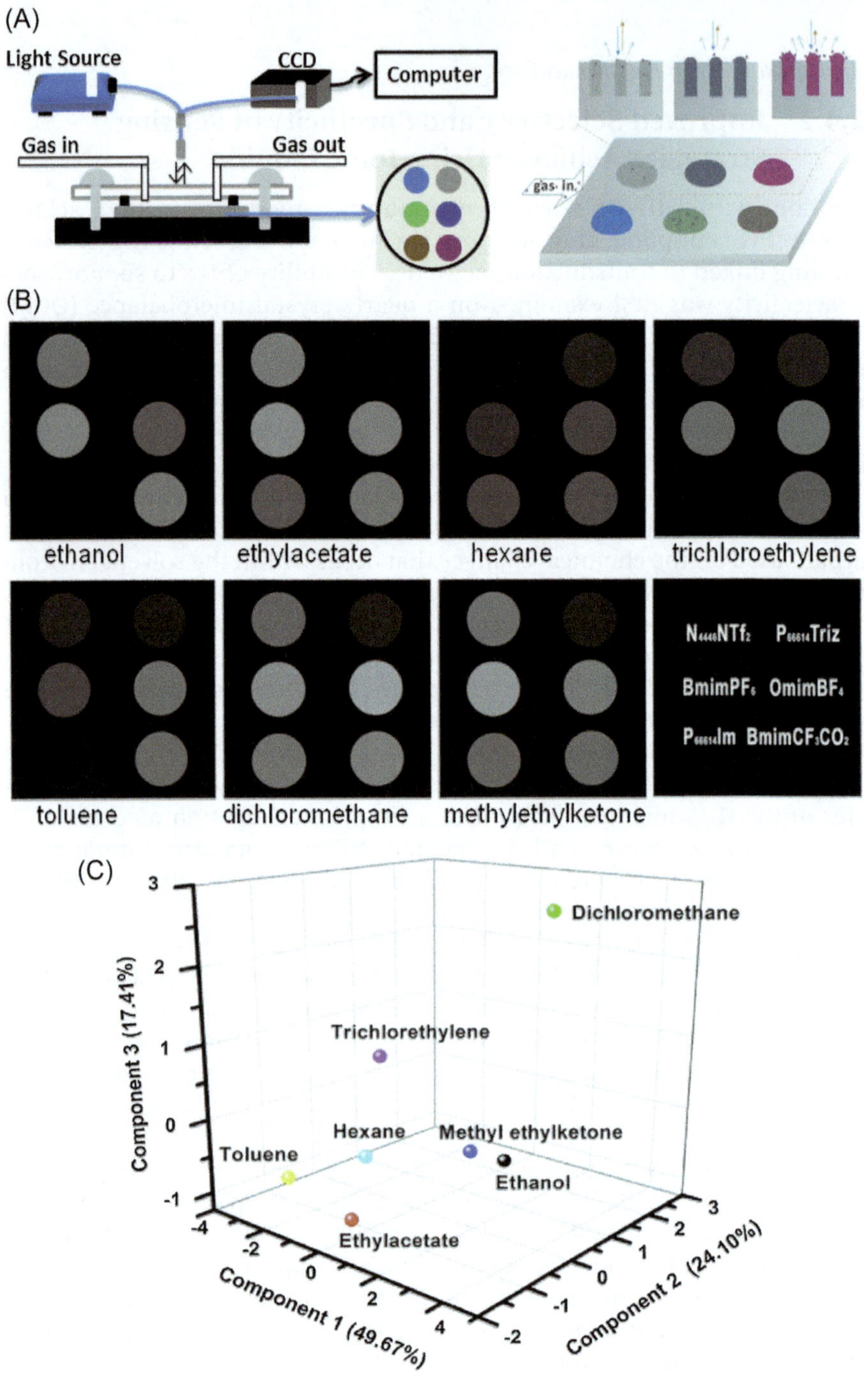

Figure 15.2 (A) Schematic of the PSi VOC detector and illustration of the array sensing mechanism. (B) Greyscale fingerprint images of the PSi VOC detector array upon exposure to a range of VOCs (ethanol, ethylacetate, hexane, trichloroethylene, toluene, dichloromethane and methylethylketone). (C) 3-dimentional principal component analysis (PCA) score plots for the PSi array for seven different VOCs. Reprinted from *Analytica Chimica Acta*, 953, H. Zhang, L. Lin, D. Liu, Q. Chen, J. Wu, Optical nose based on porous silicon photonic crystal infiltrated with ionic liquids, 71–78, Copyright (2017) with permission from Elsevier.[89]

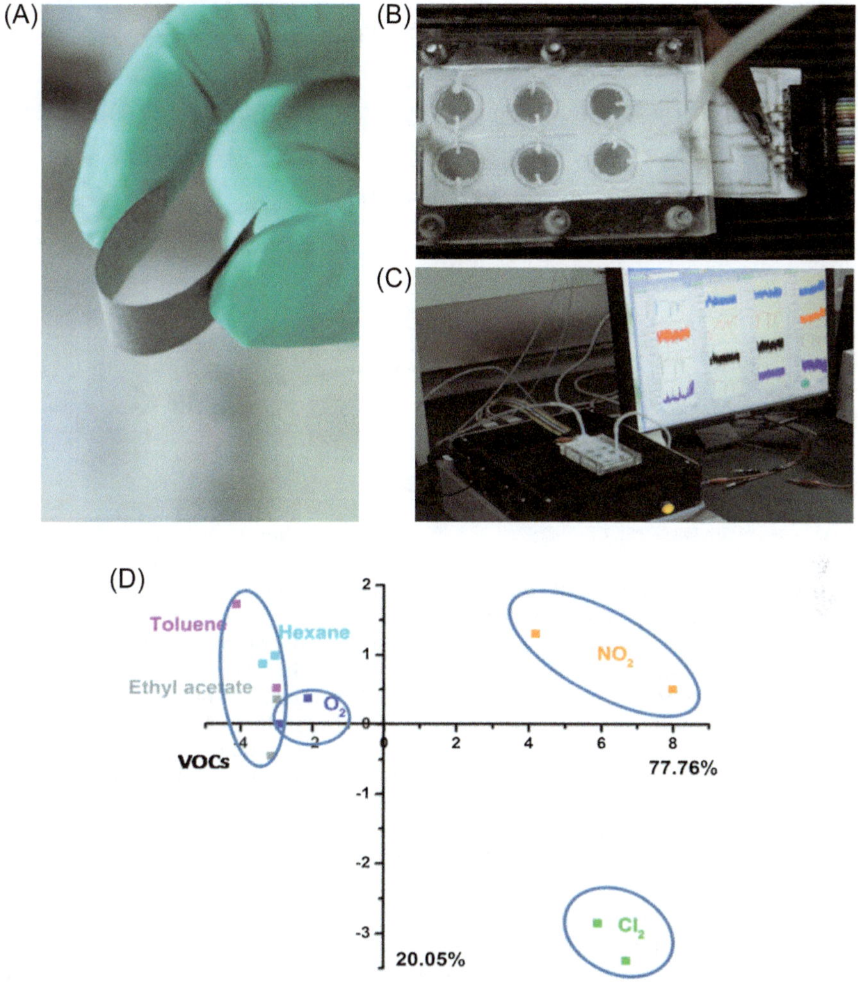

Figure 15.3 Development of (A) fully flexible paper-supported IL-reduced graphene oxide (rGO). (B) The GO–IL paper array in a gas flow chamber and (C) the performance of the sensor array e-nose system for gas detection and discrimination. (D) 2-dimensional PCA plot resulting from the normalised responses of the rGO–IL to six different gas samples. (Reproduced from ref. 90, with permission from The Royal Society of Chemistry.)

type $[Ag(L)_2][N(Tf)_2]$, where L is a long-chain alkylamine, has been shown to lead to better gas uptake, selective characteristics and complete insensitivity to water, making them particularly applicable to open architecture passive collection for environmental samples (Figure 15.4).[93]

If the application of a particular sensor aims at discrimination between, for example, materials known to be present in a complex clinical matrix, then ILs with bespoke selectivity can be designed in order to allow both analytes

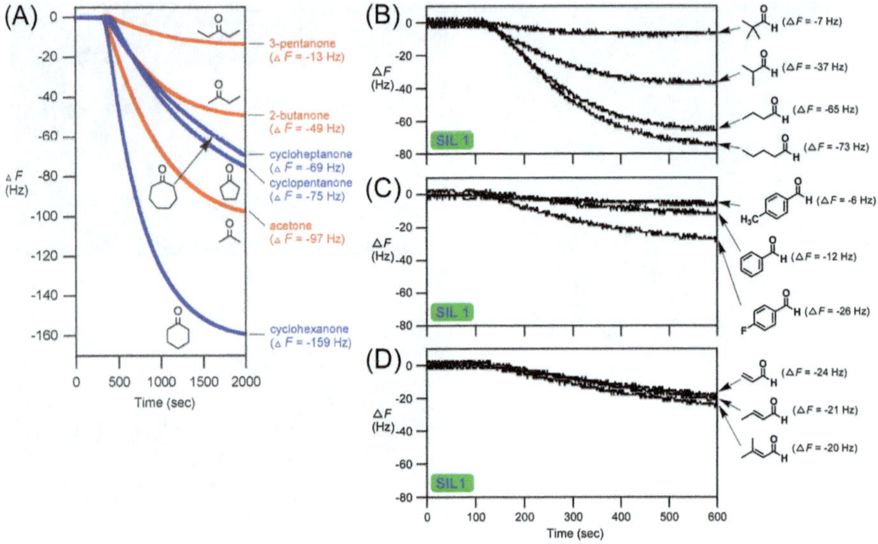

Figure 15.4 (A) Differential sensing of aldehyde and ketone gases using IL arrays. Graph shows the response of QCMs thin-coated with "SIL-2" to each of the stated VOCs at 76 ppb. (reproduced from ref. 92 with permission from The Royal Society of Chemistry). Silver-based ILs thin-coated on QCMs showing a reactivity-based response to 100 ppb volatile (B) aliphatic aldehydes (privaldehyde, isobutyraldehyde, butyraldehyde, valeraldehyde), (C) aromatic aldehydes (*p*-tolualdehyde, benzaldehyde, *p*-fluorobenzaldehyde), and (D) α,β-unsaturated aldehydes (acrolein, crotonaldehyde, 3-methylcrotonaldehyde) exhibiting no insensitivity issues in the presence of water. (Reproduced from ref. 93, with permission from The Royal Society of Chemistry.)

to be measured on the same platform. Palladium nanoparticle (NP)–GO dispersions modified with the ILs *N*-butyl-*N*-methylpyrrolidinium bis(trifluoromethylsulfonyl)imide ($[C_4C_1pyrr][N(Tf)_2]$) and *N*-butyl-*N*-methylpyrrolidinium dicyanamide ($[C_4C_1pyrr][N(CN)_2]$) exhibited specific voltammetry for glucose and ascorbic acid, respectively (Figure 15.5). Anion selection was shown to be critical to the selectivity of the sensor through cation-independent results using $[C_2C_1im]^+$-based ILs and the results from angle resolved X-ray photoelectron spectroscopy. The latter indicated that anion-rich electrode surfaces dominate the selectivity, and suggested a more favourable association of the cationic components with the π-rich graphene sheets.[94]

The majority of examples currently used in the literature restrict themselves to a narrow list of broadly classical-type ILs focusing on common imidazolium cations (mostly $[C_4C_1im]^+$). The importance of changing the anion to achieve differential specificities in IL formulations is also clear from the published work to date, as shown by the surprising levels of selectivity that have been achieved so far with only a relatively small list of anions (mainly $[OTf]^-$, $[BF_4]^-$, $[PF_6]^-$, $[ClO_4]^-$, and $[N(Tf)_2]^-$), with rare usage of more exotic combinations of anions and cations for very specific materials.

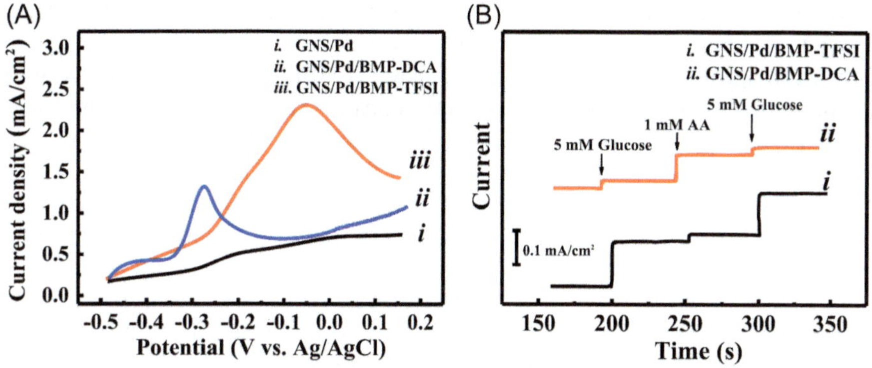

Figure 15.5 Simultaneous detection of both analytes in a mixed solution of ascorbic acid (1 mM) and glucose (5 mM) using highly specific IL formulations. (A) Linear scan voltammograms from GNS–Pd-based electrodes functionalised with BMP-DCA IL demonstrating specific measurement of ascorbic acid and BMP-TFSI for detection of glucose. (B) Amperometric dose response curves for GNS–Pd-BMP–TFSI and GNS–Pd-BMP–DCA electrodes to successive additions of 5 mM glucose, 1 mM AA, and then 5 mM glucose. (Reprinted from *Biosensors and Bioelectronics*, 89, C.-H Wang, C.-H Yang, J.-K Chang, High-selectivity electrochemical non-enzymatic sensors based on graphene/Pd nanocomposites functionalized with designated ionic liquids, 483–488, Copyright (2017) with permission from Elsevier.)[94]

This highlights a considerable bottleneck between full exploitation of this concept in a wearable platform, namely the facile identification of useful cation and anion pairs for particular molecular interactions. Limited work has been performed in this area to tackle this aspiration, including an example where anion exchange was demonstrated on tethered IL ions to generate IL microarrays.[95] However, while this strategy was successful in producing IL variants *in situ*, it was more challenging to show that an increased diversity with no rational design achieved increases in microarray performance.

In summary, predicting the solubility of ILs in a rational way and to enable selection of the most appropriate solvents for a selective array remains a substantial technical challenge for ILs. Research in multiple application areas will furnish valuable data, with developments in selective sampling and extraction continuing to provide a significant avenue for solvent understanding and subsequent design.

15.5 Progression of Ionic Liquid Sensors Towards Viable Wearable Sensor Options

In order for a sensor platform to have exploitability as a wearable sensor it must possess a number of characteristics. On a chemical level it must provide tuneable sensitivity and a relevant dynamic range to the analyte of interest, and require no consumables (including additional solvent). Previous sections have described how IL formulations might provide enhanced sensitivity and selectivity to sensor platforms. However, at the most basic level, the sensors themselves must also be physically compatible with unobtrusive integration into a format that would allow comfortable wear. Development of appropriate materials that allow the improvement of sensors of this kind has remained one of the consistent barriers to achieving portable/wearable sensing options for chemical analytes,[9,10,96] but this is a particular area where the physical properties of ILs could add real benefit to achieving wearable chemical sensors.

In this section attention is given to those sensor systems where a solvent forms a component of the system, and therefore focuses on optically transduced and electrochemical-based sensor systems. The latter is a particularly active field of research given the favourable electrochemical properties of ILs, as stated earlier.

15.5.1 Optical Systems

Within a wearable device, the sample collection medium can be part of the sensor system and as such the desired optical properties of that medium must be either negligible or such that they enhance the measured effect. Research into the optical properties of ILs has typically focused on their absorption and fluorescence behaviour. This has shown that in the UV region ILs generally exhibit a strong absorbance band at lower wavelengths coupled with a low trailing absorbance reaching up to higher wavelengths; the fluorescence intensity follows the same trend with higher intensity emissions at lower wavelengths, and lower intensities at higher wavelengths.[97,98] It should, however, be noted that in some cases careful synthesis can reduce the absorbance features found in ionic liquids.[99]

Photonic gels are generated by the suspension of colloidal crystals within a 2- or 3-dimensional gel network.[100] Control of the spatial distribution of the crystals within the gel generates a periodicity of the refractive index between the crystals thus creating a gel with an optical bandwidth. Swelling or shrinking of the gel causes the distance between the particles to shift creating a change in the refractive index and thus the optical bandwidth shifts (Figure 15.6A). Linkage of this to some form of stimulus (temperature, pH or specific chemical/biological interaction) can be used to generate a very simple transduction mechanism that has been widely exploited using conventional solvents.[101] Use of these hydrogels is in many cases impractical due to the evaporation of aqueous and organic solvents over time, and thus transition to low volatility ILs is desirable.

Early work integrating novel ILs into photonic gels showed that 1-(2-acryloyloxyhexyl)-3-methylimidazolium bromide or phosphate could be used to generate an inverse opal structure by SiO_2 microsphere templating, which could then be used to reversibly monitor relative humidity. Changing to a sodium dodecyl sulfate counter-anion ($[SDS]^-$), generated a gel with selectivity towards ethanol.[102] Hydrogels with intact microspheres were then generated with either polystyrene

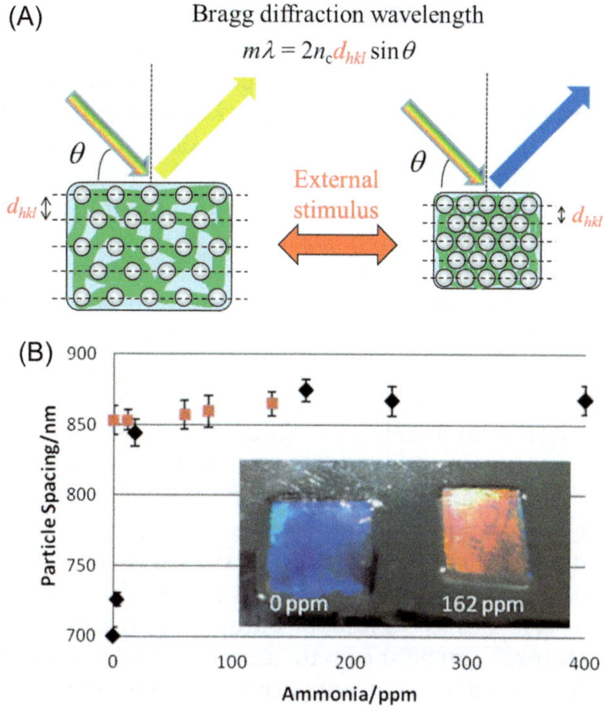

Figure 15.6 (A) Schematic describing the signalling mechanism of photonic hydrogels showing the spatial dispersion of photonic nano-colloids within a mobile matrix; associated refractive index (θ) and how stimuli induced impact on that refractive index. (Reproduced with permission from *Organic and Hybrid Photonic Crystals*, Gel-Immobilized Colloidal Photonic Crystals with Tunable Properties, 2015, 431–450, T. Kanai, © Springer International Publishing Switzerland, with permission of Springer.)[100] (B) Response of polystyrene-ethyl-guanidine perchlorate 2D photonic gel to exposure to increasing ammonia concentration (black diamonds) and the impact of decreasing ammonia concentration relative to particle spacing after exposure (red squares). Inset shows the colour change associated with ammonia exposure. (Reproduced from ref. 105, with permission from The Royal Society of Chemistry.)

or silica microspheres by replacing the water in the hydrogels with hydrophilic ionic liquids, namely 1-allyl-3-butylimidazolium bromide[103] or 1,3-diallylimidazolium bromide,[104] while maintaining the optical properties of the gel. Subsequently, it was demonstrated that the swelling of the IL gels and thus colour could be robustly controlled by tuning the ratio of hydrophilic 1,3-diallylimidazolium bromide to hydrophobic 1,3-diallylimidazolium $[N(Tf)_2]^-$.[104]

Application of these gels has since transitioned towards broader passive chemical sensing applications through the use of the novel IL ethyl-guanidine perchlorate as the supporting phase in a 2D polystyrene bead gel.[105] The gel was found to respond to both humidity and ammonia gas.

When exposed to humidity the gel shrinks reducing the spacing of the particles within the gel, which results in a visible blue shift in colour. Conversely, exposure to ammonia (NH_3) causes irreversible deprotonation of the gel polymer groups leading to swelling of the gel resulting in a visible red shift in colour (Figure 15.6B). The authors report a detection limit of 3 ppm for NH_3, which is broadly comparable to what existing metal oxide sensors can currently achieve, but would need further development to achieve the levels possible with more advanced high burden detection systems.

The potential for use of such visible indicators to indicate the presence of hazardous materials is extremely attractive. However, technical challenges still need to be addressed to transition these types of cheap gel sensors to fully wearable solutions. Most notably, all experiments within these studies were performed in a controlled environment, where humidity could be closely monitored. In a real environmental context, this will not be the case and the use of hydrophilic ILs may ultimately mean that background responses to changing humidity may become a confounding factor in their utility. Further exploitation will therefore likely be dependent on overcoming such confounding stimuli responses in air-exposed gel networks.

ILs have also shown early promise in enhancing the properties of transparent, luminescent films.[106] Films of europium(III)–nanoclay composites embedded in chitosan can be shown to luminesce under 365 nm irradiation. Replacement of acetic acid solutions with the carboxyl-functionalised ionic liquid 3-(5-carboxypentyl)-1-methylimidazolium bromide retains solubilisation of the chitosan and results in improvements in the luminescence efficiency of the film (Figure 15.7). The latter is speculated to arise from the deprotonation of acid sites on the nanoclay platelet surfaces by the IL, in turn leading to higher coordination complexes of europium and enhanced luminescence. This film has shown application as a vapour luminescent sensor with trimethylamine (Et_3N) vapour enhancing the luminescence and hydrochloric acid (HCl) vapour quenching it.

15.5.2 Electrochemical Sensors

One class of sensor platform for which ionic liquids are particularly suited is electrochemical sensors. Conventional electrolytes for use in electrochemical sensors tend to be aqueous or organic salt solutions and have very limited exploitation potential for use in wearable technologies due to evaporation of the solvent, especially in small sensor systems and in varying temperature regimes. While polymer-based electrolytes generated from conventional solvents can overcome these volatility issues, they tend to exhibit low ionic conductivity and therefore have limited exploitation in many sensor solutions.[107]

When using an IL as an electrolyte (which can be as a liquid or polymer) the important factors to consider are the electrochemical stability, electrochemical window and the conductivity. In general, ILs exhibit wide electrochemical potential windows (*i.e.* the difference between the reduction and oxidation potentials) ranging from 3.1–7 V.[108–112] These are substantially

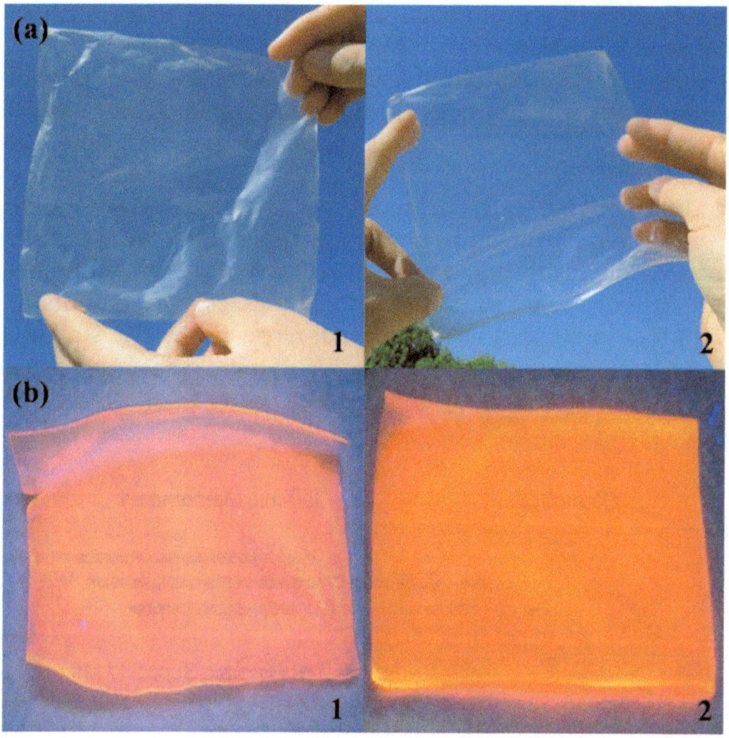

Figure 15.7 Flexible films composed of europium(III) complex–nanoclay composites within chitosan under (a) daylight conditions and (b) under 365 nm irradiation, and prepared using either (1) aqueous acetic acid or (2) carboxylate-functionalised ILs. (Reprinted with permission from ref. 106, Copyright (2016), American Chemical Society.)

higher than those found in aqueous salt solutions.[113,114] Moreover, due to the high ion density within an IL, its conductivity tends to range from 0.1 to 20 mS cm^{-1} at room temperature.[108,115,116] The intrinsically low (or near-zero) vapour pressure of ILs as solvents coupled with these electrochemical properties therefore makes them highly compatible with integrated, portable sensing technologies.

15.5.2.1 Integration of Ionic Liquids into Miniaturised Electrochemical Devices

Miniaturisation of standard sensing materials may be compatible with an extrinsic device format and this has been widely explored with ILs, particularly within the field of electrochemical sensors.[117,118] The use of ILs to enhance both the form and function of electrochemical detection platforms underpinning their application as fully integrated wearable sensing platforms is elegantly introduced through the trace detection of gases. The detection of oxygen is frequently achieved using a 'Clark' style electrode[119] where the oxygen is reduced using a platinum electrode – a process that is limited by temperature in conventional solvents.

ILs composed of imidazolium cations in combination with $[N(Tf)_2]^-$, $[B(CN)_4]^-$ and tris(pentafluoroethyl)trifluorophosphate ($[FAP]^-$) anions were selected as potential electrolytes due to their thermal stability and hydrophobicity. Each IL was applied as a liquid layer between two membranes to cover either platinum or gold electrodes and tested for detection of O_2, using multiple-potential-step-chronoamperometry (Figure 15.8).[120] In this system, the most sensitive sensing response was achieved using an Au-membrane-sensor filled with either $[C_2C_1im][B(CN)_4]$ or $[C_6C_1im][FAP]$, where limits of detection of 5 ppm and 30 ppm were achieved, respectively. The stability of the electrode systems was demonstrated with a Pt-membrane-sensor filled with $[C_2C_1im][N(Tf)_2]$ over the course of one week (Figure 15.8).

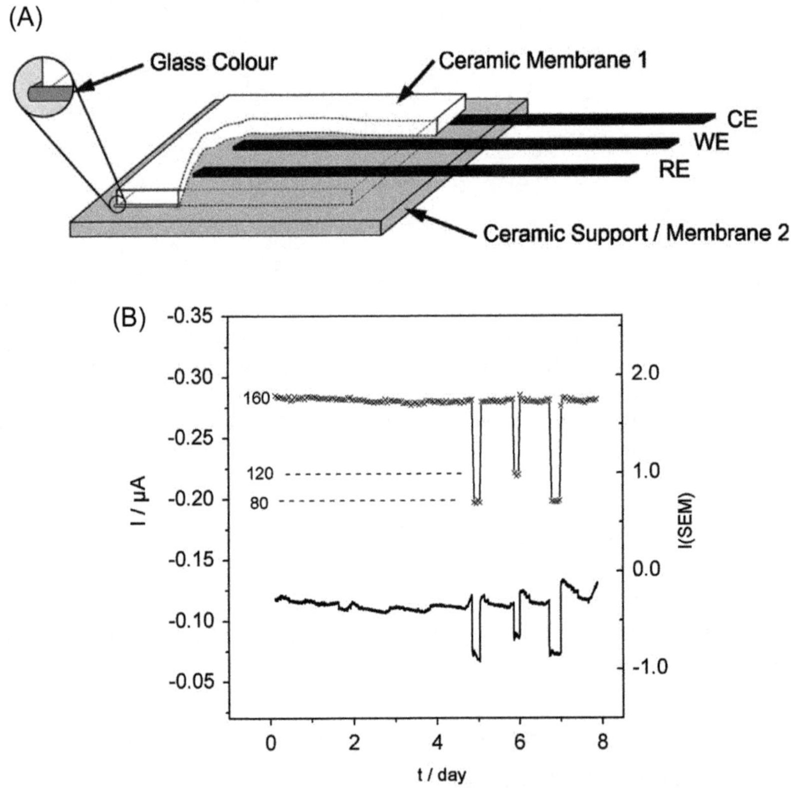

Figure 15.8 (A) Schematic of electrode format and (B) current values (black line) from continuous multiple-potential-step-chronoamperometry using the modified Clark electrode constructed with $[C_2C_1im][N(Tf)_2]$ using a Pt counter electrode. Readings taken over 8 days. T: 23 °C; $E_1 = 0$ V (20 s); $E_2 = -1.35$ V (20 s). X-symbols: controlling mass spectrometer signal. Reprinted from *Talanta*, 116, N. Baltes, F. Beyle, S. Freiner, F. Geier, M. Joos, K. Pinkwart, P. Rabenecker, Trace detection of oxygen – ionic liquids in gas sensor design, 474–481, Copyright (2013) with permission from Elsevier.[120]

Exploitation of the electrochemical characteristics of IL electrolytes can be used to further improve the performance of IL-based gas sensors. In one particularly elegant example, $[C_4C_1pyrr][N(Tf)_2]$ was used by Wang *et al.*[121] to target both methane and oxygen detection. This was possible through the use of Pt electrodes to oxidise the methane, producing CO_2 and H_2O. Incomplete oxidation of the methane also leads to generation of CO, which would result in poisoning of the Pt working electrode over time in conventional solvents (Figure 15.9A). However, exploitation of the wide electrochemical window of $[C_4C_1pyrr][N(Tf)_2]$ (~5.2 V)[109] also enables the production of oxygen superoxide radicals. These radicals could then be used to simultaneously react with CO_2 generated by the electrooxidation of methane effectively removing it from the sensor platform (Figure 15.9A).[122] The further benefit of this system is that $[C_4C_1pyrr][N(Tf)_2]$ has poor solubility for a variety of other gases increasing the overall selectivity of the system to methane over other background interferents (Figure 15.9 B and C).

The macroelectrode-based oxygen sensor upon which this sensor is based had a detection limit of 500 ppm for oxygen, good selectivity and reversibility, and showed long-term stability with only a 0.3% decrease in the peak current (at −1.2 V) in response to 20% oxygen over 60 days (during a 90-day

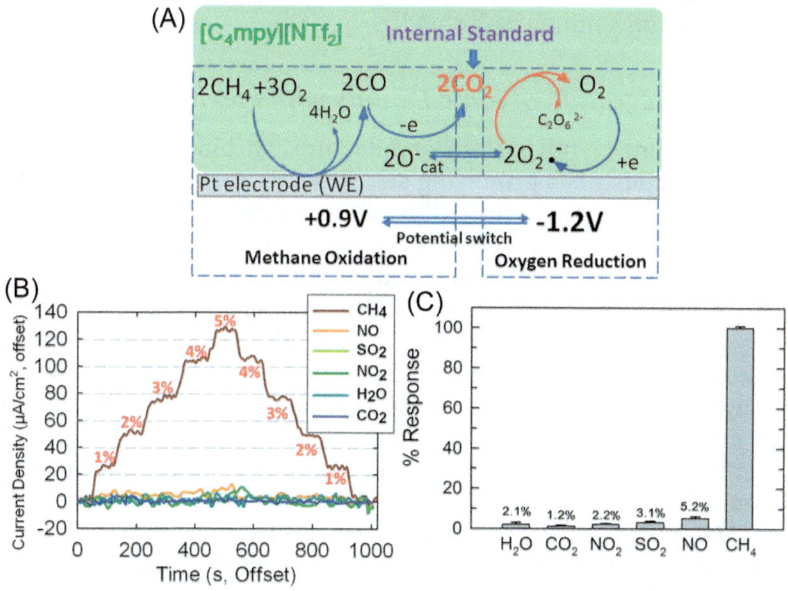

Figure 15.9 (A) Scheme describing the electrochemical coupling of methane detection to quantification of oxygen *via* the oxidation of CO generated from the incomplete oxidation of methane by electrochemically generated superoxide radicals. (B) Current density transients measured in response to increasing/decreasing concentration of methane at 0.9 V and (C) relative change in current density (%) of the sensor to 5 vol% of interferent gases in the presence of 5 vol% methane. (Reproduced from ref. 121, with permission from The Royal Society of Chemistry.)

test period).[123] In comparison, the methane-oxygen sensor exhibited the same selectivity and reversibility characteristics, along with excellent long-term stability (tested over 120 days). Quantification of methane was possible with a detection limit of 3000 ppm in air, along with confirmation of the oxygen concentration. The exploitation of IL electrochemical properties has since progressed further not only in terms of the sophistication of the transduction approach, but also in terms of integration of multi-analyte gas sensors into fully integrated miniature devices, *e.g.* the intelligent electrochemical gas analysis system (iEGAS). The iEGAS system is a fully integrated sensor array in combination with embedded microcontrollers, power (*via* an integrated supercapacitor), and algorithms designed for monitoring multiple gases of importance for health and safety.[124]

This system utilises the same single IL, $[C_4C_1pyrr][N(Tf)_2]$, as the electrolyte medium in a planar-electrode-on-semi-permeable-membrane format to minimise the diffusion times of gases to the electrode surface for analysis (8 seconds). Furthermore, this platform elegantly exploits the tuneable selectivity of IL formulations describing the use of bespoke embedded algorithms and differential applied DC bias to the IL electrolyte to achieve selectivity to four gases, namely CH_4, O_2, SO_2 and NO_2, using dual AC impedance and amperometry modes – a methodology that would be hugely challenging to achieve using conventional solvents.[125]

15.5.2.2 Integration of Ionic Liquids with Flexible Substrates

The development of full system architectures within the sensing devices described previously is a promising step forwards in terms of wearable sensors; however, these platforms still result in a rigid substrate that would be severely limited in a human-centric concept. As stated earlier, for human mounted sensors, the 'soft' nature and curvilinear surface of human tissue requires sensing devices that are capable of connecting reproducibly and robustly with the tissues beneath and to maintain this connection when subjected to regular and complex human deformations. Generally, the wearables are likely to require similarities in mechanical properties to the underlying tissues upon which they are to be mounted.

Integration of ILs into more flexible smart and functional materials is therefore a key driver within the IL field. A number of screen printed electrodes (SPEs) that use ILs as components in electrode preparation and to enhance performance have been reported, and generally indicate that use of ILs is a favourable route to low cost, readily and reproducibly fabricated, and resilient electrodes.[126,127] There are levels of sophistication where this may be achieved ranging from the simple inclusion of ILs *via* the coating of flexible electrode materials, such as the coating of a flexible interdigitated electrode (planar Pt electrode on a Teflon gas-permeable membrane) with $[C_4C_1pyrr][N(Tf)_2]$ for methane/oxygen sensing, to the printing of the IL, $[C_2C_1im][N(Tf)_2]$, as a component of a solid polymer electrolyte as part of a fully printed NO_2 sensor (Figure 15.10) on polyethylene terephthalate (PET).[121,128]

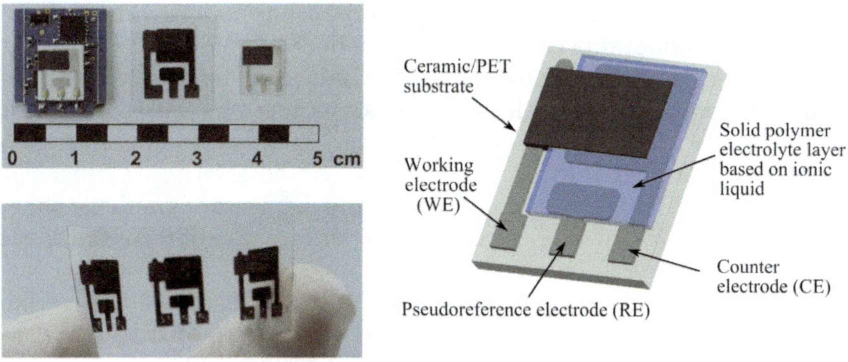

Figure 15.10 Top left: images of (from left to right) a ceramic-based NO_2 sensor with a potentiostatic circuit, a fully printed sensor on a PET-based graphite sheet and a fully printed sensor based on graphite and silver/silver chloride paste. Bottom left: images of fully printed sensors based on graphite paste. Right: Sensor topology–PET fully printed and ceramic-based. Reprinted from *Sensors and Actuators B: Chemical*, 209, P. Kuberský, T. Syrový, A. Hamáček, S. Nešpůrek, L. Syrová, Towards a fully printed electrochemical NO_2 sensor on a flexible substrate using ionic liquid based polymer electrolyte, 1084–1090, Copyright (2015) with permission from Elsevier.[128]

This printed NO_2 sensor exhibits excellent responsiveness (t_{90}/t_{10} of 70/60 s per 2 ppm step change), stability under different environmental conditions, including 10–90% relative humidity, and resolution (0.2 ppm).

The deposition of ILs on functional substrates is now being taken a step further with the full integration of these solvents into functional complexes with a variety of solid matrices, most commonly carbon nanotube (CNT)–graphene–IL composites.[94,126,127,129–136] These composite materials are relatively easily generated using simple mixing and sonication processes, and have the benefit of producing ink-like materials that can be applied to a variety of flexible electronics and can be printed onto electrodes. The fully flexible characteristics of these types of materials make them increasingly suitable for use in wearable technology due to their increased compatibility with many of the form factors required to achieve small, low visibility sensor platforms that integrate into a device or garment.[137]

Manipulation of the structures generated within electrodes through the use of different combinations of carbon supports and ILs can also enhance the overall performance of the electrode systems, increasing the conductivity and the robustness of the physical properties of the electrode mixes. These composite electrodes can frequently demonstrate greater signal outputs and thus sensitivity to their target analytes.[126,127] Judicious choice of an IL as an electrolyte within these types of composite can further be used to overcome specific challenges associated with the electrochemical detection of particular analytes. This can be a powerful tool when targeting components of complex matrices or mixtures such as clinical samples. In one specific example regarding the sensing of nicotinamide adenine dinucleotide (NADH), the use of an IL (1-(3-aminopropyl)-3-methylimidazolium bromide) as an electrolyte in a graphene–IL electrode mix was used to reduce the large overvoltage and surface fouling of the electrode, a known issue caused by NADH at electrode surfaces made with conventional substrates.[138] Furthermore, the introduction of charge through the use of an IL had the added advantage of increasing the solubility of the electrode components and dispersal of graphene sheets, leading to much improved electrochemical performance of the electrodes in general.[129]

The most sophisticated IL–substrate composites generated to date have begun to address the potential inclusion of catalytic nanoparticles (NPs).[139] Inclusion of NPs has been demonstrated to further enhance the properties of IL–CNT/graphene composites and is exemplified by the in-depth study of Wang *et al.*,[136] investigating the impact of IL–Pd-doped graphene- and carbon nanotube-based electrodes for the detection of ascorbic acid, dopamine, uric acid and glucose. They demonstrated the synergistic impact of combining ILs and Pd nanoparticles on electrode performance, substantially increasing the detection sensitivity (0.86 $\mu A\ \mu M^{-1}$ and 3.28 $\mu A\ \mu M^{-1}$ for DA with the graphene–Pd and graphene–Pd–IL electrode, respectively, with a limit of detection of 0.12 μM and 0.07 μM – Figure 15.11). Although the authors acknowledge that further investigation into the interactions of ILs with graphene sheets is warranted, they attribute the more significant positive effects of the IL inclusion in the electrode to two key reasons. First, the formation of IL layers on the graphene sheets, which leads to electrostatic repulsion and thus suppression of nano-sheet stacking (π–π interactions). Second, wrapping of the graphene nanosheets in IL layers may overcome the low conductivity of graphene in its *z*-axis (*i.e.* perpendicular to the graphene plane) and allow for improved signal transduction.[136]

Aspirations to specifically drive these substrates towards wearable/implantable applications are now occurring, commonly targeting the direct electrochemical detection of small molecules in biological fluids. In the most recent example of this, gels composed of PtAu NP-loaded graphene–CNT–ionic liquid (IL: $[C_4C_1im][BF_4]$), which had previously been shown to exhibit good conductivity and mechanical properties,[140] were printed onto graphene paper to produce flexible electrodes for the sensing of glucose in blood.[133] Optimisation of the deposition of the PtAu nanoparticles *via* a novel ultrasonic electrodeposition process (Figure 15.12A) was used to generate free-standing paper substrates (Figure 15.12B and C). The resulting electrode material is decorated with highly dense NPs with an average diameter of 20–40 nm, and X-ray diffraction (XRD) data indicates that the Pt (control), Au (control) and PtAu NPs possess a single crystallinity.

Upon investigation of the electrode materials with the redox probe $Fe(CN)_6^{3-/4-}$ a greater Faradaic current and capacitance current was shown for the graphene oxide (GO)–CNT–IL/graphene paper (GP) electrode material over plain GP, attributed to the greater surface area and porous structure. Furthermore, decoration of the 3D assembly with PtAu NPs led to an increase

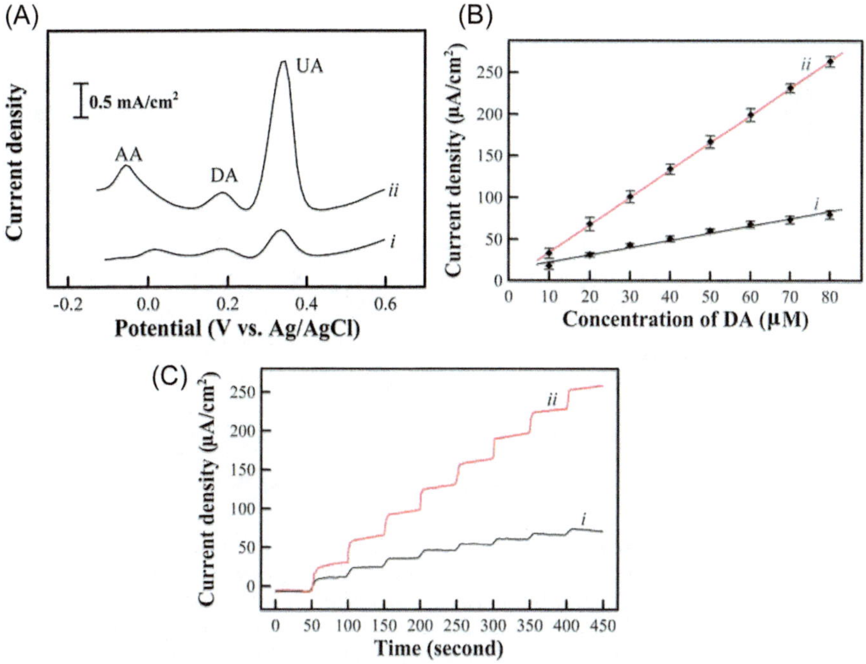

Figure 15.11 Improvement in performance of a graphene–Pd electrode through incorporation of IL 1-butyl-3-methylimidazolium hexa-fluorophosphate to generate graphene–Pd–IL electrodes. (A) Enhanced signal output demonstrated by comparison of differential pulse voltammograms from graphene–Pd and graphene–Pd–IL against a solution containing 1 mM ascorbic acid (AA), 2 mM dopamine (DA) and 50 mM uric acid (UA). (B) Calibration curves of current density *versus* dopamine concentration. (C) Chronoamperometry measurements for graphene–Pd and graphene–Pd–IL after sequential addition of dopamine (10 μM). In each case, readings from graphene–Pd are shown on curve i with graphene–Pd–IL shown on curve ii. (Reproduced from ref. 136, with permission from The Royal Society of Chemistry.)

in the peak current and a decrease in peak potential separation as observed in CV measurements, thus, claim the authors, demonstrating an increase in the electroactive surface area and accelerated electron transfer.

The materials were tested with regard to the electrochemical sensing of glucose. Initial testing demonstrated the advantage of all three NP species over the NP-free electrodes, and that the PtAu alloy NPs exhibit greater electrochemical activity and resistance to surface poisoning than the monometallic NPs. The long-term stability of the PtAu/reduced graphene oxide (rGO)–CNT–IL/GP electrode material is indicated by the retention of over 90% of the amperometric current after a period of 30 days. Comparison of 10 different PtAu/rGO–CNT–IL/GP electrodes provided a relative standard deviation (r.s.d.) of 4.2% in the amperometric current for 5 nM glucose, and ten

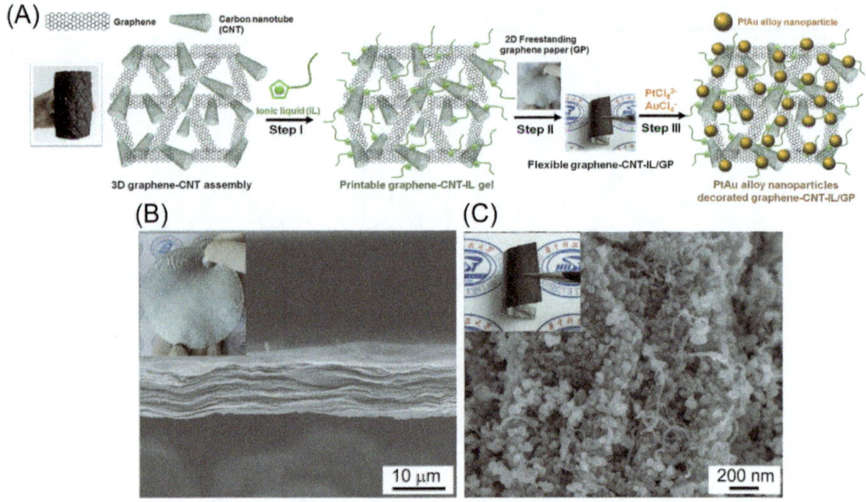

Figure 15.12 (A) Schematic of the preparation of PtAu alloy NP-decorated graphene–CNT–IL/GP electrodes reported by Duan and co-workers. Step I: 3D graphene–CNT assemblies ground with ILs to produce a printable gel. Step II: The graphene–CNT–IL gel is printed onto GP. Step III: Ultrasonic-electrodeposition of PtAu NPs onto the electrode. (B) Graphene paper at different magnifications and (C) SEM image of the PtAu/GO–CNT–IL/GP electrode material reported by Duan and co-workers showing the uniform distribution of NPs on the 3D assembly. Inset: indication of the flexibility of the electrode material. Reprinted from *Analytica Chimica Acta*, 903, W. He, Y. Sun, J. Xi, A. A. M. Abdurhman, J. Ren, H. Duan, Printing graphene-carbon nanotube-ionic liquid gel on graphene paper: Towards flexible electrodes with efficient loading of PtAu alloy nanoparticles for electrochemical sensing of blood glucose, 61–68, Copyright (2016) with permission from Elsevier.[133]

repeat measurements on a single electrode gave an r.s.d. of less than 2.9%. Addition of potential interferents, dopamine, ascorbic acid and uric acid at 1.0–5.0 mM and NaCl at 2.0–100.0 mM, resulted in changes in the response of the electrode to glucose of less than 10%.

Electrochemical sensing of glucose in diluted human serum was also claimed, with r.s.d. values of 2.21–3.58% and recoveries of 93.5–105.7%, potentially indicating the feasibility of such materials being exploited in clinical diagnosis. The authors state that these results combined indicate the good reproducibility, repeatability and stability of the electrode materials.

15.5.3 Skin-worn Chemical Sensors

Flexible organic electrodes and elastomers[141] can be applied to any number of substrates including fabric mountings[142,143] or onto the skin in the form of a tattoo.[144–146] These platforms are suitable for placement on skin surfaces

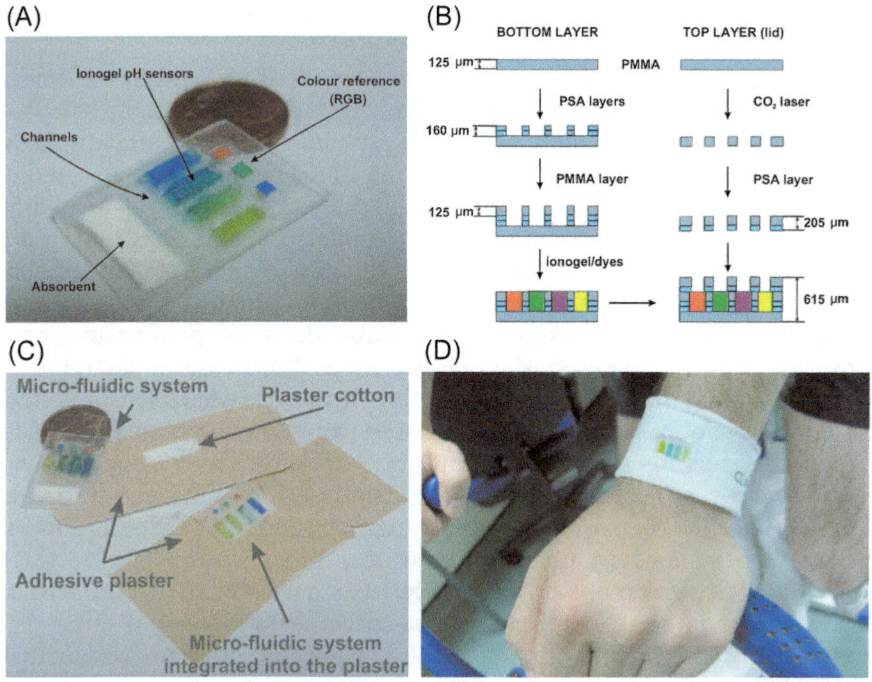

Figure 15.13 (A) Microfluidic pH sensor constructed from IL ionogel-supported pH sensitive dyes. (B) Schematic showing the construction process for the microfluidic platform. Systems for attachment of the pH sensor into (C) a bandage and (D) a wristband to achieve a prototype skin-worn device. Reprinted from *Sensors and Actuators B: Chemical*, 171–172, V. F. Curto, C. Fay, S. Coyle, R. Byrne, C. O'Toole, C. Barry, S. Hughes, N. Moyna, D. Diamond, F. Benito-Lopez, Real-time sweat pH monitoring based on a wearable chemical barcode micro-fluidic platform incorporating ionic liquids, 1327–1334, Copyright (2012) with permission from Elsevier.[143]

and in a host-based monitoring format can take advantage of biological fluids (*e.g.* sweat) to access specific biomarkers and aqueous liquids that are required for detection of specific biological events (*i.e.* enzymatic function or receptor/material binding). It is interesting to note that ILs were exploited in early forms of these wearable platforms as simple pH sensors[143] where an ionogel was used to support pH responsive gels (Figure 15.13) – a concept that had previously formed part of the ProeTEX wearable sensor system prototype[147] for health and safety monitoring of first responder communities.

Furthermore, an IL was used in one of the early skin-worn tattoo form biosensors in the form of an ionogel supported on a poly(3,3-ethylenedioxythiophene):polystyrene sulfonate (PEDOT:PSS) organic electrochemical transistor (OECT) for the detection of lactate in sweat.[148] In this example, the IL layer was composed of $[C_2C_1im][EtSO_4]$ (a water soluble IL), which was mixed with a ferrocene mediator in a solution that also contained the lactate oxidase enzymes, in a photo-cross-linked ionogel on the surface of the OECT (Figure 15.14).

A number of limitations to this system were identified within the work, including the slow response time (10 minutes). In addition, the particular form of ionogel used would have limited exploitation as a wearable sensor due to the toxicity of some of the components of the material. However, it did represent a possible starting point for ILs in these forms of sensor. Certainly the use of skin-worn sensors that exploit similar principles has moved on since this sensor was demonstrated, with the development of skin-applied sensors for a narrow group of analytes, *i.e.* ammonium, pH, sodium and lactate.[145,146] These sensors show demonstrable detection performance and consistency of function even when placed under the stresses and strains of moving skin (Figure 15.15); however, of note with this style of sensor is that it is currently not possible for this class of sensor to provide a measurement response in isolation. Instead they must be worn either with a directly connected electrochemical device or a data transmitter (shown in Figure 15.15D).

PEDOT:PSS has been used to generate highly stretchable sensor interfaces that have been shown to integrate well with skin (shown in Figure 15.15F–H).[149] Moreover, ILs combined with PEDOT:PSS have also shown properties that allow for composites with excellent electrical performance and improved stretchability. The incorporation of 1 wt% $[C_2C_1im][B(CN)_4]$ into a conductive PEDOT:PSS polymer produced a composite film with a conductivity greater than 1000 S cm^{-1}, and increased resistance up to 50% applied strain (Figure 15.16A) (pristine PEDOT:PSS has a conductivity and maximum strain of ~1 S cm^{-1} and <6%).[150] This has been achieved because inclusion of the ionic liquid causes a structural transformation of the PEDOT chains giving a fibrillar-like structure. This combination of PEDOT:PSS and IL has also shown

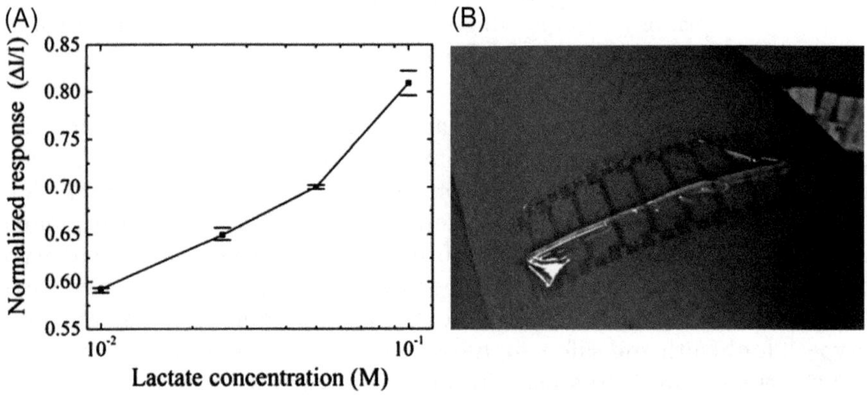

Figure 15.14 (A) Response of the IL-functionalised ionogel OECT to a change in lactate concentration (10–100 mM). (B) Illustrative positioning of the ionogel on the forearm. It should be noted that this sensor was not tested on human subjects. (Reproduced from ref. 148, with permission from The Royal Society of Chemistry.)

application as a possible actuator material, when coupled with single-walled CNTs (SWCNTs), which could allow for a wearable sensor with energy-conversion/generation built in.[151] The incorporation of IL cavities into soft elastomers has also shown application as a method of isolating a device from the skin and the stresses it can experience (Figure 15.16B).[152] An IL is used due to

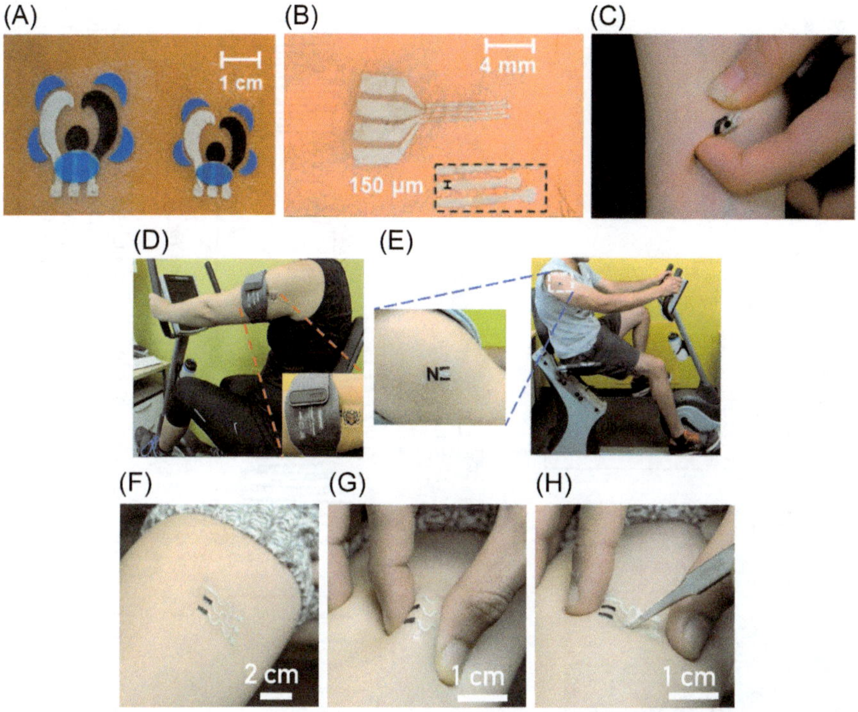

Figure 15.15 Development of organic electrode systems that can be deposited onto the skin for standard electrochemical measurements. (A) and (B) indicate the flexibility of the designs that can be printed onto the skin and (C) shows the resistance of these tattoos to twisting of the skin (Reproduced from ref. 144, with permission from The Royal Society of Chemistry). (D) Potentiometric sensor for sodium ion detection demonstrating the requirement for readout using a separate potentiostat. (Reprinted from *Biosensors and Bioelectronics*, 54, A. J. Bandodkar, D. Molinnus, O. Mirza, T. Guinovart, J. R. Windmiller, G. Valdés-Ramírez, F. J. Andrade, M. J. Schöning, J. Wang, Epidermal tattoo potentiometric sodium sensors with wireless signal transduction for continuous non-invasive sweat monitoring, 603–609, Copyright (2014), with permission from Elsevier).[145] (E) Amperometric detection of lactate production during exercise using epidermal mounted temporary tattoo sensor platforms (Reprinted with permission from ref. 146, Copyright (2013), American Chemical Society). (F – H) Deformation resilience of PEDOT-based transistors deposited onto skin. (Reproduced with permission from A. J. Bandodkar, R. Nuñez-Flores, W. Jia, J. Wang, All-Printed Stretchable Electrochemical Devices, *Advanced Materials*, John Wiley and Sons, Copyright (2016) Wiley-VCH Verlag GmbH &Co. KGaA, Weinheim).[149]

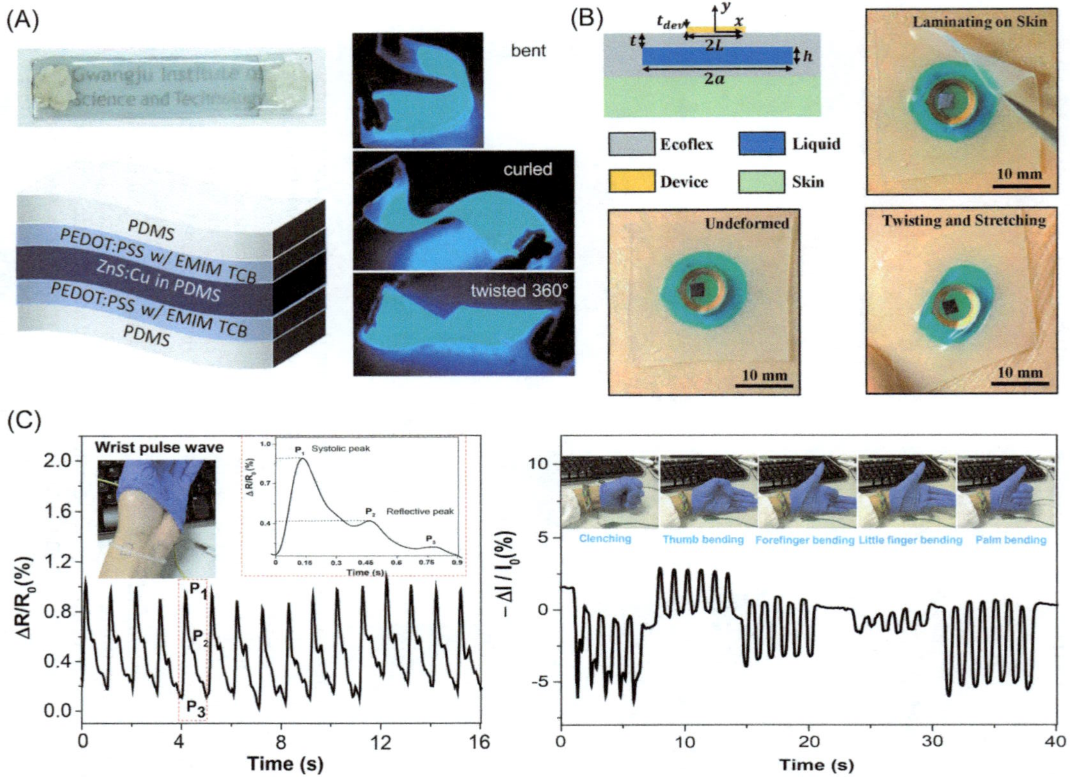

Figure 15.16 Examples of the use of ILs to enhance the properties of stretchable sensor substrates. (A) Incorporation of 1-ethyl-3-methylimidazolium tetracyanoborate into (PEDOT:PSS) to generate a polymer with a high conductivity of >1000 S cm^{-1} with stable performance at tensile strains. (Reprinted with permission from ref. 150, Copyright (2017) American Chemical Society.) (B) Use of ionic liquid-filled elastomer cavities ($[C_2C_1im][EtSO_4]$) as a strain isolating support to ensure full connectivity of inflexible sensor materials to the skin. (Reproduced with permission from Y. Ma, M. Pharr, L. Wang, J. Kim, Y. Liu, Y. Xue, R. Ning, X. Wang, H. U. Chung, X. Feng, J. A. Rogers, Y. Huang, Soft Elastomers with Ionic Liquid-Filled Cavities as Strain Isolating Substrates for Wearable Electronics, *Small*, John Wiley and Sons, Copyright (2016) Wiley-VCH Verlag GmbH &Co. KGaA, Weinheim).[152] (C) Real-time monitoring of wrist pulse (left) and hand gestures (right), using ionic liquid-filled rubber band sensors. (Reproduced from ref. 153, with permission from The Royal Society of Chemistry.)

its negligible vapour pressure and in this case, its negligible permeability through the elastomer.

This approach of using IL-filled cavities has been applied to microchannels to form IL-based wearable rubber band-like sensors.[153] The incorporation of $[C_8C_1im]Cl$ into the microchannels between two layers of Ecoflex produced a rubber band-like stretchable sensor which had durability under low and high strains (5% and 100% respectively, for >50 000 cycles), excellent long-term stability, wide strain detectability (0.1%–500%) and maintained conductivity after being stretched up to 500%. The authors targeted the design of flexible sensors, and were able to identify wrist pulses when the sensor was attached to the human wrist. This potentially offers application to chemical detection and exemplifies how ILs could improve integration of sensors directly with the human form (Figure 15.16C).

Given the obvious compatibility of ILs with these formats, it is curious that IL skin-worn sensors have not appeared at greater densities in the literature. While the $[C_8C_1im]Cl$ IL selected for use in the Ecoflex elastomer described above was chosen based on its putative biocompatibility,[152] possible toxicity issues must be considered. These issues have yet to be tackled or tested for ILs with this application in mind.[148] More generally while it is considered likely that this approach will continue to develop for biochemical targets, current sensors of this type do not currently provide a comparable performance in terms of reliability in comparison to traditional solid state electrode materials.[154] In addition, the breadth of targets is currently very limited by the need to use targets with an innate electrochemical profile (*i.e.* those with known characteristic electrochemical signatures).[142,144] Furthermore, the broad utility of skin sensors is likely to be limited by the presence of a clinically relevant analyte in sweat.

15.5.4 *In situ* Environmental Detection Using Paper-based Sensors

Initial work incorporating ILs into paper substrates began with printing of electrodes onto filter paper substrates with carbon wax inks. These inks were modified by Co(II) phthalocyanine to facilitate thiol-based detection of 1-butanethiol. An IL electrolyte was then added by simply soaking hydrophobic imidazolium ILs ($[C_4C_1im][N(Tf)_2]$, $[PF_6]^-$ and $[BF_4]^-$) into the paper to generate the functionalised electrodes.[155] In subsequent studies, improvements to the electrochemical performance of this system were made by changing the anion to a basic acetate, thus reducing the potentials required to perform oxidation of 1-butanethiol and phenol, and removing reactions from interfering species.[156]

Paper-based devices have since progressed from this point finding particular application in the analysis of analytes that do not easily present as mobile vapour or gas phases in the environment. A good example of this application is detection of explosives; many of which may be present as deposited solids, with low vapour pressures.[157] Trinitrotoluene (TNT), and associated nitroaromatic-based materials, have emerged as model analytes for the development of potential portable and future wearable solutions for explosives, and ILs have developed as promising electrolytes for *in situ* detection techniques to allow these materials to be effectively sampled and analysed in the field. Paper-based sensors have demonstrated good utility in this application, being resilient to deformation and thus capable of sampling across surface types likely to be found in real world scenarios.

The hydrophobic ionic liquids $[C_4C_1im]^+$ and $[C_4C_1pyrr]^+$ with $[N(Tf)_2]^-$ and $[FAP]^-$ anions have been characterised for their behaviour as electrolytes for TNT and dinitrotoluene (DNT) using thin layers of each IL on glassy carbon electrodes.[158] This work indicated that the IL layer could act as a pre-concentrator, preferentially pulling the TNT and DNT from the gas phase, leading to detection at estimated concentrations of 0.27 and 2.05 ppm in the gas phase for TNT and DNT, respectively. ILs $[C_4C_1im][N(Tf)_2]$, $[C_8C_1im][PF_6]$ and $[C_4C_1im][PF_6]$ have since formed the basis of novel surface sampling paper-style sensors for *in situ* analysis of DNT from surfaces.[159] These sensors utilise gold–PVdF-sandwich electrodes infiltrated with IL as an electrolyte. These materials have been shown to be effective at direct sampling and analysis of DNT from surfaces, *e.g.* plant leaves and knife blades (Figure 15.17) with a detection limit as low as 0.33 ng mm^{-2}. Technical challenges do still, however, remain in the transition of this system to real life scenarios, the most pressing being the requirement to purge each functionalised paper with nitrogen prior to electrochemical analysis.

The enhanced properties of graphene nanosheet/CNT electrodes functionalised with a functionalised IL (1-(3-aminopropyl)-3-methylimidazolium bromide) *versus* standard IL–CNT composites on glassy carbon electrodes (GCEs) and bare GCEs has previously been demonstrated for environmental

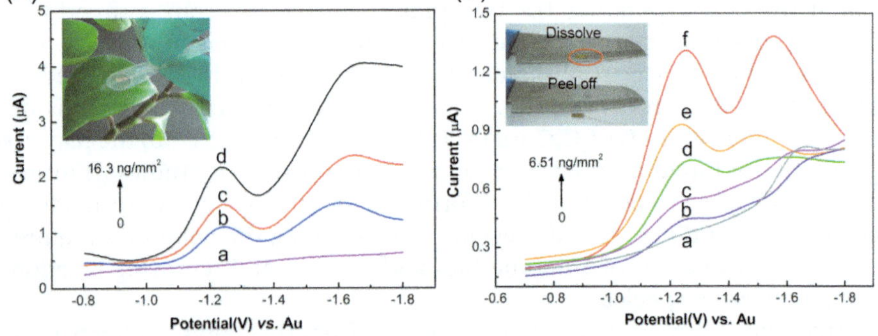

Figure 15.17 Demonstration of the detection of DNT deposited on (A) a plant leaf and (B) a knife blade using the paper-like smart sensor to solubilise deposited solid. (Reprinted with permission from ref. 159, Copyright (2014) American Chemical Society.)

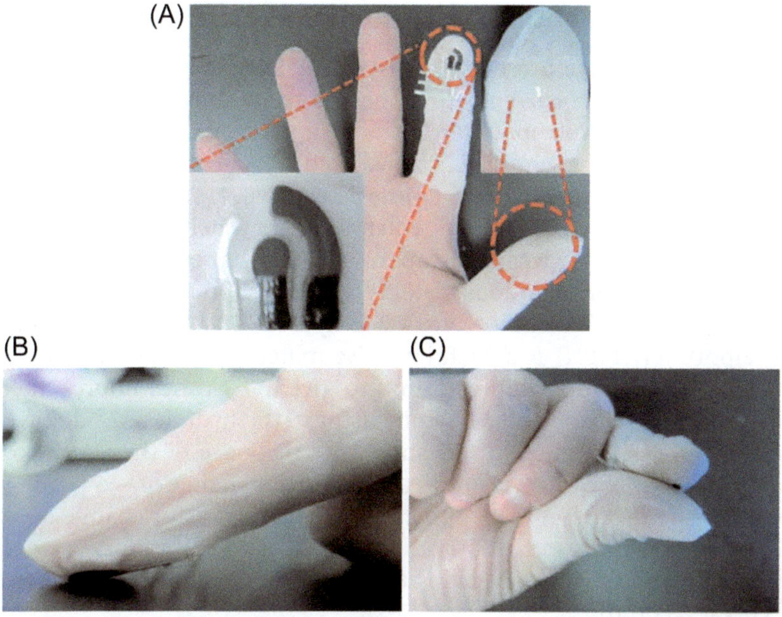

Figure 15.18 Two-component printed electrochemical cell showing (A) a printed electrode (forefinger) and solid IL-based solvent (thumb), (B) sampling of a surface onto the electrode and (C) completing the electrochemical circuit for analysis by voltammetry. (Reproduced from ref. 160, with permission from The Royal Society of Chemistry.)

sampling and analysis of TNT in ground water, tap water and lake water, without any pre-filtering or pre-treatment.[134]

In situ detection of DNT in gunshot residue has also been applied to glove-worn options. In this example, electrochemical sensors were printed from organic materials functionalised with a PEG-diacrylate (PEDGA) $[C_2C_1im][BF_4]$ ionogel on plastic/nitrile gloves to generate a contact-based detection technology for firearms residue.[160] This sensor exploits IL-based solid electrolytes (printed on the thumb) as the solvent for performing voltammetry-based electrochemical analysis using an organic printed electrode (printed on the index finger) (Figure 15.18A). Analysis is performed by sampling the surface with the electrode, then bringing the electrode into contact with the IL solvent to complete the circuit (Figure 15.18B and 15.18C). The flexibility of the electrode means that the electrochemical response can be measured with the hands in a variety of positions and while under strain.

It should be noted that the printed electrochemical sensors described here have no current intrinsic connectivity to a power/measurement source in the form pictured. This represents a current limitation to fully wearable formats, such as the glove. Connection to a potentiostat with data transmission capability such as that described for tattoo-style sensors would therefore be required for measurement to occur.[161]

15.5.5 Environmental Detection of Vapours

The sensor platforms closest to delivery of realistic solutions for testing in open environments are currently those associated with passive collection and detection of VOCs in the vapour phase. Ionic liquids form a component part of several of these systems. Zhu *et al.*,[162] have used Chemically Actuated Resonant Devices (CARDs) generated from chemoresistive SWCNTs as a platform for their system (Figure 15.19A). This work, which builds on their previous studies,[163] developed a concept for wearable vapour sensing that currently represents a realistic dosimeter for hazardous materials. This sensor uses a SWCNT/IL composite (IL: [C_4C_1im]Cl) combined with 2-(2-hydroxy-1,1,1,3,3,3-hexafluoropropyl)-1-naphthol (HFIPN) as a chemiresistive material. The addition of HFIPN to the SWCNT/IL composite was found to give a 3.3 times improvement in the irreversible response of the chemiresistor to the hydrolysis of diethylchlorophosphate (DCP). Figure 15.19B shows the response of the CARD to DCP vapour. At a 2.4 ppm DCP exposure, a chemiresistor change of ~65% min^{-1} was achieved, while detection at 28 ppb caused a chemiresistor change of ~0.4% min^{-1}. The CARD was exposed until gain readout saturation was reached, which is represented by the plateau at approximately ΔGain ≈ −1.9 dB (ΔGain ≈ −1.5 dB at 28 ppm).

The use of the CARD as a triggered smartphone binary switch was also demonstrated through repeated exposure to DCP where it successfully switched on when the protective action criterion (PAC-1) time-weighted average (TWA) of 10 ppb was exceeded. Selectivity of the sensor was established against a range of possible interferents. While not tested concurrently with the target materials, the sensor responses for the targets were at least an order of magnitude larger (Figure 15.19C), which implies that a selective response could be possible. It should be noted that the authors stringently defined the current utility of the existing platform in a practical sense and highlight the areas where substantial development is required to transition this platform towards practical chemical sensing.

15.6 Prospects for the Future of Ionic Liquids in Smart Chemical Sensors

The span of technologies, applications and devices, both optically driven and electrochemically monitored, that have been described here illustrate the influence that ILs could have in translating sensing technologies from the bench to the field or clinic into a future man-worn format. The inherent physical characteristics of these solvents make them favourable, in general, for use in liquid-based sensors and thus demonstrate their compatibility with man-worn sensor systems. In addition, inclusion of these solvents in sensor substrates, for example, in graphene-based and NP-containing electrodes, has been shown to both enhance performance and impart solvation-based selectivity towards particular analytes in sensor platforms that could be compatible with wearable sensors in future iterations.

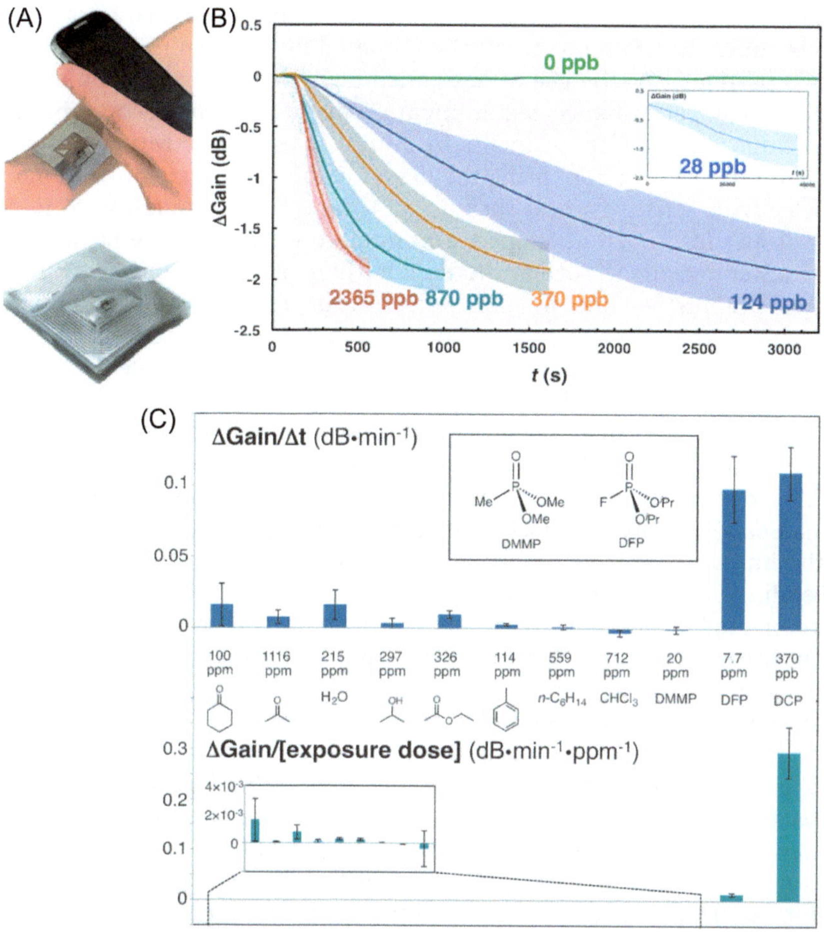

Figure 15.19 Prototype chemical dosimeter for DCP. (A) Passive sensor based on a single-walled carbon nanotube (SWCNT) chemically actuated resonant device (CARD), applied as a single-use skin-worn sticker read by a mobile phone. (B) Measurement of the saturation response of the SWCNT CARD after exposure to different concentrations of DCP, showing reduced gain associated with concentrations below 100 ppb (inset 28 ppb). (C) Sensor response curves to different concentrations of DCP as a function of time (upper) and dose (lower). (Reproduced with permission from R. Zhu, J. M. Azzarelli, T. M. Swager, Wireless Hazard Badges to Detect Nerve-Agent Simulants, *Angewandte Chemie International Edition*, John Wiley and Sons, Copyright (2016) Wiley-VCH Verlag GmbH &Co. KGaA, Weinheim).[162]

Within these examples it is clear to see that a number of IL families are commonly used. This is likely to be due to the easy availability of some of these materials from commercial providers as well as an existing understanding of behaviour and proven enhanced performance from previous studies. However, it also indicates that there is a very large number of ILs that have yet to be investigated in this context. Significant further performance enhancements could therefore potentially be garnered by broadening these studies and in particular by the use of design-led task-specific IL selection specific to particular performance requirements. This is likely to become increasingly important as improvements in overall sensor sensitivity continue to be reported and therefore bring the need for commensurate improvements in selectivity, sensitivity and specificity. Although understanding of IL solvent properties is well developed, and numerous models have been shown to be advantageous in solvent prediction, there are still additional gains to be made by further fundamental work in this area. This will undoubtedly furnish data that can be exploited positively in future sensor designs. Similar arguments may also be made for understanding the positive effects of ILs on the ordering and assembly of materials in composite electrodes.

Test criteria to enable development of sensors beyond controlled proof-of-concept studies performed in controlled laboratory environments need to be developed in parallel with the sensor platforms. The resilience of these classes of materials to environmental exposure is currently poorly validated and many of the immature sensors reported to date may be subject to baseline shifts depending on the natural background of materials in the environment. Furthermore, to be applicable to sensors on which human health might depend, a sensor would need to conform to key requirements regarding the speed of response and sensitivity required to have any utility in response to high levels of toxic challenges. In short, the sensor system must be compliant with the needs of the analyte of interest. In this context, it is clear that there are substantial technical challenges ahead to transition some of the technologies described here to real world solutions for high impact sensing requirements in public health. However, the diversity of new technologies reported in the literature exploiting the unique and tailorable properties of ILs is encouraging and the broad options available in terms of IL enabled sensors (active surface-based sampling through to full passive vapour sensors and dosimeters) implies that this will be a strength of systems developed including these solvents in the future.

References

1. S. Park, K. Chung and S. Jayaraman, in *Wearable Sensors: Fundamentals, Implementation and Applications*, ed. E. Sazonov and M. R. Neuman, Academic Press, Amsterdam, 1st edn, 2014, ch. 1.1, pp. 1–23.
2. M. D. Steinberg, P. Kassal and I. M. Steinberg, *Electroanalysis*, 2016, **28**, 1149.
3. G. Andreoni, C. Standoli and P. Perego, *Sensors*, 2016, **16**, 769.
4. P. Düking, A. Hotho, H.-C. Holmberg, F. K. Fuss and B. Sperlich, *Front. Physiol.*, 2016, **7**, 1.
5. M. H. Iqbal, A. Aydin, O. Brunckhorst, P. Dasgupta and K. Ahmed, *J. R. Soc. Med.*, 2016, **109**, 372.
6. P. Kumari, L. Mathew and P. Syal, *Biosens. Bioelectron.*, 2017, **90**, 298.

7. S. Coyle, V. F. Curto, F. Benito-Lopez, L. Florea and D. Diamond, in *Wearable Sensors: Fundamentals, Implementation and Applications*, ed. E. Sazonov and M. R. Neuman, Elsevier Academic Press, Amsterdam, 1st edn, 2014, ch. 2.1, pp. 65–83.
8. R. A. Potyrailo, N. Nagraj, C. Surman, H. Boudries, H. Lai, J. M. Slocik, N. Kelley-Loughnane and R. R. Naik, *TrAC, Trends Anal. Chem.*, 2012, **40**, 133.
9. R. A. Potyrailo, *Chem. Rev.*, 2016, **116**, 11877.
10. A. J. Bandodkar, I. Jeerapan and J. Wang, *ACS Sens.*, 2016, **1**, 464.
11. Y. Guo, T. Wang, F. Chen, X. Sun, X. Li, Z. Yu, P. Wan and X. Chen, *Nanoscale*, 2016, **8**, 12073.
12. T. Wang, Y. Guo, P. Wan, H. Zhang and X. Chen, *Small*, 2016, **12**, 3748.
13. H. H. Huang, M. Sakhdari, M. Hajizadegan, A. Shahini, D. Akinwande and P. Chen, *Appl. Phys. Lett.*, 2016, **108**, 173503.
14. B. Cho, A. R. Kim, D. J. Kim, H. Chung, S. Y. Choi, J.-D. Kwon, S. W. Park, Y. Kim, B. H. Lee, K. H. Lee, D. Kim, J. Nam and M. G. Hahm, *ACS Appl. Mater. Interfaces*, 2016, **8**, 19635.
15. T. Sekitani, H. Nakajima, H. Maeda, T. Fukushima, T. Aida, K. Hata and T. Someya, *Nat. Mater.*, 2009, **8**, 494.
16. S. Choi, H. Lee, R. Ghaffari, T. Hyeon and D. Kim, *Adv. Mater.*, 2016, **28**, 4203.
17. M. Novell, M. Parrilla, G. A. Crespo, F. X. Rius and F. J. Andrade, *Anal. Chem.*, 2012, **84**, 4695.
18. S. R. Das, Q. Nian, A. A. Cargill, J. A. Hondred, S. Ding, M. Saei, G. J. Cheng and J. C. Claussen, *Nanoscale*, 2016, **8**, 15870.
19. B. Saha, S. Baek and J. Lee, *ACS Appl. Mater. Interfaces*, 2017, **9**, 4658.
20. S. Yao and Y. Zhu, *Adv. Mater.*, 2015, **27**, 1480.
21. W. Weng, P. Chen, S. He, X. Sun and H. Peng, *Angew. Chem., Int. Ed.*, 2016, **55**, 6140.
22. T. Wang, M. Farajollahi, Y. S. Choi, I. Lin, J. E. Marshall, N. M. Thompson, S. Kar-Narayan, J. D. W. Madden and S. K. Smoukov, *Interface Focus*, 2016, **6**, 20160026.
23. Y. Ju Yun, W. G. Hong, N.-J. Choi, B. Hoon Kim, Y. Jun and H.-K. Lee, *Sci. Rep.*, 2015, **5**, 10904.
24. A. K. Yetisen, H. Qu, A. Manbachi, H. Butt, M. R. Dokmeci, J. P. Hinestroza, M. Skorobogatiy, A. Khademhosseini and S. H. Yun, *ACS Nano*, 2016, **10**, 3042.
25. J. Di, X. Zhang, Z. Yong, Y. Zhang, D. Li, R. Li and Q. Li, *Adv. Mater.*, 2016, **28**, 10529.
26. R. Zhang, Y. Chen and R. Montazami, *Materials (Basel)*, 2015, **8**, 2735.
27. Y. Lu, S. K. Das, S. S. Moganty and L. A. Archer, *Adv. Mater.*, 2012, **24**, 4430.
28. S. Uhl, E. Laux, T. Journot, J. Brossard, L. Jeandupeux and H. Keppner, *J. Energy Challenges Mech.*, 2016, **3**, 42.
29. C. F. Poole and S. K. Poole, *J. Chromatogr. A*, 2010, **1217**, 2268.

30. R. Liu, J. Liu, Y. Yin, X. Hu and G. Jiang, *Anal. Bioanal. Chem.*, 2009, **393**, 871.
31. T. D. Ho, C. Zhang, L. W. Hantao and J. L. Anderson, *Anal. Chem.*, 2014, **86**, 262.
32. N. V. Plechkova and K. R. Seddon, *Chem. Soc. Rev.*, 2008, **37**, 123.
33. I. J. Villar-Garcia, K. R. J. Lovelock, S. Men and P. Licence, *Chem. Sci.*, 2014, **5**, 2573.
34. M. G. Freire, P. J. Carvalho, A. M. S. Silva, L. M. N. B. F. Santos, L. P. N. Rebelo, I. M. Marrucho and J. A. P. Coutinho, *J. Phys. Chem. B*, 2009, **113**, 202.
35. M. G. Freire, P. J. Carvalho, R. L. Gardas, I. M. Marrucho, L. M. N. B. F. Santos and J. A. P. Coutinho, *J. Phys. Chem. B*, 2008, **112**, 1604.
36. T. Kakiuchi, *Anal. Sci.*, 2008, **24**, 1221.
37. D. W. Armstrong, L. He and Y. S. Liu, *Anal. Chem.*, 1999, **71**, 3873.
38. M. H. Abraham, *Chem. Soc. Rev.*, 1993, **22**, 73.
39. T. O. Kollie, C. F. Poole, M. H. Abraham and G. S. Whiting, *Anal. Chim. Acta*, 1992, **259**, 1.
40. M. N. Kobrak, *Green Chem.*, 2008, **10**, 80.
41. S. Bruzzone, M. Malvaldi and C. Chiappe, *Phys. Chem. Chem. Phys.*, 2007, **9**, 5576.
42. E. D. Bates, R. D. Mayton, I. Ntai and J. H. Davis, *J. Am. Chem. Soc.*, 2002, **124**, 926.
43. Á. Pérez-Salado Kamps, D. Tuma, J. Xia and G. Maurer, *J. Chem. Eng. Data*, 2003, **48**, 746.
44. P. Scovazzo, J. Kieft, D. A. Finan, C. Koval, D. DuBois and R. Noble, *J. Membr. Sci.*, 2004, **238**, 57.
45. J. Jacquemin, P. Husson, V. Majer and M. F. Costa Gomes, *Fluid Phase Equilib.*, 2006, **240**, 87.
46. Y. Chen, S. Zhang, X. Yuan, Y. Zhang, X. Zhang, W. Dai and R. Mori, *Thermochim. Acta*, 2006, **441**, 42.
47. M. J. Muldoon, S. N. V. K. Aki, J. L. Anderson, J. K. Dixon and J. F. Brennecke, *J. Phys. Chem. B*, 2007, **111**, 9001.
48. G. Hong, J. Jacquemin, M. Deetlefs, C. Hardacre, P. Husson and M. F. Costa Gomes, *Fluid Phase Equilib.*, 2007, **257**, 27.
49. M. B. Shiflett and A. Yokozeki, *J. Phys. Chem. B*, 2007, **111**, 2070.
50. N. M. Yunus, M. I. A. Mutalib, Z. Man, M. A. Bustam and T. Murugesan, *Chem. Eng. J.*, 2012, **189–190**, 94.
51. J. Jacquemin, M. F. Costa Gomes, P. Husson and V. Majer, *J. Chem. Thermodyn.*, 2006, **38**, 490.
52. J. L. Anderson, J. K. Dixon and J. F. Brennecke, *Acc. Chem. Res.*, 2007, **40**, 1208.
53. J. L. Anthony, E. J. Maginn and J. F. Brennecke, *J. Phys. Chem. B*, 2002, **106**, 7315.
54. J. L. Anthony, J. L. Anderson, E. J. Maginn and J. F. Brennecke, *J. Phys. Chem. B*, 2005, **109**, 6366.
55. D. Camper, J. Bara, C. Koval and R. Noble, *Ind. Eng. Chem. Res.*, 2006, **45**, 6279.

56. Q. Gan, D. Rooney, M. Xue, G. Thompson and Y. Zou, *J. Membr. Sci.*, 2006, **280**, 948.
57. P. Scovazzo, D. Havard, M. McShea, S. Mixon and D. Morgan, *J. Membr. Sci.*, 2009, **327**, 41.
58. J. Kumełan, Á. P.-S. Kamps, I. Urukova, D. Tuma and G. Maurer, *J. Chem. Thermodyn.*, 2005, **37**, 595.
59. Z. Lei, C. Dai and B. Chen, *Chem. Rev.*, 2014, **114**, 1289.
60. C. L. Arthur and J. Pawliszyn, *Anal. Chem.*, 1990, **62**, 2145.
61. M. Andre, J. Loidl, G. Laus, H. Schottenberger, G. Bentivoglio, K. Wurst and K. H. Ongania, *Anal. Chem.*, 2005, **77**, 702.
62. F. Liu and Y. Jiang, *J. Chromatogr. A*, 2007, **1167**, 116.
63. G. Laus, M. Andre, G. Bentivoglio and H. Schottenberger, *J. Chromatogr. A*, 2009, **1216**, 6020.
64. M. Ni, T. Sun, L. Zhang, Y. Liu, M. Xu and Y. Jiang, *J. Chromatogr. B*, 2014, **945–946**, 60.
65. G. Von Wald, D. Albers, H. Cortes and T. McCabe, *J. Chromatogr. A*, 2008, **1201**, 15.
66. D. Ge and H. K. Lee, *J. Chromatogr. A*, 2012, **1229**, 1.
67. X. Ma, M. Huang, Z. Li and J. Wu, *J. Hazard. Mater.*, 2011, **194**, 24.
68. Y. Zhang and H. K. Lee, *Anal. Chim. Acta*, 2012, **750**, 120.
69. J.-F. Peng, J.-F. Liu, X.-L. Hu and G.-B. Jiang, *J. Chromatogr. A*, 2007, **1139**, 165.
70. D. W. Armstrong, L. Zhang, L. He and M. L. Gross, *Anal. Chem.*, 2001, **73**, 3679.
71. Y. L. Li and M. L. Gross, *J. Am. Soc. Mass Spectrom.*, 2004, **15**, 1833.
72. J. A. Crank and D. W. Armstrong, *J. Am. Soc. Mass Spectrom.*, 2009, **20**, 1790.
73. P. Sun and D. W. Armstrong, *Anal. Chim. Acta*, 2010, **661**, 1.
74. J. L. Anderson and D. W. Armstrong, *Anal. Chem.*, 2005, **77**, 6453.
75. Q. Zhao and J. L. Anderson, *J. Sep. Sci.*, 2010, **33**, 79.
76. C. Yao and J. L. Anderson, *J. Chromatogr. A*, 2009, **1216**, 1658.
77. Q. Q. Baltazar, S. K. Leininger and J. L. Anderson, *J. Chromatogr. A*, 2008, **1182**, 119.
78. J. L. Anderson and D. W. Armstrong, *Anal. Chem.*, 2003, **75**, 4851.
79. S. Risticevic, H. Lord, T. Górecki, C. L. Arthur and J. Pawliszyn, *Nat. Protoc.*, 2010, **5**, 122.
80. B. Bojko, E. Cudjoe, G. A. Gómez-Ríos, K. Gorynski, R. Jiang, N. Reyes-Garcés, S. Risticevic, É. A. S. Silva, O. Togunde, D. Vuckovic and J. Pawliszyn, *Anal. Chim. Acta*, 2012, **750**, 132.
81. T. D. Ho, A. J. Canestraro and J. L. Anderson, *Anal. Chim. Acta*, 2011, **695**, 18.
82. F. Bianchi, F. Bisceglie, S. Dugheri, G. Arcangeli, V. Cupelli, E. del Borrello, L. Sidisky and M. Careri, *J. Chromatogr. A*, 2014, **1331**, 1.
83. C. Liang, C. Y. Yuan, R. J. Warmack, C. E. Barnes and S. Dai, *Anal. Chem.*, 2002, **74**, 2172.
84. X. Jin, L. Yu, D. Garcia, R. X. Ren and X. Zeng, *Anal. Chem.*, 2006, **78**, 6980.

85. I. Levitsky, *Sensors*, 2015, **15**, 19968.
86. Y. Shang, H. Zhang, X. Wang and J. Wu, *Chem.–Eur. J.*, 2011, **17**, 13400.
87. S. Trivedi, S. G. Coombs, D. V. Wagle, N. Bhawawet, G. A. Baker and F. V. Bright, *Chem.–Eur. J.*, 2016, **22**, 11677.
88. H. Zhang, X. Wang and J. Wu, *Small*, 2012, **8**, 3775.
89. H. Zhang, L. Lin, D. Liu, Q. Chen and J. Wu, *Anal. Chim. Acta*, 2017, **953**, 71.
90. X. Zhu, D. Liu, Q. Chen, L. Lin, S. Jiang, H. Zhou, J. Zhao and J. Wu, *Chem. Commun.*, 2016, **52**, 3042.
91. M.-C. Tseng and Y.-H. Chu, *Chem. Commun.*, 2010, **46**, 2983.
92. Y.-L. Liu, M.-C. Tseng and Y.-H. Chu, *Chem. Commun.*, 2013, **49**, 2560.
93. H.-Y. Li, T.-H. Hsu, C.-Y. Chen, M.-C. Tseng and Y.-H. Chu, *Analyst*, 2015, **140**, 6245.
94. C.-H. Wang, C.-H. Yang and J.-K. Chang, *Biosens. Bioelectron.*, 2017, **89**, 483.
95. W. Zhu, W. Li, H. Yang, Y. Jiang, C. Wang, Y. Chen and G. Li, *Chemistry*, 2013, **19**, 11603.
96. A. J. Bandodkar and J. Wang, *Trends Biotechnol.*, 2014, **32**, 363.
97. A. Paul, P. K. Mandal and A. Samanta, *J. Phys. Chem. B*, 2005, **109**, 9148.
98. S. Cha, T. Shim, Y. Ouchi and D. Kim, *J. Phys. Chem. B*, 2013, **117**, 10818.
99. P. Nockemann, K. Binnemans and K. Driesen, *Chem. Phys. Lett.*, 2005, **415**, 131.
100. T. Kanai, in *Organic and Hybrid Photonic Crystals*, ed. D. Comoretto, Springer International Publishing, Cham, 2015, pp. 431–450.
101. Z. Cai, N. L. Smith, J.-T. Zhang and S. A. Asher, *Anal. Chem.*, 2015, **87**, 5013.
102. J. Huang, C. Tao, Q. An, C. Lin, X. Li, D. Xu, Y. Wu, X. Li, D. Shen and G. Li, *Chem. Commun.*, 2010, **46**, 4103.
103. S. Furumi, T. Kanai and T. Sawada, *Adv. Mater.*, 2011, **23**, 3815.
104. T. Kanai, S. Yamamoto and T. Sawada, *Macromolecules*, 2011, **44**, 5865.
105. N. L. Smith, Z. Hong and S. A. Asher, *Analyst*, 2014, **139**, 6379.
106. D. Yang, Y. Wang, L. He and H. Li, *ACS Appl. Mater. Interfaces*, 2016, **8**, 19709.
107. J.-H. Shin, W. A. Henderson and S. Passerini, *Electrochem. Commun.*, 2003, **5**, 1016.
108. *Electrochemical Aspects of Ionic Liquids*, ed. H. Ohno, John Wiley & Sons, Inc., Hoboken, NJ, USA, 2nd edn, 2011.
109. M. Hayyan, F. S. Mjalli, M. A. Hashim, I. M. AlNashef and T. X. Mei, *J. Ind. Eng. Chem.*, 2013, **19**, 106.
110. P. A. Z. Suarez, V. M. Selbach, J. E. L. Dullius, S. Einloft, C. M. S. Piatnicki, D. S. Azambuja, R. F. de Souza and J. Dupont, *Electrochim. Acta*, 1997, **42**, 2533.
111. N. V. Ignat'ev, U. Welz-Biermann, A. Kucheryna, G. Bissky and H. Willner, *J. Fluorine Chem.*, 2005, **126**, 1150.
112. D. Bejan, N. Ignat'ev and H. Willner, *J. Fluorine Chem.*, 2010, **131**, 325.

113. L. Suo, O. Borodin, T. Gao, M. Olguin, J. Ho, X. Fan, C. Luo, C. Wang and K. Xu, *Science*, 2015, **350**, 938.
114. H. Kim, J. Hong, K.-Y. Park, H. Kim, S.-W. Kim and K. Kang, *Chem. Rev.*, 2014, **114**, 11788.
115. *Ionic Liquids in Synthesis*, ed. P. Wasserscheid and T. Welton, Wiley-VCH Verlag GmbH & Co. KGaA, Weinheim, Germany, 2nd edn, 2008.
116. M. Galiński, A. Lewandowski and I. Stepniak, *Electrochim. Acta*, 2006, **51**, 5567.
117. J. Gębicki, A. Kloskowski, W. Chrzanowski, P. Stepnowski and J. Namiesnik, *Crit. Rev. Anal. Chem.*, 2016, **46**, 122.
118. G. Korotcenkov, *Handbook of Gas Sensor Materials: Properties, Advantages and Shortcomings for Applications Volume 2: New Trends and Technologies*, Springer, New York, NY, 2014, pp. 121–130.
119. L. C. Clark Jr, R. Wolf, D. Granger and Z. Taylor, *J. Appl. Physiol.*, 1953, **6**, 189.
120. N. Baltes, F. Beyle, S. Freiner, F. Geier, M. Joos, K. Pinkwart and P. Rabenecker, *Talanta*, 2013, **116**, 474.
121. Z. Wang, M. Guo, G. A. Baker, J. R. Stetter, L. Lin, A. J. Mason and X. Zeng, *Analyst*, 2014, **139**, 5140.
122. Z. Wang and X. Zeng, *J. Electrochem. Soc.*, 2013, **160**, H604.
123. Z. Wang, P. Lin, G. A. Baker, J. Stetter and X. Zeng, *Anal. Chem.*, 2011, **83**, 7066.
124. H. Li, X. Mu, Y. Yang and A. J. Mason, *IEEE Sens. J.*, 2014, **14**, 3391.
125. X. Wang, Y. He, L. Lin, F. Zeng and T. Luan, *Sci. Total Environ.*, 2014, **470–471**, 263.
126. J. Ping, J. Wu and Y. Ying, *Electrochem. Commun.*, 2010, **12**, 1738.
127. J. Ping, Y. Wang, K. Fan, J. Wu and Y. Ying, *Biosens. Bioelectron.*, 2011, **28**, 204.
128. P. Kuberský, T. Syrový, A. Hamáček, S. Nešpůrek and L. Syrová, *Sens. Actuators, B*, 2015, **209**, 1084.
129. C. Shan, H. Yang, D. Han, Q. Zhang, A. Ivaska and L. Niu, *Biosens. Bioelectron.*, 2010, **25**, 1504.
130. M. Du, T. Yang, S. Ma, C. Zhao and K. Jiao, *Anal. Chim. Acta*, 2011, **690**, 169.
131. P. Jing, X. Zhang, Z. Wu, L. Bao, Y. Xu, C. Liang and W. Cao, *Talanta*, 2015, **141**, 41.
132. Y. Sun, H. Zheng, C. Wang, M. Yang, A. Zhou and H. Duan, *Nanoscale*, 2016, **8**, 1523.
133. W. He, Y. Sun, J. Xi, A. A. M. Abdurhman, J. Ren and H. Duan, *Anal. Chim. Acta*, 2016, **903**, 61.
134. S. Guo, D. Wen, Y. Zhai, S. Dong and E. Wang, *Biosens. Bioelectron.*, 2011, **26**, 3475.
135. A. Wong, T. A. Silva, F. C. Vicentini and O. Fatibello-Filho, *Talanta*, 2016, **161**, 333.
136. C.-H. Wang, C.-H. Wu, J.-W. Wu, M.-T. Lee, J.-K. Chang, M.-D. Ger and C.-L. Sun, *Analyst*, 2013, **138**, 576.

137. X. Wang and J. Hao, *Sci. Bull.*, 2016, **61**, 1281.
138. J. Wang, L. Angnes and T. Martinez, *Bioelectrochem. Bioenerg.*, 1992, **29**, 215.
139. Z. He and P. Alexandridis, *Adv. Colloid Interface Sci.*, 2017, **244**, 54.
140. Y. Sun, Y. Cheng, K. He, A. Zhou and H. Duan, *RSC Adv.*, 2015, **5**, 10178.
141. C. Liao, M. Zhang, M. Y. Yao, T. Hua, L. Li and F. Yan, *Adv. Mater.*, 2015, **27**, 7493.
142. P. Kuberský, J. Altšmíd, A. Hamáček, S. Nešpůrek and O. Zmeškal, *Sensors*, 2015, **15**, 28421.
143. V. F. Curto, C. Fay, S. Coyle, R. Byrne, C. O'Toole, C. Barry, S. Hughes, N. Moyna, D. Diamond and F. Benito-Lopez, *Sens. Actuators, B*, 2012, **171–172**, 1327.
144. J. R. Windmiller, A. J. Bandodkar, G. Valdés-Ramírez, S. Parkhomovsky, A. G. Martinez and J. Wang, *Chem. Commun.*, 2012, **48**, 6794.
145. A. J. Bandodkar, D. Molinnus, O. Mirza, T. Guinovart, J. R. Windmiller, G. Valdés-Ramírez, F. J. Andrade, M. J. Schöning and J. Wang, *Biosens. Bioelectron.*, 2014, **54**, 603.
146. W. Jia, A. J. Bandodkar, G. Valdés-Ramírez, J. R. Windmiller, Z. Yang, J. Ramírez, G. Chan and J. Wang, *Anal. Chem.*, 2013, **85**, 6553.
147. D. Curone, E. L. Secco, A. Tognetti, G. Loriga, G. Dudnik, M. Risatti, R. Whyte, A. Bonfiglio and G. Magenes, *IEEE Trans. Inf. Technol. Biomed.*, 2010, **14**, 694.
148. D. Khodagholy, V. F. Curto, K. J. Fraser, M. Gurfinkel, R. Byrne, D. Diamond, G. G. Malliaras, F. Benito-Lopez and R. M. Owens, *J. Mater. Chem.*, 2012, **22**, 4440.
149. A. J. Bandodkar, R. Nuñez-Flores, W. Jia and J. Wang, *Adv. Mater.*, 2015, **27**, 3060.
150. M. Y. Teo, N. Kim, S. Kee, B. S. Kim, G. Kim, S. Hong, S. Jung and K. Lee, *ACS Appl. Mater. Interfaces*, 2017, **9**, 819.
151. N. Terasawa and K. Asaka, *Langmuir*, 2016, **32**, 7210.
152. Y. Ma, M. Pharr, L. Wang, J. Kim, Y. Liu, Y. Xue, R. Ning, X. Wang, H. U. Chung, X. Feng, J. A. Rogers and Y. Huang, *Small*, 2016, 1602954.
153. Y. Wang, S. Gong, S. J. Wang, G. P. Simon and W. Cheng, *Mater. Horiz.*, 2016, **3**, 208.
154. P. Kuberský, A. Hamáček, S. Nespůrek, R. Soukup and R. Vik, *Sens. Actuators, B*, 2013, **187**, 546.
155. N. Dossi, R. Toniolo, A. Pizzariello, E. Carrilho, E. Piccin, S. Battiston and G. Bontempelli, *Lab Chip*, 2012, **12**, 153.
156. R. Toniolo, N. Dossi, A. Pizzariello, A. Casagrande and G. Bontempelli, *Anal. Bioanal. Chem.*, 2013, **405**, 3571.
157. B. C. Dionne, D. P. Rounbehler, E. K. Achter, J. R. Hobbs and D. H. Fine, *J. Energ. Mater.*, 1986, **4**, 447.
158. C. Xiao, A. Rehman and X. Zeng, *Anal. Chem.*, 2012, **84**, 1416.
159. J. Wang, W. Jin, X. Zhang, C. Hu, Q. Luo, Y. Lin and S. Hu, *Anal. Chem.*, 2014, **86**, 8383.

160. A. J. Bandodkar, A. M. O'Mahony, J. Ramírez, I. A. Samek, S. M. Anderson, J. R. Windmiller and J. Wang, *Analyst*, 2013, **138**, 5288.
161. M. D. Steinberg, P. Kassal, I. Kereković and I. M. Steinberg, *Talanta*, 2015, **143**, 178.
162. R. Zhu, J. M. Azzarelli and T. M. Swager, *Angew. Chem., Int. Ed.*, 2016, **55**, 9662.
163. J. M. Azzarelli, K. A. Mirica, J. B. Ravnsbæk and T. M. Swager, *Proc. Natl. Acad. Sci.*, 2014, **111**, 18162.

CHAPTER 16

Ionic Electrochemical Actuators

A. MAZIZ*[a], A. SIMAITE[b] AND C. BERGAUD*[a]

[a]LAAS-CNRS, Université de Toulouse, CNRS, 7 Avenue du Colonel Roche, F-31031, Toulouse, France; [b]Mechanical, Materials, Aerospace Engineering Department, Illinois Institute of Technology, IL 60616, Chicago, USA
*E-mail: maziz@laas.fr, bergaud@laas.fr

16.1 Introduction

There is growing interest in robots working with or in the close vicinity of humans. Making them compliant is one of the biggest challenges. For the next generation of robots, adapted control strategies will need to be combined with lightweight materials, actuators, sensors and power sources. For example, traditional actuators have low power density, power-to-mass ratio and efficiency. These and other properties are limiting the creation of dynamic micro-robots. Furthermore, their cost, bulkiness and lack of integrated tactile or force feedback are limiting their applications in medicine.

Even though very limited at the moment, electroactive polymers (EAPs) have the potential to replace prevalent actuators, where a compliant muscle-like response is desirable. EAP actuators are categorized into two classes: electric actuators mainly based on the use of dielectric elastomers (DEs) and ionic actuators mainly based on ionic polymer–metal composites (IPMCs) and conducting polymer actuators (CPAs).

DEAs are typically made of a passive elastomer film (*e.g.* silicone, polyurethane, polyacrylate) that is sandwiched between two compliant electrodes. When a voltage is applied, the electrostatic pressure between the electrodes

Smart Materials No. 29
Polymerized Ionic Liquids
Edited by Ali Eftekhari

Published by the Royal Society of Chemistry, www.rsc.org

(Maxwell stress), arising due to the Coulomb forces, compresses the thickness and expands the area of the elastomer in-between. DE actuators come in a variety of shapes and sizes, driving *e.g.* small robots.[1,2] Nevertheless, DEAs require high electric fields (150 MV m^{-1}) and, consequently, voltages up to 5 kV; therefore their application in medicine and other fields that require contact with humans is very limited.[3,4]

Ionic EAP actuators are most promising for biomedical applications, where biocompatibility, compactness and accurate positioning are essential. They are driven by low potentials, typically 1–2 V, but need an ion source/sink. This can be an ionic liquid embedded in a polymer matrix making the technology work in air.[5] Compared to DEAs, they produce low forces and moderate strains. However, as ionic EAP actuators are driven by low potentials, they operate silently. Moreover, being polymers, they are lightweight, potentially cheap to mass fabricate and can be used in the vicinity of human beings, *i.e.* for applications in medical devices.[6–8] Another major advantage of ionic EAP actuators is their compactness, which reduces the overall weight of the devices and simplifies their handling and use in various environments making them especially interesting for applications in soft robotics.[9–11] Nevertheless, the development of applicable devices is held back by their low efficiency and the absence of tactile, force or position feedback. Another feature limiting the development of functional devices is their fast performance deterioration, often due to the evaporation of the solvent. A major breakthrough in the development of actuators was the discovery of ionic liquids used as an electrolyte, which prolonged the lifetime of the actuators by several orders of magnitude.[12,13]

A variety of other emerging ionic EAP actuator technologies have also been reported, including ionic gels, ionic polymer–metal composites (IPMCs) and more recently, carbon nanomaterial-based composites. Several types of ionic electromechanical actuators were investigated during the last decades to design artificial muscles for prosthetics (hands and arms) or soft robotics (grippers, manipulators, *etc.*)[10,11,14] but also to develop optical microsystems[15] and in the biomedical field (minimally invasive surgical and diagnostic tools).[6–8]

16.1.1 Ionic Gels

Active hydrogels are three-dimensional polymer networks in which the voids are filled with water. They can undergo a volume change in response to an environmental stimulus and already have various applications in biotechnology and medicine.[16–18] As an emerging method for producing artificial muscles, ionic gels could potentially match the force and the energy density of a skeletal muscle at low voltages. In active ionic gels, the reversible volume or shape change is a response to a chemical reaction, *e.g.* pH oscillations[19,20] (Figure 16.1) through diffusion of a reactive substrate into the gel.[21] Because the flow of the solvent into the gel matrix is a diffusion-controlled process, the response time of ionic gels is limited to several seconds. Nevertheless,

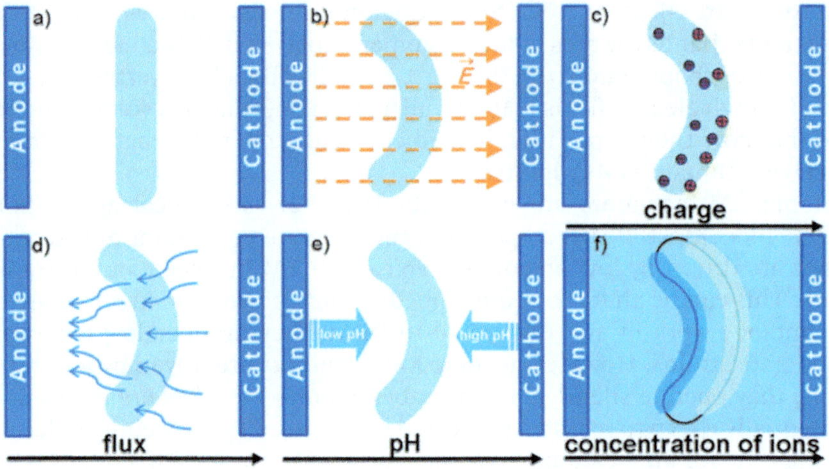

Figure 16.1 Actuation of an electroresponsive polyelectrolyte gel placed in salt solution and schematic illustration of historically proposed models to explain electroactuation. Rod-like gel (a) before and (b) after applying electric potential. Electroactuation mechanisms: (c) Coulomb mechanism, (d) electroosmosis mechanism, (e) electrochemical mechanism, and (f) dynamic enrichment/depletion mechanism (darker and lighter colours at the gel/solution boundaries represent ion accumulation and depletion at the anode and cathode side, respectively). Reproduced from ref. 19 with permission from The Royal Society of Chemistry.

fast gels could be made of thin films but at the expense of the force.[22] In order to make active gels more practical, the speed of their stimuli responsiveness and their weak mechanical properties need to be improved.[23,24] Furthermore, robust electroding techniques need to be developed that will not damage the gel surface.

16.1.2 Ionic Polymer–Metal Composites

IPMCs are made of a semipermeable polymer membrane sandwiched between two metal electrodes. These membranes are usually ion-conducting polyelectrolytes, having a backbone chain of perfluorinated alkane terminated with ionic groups, *e.g.* sulfonate (SO_3^-) for Nafion® or carboxylate (COO^-) for Flemion®. These ion groups form hydrophilic cluster networks, where the solvent and the mobile counter-ions accumulate. Once an electric field is applied, the cations along with the solvent move towards one of the electrodes, causing swelling and a fast bending motion (Figure 16.2) (more sophisticated actuation mechanisms are reported in ref. 25–28). Factors influencing the electrochemomechanical response of the actuators, such as (1) the chemical composition and structure of the polymer,[29,30] (2) the morphology of the electrodes,[31] (3) the nature of the mobile ions,[30,32,33] (4) the solvent saturation,[34] *etc.*, are rather well

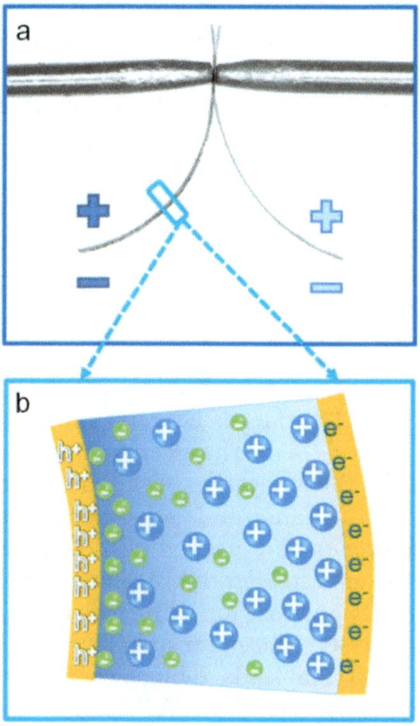

Figure 16.2 (a) Optical image of the bending actuation of an IPMC actuator under AC electric signals. (b) Schematic of the bending actuation mechanism. Reproduced with permission from ref. 26.

understood and are extensively reviewed.[35–37] Due to their low actuation voltage and relatively large bending, IPMCs have potential applications in underwater robotics (*e.g.* grippers,[38] swimming devices[39–41]) and medicine (*e.g.* microgrippers,[42] steerable catheters).[7]

When used with aqueous solvents, one of their drawbacks is back-relaxation – the actuation is followed by a slow relaxation due to the water diffusion out from the cation-rich area. In addition, IPMCs are sensitive to dehydration and hydrolysis above 1.23 V, and tend to drift in position or get permanently deformed when a direct current is applied. Furthermore, the lifetime of IPMCs is rather limited,[43] primarily because of:

- Gradual water evaporation during cycling.[44,45] The evaporation of water in air could be avoided if ionic liquids were used as the electrolytes[32] but at the cost of the response speed and strain. Alternatively, devices could be encapsulated in *e.g.* parylene,[46,47] polydimethylsiloxane (PDMS)[48] or other materials.[49,50] Nevertheless, because of encapsulation, the actuator stiffness increases and this leads to a decrease of the deflection amplitude.

- Poor adhesion of the metal electrodes to the membrane. In order to ensure sufficient adhesion between the membrane and the electroplated metals, expensive and time-consuming surface roughening techniques are required.[51,52] Therefore, alternative adhesion improvement methods, *e.g.* polymer coating[53,54] and nano-powder casting,[55] have been investigated.
- Damage of the electrode surface.[44,56] In order to decrease the electrode cracking and degradation during bending cycles[54,57] development of fabrication techniques, *e.g.* introducing novel electrodes,[58,59] is necessary.

16.1.3 Carbon Nanotubes

The most recent ionic actuator type is based on carbon nanotubes (CNTs) and their composites suspended in electrolytes[60] or ILs.[61] When a voltage is applied (1–4 V), the CNT surface is charged and the electrolytes form an electric double layer around them. The electrostatic repulsion of the charges on the nanotubes causes the elongation of the carbon–carbon bonds, which consequently elongates the nanotube.[60] Due to their stiffness, networks of entangled nanotubes or yarns are needed to induce macroscopic deformations, and even then, strains smaller than 1% are achievable.[60] On the other hand, due to the porosity and fast ion diffusion, the strain rate of the CNT actuators can reach 19% s^{-1} (<10 ms response time).[62] Furthermore, their high elastic modulus (640 GPa) leads to huge work densities of 108 J m^{-3}.

The discovery of the 'bucky gel'[63] was followed by its application in actuators.[64] 'Bucky gel' actuators are similar to IPMCs in their trilayer structure: the polymer impregnated with the ionic liquid in-between two electrodes made of a gelatinous ionic liquid containing single-walled carbon nanotubes (SWCNTs). Nevertheless, the motion is caused by charging of the SWCNTs and subsequently the ion transport within and to the gel. The main advantages of the 'bucky gel' actuators are their facile fabrication (layer-by-layer casting) and fast actuation times (responding to up to 100 Hz) (Figure 16.3). Nevertheless, just like SWCNT actuators, they are difficult and expensive to

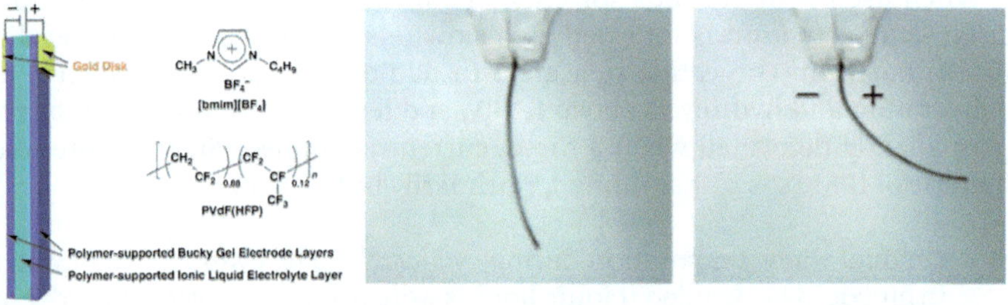

Figure 16.3 (Left) Schematic configuration of a bucky-gel-actuator; (Right) bending motion of the corresponding actuator (3.5 V, 0.01 Hz). Reproduced from T. Fukushima and T. Aida, Ionic Liquids for Soft Functional Materials with Carbon Nanotubes, *Chemistry - A European Journal*,[70] John Wiley and Sons, © 2007 WILEY-VCH Verlag GmbH & Co. KGaA, Weinheim.

mass-produce.[65–67] As a cheaper alternative, carbon black and carbon fiber mixtures with ILs are being investigated.[68] However, the strains remain low, less than 0.01% for a frequency of 100 Hz.[69]

16.1.4 Conducting Polymers

CPs are also a class of ionic electroactive materials widely investigated for the development of lightweight, noiseless, high stress, large strain and low driving voltage actuators. In CPs, ion transport occurs during the electrochemical oxidation and reduction of the polymer. This redox change results in a change of material properties such as conductivity,[71] color,[72] wettability[73] and shape or volume.[74] Indeed, when the CP is oxidized or reduced electrochemically, ions and solvent molecules are inserted or expelled from the CP in order to ensure the overall electroneutrality leading to a variation of the CP volume. This behavior makes them particularly interesting in the family of electroactive polymers for actuator applications.[4] Moreover, considering that they are biocompatible, down-sizable and can present large deformations, they are the most promising candidates for applications in medicine and implantable devices.[75]

It is also worth mentioning the growing interest in using composite material electrodes along with conducting polymers to improve the electromechanical properties of ionic conducting polymer actuators (CPAs). Alternatives rely on the use of carbon-conducting polymer composites[76,77] or the use of CNT mixtures with nanoparticles.[78] Torop *et al.* described an *in situ* synthesis of nanoporous carbide-derived carbon (CDC) powder and pyrrole monomers using electrochemical polymerization.[76] They have demonstrated that for iEAP actuators, the amount of swelling per injected charge is double for PPy-CDC films compared to non-modified PPy actuators.

16.2 Volume Change in Ionic Conducting Polymers

As mentioned above, CPs are reactive materials under a flow of electric current. An electrical energy input changes the electrochemical state of the conjugated polymer and is followed by several simultaneous and subsequent processes, *e.g.* volume changes of the material. Indeed, when the CP is oxidized or reduced electrochemically, ions and solvent molecules are inserted or expelled from the CP in order to ensure the overall electroneutrality. In 1995, R. Gandhi *et al.* summarized the different volume variations of a material in two different redox reactions and the accompanying ion flows (Figure 16.4).[79]

A complete volumetric expansion mechanism will depend essentially on the nature and size of the ions involved, and on their solvation state. In a general sense, the ion flow and redox reactions can be described as follows. For a polymer P doped with small or medium-sized mobile dopant anions (A^-), such as perchlorate (ClO_4^-) anions, in contact with an electrolyte containing both mobile cations and anions, the redox reaction is mainly controlled by

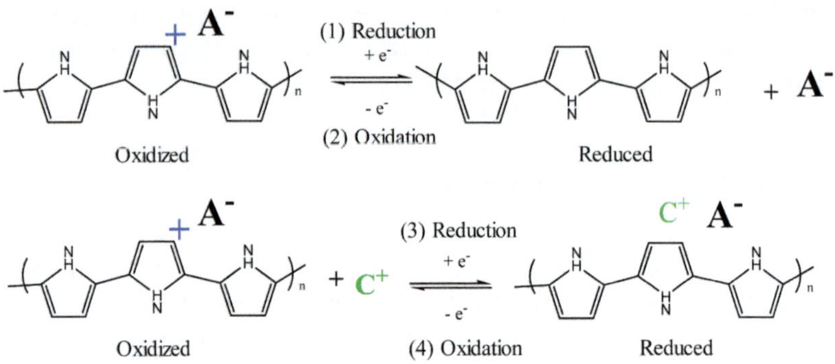

Figure 16.4 Molecular mechanisms of volume variation in conducting polymers.

mechanisms (1) and (2) (Figure 16.4). That is, when reducing the polymer to its neutral state, anions A^- are expelled and when oxidizing the polymer, anions are inserted into the polymer matrix in order to compensate for the charge imbalance.[80–83] On the other hand, for a polymer P doped with large immobile dopant anions A^-, such as dodecylbenzenesulphonate (DBS^-), poly(styrene sulfonate) (PSS) or trifluoromethanesulfonate ($CF_3SO_3^-$) anions, in contact with an electrolyte containing small mobile cations C^+, the redox reaction is mainly controlled by mechanisms (3) and (4) (Figure 16.4). That is, cations C^+ are inserted when the polymer is reduced and expelled when the polymer is oxidized.[84–86] In the former case, the volume typically expands in the oxidized state, *i.e.* when a positive potential is applied, and in the latter case the volume of the polymer expands in the reduced state, *i.e.* when a negative potential is applied. In the former case, however, there may be two moving species because not only redox reactions 1 and 2 (Figure 16.4) occur, but reactions 3 and 4 (Figure 16.4) may also occur, which can lead to a "twitching" or "salt draining behavior", *i.e.* the neutral pairs of ions could form and move out from the polymer, leading to a decrease in expansion.[87] Therefore, CPs doped with large immobile anions, such as DBS or PSS, have frequently been employed since they provide a smooth motion (with only cations as the moving species), stability, and long life-time. Ions in the electrolyte are usually solvated and their actual volume and mobility depend on the solvation shell. Therefore, solvents can be critical in determining the mobile species. Furthermore, osmotic effects that lead to solvent molecules moving in and out of the polymer were also shown to play a significant role in actuation.[13] In addition, conformational changes and coulombic repulsion of the polymer chains may also contribute to the volume change.[88,89]

16.3 Synthesis of Conducting Polymers

The properties of conducting polymers are strongly dependent on their synthesis conditions. To achieve a good conductivity, it is required that the conjugated nature (Figure 16.5) of the monomer is conserved in the repeating

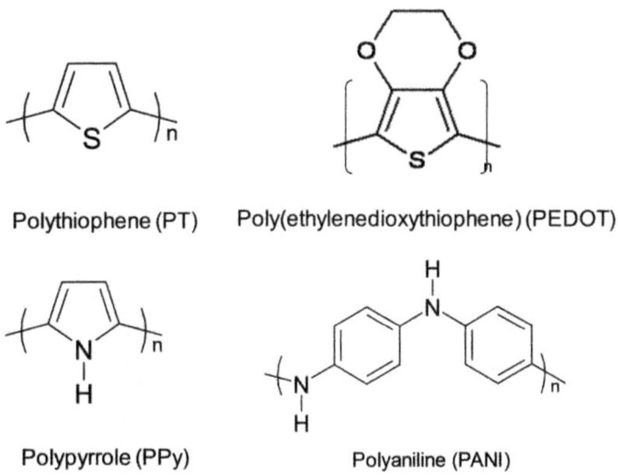

Figure 16.5 The chemical structures of some common conjugated polymers used as actuator materials.

unit during the polymerization process, which strongly limits both the choice of monomer and the choice of the deposition pathway. The polymerization of designated conducting polymer monomers can be carried out in many ways. However, the most widely used technique is oxidative coupling involving the oxidation of monomers *via* chemical or electrochemical routes.[90–93]

Its simplicity, high selectivity and reproducibility have made electrochemical synthesis the preferred method to synthesize conducting polymers for actuator applications. It has the advantage of producing the material on an electrode facilitating further analysis, and hence eliminating processability problems. Moreover, the electrochemical route has the advantage of allowing good control of the polymer thickness, morphology and degree of polymer doping by mastering the synthesis parameters, such as the quantity of charge injected during the deposition process. The electrosynthesis is usually achieved using a two- or three-electrode configuration through the electro-oxidation of a given monomer (*e.g.* pyrrole or 3,4-ethylenedioxythiophene) in a solution containing a supporting electrolyte (Figure 16.6). This may be an inert organic solvent, such as propylene carbonate with *e.g.* bis(trifluoromethane)sulfonimide lithium (LiTFSI), an aqueous electrolyte, typically sodium dodecylbenzenesulfonate (NaDBS), or an ionic liquid. Indeed, Pringle *et al.* reported the use of the ionic liquids 1-butyl-3-methylimidazolium hexafluorophosphate, 1-ethyl-3-methylimidazolium bis(trifluoromethanesulfonyl) amide and *N*,*N*-butylmethylpyrrolidinium bis(trifluoromethanesulfonyl) amide, both as the growth medium and as an electrolyte for the electrochemical cycling of polypyrrole films. The use of the ionic liquid as the growth medium results in significantly altered film morphologies and enhanced electrochemical activities compared to films grown in conventional molecular solvent/electrolyte systems.[94]

As mentioned, CPs are deposited on conducting substrates, such as gold, stainless steel, platinum, indium-tin oxide (ITO)-coated glass or more recently CP-coated textiles.[95] The electrochemical polymerization process can be

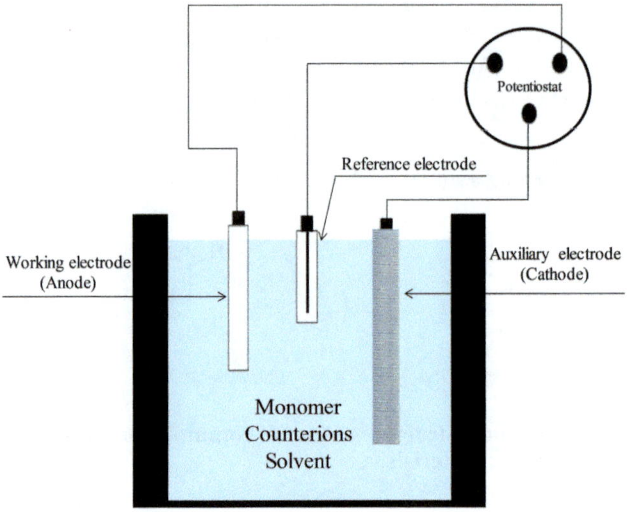

Figure 16.6 A three-electrode electrochemical polymerization cell. The sample, for instance a stainless steel substrate, is connected as the working electrode.

performed using different ways of applying the current or the potential to the conducting substrate. The simplest electrochemical methods include potentiostatic or galvanostatic methods (constant potential and constant current, respectively), but potentiodynamic methods have also been employed.[12,76,96]

In some applications, it is either impractical or undesirable to synthesize the polymer on a conducting surface. In these cases, CPs can often also be synthesized using chemical oxidative polymerization. Distinct advantages of the chemical route over electrochemical synthesis are that there is no need for electrochemical instruments such as a potentiostat and that non-conductive surfaces can also be coated. Moreover, chemical polymerization is known to be a simple and fast method and is recommended if large amounts of CP are needed. In typical chemical polymerization, an oxidizing reagent such as iron(III) chloride is used in aqueous or non-aqueous solvents[97,98] to produce a polymer in the conducting form. Although other oxidizing agents can be used as well,[99] iron(III) chloride was found to be one of the most efficient oxidants for pyrrole or EDOT polymerization because of the high polymerization yield. In chemical synthesis, however, it is difficult to gain direct control over the polymer deposition, which results in a lack of accuracy and reproducibility.

An alternative chemical synthesis method to conventional solution-based CP deposition is the so-called vapour phase polymerization (VPP) technique.[100–104] There is much interest in this method because it provides high control over CP film thickness, uniformity and density. Typically, the formation of CPs is carried out directly on the surface of the substrate in a two-step process (Figure 16.7). First, the oxidant such as iron tosylate in *n*-butanol is

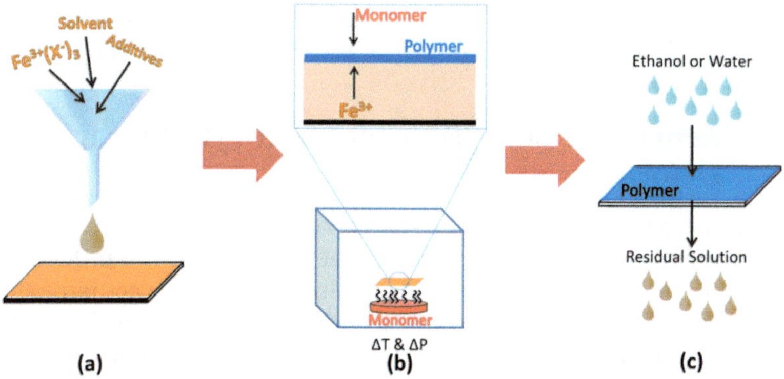

Figure 16.7 The VPP process involving (a) deposition of the oxidant solution (onto the substrate, black) commonly containing an Fe^{3+} salt in a solvent (typically an alcohol such as ethanol) with possible additives, (b) exposure of said oxidant to monomer vapour at a given temperature (T) and pressure (P), where the oxidant/monomer is transported/condensed at the interface to initiate polymerization, and (c) washing away excess oxidant and monomer to yield an ICP thin film. Reprinted from *Progress in Materials Science*, 86, R. Brooke, P. Cottis, P. Talemi, M. Fabretto, P. Murphy, D. Evans, Recent advances in the synthesis of conducting polymers from the vapour phase, 127–146, Copyright (2017) with permission from Elsevier.[111]

applied on the substrate using solvent coating processes and then the coated surface is exposed to a reactive CP monomer vapour. PEDOT films have been reported to have conductivities as high as approximately 3400 S cm^{-1} for a thickness of 60 nm, which is equivalent to commercially available indium tin oxide (ITO).[105] The ability to fabricate CP thin films directly onto substrates opened up the opportunity to integrate them into practical devices. For example, Maziz *et al.* have recently employed VPP PEDOT in ionic actuator devices and electroactive fabrics for robotics.[95,106]

CP processing using commonly employed deposition methods such as spin-coating, solvent-casting or ink-jet printing is virtually impossible since these CPs are relatively insoluble in most solvents. To overcome this problem, alternative synthetic routes have been proposed that involve the attachment of soluble functional groups to the polymer or doping with stabilizing polyelectrolytes.[107–110] An aqueous dispersion of PEDOT:PSS is commercially available under the brand names Baytron P or Clevios P.

16.4 Ionic Electromechanical Actuators Based on Conducting Polymers

As previously mentioned, CPs can undergo dimensional changes in response to an electric charge, and consequently they are capable of transducing electrochemical energy directly into mechanical work. The latter can be converted into a variety of forms (Figure 16.8), *i.e.* bending, linear or out-of-plane deformation according to the different device architectures reviewed below.

16.4.1 Actuators Immersed in an Electrolyte: Linear Deformation

The easiest way to design an electromechanical actuator system based on CPs is to shape the active material as a freestanding film to induce a linear deformation (Figure 16.9). Freestanding film, wire or fibre actuators provide access to the intrinsic actuator characteristics, such as strain, force and generated stress. In all cases, the characterizations of the system are done in an electrochemical cell allowing both isotonic and isometric measurements, including strain, force and stress under electrical stimulation.[112] PPy is the most studied CP in a linear configuration since it is quite easy to prepare PPy films by an electrochemical synthesis process. The typical values for strain and stress are around 2% and 5 MPa, respectively.[113,114] Nonetheless, PPy actuators prepared using specific conditions *e.g.* by varying the electrochemical synthesis conditions and/or the choice of the dopant, can show large strains of up to 30%.[115–118] However, for these systems, the electrochemomechanical stability remains a critical issue for practical applications. A PPy actuator film doped with TFSI generated 24.5% strain in the first cycle, whereas only 12.4% was measured in the 7th cycle.[117] The reduction

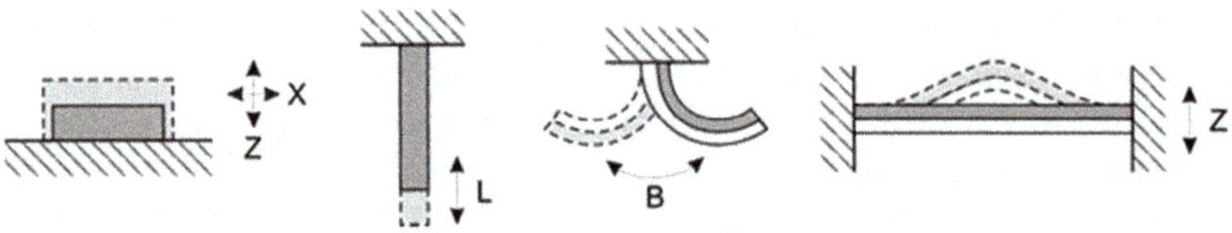

Figure 16.8 Different actuation modes used for conducting polymer actuators: bulk or perpendicular expansion, linear strain, bending bilayer, and buckling sheet.

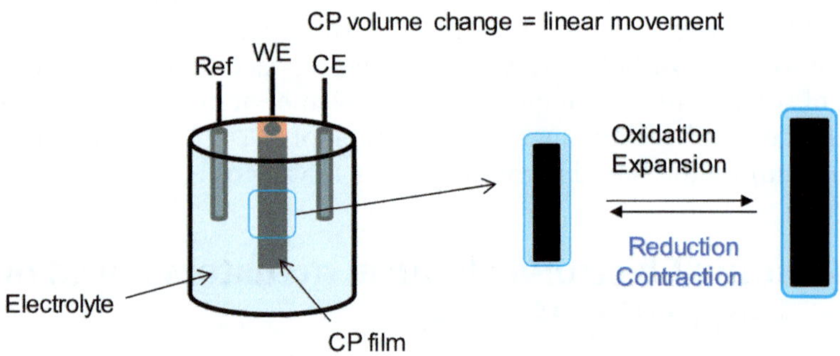

Figure 16.9 Architectures of CPAs: linear deformation actuators immersed in an electrolyte.

in performance might be caused by parasitic reactions, *i.e.* over oxidation – side reactions of dissolved oxygen in driving electrolyte solutions leading to structural changes along the PPy backbone and hence to a lower capacity to generate movement after several cycles. In this context, Kaneto *et al.* reported the use of ILs as the electrolytic solution, which results in a significant increase in stability with increasing cycle number; however, in such cases, the strain is less than 5% and the ability of the PPy film to undergo large deformation is hampered.[119]

Even though CPA actuators generate high stress, the exerted forces are typically low. The actuation mechanism of CPAs is dominated by mass transfer, including ions and solvents into the polymer. Therefore, the actuation speed will be determined by the diffusion rate and diffusion distance.[13,84,120–122] To keep a moderate actuation speed, only thin layers or fibres of CPs are used, and hence, the exerted forces remain rather small. Higher actuation forces were demonstrated by assembling CP fibers into macrodevices that increase their thickness and/or cross-sectional area (Figure 16.10).[112,122] The prototype shown in Figure 16.10(A) is a CPA assembled with bundles of PPy deposited on nickel-plated stainless steel coils. This actuator configuration lifted 22 kg when immersed in aqueous $NaBF_4$ solution. Maziz *et al.* proposed another approach by developing textile actuators. Advanced textile technology allows

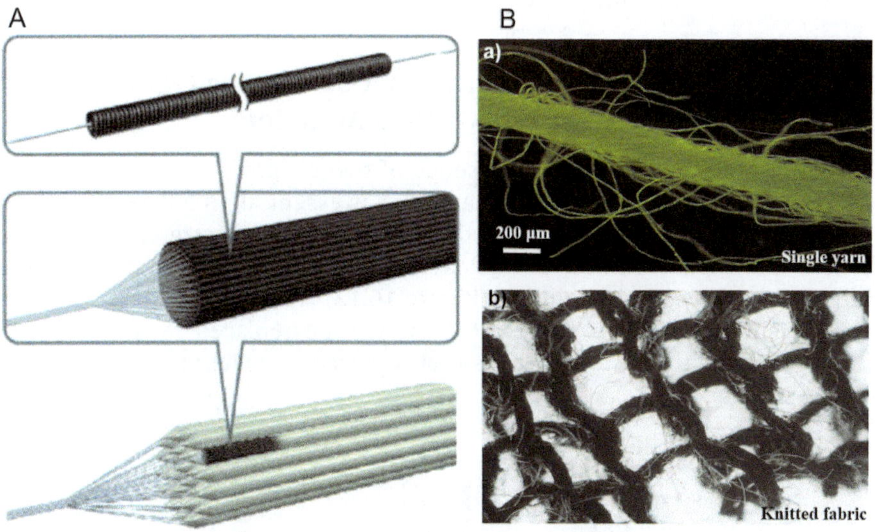

Figure 16.10 (A): Schematic diagrams of some bundles of PPy–metal coil actuators. One piece of PPy–metal coil composite exhibits only 0.20 N force, but a bundle and a bundle of bundles, *etc.* can create huge force (reprinted from *Synthetic Metals*, 146, S. Hara, T. Zama, W. Takashima and K. Kaneto, Polypyrrole-metal coil composite actuators as artificial muscle fibres, 47–55, Copyright (2004), with permission from Elsevier).[123] (B): Textile actuators. (a) Lyocell-based single yarn. (b) CP-coated Lyocell-based knitted fabric.

rational parallel assembly of fibres/yarns, and this increases the total force of the actuators (Figure 16.10).[96] Soft actuators were produced from cellulose yarns assembled into fabrics and coated with CPs. When immersed in propylene carbonate with LiTFSI, the generated force scaled linearly with the number of yarns in the fabric, while conserving the strain and keeping the advantages of single, thin yarns, that is, a high surface-to-volume ratio.

16.4.2 Bilayer Bending Actuators

The volume variation during the redox process can also be used to get out-of-plane movement by designing the actuator in a bilayer configuration. Bilayer actuators are laminates of conducting polymers and passive or "non-volume changing" thin films, in which the relative expansion or contraction of the conducting polymer with respect to the other layer leads to a bending of the structure (Figure 16.11). In these structures, the CP film is connected as a working electrode in a three-electrode configuration cell in the presence of the electrolytic solution. The volume increase due to the oxidation (or reduction) of the CP layer generates a stress gradient at the CP/passive layer interface causing the bending of the actuator. The first design of bilayer actuators was proposed by Baughman *et al.*[60,124] Since then, these types of actuators have been extensively studied by T.F. Otero *et al.*,[120,125,126] Q. Pei *et al.*,[84,127] K. Kaneto *et al.*[128,129] and E. Smela *et al.*[130–132]

16.4.3 Trilayer Bending Actuators: Use of Ionic Liquids as Electrolytes for Air Working Actuators

For CPAs operating in air, the presence of two electroactive electrodes and an ion reservoir from which ions can move is essential. In a three-layer configuration, active CPs are laminated with an inactive substrate. The relative differential expansion between CP layers generates a stress gradient at the interface and results in bending (Figure 16.12). To operate trilayer CPAs in air, the substrate between the CP layers can be replaced by a membrane impregnated with an electrolyte as first reported by MacDiarmid in 1994.[133]

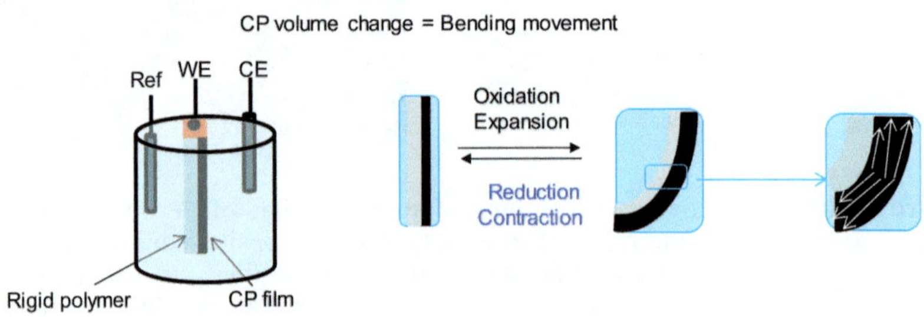

Figure 16.11 Architectures of CPAs: bilayer bending actuators immersed in an electrolyte. Reprinted with permission from ref. 146. Copyright 2015 American Chemical Society.

The configuration was based on a piece of paper, acting as an ion reservoir membrane, two PANI electrodes and hydrochloric acid as an electrolyte. Thereafter, more classical solid polymer electrolytes (SPEs) have been used as the ion reservoir for the fabrication of trilayer actuators. In this case, requirements of SPEs are much more difficult to fulfil compared to other solid-state electrochemical devices, such as batteries or supercapacitors. The stringent requirements for a SPE to be used in an electromechanical actuator include adequate ionic conductivity and electrical stability. It must also be flexible, present good mechanical properties, and maintain good adhesion with the electroactive layers during bending. SPEs are based on either poly(epichlorohydrin-*co*-ethylene oxide),[134] nitrile butadiene rubber,[135] or polyethylene oxide (PEO),[136] or elaborated using more complex macromolecular architectures, such as an interpenetrating polymer network (IPN). SPE IPNs are based on PEO and either polycarbonate,[136] polybutadiene[137] polytetrahydrofurane[138] or nitrile butadiene rubber.[139,140] In SPE IPNs, one polymer network is really used as the SPE, *i.e.* the PEO partner, whereas the second one provides the required mechanical properties for the device. Porous polyvinylidene fluoride (PVDF) membranes have also been used, with the porosity of the membrane being filled by the electrolytic solution.[141]

Table 16.1 shows the molecular structures of the most widely employed polymer electrolytes for the development of CP actuators.

In most of these approaches, the use of aqueous or organic electrolytic solutions strongly limits the potential applications of the ionic actuators. Aqueous or organic electrolytic solutions evaporate during actuation, decreasing the ionic conductivity within the membrane and ultimately the actuators stop working due to the lack of ionic movement. Moreover, water electrolysis severely restricts the actuation voltage to less than 1.23 V at room temperature, which limits the strain that the actuators can generate.[144] Various polymer-based encapsulating layers were investigated in an attempt to minimize or eliminate the electrolyte evaporation, but none of them was successful.[134,137,145]

A major breakthrough in the development of actuators was the discovery of room temperature ILs, also known as molten salts, used as an electrolyte. ILs have excellent physical and electrochemical properties, such as high ionic conductivity, non-volatility and larger electrochemical window.

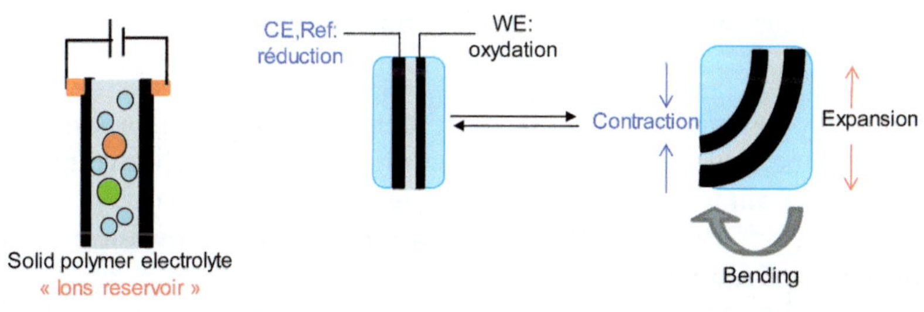

Figure 16.12 Architectures of CPAs: trilayer bending actuators operating in air.

Table 16.1 Molecular structures of the most widely employed solid polymer electrolytes for conducting polymer actuator applications.

SPE Membrane	Molecular structure	Electrode	Electrolyte	Ref.
Poly(epichlorohydrin-*co*-ethylene oxide)	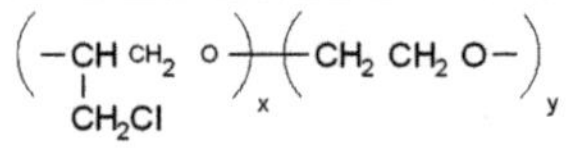	PPy	$LiClO_4$/water solution	133
Nitrile butadiene rubber	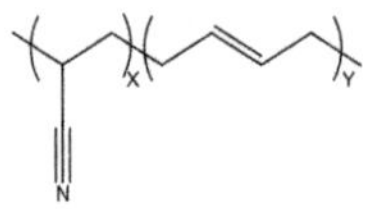	PEDOT	EMITFSI, $BMMPF_6$ $BMIMBF_4$, BMIMTFSI	134
Branched polyethylene oxide (PEO)	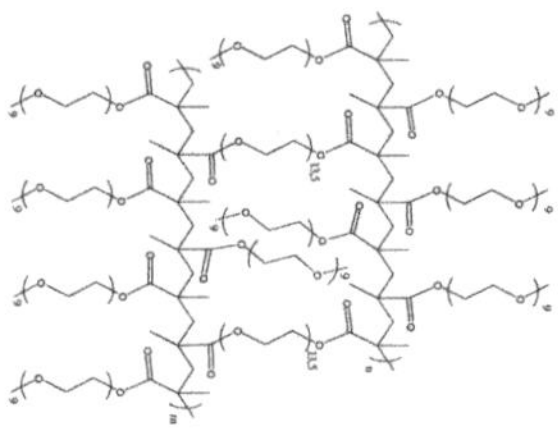	PEDOT	$LiCLO_4$/water solution	135
Poly(vinylidene fluoride) (PVdF)	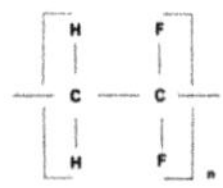	PPy	LiTFSI/PC solution	136,137
Poly (vinylidene fluoride-*co*-hexafluoropropylene) (PVDF-*co*-HFP)	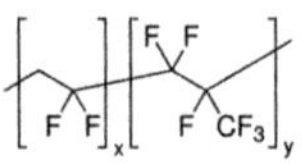	PPy	LiTFSI/PC solution	137
Poly(methyl methacrylate) (PMMA)	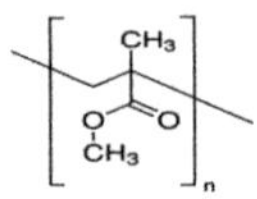	PPy	Tetraethylammoniumhexafluorophosphate (TEAH°/PC solution)	114
Polycarbonate[a]	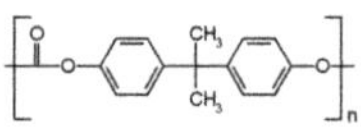	PEDOT	$LiCLO_4$/water solution	135
Polybutadiene[a]	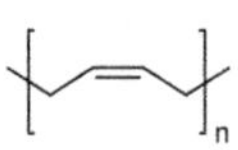	PEDOT	EMITFSI	138
Polytetrahydrofuran[a]	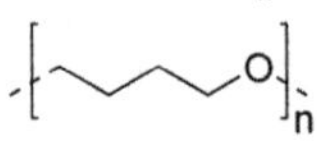	PEDOT	EMITFSI	139

[a]The SPE was associated with a PEO partner as an IPN architecture.

Non-volatility prolongs the lifetime of the actuators by several orders of magnitude.[12,32] Due to their larger electrochemical window, higher voltages can be applied without leading to an irreversible electrochemical deterioration of the actuators.[32,140] Moreover, ILs can be tailored to exhibit specific properties for various applications.[28]

Table 16.2 shows the molecular structures of the most widely employed ILs as electrolytes for the development of CPAs. W. Lu *et al.* used ILs for the first

Table 16.2 Molecular structures of the most widely employed ILs as electrolytes for CP actuator applications.

Name	Abbreviation	Molecular Structure	Ref.
1-Butyl-3-methylimidazolium bis(trifluoromethyl sulfonyl)imide	BMITFSI	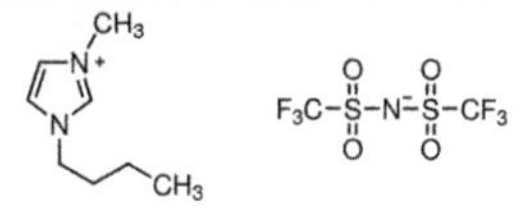	134
1-Ethyl-3-methylimidazolium bis(trifluoromethylsulfonyl)imide	EMI TFSI	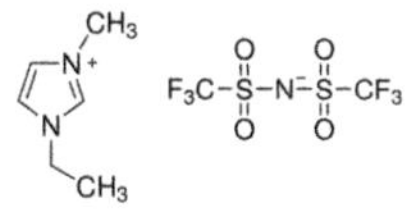	145
1-Butyl-3-methylimidazolium hexafluorophosphate	BMIM PF_6	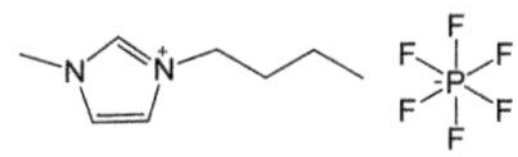	141
1-Butyl-3-methyl imidazolium tetrafluoroborate	$BMIMBF_4$	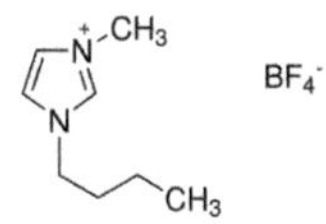	141
1-Butyl-1-methylpyrrolidinium bis(trifluoromethylsulfonyl)imide	BMP TFSI	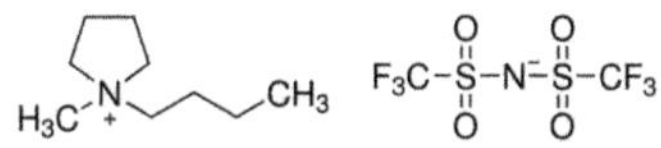	119
Methyl-trioctylammonium bis(trifluoromethylsulfonyl)imide	MTOATFSI	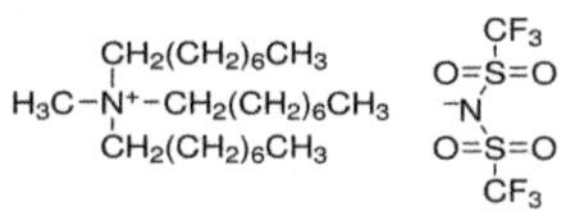	119
Triethylsulfonium bis(trifluoromethylsulfonyl) imide	TESTFSI	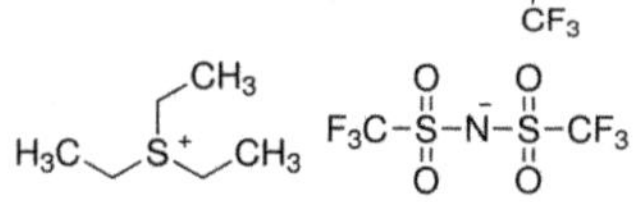	119

time in 2002 for actuator applications.[141] The polymer-in-ionic liquid electrolytes were prepared by using PANI and PPy both with BMIM(PF_6) and EMI-MTFSI. The electrochemomechanical actuators showed a maximum strain of ±2% when a potential of ±2 V was applied. The stability of this system was evaluated by subjecting the actuators to continuous pulsing for more than 3600 cycles without degradation.[5,142] The cyclic lifetime and stability of the trilayer actuators were significantly enhanced compared to those containing propylene carbonate-based electrolytes.[5,138] The presence of ILs in polymer electrolytes has also proved to enhance the switching speeds and to increase the amplitude of actuation.[143] The types of ILs play a critical role in the actuating properties of ionic polymer-based actuators. Mattes and Lu observed enhanced electroactivity and mechanical actuation in a $BMIMBF_4$ electrolyte after doping PANI with CF_3SO_3H, suggesting a strong interaction between the $CF_3SO_3^-$ dopant of the conducting polymer and the BMIM cation. This actuator showed an excellent lifetime of over 100000 cycles between +2.5 V and −2.5 V at 2 Hz.[144]

Another major drawback for the fabrication of very long-life air working actuators is the delamination at the interface between the conducting polymer layers and the flexible SPE.[151] The delamination problem has been overcome *via* three different strategies (Figure 16.13). First, a porous

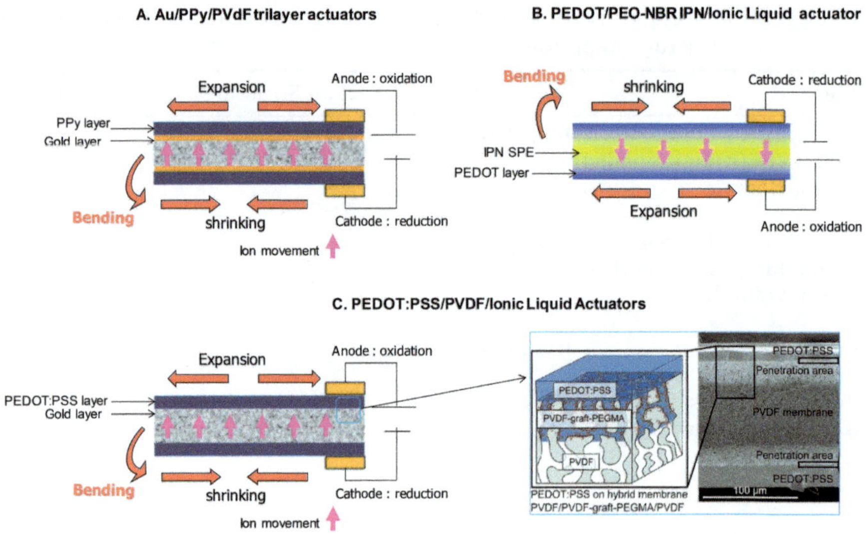

Figure 16.13 Air-operating trilayer systems based on CPs. (A) Au/PPy/PVDF actuators: a porous PVDF membrane onto which two PPy electrodes are electrochemically deposited. (B) PEDOT/PEO-NBR IPN actuators: the trilayer system is shaped as a conducting IPN, where two PEDOT electrodes are chemically embedded through the thickness of the IPN. (C) PEDOT:PSS/PVDF-*graft*-PEGMA actuators: the PEDOT:PSS water solution fills the pores of the functionalized PVDF membrane and a mixing layer between PEDOT:PSS and PVDF is created. Reprinted with permission from ref. 146. Copyright (2015) American Chemical Society.

commercial PVDF membrane was used as the electrolyte storage layer and electrochemical and/or chemical deposition of the CP working material was performed.[142,152] The CP is anchored in the pores of the PVDF membrane, which creates an interfacial layer between the membrane and the electrode. The second strategy was proposed by designing actuators such as conducting IPNs where a CP is embedded in an IPN SPE. These two architectures, combined with ILs as electrolytes, allow the cyclic life-time and the electrochemical stability of actuators to be enhanced. Operations in air of up to 3600 (1 Hz) to 106 (10Hz) cycles were reported.[150] These performance advantages were obtained by using the environmentally stable, room temperature ionic liquid 1-ethyl-3-methylimidazolium bis(trifluoromethanesulfonyl) imide (EMITFSI) (Table 16.2). C. Bergaud *et al.* proposed another approach by developing hybrid ion-storing membranes.[146,154] The wetting properties of the surface of a PVDF membrane were improved using argon-plasma-induced surface polymerization of poly(ethylene glycol) monomethyl ether methacrylate (PEGMA). Hybrid membranes were created with hydrophilic

PVDF-*graft*-PEGMA outer surfaces and hydrophobic bulk. The width of each layer is controlled by spray coating, as this technique allows for the deposition of the reaction precursor to a certain depth. Subsequently, a PEDOT:PSS water solution fills the pores of the functionalized part of the membrane and a mixing layer between PEDOT:PSS and PVDF is created. Finally, PEDOT:PSS/PVDF-*graft*-PEGMA/PEDOT:PSS actuators are made by simple solution casting. They are capable of producing high strains of 0.6% in the EMITFSI ionic liquid, and show no signs of delamination after more than 150 h or 10^4 actuation cycles at 0.1 Hz.

16.4.4 Creeping Effects

Creeping in CPAs is observed as a continuous displacement of the movement range.[147] It originates from the incomplete recovery of the initial position on consecutive actuation cycles. There are two identified reasons causing such a behavior: (1) mechanical creeping, *i.e.* cycling-induced solvent swelling as suggested by Melling *et al.*,[148] and (2) chemical creeping, *i.e.* unbalanced charging as characterized by Kaneto *et al.*[149,150] In the first case, cycling opens the polymer matrix so that with each cycle the ions diffuse faster and therefore deeper into the film. Consequently, if present, more solvent can be irreversibly accommodated in-between the polymer chains.[148,149,151,152] Concerning chemical creeping in aqueous solutions, faster actuation rates during both reduction[148] and oxidation[153] were previously observed. This was attributed to the higher electronic conductivity in the oxidized state and the higher ion conductivity because of the more expanded, open and swollen polymer. For polypyrrole (PPy) actuators operating in propylene carbonate, J. D. Madden *et al.* have shown a correlation between a reduction in strain amplitude and a decrease in charge transfer that could suggest an irreversible degradation of the polymer.[147] Kaneto *et al.* have compared the swelling of PPy films in ILs and mixed solutions of ILs and propylene carbonate (PC).[149] The electrochemomechanical strain (ECMS) was studied as a function of the stress induced by applied tensile loads. It was shown that if ILs are used, creeping only takes place at high tensile stresses. However, in a mixed solution of IL and PC, a decrease of the ECMS was rapidly observed at lower stresses after a few electrochemical cycles. This was explained as a loss of the electrochemical activity and the conductivity of the PPy films. Recently, the use of carbon nanotubes spray-coated onto the surface of a PEDOT:PSS actuator was shown to significantly reduce creeping by avoiding charge accumulation in the polymer film (Figure 16.14). This led to a significantly improved performance during cycling actuation.[154] Modeling conductive polymer actuators as viscoelastic materials could be useful to more accurately predict creeping. An advanced mathematical model was, for example, proposed by Nguyen *et al.* for this purpose for trilayer conjugated polymer actuators.[173]

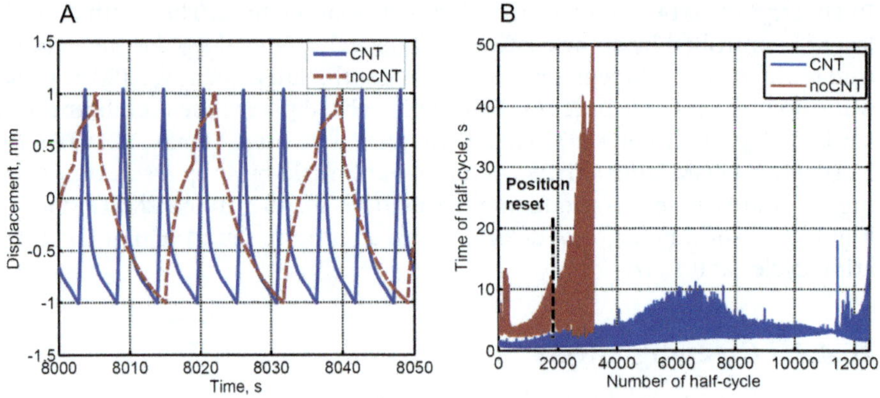

Figure 16.14 (A) Bending profiles *versus* time for actuators with (blue dotted line) and without (red line) a CNT layer. The voltage polarity (±1.5 V) was changed every time the actuator reached the ±1 mm set value. (B) Time needed for the tip of the actuator to move from its initial position to ±1 mm *vs.* time for the CNT (blue) and simple (red) actuators. ±1.5 V was used to change the direction of bending as soon as the actuator tip reached the set value. For the actuator without a CNT layer, the value was reset to −0.8 mm and +1.2 mm due to the large creeping (shown as the dotted line).

16.5 Interfacing and Actuation

The lack of proper electrical connections and interfacing systems to CPAs working in air has been a major obstacle hindering further progress of these smart devices into practical applications. Macroscopic electrical connectors such as Kelvin clips, alligator clips, or home-made tweezers[12,139,155–157] have been used to connect the two CP electrodes in the trilayer structure into a functional electrical device. Therefore, designing a complex device with several individually controlled trilayer actuators operating in air remains a major issue. Jager *et al.* developed a new interfacing method for PPy trilayer actuators that can be individually controlled.[158] The millimetre-sized PPy actuators were first patterned on a PVDF membrane to form an actuator unit comprising three individually controllable actuators. The new interface was based on a commercial flexible printed circuit board (FPCB) that comprised the electronic circuit into which the actuator unit was embedded by simply inserting and folding the FPCB (Figure 16.15). The contact pads were automatically aligned to the actuator unit. However, this methodology is not suitable for microscale engineering since it uses thick PVDF membranes. In addition, the substrates need to be flipped during the fabrication process in order to pattern both electrodes making the alignment difficult across an optically opaque layer. For these reasons, A. Maziz *et al.* and A. Khaldi *et al.* developed bottom-up microfabrication processes to microfabricate CP actuators on the substrates on which they ultimately operate.[106,159] The bending microactuators, which use ILs as electrolytes, were fabricated by sequentially

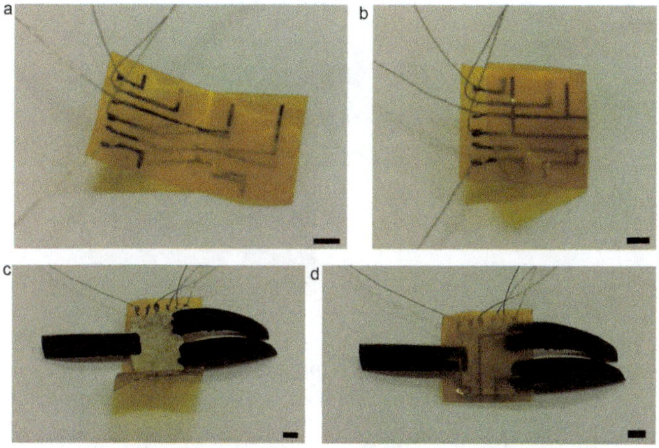

Figure 16.15 An interface based on flexible printed circuit boards (FPCBs) for contacting CP actuators in the open (a) and closed (b) position. The actuator unit is placed in the open FPCB interface with the actuators aligned to the Cu leads (c). Thereafter, the FPCB interface is folded around the actuator unit to provide contact to both sides of the actuators (d). Scale bars: 2 mm. LiTFSI/propylene carbonate was used as the supporting electrolyte. Reprinted from *Sensors and Actuators B: Chemical*, **183**, E. W. H. Jager, N. Masurkar, N. F. Nworah, B. Gaihre, G. Alici and G. M. Spinks, Patterning and electrical interfacing of individually controllable conducting polymer microactuators, 283–289, Copyright (2013) with permission from Elsevier.[158].

stacking layers using a layer-by-layer polymerization of CP electrodes and various SPEs (Figure 16.16). In addition, the process design integrates contact pads. This has been an issue for small-scale conducting polymer actuators.

16.6 Applications

CPAs have been employed for a wide range of applications.[78,160–166] Their use with ILs as electrolytes has led to fewer applicative developments, among which we can find optical devices, microgrippers and bioinspired devices.[15,167,168] Nonetheless, the fact that iEAP actuators are compatible with harsh space environments may lead to new potential applications.[169] In addition, as ILs have become the "green alternative" of volatile organic solvents for pharmaceutical synthesis,[164] biocompatible devices based on iEAP actuators may spark renewed interest and open new challenging issues in the near future for biomedical applications.

Optical MEMS: The development of autofocus lens actuators using CPs has been reported by Ikushima *et al.* The design was based on the requirements of standard camera module dimensions for cellular phones.[15] The whole system comprises an image sensor and a lens holder with two stages of four actuators and separators as shown in Figure 16.17. This configuration allows the lens holder to be moved smoothly along the optical axis. Three-layer actuators were

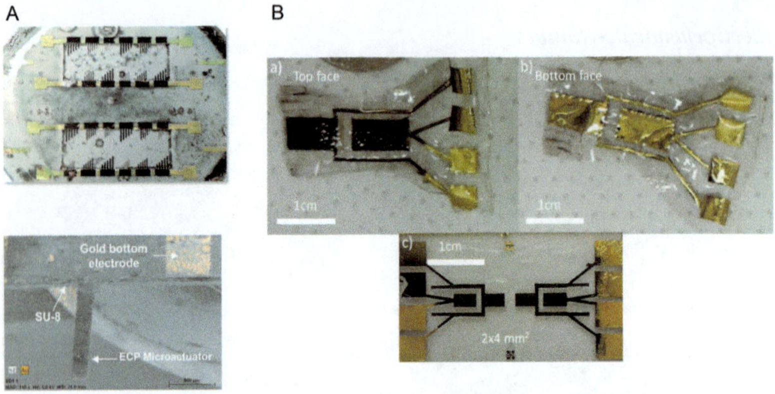

Figure 16.16 (A) Top-down fabricated PEDOT-PEO-NBR semi-IPN microactuators on flexible substrates. EMITFSI was used as the supporting electrolyte. Reprinted with permission from ref. 106. Copyright 2016 American Chemical Society. (B) Bottom-up fabricated PPy-PVdF actuator devices (or "fingers") with individually addressable PPy actuators. (a) top face and (b) bottom face of a device comprising two 5×10 mm^2 sized PPy actuators and (c) top face of two devices each comprising two 2×4 mm^2 sized PPy actuators. LiTFSI/propylene carbonate was used as the supporting electrolyte. Reprinted from *Sensors and Actuators B: Chemical*, 230, A. Khaldi, A. Maziz, G. Alici, G. M. Spinks, E. W. H. Jager, Bottom-up microfabrication process for individually controlled conjugated polymer actuators, 818–824, Copyright 2016 with permission from Elsevier.[159]

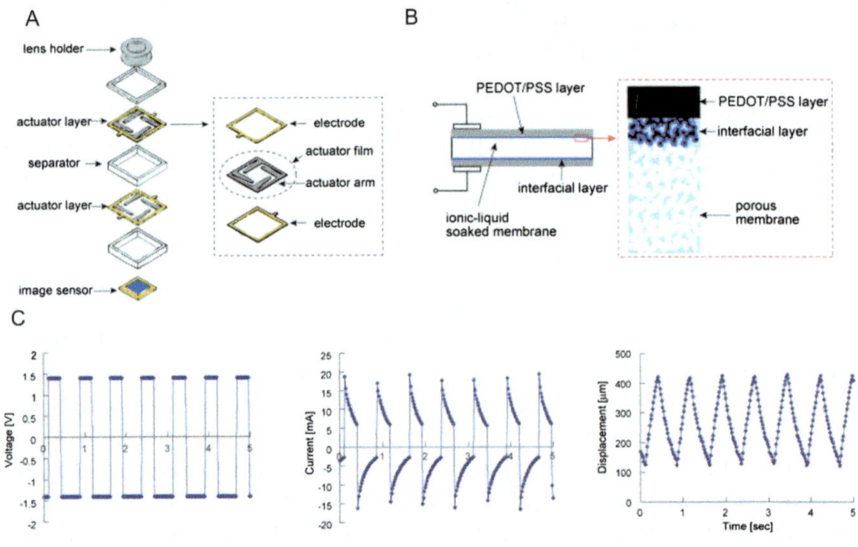

Figure 16.17 (A) Micro-autofocus lens actuator that uses bending CP actuators. (B) The bending-type CP actuator produced by hydrophilic treatment of a porous PVDF membrane and PEDOS:PSS casting. (C) Changes in voltage, current, and displacement over time after elapse of 500,000 s since the start of the actuation. On the horizontal time axis, the clock time at which 500,000 s has elapsed since the start of the actuation is defined as "0." Reprinted from *Synthetic Metals*, 160 (17–18), K. Ikushima, S. John, A. Ono, S. Nagamitsu, PEDOT/PSS bending actuators for autofocus micro lens applications, 1877–1883, Copyright 2010 with permission from Elsevier.[15]

used made of a porous membrane in PDVF incorporating an IL and sandwiched between two PEDOT:PSS conductive layers. In order to enhance the electrical and mechanical properties of the actuators, PEO was added to the PEDOT:PSS. The authors have shown that, an increase in the PEO concentration leads to a decrease of the elastic modulus, higher elongation at break and higher electrical conductivity. The surface of the hydrophobic PVDF membrane was made hydrophilic by grafting polyethylene glycol methacrylate (PEGMA) under UV irradiation to favour the penetration of the PEDOT:PSS within a few microns on both sides of the porous membrane. The actuation tests were performed for more than a million cycles with an actuation frequency of around 1 Hz. It was shown that despite some remaining creeping effects, the operating device met the requirements for the targeted application.

Festin *et al.* used trilayer actuators in ILs to design an autonomous biomimetic vision system composed of a visual perception board, a micro-imager and a PEO/NBR/PEDOT actuator in EMITFSI IL.[167] The micro-imager position was controlled by a closed loop of the actuator with the visual perception board as shown in Figure 16.18. The device was tested in a tracking object experiment showing robust tracking at an angular speed of 4° s^{-1} for a maximum movement of 20 degrees.

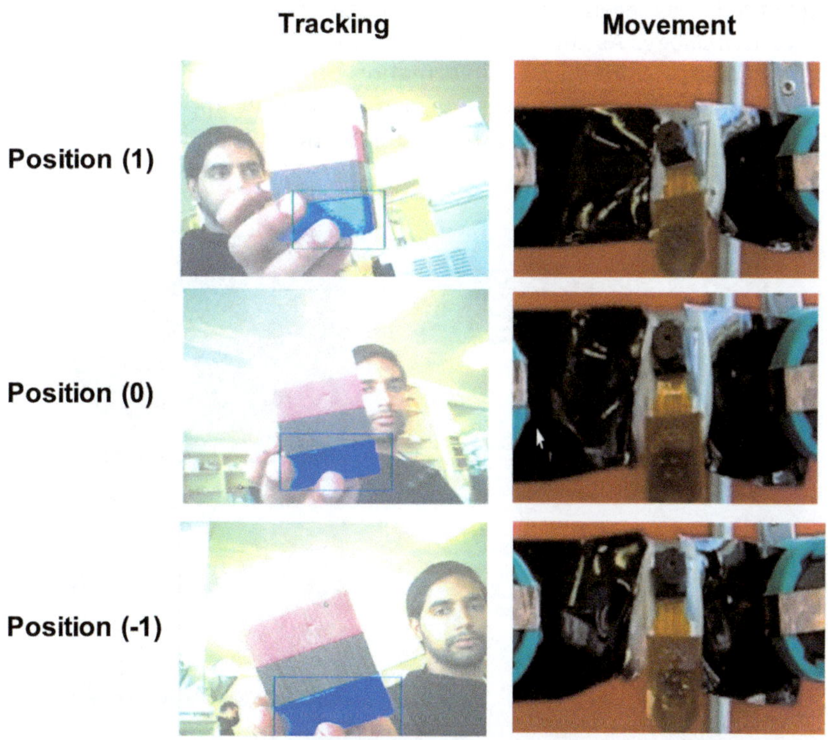

Figure 16.18 Resulting visual target tracking pictures (left) and corresponding device (right) in three different positions, (1), (0) and (−1).

Bio-inspired devices: Using microfabrication, Khaldi and co-workers have used the trilayer configuration to fabricate bio-mimicking insect wings. Using SU8, they fabricated artificial wings on the surface of which they attached IPN microactuators to operate them. They showed that the IPN microactuator has enough force to put a wing in motion at a frequency of 24 Hz with a bending amplitude of 20 degrees.[168] Figure 16.19(A) shows an IPN microactuator moving as an artificial wing with a bending amplitude of 60° at 16 Hz demonstrating the potential of CP actuators to power micro-UAVs (unmanned aerial vehicles or drones). Micro-UAVs would need a higher flapping wing frequency to lift; however, this preliminary result is very encouraging. The authors used the same trilayer configuration to fabricate standalone gripper-like microactuators. The actuator with its three fingers was electrically stimulated using an input voltage of 4 V at 0.05 Hz. The displacement response of the actuator could be observed inside a SEM, *i.e.*, in a vacuum environment, since EMITFSI IL does not evaporate. A large displacement amplitude of 475 μm was measured as presented in Figure 16.19(B). This result shows the great interest of this innovative material for creating micro-tools and bio-inspired devices. Further development related to downscaling the size of the wing, as well as increasing the resonance frequency of the system is currently on going. For instance, a drastic increase of the resonance frequency to approximately 1 kHz was recently presented by Maziz *et al.*[156]

Space technology: As already mentioned, CP actuators are light materials and generally exhibit large deformation under low voltage stimulation. These properties make them very attractive for space applications.[169] However, the space environment is hostile for polymeric materials. High vacuum

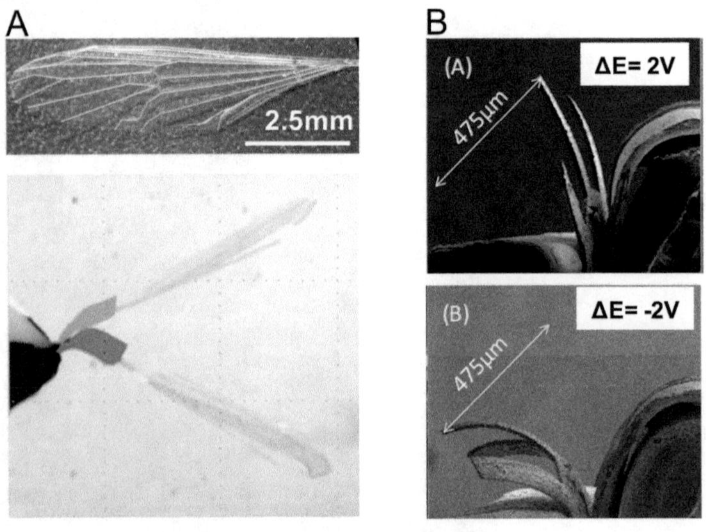

Figure 16.19 (A) An artificial wing made of SU8; (B) flapping SU8 wing movement at a resonant frequency of 16 Hz, under an applied voltage of ±2 V and with a bending angle of 60°.

conditions prohibit the presence of low molecular weight volatile compounds in the system. In addition, under space vacuum, ionic actuators may be less efficient due to the evaporation of the electrolyte. Vidal *et al.* have studied the behavior of all-polymer ionic conducting IPN actuators after aging under space-hazardous environmental factors in laboratory conditions.[167,169] The IL EMITFSI was used as an electrolyte to avoid any evaporation under high vacuum. Actuators have been submitted to thermal cycling tests under vacuum and to ionizing gamma radiation. Despite a small decrease in performance after ageing, the authors observed that conducting IPN actuators remain functional and largely maintain their actuation properties.

CPs have also been employed in iEAP actuators with various organic electrolytes for a wider range of applications, including microfluidics, among others. Wu *et al.* employed the trilayer configuration to design a microfluidic pumping system.[170] The configuration, called "The Tube-In-Tube Actuators Node, (TITAN)", uses the electromechanical actuation properties of PPy to induce fluid movement through an inner channel in a single unit (0.25 M of tetrabutylammonium hexafluorophosphate/propylene carbonate was used as the supporting electrolyte). Series connection of these PPy units and appropriate electrical connection/stimulation induces a novel peristaltic action that enables fluids to be pumped in a predetermined direction. Flow rates of up to 2.5 μl min^{-1} can be achieved against a back pressure of 50 mbar, which is sufficient to enable fluid movement in a glass capillary channel (diameter: 266 μm, length: 12.8 cm). The use of ILs as electrolytes could be beneficially explored with the same approaches.

16.7 Conclusions and Challenges

In this chapter, advancements in creating iEAP actuators have been presented, with a particular focus on CP technology. During the last two decades, CP actuator technology has shown tremendous progress from concept to a plethora of devices in the field of iEAP. They can be fabricated using a variety of methods and technologies and integrated into functional devices. The CP actuator technology is both versatile and exciting, and it is an enabling technology for many applications including biomimetic robotics, optical displays, bioinspired systems, space technology, microfluidic systems, *etc.* This large number of possible applications can be attributed to the unique properties of CP actuators, such as compliance, lightness, silent operation, low driving voltage, biocompatibility, simple structure and ability to be miniaturized.

The use of ILs as a supporting electrolyte provides an interesting alternative electrolyte for iEAP actuators. The wide electrochemical potential window provides an added degree of stability, enabling thousands of cycles of actuation. It would be very interesting to consider the use of polymeric/polymerized ionic liquids or poly(ionic liquid)s (PILs) in iEAP actuators. PILs refer to a special type of polyelectrolytes that carry an IL species in each of the repeating units. The major advantages of using a PIL instead of an IL in iEAP actuators would be an enhanced mechanical stability, improved processability, durability, and spatial controllability over the IL species. Therefore, the combination of the unique properties of ILs with the macromolecular architecture may be a novel way of creating new properties and functions in iEAP actuators.

Furthermore, a number of CP materials are also known to function as sensors for detecting motion[170,171] or other ambient parameters.[172] However, although there have been many advanced demonstrations, prototypes, and application proposals, to the best of our knowledge, commercial products have not been introduced to the market yet. Several challenges still lie ahead. For instance, although a comprehensive understanding of the electrochemical nature of CP actuators has already been experimentally validated, it has not yet been successfully implemented for controlling these actuators.

The development of various system identification and validation techniques for both linear and non-linear modelling must be considered to better understand the physical, chemical and mechanical mechanisms underlying the behaviour of the whole system.[173,174] In the case of iEAP actuators, large time constants, unknown systems dynamics and positional drift are the main control challenges, but closed-loop[175] and feedforward[176] control strategies have been suggested. Closed-loop control usually requires laser displacement or other sensors that are bulky and troublesome to implement.[176,177] Environment-dependent behaviour is challenging for feedforward controllers.[43] The design of self-sensing actuators is therefore one of the main current research axes.[178,179] In this case, the implementation of closed-loop control algorithms for actuator control is linked to integrated sensors to get feedback on its state of deformation or stress. Novel designs should therefore be developed to integrate both the actuating and sensing functions in the iEAP transducers to make them easy to implement in a system usable for practical applications.[180] The possibility to tune the stiffness is also an important feature for artificial muscles, where a variable compliance of the device is required during real-time feedback-loop operation.

One of the weak points of iEAP actuators is the low forces they can generate. The challenge here lies in combining high forces while having large displacements. Stacking of trilayer actuators driven by CPs in order to combine large force and reasonable deflection is an interesting strategy both at the macro- and at the micro-scale.[181]

The design of more complex 3D structures from 2D materials by folding could lead to more sophisticated multifunctional systems. Origami actuator design through topology optimization has been proposed by Fuchi *et al.*[182] The design complexity in terms of shape but also regarding the optimization of function could be tuned by changing the number of fold lines. An origami-inspired approach was also used to design worm robots with folding induced by electrothermal actuation.[183] It may be possible to design reconfigurable electronic flexible devices using iEAP actuators combined with other actuation schemes.[184] Bio-inspired adaptive structures and materials are also interesting alternatives where iEAP could help foster improved performances for future applications.[185]

References

1. P. Brochu and Q. Pei, *Macromol. Rapid Commun.*, 2010, **31**, 10–36.
2. F. Carpi, R. Kornbluh, P. Sommer-Larsen and G. Alici, *Bioinspiration Biomimetics*, 2011, **6**, 045006.
3. P. G. Madden, J. D. Madden, P. A. Anquetil, N. A. Vandesteeg and I. W. Hunter, *IEEE J. Oceanic Eng.*, 2004, **29**, 696–705.
4. T. Mirfakhrai, J. D. Madden and R. H. Baughman, *Mater. Today*, 2007, **10**, 30–38.
5. D. Z. Zhou, G. M. Spinks, G. G. Wallace, C. Tiyapiboonchaiya, D. R. MacFarlane, M. Forsyth and J. Z. Sun, *Electrochim. Acta*, 2003, **48**, 2355–2359.
6. N. K. Guimard, N. Gomez and C. E. Schmidt, *Prog. Polym. Sci.*, 2007, **32**, 876–921.
7. S. Ruiz, B. Mead, V. Palmre, K. J. Kim and W. Yim, *Smart Mater. Struct.*, 2014, **24**, 015007.
8. P. Tozzi, *Swiss Med. Wkly.*, 2011, **141**, w13311.
9. H. Choi, K. Jung, J. Koo and J. Nam, in *Electroactive Polymers for Robotic Applications*, Springer, 2007, pp. 49–90.
10. F. Ilievski, A. D. Mazzeo, R. F. Shepherd, X. Chen and G. M. Whitesides, *Angew. Chem.*, 2011, **123**, 1930–1935.
11. B. Tondu, *Artificial Muscles for Humanoid Robots*, INTECH Open Access Publisher, 2007.
12. R. Temmer, A. Maziz, C. Plesse, A. Aabloo, F. Vidal and T. Tamm, *Smart Mater. Struct.*, 2013, **22**, 104006.
13. L. Bay, T. Jacobsen, S. Skaarup and K. West, *J. Phys. Chem. B*, 2001, **105**, 8492–8497.
14. K. J. Kim and S. Tadokoro, *Artificial Muscles and Sensors*, 2007.
15. K. Ikushima, S. John, A. Ono and S. Nagamitsu, *Synth. Met.*, 2010, **160**, 1877–1883.
16. A. Döring, W. Birnbaum and D. Kuckling, *Chem. Soc. Rev.*, 2013, **42**, 7391–7420.
17. P. Bawa, V. Pillay, Y. E. Choonara and L. C. Du Toit, *Biomed. Mater.*, 2009, **4**, 022001.
18. E. Cabane, X. Zhang, K. Langowska, C. G. Palivan and W. Meier, *Biointerphases*, 2012, **7**, 9.
19. P. Glazer, M. van Erp, A. Embrechts, S. Lemay and E. Mendes, *Soft Matter*, 2012, **8**, 4421–4426.
20. S. Maeda, Y. Hara, R. Yoshida and S. Hashimoto, *Int. J. Mol. Sci.*, 2010, **11**, 52–66.
21. V. Labrot, P. De Kepper, J. Boissonade, I. Szalai and F. Gauffre, *J. Phys. Chem. B*, 2005, **109**, 21476–21480.
22. S. Nakamaru, S. Maeda, Y. Hara and S. Hashimoto, *J. Phys. Chem. B*, 2009, **113**, 4609–4613.
23. A. B. Imran, T. Seki and Y. Takeoka, *Polym. J.*, 2010, **42**, 839–851.
24. M. Bassil, J. AL Moussawel, M. Ibrahim, G. Azzi and M. El Tahchi, *J. Appl. Polym. Sci.*, 2014, **131**, 22.

25. W. Hong, C. Meis, J. R. Heflin and R. Montazami, *Sens. Actuators, B*, 2014, **205**, 371–376.
26. Y. Liu, C. Lu, S. Twigg, M. Ghaffari, J. Lin, N. Winograd and Q. Zhang, *Sci. Rep.*, 2013, **3**, 973.
27. S. Nemat-Nasser, *J. Appl. Phys.*, 2002, **92**, 2899–2915.
28. S. Nemat-Nasser and J. Y. Li, *J. Appl. Phys.*, 2000, **87**, 3321–3331.
29. C. Jo, D. Pugal, I.-K. Oh, K. J. Kim and K. Asaka, *Prog. Polym. Sci.*, 2013, **38**, 1037–1066.
30. O. Kim, S. Y. Kim, B. Park, W. Hwang and M. J. Park, *Macromolecules*, 2014, **47**, 4357–4368.
31. R. Tiwari and E. Garcia, *Smart Mater. Struct.*, 2011, **20**, 083001.
32. M. D. Bennett and D. J. Leo, *Sens. Actuators, A*, 2004, **115**, 79–90.
33. S. Nemat-Nasser and Y. Wu, *J. Appl. Phys.*, 2003, **93**, 5255–5267.
34. T. Kobayashi and M. Omiya, *Smart Mater. Struct.*, 2012, **21**, 105031.
35. V. De Luca, P. Digiamberardino, G. Di Pasquale, S. Graziani, A. Pollicino, E. Umana and M. G. Xibilia, *J. Polym. Sci., Part B: Polym. Phys.*, 2013, **51**, 699–734.
36. K. J. Kim and M. Shahinpoor, *Smart Mater. Struct.*, 2003, **12**, 65.
37. M. Shahinpoor and K. J. Kim, *Smart Mater. Struct.*, 2001, **10**, 819.
38. S. Ford, G. Macias and R. Lumia, *Smart Mater. Struct.*, 2014, **24**, 025015.
39. Z. Chen, T. I. Um and H. Bart-Smith, *Sens. Actuators, A*, 2011, **168**, 131–139.
40. B. Kim, D.-H. Kim, J. Jung and J.-O. Park, *Smart Mater. Struct.*, 2005, **14**, 1579.
41. V. Palmre, J. J. Hubbard, M. Fleming, D. Pugal, S. Kim, K. J. Kim and K. K. Leang, *Smart Mater. Struct.*, 2012, **22**, 014003.
42. G.-H. Feng and W.-L. Huang, *Mater. Sci. Eng., C*, 2014, **45**, 241–249.
43. A. Punning, I. Must, I. Põldsalu, V. Vunder, R. Temmer, K. Kruusamäe, F. Kaasik, J. Torop, P. Rinne and T. Lulla, *J. Intell. Mater. Syst. Struct.*, 2014, **25**, 2267–2275.
44. K. Park, M.-K. Yoon, S. Lee, J. Choi and M. Thubrikar, *Smart Mater. Struct.*, 2010, **19**, 075002.
45. P. Kiwon, *J. Korean Phys. Soc.*, 2011, **59**, 3401.
46. S. J. Kim, I. T. Lee, H.-Y. Lee and Y. H. Kim, *Smart Mater. Struct.*, 2006, **15**, 1540.
47. H. Lei, W. Li and X. Tan, *Sens. Actuators, A*, 2014, **217**, 1–12.
48. S.-A. Tsai, H.-C. Wei and G.-D. J. Su, *Appl. Opt.*, 2012, **51**, 8315–8323.
49. B. Akle and D. J. Leo, *Smart Mater. Struct.*, 2004, **13**, 1081.
50. J. Barramba, J. Silva and P. C. Branco, *Sens. Actuators, A*, 2007, **140**, 232–238.
51. L. Chang, K. Asaka, Z. Zhu, Y. Wang, H. Chen and D. Li, *J. Appl. Phys.*, 2014, **115**, 244901.
52. S. J. Kim, I. T. Lee and Y. H. Kim, *Smart Mater. Struct.*, 2007, **16**, N6.
53. K. J. Kim and M. Shahinpoor, *Polymer*, 2002, **43**, 797–802.
54. F. Cellini, A. Grillo and M. Porfiri, *Appl. Phys. Lett.*, 2015, **106**, 131902.

55. C.-K. Chung, P. Fung, Y. Hong, M.-S. Ju, C.-C. K. Lin and T. Wu, *Sens. Actuators, B*, 2006, **117**, 367–375.
56. X. Yan, Z. Gang, Z. Yuming and S. Zhuangzhi, *Ionics*, 2015, **21**, 1089–1094.
57. V. Palmre, D. Pugal, K. J. Kim, K. K. Leang, K. Asaka and A. Aabloo, *Sci. Rep.*, 2014, **4**, 6176.
58. Q. He, M. Yu, X. Yang, K. J. Kim and Z. Dai, *Smart Mater. Struct.*, 2015, **24**, 065026.
59. J. Kim, J.-H. Jeon, H.-J. Kim, H. Lim and I.-K. Oh, *ACS Nano*, 2014, **8**, 2986–2997.
60. R. H. Baughman, C. X. Cui, A. A. Zakhidov, Z. Iqbal, J. N. Barisci, G. M. Spinks, G. G. Wallace, A. Mazzoldi, D. De Rossi, A. G. Rinzler, O. Jaschinski, S. Roth and M. Kertesz, *Science*, 1999, **284**, 1340–1344.
61. J. Barisci, G. Wallace, D. MacFarlane and R. Baughman, *Electrochem. Commun.*, 2004, **6**, 22–27.
62. J. D. Madden, J. N. Barisci, P. A. Anquetil, G. M. Spinks, G. G. Wallace, R. H. Baughman and I. W. Hunter, *Adv. Mater.*, 2006, **18**, 870–873.
63. T. Fukushima, A. Kosaka, Y. Ishimura, T. Yamamoto, T. Takigawa, N. Ishii and T. Aida, *Science*, 2003, **300**, 2072–2074.
64. T. Fukushima, K. Asaka, A. Kosaka and T. Aida, *Angew. Chem., Int. Ed.*, 2005, **44**, 2410–2413.
65. J. Lee and T. Aida, *Chem. Commun.*, 2011, **47**, 6757–6762.
66. C. Li, E. T. Thostenson and T.-W. Chou, *Compos. Sci. Technol.*, 2008, **68**, 1227–1249.
67. M. Tunckol, J. Durand and P. Serp, *Carbon*, 2012, **50**, 4303–4334.
68. N. Terasawa and I. Takeuchi, *Electrochim. Acta*, 2014, **123**, 340–345.
69. N. Terasawa and I. Takeuchi, *Sens. Actuators, B*, 2010, **145**, 775–780.
70. T. Fukushima and T. Aida, *Chem.–Eur. J.*, 2007, **13**, 5048–5058.
71. C. K. Chiang, C. R. Fincher, Y. W. Park, A. J. Heeger, H. Shirakawa, E. J. Louis, S. C. Gau and A. G. MacDiarmid, *Phys. Rev. Lett.*, 1977, **39**, 1098–1101.
72. P. J. S. Foot and R. Simon, *J. Phys. D: Appl. Phys.*, 1989, **22**, 1598–1603.
73. J. Isaksson, C. Tengstedt, M. Fahlman, N. Robinson and M. Berggren, *Adv. Mater.*, 2004, **16**, 316–320.
74. R. H. Baughman, L. W. Shacklette, R. L. Elsenbaumer, E. J. Plichta and C. Becht, in *Molecular Electronics*, ed. P. I. Lazarev, Kluwer Academic Publishers, Dordrecht, 1991, pp. 267–289.
75. S. A. Wilson, R. P. J. Jourdain, Q. Zhang, R. A. Dorey, C. R. Bowen, M. Willander, Q. U. Wahab, M. Willander, M. A. H. Safaa, O. Nur, E. Quandt, C. Johansson, E. Pagounis, M. Kohl, J. Matovic, B. Samel, W. van der Wijngaart, E. W. H. Jager, D. Carlsson, Z. Djinovic, M. Wegener, C. Moldovan, R. Iosub, E. Abad, M. Wendlandt, C. Rusu and K. Persson, *Mater. Sci. Eng., R*, 2007, **56**, 1–129.
76. J. Torop, A. Aabloo and E. W. Jager, *Carbon*, 2014, **80**, 387–395.
77. K. Surana, P. K. Singh, B. Bhattacharya, C. Verma and R. Mehra, *Ceram. Int.*, 2015, **41**, 5093–5099.

78. T. Sugino, K. Kiyohara, I. Takeuchi, K. Mukai and K. Asaka, *Carbon*, 2011, **49**, 3560–3570.
79. M. R. Gandhi, P. Murray, G. M. Spinks and G. G. Wallace, *Synth. Met.*, 1995, **73**, 247–256.
80. C. Lopez, M. M. Viegas, G. Bidan and E. Vieil, *Synth. Met.*, 1994, **63**, 73–78.
81. J. Reynolds, M. Pyo and Y.-J. Qiu, *Synth. Met.*, 1993, **55**, 1388–1395.
82. Y. J. Qiu and J. R. Reynolds, *Polym. Eng. Sci.*, 1991, **31**, 417–421.
83. K. Naoi, M. Lien and W. H. Smyrl, *J. Electrochem. Soc.*, 1991, **138**, 440–445.
84. Q. Pei and O. Inganäs, *J. Phys. Chem.*, 1992, **96**, 10507–10514.
85. V. M. Schmidt, C. Barbero and R. Kötz, *J. Electroanal. Chem.*, 1993, **352**, 301–307.
86. T. Matencio, M. A. De Paoli, R. C. D. Peres, R. M. Torresi and S. I. Cordoba de Torresi, *Synth. Met.*, 1995, **72**, 59–64.
87. Q. Pei and O. Inganäs, *J. Phys. Chem.*, 1993, **97**, 6034–6041.
88. T. F. Otero, H. Grande and J. Rodriguez, *Electrochim. Acta*, 1996, **41**, 1863–1869.
89. T. Otero and J. Padilla, *J. Electroanal. Chem.*, 2004, **561**, 167–171.
90. T. Chao and J. March, *Polym. Chem.*, 1988, **26**, 743–753.
91. A. F. Diaz and J. Bargon, in *Handbook of Conducting Polymers*, ed. T. A. Skotheim, Marcel Dekker, Inc., New York, 1986, vol. 1, pp. 81–115.
92. S. Machida, S. Miyata and A. Techagumpuch, *Synth. Met.*, 1989, **31**, 311–318.
93. S. Rapi, V. Bocchi and G. Gardini, *Synth. Met.*, 1988, **24**, 217–221.
94. J. M. Pringle, J. Efthimiadis, P. C. Howlett, J. Efthimiadis, D. R. MacFarlane, A. B. Chaplin, S. B. Hall, D. L. Officer, G. G. Wallace and M. Forsyth, *Polymer*, 2004, **45**, 1447–1453.
95. A. Maziz, A. Concas, A. Khaldi, J. Stålhand, N.-K. Persson and E. W. Jager, *Sci. Adv.*, 2017, **3**, e1600327.
96. V. Castagnola, C. Bayon, E. Descamps and C. Bergaud, *Synth. Met.*, 2014, **189**, 7–16.
97. K. K. Kanazawa, A. Diaz, R. H. Geiss, W. D. Gill, J. F. Kwak, J. A. Logan, J. F. Rabolt and G. B. Street, *J. Chem. Soc., Chem. Commun.*, 1979, 854–855.
98. R.-I. Sugimoto, S. Takeda, H. Gu and K. Yoshino, *Chem. Express*, 1986, **1**, 635–638.
99. M. Castillo-Ortega, M. Inoue and M. Inoue, *Synth. Met.*, 1989, **28**, 65–70.
100. A. Mohammadi, M.-A. Hasan, B. Liedberg, I. Lundström and W. Salaneck, *Synth. Met.*, 1986, **14**, 189–197.
101. J. Kim, E. Kim, Y. Won, H. Lee and K. Suh, *Synth. Met.*, 2003, **139**, 485–489.
102. B. Winther-Jensen and K. West, *Macromolecules*, 2004, **37**, 4538–4543.
103. B. Winther-Jensen, D. W. Breiby and K. West, *Synth. Met.*, 2005, **152**, 1–4.
104. M. Fabretto, K. Zuber, C. Hall and P. Murphy, *Macromol. Rapid Commun.*, 2008, **29**, 1403–1409.
105. M. V. Fabretto, D. R. Evans, M. Mueller, K. Zuber, P. Hojati-Talemi, R. D. Short, G. G. Wallace and P. J. Murphy, *Chem. Mater.*, 2012, **24**, 3998–4003.

106. A. Maziz, C. Plesse, C. Soyer, E. Cattan and F. Vidal, *ACS Appl. Mater. Interfaces*, 2015, **8**, 1559–1564.
107. Y. Cao, P. Smith and A. J. Heeger, *Synth. Met.*, 1992, **48**, 91–97.
108. G. Heywang and F. Jonas, *Adv. Mater.*, 1992, **4**, 116–118.
109. F. Hide, M. A. Diaz-Garcia, B. J. Schwartz, M. R. Andersson, Q. Pei and A. J. Heeger, *Science*, 1996, **273**, 1833–1836.
110. M. Kobayashi, N. Colaneri, M. Boysel, F. Wudl and A. Heeger, *J. Chem. Phys.*, 1985, **82**, 5717–5723.
111. R. Brooke, M. Fabretto, P. Murphy, D. Evans, P. Cottis and P. Talemi, *Prog. Mater. Sci.*, 2017, **86**, 127–146.
112. G. M. Spinks, T. E. Campbell and G. G. Wallace, *Smart Mater. Struct.*, 2005, **14**, 406–412.
113. G. M. Spinks, L. Liu, G. G. Wallace and D. Zhou, *Adv. Funct. Mater.*, 2002, **12**, 437–440.
114. J. D. Madden, R. A. Cush, T. S. Kanigan and I. W. Hunter, *Synth. Met.*, 2000, **113**, 185–192.
115. L. Bay, K. West, P. Sommer-Larsen, S. Skaarup and M. Benslimane, *Adv. Mater.*, 2003, **15**, 310–313.
116. S. Hara, T. Zama, W. Takashima and K. Kaneto, *J. Mater. Chem.*, 2004, **14**, 1516–1517.
117. S. Hara, T. Zama, W. Takashima and K. Kaneto, *Synth. Met.*, 2006, **156**, 351–355.
118. J. Foroughi, G. M. Spinks and G. G. Wallace, *Sens. Actuators, B*, 2011, **155**, 278–284.
119. K. Yamato, K. Tominaga, W. Takashima and K. Kaneto, *Synth. Met.*, 2009, **159**, 839–842.
120. T. F. Otero, J. Rodriguez, E. Angulo and C. Santamaria, *Synth. Met.*, 1993, **57**, 3713–3717.
121. M. J. M. Jafeen, M. A. Careem and S. Skaarup, *Ionics*, 2010, **16**, 1–6.
122. J. G. Martinez, T. F. Otero and E. W. H. Jager, *Langmuir*, 2014, **30**, 3894–3904.
123. S. Hara, T. Zama, W. Takashima and K. Kaneto, *Synth. Met.*, 2004, **146**, 47–55.
124. R. Baughman, L. Shacklette, R. Elsenbaumer, E. Plichta and C. Becht, in *Conjugated Polymeric Materials: Opportunities in Electronics, Optoelectronics, and Molecular Electronics*, Springer, 1990, pp. 559–582.
125. T. F. Otero and J. M. Sansieña, *Adv. Mater.*, 1998, **10**, 491–494.
126. T. F. Otero and J. M. Sansiñena, *Bioelectrochem. Bioenerg.*, 1997, **42**, 117–122.
127. Q. Pei and O. Inganäs, *Solid State Ionics*, 1993, **60**, 161–166.
128. A. MacDiarmid, K. Kaneto, H. Saito and Y. Min, Abstracts of Papers of The Amercian Chemical Society, 1994.
129. M. Kaneko and K. Kaneto, *React. Funct. Polym.*, 1998, **37**, 155–161.
130. E. Smela, O. Inganäs and I. Lundström, *Science*, 1995, **268**, 1735–1738.
131. E. Smela, *J. Micromech. Microeng.*, 1999, **9**, 1–18.
132. E. Smela, M. Kallenbach and J. Holdenried, *J. Microelectromech. Syst.*, 1999, **8**, 373–383.

133. A. G. MacDiarmid, K. Kaneto, H. Saito and Y. Min, *Polym. Mater. Sci. Eng.*, 1994, **71**, 713.
134. M. Cho, H. Seo, J. Nam, H. Choi, J. Koo, K. Song and Y. Lee, *Sens. Actuators, B*, 2006, **119**, 621–624.
135. F. Vidal, J. F. Popp, C. Plesse, C. Chevrot and D. Teyssie, *J. Appl. Polym. Sci.*, 2003, **90**, 3569–3577.
136. Y. Wu, G. Alici, G. M. Spinks and G. G. Wallace, *Synth. Met.*, 2006, **156**, 1017–1022.
137. R. Temmer, I. Must, F. Kaasik, A. Aabloo and T. Tamm, *Sens. Actuators, B*, 2012, **166**, 411–418.
138. C. Plesse, F. Vidal, H. Randriamahazaka, D. Teyssié and C. Chevrot, *Polymer*, 2005, **46**, 7771–7778.
139. C. Plesse, A. Khaldi, Q. Wang, E. Cattan, D. Teyssié, C. Chevrot and F. Vidal, *Smart Mater. Struct.*, 2011, **20**.
140. Y. Liu, Q. Gan, S. Baig and E. Smela, *J. Phys. Chem. C*, 2007, **111**, 11329–11338.
141. W. Lu, A. G. Fadeev, B. Qi, E. Smela, B. R. Mattes, J. Ding, G. M. Spinks, J. Mazurkiewicz, D. Zhou and G. G. Wallace, *Science*, 2002, **297**, 983–987.
142. J. Ding, D. Z. Zhou, G. Spinks, G. Wallace, S. Forsyth, M. Forsyth and D. MacFarlane, *Chem. Mater.*, 2003, **15**, 2392–2398.
143. M. J. Park, I. Choi, J. Hong and O. Kim, *J. Appl. Polym. Sci.*, 2013, **129**, 2363–2376.
144. W. Lu and B. R. Mattes, *Synth. Met.*, 2005, **152**, 53–56.
145. F. Vidal, C. Plesse, H. Randriamahazaka, D. Teyssie and C. Chevrot, *Mol. Cryst. Liq. Cryst.*, 2006, **448**, 95/[697]–102/[704].
146. A. Simaite, B. Tondu, P. Souères and C. Bergaud, *ACS Appl. Mater. Interfaces*, 2015, **7**, 19966–19977.
147. J. D. Madden, D. Rinderknecht, P. A. Anquetil and I. W. Hunter, *Sens. Actuators, A*, 2007, **133**, 210–217.
148. D. Melling, S. Wilson and E. W. Jager, *Smart Mater. Struct.*, 2013, **22**, 104021.
149. K. Kaneto, T. Shinonome, K. Tominaga and W. Takashima, *Jpn. J. Appl. Phys.*, 2011, **50**, 091601.
150. K. Kaneto, H. Suematsu and K. Yamato, *Bioinspiration Biomimetics*, 2008, **3**, 035005.
151. T. F. Otero, H. J. Grande and J. Rodriguez, *J. Phys. Chem. B*, 1997, **101**, 3688–3697.
152. X. Wang and E. Smela, *J. Phys. Chem. C*, 2008, **113**, 359–368.
153. E. Smela and N. Gadegaard, *Adv. Mater.*, 1999, **11**, 953–957.
154. A. Simaite, A. Delagarde, B. Tondu, P. Souères, E. Flahaut and C. Bergaud, *Nanotechnology*, 2016, **28**, 025502.
155. G. Alici, V. Devaud, P. Renaud and G. Spinks, *J. Micromech. Microeng.*, 2009, **19**, 025017.
156. A. Maziz, C. Plesse, C. Soyer, C. Chevrot, D. Teyssié, E. Cattan and F. Vidal, *Adv. Funct. Mater.*, 2014, **24**, 4851–4859.
157. N. Festin, A. Maziz, C. Plesse, D. Teyssié, C. Chevrot and F. Vidal, *Smart Mater. Struct.*, 2013, **22**, 104005.

158. E. W. H. Jager, N. Masurkar, N. F. Nworah, B. Gaihre, G. Alici and G. M. Spinks, *Sens. Actuators, B*, 2013, **183**, 283–289.
159. A. Khaldi, A. Maziz, G. Alici, G. M. Spinks and E. W. Jager, *Sens. Actuators, B*, 2016, **230**, 818–824.
160. E. D. Daneshvar and E. Smela, *Adv. Healthcare Mater.*, 2014, **3**, 1026–1035.
161. A. Gelmi, M. Ljunggren, M. Rafat and E. W. H. Jager, in *NanoMed, International Conference on Nanotechnology in Medicine*, Royal Free Hospital, University College London, UK, 2014.
162. K. Svennersten, M. Berggren, A. Richter-Dahlfors and E. W. H. Jager, *Lab Chip*, 2011, **11**, 3287–3293.
163. E. W. H. Jager, O. Inganäs and I. Lundström, *Science*, 2000, **288**, 2335–2338.
164. T. Siódmiak, M. Piotr Marszall and A. Proszowska, *Mini-Rev. Org. Chem.*, 2012, **9**, 203–208.
165. C. Immerstrand, K. H. Peterson, K.-E. Magnusson, E. Jager, M. Krogh, M. Skoglund, A. Selbing and O. Inganäs, *MRS Bull.*, 2002, **27**, 461–464.
166. K. K. C. Lee, N. R. Munce, T. Shoa, L. G. Charron, G. A. Wright, J. D. Madden and V. X. D. Yang, *Sens. Actuators, A*, 2009, **153**, 230–236.
167. N. Festin, C. Plesse, C. Chevrot, D. Teyssié, P. Pirim and F. Vidal, *SPIE Smart Structures and Materials+Nondestructive Evaluation and Health Monitoring*, 2011, pp. 79760K-79760K-8.
168. A. Khaldi, C. Plesse, C. Soyer, E. Cattan, F. Vidal, C. Chevrot and D. Teyssié, *ASME 2011 International Mechanical Engineering Congress and Exposition, IMECE 2011*, Denver, USA, 2011.
169. A. Punning, K. J. Kim, V. Palmre, F. Vidal, C. Plesse, N. Festin, A. Maziz, K. Asaka, T. Sugino and G. Alici, *Sci. Rep.*, 2014, **4**, 6913.
170. Y. Wu, G. Alici, J. D. Madden, G. M. Spinks and G. G. Wallace, *Adv. Funct. Mater.*, 2007, **17**, 3216–3222.
171. N. Festin, C. Plesse, C. Chevrot, D. Teyssié, L. Josselin, P. Pirim and F. Vidal, *Procedia Comput. Sci.*, 2011, **7**, S4–S7.
172. J. G. Martinez and T. F. Otero, *J. Phys. Chem. B*, 2012, **116**, 9223–9230.
173. C. H. Nguyen, G. Alici and G. Wallace, *IEEE ASME Trans. Mechatron.*, 2014, **19**, 1279–1288.
174. X. Wang, G. Alici and X. Tan, *Smart Mater. Struct.*, 2013, **23**, 025015.
175. Y. Fang, X. Tan and G. Alici, *IEEE Trans. Control Syst. Technol.*, 2008, **16**, 600–612.
176. S. W. John, G. Alici and C. D. Cook, *IEEE ASME Trans. Mechatron.*, 2010, **15**, 149–156.
177. Q. Yao, G. Alici and G. M. Spinks, *Sens. Actuators, A*, 2008, **144**, 176–184.
178. T. F. Otero and J. G. Martinez, *Prog. Polym. Sci.*, 2015, **44**, 62–78.
179. K. Kruusamäe, A. Punning, A. Aabloo and K. Asaka, *Actuators*, 2015, **4**, 17–38.
180. A. Simaite, F. Mesnilgrente, B. Tondu, P. Souères and C. Bergaud, *Sens. Actuators, B*, 2016, **229**, 425–433.
181. M. Farajollahi, S. E. Takallo, V. Woehling, A. Fannir, C. Plesse, F. Vidal, F. Sassani and J. D. Madden, *SPIE Smart Structures and Materials+Nondestructive Evaluation and Health Monitoring*, 2015.

182. K. Fuchi, P. R. Buskohl, G. Bazzan, M. F. Durstock, G. W. Reich, R. A. Vaia and J. J. Joo, *J. Mech. Des.*, 2015, **137**, 091401.
183. C. D. Onal, R. J. Wood and D. Rus, *IEEE ASME Trans. Mechatron.*, 2013, **18**, 430–438.
184. S. Felton, M. Tolley, E. Demaine, D. Rus and R. Wood, *Science*, 2014, **345**, 644–646.
185. S. Li and K. Wang, *Bioinspiration Biomimetics*, 2016, **12**, 011001.

CHAPTER 17

Capturing CO_2 with Poly(Ionic Liquid)s

S. EINLOFT*[a,b], F. L. BERNARD[a,b] AND F. DALLA VECCHIA[a]

[a]Pontifical Catholic University of Rio Grande do Sul – PUCRS, Post-Graduation Program in Materials Engineering and Technology, Av. Ipiranga, 6681 Partenon, Porto Alegre/RS, 90619-900, Brazil; [b]Pontifical Catholic University of Rio Grande do Sul – PUCRS, School of Chemistry, Av. Ipiranga, 6681 Partenon, Porto Alegre / RS, 90619-900, Brazil
*E-mail: einloft@pucrs.br

17.1 Introduction

The burning of fossil fuels is the largest anthropogenic contributor to climate change due to the huge carbon dioxide (CO_2) emissions involved.[1–4] Reduction of CO_2 emissions from flue gas is imperative to mitigate environmental impacts.[1,4,5] There are different commercial technologies used to capture CO_2 from stationary energy sources: post-combustion, pre-combustion and oxy-fuel.[2,5–8]

Separation and compression of CO_2 can help global warming prevention but also provides CO_2 for other processes such as carbon capture and utilization (CCU) and enhanced oil recovery (EOR).[2,3,9] On the other hand, CO_2 is an abundant non-toxic resource that can be used in the food and chemical industries, such as in urea and fertilizer production, dry ice production or even in the supercritical state as a solvent.[2]

Smart Materials No. 29
Polymerized Ionic Liquids
Edited by Ali Eftekhari

Published by the Royal Society of Chemistry, www.rsc.org

The development of new and the improvement of existing materials for designing new solvents/sorbents for CO_2 separation is an urgent priority and significant goal in this field. Room temperature ionic liquids (RTIL) have been highlighted as the next generation of solvents for selective CO_2 separation.[6,10–14] Several RTILs present unique physicochemical properties, such as high thermal stability, negligible vapor pressure, non-flammability, and a myriad of possible combinations of cations and anions resulting in compounds with different properties.[8,10,12,14–16] High viscosity and production costs are seen as drawbacks for RTILs.[10,11] Poly(ionic liquid)s (PILs) appear to be an alternative to RTILs and promising materials for CO_2 capture.[15,17–21] Polymerized ionic liquids or PILs represent an emerging subclass of polyelectrolytes, where each repeating unit is ionic and connected through a polymeric backbone forming a macromolecular structure.[19,20] PILs combine the useful features of RTILs with the good mechanical stability, processing and tunable macromolecular design of polymeric materials.[14,20,22] PILs present higher CO_2 sorption capacities and sorption/desorption velocities than RTILs.[9,14,18,23,24] PILs, materials with smart designs, can be used for CO_2 separation from flue gas (CO_2/N_2) and natural gas purification (CO_2/CH_4).[14,17]

The aim of this chapter is to present a concise overview of PILs described in the literature. Furthermore, their synthesis routes, as well as the influence of the PIL backbone, anion type and modification on the CO_2 sorption capacity, will be discussed.

17.2 Carbon Capture Technologies

The most discussed technological routes for CO_2 capture from facilities using fossil fuels or biomass for power generation are post-combustion, pre-combustion and oxy-combustion.[2,5–8]

Post-combustion involves the removal of CO_2 from flue gases produced after burning fossil fuels or biomass in the presence of air[2,5,6,8,16] (Figure 17.1).

The gaseous mixture, *i.e.* the products of combustion, is usually at atmospheric pressure and high temperatures (320–400 K),[2] and contains a low CO_2 concentration (3–15% for coal facilities, 7–8% for gas facilities).[2,7,8] Despite the unfavorable conditions of flue gases from post-combustion for CO_2 capture, this process uses low cost technologies.[2,5,6] Processes based on

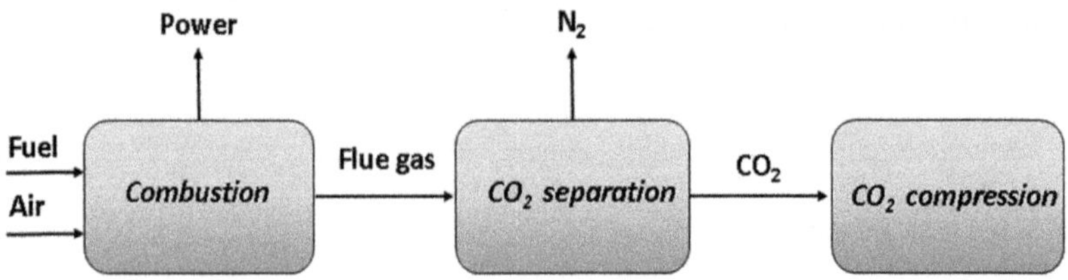

Figure 17.1 Principle of post-combustion CO_2 capture.

chemical absorption are the most suitable.[7] Aqueous solutions of alkanolamines such as monoethanolamine (MEA) are the benchmark solvents for CO_2 removal in post-combustion.[8,25–27] However, there are some drawbacks for the commercial use of these amine solutions, including formation of corrosive by-products, amine loss (requiring solvent replacement), the need for large volume absorbers and the energy penalty during the regeneration step.[28,29] Other separation technologies can also be employed, such as membrane separation, cryogenic separation and adsorption in solids.[5,8]

Pre-combustion involves primarily the reaction of fuel with oxygen or steam to produce a gas mixture containing mainly CO and H_2 (syngas) (Figure 17.2),[2,5,8] which reacts in a water-gas shift reactor producing more H_2 and a stream with a high partial pressure and concentration of CO_2 (15 to 60% volume).[8] CO_2 and H_2 are separated and H_2 used as a fuel.[5,6,8,16] The high pressure and concentration of CO_2 related to post-combustion capture allows for cost reduction in the loading and compression steps.[6] In addition, the use of more efficient separation technologies with a lower energy penalty required for the regeneration step (physical sorbents) is possible.[5,6] The main disadvantage of this system is the high cost required for the generating facility.[5,6]

In oxy-combustion (Figure 17.3), the oxygen is first separated from air in a separation unit.[6,8] The fuel is burned with O_2 and the resulting combustion gas is composed mainly of CO_2 (>80% volume) and water.[2,5,6,8,16] The great advantage of oxy-fuel is the high partial pressure of CO_2 allowing easy separation from the flue gas.[2,8] The main drawback of this technology is related to the large amount of oxygen needed, increasing costs and requiring high energy consumption.[2,5,6]

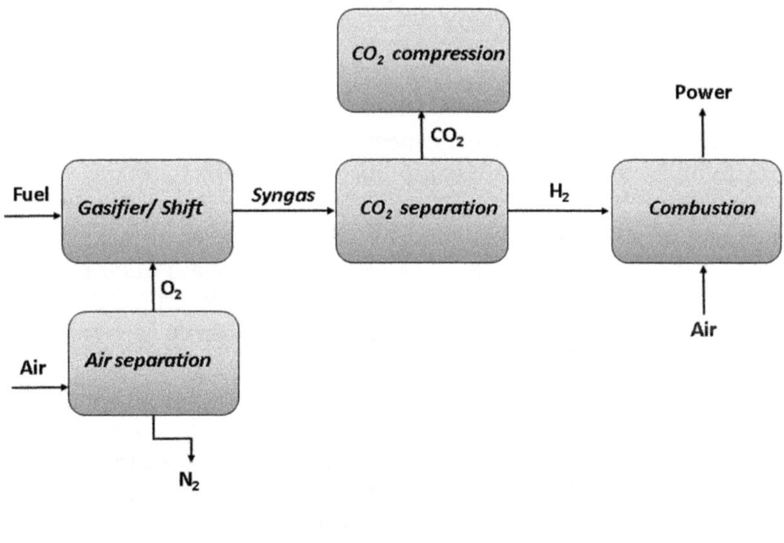

Figure 17.2 Principle of pre-combustion CO_2 capture.

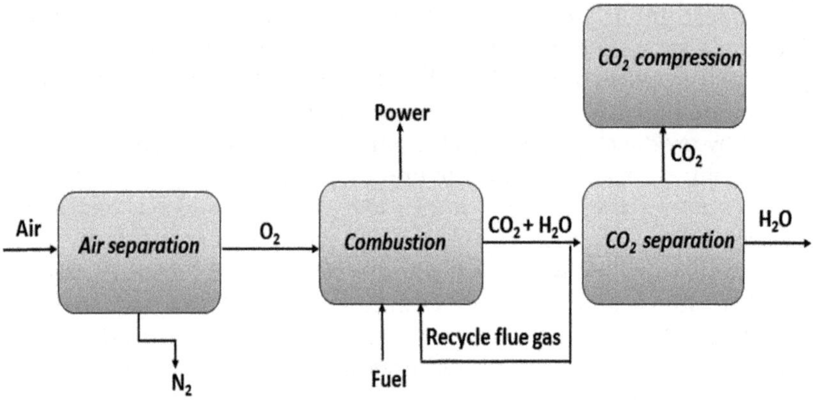

Figure 17.3 Principle of oxy-combustion CO_2 capture.

17.3 Ionic Liquids (ILs)

Ionic liquids (ILs) composed of a combination of an organic cation and an inorganic or organic anion (Figure 17.4) are usually liquid at room temperature having a melting point below 100 °C.[6,13,14,22,30,31]

ILs exhibit useful properties, such as negligible vapor pressure, non-flammability, high thermal stability, and tenability.[6,10,13,30] ILs can be divided into two broad categories: room-temperature ionic liquids (RTILs), considered conventional ILs, and task-specific ionic liquids (TSILs), also called functionalized ILs. RTILs show the typical behavior of a physical solvent, and gas solubility in RTILs increases with gas pressure (for gas pressures up to 10 bar).[13,14] According to recent advances, certain ILs are potential solvents for acidic gas capture, such as CO_2.[6,8,10,11,14] Most RTILs are selective for CO_2 absorption in gas mixtures,[13,32] yet the CO_2 absorption capacity needs to be improved.[13]

RTILs based on alkyl-imidazolium cations are the most investigated for use in gas absorption processes, especially for CO_2.[12,14,33–36] Several studies seeking to understand the high solubility of CO_2 in alkyl-imidazolium-based ionic liquids are described elsewhere.[14,34,35,37,38] The anion–CO_2 interaction force is one of the most highlighted factors. However, to a lesser extent, changes in the imidazolium cation involving side chain alkyl groups may also influence the CO_2 solubility.[11,14,33,38] It was also described that fluorination of the anion and cation can contribute significantly to increasing CO_2 solubility in these compounds.[11,34,38–40]

RTILs presents some drawbacks for implementation in CO_2 capture processes, including a relatively low CO_2 sorption capacity, poor selectivity in relation to hydrocarbons (being a problem mainly in natural gas processing), necessity of high circulation rates, and high viscosity. In addition, ILs show little economical competitiveness, due to the high cost of their synthesis.[10,11]

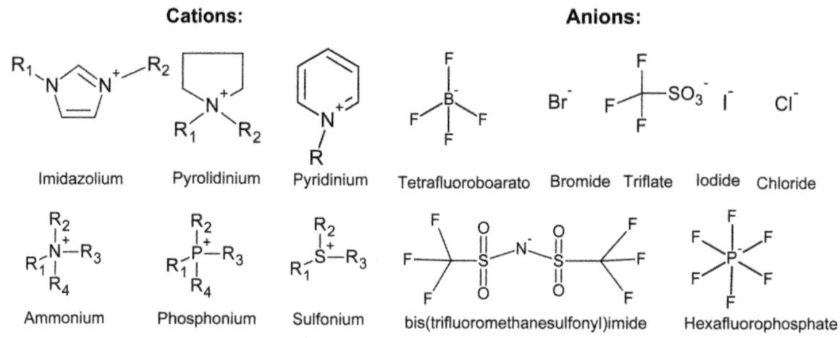

Figure 17.4 Examples of cation and anion structures typically found in ionic liquids.

17.4 Poly(Ionic Liquid)s

Tang and co-workers[24] first highlighted the increased CO_2 sorption ability when turning ionic liquid monomers into their polymeric form as poly(ionic liquid)s (PILs). For instance, ammonium-based PILs absorb 6.0 to 7.6 times more CO_2 relative to the corresponding RTIL monomers. PILs also present faster CO_2 sorption/desorption capacity compared to RTILs. Furthermore, the sorption/desorption process is completely reversible indicating that PILs are more suitable candidates for industrial CO_2 capture relative to IL monomers.

Several factors can affect the CO_2 sorption capacity of PILs, namely alkyl chain length, cation, anion, porosity, polymer backbone, cross-linking and molecular weight.[15,17,22] PIL syntheses are generally performed *via* the following routes: direct IL monomer polymerization, condensation and polymer modification.[19,20,22,41] The definition of the synthesis method is directly linked to the target poly(ionic liquid). Each route can produce PILs with different characteristics presenting advantages and limitations, directly affecting the resulting material properties. PILs of different structure and backbone are produced. Different architectures, such as linear, star-shaped or hyperbranched, can also be designed.[20]

17.4.1 Poly(Ionic Liquid) Syntheses

17.4.1.1 Via *IL Monomer Radical Polymerization*

The polymerizable unit in the cation is the dominant form.[19,42] It is worth mentioning the myriad of possible combinations of cations and anions resulting in a platform of options to design cationic PILs with desired properties. Figure 17.5 presents the main cationic PILs described in the literature.

These PILs are mainly synthesized by direct radical polymerization of IL monomers.[15,19,20] This method presents great tolerance towards impurities and moisture[19] favoring PIL production in contrast with other techniques sensitive to those components. PILs based on polymerizable monomers of

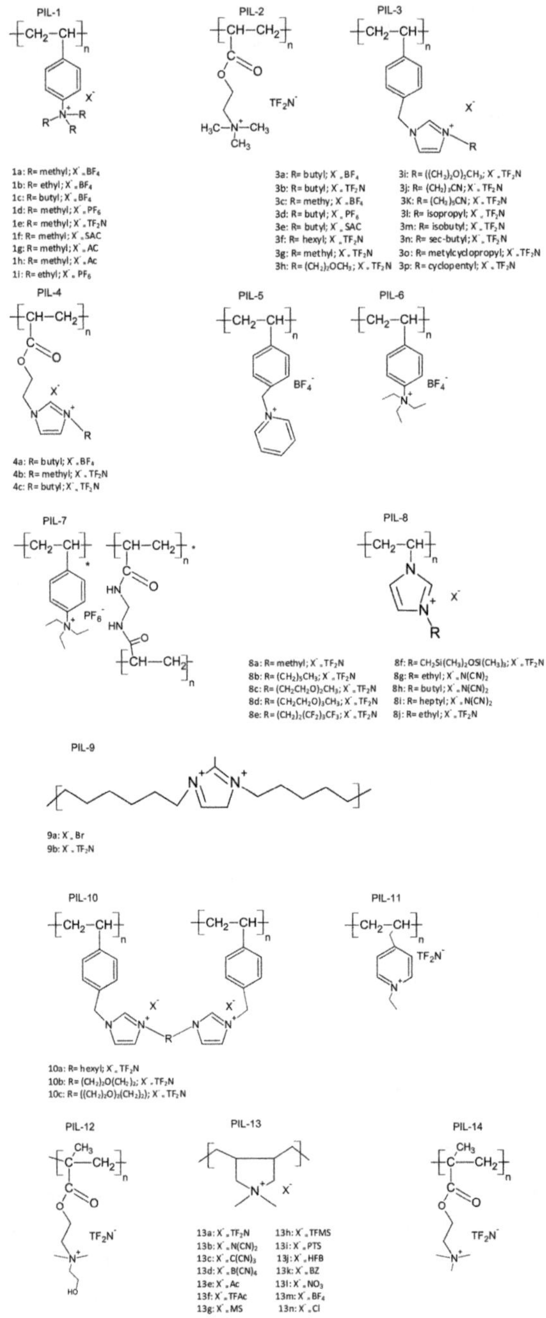

Figure 17.5 Chemical structures of various cationic PILs obtained by direct radical polymerization.

acryloyl or styrene imidazolium with diverse functional substituents are the most described in the literature.[17,20,42] Other cations based on tetraalkyl ammonium and phosphonium, and pyridinium have also been used to obtain PILs.[9,23,43,44]

17.4.1.2 Via *Condensation Polymerization and Polymer Modification*

The use of condensation polymerization for PIL synthesis has received little attention when compared to radical polymerization. This polymerization process could enrich the PIL classes producing new materials for CO_2 separation. PIL syntheses are usually performed by condensation of the respective ionic monomers.[41] Cationic PILs can also be synthesized *via* polymer modification. Polyethylene oxide is obtained by this route.[45] The synthesis of poly[(1-butyl imidazolium-3) methyl ethylene oxide tetrafluoroborate] (P[BIEO][BF_4]) *via* the modification of polyethylene oxide was described by Tang *et al.*[45] The formation of the poly(ethylene oxide) backbone was carried out from poly(epichlorohydrin) with the addition of aluminum isopropoxide, *N*-butylimidazole and sodium tetrafluoroborate ($NaBF_4$) (Figure 17.6).

PILs based on polybenzimidazoles (PBIs) have been described in the literature.[46,47] PBIs were synthesized *via* a solution polycondensation method using polyphosphoric acid (PPA) as a solvent and condensation agent. Systematic variation of the PBI family based on a tetraamine (3,3-diaminobenzidine) and different carboxylic acids was described elsewhere.[48] Aiming to obtain PILs based on a strong film-forming capacity, the group performed *N*-quaternarization with a methyl group, followed by iodine exchange with various promising anions (Figure 17.7).[46] Figure 17.7 and 17.8 show the PBI-based PILs used for CO_2 separation.[47]

Preparation of functional imidazolium-based ILs with symmetrical ester and hydroxyl groups was described elsewhere (Figure 17.9). These monomers were used to prepare PILs by condensation polymerization using stannous octoate as a catalyst with equal moles of IL.[49]

An efficient method for transforming polyimides (PIs), containing benzimidazole or quinuclidine in their structure, into PILs has been reported elsewhere (Figure 17.10).[41] Benzimidazole and quinuclidine were incorporated into the polymer chain through their respective diamines

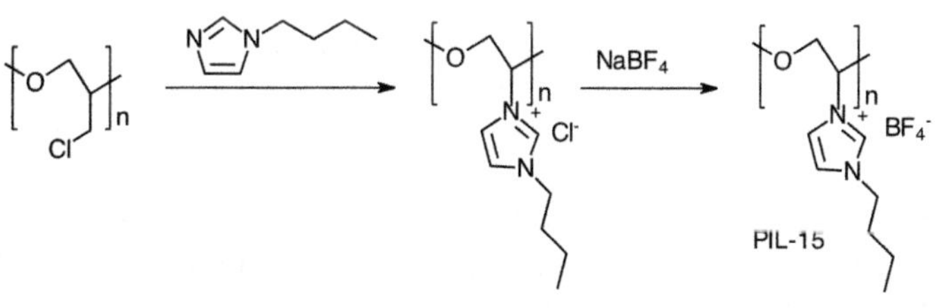

Figure 17.6 Synthesis of P[BIEO][BF_4].

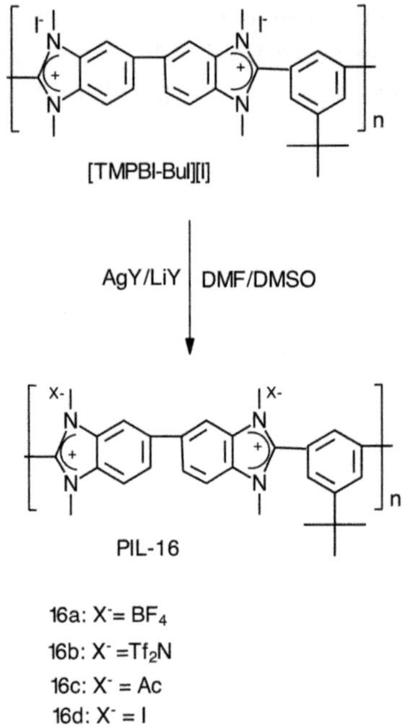

Figure 17.7 Scheme of the synthesis steps of PBI-based PILs.

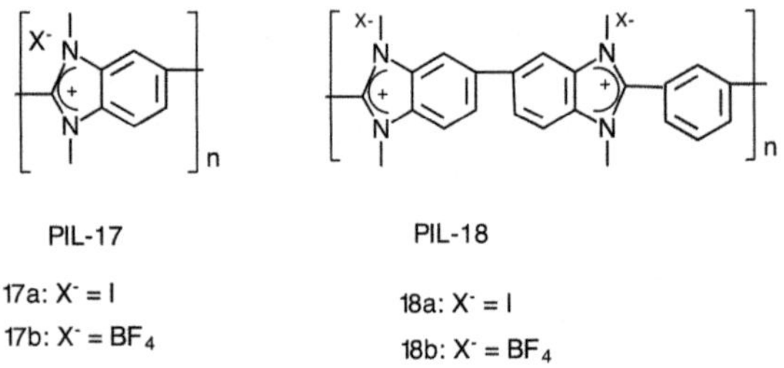

Figure 17.8 Chemical structures of PILs based on PBI.

(5(6)-amino-2-(4′-aminophenyl) benzimidazole (ABIZ) and 3,3-bis(4-aminophenyl) quinuclidine (AQ)). 4,4′-(Hexafluoroisopropylidene) diphthalic anhydride (HFDA) was chosen as the pairing monomer for condensation. The polycondensation reaction was performed in *m*-cresol at 180 °C, resulting in PIs with high molecular weight. PI with benzimidazole was initially deprotonated and *N*-alkylated with bromoethane, quaternized with iodoethane, and finally subjected to metathesis ion exchange reaction with LiTFSI. PI bearing the quinuclidine fragment was exposed to quaternization with iodoethane and underwent metathesis ion exchange reaction with LiTFSI.

Synthesis of versatile and inexpensive polyurethane-based PILs designed for CO_2 capture was described in the literature.[21,50,51] Two different approaches were used for the synthesis of these materials (Figure 17.11). The first

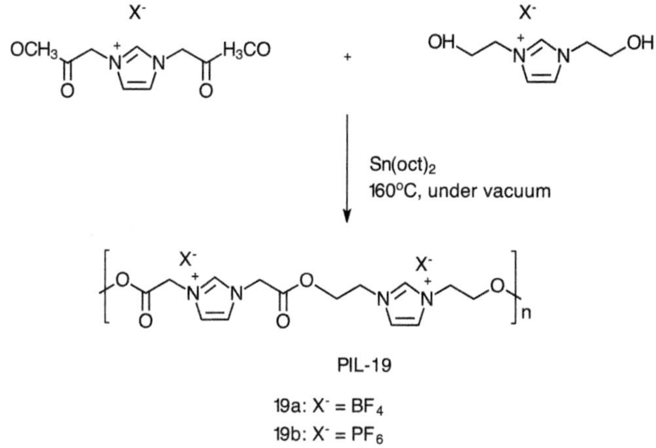

Figure 17.9 Scheme of the synthesis steps of PILs obtained from imidazolium-based ILs with symmetrical ester and hydroxyl groups.

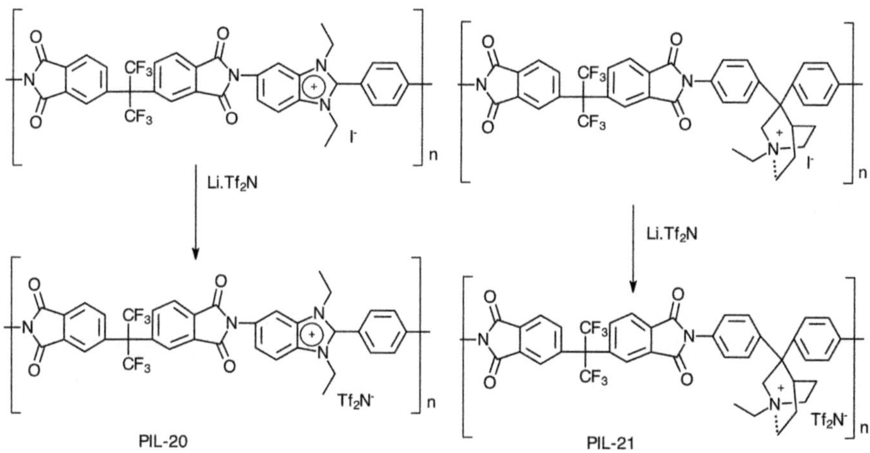

Figure 17.10 Scheme of the transformation of polyimides (PIs) containing benzimidazole or quinuclidine in their structure into PILs.

approach produces anionic PILs with the introduction of imidazolium, phosphonium or ammonium as a cation.[21,50] PILs were obtained using dibutyl tin dilaurate (IV) (DBTDL) as a catalyst, hexamethylene diisocyanate (HDI) as the diisocyanate, dimethylol butanoic acid (DMBA) or dimethylol propanoic acid (DMPA) as the diol, and polytetramethylene glycol (MM = 2000 or 1000 g mol^{-1}) (PTMG) as the polyol. The ion exchange step was performed using the following ionic liquids: (butyl methyl imidazolium chloride) [bmim][Cl], (dimethylbutyl methyl imidazolium chloride) [dmbmim][Cl], tetrabutylammonium bromide (TBAB) or tetrabutylphosphonium bromide (TBPB) acting as cations for the anionic polyurethanes.[21,50] The second approach produces

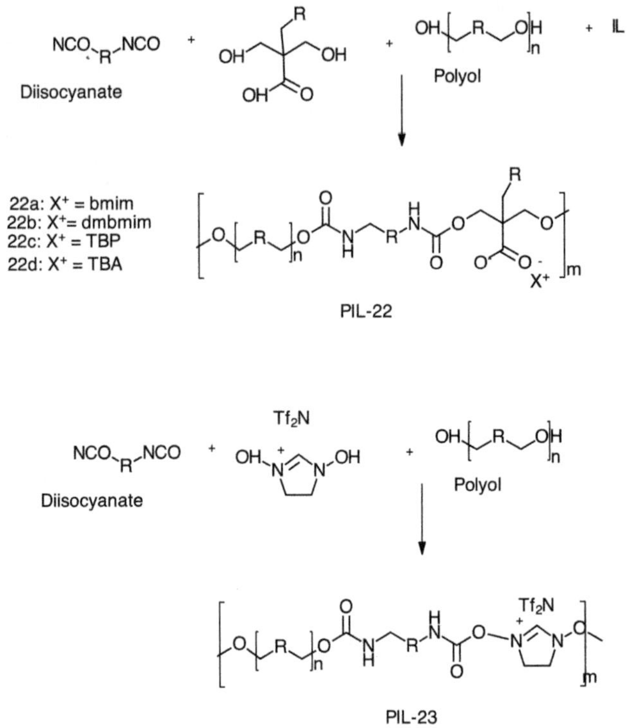

Figure 17.11 Scheme of typical syntheses to obtain PU-based PILS. PIL-22, anionic PU; PIL-23, cationic PU.

cationic polyurethanes. For polycation synthesis, [dhim][Tf_2N] (bis(trifluoromethylsulfonyl) imide-dihidroxiimidazol-1,3) was used as the diol.[50] The use of anionic PU chains to produce PILs represents an easy way to obtain different materials.

17.5 Performance of PILs Synthesized by Direct Radical Polymerization of IL Monomers in CO_2 Capture and Separation

Tang and co-workers[18,23,24] described the use of imidazolium-based PILs obtained *via* radical polymerization as CO_2 sorbents. The synthesis and sorption evaluation of three PILs were first reported:[18] poly[1-(*p*-vinylbenzyl)-3-butylimidazolium tetrafluoroborate] (PVBIT) (PIL-3a), poly[1-(*p*-vinylbenzyl)-3-butylimidazolium hexafluorophosphate] (PVBIH) (PIL-3d) and poly[2-(methacryloyloxy)ethyl-3-butylimidazolium tetrafluoroborate] (PBIMT) (PIL-4a). The PIL structures are shown in Figure 17.1. The CO_2 sorption capacity of these materials was higher than that of the ionic liquid [bmim][BF_4]. In addition, these PILs required only a few minutes (4 min for

PIL-3a and PIL-3d and 3 min for PIL-4a) to reach 90% of their CO_2 sorption capacity and about 30 minutes to reach equilibrium, whereas for [bmim][BF_4] more than 400 min was needed to reach its total sorption capacity.[18] The PVBIT (PIL-3a) and PBIMT (PIL-4a) PILs were submitted to four sorption/desorption cycles and no variation in sorption capacity was observed. Desorption was complete after 15 minutes under vacuum, suggesting that the process is completely reversible.[18] CO_2 sorption in the PILs is very selective, since no mass gain was observed when PVBIT (PIL-3a) was exposed to N_2 or O_2 under similar conditions.[18] These results aroused huge interest in the scientific community and stimulated research on the use of PILs for CO_2 capture.

Several studies have been conducted on the use of cationic PILs synthesized *via* direct radical polymerization of monomers in CO_2 capture.[9,43,52–60] These PILs offer an extremely versatile and adjustable platform for developing new sorbents for CO_2 capture.[14,17,20] PILs exhibiting CO_2 sorption performances higher than those of conventional ILs have been described elsewhere.[9,18,23,24] Table 17.1 presents the CO_2 sorption performance of several cationic PILs synthesized *via* direct radical polymerization.

Experimental results regarding the CO_2 sorption capacity of PILs have highlighted that the effects of the anion and cation on CO_2 sorption are different when compared to those of conventional RTILs. In contrast to conventional RTILs, the cation plays a major role in CO_2 sorption, whereas the anion plays a secondary role.[23,43]

The influence of the cation, anion and backbone structure of PILs on their CO_2 solubility, permeability, selectivity and thermal properties has been continuously evaluated.[23,43,52,60]

17.5.1 The Effect of the Cation, Anion and Backbone Structure on CO_2 Sorption

To our knowledge, the literature describes only experimental data for CO_2 sorption and/or selectivity for cationic PILs. Factors such as the anion, cation and polymer backbone affect the CO_2 sorption capacity of cationic PILs. However, the cation type exerts a greater influence relative to the other factors.[24]

PILs for CO_2 capture and separation have been synthesized using mainly acryloyl[22,24,43,45,57,61] and styrene[9,23,43,54,57,60] as the main polymeric chain structure. However, other structures have also been explored such as vinyl,[58,61–64] diallyldimethylammonium[44,52,61,65] and ethylene oxide[45] (see Figure 17.1).

The CO_2 sorption capacity of cationic PILs with different structures in their main chain (styrene (PIL-3a) and acryloyl (PIL-4a)) has been investigated.[24] PIL-3a presented a sorption capacity about 20% (Table 17.1, entry 33, 2.27 mol%) higher than that of PIL-4a (Table 17.1, entry 57, 1.80 mol%). Similar behavior was observed for the alkylammonium cation styrene PIL-1a (Table 17.1, entry 1, 10.22 mol%) and the acryloyl-based PIL PIL-2 (Table 17.1, entry 24, 7.99 mol%). Styrene-based PILs provide greater CO_2 sorption capacity than acryloyl-based PILs, probably due to their more rigid structure[43]

Table 17.1 CO_2 sorption performance of various cationic PILs synthesized *via* direct radical polymerization.

PIL	Entry	Experimental conditions (bar–°C)	CO_2 Loading			N_2 Loading	CH_4 Loading	Permeability (Barrers)[a]			Permselectivity (dimensionless)		Ref.
			mol%	mg g^{-1}	cm^3(STP)/cm^3 atm	cm^3(STP)/cm^3 atm	cm^3(STP)/cm^3 atm	CO_2	N_2	CH_4	CO_2/CH_4	CO_2/N_2	
PIL-1a	1	0.79–22	10.22	–	–	ND	–	–	–	–	–	–	24
	2	12.15–22	44.80	–	–	–	–	–	–	–	–	–	24
	3	0.79–22	10.22	–	–	ND	–	–	–	–	–	–	23
	4	1.01–22	–	–	10.68	ND	–	–	–	–	–	–	43
	5	1.00–25	–	–	7.54	–	–	–	–	–	–	–	9
	6	10.00–25	–	–	31.77	–	–	–	–	–	–	–	9
	7	1.00–50	–	–	3.42	–	–	–	–	–	–	–	9
	8	10.00–50	–	–	20.09	–	–	–	–	–	–	–	9
	9	1.00–75	–	–	1.42	–	–	–	–	–	–	–	9
	10	10.00–75	–	–	11.61	–	–	–	–	–	–	–	9
	11	1.01–35	–	–	4.86	0.01	–	–	–	–	–	–	52
PIL-1b	12	0.79–22	4.85	–	–	ND	–	–	–	–	–	–	23
	13	1.01–22	–	–	5.90	ND	–	–	–	–	–	–	43
PIL-1c	14	0.79–22	3.11	–	–	ND	–	–	–	–	–	–	23
PIL-1d	15	0.79–22	10.66	–	–	ND	–	–	–	–	–	–	23
	16	1.01–22	–	–	8.95	ND	–	–	–	–	–	–	43
	17	1.01–25	–	770	–	–	–	–	–	–	–	70	56
PIL-1e	18	0.79–22	2.85	–	–	ND	–	–	–	–	–	–	23
	19	1.01–22	–	–	2.59	ND	–	–	–	–	–	–	43
PIL-1f	20	0.79–22	2.67	–	–	ND	–	–	–	–	–	–	23
PIL-1g	21	1.01–35	–	–	8.34	0.10	–	–	–	–	–	–	52
PIL-1h	22	1.01–35	–	–	2.48	0.15	–	–	–	–	–	–	52
PIL-1i	23	1.01–25	–	10.36	–	–	–	–	–	–	–	–	60

PIL-2	24	0.79–22	7.99	–	–	ND	–	–	–	–	–	–	24
	25	0.79–22	7.99	–	–	ND	–	–	–	–	–	–	23
	26	1.01–22	–	–	8.13	ND	–	–	–	–	–	–	43
	27	1.00–25	–	–	7.01	–	–	–	–	–	–	–	9
	28	10.00–25	–	–	27.33	–	–	–	–	–	–	–	9
	29	1.00–50	–	–	3.50	–	–	–	–	–	–	–	9
	30	10.00–50	–	–	18.09	–	–	–	–	–	–	–	9
	31	1.00–75	–	–	1.68	–	–	–	–	–	–	–	9
	32	10.00–75	–	–	11.32	–	–	–	–	–	–	–	9
PIL-3a	33	0.79–22	2.27	–	–	ND	–	–	–	–	–	–	24
	34	0.79–22	2.22	3.05	–	ND	–	–	–	–	–	–	18
	35	0.79–22	2.27	–	–	ND	–	–	–	–	–	–	45
PIL-3b	36	0.79–22	2.27	–	–	ND	–	–	–	–	–	–	24
	37	0.79–22	2.23	–	–	ND	–	–	–	–	–	–	45
	38	2.02–20	–	–	4.40	ND	0.55	20.00	0.67	0.91	22	30	57
	39[a]	10.0–20	–	–	–	–	–	27.69	–	0.96	29	–	54
PIL-3c	40	0.79–22	3.05	–	–	ND	–	–	–	–	–	–	45
	41	1.01–22	–	–	3.99	ND	–	–	–	–	–	–	43
PIL-3d	42	0.79–22	2.75	3.22	–	ND	–	–	–	–	–	–	18
	43	0.79–22	2.80	–	–	ND	–	–	–	–	–	–	45
PIL-3e	44	0.79–22	1.55	–	–	ND	–	–	–	–	–	–	45
PIL-3f	45	2.02–20	–	–	3.90	0.10	0.57	32.00	1.40	2.30	17	28	57
	46	10.0–20	–	–	–	–	–	28.6	–	1.69	17	–	54
PIL-3g	47	2.02–20	–	–	4.00	ND	0.21	9.20	0.29	0.24	39	32	57
	48	10.0–20	–	–	–	–	–	7.38	–	0.18	43	–	54
PIL-3h	49	2.02–22	–	–	–	–	–	16.0	0.39	0.48	33	41	58
PIL-3i	50	2.02–22	–	–	–	–	–	22.0	0.50	0.74	29	44	58
PIL-3j	51	2.02–22	–	–	–	–	–	4.10	0.11	0.11	37	37	58
PIL-3k	52	2.02–22	–	–	–	–	–	8.20	0.21	0.28	30	40	58
PIL-3l	53	3.04–20	–	–	–	–	–	10.40	0.33	0.35	30	31	55
PIL-3n	54	3.04–20	–	–	–	–	–	13.60	0.39	0.55	25	35	55
PIL-3o	55	3.04–20	–	–	–	–	–	7.94	0.24	0.32	25	33	55
PIL-3p	56	3.04–20	–	–	–	–	–	6.65	0.19	0.25	27	34	55
PIL-4a	57	0.79–22	1.80	–	–	ND	–	–	–	–	–	–	24
	58	0.79–22	1.77	2.41	–	ND	–	–	–	–	–	–	18
	59	0.79–22	1.78	–	–	ND	–	–	–	–	––	–	45
PIL-4b	60	2.02–20	–	–	3.60	ND	0.17	7.00	0.23	0.19	37	31	57
PIL-4c	61	2.02–20	–	–	4.50	ND	0.57	22.00	0.71	2.30	22	30	57
PIL-5	62	1.01–22	–	–	5.62	ND	–	–	–	–	–	–	43

(continued)

Table 17.1 (*continued*)

PIL	Entry	Experimental conditions (bar–°C)	CO_2 Loading			N_2 Loading	CH_4 Loading	Permeability (Barrers)[a]			Permselectivity (dimensionless)		Ref.
			mol%	mg g^{-1}	cm^3(STP)/ cm^3 atm	cm^3(STP)/ cm^3 atm	cm^3(STP)/ cm^3 atm	CO_2	N_2	CH_4	CO_2/CH_4	CO_2/N_2	
PIL-6	63	1.01–22	–	–	4.77	ND	–	–	–	–	–	–	43
PIL-7	64	1.01–25	–	14.04	–	–	–	–	–	–	–	–	60
PIL-8a	65	2.02–22	–	–	3.30	–	–	4.80	–	–	40	29	63
PIL-8b	66	2.02–22	–	–	3.80	–	–	69.0	–	–	9.9	17	63
PIL-8c	67	2.02–22	–	–	3.20	–	–	14.00	–	–	32	32	63
PIL-8d	68	2.02–22	–	–	3.90	–	–	26.00	–	–	25	34	63
PIL-8e	69	2.02–22	–	–	3.90	–	–	69.00	–	–	14	11	63
PIL-8f	70	2.02–22	–	–	3.50	–	–	130.0	–	–	8.7	14	63
PIL-8g	71	10.0–35	–	–	0.48	–	–	0.09	–	–	–	–	64
PIL-8h	72	10.0–35	–	–	0.73	0.03	–	4.24	0.16	–	–	26	64
PIL-8i	73	10.0–35	–	–	1.08	0.05	–	33.50	1.60	–	–	21	64
PIL-9a	74	4.04–22	–	–	0.45	–	–	0.13	ND	ND	N–D	ND	53
PIL-9b	75	2.02–22	–	–	0.79	–	–	5.30	0.22	0.26	20	24	53
PIL-10a	76	2.02–20	–	–	2.80	–	–	4.40	0.20	0.16	27	22	59
PIL-10b	77	2.02–20	–	–	3.80	–	–	4.00	0.14	0.12	32	28	59
PIL-10c	78	2.02–20	–	–	3.50	–	–	3.80	0.16	0.13	28	23	59
PIL-13a	79	1.00–21	–	–	1.40	0.02	0.02	5.09	0.23	0.18	28	22	44
	80	20.00–35	–	–	15.10	1.14	–	–	–	–	–	–	52
PIL-13e	81	20.00–35	–	–	78.72	0.70	–	–	–	–	–	–	52
PIL-13f	82	20.00–35	–	–	12.62	2.26	–	–	–	–	–	–	52
PIL-13g	83	20.00–35	–	–	5.01	3.64	–	–	–	–	–	–	52
PIL-13h	84	20.00–35	–	–	19.45	7.66	–	–	–	–	–	–	52
PIL-13i	85	20.00–35	–	–	5.02	0.66	–	–	–	–	–	–	52
PIL-13j	86	20.00–35	–	–	21.12	1.22	–	–	–	–	–	–	52
PIL-13k	87	20.00–35	–	–	27.48	0.66	–	–	–	–	–	–	52
PIL-13l	88	20.00–35	–	–	8.42	1.74	–	–	–	–	–	–	52
PIL-13m	89	20.00–35	–	–	17.01	1.26	–	–	–	–	–	–	52
PIL-13n	90	20.00–35	–	–	2.45	1.06	–	–	–	–	–	–	52

[a]Permeability in Barrers (10^{-10} cm^3(STP) cm/cm^2 s cm Hg).

(Table 17.1, entries 4 and 26). Styrene and acryloyl monomers show no CO_2 absorption due to their crystalline structures.[24]

The cation plays a decisive role in CO_2 solubility in cationic PILs obtained *via* radical polymerization. The effect of cation type and changes involving the side chain alkyl groups of the cation in different backbone structures has been investigated.[24,43,45] The CO_2 sorption is higher in the tetraalkylammonium cation PIL-1a (Table 17.1, entry 1, 10.22 mol%) when compared to the imidazolium cation PIL-3a (Table 17.1, entry 33, 2.27 mol%).[24] According to Tang and co-workers, this result is probably related to the high density of positive charge in the tetraalkylammonium cation resulting in a strong interaction with CO_2. On the other hand, the positive imidazolium charge is delocalized.[24]

It has been described elsewhere[43] that cross-linking in cationic PILs reduces the CO_2 sorption capacity. Although cross-linking increases the microvoid volume due to the stiffness increase, it may hinder the PIL/CO_2 interaction.[43] The insertion of cross-links into the tetraalkylammonium cation styrene-based PIL-1a structure, with BF_4 as the anion, resulted in a slight reduction in CO_2 solubility.[43]

Yu *et al.*[60] observed distinct behavior when they synthesized *N*,*N*-methylenebisacrylamide-cross-linked poly(4-vinylbenzyltriethylammonium hexafluorophosphate), (MBA-cross-linked-P[VBTEA][PF_6]) (PIL-7a) and compared the CO_2 sorption results with those of P[VBTEA][PF_6] (PIL-1i). In this case, when cross-linking was inserted into the PIL-1i structure, the CO_2 sorption capacity was increased from 10.36 mg g^{-1} to 14.04 mg g^{-1}. The cross-linking allowed an increase in the specific surface and pore formation.[60] The reuse of PIL-7a was evaluated in sorption/desorption cycles and the reduction in CO_2 sorption capacity was <1% after four cycles.[60]

In cationic PILs, longer alkyl side chains in the cations hinder CO_2 sorption due to steric effects, decreasing the sorption capacity.[43,45] The substitution of butyl groups (PIL-3a) for methyl groups (PIL-3c) in the imidazolium cation resulted in an increase in CO_2 sorption capacity of approximately 34%. The values increased from 2.27 mol% (PIL-3a) to 3.05 mol% (PIL-3c)[45] (Table 17.1, entry 35 and 40). In PILs containing ammonium cations, the CO_2 sorption capacity also decreases with the introduction of long side chains into the cation. Replacement of methyl (PIL-1a) by ethyl (PIL-1b) and butyl (PIL-1c) reduced the CO_2 solubility by nearly 53% and 70%, respectively[23] (see Table 17.1, entries 3, 12 and 14).

Experimental CO_2 sorption data obtained for styrene-based PILs with BF_4 as the anion and different cations (ammonium (PIL-1b), pyridinium (PIL-5), phosphonium (PIL-6) and imidazolium (PIL-3c)) indicated that CO_2 sorption capacity decreases in the following order: ammonium > pyridinium > phosphonium > imidazolium[43] (see Table 17.1, entries 13, 62, 63 and 41). In addition, as observed in conventional ILs, sorption capacity increases with increasing pressure.[43]

In general, CO_2 permeability tends to increase with growing pressure and temperature. This behavior is due to the increase in diffusion at higher temperatures and pressures. On the other hand, selectivity decreases with growing temperature as well.[54]

The structure of the polymer backbone exerts a considerable effect on CO_2 permeability and selectivity. Styrene-based PILs (PIL-3g, Table 17.1, entry 47, 9.2 PCO_2)[57] presented higher selectivity to CO_2 when compared to acrylate- (PIL-4b, Table 17.1, entry 60, 7.0 PCO_2)[57] and vinyl-based PILs (PIL-8a, Table 17.1, entry 65, 4.8 PCO_2).[63] Styrene- and acrylate-based PILs are more selective to CO_2 relative to vinyl-based PILs.[57,63]

Cationic PILs demonstrate greater selectivity to CO_2 when compared to RTILs.[56] The nature and length of the alkyl substituents of the cation may influence the selectivity and permeability of the PILs.[54,58,63] An increase in alkyl chain length or the introduction of polar substituents on the cation can improve the separation of CO_2 from CO_2/N_2 and CO_2/CH_4 gas streams.[57,58,63] The disiloxane substituent on the cation can promote increased permeability.[63] The presence of polar groups such as oligo(ethylene glycol) (PIL-3h and PIL-3i, Table 17.1, entries 49 and 50) as substituents on the imidazolium cation produces more permeable membranes than nitrile-terminated functional groups (PIL-3j and PIL-3k, Table 17.1, entries 51 and 52) and alkyl substituents (PIL-3g, PIL-3b and PIL-3f, Table 17.1, entries 47, 38 and 45).[57,58] In addition, an increase of alkyl chain length between polar groups may also contribute to improved selectivity and permeability.[58]

Introduction of cross-linked monomers into cationic PILs shows no significant selectivity influence, but it can provide a significant reduction in the permeability of these materials, making them poor candidates for high-performance gas separation.[59] The introduction of polar groups or an increase in the alkyl chain length between polar groups of the cation substituents seems to have little effect on the permeability in the presence of cross-linked structures. The PIL with a hexyl group (PIL-10a, Table 17.1, entry 76) was slightly more permeable than those with polar substituents (PIL-10b and PIL-10c, Table 17.1, entries 77 and 78).[59] PILs with branched (PIL-3l and PIL-3n, Table 17.1, entries 53 and 54) or cyclic substituents (PIL-3o and PIL-3p, Table 17.1, entries 55 and 56)[55] in the imidazolium cation exhibited selectivities for CO_2/N_2 and CO_2/CH_4 greater than those of PILs with linear alkyl chain substituents (PIL-3g, PIL-3b and PIL-3f, Table 17.1, entries 47, 38 and 45).[57] Selectivity improvements associated with branched and cyclic functionalities are due to smaller free fractional volumes (FFVs).[55] Although the nature of the anion exerts less influence on the use of PILs as sorbents for CO_2 capture, it is still a parameter that deserves attention and must be taken into consideration. The use of various inorganic (Br, BF_4, PF_6, Tf_2N *etc.*) and organic (Sac, Bz, Ac *etc.*) anions has been reported in the literature for obtaining cationic PILs (Figure 17.5). Different from the trend observed in RTILs, where studies indicate that the presence of fluorinated atoms in the anions increases the CO_2 sorption capacity,[35,37] surprisingly, for PILs, fluorinated anions show no direct impact on CO_2 sorption.[23,43,45] This behavior was evidenced in sorption tests carried out on ammonium-based PILs, synthesized with different fluorinated inorganic anions (PF_6 (PIL-1d), BF_4 (PIL-1a), and Tf_2N (PIL-1e)) and the non-fluorinated organic anion Sac (PIL-1f). The CO_2 sorption capacity obtained for the PIL containing the Tf_2N anion (PIL-1e, Table 17.1, entry 18, 2.85 mol%) was much lower than that of the PILs with PF_6 (PIL-1d, Table 17.1, entry 15, 10.66 mol%), and BF_4 (PIL-1a, Table 17.1, entry 3, 10.22 mol%) and similar to that achieved with the non-fluorinated anion Sac (PIL-1f, Table 17.1, entry 20, 2.67 mol%).[23] Studies with imidazolium-based PILs showed similar behavior, with a CO_2 sorption capacity of 2.80 mol% for PIL-3d (Table 17.1, entry 43), 3.05 mol% for PIL-3c (Table 17.1, entry 40), 2.23 mol% for PIL-3b (Table 17.1, entry 37) and 1.55 mol% for PIL-3e (Table 17.1, entry 44).[45]

In the same way that the anion can influence the CO_2 solubility, it can also impact the permeability and selectivity.[52,63] Comparison of the selectivity and permeability values obtained for PIL-9a and PIL-9b (Table 17.1, entries 74 and 75) showed that the substitution of an inorganic anion with low molar mass (Br) with an anion of sulfonamide class (Tf_2N) increased the CO_2 permeability and selectivity.[53] A number of diallyldimethylammonium-based cationic PILs (PIL-13) with different carboxylate (Ac, TFAc, HFB, BZ), sulfonate (MS, PTS, TFMS), sulfonamide (Tf_2N), and inorganic (Cl, NO_3, BF_4) anions were investigated regarding their CO_2 sorption and selectivity capacity.[52] Among all PILs evaluated, the sample with the carboxylate anion Ac (PIL-13e) exhibited the highest CO_2 solubility (Table 17.1, entry 81) and selectivity in relation to N_2 (SCO_2/SN_2 of 114.3).[52]

As observed in RTILs, changes in the anion basicity of PILs affect the CO_2 sorption and selectivity.[52] The CO_2 solubility for PILs with carboxylate anions increased in the following order: [TFAc] < [HFB] < [Bz] < [Ac] (Table 17.1, entries 82, 86, 87 and 81) according to the increase in basicity (pKa of the conjugate acid). This behavior is similar to that of RTILs, where increasing basicity of the anion tends to increase the CO_2 solubility.[52]

For PILs containing sulfonated anions, the basicity of the anion cannot be considered the only factor responsible for the increased sorption and selectivity of CO_2. Other properties such as FFV and density can also have an influence. SCO_2 showed the following variation: [TFMS] > [PTS] ≈ [MS]; and the basicity values (pKa of the conjugate acid) were: [TFMS] = −13, [PTS] = −2.8, [MS] = −2.0.[52]

In inorganic anions such as Cl, NO_3 and BF_4 the increase in the anion molar mass improves the CO_2 solubility (Table 17.1, entries 90, 88 and 89), SCO_2 and SCO_2/SN_2.[52]

17.6 Performance of PILs Synthesized by Condensation Polymerization and Polymer Modification in CO_2 Capture and Separation

Studies reporting the use of PILs synthesized *via* condensation polymerization and polymer modification for CO_2 capture and separation are still scarce relative to those synthesized by radical polymerization. However, these strategies have proved to be a promising route to obtain cationic and anionic

PILs. PILs obtained by this route are mainly based on poly(ethylene oxide),[45] polyurethane[21,50,51] polybenzimidazoles,[46,47] polyimides[41] and imidazolium containing ester and hydroxyl groups.[49] Changes in the structure of cationic[41,46,47,49,50] or anionic[21,50,51] PILs can impact heavily on the CO_2 sorption capacity. The main results described in the literature will be discussed in this section and are presented in Table 17.2.

Tang *et al.*[45] were among the first to design and synthesize cationic PILs (PIL-15) (Figure 17.6) for CO_2 capture. The PIL was obtained from direct modification of the polymer (poly(epichlorohydrin)). The CO_2 sorption capacity of PIL-15 was compared to that of imidazolium-based PILs synthesized by radical polymerization, containing the same anion (BF_4) but a different structure in the main chain (styrene (PIL-3a) and acryloyl (PIL-4a)). PIL-15 showed lower CO_2 sorption capacity (Table 17.2, entry 1, 1.06 mol%) relative to PIL-3a (2.27 mol%) and PIL-4a (1.78 mol%) (Table 17.1, entries 35 and 59) presenting styrene and acryloyl groups, respectively, in their structures.

The effect of the anion on the ability to solubilize CO_2 was investigated. Cationic PILs based on imidazolium containing symmetrical ester and OH groups synthesized with PF_6 (PIL-19b) and BF_4 (PIL-19a) anions were evaluated.[49] The PF_6 anion resulted in a higher CO_2 sorption capacity than BF_4 (Table 17.2, entries 12 and 11). The higher sorption value was attributed to the uneven morphology of PIL-19b, resulting in an improved superficial area and a higher CO_2 sorption capacity.[49]

The CO_2 sorption capacity of PIL-19a/b was superior to that of a number of ILs: 1-*n*-butyl-3-methylimidazolium hexafluorophosphate ([Bmim]PF_6) (2.9 mol%), 1,3-bis(2-methoxy-2-oxyl ethyl) imidazolium hexafluorophosphate ([EEIm]PF_6) (2.1 mol%), 1,3-bis(2-methoxy-2-oxyl ethyl) imidazolium tetrafluoroborate ([EEIm]BF_4) (1.0 mol%) and 1,3-bis(2-hydroxyl ethyl) imidazolium tetrafluoroborate ([HHIm]BF_4) (1.6 mol%). The CO_2 sorption capacity was inferior only to that of 1,3-bis-(2-hydroxyl ethyl) imidazolium hexafluorophosphate ([HHIm]PF_6) (10 mol%).[49] The CO_2 sorption kinetics of the PILs (19a/b) were also evaluated and compared to those of RTILs. The results indicated that the fundamental time to reach the sorption equilibrium was higher for PILs than for most RTILs, except for ([Bmim]PF_6) and ([EEIm]PF_6).[49]

The variation of the anion affects the PIL's solubility, selectivity and permeability.[47] Three different anions, BF_4 (PIL-16a), Tf_2N (PIL-16b) and Ac (PIL-16c) (Figure 17.7), were inserted into cationic polybenzimidazole-based PILs.[47] Anion variation in these PILs caused significant changes in the polymer appearance and stability. Unlike the BF_4 (PIL-16a) and Tf_2N (PIL-16b) PILs, which formed amber and pale yellow resistant films, respectively, the Ac anion (PIL-16c) formed no resistant membrane. The PILs (PIL-16a–c) exhibited excellent CO_2 separation performance (Table 17.2, entries 3, 4 and 5). In addition, the CO_2 sorption values of these membranes were about two to three times higher than the values obtained with conventional polysulfone (PSF) and polycarbonate (PC).[47]

CO_2 solubility as well as permeability and selectivity data for cationic polybenzimidazole-based PILs (PIL-16a–c)[47] demonstrated that the anion had a

Table 17.2 CO_2 sorption performance of various PILs synthesized *via* condensation polymerization and polymer modification.

PIL	Entry	Experimental condition (bar–°C)	CO_2 Loading mol%	CO_2 Loading mg g^{-1}	CO_2 Loading cm^3 (STP)/cm^3 atm	N_2 Loading cm^3 (STP)/cm^3 atm	CH_4 Loading cm^3 (STP)/cm^3 atm	Permeability (Barrers)[a] CO_2	Permeability (Barrers)[a] N_2	Permeability (Barrers)[a] CH_4	Permselectivity (dimensionless) CO_2/CH_4	Permselectivity (dimensionless) CO_2/N_2	Ref.
PIL-15	1	0.79–22	1.06	–	–	–	–	–	–	–	–	–	45
PIL-16a	2	10.00–35	–	–	38.20	–	–	–	–	–	–	–	46
	3	20.00–35	–	–	49.10	7.53	11.69	4.78	–	–	15.9	7.9	47
PIL-16b	4	20.00–35	–	–	36.90	5.74	6.82	17.90	–	–	14.4	7.6	47
PIL-16c	5	20.00–35	–	–	59.00	4.72	6.57	ND	–	–	ND	ND	47
PIL-16d	6	10.00–35	–	–	33.60	–	–	–	–	–	–	–	46
PIL-17a	7	10.00–35	–	–	14.50	–	–	–	–	–	–	–	46
PIL-17b	8	10.00–35	–	–	30.50	–	–	–	–	–	–	–	46
PIL-18a	9	10.00–35	–	–	25.90	–	–	–	–	–	–	–	46
PIL-18b	10	10.00–35	–	–	34.50	–	–	–	–	–	–	–	46
PIL-19a	11	0.86–25	4.50	–	–	–	–	–	–	–	–	–	49
PIL-19b	12	0.86–25	4.20	–	–	–	–	–	–	–	–	–	49
PIL-20	13	1.00–20	–	–	–	–	–	28.75	7.23	7.25	3.9	4.0	41
PIL-21	14	1.00–20	–	–	–	–	–	–	–	–	–	–	41
PIL-22a	15	0.82–30	–	9.40	–	–	–	–	–	–	–	–	21
PIL-22b	16	10.00–25	66.10	–	–	–	–	–	–	–	–	–	50
PIL-22c	17	0.82–30	–	15.70	–	–	–	–	–	–	–	–	21
PIL-22d	18	0.82–30	–	16.10	–	–	–	–	–	–	–	–	21
PIL-23	19	5.00–25	56.30	–	–	–	–	–	–	–	–	–	50

[a]Permeability in Barrers (10^{-10} cm^3(STP) cm/cm^2 s cm Hg).

strong effect on the CO_2 sorption capacity. PIL-16b presented a greater CO_2 permeability than PIL-16a (Table 17.2, entries 4 and 3). This result is probably due to greater gas diffusion resulting from the elevated chain mobility provided by the Tf_2N anion.[47] The $S(CO_2)/S(N_2)$ and $S(CO_2)/S(CH_4)$ selectivity, as well as $S(CO_2)$, increased with the anions in the following order: $Tf_2N < BF_4 < Ac$. The CO_2 sorption capacity of these PILs followed the increasing basicity order of the anions (pKa value of their conjugate acids $[Tf_2N]$ −4 < $[BF_4]$ −0.44 < [Ac] 4.75).[47] At high pressure levels, PIL-16a and PIL-16c presented a higher $S(CO_2)$ than other PILs described elsewhere, such as P[VBTMA][BF4].[52] Such behavior demonstrated the advantage of the use of aromatic structures for cationic PIL synthesis.[47]

Bhavsar *et al.*[46] also studied the influence of the anion on the CO_2 sorption capacity of cationic polybenzimidazole-based PILs. PILs (PIL-16a, PIL-16d, PIL-17a/b and PIL-18a/b) (Figure 17.7 and 17.8) were obtained using two different anions (I and BF_4) and three distinct cations. Their sorption capacities were evaluated. The experimental data revealed an increase in $S(CO_2)$ and CO_2 sorption capacity when the I anion was replaced with BF_4 for all tested cations (Table 17.2, entries 2, 6, 7, 8, 9 and 10).

The effect of structural variation of the cationic chain of polyimide-based PILs on CO_2 separation has been studied.[41] Polyimides (PIs) usually present high thermal stability and chemical resistance. Cationic PI-based PILs with two different groups in their structure, benzimidazole (PIL-20) and quinuclidine (PIL-21) (Figure 17.10) both with the Tf_2N cation, were developed. N_2, CH_4 and CO_2 permeability at 20 °C and 35 °C were also evaluated.[41] Membranes of these materials were prepared. PIL-21 produced a brittle membrane that broke during the gas permeation experiments. PIL-20 showed the following permeability tendency: $PCO_2 \gg PCH_4 > PN_2$ at both temperatures (Table 17.2, entry 13). The CO_2 permeability values were 2× higher than those of CH_4 and N_2.[41]

CO_2 sorption capacity evaluation in polyurethane-based PILs, both cationic[50] and anionic,[21,50,51] has been described. Anionic PILs presented higher CO_2 sorption capacity than cationic PILs.[50] PIL-22a (Figure 17.11) containing imidazolium as the counter-cation presented a higher CO_2 sorption capacity than the polycation PIL-23 (Figure 17.11) containing imidazolium in the backbone and $[Tf_2N]$ as the counter-anion.[50] Polyanion PIL-22a had a higher CO_2 sorption capacity compared to PIL $[VBTMA][BF_4]$ (PIL-1a),[24] MDEA 50 wt% and $[bmim][TF_2N]$. At 20 bar and 25 °C, the CO_2 sorption capacities for $[bmim][Tf_2N]$ and MDEA were 56% and 51%, respectively, when compared to that of PIL-22a.[50] PIL-22a (69.0 mol% of CO_2, at 10 bar and 25 °C) also possessed a higher CO_2 sorption capacity than $[VBTMA][BF_4]$ (PIL-1a) (approximately 39 mol% at 10.13 bar and 22 °C).[50] Changes in the main chain of anionic polyurethane-based PILs were investigated.[50] The increment of polyether groups attained by increasing the molecular weight of polytetramethylene glycol (PTMG) from 1000 to 2000 (PIL-22a) resulted in a 71% improvement in the CO_2 sorption capacity at 20 bar. The ether group can strengthen the physical forces between the PIL and CO_2. Distance increase between the ionic sites can improve the PIL–CO_2 interaction increasing the CO_2 sorption capacity.[50]

The effect of the counter-cation on the CO_2 sorption capacity of anionic polyurethane-based PILs was evaluated.[21,50] It is well known that the introduction of branching in the alkyl side chain of RTILs improves the CO_2 solubility.[66] Anionic polyurethane-based PILs did not show a similar trend. PIL-22a (Table 17.2, entry 15) with the counter-cation $[bmim]^+$(linear) showed a slightly higher sorption capacity than PIL-22b (Table 17.2, entry 16) containing $[dmbmim]^+$(branched). This behavior can be associated with the steric hindrance caused by the branched chain.[50]

Three counter-cations were evaluated in the CO_2 sorption capacity of anionic polyurethane-based PILs,[21] imidazolium (PIL-22a, Table 17.2, entry 15), phosphonium (PIL-22c, Table 17.2, entry 17) and ammonium (PIL-22d, Table 17.2, entry 18). The highest values of CO_2 sorption were achieved with PILs containing tetrabutylammonium and tetrabutylphosphonium as counter-cations. Simulation studies revealed that the interaction of deprotonated carboxylic groups (COO^-) with CO_2 is very important. Tetrabutylammonium and tetrabutylphosphonium cations exhibit weaker coordination with carboxylic groups (COO^-) than imidazolium, promoting electrostatic binding of the CO_2 molecule and the carboxyl group.[21]

17.7 Composites (PIL–ILs)

The formation of composites using different poly(ionic liquid)s and ionic liquids (PIL–ILs) has also been reported in the literature.[44,53,61,63–65,67] The results (presented in Table 17.3) demonstrated that PIL–IL composites are promising materials for CO_2 separation from gas streams, since their properties can be improved with the incorporation of a non-polymerizable IL ("free" ionic liquid).[44,61,63,65,67]

RTILs containing the anion bis(trifluoromethylsulfonyl)imide (Tf_2N) have been combined with PILs in order to obtain PIL–IL composites.[44,53,58,63] The incorporation of 20 wt% of 1-butyl-1-methylpyrrolidiniumbis(tri-fluoromethylsulfonyl)imide ([pyr14][Nf_2T]) into PIL-13a increased the CO_2 permeability by up to 391% and the CO_2/N_2 selectivity by up to 44%[44] (Table 17.3, entry 16).

Vinyl-based PILs, mainly functionalized with *n*-hexyl (PIL-8b) and disiloxane (PIL-8f), are good candidates to contribute to the CO_2 permeability and selectivity when combined with free RTILs.[63] Significant improvements in permeability and selectivity were observed when 20 mol% of the 1-ethyl-3-methylimidazolium-bis-(trifluoromethylsulfonyl) imide ([emim][Tf_2N]) ionic liquid was added to PIL-8b and PIL-8f (Table 17.3, entries 1 and 2). For example, the CO_2 permeability of PIL-8b increased from 69 to 105 Barrers with the addition of free RTIL, whereas the PCO_2/PN_2 selectivity increased from 17 to 21. Similarly, it was found that the permeability of PIL-8f increased from 130 to 190 Barrers and the PCO_2/PN_2 selectivity increased from 14 to 19.[63] The incorporation of free RTILs could increase the selectivity relative to neat PILs.[63]

Incorporation of 20 mol% of [emim][Tf_2N] into a styrene-based PIL (PIL-3g) resulted in a hybrid stable composite.[58] The permeability of several gases (CO_2, N_2 and CH_4) increased up to 300–600% when free RTIL was incorporated into the PIL matrix. However, it should be highlighted that the permeability

Table 17.3 CO_2 sorption performance of PIL–IL composites.

Composite (PIL + IL)	Entry	Experimental Condition [bar–°C]	Loading [cm^3 (STP)/cm^3 atm]			Permeability [Barrers][a]			Permselectivity [dimensionless]		Ref.
			CO_2	N_2	CH_4	CO_2	N_2	CH_4	CO_2/CH_4	CO_2/N_2	
PIL-8b + 20 mol% [emim][Tf$_2$N]	1	2.02–22	–	–	–	105.0	–	–	12.0	21.0	63
PIL-8f + 20 mol% [emim][Tf$_2$N]	2	2.02–22	–	–	–	190.0	–	–	11.0	19.0	63
PIL-8g – [emim][dca] (1 : 1)	3	10.00–35	1.04	0.021	–	62.1	1.03	–	–	60.3	64
PIL-8g – [emim][BF$_4$] (1 : 1)	4	10.00–35	1.05	0.026	–	36.0	0.79	–	–	45.6	64
PIL-8g – [emim][B(CN)4] (1 : 1)	5	10.00–35	1.66	0.046	–	99.8	2.23	–	–	44.8	64
PIL-8h – [emim][dca] (1 : 1)	6	10.00–35	1.16	0.026	–	118.8	2.40	–	–	49.5	64
PIL-8h – [emim][BF$_4$] (1 : 1)	7	10.00–35	1.21	0.032	–	74.8	1.79	–	–	41.8	64
PIL-8h – [emim][B(CN)4] (1 : 1)	8	10.00–35	1.79	0.05	–	184.7	4.67	–	–	40.0	64
PIL-8i – [emim][dca] (1 : 1)	9	10.00–35	1.35	0.03	–	138.0	3.28	–	–	42.0	64
PIL-8i – [emim][BF$_4$] (1 : 1)	10	10.00–35	1.32	0.03	–	89.7	2.35	–	–	38.2	64
PIL-8i – [emim][B(CN)4] (1 : 1)	11	10.00–35	1.99	0.05	–	230.0	6.42	–	–	35.8	64
PIL-8j + 10 wt% [C$_2$mim][Tf$_2$N]	12	1.00–20	1.29	0.04	0.09	14.9	0.63	0.59	25.1	23.7	65
PIL-8j + 20 wt% [C$_2$mim][Tf$_2$N]	13	2.02–22	1.20	–	–	19.0	0.83	1.10	17.0	23.0	53
PIL-11a + 10 wt% [C$_2$Py][Tf$_2$N]	14	1.00–20	1.10	0.03	0.06	20.4	1.01	0.81	25.1	20.1	65
PIL-12a + 10 wt% [N$_{1122}$OH]	15	1.00–20	1.81	0.03	0.06	3.6	0.11	0.08	43.0	34.7	65
PIL-13a + 20 wt% pyr$_{14}$[Tf$_2$N]	16	1.00–21	1.75	0.04	0.10	25.0	0.78	0.98	25.5	32.0	44
PIL-13a + 40 wt% pyr$_{14}$[Tf$_2$N]	17	1.00–21	2.07	0.05	0.15	105.5	3.49	5.65	18.7	30.3	44
PIL-13a + 10 wt% [pyrr$_{14}$][Tf$_2$N]	18	1.00–20	1.38	0.02	0.03	11.5	0.39	0.40	28.3	28.9	65
PIL-13b + 20 wt% [C$_2$mim][N(CN)2]	19	1.00–20	1.02	0.01	–	3.7	0.08	–	–	–	61
PIL-13c + 20 wt% [C$_2$mim][C(CN)3]	20	1.00–20	1.02	0.01	–	7.9	0.17	–	–	–	61
PIL-13c + 40 wt% [C$_2$mim][C(CN)3]	21	1.00–20	1.25	0.01	–	142.3	56.4	–	–	–	61
PIL-14a + 10 wt% [N$_{4111}$][Tf$_2$N]	22	1.00–20	1.94	0.03	0.02	9.4	0.36	0.08	31.5	25.8	65
PIL-20a + 50 wt% [1-Me-3-Et im][TFSI]	23	1.00–20	–	–	–	69.8	2.81	2.86	24.5	34.0	41
PIL-20a + 50 wt% [1-Me-3-Et im][TFSI]	24	1.00–35	–	–	–	102.6	5.64	5.66	18.2	27.5	41
PIL-21a + 50 wt% [1-Me-3-Et im][TFSI]	25	1.00–35	–	–	–	84.4	4.03	4.07	20.7	26.0	41

[a]Permeability in Barrers (10^{-10} cm^3(STP) cm/cm^2 s cm Hg).

values presented for N_2 and CH_4 are extremely low relative to that of CO_2.[58] The CO_2/N_2 selectivity increased with the incorporation of 20 mol% [emim][TF$_2$N] into PIL-3g. The selectivity could result from the increase of CO_2 diffusion. Surprisingly, the CO_2/CH_4 selectivity decreased with IL incorporation.[58]

Incorporation of 20 wt% of [C_6mim][Tf2N] into a PIL-9b matrix increased the CO_2, CH_4 and N_2 permeability up to 360%, 430% and 380%, respectively, with little or no decrease in the CO_2/CH_4 selectivity relative to neat PIL-9b.[53]

PIL–IL composites produced by the combination of poly(ionic liquid)s PIL-8j, PIL-11, PIL-12, PIL-13a, and PIL-14 with ionic liquids of similar structure were investigated concerning permeability and selectivity. The properties listed in Table 17.3 evidenced that the polycation structure plays an important role in the permeation properties and in the performance improvement of PIL–IL membranes in CO_2 separation.[65]

In PIL–IL composites, the PIL counter-anion also plays an important role in CO_2 permeability and selectivity.[61] Studies carried out with some cationic diallyldimethylammonium-based PILs and different nitrile-functionalized anions (PIL-13b–d) (Table 17.3, entries 19–21) indicated that the number of nitrile groups present in the anion effects the permeability and selectivity.[61]

Three free ILs (1-ethyl-3-methylimidazolium dicyanamide ([emim][dca]), 1-ethyl-3-methylimidazolium tetracyanoborate ([emim][B(CN)$_4$]) and 1-ethyl-3-methylimidazolium tetrafluoroborate ([emim][BF$_4$])) were incorporated into three vinyl-based PILs containing dicyanamide as the anion and imidazolium as the cation (PIL-8g–i). The results revealed that the incorporation of a RTIL into the PIL matrix increased the CO_2 permeability and CO_2/N_2 selectivity[64] (Table 17.3, entries 3–11).

Membranes of PIL-20 and PIL-21 with incorporation of 50% 1-ethyl-3-methylimidazolium bis-(trifluoromethylsulfonyl)imide [1-Me-3-Etim] TFSI IL were obtained. The permeability results for the neat PILs and composites reveal the same tendency ($PCO_2 \gg PCH_4 > PN_2$). The permeability values of CO_2 were 1× to 2× higher than those of CH_4 and N_2[41] for neat PILs. The incorporation of 50 wt% of IL into the PIL-20 matrix increased the CO_2 permeability up to 164% at 30 °C.[41] The presence of a quinuclidine group in the PIL-21–50% IL matrix composite increased the CO_2, CH_4 and N_2 permeability by 21, 42 and 57%, respectively, relative to the composite obtained with PIL-20 as the matrix (with a benzimidazole group in the polymeric chain). The presence of the quinuclidine group, which is more voluminous than benzimidazole, contributes to the development of a less packed membrane with less resistance to gas diffusion and consequently greater gas permeability.[41]

Acknowledgements

Support from the EU 6 Multihybrids (IP 026685-2) and Nanofire (NMP3-CT 2004-505637) projects, the Hungarian Research Foundation (OTKA T049121), the Fund of the European Union and Hungarian state (GVOP/3.1.1.-2004-0531/3.0), and the Public Benefit Association of Sciences and Sport of the Budapest University of Technology and Economics is acknowledged.

References

1. S. A. Montzka, E. J. Dlugokencky and J. H. Butler, *Nature*, 2011, **476**, 43.
2. M. Songolzadeh, M. Soleimani, M. Takht Ravanchi and R. Songolzadeh, *Sci. World J.*, 2014, **2014**, 828131.
3. T. Bruhn, H. Naims and B. Olfe-Kräutlein, *Environ. Sci. Policy*, 2016, **60**, 38.
4. M. Wang, A. S. Joel, C. Ramshaw, D. Eimer and N. M. Musa, *Appl. Energy*, 2015, **158**, 275.
5. A. A. Olajire, *Energy*, 2010, **35**, 2610.
6. J. D. Figueroa, T. Fout, S. Plasynski, H. McIlvried and R. D. Srivastava, *Int. J. Greenhouse Gas Control*, 2008, **2**, 9.
7. M. K. Mondal, H. K. Balsora and P. Varshney, *Energy*, 2012, **46**, 431.
8. J. M. Ketzel, R. Igleisias and S. Einloft, in *Handbook of Climate Change Mitigation and Adaptation*, ed. M. L. Wei-Yin Chen and Toshio Suzuki, 2015, pp. 1–40.
9. A. Blasig, J. Tang, X. Hu, Y. Shen and M. Radosz, *Fluid Phase Equilib.*, 2007, **256**, 75.
10. S. D. Kenarsari, D. Yang, G. Jiang, S. Zhang, J. Wang, A. G. Russell, Q. Wei and M. Fan, *RSC Adv.*, 2013, **3**, 22739.
11. M. Hasib-ur-Rahman, M. Siaj and F. Larachi, *Chem. Eng. Process. Process Intensif.*, 2010, **49**, 313.
12. J. F. Brennecke and B. E. Gurkan, *J. Phys. Chem. Lett.*, 2010, **1**, 3459.
13. Z. Dai, R. D. Noble, D. L. Gin, X. Zhang and L. Deng, *J. Membr. Sci.*, 2016, **497**, 1.
14. E. I. Privalova, P. Mäki-Arvela, D. Y. Murzin and J. P. Mikkhola, *Russ. Chem. Rev.*, 2012, **81**, 435.
15. M. Sadeghpour, R. Yusoff and M. K. Aroua, *Rev. Chem. Eng.*, 2017, **33**, 183–200.
16. P. Markewitz, W. Kuckshinrichs, W. Leitner, J. Linssen, P. Zapp, R. Bongartz, A. Schreiber and T. E. Müller, *Energy Environ. Sci.*, 2012, **5**, 7281.
17. S. Zulfiqar, M. I. Sarwar and D. Mecerreyes, *Polym. Chem.*, 2015, 6435.
18. J. Tang, W. Sun, H. Tang, M. Radosz and Y. Shen, *Macromolecules*, 2005, **38**, 2037.
19. J. Yuan and M. Antonietti, *Polymer (Guildf.)*, 2011, **52**, 1469.
20. J. Yuan, D. Mecerreyes and M. Antonietti, *Prog. Polym. Sci.*, 2013, **38**, 1009.
21. F. L. Bernard, B. B. Polesso, F. W. Cobalchini, A. J. Donato, M. Seferin, R. Ligabue, V. V. Chaban, J. F. do Nascimento, F. Dalla Vecchia and S. Einloft, *Polymer (Guildf.)*, 2016, **102**, 199.
22. L. C. Tome and I. M. Marrucho, *Chem. Soc. Rev.*, 2016, **45**, 2785.
23. J. Tang, H. Tang, W. Sun, M. Radosz and Y. Shen, *Polymer (Guildf.)*, 2005, **46**, 12460.
24. J. Tang, H. Tang, W. Sun, H. Plancher, M. Radosz and Y. Shen, *Chem. Commun. (Camb.)*, 2005, 3325.
25. H. Yang, Z. Xu, M. Fan, R. Gupta, R. B. Slimane, A. E. Bland and I. Wright, *J. Environ. Sci.*, 2008, **20**, 14.
26. T. T. N. Bachelor and P. Toochinda, *Environ. Technol.*, 2012, **33**, 2645.

27. M. Ramdin, A. Amplianitis, S. Bazhenov, A. Volkov, V. Volkov, T. J. H. Vlugt and T. W. de Loos, *Ind. Eng. Chem. Res.*, 2014, **53**, 15427.
28. S. Seo, L. D. Simoni, M. Ma, M. A. DeSilva, Y. Huang, M. A. Stadtherr and J. F. Brennecke, *Energy Fuels*, 2014, **28**, 5968.
29. M. Kanniche, R. Gros-Bonnivard, P. Jaud, J. Valle-Marcos, J. M. Amann and C. Bouallou, *Appl. Therm. Eng.*, 2010, **30**, 53.
30. J. S. Wilkes, *Green Chem.*, 2002, **4**, 73–80.
31. Z. Xue, Z. Zhang, J. Han, Y. Chen and T. Mu, *Int. J. Greenhouse Gas Control*, 2011, **5**, 628.
32. J. L. Anthony, E. J. Maginn and J. F. Brennecke, *J. Phys. Chem. B*, 2002, **106**, 7315.
33. L. A. Blanchard, Z. Gu and J. F. Brennecke, *J. Phys. Chem. B*, 2001, **105**, 2437.
34. J. L. Anthony, J. L. Anderson, E. J. Maginn and J. F. Brennecke, *J. Phys. Chem. B*, 2005, **109**, 6366.
35. C. Cadena, J. L. Anthony, J. K. Shah, T. I. Morrow, J. F. Brennecke and E. J. Maginn, *J. Am. Chem. Soc.*, 2004, **126**, 5300.
36. J. L. Anthony, E. J. Maginn and J. F. Brennecke, *J. Phys. Chem. B*, 2002, **106**, 7315.
37. S. N. V. K. Aki, B. R. Mellein, E. M. Saurer and J. F. Brennecke, *J. Phys. Chem. B*, 2004, **108**, 20355.
38. S. Babamohammadi, A. Shamiri and M. K. Aroua, *Rev. Chem. Eng.*, 2015, **31**, 383.
39. M. J. Muldoon, S. N. V. K. Aki, J. L. Anderson, J. K. Dixon and J. F. Brennecke, *J. Phys. Chem. B*, 2007, **111**, 9001.
40. S. N. V. K. Aki, B. R. Mellein, E. M. Saurer and J. F. Brennecke, *J. Phys. Chem. B*, 2004, **108**, 20355.
41. A. S. Shaplov, S. M. Morozova, E. I. Lozinskaya, P. S. Vlasov, A. S. L. Gouveia, L. C. Tomé, I. M. Marrucho and Y. S. Vygodskii, *Polym. Chem.*, 2016, **7**, 580.
42. D. Mecerreyes, *Prog. Polym. Sci.*, 2011, **36**, 1629.
43. J. Tang, Y. Shen, M. Radosz and W. Sun, *Ind. Eng. Chem. Res.*, 2009, **48**, 9113.
44. L. C. Tomé, D. Mecerreyes, C. S. R. Freire, L. P. N. Rebelo and I. M. Marrucho, *J. Membr. Sci.*, 2013, **428**, 260.
45. J. Tang, H. Tang, W. Sun, M. Radosz and Y. Shen, *J. Polym. Sci., Part A: Polym. Chem.*, 2005, **43**, 5477.
46. R. S. Bhavsar, S. Kumbharkar, A. S. Rewar and U. K. Kharul, *Polym. Chem.*, 2014, 4083.
47. S. C. Kumbharkar, R. S. Bhavsar and U. K. Kharul, *RSC Adv.*, 2014, **4**, 4500.
48. S. C. Kumbharkar, M. N. Islam, R. A. Potrekar and U. K. Kharul, *Polymer (Guildf.)*, 2009, **50**, 1403.
49. Y. B. Xiong, H. Wang, Y. J. Wang and R. M. Wang, *Polym. Adv. Technol.*, 2012, **23**, 835.
50. T. O. Magalhaes, A. S. Aquino, F. Dalla Vecchia, F. L. Bernard, M. Seferin, S. C. Menezes, R. Ligabue and S. Einloft, *RSC Adv.*, 2014, **4**, 18164.

51. M. Fernández, L. Á. Carreño, F. Bernard, R. Ligabue and S. Einloft, *Macromol. Symp.*, 2016, **368**, 98.
52. R. S. Bhavsar, S. C. Kumbharkar and U. K. Kharul, *J. Membr. Sci.*, 2012, **389**, 305.
53. T. K. Carlisle, J. E. Bara, A. L. Lafrate, D. L. Gin and R. D. Noble, *J. Membr. Sci.*, 2010, **359**, 37.
54. K. Simons, K. Nijmeijer, J. E. Bara, R. D. Noble and M. Wessling, *J. Membr. Sci.*, 2010, **360**, 202.
55. W. J. Horne, M. A. Andrews, M. S. Shannon, K. L. Terrill, J. D. Moon, S. S. Hayward and J. E. Bara, *Sep. Purif. Technol.*, 2014, **155**, 89.
56. S. Supasitmongkol and P. Styring, *Energy Environ. Sci.*, 2010, **3**, 1961.
57. J. E. Bara, S. Lessmann, C. J. Gabriel, E. S. Hatakeyama, R. D. Noble and D. L. Gin, *Ind. Eng. Chem. Res.*, 2007, **46**, 5397.
58. J. E. Bara, C. J. Gabriel, E. S. Hatakeyama, T. K. Carlisle, S. Lessmann, R. D. Noble and D. L. Gin, *J. Membr. Sci.*, 2008, **321**, 3.
59. J. E. Bara, E. S. Hatakeyama, C. J. Gabriel, X. Zeng, S. Lessmann, D. L. Gin and R. D. Noble, *J. Membr. Sci.*, 2008, **316**, 186.
60. G. Yu, Q. Li, N. Li, Z. Man, C. Pu, C. Asumana and X. Chen, *Polym. Eng. Sci.*, 2014, 2.
61. L. C. Tomé, M. Isik, C. S. R. Freire, D. Mecerreyes and I. M. Marrucho, *J. Membr. Sci.*, 2015, **483**, 155.
62. L. C. Tomé, 2014.
63. T. K. Carlisle, E. F. Wiesenauer, G. D. Nicodemus, D. L. Gin and R. D. Noble, *Ind. Eng. Chem. Res.*, 2013, **52**, 1023.
64. P. Li, D. R. Paul and T.-S. Chung, *Green Chem.*, 2012, **14**, 1052.
65. L. C. Tomé, A. S. L. Gouveia, C. S. R. Freire, D. Mecerreyes and I. M. Marrucho, *J. Membr. Sci.*, 2015, **486**, 40.
66. H. W. Pennline, D. R. Luebke, K. L. Jones, C. R. Myers, B. I. Morsi, Y. J. Heintz and J. B. Ilconich, *Fuel Process. Technol.*, 2008, **89**, 897.
67. J. E. Bara, E. S. Hatakeyama, D. L. Gin and R. D. Noble, *Polym. Adv. Technol.*, 2008, **19**, 1415.

CHAPTER 18

Ionic Liquid-based Polymers and Crystals for Dye-sensitized Solar Cells

CHUAN-PEI LEE[a] AND KUO-CHUAN HO*[a,b]

[a]Department of Chemical Engineering, National Taiwan University, Taipei 10617, Taiwan; [b]Institute of Polymer Science and Engineering, National Taiwan University, Taipei 10617, Taiwan
*E-mail: kcho@ntu.edu.tw

18.1 Introduction to Solar Energy & Dye-sensitized Solar Cells

For a long time, the utilization of non-renewable energy resources from fossil fuels has resulted in the pollution of the Earth, such as air pollution, carbon dioxide emission and greenhouse effects, which has significantly affected our health and the environment.[1] Therefore, researchers around the world have dedicated tremendous effort to exploring alternative energy sources (*i.e.*, sunlight, wind, hydro, rain, waves and geothermal heat *etc.*) that are sustainable, safe and environmentally friendly.[2] Among the alternative energy sources, solar energy is the most promising because of its abundance, cleanliness, safety and higher economic value. It is remarkable that a mere 10 min of solar irradiation onto the Earth's surface is equal to the total yearly human energy consumption.[3] Moreover, the direct conversion of solar light into electrical energy by photovoltaic cells is of particular interest due to it

Smart Materials No. 29
Polymerized Ionic Liquids
Edited by Ali Eftekhari

Published by the Royal Society of Chemistry, www.rsc.org

being very effective among all the available renewable energy resources. Currently, the most commonly used photovoltaic cells are made of silicon (Si), cadmium telluride (CdTe), copper indium selenide/sulphide (CIS), or multi-junction-based materials, *etc.*[4] However, expensive and complex fabrication processes are needed for the above-mentioned solar cells, and therefore their usage and applications are limited. As a result, the dominance of these kinds of inorganic solid-state junction devices in the solar cell market is now being challenged by the emergence of third generation solar cells based on interpenetrating network structures, such as dye-sensitized solar cells (DSSCs);[5] these third generation solar cells are designed to cost less and are based on cheap, simple, and easy fabrication processes.

In 1991, O'Regan and Grätzel made a breakthrough of ~7% cell efficiency in a DSSC, thus stimulating the intensive investigation of these types of solar cells, which are considered to be one of the most promising renewable power sources since they do not rely on expensive or energy-intensive processing methods and can be printed onto flexible substrates using roll-to-roll methods.[5] In addition, as compared to conventional Si solar cells, the performance of DSSCs is less affected by the light intensity and less sensitive to the angle of incidence (AOI) of solar light,[6] which makes DSSCs the most efficient photo-to-electron conversion devices under indoor and low-level outdoor lighting applications,[6] such as in smart homes and intelligent buildings, with an average power consumption in the range of microwatts, including sensors, chargers, and standby power. Figure 18.1a shows a statistical graph of the number of publications related to DSSCs from 1991 to 2016. Obviously, the statistical data show that the number of reports on DSSCs surged drastically up to 2014, since DSSCs drew much attention after a cell efficiency of DSSCs higher than 10% was reported in 2000.[7] In 2015, a DSSC with the co-sensitization of two organic dyes and with a Co(II/III) tris(phenanthroline)-based liquid electrolyte showed an excellent cell performance of 14.30%, which is the highest efficiency recorded to date.[8]

Figure 18.1b shows the typical structure of a traditional DSSC device, which is composed of a working electrode with a dye-sensitized mesoporous titanium dioxide (TiO_2) nanocrystalline film for the absorption of solar light, a counter electrode with a platinum (Pt) layer for the collection of electrons and a redox electrolyte, sandwiched between the two electrodes. The usual choice for dye sensitizers are the ruthenium bipyridyl derivatives (*i.e.*, N719 and N3 *etc.*, see Figure 18.1c). The electrolyte mostly contains the iodide/triiodide (I^-/I_3^-) redox couple, which is obtained by mixing iodine (I_2) and inorganic or organic iodides in suitable non-aqueous solvents. The solar-to-electricity conversion mechanism of the DSSCs is shown in Figure 18.1d. Upon absorption of light, an electron is injected from the metal-to-ligand charge transfer excited state of the dye into the conduction band of the TiO_2. The rate of this electron injection reaction is ultrafast, typically occurring on the order of hundreds of femtoseconds to tens of picoseconds.[9] The injected electron percolates through the TiO_2 film, and is thought to move by a "hopping" mechanism driven by a chemical diffusion gradient (rather than

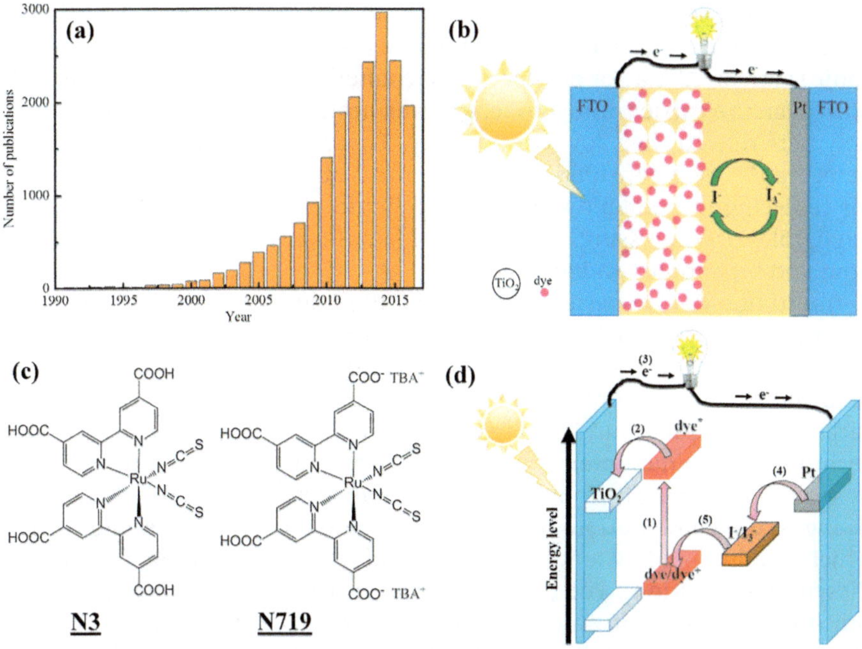

Figure 18.1 (a) A statistical graph of the number of publications related to DSSCs from 1991 to 2016. Data were obtained from Scopus (using title, abstract and keyword searches for all types of literature: dye solar cell). (b) Schematic sketch of a DSSC. (c) The structures of ruthenium bipyridyl derivatives N3 and N719. (d) The basic sequence of events in a DSSC.

an electric field), before being collected at a transparent conductive substrate of fluorine-doped tin oxide glass (SnO_2:F), on which the TiO_2 film is formed. After passing through an external circuit, the electron is reintroduced into the solar cell at the Pt counter electrode, where triiodide ions are reduced to iodide ions. Immediately, the iodide ions then regenerate the oxidized dye, thereby completing the circuit with no net chemical change.

18.2 Toward All/Quasi-solid-state Dye-sensitized Solar Cells *via* Ionic Liquid Electrolytes

Although the DSSCs have reached an attractive efficiency record of 14.30%,[8] they still use a highly volatile organic solvent for preparing the electrolytes. Unfortunately, DSSCs using organic solvent-based liquid electrolytes face challenges in the integration of large area modules, cell sealing, evaporation of organic solvent, and precipitation of salts at low temperature; thus, long-term stability becomes the major problem for this type of cell. Accordingly, organic hole-conducting materials,[10] p-type inorganic semiconductors,[11,12] ionic gel electrolytes with a polymer[13,14] or a gelator,[15] and room temperature ionic liquid (IL)-based electrolytes[16–21] were recently investigated as all/quasi-solid-state electrolytes for DSSCs. In these cases, imperfect filling of the dye–sensitized mesoporous TiO_2 nanocrystalline film by the p-type inorganic semiconductors or polymers has resulted in low efficiency for the cells. Moreover, the carrier diffusion length was limited in the case of conducting polymers due to their low conductivity. Therefore, solvent-free room-temperature ILs were considered to be the most attractive electrolyte materials for replacing the organic solvent-based liquid electrolytes because of their negligible vapor pressure, high thermal stability, wide electrochemical window, and high ionic conductivity.[16–21]

Traditional ILs are basically composed of organic ions that may undergo almost unlimited structural variations because of the easy preparation of a large variety of their components; thus, various kinds of salts can be added to formulate an ionic liquid that has the desired properties for a given application. In general, room-temperature ILs are low-temperature molten salts with melting points lower than ambient temperature, that is, liquids composed of ions only. The molten salts are characterized by weak interactions, owing to the combination of a large cation and a charge-delocalized anion, which results in a low tendency to crystallize due to flexibility (anion) and dissymmetry (cation). For the application of room-temperature ILs as solvent-free electrolytes in efficient DSSCs, Wang *et al.* first prepared a binary-IL electrolyte that consisted of 1-methyl-3-propyl imidazolium iodide (PMII), 1-methyl-3-ethylimidazolium dicyanamide (EMIDCN), and lithium iodide (LiI) for a DSSC to obtain a cell efficiency of 6.6% at full solar irradiation.[16] Soon after, they developed a new binary-IL electrolyte composed of PMII and 1-ethyl-3-methylimidazolium thiocyanate (EMINCS), and a cell efficiency of 7.00% was obtained.[17] In addition, they also synthesized an iodine-free $SeCN^-/(SeCN)_3^-$-based IL electrolyte by incorporation of a new room-temperature IL, 1-ethyl-3-methylimidazolium selenocyanate (EMISeCN), and a high efficiency of 7.5% was achieved.[18] In the same group, Kuang *et al.* further reported on a new record of 7.6% cell efficiency under full sunlight irradiation using a 1-ethyl-3-methyl-imidazolium tetracyanoborate ($EMIB(CN)_4$)-based binary-IL electrolyte.[22] However, room-temperature ILs are still liquid-state under ambient conditions; thus, the fluidity and potential leakage of room-temperature IL-based electrolytes during long-term operation or under high temperature conditions is still unavoidable.[4] To overcome this problem, solid-state ILs have been intensively investigated as all/quasi-solid-state electrolytes for DSSCs recently. This chapter reviews some of the key research on the topic of solid-state ILs using polymeric ILs or IL crystals for DSSCs, and some of the relevant literature has also been briefly introduced.

18.2.1 Polymeric Ionic Liquids for Solid-state Dye-sensitized Solar Cells

To further enhance the long-term stability of DSSCs with IL-based electrolytes, a new class of ILs was designed to prevent the leakage and volatilization of the electrolyte. Recently, polymeric ILs were designed by incorporating a polymer chain into the IL moieties as a gelator for the solidification of IL-based electrolytes.[23,24] These electrolytes possess high mechanical strength, and even show good ionic conductivity without liquid components due to their specific functions, such as transport of target ions, and specific polar environment.[25] For the application of polymeric ILs in DSSCs, Wang *et al.* synthesized a group of solid-state IL polymers, poly(1-alkyl-3-(acryloyloxy) hexylimidazolium iodide) (PAAII), with four different alkyl side chains (*i.e.*, methyl, ethyl, propyl, and butyl).[26–28] Among all the PAAII polymeric ILs, the one with an ethyl side chain, namely poly(1-ethyl-3-(acryloyloxy)hexylimidazolium iodide) (PEAII; Figure 18.2), shows the highest ionic conductivity (3.63×10^{-4} S cm^{-1}). Due to the conjugation effect of the imidazolium ring and the steric hindrance of the polymer backbone, the attraction between the cation and the iodide anion was weak, resulting in easy diffusion of iodide anions in the PEAII electrolyte; thus, the electrons can be quickly transported from the counter electrode to the oxidized dyes by diffusion of I^- ions. Using PEAII as the electrolyte, the corresponding DSSC exhibits a cell efficiency of 5.29% under AM 1.5G illumination of 100 mW cm^{-2}, and maintains about 85% of its initial efficiency after a long-term stability test for 1000 h without sealing. Fang *et al.* synthesized an acidic IL polymer, P[((3-(4-vinylpyridine) propanesulfonic acid) iodide)-*co*-(acrylonitrile)] (P–HI; Figure 18.2), for use as an electrolyte in DSSCs.[29]

The polymer P–HI contains sulfonic acid groups that exhibit electrostatic forces with ILs to form a homogeneous and continuous framework, which enhances transportation of redox couples in the electrolyte. A DSSC with a novel IL electrolyte containing P–HI, *N*-methylbenzimidazole (NMB), guanidinium thiocyanate (GuNCS), 1-hexyl-3-methylimidazolium iodide (HMII) and 1-allyl-3-methylimidazoliumiodide (AMII) achieved an excellent cell efficiency of 6.95% under AM 1.5G illumination of 100 mW cm^{-2}. Chi *et al.* synthesized a polymeric IL, namely poly((1-(4-ethenylphenyl)methyl)-3-butyl-imidazolium iodide) (PEBII; Figure 18.2), as the electrolyte of DSSCs.[30] Here, the (4-ethenylphenyl)methyl functional group was anchored onto 1-butylimidazolium iodide and served as the self-polymerization site in PEBII. Due to the strong π–π stacking interaction between the benzene groups, PEBII exhibits a high conductivity of 2.0×10^{-4} S cm^{-1} at 25 °C. For application in DSSCs, the PEBII electrolyte was able to deeply penetrate into the mesoporous TiO_2 film; thereby, the charge transfer at the photoanode/electrolyte interface would be significantly improved. A DSSC using PEBII electrolyte shows a good cell efficiency of 5.93%, measured under AM 1.5G illumination of 100 mW cm^{-2}. With the incorporation of their newly designed photoanode containing double-layer mesoporous TiO_2 beads, the corresponding DSSCs with PEBII electrolyte achieved a higher cell efficiency of up to 6.70%.[31] Chen *et al.* designed a novel bis-imidazolium-based polymeric IL, namely poly(1-butyl-3-(1-vinylimidazolium-3-hexyl)-imidazolium bis(trifluoromethanesulfonyl)imide) (poly[BVIm][HIm][TFSI]; Figure 18.2),[32] which was synthesized and utilized to solidify an IL-based electrolyte composed of 1-ethyl-3-methylimidazolium (EMII), PMII, EMINCS, I_2,

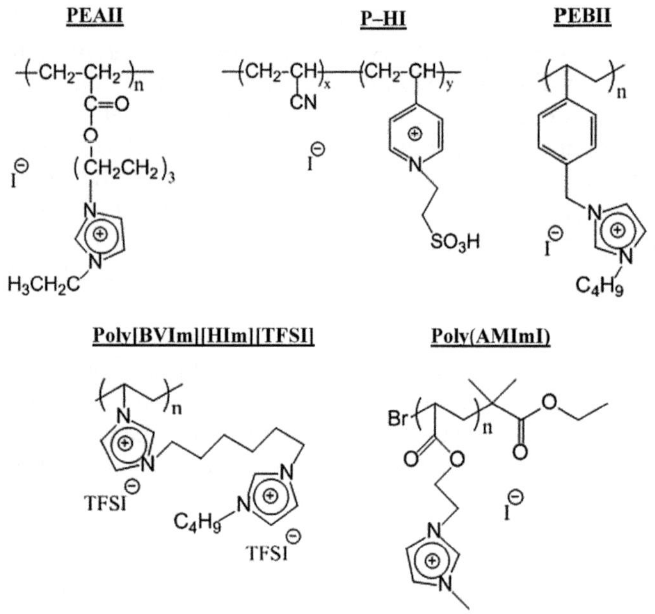

Figure 18.2 The molecular structures of polymeric ILs used as electrolytes in DSSCs.

guanidinium thiocyanate (GuSCN) and *N*-butylbenzimidazole (NBB) for DSSCs. They found that the as-prepared electrolyte possesses good conductivity due to the charge transport networks formed in the electrolyte *via* the π–π stacked imidazolium rings. The DSSC using this poly[BVIm][HIm][TFSI]-based electrolyte yielded a cell efficiency of 5.92% under AM 1.5G illumination of 100 mW cm^{-2}, and it retained about 96% of the initial efficiency after 1200 h of continuous light-soaking testing at 60 °C. Chang *et al.* synthesized a novel polymeric IL, poly(1-(2-acryloyloxy-ethyl)-3-methyl-imidazol-1-ium iodide) (poly(AMImI); Figure 18.2), by using an atom transfer radical polymerization method.[33] The monomeric AMImI possesses a 2-acryloyloxy-ethyl functional group anchored onto the 1-methylimidazolium iodide IL, and the 2-acryloyloxy-ethyl functional groups served as the self-polymerization site. A DSSC with a poly(AMImI)-based electrolyte containing *N*-methylbenzimidazole (NMB), GuSCN, and I_2 reached an efficiency of 1.16%. After introducing multi-walled carbon nanotubes (MWCNTs) into the poly(AMImI)-based electrolyte, the corresponding DSSC achieved a 3.55% cell efficiency.

In our group, Chang *et al.* synthesized a novel polymeric IL, poly(oxyethylene)-imide imidazolium iodide (POEI–II; Figure 18.3), for application as a quasi-solid-state electrolyte in DSSCs.[34] As shown in Figure 18.3; two 1-butylimidazolium iodide ILs are anchored at both ends of a poly(oxyethylene)-segmented oligo(imide) (POEI) monomer to construct the POEI–II IL. The POEI functional groups possess aromatic imides and hydrophilic POE segments with multiple dipole–dipole interaction sites, which endow

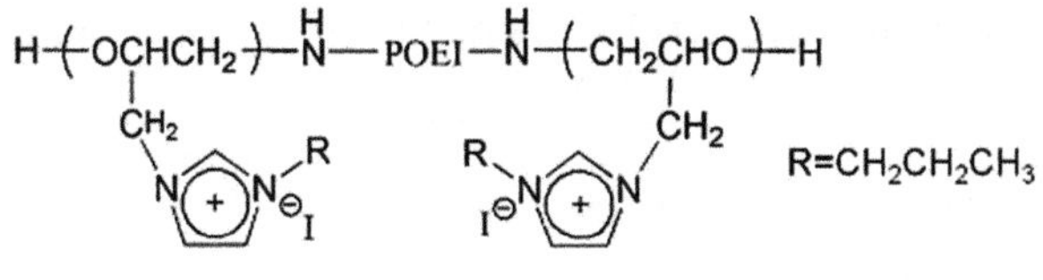

Figure 18.3 The structure of poly(oxyethylene)-imide imidazolium iodide (POEI–II).

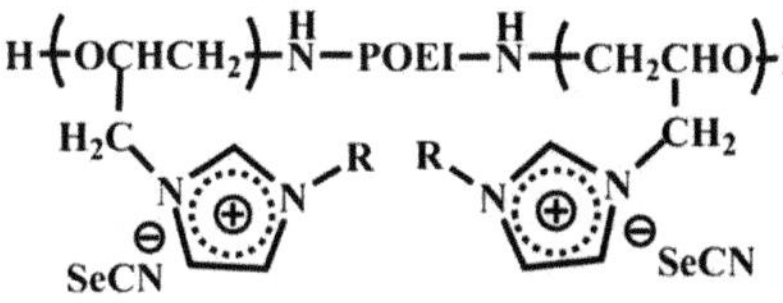

Figure 18.4 The structure of poly(oxyethylene)-imide–imidazolium selenocyanate (POEI–IS).

POEI–II with high solubility in aqueous solutions as well as in organic solvents. The POE segment in the POEI–II could chelate lithium cations (Li^+) within the electrolyte to improve the open-circuit-voltage (V_{OC}) value of a DSSC and enable the strong inner π–π and long-pair–π electron interactions to enhance the ionic conductivity and the diffusivity of POEI–II. Consequently, a DSSC with the quasi-solid-state POEI–II gel electrolyte reached a high cell efficiency of 7.19%. Furthermore, MWCNTs were incorporated into the quasi-solid-state POEI–II gel electrolyte as an extended electron transfer material (EETM) to facilitate charge transfer from the counter electrode to the redox mediator, which benefited the dye regeneration more efficiently. Meanwhile, the POE segments on POEI–II could prevent the MWCNTs from aggregating, which made the well-dispersed MWCNTs largely exposed to I^-/I_3^- redox mediators and thereby gave more charge transfer active sites. The highest cell efficiency of 7.65% was achieved by using the MWCNTs/POEI–II gel electrolyte, and the corresponding DSSC showed an unfailing performance after the at-rest long-term stability test for 1000 h at 50 °C. Soon after, based on a similar design concept, our group (Lin *et al.*) further reported an iodide-free polymeric IL, namely poly(oxyethylene)-imide–imidazolium selenocyanate (POEI–IS; see Figure 18.4), for use as an electrolyte in DSSCs.[35] POEI–IS possesses several functions: (a) it acts as a gelling agent for the electrolyte of the DSSC, (b) it possesses a redox mediator of $SeCN^-$, which is aimed at forming a $SeCN^-/(SeCN)_3^-$ redox couple with a more positive redox potential than that of traditional I^-/I_3^-, (c) it chelates the potassium cations through the lone pair electrons of the oxygen atoms of its poly(oxyethylene)-imide–imidazolium (POEI–I) segments, and (d) it obstructs the recombination of photoinjected electrons with $(SeCN)_3^-$ ions in the electrolyte through its POEI–I segments. Thus, POEI–IS prolongs the stability of the corresponding DSSC and endows the cell with a high V_{OC}. A DSSC with a POEI–IS gel electrolyte exhibits a high V_{OC} of 825 mV and a high power conversion efficiency of 8.18%, and it also

Table 18.1 Partial literature results reported for DSSCs using all/quasi-solid-state electrolytes with polymeric ILs.

Polymeric ILs	Electrolyte composition	Efficiency (η)	Durability (% to the initial η)	Ref.
PEAII	PEAII	5.29%	85%; after 1000 h	26–28
P–HI	P–HI; HMII; AMII; NMB; GuSCN	6.95%	N.A.	29
PEBII	PEBII	6.70%	N.A.	30 and 31
Poly[BVIm][HIm][TFSI]	Poly[BVIm][HIm][TFSI]; EMII; PMII; EMISCN; I_2; GuSCN; NBB	5.92%	96%; after 1200 h	32
Poly(AMImI)	poly(AMImI); AMBImI-modified MWCNTs; I_2; NMBI; GuSCN; ACN	3.55%	N.A.	33
POEI–II	POEI–II; MWCNTs; LiI; I_2; tBP; ACN/MPN	7.65%	100%; after 1000 h	34
POEI–IS	POEI–IS; KSeCN; $(SeCN)_2$; ACN	8.18%	95%; after 1000 h	35

retained up to 95% of its initial efficiency after an at-rest long-term stability test for 1000 h.

Table 18.1 is a partial list of the performance of DSSCs using the above polymeric ILs in the all/quasi-solid-state electrolytes.

18.2.2 Ionic Liquid Crystals for Solid-state Dye-sensitized Solar Cells

Via chemical structure design, some ILs could exhibit melting points higher than room temperature; thereby, they are solid crystals under ambient conditions, which could significantly eliminate the problems of the fluidity and potential leakage of room-temperature ILs in a DSSC device during long-term operation. Yamanaka *et al.* reported a new concept for enhancing the conductivity of IL-based electrolytes by employing an IL crystal as a constituent of an electrolyte[36], which forms a self-assembled structure to promote the charge exchange reaction due to the locally increased concentrations of I_2 and I_3^-[37,38]. They utilized 1-dodecyl-3-methylimidazolium iodide (C12MImI; Figure 18.5) as the IL crystal, which provides a self-assembled structure of the imidazolium cations like a solid, while maintaining the molecular dynamics like a liquid[36]. The IL crystal, C12MImI, with the smectic A phase has a bilayer structure of interdigitated alkyl chains of the imidazolium cations, and I_2 and I_3^- would be localized between the smectic A layers, whereas the locally high concentrations of I_2 and I_3^- would promote the charge exchange reaction. A few examples of IL crystals with the smectic A phase, such as imidazolium-based ILs consisting of cations with alkyl chains of C12–C18 and anions of hexafluorophosphate or bromide, have been reported.[39,40] Their study shows for the first time that imidazolium-based ILs with an iodide

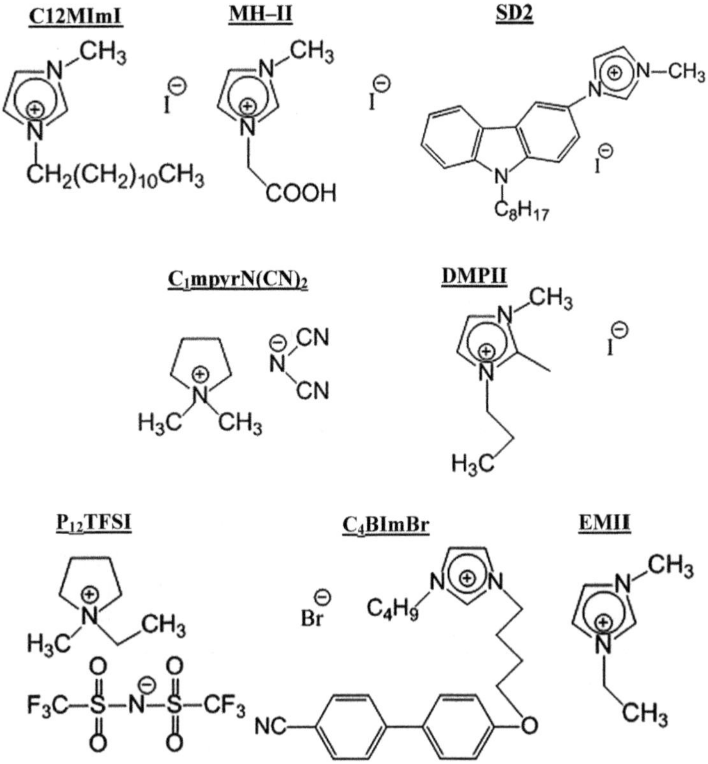

Figure 18.5 The molecular structures of IL crystals used for the electrolytes of DSSCs.

counter-anion and with alkyl chains longer than C12 also exhibit a smectic A phase and that the liquid crystalline nature is preferable in terms of the hole transport layer in DSSCs. The DSSC with C12MImI/I_2 electrolyte exhibited a cell efficiency of 2.30% under AM 1.5G illumination of 100 mW cm^{-2}. Zhao *et al.* reported a solid-state DSSC utilizing an imidazolium-based IL crystal, 1-methyl-3-hydroxyethyl-imidazolium iodide (MH–II; Figure 18.5), as the charge transfer layer, 1-methyl-3-propylimidazolium tetra-fluoroborate (MP–BF_4) as a crystal growth inhibitor, lithium bis-trifluoro-methanesulfonylimide (Li[$(CF_3SO_2)_2N$]) as a charge transport enhancer, and 4-*tert*-butylpyridine (tBP) as a carrier recombination inhibitor; the corresponding cell showed a cell efficiency of 3.10% under AM 1.5G illumination of 100 mW cm^{-2},[41] and the cell efficiency remained at 60% of the initial value after 30 days at room temperature without any sealing and protection from ambient conditions. Midya *et al.* have designed and synthesized a new class of solid-state carbazole-imidazolium-based ILs as electrolytes for DSSCs;[42] carbazole was used as the hole conductor because polyvinyl carbazole had already been employed successfully in a solid-state DSSC as a hole conductor.[43] In their study, the carbazole-imidazolium-based ILs with SCN^- anions (SD1) or I^- anions (SD2) were synthesized and applied in an all-solid-state DSSC.[42] They found that the electrolyte containing SD2 and I_2 can provide dual channels for hole/triiodide transportation. Thus, the DSSC with the electrolyte containing SD2 (Figure 18.5), I_2, $Li[(CF_3SO_2)_2N]$, tBP and $EMIB(CN)_4$ shows a higher cell efficiency (2.85%) than the cell with the electrolyte containing SD1 (1.43%).[43] Armel *et al.* have developed an organic ionic plastic crystal, *N*,*N*-dimethylpyrrolidinium dicyanamide ($C_1mpyrN(CN)_2$; Figure 18.5), as a new class of solid-state electrolyte for DSSCs.[44] The DSSCs with solid-state electrolytes containing $C_1mpyrN(CN)_2$, EMII, lithium iodide (LiI), I_2 and NMB achieved a good cell efficiency of 5.10% under AM 1.5G illumination of 100 mW cm^{-2}. Chen *et al.* have developed a solid-state IL, 1,2-dimethyl-3-propylimidazolium iodide (DMPII; Figure 18.5), as a charge transfer intermediate (CTI) for use as an electrolyte in DSSCs.[45] Meanwhile, potassium iodide (KI) and polyethylene oxide (PEO, MW = 100 000) were incorporated into the DMPII-based CTI as the charge transfer auxiliary agent and the crystal growth inhibitor, respectively. An optimal cell efficiency of 5.87% was obtained for the DSSC fabricated with the solid-state electrolyte of DMPII/KI/PEO.

Recently, Li *et al.* also utilized an organic ionic plastic crystal, 1-ethyl-1-methylpyrrolidinium bis(tri-fluoromethane sulfonyl)imide ($P_{12}TFSI$; Figure 18.5), as the electrolyte for solid-state DSSCs.[46] $P_{12}TFSI$ is a well-known organic ionic plastic crystal, which exhibits a broad solid–solid transformation at 14 °C and a melting point at 86 °C.[47–50] The DSSC with an electrolyte containing $P_{12}TFSI$ and PMII yielded a cell efficiency of 3.92% under AM 1.5G illumination of 100 mW cm^{-2}. Further addition of LiI and NBB to the $P_{12}TFSI$/PMII electrolyte enhanced the cell efficiency up to 4.78%. The DSSC using $P_{12}TFSI$/PMII/LiI/NBB electrolyte retained 90% of the initial value after 50 days under ambient conditions without further sealing. Cao-Cen *et al.* have synthesized the *N*-4-(4-cyanobiphenyl-40-oxy)-butyl-3-butylimidazolium bromide (C_4BImBr; Figure 18.5) IL crystal and applied it as the electrolyte for DSSCs.[51] It has been reported that biphenyl and 4-cyanobiphenyl are important core units for mesogenic molecules.[52–55] Cyanobiphenyl-functionalized compounds can act as a kind of low light-transmitting and high light-scattering material and obtained a high light-harvesting efficiency when they were employed as an electrolyte placed at the back of the working electrode of a DSSC. The DSSC with the all-solid-state $C_4BImBr/EMII/I_2$ electrolyte achieved a good cell efficiency of 4.45% under AM 1.5G illumination of 100 mW cm^{-2} because of the enhanced light-harvesting capability of the electrolyte containing C_4BImBr. To further improve the cell efficiency, PMII was added into the $C_4BImBr/EMII/I_2$ electrolyte as a crystal growth inhibitor. The thus fabricated DSSC with $C_4BImBr/EMII/I_2$/PMII electrolyte exhibited an enhanced cell efficiency of 5.07%, and possessed a good long-term stability (*i.e.*, 5% decay) during an accelerated aging test (*i.e.*, one sun light soaking) for 1000 h at 25 °C.

In our group, Lee *et al.* have fabricated all-solid-state DSSCs with a hybrid SWCNT–binary charge transfer intermediate (CTI) consisting of single-walled carbon nanotubes (SWCNTs), EMII and PMII.[56] Here, the EMII IL crystal (Figure 18.5) was employed as the CTI to fabricate all-solid-state DSSCs; the

SWCNTs were incorporated into the CTI as extended electron transfer materials (EETMs; Figure 18.6), which can reduce the charge diffusion length and serve simultaneously as a catalyst for the electrochemical reduction of I_3^-. An all-solid-state DSSC with this hybrid SWCNTs/EMII electrolyte showed a higher cell efficiency (1.88%) than that containing bare EMII alone (0.41%). To further improve the cell efficiency, the PMII IL was used as a co-charge transfer intermediate and crystal growth inhibitor (Figure 18.7), simultaneously. An improved cell efficiency of 3.49% was thus obtained by using the hybrid SWCNTs/EMII/PMII electrolyte; the durability of the corresponding solid-state DSSCs was studied at room temperature and was found to be far superior to that of a cell with an organic solvent electrolyte. On the other hand, based on a similar concept, Lee *et al.* also developed a composite electrolyte comprising two ILs and a carbon material to fabricate a solid-state DSSC;[57] the ILs were EMII and 1-ethyl-3-methylimidazolium tetrafluoroborate ($EMIBF_4$), and the carbon materials were carbon black (CB), multi-walled carbon nanotubes (MWCNTs), and single-walled carbon nanotubes (SWCNTs). A cell efficiency of 0.41% was achieved by using the bare EMII as the CTI; an efficiency of 2.52% was achieved for a solid-state DSSC with the incorporation of carbon black (CB) in the EMII. To further improve the cell efficiency, they utilized $EMIBF_4$, a crystal growth inhibitor, as an additive to the CB/EMII electrolyte. A cell efficiency of 3.09% was obtained using an electrolyte containing the CB and the binary CTI (*i.e.*, EMII/$EMIBF_4$). When the CB was replaced with MWCNTs or SWCNTs, the cell efficiency could be further improved to 3.53% and 4.01%, respectively. The long-term durability of the DSSC with SWCNTs/EMII/$EMIBF_4$ was found to be far superior to that of the cell with an organic solvent electrolyte, and in fact the durability was uninterrupted for at least 1000 h.

Table 18.2 is a partial list of the performance of DSSCs using the above-mentioned IL crystals in the all/quasi-solid-state electrolytes.

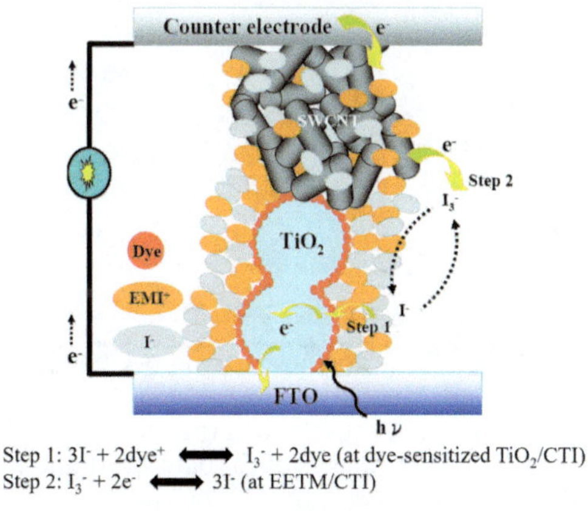

Figure 18.6 A schematic illustration of the charge transport processes in the DSSC with a hybrid SWCNTs/EMII electrolyte.

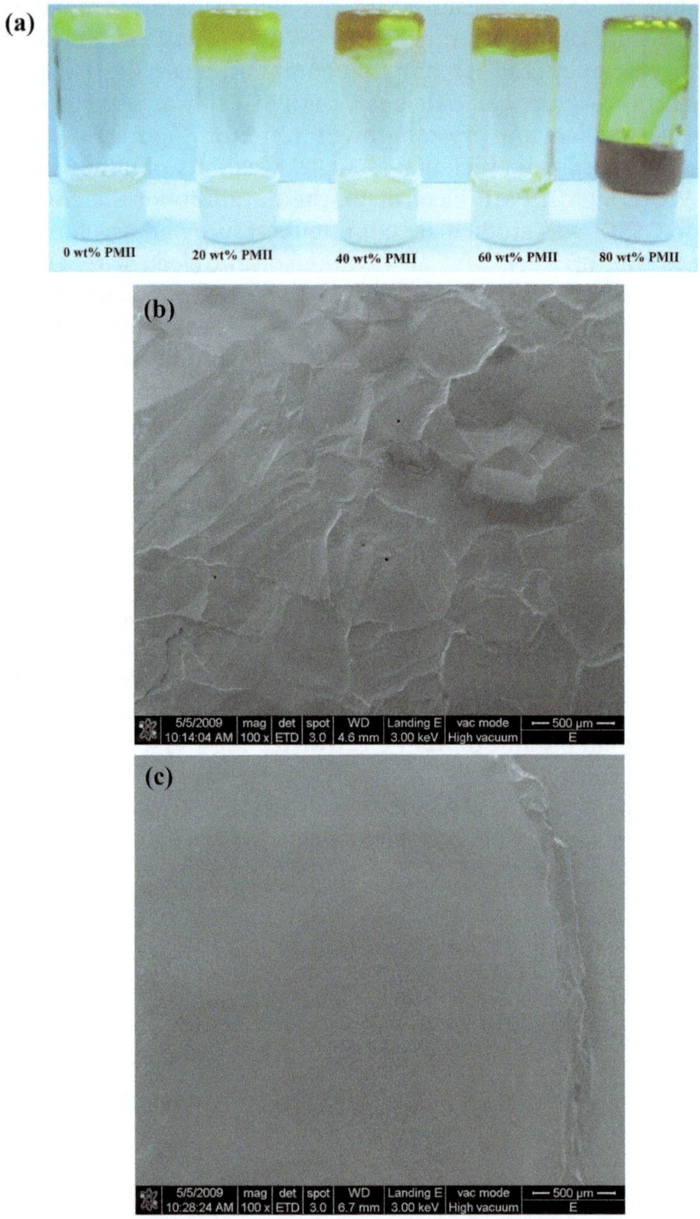

Figure 18.7 (a) Pictures of the EMII CTI with the addition of different weight percents of PMII. SEM images of (b) bare EMII and (c) EMII/PMII (w/w = 40/60).

Table 18.2 Partial literature results reported for DSSCs using all/quasi-solid-state electrolytes with IL crystals.

IL crystals	Electrolyte composition	Cell efficiency (η)	Durability (% to the initial η)	Ref.
C12MImI	C12MImI; I_2	2.30%	N.A.	36
MH–II	MH-II; MP-BF_4; Li[$(CF_3SO_2)_2N$]; tBP	3.10%	60%; after 30 days	41
SD2	SD2; I_2; Li[$(CF_3SO_2)_2N$]; tBP; EMIB$(CN)_4$	2.85%	N.A.	42
C_1mpyrN$(CN)_2$	C_1mpyrN$(CN)_2$; EMII; LiI; I_2; NMB	5.10%	N.A.	44
DMPII	DMPII; KI; PEO	5.87%	N.A.	45
P_{12}TFSI	P_{12}TFSI; PMII; LiI; NBB	4.78%	90%; after 50 days	46
C_4BImBr	EMII; I_2; C_4BImBr; PMII	5.07%	95%; after 1000 h	51
EMII	EMII; PMII; SWCNTs	3.49%	100%; after 1000 h	56
	EMII; EMIBF_4; SWCNTs	4.01%	100%; after 1000 h	57

18.3 Summary

In this chapter, we mainly review some of the key research on the topic of polymeric ILs or IL crystals for use as all/quasi-solid-state electrolytes in DSSCs, and some of the relevant literature is also selected for comparison. In the topic of polymeric ILs, our group designed and synthesized a novel polymeric IL, POEII, for preparing a quasi-solid-state electrolyte for DSSCs. The POEII acted simultaneously as a redox mediator and a gelator; its POE segment could chelate Li^+ ions within the electrolyte to improve the cell voltage and enable strong inner π–π and long-pair–π electron interactions to enhance the ionic conductivity and the diffusivity of the redox couple. Meanwhile, the POE segments could prevent the MWCNTs from aggregating, which meant that the well-dispersed MWCNTs were largely exposed to the I^-/I_3^- redox mediators. MWCNTs were incorporated into the POEII-based electrolyte as extended electron transfer materials to facilitate charge transfer from the counter electrode to the redox mediator. The highest cell efficiency of 7.65% was achieved by using the MWCNTs/POEII-based electrolyte and showed an unfailing durability for greater than 1000 h under 50 °C. On the basis of a similar design concept, our group further designed an iodide/iodine ion-free polymeric IL, POEI–IS, containing the redox mediator $SeCN^-$, which is aimed at forming a $SeCN^-/(SeCN)_3^-$ redox couple with a more positive redox potential than that of traditional I^-/I_3^-. Thus, the POEI–IS-based electrolyte endows the corresponding DSSC with a high V_{OC} (825 mV). The DSSC with the POEI–IS-based electrolyte exhibits a high excellent cell efficiency of 8.18%, and it also retained up to 95% of its initial efficiency after an at-rest long-term stability test for 1000 h. On the topic of IL crystals, our group developed an efficient all-solid-state DSSC using a hybrid SWCNTs//EMII/PMII electrolyte without the addition of iodine and tBP, and achieved a high efficiency of 3.49%; moreover, the cell also showed an excellent durability at room temperature for 1000 h. Here, SWCNTs and the EMII IL crystal were employed as the EETM and solid-state CTI, respectively; the PMII IL acts as a co-CTI and crystal growth inhibitor, simultaneously. By using a similar electrolyte composite, our DSSCs with all-solid-state electrolytes containing the binary CTI of EMII/$EMIBF_4$ and different carbon materials of CB, MWCNTs or SWCNTs exhibited cell efficiencies of 3.09%, 3.53% and 4.01%, respectively. The long-term durability of the DSSC with SWCNTs/EMII/$EMIBF_4$ was found to be uninterrupted for at least 1000 h.

To date, most DSSCs with all/quasi-solid-state IL-based electrolytes still show relatively low efficiency as compared to those with traditional liquid electrolytes. Although the all/quasi-solid-state IL-based electrolytes endow the DSSCs with superior long-term stability, the cell efficiency indeed needs to be further improved because both stability and cell conversion efficiency are the two essential criteria for a good DSSC. From our perspective, we believe that in the future, iodide/iodine-free polymeric IL electrolytes will be the most promising electrolytes in quasi-solid-state DSSCs due to the absence of iodide/iodine components in the DSSC, which will enhance the cell durability, photocurrent and electron life-time, and retard the photocorrosion effects of iodine radicals on the electrodes and sealing materials.

Acknowledgement

This work was supported in part by the Ministry of Science and Technology (MOST) of Taiwan.

References

1. N. Armaroli and V. Balzani, *Angew. Chem., Int. Ed.*, 2007, **46**, 52–66.
2. D. G. Nocera, *Inorg. Chem.*, 2009, **48**, 10001–10017.
3. J. H. Yum, E. Baranoff, S. Wenger, M. K. Nazeeruddin and M. Grätzel, *Energy Environ. Sci.*, 2011, **4**, 842–857.
4. M. S. Su'ait, M. Y. A. Rahman and A. Ahmad, *Sol. Energy*, 2015, **115**, 452–470.
5. B. O'Regan and M. Grätzel, *Nature*, 1991, **353**, 737–740.
6. C. P. Lee, C. A. Lin, T. C. Wei, M. L. Tsai, Y. Meng, C. T. Li, K. C. Ho, C. I. Wu, S. P. Lau and J. H. He, *Nano Energy*, 2015, **18**, 109–117.
7. A. Hagfeldt and M. Grätzel, *Acc. Chem. Res.*, 2000, **33**, 269–277.
8. K. Kakiage, Y. Aoyama, T. Yano, K. Oya, J.-i. Fujisawa and M. Hanaya, *Chem. Commun.*, 2015, **51**, 15894–15897.
9. M. Grätzel, *Nature*, 2001, **414**, 338–344.
10. U. Bach, D. Lupo, P. Comte, J. E. Moser, F. Weissortel, J. Salbeck, H. Spreitzer and M. Grätzel, *Nature*, 1998, **395**, 583–585.
11. G. R. A. Kumara, A. Konno, K. Shiratsuchi, J. Tsukahara and K. Tennakone, *Chem. Mater.*, 2002, **14**, 954–955.
12. G. Hodes and D. Cahen, *Acc. Chem. Res.*, 2012, **45**, 705–713.
13. C. P. Lee, L. Y. Lin, R. Vittal and K. C. Ho, *J. Power Sources*, 2011, **196**, 1665–1670.

14. J. H. Wu, S. C. Hao, Z. Lan, J. M. Lin, M. L. Huang, Y. F. Huang, L. Q. Fang, S. Yin and T. Sato, *Adv. Funct. Mater.*, 2007, **17**, 2645–2652.
15. W. Kubo, K. Murakoshi, T. Kitamura, S. Yoshida, M. Haruki, K. Hanabusa, H. Shirai, Y. Wada and S. Yanagida, *J. Phys. Chem. B*, 2001, **105**, 12809–12815.
16. P. Wang, S. M. Zakeeruddin, J. E. Moser and M. Grätzel, *J. Phys. Chem. B*, 2003, **107**, 13280–13285.
17. P. Wang, S. M. Zakeeruddin, R. Humphry-Baker and M. Grätzel, *Chem. Mater.*, 2004, **16**, 2694–2696.
18. P. Wang, S. M. Zakeeruddin, J. E. Moser, R. Humphry-Baker and M. Grätzel, *J. Am. Chem. Soc.*, 2004, **126**, 7164–7165.
19. C. P. Lee, K. M. Lee, P. Y. Chen and K. C. Ho, *Sol. Energy Mater. Sol. Cells*, 2009, **93**, 1411–1416.
20. C. P. Lee, P. Y. Chen, R. Vittal and K. C. Ho, *J. Mater. Chem.*, 2010, **20**, 2356–2361.
21. C. P. Lee, J. D. Peng, D. Velayutham, J. Chang, P. W. Chen, V. Suryanarayanan and K.-C. Ho, *Electrochim. Acta*, 2013, **114**, 303–308.
22. D. Kuang, C. Klein, Z. Zhang, S. Ito, J. E. Moser, S. M. Zakeeruddin and M. Grätzel, Stable, *Small*, 2007, **3**, 2094–2102.
23. H. Ohno, *Electrochim. Acta*, 2001, **46**, 1407–1411.
24. M. Yoshizawa and H. Ohno, *Electrochim. Acta*, 2001, **46**, 1723–1728.
25. W. Ogihara, S. Washiro, H. Nakajima and H. Ohno, *Electrochim. Acta*, 2006, **51**, 2614–2619.
26. G. Wang, L. Wang, S. Zhuo, S. Fang and Y. Lin, *Chem. Commun.*, 2011, **47**, 2700–2702.
27. G. Wang, S. Zhuo, L. Wang, S. Fang and Y. Lin, *Sol. Energy*, 2012, **86**, 1546–1551.
28. G. Wang, S. Zhuo and Y. Lin, *J. Appl. Polym. Sci.*, 2013, **127**, 2574–2580.
29. Y. Fang, W. Xiang, X. Zhou, Y. Lin and S. Fang, *Electrochem. Commun.*, 2011, **13**, 60–63.
30. W. S. Chi, J. K. Koh, S. H. Ahn, J.-S. Shin, H. Ahn, D. Y. Ryu and J. H. Kim, *Electrochem. Commun.*, 2011, **13**, 1349–1352.
31. D. K. Roh, J. A. Seo, W. S. Chi, J. K. Koh and J. H. Kim, *J. Mater. Chem.*, 2012, **22**, 11079–11085.
32. X. Chen, J. Zhao, J. Zhang, L. Qiu, D. Xu, H. Zhang, X. Han, B. Sun, G. Fu, Y. Zhang and F. Yan, *J. Mater. Chem.*, 2012, **22**, 18018–18024.
33. Y. H. Chang, P. Y. Lin, S. R. Huang, K. Y. Liu and K. F. Lin, *J. Mater. Chem.*, 2012, **22**, 15592–15598.
34. L. Y. Chang, C. P. Lee, C. T. Li, M. H. Yeh, K. C. Ho and J. J. Lin, *J. Mater. Chem. A*, 2014, **2**, 20814–20822.
35. Y. F. Lin, C. T. Li, C. P. Lee, Y. A. Leu, Y. Ezhumalai, R. Vittal, M. C. Chen, J. J. Lin and K. C. Ho, *ACS Appl. Mater. Interfaces*, 2016, **8**, 15267–15278.
36. N. Yamanaka, R. Kawano, W. Kubo, T. Kitamura, Y. Wada, M. Watanabe and S. Yanagida, *Chem. Commun.*, 2005, 740–742.
37. W. Kubo, K. Murakoshi, T. Kitamura, S. Yoshida, M. Haruki, K. Hanabusa, H. Shirai, Y. Wada and S. Yanagida, *J. Phys. Chem. B*, 2001, **105**, 12809–12815.

38. R. Kawano and M. Watanabe, *Chem. Commun.*, 2003, 330–331.
39. C. M. Gordon, J. D. Holbrey, A. R. Kennedy and K. R. Seddon, *J. Mater. Chem.*, 1998, **8**, 2627–2636.
40. A. E. Bradley, C. Hardacre, J. D. Holbrey, S. Johnston, S. E. J. McMath and M. Nieuwenhuyzent, *Chem. Mater.*, 2002, **14**, 629–635.
41. Y. Zhao, J. Zhai, J. He, X. Chen, L. Chen, L. Zhang, Y. Tian, L. Jiang and D. Zhu, *Chem. Mater.*, 2008, **20**, 6022–6028.
42. A. Midya, Z. Xie, J. X. Yang, Z. K. Chen, D. J. Blackwood, J. Wang, S. Adams and K. P. Loh, *Chem. Commun.*, 2010, **46**, 2091–2093.
43. N. Ikeda and T. Miyasaka, *Chem. Commun.*, 2005, 1886–1888.
44. V. Armel, M. Forsyth, D. R. MacFarlane and J. M. Pringle, *Energy Environ. Sci.*, 2011, **4**, 2234–2239.
45. J. Chen, T. Peng, K. Fan and J. Xia, *J. Mater. Chem.*, 2011, **21**, 16448–16452.
46. Q. Li, X. Chen, J. Zhao, L. Qiu, Y. Zhang, B. Sun and F. Yan, *J. Mater. Chem.*, 2012, **22**, 6674–6679.
47. D. R. MacFarlane and M. Forsyth, *Adv. Mater.*, 2001, **13**, 957–966.
48. Q. Li, J. Zhao, B. Sun, B. Lin, L. Qiu, Y. Zhang, X. Chen, J. Lu and F. Yan, *Adv. Mater.*, 2012, **24**, 945–950.
49. P. Wang, Q. Dai, S. M. Zakeeruddin, M. Forsyth, D. R. MacFarlane and M. Grätzel, *J. Am. Chem. Soc.*, 2004, **126**, 13590–13591.
50. Z. Chen, H. Yang, X. Li, F. Li, T. Yi and C. Huang, *J. Mater. Chem.*, 2007, **17**, 1602–1607.
51. H. Cao-Cen, J. Zhao, L. Qiu, D. Xu, Q. Li, X. Chen and F. Yan, *J. Mater. Chem.*, 2012, **22**, 12842–12850.
52. N. Sebastián, M. R. De La Fuente, D. O. López, M. A. Pérez-Jubindo, J. Salud, S. Diez-Berart and M. B. Ros, *J. Phys. Chem. B*, 2011, **115**, 9766–9775.
53. G. Vijayakumar, M. J. Lee, M. Song, S. H. Jin, J. W. Lee, C. W. Lee, Y. S. Gal, H. J. Shim, Y. Kang, G. W. Lee, K. Kim, N. G. Park and S. Kim, *Macromol. Res.*, 2009, **17**, 963–968.
54. K. Binnemans, *Chem. Rev.*, 2005, **105**, 4148–4204.
55. R. K. Lammi, K. P. Fritz, G. D. Scholes and P. F. Barbara, *J. Phys. Chem. B*, 2004, **108**, 4593–4596.
56. C. P. Lee, L. Y. Lin, P. Y. Chen, R. Vittal and K. C. Ho, *J. Mater. Chem.*, 2010, **20**, 3619–3625.
57. C. P. Lee, M. H. Yeh, R. Vittal and K. C. Ho, *J. Mater. Chem.*, 2011, **21**, 15471–15478.

Subject Index

ABIL. *see* amino-functionalized basic IL (ABIL)
ABSs. *see* aqueous biphasic systems (ABSs)
abundant polysaccharide ion gels, 321–327
actuators, and redox-active ILs, 251–252
agarose, 306–307
alga fractionation, and SPSs, 166–167
amino-functionalized basic IL (ABIL), 57
(aminopropyl)trialkylsilane, 157
angle of incidence (AOI), 516
anti-static PVDF/IL
  –CNT nanocomposites, 286–287
  miscible films, 281–284
  nanofibrous films, 284–286
AOI. *see* angle of incidence (AOI)
aqueous biphasic systems (ABSs), 118
ATR-FTIR. *see* attenuated total reflectance-Fourier transform infrared (ATR-FTIR) analysis
attenuated total reflectance-Fourier transform infrared (ATR-FTIR) analysis, 157

battery applications, and wearable energy storage for, 389–393
bilayer bending actuators, 468
bio-inspired devices, 478
block-like copolymers, 291–296

carbide-derived carbon (CDC), 461
carbon-based supports, SPILs on, 56–57
carbon capture
  composites (PIL–ILs), 509–511
  description, 490–492
  and ionic liquids (ILs), 492–493
  overview, 489–490
  and poly(ionic liquid)s (PILs)
    cation, anion and sorption, 499–505
    condensation polymerization and polymer modification, 495–498
    and separation, 505–509
carbon (CO_2) capture
  switchable polarity liquids for, 145–146, 162–166
    and (aminopropyl) trialkylsilane, 157
    attenuated total reflectance-Fourier transform infrared (ATR-FTIR) analysis, 157
    fractionation of alga, 166–167
    Henry's constant, 158–159
    NMR analysis, 159–160
    (trialkoxysilyl)propylamines and, 156–157
    wood fractionation, 167–174

carbon nanotubes (CNTs), 460–461
anti-static PVDF/IL-nano-composites, 286–287
cationic and anionic PILs
CO_2 sorption, 96–103
electroconductivity, 84–95
overview, 83–84
sensors, 103–105
thermoresponsive materials, 105–108
CB nanocomposites, 296–299
CDC. *see* carbide-derived carbon (CDC)
cellulose, 307–308
acetate, dissolution of, 268–270
cetyltrimethylammonium bromide (CTABr), 72
chain transfer agents (CTAs), 108
chemically cross-linked polymer scaffolds, 392–393
chitin and chitosan, 308–310
CNTs. *see* carbon nanotubes (CNTs)
co-condensation, 72–74
COF. *see* covalent organic framework (COF)
colloidal ionic liquid gels, 400–402
colloidal scaffolds, 387–388
conducting polymer actuators (CPAs), 461
conducting polymers (CPs), 461
bilayer bending actuators, 468
creeping effects, 473–474
linear deformation, 466–468
synthesis of, 462–465
trilayer bending actuators, 468–473
volume change in, 461–462
copolymerization, and thermo-responsive PIL nanogels, 206–221
CO_2 release, and switchable polarity liquids, 149–150
CO_2 separation/solubility, for ILs and PILs
factors affecting, 365–368
selective solution, 364–365
supported ILs, 368–372
CO_2 sorption, 96–103
covalent organic framework (COF), 45
CPAs. *see* conducting polymer actuators (CPAs)
creeping effects, 473–474
crystallization, and polymer doping, 267–268
CTABr. *see* cetyltrimethylammonium bromide (CTABr)
CTAs. *see* chain transfer agents (CTAs)

device assembly, and wearable energy storage
colloidal ionic liquid gels, 400–402
durability and comfort, 405–407
polymer-supported, 395–400
and safety, 407–408
dielectric PVDF/IL composites
block-like copolymers, 291–296
–CB nanocomposites, 296–299
film formation, 288–290
nanostructured composites, 290–291
direct synthesis, of polymeric networks, 40–48
doping, polymers
description, 264–265
overview, 262–263
and permeation properties
gas diffusivity, 272–274
gas separation, 274–275
gas solubility, 270–271
on T_g depression
estimation of, 266
Gordon–Taylor equation, 265–266
on T_m depression
crystallization, 267–268
dissolution of cellulose acetate, 268–270
polymer crystallinity, 268

DSSCs. *see* dye-sensitized solar cells (DSSCs)
dye-sensitized solar cells (DSSCs)
overview, 515–517
solid-state
ionic liquid crystals for, 522–527
overview, 517–518
polymeric ionic liquids for, 518–522

electrochemical polymerization, 12
electrochemical sensors, 430–438
electrochemical stability window, 389
electrochemistry, and ILs
electrodeposition, 226–227
electropolymerization, 227–228
electrochromism, 357–363
electroconductivity, 84–95
electrodeposition, 226–227
electro/magneto-responsive smart fluids, 182–184
electropolymerization, 227–228
electro-responsive electrorheological smart fluids
based on ILs, 187–191
based on PILs, 191–198
description, 184–187
environmental detection, of vapours, 446
enzymatic polymerization, 12–13

fiber and yarn production, 402–403
film formation, and dielectric PVDF/IL composites, 288–290
free radical polymerization, 36–40
and ionic liquids (ILs)
controlled, 6–9
conventional, 2–6
freestanding device incorporation, 404
Friedel–Craft reaction, 47
fuel cells, and redox-active ILs, 250–251

gas diffusivity, 272–274
gas separation, 274–275
gas solubility, 270–271
gel rheology, 389
Gordon–Taylor equation, 265–266
group transfer polymerizations, 9–10

hard templating, 32–33
Henry's constant, 158–159
horse radish peroxidase (HRP), 135
HRP. *see* horse radish peroxidase (HRP)
hybrid porous IL materials
metal–organic frameworks (MOFs), 64–70
silica, 70–77
from co-condensation, 72–74
overview, 70–72
from self-condensation, 74–77
hydrocolloid polysaccharide ion gels, 327–332

ILs. *see* ionic liquids (ILs)
inorganic colloidal scaffolds, 393
inorganic oxide supports, SPILs on, 59–60
International Union of Pure and Applied Chemistry (IUPAC), 25
ionic conducting polymers, volume change in, 461–462
ionic conductivity, 388–389
ionic electrochemical actuators
applications, 475–479
bio-inspired devices, 478
optical MEMS, 475–477
space technology, 478–479
carbon nanotubes (CNTs), 460–461
challenges, 479–480
conducting polymers (CPs), 461
bilayer bending actuators, 468

ionic electrochemical actuators (*continued*)
creeping effects, 473–474
linear deformation, 466–468
synthesis of, 462–465
trilayer bending actuators, 468–473
volume change in, 461–462
interfacing and actuation, 474–475
ionic gels, 457–458
ionic polymer–metal composites (IPMCs), 458–460
overview, 456–457
volume change in ionic conducting polymers, 461–462
ionic gels, 457–458
ionic liquid crystals, 522–527
ionic liquid gels, and wearable energy storage
for battery applications, 389–393
chemically cross-linked polymer scaffolds, 392–393
colloidal scaffolds, 387–388
considerations for, 404–408
deposition methods, 388
electrochemical stability window, 389
inorganic colloidal scaffolds, 393
ionic conductivity, 388–389
linear/chemically cross-linked polymers, 385–386
lithium ion-containing electrolytes, 386–387
mechanical properties/gel rheology, 389
molecular scaffolds, 388
overview, 384–385
physically cross-linked polymer scaffolds, 390–392
for supercapacitors, 394–395
in textile energy storage, 402–404
fiber and yarn production, 402–403
freestanding device incorporation, 404
textile printing, 403–404
triblock copolymers, 386
ionic liquid sensors. *see* wearable chemical sensors, ILs in
ionic liquids (ILs). *see also* poly(ionic liquid)s (PILs)
applications of, 13–14
and carbon capture, 492–493
and electrochemical polymerization, 12
electro-responsive electrorheological smart fluids based on, 187–191
and enzymatic polymerization, 12–13
free radical polymerization
controlled, 6–9
conventional, 2–6
in ionic and group transfer polymerizations, 9–10
overview, 1–2
and PVDF (*see* poly(vinylidene fluoride) (PVDF), and ILs)
redox-active (*see* redox-active immobilized ionic liquids)
and ring opening polymerization, 10–11
and smart polymeric hydrogels
agarose, 306–307
cellulose, 307–308
chitin and chitosan, 308–310
description, 311–314
overview, 304–306
silk fibroin (SF), 310–311
xanthan gum, 311
as solvents for grafting from biopolymers, 13
and step-growth polymerization, 12

and transition metal-catalyzed polymerizations, 11
transport properties
and CO_2 separation/ solubility, 365–372
description, 348–356
and electrochromism, 357–363
factors affecting, 365–368
overview, 342–346
and physicochemical properties, 346–348
selective solution, 364–365
in wearable chemical sensors
benefits of, 418–419
description, 417–418
electrochemical sensors, 430–438
environmental detection of vapours, 446
optical systems, 428–430
overview, 416–417
paper-based sensors, 443–445
research on, 446–448
selective sampling, 420–422
selectivity and specificity of, 423–427
skin-worn chemical sensors, 438–443
ionic polymerizations, 9–10
ionic polymer-metal composites (IPMCs), 458–460
IPMCs. *see* ionic polymer-metal composites (IPMCs)
IUPAC. *see* International Union of Pure and Applied Chemistry (IUPAC)

LCST. *see* lower critical solution temperaturc (LCST)
Li-ion batteries, and redox-active ILs, 248–250
linear/chemically cross-linked polymers, 385–386
linear deformation, 466–468
lithium ion-containing electrolytes, 386–387
lower critical solution temperature (LCST), 117, 204

mechanical properties, of ionic liquid gels, 389
metal catalysts, SPILs on, 55–56
metal–organic frameworks (MOFs)
hybrid porous IL materials, 64–70
supported porous ionic liquids (SPILs) on, 57–59
miscible films, anti-static PVDF/IL, 281–284
MOFs. *see* metal–organic frameworks (MOFs)
molecular scaffolds, 388
multiwalled carbon nanotubes (MWCNTs), 56
MWCNTs. *see* multiwalled carbon nanotubes (MWCNTs)

nanofibrous films, 284–286
nanostructuration, and redox-active ILs, 243–245
nanostructured composites, and dielectric PVDF/IL, 290–291
NLDFT. *see* nonlocal density functional theory (NLDFT)
NMR. *see* nuclear magnetic resonance (NMR)
nonlocal density functional theory (NLDFT), 70
nuclear magnetic resonance (NMR), 159–160

optical MEMS, 475–477
optical systems, and wearable chemical sensors, 428–430

PAA. *see* poly(acrylic acid) (PAA)
paper-based sensors, 443–445
PEGS. *see* poly(ethylene glycol)-tailed sulfonate (PEGS)

periodic mesoporous organosilica (PMOs), 24, 72
permeation, and polymer doping
gas diffusivity, 272–274
gas separation, 274–275
gas solubility, 270–271
PET. *see* polyethylene terephthalate (PET)
physically cross-linked polymer scaffolds, 390–392
physicochemical properties
of ILs and PILs
density, 347–348
overview, 346–347
and redox-active ILs, 241–243
PILs. *see* poly(ionic liquid)s (PILs)
PMOs. *see* periodic mesoporous organosilica (PMOs)
PNIPAM. *see* poly(*N*-isopropyl-acrylamide) (PNIPAM)
poly(acrylic acid) (PAA), 48
polyelectrolyte complexation, 48–51
poly(ethylene glycol)-tailed sulfonate (PEGS), 61
polyethylene terephthalate (PET), 203
poly(ionic liquid)s (PILs). *see also* ionic liquids (ILs)
and carbon capture
cation, anion and sorption, 499–505
condensation polymerization and polymer modification, 495–498
and separation, 505–509
electro-responsive electrorheological smart fluids based on, 191–198
and polysaccharide ion gels
composites preparation, 335–338
film preparation, 334–335
overview, 332–334
porous
hard templating, 32–33
soft templating, 33–35
and redox-active immobilized ionic liquids
actuators, 251–252
energy applications, 247–252
fuel cells, 250–251
Li-ion batteries, 248–250
nanostructuration, 243–245
physicochemical properties, 241–243
sensors, 252–255
solid state dye-sensitized solar cells (SS-DSSCs), 251
supercapacitors, 247–248
switchable devices, 245–247
synthetic route and structure, 238–241
for solid-state dye-sensitized solar cells, 518–522
thermoresponsive, 125–134
applications of, 134–139
thermo-responsive nanogels
and copolymerization, 206–221
description, 204–206
overview, 202–203
transport properties
and CO_2 separation/solubility, 365–372
description, 348–356
and electrochromism, 357–363
factors affecting, 365–368
overview, 342–346
and physicochemical properties, 346–348
selective solution, 364–365
polymer crystallinity, 268
polymer doping
description, 264–265
overview, 262–263

and permeation properties
gas diffusivity, 272–274
gas separation, 274–275
gas solubility, 270–271
on T_g depression
estimation of, 266
Gordon–Taylor equation, 265–266
on T_m depression
crystallization, 267–268
dissolution of cellulose acetate, 268–270
polymer crystallinity, 268
polymeric supports, 55
polymerizations
electrochemical, 12
enzymatic, 12–13
group transfer, 9–10
ionic, 9–10
ring opening, 10–11
step-growth, 12
transition metal-catalyzed, 11
polymer supported ionic liquids (PSILs), 55
polymer-supported wearable energy storage, 395–400
poly(*N*-isopropylacrylamide) (PNIPAM), 127
polysaccharide ion gels
abundant, 321–327
hydrocolloid, 327–332
overview, 319–321
and PILs
composites preparation, 335–338
film preparation, 334–335
overview, 332–334
poly(vinylidene fluoride) (PVDF), and ILs
anti-static
–CNT nanocomposites, 286–287
miscible films, 281–284
nanofibrous films, 284–286
dielectric composites
block-like copolymers, 291–296
–CB nanocomposites, 296–299
film formation, 288–290
nanostructured composites, 290–291
overview, 280–281
porosity/porous ionic liquid materials
definition of, 24–28
description, 28–30
hybrid materials
metal–organic frameworks (MOFs), 64–70
silica, 70–77
overview, 23–24
porous PILs
hard templating, 32–33
soft templating, 33–35
supported porous ionic liquids (SPILs)
on carbon-based supports, 56–57
on inorganic oxide supports, 59–60
on metal catalysts, 55–56
overview, 51–55
on polymeric supports, 55
porous liquids, 61–63
on porous metal–organic frameworks (MOFs), 57–59
supramolecular, 77
templating-free methodologies
direct synthesis of polymeric networks, 40–48
free radical polymerization, 36–40
overview, 35–36
polyelectrolyte complexation, 48–51

porous liquids, 61–63
porous PILs
hard templating, 32–33
soft templating, 33–35
PSILs. *see* polymer supported ionic liquids (PSILs)
PVDF. *see* poly(vinylidene fluoride) (PVDF)

redox-active immobilized ionic liquids
description, 228–230
and electrochemistry
electrodeposition, 226–227
electropolymerization, 227–228
overview, 225–226, 230–231
and PILs
actuators, 251–252
energy applications, 247–252
fuel cells, 250–251
Li-ion batteries, 248–250
nanostructuration, 243–245
physicochemical properties, 241–243
sensors, 252–255
solid state dye-sensitized solar cells (SS-DSSCs), 251
supercapacitors, 247–248
switchable devices, 245–247
synthetic route and structure, 238–241
self-assembled monolayers (SAMs), 231–237
ring opening polymerization, 10–11

safety, and device assembly, 407–408
SAMs. *see* self-assembled monolayers (SAMs)
scaffolds
chemically cross-linked polymer, 392–393
colloidal, 387–388
inorganic colloidal, 393
molecular, 388
physically cross-linked polymer, 390–392
SCILL. *SEE* solid catalyst with an ionic liquid layer (SCILL)
self-assembled monolayers (SAMs), 231–237
self-condensation, 74–77
sensors
cationic and anionic PILs, 103–105
and redox-active ILs, 252–255
SF. *see* silk fibroin (SF)
“Ship-In-Bottle” (SIB) approach, 59
SIB. *see* “Ship-In-Bottle” (SIB) approach
SILCs. *see* supported ionic liquid catalysts (SILCs)
silica-ILs, 70–77
from co-condensation, 72–74
overview, 70–72
from self-condensation, 74–77
silk fibroin (SF), 310–311
SILMs. *see* supported ionic liquid membranes (SILMs)
SILs. *see* supported ionic liquids (SILs)
single-walled carbon nanotubes (SWCNTs), 57
skin-worn chemical sensors, 438–443
smart chemical sensors. *see* wearable chemical sensors, ILs in
smart fluids
electro/magneto-responsive, 182–184
electro-responsive electrorheological
based on ILs, 187–191
based on PILs, 191–198
description, 184–187
overview, 180–182

smart polymeric hydrogels, and ILs
agarose, 306–307
cellulose, 307–308
chitin and chitosan, 308–310
description, 311–314
overview, 304–306
silk fibroin (SF), 310–311
xanthan gum, 311
soft templating, 33–35
solid catalyst with an ionic liquid layer (SCILL), 55
solid-phase microextraction (SPME), 104
solid-state dye-sensitized solar cells
ionic liquid crystals for, 522–527
overview, 517–518
polymeric ionic liquids for, 518–522
solid state dye-sensitized solar cells (SS-DSSCs), 251
Sonogashira coupling, 41
space technology, and ionic electrochemical actuators, 478–479
specific surface area (SSA), 26
SPME. *see* solid-phase microextraction (SPME)
SPSs. *see* switchable polarity solvents (SPSs)
SRPs. *see* stimuli-responsive polymers (SRPs)
SSA. *see* specific surface area (SSA)
SS-DSSCs. *see* solid state dye-sensitized solar cells (SS-DSSCs)
step-growth polymerization, 12
stimuli-responsive polymers (SRPs), 204
supercapacitors
and ionic liquid gels, 394–395
and redox-active ILs, 247–248
supported ionic liquid catalysts (SILCs), 51
supported ionic liquid membranes (SILMs), 55
supported ionic liquids (SILs), 51
supported porous ionic liquids (SPILs)
on carbon-based supports, 56–57
on inorganic oxide supports, 59–60
on metal catalysts, 55–56
overview, 51–55
on polymeric supports, 55
porous liquids, 61–63
on porous metal–organic frameworks (MOFs), 57–59
supramolecular ILs, 77
SWCNTs. *see* single-walled carbon nanotubes (SWCNTs)
switchable devices, and redox-active ILs, 245–247
switchable hydrophobicity/hydrophilicity
depending on temperature, 118–125
overview, 117–118
thermoresponsive PILs, 125–134
applications of, 134–139
switchable polarity solvents (SPSs)
CO_2 release, 149–150
for CO_2/SO_2 capture, 145–146, 162–166
and (aminopropyl) trialkylsilane, 157
attenuated total reflectance-Fourier transform infrared (ATR-FTIR) analysis, 157
fractionation of alga, 166–167
Henry's constant, 158–159
NMR analysis, 159–160
(trialkoxysilyl) propylamines and, 156–157
wood fractionation, 167–174

switchable polarity solvents (SPSs) (*continued*)
from DBU, alcohols and CO_2, 146–147
DBU bicarbonate, 153–155
from glycerol, 151–153
overview, 143–145
as reaction and separation media, 174–176
recyclability of SPSs, 150
and silylamines
recyclability of, 161–162
regeneration of, 161
from TMG, alcohols and CO_2, 147–149
synthesis, of conducting polymers, 462–465

templating
-free methodologies
direct synthesis of polymeric networks, 40–48
free radical polymerization, 36–40
overview, 35–36
polyelectrolyte complexation, 48–51
hard, 32–33
soft, 33–35
textile energy storage, 402–404
fiber and yarn production, 402–403
freestanding device incorporation, 404
textile printing, 403–404
textile printing, 403–404
T_g depression, and polymer doping
estimation of, 266
Gordon–Taylor equation, 265–266
thermoresponsive materials, 105–108
thermo-responsive poly(ionic liquid) nanogels
and copolymerization, 206–221
description, 204–206
overview, 202–203
T_m depression, and polymer doping
crystallization, 267–268
dissolution of cellulose acetate, 268–270
polymer crystallinity, 268
transition metal-catalyzed polymerizations, 11
transport properties, for ILs and PILs
and CO_2 separation/solubility
factors affecting, 365–368
selective solution, 364–365
supported ILs, 368–372
description, 348–356
and electrochromism, 357–363
overview, 342–346
and physicochemical properties
density, 347–348
overview, 346–347
trilayer bending actuators, 468–473

UCST. *see* upper critical solution temperature (UCST)
upper critical solution temperature (UCST), 118–119

volume change, in CPs, 461–462

wearable chemical sensors, ILs in
benefits of, 418–419
description, 417–418
overview, 416–417
progression of
electrochemical sensors, 430–438
environmental detection of vapours, 446
optical systems, 428–430
paper-based sensors, 443–445

skin-worn chemical sensors, 438–443
research on, 446–448
selective solvation of
overview, 419–420
selective sampling, 420–422
selectivity and specificity of, 423–427
wearable energy storage, and ILs
description, 382–383
device assembly techniques
colloidal ionic liquid gels, 400–402
durability and comfort, 405–407
polymer-supported, 395–400
and safety, 407–408
device selection based on electrical properties, 405
ionic liquid gels
for battery applications, 389–393
chemically cross-linked polymer scaffolds, 392–393
colloidal scaffolds, 387–388
considerations for, 404–408
deposition methods, 388
electrochemical stability window, 389
inorganic colloidal scaffolds, 393
ionic conductivity, 388–389
linear/chemically cross-linked polymers, 385–386
lithium ion-containing electrolytes, 386–387
mechanical properties/ gel rheology, 389
molecular scaffolds, 388
overview, 384–385
physically cross-linked polymer scaffolds, 390–392
for supercapacitors, 394–395
in textile energy storage, 402–404
triblock copolymers, 386
overview, 381–382
wood fractionation, 167–174